# HANDBOOK OF POLYPROPYLENE AND POLYPROPYLENE COMPOSITES

# PLASTICS ENGINEERING

*Founding Editor*

**Donald E. Hudgin**

Professor
Clemson University
Clemson, South Carolina

1. Plastics Waste: Recovery of Economic Value, *Jacob Leidner*
2. Polyester Molding Compounds, *Robert Burns*
3. Carbon Black-Polymer Composites: The Physics of Electrically Conducting Composites, *edited by Enid Keil Sichel*
4. The Strength and Stiffness of Polymers, *edited by Anagnostis E. Zachariades and Roger S. Porter*
5. Selecting Thermoplastics for Engineering Applications, *Charles P. MacDermott*
6. Engineering with Rigid PVC: Processability and Applications, *edited by I. Luis Gomez*
7. Computer-Aided Design of Polymers and Composites, *D. H. Kaelble*
8. Engineering Thermoplastics: Properties and Applications, *edited by James M. Margolis*
9. Structural Foam: A Purchasing and Design Guide, *Bruce C. Wendle*
10. Plastics in Architecture: A Guide to Acrylic and Polycarbonate, *Ralph Montella*
11. Metal-Filled Polymers: Properties and Applications, *edited by Swapan K. Bhattacharya*
12. Plastics Technology Handbook, *Manas Chanda and Salil K. Roy*
13. Reaction Injection Molding Machinery and Processes, *F. Melvin Sweeney*
14. Practical Thermoforming: Principles and Applications, *John Florian*
15. Injection and Compression Molding Fundamentals, *edited by Avraam I. Isayev*
16. Polymer Mixing and Extrusion Technology, *Nicholas P. Cheremisinoff*
17. High Modulus Polymers: Approaches to Design and Development, *edited by Anagnostis E. Zachariades and Roger S. Porter*
18. Corrosion-Resistant Plastic Composites in Chemical Plant Design, *John H. Mallinson*
19. Handbook of Elastomers: New Developments and Technology, *edited by Anil K. Bhowmick and Howard L. Stephens*
20. Rubber Compounding: Principles, Materials, and Techniques, *Fred W. Barlow*
21. Thermoplastic Polymer Additives: Theory and Practice, *edited by John T. Lutz, Jr.*
22. Emulsion Polymer Technology, *Robert D. Athey, Jr.*
23. Mixing in Polymer Processing, *edited by Chris Rauwendaal*
24. Handbook of Polymer Synthesis, Parts A and B, *edited by Hans R. Kricheldorf*

25. Computational Modeling of Polymers, *edited by Jozef Bicerano*
26. Plastics Technology Handbook: Second Edition, Revised and Expanded, *Manas Chanda and Salil K. Roy*
27. Prediction of Polymer Properties, *Jozef Bicerano*
28. Ferroelectric Polymers: Chemistry, Physics, and Applications, *edited by Hari Singh Nalwa*
29. Degradable Polymers, Recycling, and Plastics Waste Management, *edited by Ann-Christine Albertsson and Samuel J. Huang*
30. Polymer Toughening, *edited by Charles B. Arends*
31. Handbook of Applied Polymer Processing Technology, *edited by Nicholas P. Cheremisinoff and Paul N. Cheremisinoff*
32. Diffusion in Polymers, *edited by P. Neogi*
33. Polymer Devolatilization, *edited by Ramon J. Albalak*
34. Anionic Polymerization: Principles and Practical Applications, *Henry L. Hsieh and Roderic P. Quirk*
35. Cationic Polymerizations: Mechanisms, Synthesis, and Applications, *edited by Krzysztof Matyjaszewski*
36. Polyimides: Fundamentals and Applications, *edited by Malay K. Ghosh and K. L. Mittal*
37. Thermoplastic Melt Rheology and Processing, *A. V. Shenoy and D. R. Saini*
38. Prediction of Polymer Properties: Second Edition, Revised and Expanded, *Jozef Bicerano*
39. Practical Thermoforming: Principles and Applications, Second Edition, Revised and Expanded, *John Florian*
40. Macromolecular Design of Polymeric Materials, *edited by Koichi Hatada, Tatsuki Kitayama, and Otto Vogl*
41. Handbook of Thermoplastics, *edited by Olagoke Olabisi*
42. Selecting Thermoplastics for Engineering Applications: Second Edition, Revised and Expanded, *Charles P. MacDermott and Aroon V. Shenoy*
43. Metallized Plastics: Fundamentals and Applications, *edited by K. L. Mittal*
44. Oligomer Technology and Applications, *Constantin V. Uglea*
45. Electrical and Optical Polymer Systems: Fundamentals, Methods, and Applications, *edited by Donald L. Wise, Gary E. Wnek, Debra J. Trantolo, Thomas M. Cooper, and Joseph D. Gresser*
46. Structure and Properties of Multiphase Polymeric Materials, *edited by Takeo Araki, Qui Tran-Cong, and Mitsuhiro Shibayama*
47. Plastics Technology Handbook: Third Edition, Revised and Expanded, *Manas Chanda and Salil K. Roy*
48. Handbook of Radical Vinyl Polymerization, *Munmaya K. Mishra and Yusuf Yagci*
49. Photonic Polymer Systems: Fundamentals, Methods, and Applications, *edited by Donald L. Wise, Gary E. Wnek, Debra J. Trantolo, Thomas M. Cooper, and Joseph D. Gresser*
50. Handbook of Polymer Testing: Physical Methods, *edited by Roger Brown*
51. Handbook of Polypropylene and Polypropylene Composites, *edited by Harutun G. Karian*
52. Polymer Blends and Alloys, *edited by Gabriel O. Shonaike and George P. Simon*
53. Star and Hyperbranched Polymers, *edited by Munmaya K. Mishra and Shiro Kobayashi*
54. Practical Extrusion Blow Molding, *edited by Samuel L. Belcher*

55. Polymer Viscoelasticity: Stress and Strain in Practice, *Evaristo Riande, Ricardo Díaz-Calleja, Margarita G. Prolongo, Rosa M. Masegosa, and Catalina Salom*
56. Handbook of Polycarbonate Science and Technology, *edited by Donald G. LeGrand and John T. Bendler*
57. Handbook of Polyethylene: Structures, Properties, and Applications, *Andrew J. Peacock*
58. Polymer and Composite Rheology: Second Edition, Revised and Expanded, *Rakesh K. Gupta*
59. Handbook of Polyolefins: Second Edition, Revised and Expanded, *edited by Cornelia Vasile*
60. Polymer Modification: Principles, Techniques, and Applications, *edited by John J. Meister*
61. Handbook of Elastomers: Second Edition, Revised and Expanded, *edited by Anil K. Bhowmick and Howard L. Stephens*
62. Polymer Modifiers and Additives, *edited by John T. Lutz, Jr., and Richard F. Grossman*
63. Practical Injection Molding, *Bernie A. Olmsted and Martin E. Davis*
64. Thermosetting Polymers, *Jean-Pierre Pascault, Henry Sautereau, Jacques Verdu, and Roberto J. J. Williams*
65. Prediction of Polymer Properties: Third Edition, Revised and Expanded, *Jozef Bicerano*
66. Fundamentals of Polymer Engineering: Second Edition, Revised and Expanded, *Anil Kumar and Rakesh K. Gupta*
67. Handbook of Polypropylene and Polypropylene Composites: Second Edition, Revised and Expanded, *edited by Harutun G. Karian*
68. Handbook of Plastics Analysis, *edited by Hubert Lobo and Jose Bonilla*

**Additional Volumes in Preparation**

# HANDBOOK OF POLYPROPYLENE AND POLYPROPYLENE COMPOSITES

Second Edition, Revised and Expanded

edited by
## Harutun G. Karian
*RheTech, Inc.*
*Whitmore Lake, Michigan, U.S.A.*

MARCEL DEKKER, INC.　　　　NEW YORK · BASEL

**Library of Congress Cataloging-in-Publication Data**
A catalog record for this book is available from the Library of Congress.

**ISBN: 0-8247-4064-5**

This book is printed on acid-free paper.

**Headquarters**
Marcel Dekker, Inc.
270 Madison Avenue, New York, NY 10016
tel: 212-696-9000; fax: 212-685-4540

**Eastern Hemisphere Distribution**
Marcel Dekker AG
Hutgasse 4, Postfach 812, CH-4001 Basel, Switzerland
tel: 41-61-260-6300; fax: 41-61-260-6333

**World Wide Web**
http://www.dekker.com

The publisher offers discounts on this book when ordered in bulk quantities. For more information, write to Special Sales/Professional Marketing at the headquarters address above.

**Copyright © 2003 by Marcel Dekker, Inc. All Rights Reserved.**

Neither this book nor any part may be reproduced or transmitted in any form or by any means, electronic or mechanical, including photocopying, microfilming, and recording, or by any information storage and retrieval system, without permission in writing from the publisher.

Current printing (last digit):
10 9 8 7 6 5 4 3 2 1

**PRINTED IN THE UNITED STATES OF AMERICA**

# PREFACE

Since the publication of the first edition of this handbook in March 1999, there have been significant changes in the manufacture of polypropylene resin with the consolidation of many companies in order to better utilize raw material and technology resources: BP (BP + Amoco), Basell (BASF + Shell + Montell), ExxonMobil Chemical Company (Exxon + Mobil) and Chevron Phillips (Chevron + Phillips) Chemical Company. Most recently, Dow Chemical Company is striving to be a major producer of polypropylene resin. In addition, Dow has acquired Union Carbide business that includes impact modifier technology. Similarly, Crompton Corporation includes Uniroyal and Aristech for the manufacture of maleated polypropylene used as chemical coupling agent in glass fiber reinforced polypropylene.

With the rapid growth of TPO applications, the automotive industry is pushing for cost effective replacement of polycarbonate, ABS and PPO/PS into molded parts having molded-in-color and scratch-mar resistant characteristics. Likewise there are joint ventures in order to meet marketing demands, e.g. GM-Basell to develop nanocomposites in polypropylene resin.

Associated with increased interest in utilizing polypropylene technology, the first edition of the Handbook has been well received worldwide. Consequently, I have been asked by Russell Dekker to be Editor for a revised and expanded second edition. This request provides an opportunity to include more information concerning options to make polypropylene composites that better suit marketing requirements. A number of modifications have been made to several existing chapters (1, 3, 6, 7 and 8) of the first edition, along with the addition of six new chapters (15–20) to the second edition of the Handbook.

In Chapter 1, global trends of polypropylene usage are described in light of recent economic slow-down. The development of specialty products is one area of increased activity.

Chapter 3 includes new data regarding reinforcement of polypropylene using a new grade of OCF glass fiber. Wood filled highly crystalline polypropylene using chemically coupled polypropylene is described as having enhanced mechanical properties. A key addition to Chapter 3 summarizes recent development of nanocomposites using exfoliated clay treated with maleated polypropylene.

Chapter 6 has been rewritten to provide new insights into experimental techniques in impact testing to better characterize anticipated end-use impact behavior of polypropylene based materials.

Chapter 7 is updated to include new metallocene technology with the growth of TPO applications to meet increasing end product demands.

Chapter 8 has been expanded to include recent developments in surface modification of talc to replace PVC and engineering thermoplastics by talc filled materials with molded in color. The section on surface modifiers for talc filler includes description of a new grade of talc called R-Talc that improves scratch resistance and impact properties of TPO composites. Zero Force technology is cited as a recent break-through in talc manufacture to effectively compact fine grades of talc into easy to feed granules for enhanced processability via compounding extrusion.

Chapter 15 describes recent advancements in surface treatment of mica to suit hybridization with glass fiber reinforcement. This combination yields composites with enhanced mechanical properties and minimum warpage of mold parts.

Chapter 16 provides up-to-date technology of high purity submicrometer talc filler with lamellar microstructure. By compacting the fine talc grade into densified granules for ease of processability, enhancement of mechanical properties are practically attained for polypropylene composites.

Chapter 17 describes automotive applications for polypropylene and polypropylene composites with the utilization of new technology, e.g. exfoliated clay nanocomposites.

*Preface*

The utilization of wollastonite fibers to reinforce polypropylene is given in Chapter 18. This avenue to interphase design has particular merit; since it features inherent mar-scratch resistance combined with mechanical properties attributed to high aspect ratio fibers.

Chapter 19 provides fundamental description of mold shrinkage behavior for polypropylene composites. Shrinkage is described as a combined function of material characteristics, process conditions for injection molding, and mold design parameters. Hence, notions of single valued shrinkage are replaced by a range of values depending on the operating window for a given molded product design.

Finally, Chapter 20 provides an update on developments of nanocomposite concentrates to enhance the compounding of materials with enhanced mechanical and thermal properties.

*Harutun G. Karian*

# CONTENTS

*Preface*                                                                                           *iii*

1. **Global Trends for Polypropylene**                                                              1
   *Michael J. Balow*

2. **Polypropylene: Structure, Properties, Manufacturing Processes, and Applications**              11
   *William J. Kissel, James H. Han, and Jeffrey A. Meyer*

3. **Chemical Coupling Agents for Filled and Grafted Polypropylene Composites**                     35
   *Darilyn Roberts and Robert C. Constable*

4. **Stabilization of Flame-Retarded Polypropylene**                                                81
   *Robert E. Lee, Donald Hallenbeck, and Jane Likens*

5. **Recycling of Polypropylene and Its Blends: Economic and Technology Aspects**                   113
   *Akin A. Adewole and Michael D. Wolkowicz*

6. **Impact Behavior of Polypropylene, Its Blends and Composites** 155
   *Josef Jancar*

7. **Metallocene Plastomers as Polypropylene Impact Modifiers** 221
   *Thomas C. Yu and Donald K. Metzler*

8. **Talc in Polypropylene** 281
   *Richard J. Clark and William P. Steen*

9. **Glass Fiber-Reinforced Polypropylene** 311
   *Philip F. Chu*

10. **Functionalization and Compounding of Polypropylene Using Twin-Screw Extruders** 383
    *Thomas F. Bash and Harutun G. Karian*

11. **Engineered Interphases in Polypropylene Composites** 413
    *Josef Jancar*

12. **Mega-Coupled Polypropylene Composites of Glass Fibers** 465
    *Harutun G. Karian*

13. **Characterization of Long-Term Creep-Fatigue Behavior for Glass Fiber-Reinforced Polypropylene** 517
    *Les E. Campbell*

14. **Mica Reinforcement of Polypropylene** 543
    *Levy A. Canova*

15. **Use of Coupled Mica Systems to Enhance Properties of Polypropylene Composites** 593
    *Joseph Antonacci*

16. **Performance of Lamellar High-Purity Submicrometer and Compacted Talc Products in Polypropylene Compounds** 617
    *Wilhelm Schober and Giovanni Canalini*

17. **Automotive Applications for Polypropylene and Polypropylene Composites** 641
    *Brett Flowers*

| 18. | **Wollastonite-Reinforced Polypropylene**<br>*Roland Beck, Dick Columbo, and Gary Phillips* | 651 |
|---|---|---|
| 19. | **Part Shrinkage Behavior of Polypropylene Resins and Polypropylene Composites**<br>*Harutun G. Karian* | 675 |
| 20. | **Polypropylene Nanocomposite**<br>*Guoqiang Qian and Tie Lan* | 707 |

*Index*     *729*

# CONTRIBUTORS

*Akin A. Adewole*   Basell USA, Inc., Elkton, Maryland, U.S.A.

**Joseph Antonacci**   Suzorite Mica Products Inc., Boucherville, Quebec, Canada

**Michael J. Balow**   Basell Polyolefins USA, Inc., Lansing, Michigan, U.S.A.

**Thomas F. Bash**   Ametek Westchester Plastics, Nesquehoning, Pennsylvania, U.S.A.

**Roland Beck**   Nyco Sales, Calgary, Alberta, Canada

**Les E. Campbell**   Owens Corning Fiberglas, Anderson, South Carolina, U.S.A.

**Giovanni Canalini**   Superlab S.r.l., Italy

**Levy A. Canova**   Franklin Industrial Minerals, Kings Mountain, North Carolina, U.S.A.

**Philip F. Chu**   Saint-Gobain Vetrotex America, Wichita Falls, Texas, U.S.A.

**Richard J. Clark**   Luzenac America, Englewood, Colorado, U.S.A.

**Dick Columbo**   Nyco Sales, Calgary, Alberta, Canada

**Robert C. Constable**   BRG Townsend, Mt. Olive, New Jersey, U.S.A.

**Brett Flowers**   General Motors Corporation, Pontiac, Michigan, U.S.A.

**Donald Hallenbeck**   Great Lakes Chemical Corporation, West Lafayette, Indiana, U.S.A.

**James H. Han**   BP Amoco Polymers, Inc., Alpharetta, Georgia, U.S.A.

**Josef Jancar**   Technical University Brno, Purkynova, Brno, Czech Republic

**Harutun G. Karian**   RheTech, Inc., Whitmore Lake, Michigan, U.S.A.

**William J. Kissel**   BP Amoco Polymers, Inc., Alpharetta, Georgia, U.S.A.

**Tie Lan**   Nanocor, Inc., Arlington Heights, Illinois, U.S.A.

**Robert E. Lee**   Great Lakes Chemical Corporation, West Lafayette, Indiana, U.S.A.

**Jane Likens**   Great Lakes Chemical Corporation, West Lafayette, Indiana, U.S.A.

**Donald K. Metzler**   ExxonMobil Chemical Company, Houston, Texas, U.S.A.

**Jeffrey A. Meyer**   BP Amoco Corporation, Naperville, Illinois, U.S.A.

**Gary Philips**   Nyco Sales, Calgary, Alberta, Canada

**Guoqiang Qian**   Nanocor, Inc., Arlington Heights, Illinois, U.S.A.

**Darilyn Roberts**   Crompton Corporation, Middlebury, Connecticut, U.S.A.

**Wilhelm Schober**   HiTalc Marketing and Technology GmbH, Schoconsult, GmbH, Austria

**William P. Steen**   Luzenac America, Englewood, Colorado, U.S.A.

**Michael D. Wolkowicz**   Basell USA, Inc., Elkton, Maryland, U.S.A.

**Thomas C. Yu**   ExxonMobil Chemical Company, Baytown, Texas, U.S.A.

# 1

# Global Trends for Polypropylene

**Michael J. Balow**
Basell Polyolefins USA, Inc., Lansing, Michigan, U.S.A.

## 1.1 INTRODUCTION

Polypropylene (PP) underwent phenomenal growth in production and use throughout the world during the latter half of the 20th century. From the early 1960s until the oil crisis of the early 1970s, the growth rate was nearly 25% annually. During the period from about 1974 through 1999, the rate of consumption increased between 7% and 12% annually. This high growth rate of PP consumption has required that production capacity keep up with the growing demand. Though installed-capacity utilization has typically ranged from approximately 85% to nearly 98% during peak demand periods, for the most part, supply and demand have remained in balance. However during the latter half of the 1990s, gross national product (GNP) was increasing at a record pace and many companies invested in new polyolefin capacity. With the average annual growth rate (AAGR) approaching 12% in 1999, there was continued expectation of high demand. However, a slowdown in the economy began during the latter half of 1999. This slowdown, coupled with a significant increase in new plants and debottlenecking of existing plants, created a very significant supply imbalance position.

A typical world scale polyolefins polymerization plant has a life expectancy of more than 25 years. The initial investment in a PP plant is about $150 MM, and operational efficiencies are typically only achieved when operating at >85% of nameplate capacity. This situation has led to a serious structural problem for the industry.

Changes in the demand patterns have had an impact on the industry. Growth of any commodity should ideally be based on demand factors. Polypropylene is used in a great number of end-use applications. Due to the relatively low investment for manufacturing processes that utilize PP, the consumers of PP can often move quickly into areas where labor costs are low. In addition, transportation costs for the polymer between the production site and the consumer can be a significant contributor to the cost of the polymer. These factors, combined with the increased consumption of gasoline in developed regions, have led to monomer being readily available in large quantities at refiners in developed countries. However, some of the largest growth rate for PP is in less developed areas.

Today PP still accounts for the largest consumption of propylene monomer, followed by acrylonitrile, oxo chemicals, and propylene oxide. Given the choices of downstream derivatives for propylene, many producers have considered PP as one of the easiest means for dealing with excess supply of the monomer, especially in the case of refinery operations. Many of the major producers of polypropylene are also refineries, at least in North America and Europe. Without adequate gas pipeline infrastructure found predominately in the gulf coast of North America and the Western coast of western Europe, it is not very practical to transport the monomer large distances by overland or sea. It is therefore likely that into the 21st century we will see most capacity increases for PP move either to refinery locations in developed countries or to locations much closer to the source of the oil (especially the Middle East). Another complicating factor is tariff protection for local PP production. Many regions are concerned about protecting their investments in developing countries. Often tariffs do not apply to the articles made from PP.

These combinations of factors have led to a very serious and unprecedented problem of overcapacity as we enter the 21st century. Although we have noted a reduction in the new plant announcements in 2001 and 2002, it will take a considerable period of time, perhaps as long as 2004, until the supply–demand balance is corrected.

Another factor for concern, as mentioned above, is that the production of commodity plastic parts is largely moving to third world countries. The most rapid growth rate for polypropylene consumption is Africa and the Middle East, Asia Pacific, and Eastern Europe. These regions are all increasing consumption rates faster than Japan or Western Europe or North America where the major production is located.

# Global Trends for Polypropylene

Intermaterial competition versus other commodity plastics is also working strongly in favor of selecting PP. The other major competitors in the commodity market are high-density polyethylene (HDPE), polystyrene (PS), and polyvinyl chloride resins. These are all less favorable on a cost per volume basis. Experts predict that the growth rates of PP could be as high as 8.3% annually. This growth will be spurred by continued expectations of relatively low cost, improvements in performance due to new catalyst introductions, more efficient catalysts, and strong growth in developing countries based on the Western world average per capita consumption.

## 1.2 REGIONAL ISSUES

There are many economic issues that influence regional consumption (Fig. 1.1) and growth rates (Fig. 1.2). The various contributing factors are given below for each geographical region.

With a move to a single currency, Western European producers will have to strive to remain price competitive. Lower labor costs in Eastern Europe will increasingly attract conversion operations. Eventually the Middle East will put significant pressure on this region. Today we see industry consolidation taking

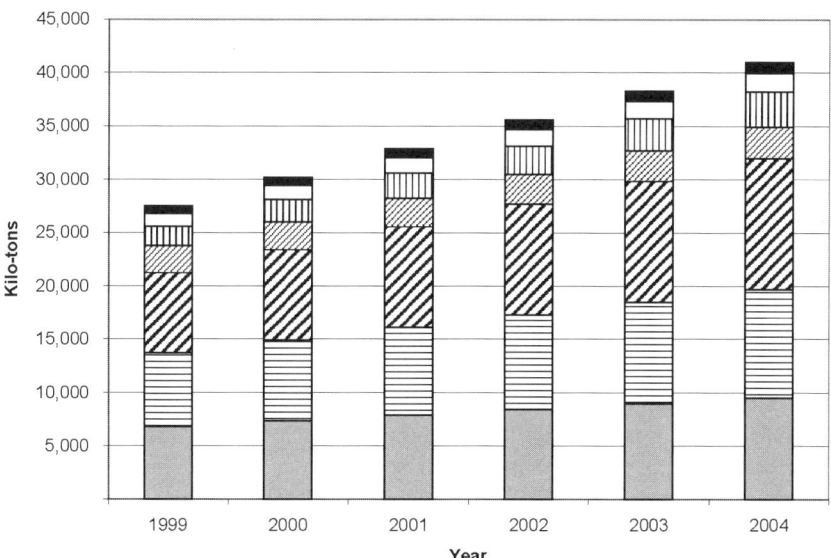

FIGURE 1.1 Projected polypropylene consumption by region. ■, Eastern Europe; □, South America; ▥, Africa and Middle East; ▨, Japan; ▰, Asia Pacific; ☰ North America; ▪, Western Europe.

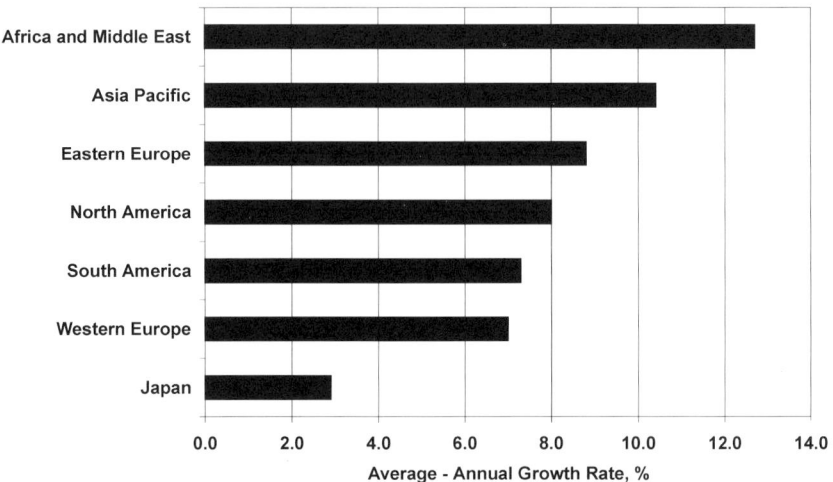

FIGURE 1.2  Regional growth rates for years 1999–2004.

place between Middle Eastern and European producers. Improvements in catalysis will lead to innovations that will further broaden PP competitiveness. In this area as well, PP capacity will outpace consumption. Exports back to the Middle East and Africa will help support this overproduction. Polypropylene consumption and trends in recycling in Western Europe are being closely followed.

In North America, PP producers are suffering from low margins. Leading producers have strived to reduce costs to cope with this trend, but industry consolidation will continue. From a market perspective, PP still has significant inroads left in the automotive industry, especially in interiors. The North American Free Trade Agreement (NAFTA) has removed the borders between Canada, Mexico, and United States. As this occurs, more production is shifting to Mexico, and eventually PP production is likely to increase there as transportation costs increase.

China recently announced its intention to increase its PP production capacity. Today it is the single largest importing nation. Still the per capita use is far below that of the rest of the world, so we can anticipate China will be a net importer for some time. The inhibiting factor to this occurring is a need to develop a more effective transportation infrastructure. Transportation of goods into and out of China's interior is expensive. China already has approximately 65 producers. However, many are small and older plants. They use the propylene feedstock from local refineries that have very limited alternative uses for the monomer.

The economy in the Asia Pacific region has just been through a very difficult time. This has resulted in industry rationalization and joint venture arrangements with Western producers. Consequently, PP manufacture has had slower than expected growth. China's import needs are the single largest driving force for newer capacity in the region. Therefore, the stability and continued growth of the Chinese market is the key to PP growth in the Asia Pacific region. India's growth in local production is allowing increased conversion industry to take place there. The ASEAN free trade area (Malaysia, Indonesia, Singapore, Vietnam, Thailand) represents a big step in creating a common market for the six member countries. Interregional trade of PP and it converted products will increase at the expense of import from other regions.

Eastern and Central European plants are being designed for a planned recovery of the economies. Historically, they have not been competitive on a world market scale. However, often growth is inhibited by lack of hard currency needed for these investments. In addition, feedstock alignment is difficult. Low labor rates are attractive but political instability of the region keeps the risk high.

Steady growth will continue in South America and Mercosur market countries where tariffs are imposed to protect local markets. Mercosur is the Common Market of the South, which includes Argentina, Brazil, Paraguay, and Uruguay. These tariff measures have resulted in higher prices and limited growth. Investments in these regions are primarily made to upgrade quality. Local economy is more or less sized to meet the local consumer needs. Demand is focused on meeting packaging and household demands. Free trade agreements as in Mercosur have created a greater homogeneous market size and make investment in the region more attractive. Currency stability in this region will continue to be an important issue.

## 1.3 GROWTH IN CONSUMPTION BY APPLICATION TYPES

Figure 1.3 depicts PP consumption of end-use products manufactured by a variety of conversion processes. Fibers are expected to grow at or near the average with significant gains in spun-bonded fibers. Injection molding will continue to grow as PP displaces other materials. Sheet application will grow in two predominate areas. Soft PP will grow to displace some rubber applications, and rigid and transparent PP will penetrate PS in the extrusion/thermoforming area. Lastly, PP will continue to grow in some regions at the expense of engineering resins such as PC/ABS, nylon, or modified polyphenylene oxide (PPO), polyurethane in application such as wheel covers, instrument panels, body side molding, and headlines.

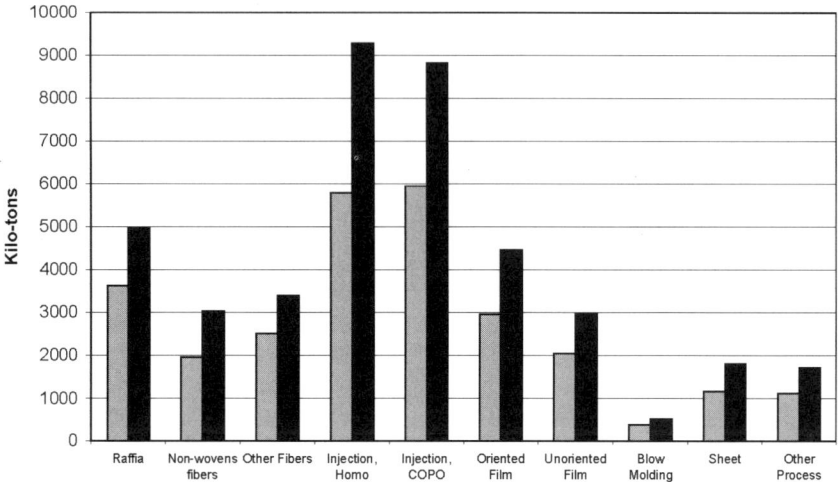

**FIGURE 1.3** Polypropylene consumption by conversion process. ■, Year 2000; ☐, Year 2005.

## 1.4 SUPPLIERS AND CONSOLIDATION

Since the mid-1990s, we have seen significant consolidation of the PP industry. This trend has been true in all major producing regions of Japan, Western Europe, and North America. This consolidation has been led by the interests of being a global supplier, interests in having critical mass for a sizable business, lower margins, increased customer service expectation as PP moves into more nontraditional markets, and a strong interest in having reliable outlets for the propylene monomer. Often key synergies have been created by these consolidations. In addition, as margins decrease, older plants can no longer remain cost competitive. Even high-efficiency catalyst processes that have been installed in just 20 plants do not have broad product capabilities (e.g., can only produce homopolymer). These upgraded plants can be obsolete, even before their expected lifetime runs out.

Homopolymer and copolymer production for different geographical locations are given in Fig. 1.4. The location of production facilities has become an important competitive advantage for shipping finished product or procuring monomer. Also locations near deep-sea ports can have a significant advantage to export product, especially in a local economic slowdown.

## 1.5 RECYCLING

In the face of huge capacity increases another issue is starting to take shape. Reuse of PP is also on the increase. Some recycle streams of PP are now fully

# Global Trends for Polypropylene

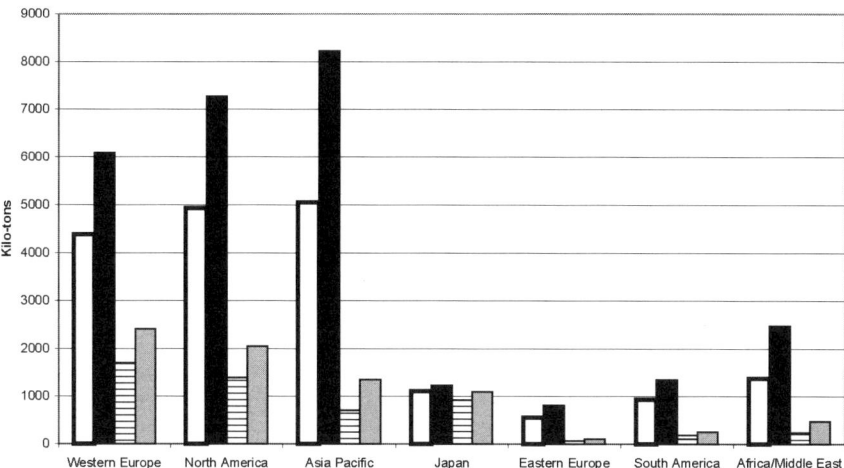

**FIGURE 1.4** Homopolymer and copolymer production by region in years 2000 and 2005. ☐, Homopolymer 2000; ≣, copolymer 2000; ■, homopolymer 2005; ▨, copolymer 2005.

integrated into the supply chain. These include postconsumer battery recycling and packaging, postindustrial fiber reclaim, and defective painted thermoplastic olefins from molding/painting operations. This trend will most likely significantly increase as the end-of-life vehicle requirements, now in place for European car producers, come into full reality. The recycling of PP tends to be very difficult in other applications; since the material is not often used without some kind of secondary operation. These include painting in automotive, metallization or multilayer structures in packaging, and, most frequently, pigmentation that limits secondary uses often to black colors.

Polypropylene itself remains a very viable fuel for burning when recycled resin is depleted of its usefulness due to ultimate degradation resulting from multiple extrusion passes. However, these same issues mentioned above are also complications for burning. Problems associated with contaminants in the plastic can be observed in some countries that have tended to use mixed reclaimed plastics for fuel in cement kilns, e.g., particularly in Asia.

## 1.6 SPECIALTY POLYPROPYLENE

As the PP market continues to suffer from poor return on investment for producers, most producers are seriously considering conversion to specialty products. In Europe and Asia, most of the specialty products are produced by the manufacturers of the PP itself. There seems to have been a different strategy in North America. In that region, most of the specialty PP products have been produced by compounders and distributors. While this trend tends to be effective in regional markets, the cost to users of these products on a global basis can go up. In addition, as these products are used in more specialty applications, regional compounding companies frequently lack the capability to support market demands globally. Furthermore, some of the specialty items have reached sufficient volume to allow alternative processes to be more cost-effective in order to reduce manufacturing costs. However, this trend could change as pressures to recycle postconsumer PP back into main market stream continues to grow.

Some of the specialty products include filled and reinforced PP with new, higher performance additives such as nanocomposites. Postreactor modifications, such as grafting with functional monomers, has reached significant magnitude that the bulk polymer properties in the polymer matrix are being changed. This enhancement may surpass simple adhesive characteristics between polymer matrix and filler–glass fiber reinforcement that are attributed to functional groups like maleic anhydride and acrylic acid, described in Chapter 3 of this book. Also irradiation techniques used to branch PP are widely known; however, investment costs are very significant.

Another driving force for PP in specialty markets is the desire to have all one polymer type present in an article. For example, the total recycling of automotive parts can avert the need for land fills, which seriously impact the environment. This has the significant benefit of allowing easier reuse at the end of life. Increase in the use of specialty PP material is extending beyond automotive applications. This broad-based market growth of specialty products can be seen in Europe where new PP wire coating applications have originated outside the automotive industry.

## 1.7 CONCLUSIONS

The challenges that lie ahead for the PP industry will be significant. However, products based on this polymer resin, borne out of a relatively simply, widely available monomer, have proven to be very versatile in all-around performance. Therefore, we can expect that PP resins will continue to be one of the major choices of raw materials of construction for mankind well into the future.

## REFERENCES

1. J Hoffman. Polypropylene beginning to show signs of strength. Chemical Market Reporter, March 11, 2002, p. 8.
2. R Brown. Polypropylene producers seek 5-cent increase. Chemical Market Reporter, January 28, 2002, p. 6.

3. J Hoffman. Petrochemicals outlook cloudy for the near term. Chemical Market Reporter, January 7, 2002, p. 6.
4. F Esposito. PS, PP prices manage to rise since Feb. 1. Plastics News, March 18, 2002, p. 4.
5. R Brown. Commodity resins face uncertainty. Chemical Market Reporter, January 7, 2002, p. 4.
6. AMI. European plastics demand keeps on rising. European Plastics Industry Report, June 20, 2000.
7. C Platz. Opportunities and challenges in polypropylene. Polyolefins International Conference, Houston, February 26, 2001.
8. Basell. Polyolefins Design Manual, October 2001.
9. Polypropylene Annual Report 2000, Phillip Townsend and Associates.
10. Phillip Townsend and Associates. Global polyolefins consumption better than expected. The Townsend Profile, Vol. 56, June 2001.

# 2

# Polypropylene: Structure, Properties, Manufacturing Processes, and Applications

**William J. Kissel and James H. Han**
BP Amoco Polymers, Inc., Alpharetta, Georgia, U.S.A.

**Jeffrey A. Meyer**
BP Amoco Corporation, Naperville, Illinois, U.S.A.

## 2.1 TYPES OF POLYPROPYLENE

Polypropylene (PP) is a thermoplastic material that is produced by polymerizing propylene molecules, which are the monomer units, into very long polymer molecule or chains. There are a number of different ways to link the monomers together, but PP as a commercially used material in its most widely used form is made with catalysts that produce crystallizable polymer chains. These give rise to a product that is a semicrystalline solid with good physical, mechanical, and thermal properties. Another form of PP, produced in much lower volumes as a byproduct of semicrystalline PP production and having very poor mechanical and thermal properties, is a soft, tacky material used in adhesives, sealants, and caulk products. The above two products are often referred to as "isotactic" (crystallizable) PP (i-PP) and "atactic" (noncrystallizable) PP (a-PP), respectively.

As is typical with most thermoplastic materials, the main properties of PP in the melt state are derived from the average length of the polymer chains and the breadth of the distribution of the polymer chain lengths in a given product. In the solid state, the main properties of the PP material reflect the type and amount of crystalline and amorphous regions formed from the polymer chains.

Semicrystalline PP is a thermoplastic material containing both crystalline and amorphous phases. The relative amount of each phase depends on structural and stereochemical characteristics of the polymer chains and the conditions under which the resin is converted to final products such as fibers, films, and various other geometric shapes during fabrication by extrusion, thermoforming, or molding.

Polypropylene has excellent and desirable physical, mechanical, and thermal properties when used in room-temperature applications. It is relatively stiff and has a high melting point, low density, and relatively good resistance to impact. These properties can be varied in a relatively simple manner by altering the chain regularity (tacticity) content and distribution, the average chain lengths, the incorporation of a comonomer such as ethylene into the polymer chains, and the incorporation of an impact modifier into the resin formulation.

The following notation is used in this chapter. Polypropylene containing only propylene monomer in the semicrystalline solid form is referred to as homopolymer PP (HPP), and we use this to mean the i-PP form. Polypropylene containing ethylene as a comonomer in the PP chains at levels in about the 1–8% range is referred to as random copolymer (RCP). HPP containing a commixed RCP phase that has an ethylene content of 45–65% is referred to as an impact copolymer (ICP). Each of these product types is described below in more detail.

### 2.1.1 Homopolymer

Homopolymer PP is the most widely used polypropylene material in the HPP, RCP, and ICP family of products. It is made in several different reactor design using catalysts that link the monomers together in a stereospecific manner, resulting in polymer chains that are crystallizable. Whether they crystallize and to what extent depends on the conditions under which the entangled mass of polymer chains transitions from the melt to the solid state or how a heat-softened solid PP material is strained during a further fabrication procedure like fiber drawing.

Homopolymer PP is a two-phase system because it contains both crystalline and noncrystalline regions. The noncrystalline, or amorphous, regions are composed of both isotactic PP and atactic PP. The isotactic PP in the amorphous regions is crystallizable, and it will crystallize slowly over time up to the limit that entanglement will allow. The extent of crystallization after the initial fabrication step of converting PP pellets or powder to a molded article will slowly increase over time, as will the stiffness. A widely accepted model of HPP morphology likens the solid structure to a system consisting of pieces of stiff cardboard linked together by strands of softer material. In the areas represented by flat pieces of cardboard, PP polymer chains weave up and down into close-packed arrays called crystallites ("little crystals"), which are called lamella by morphologists. The soft strands linking the pieces of stiff cardboard are polymer chains that exit one crystallite, enter another, and then begin weaving up and down in another crystallite. The crystallizability of the chains is one factor that determines how thick the crystallites will be, and the thickness of the crystallites determines how much heat energy is required to melt them (the melting temperature). A typical HPP has an array of crystallites from thick ones to very thin ones, and these manifest themselves as an array of melting points.

Homopolymer PP is marketed mainly by melt flow rate (MFR) and additive formulation into fiber, film, sheet, and injection molding applications. Melt flow rate is an indicator of the weight-average molecular weight as measured by the ASTM or ISO MFR test method.

### 2.1.2 Random Copolymer

Random copolymers are ethylene/propylene copolymers that are made in a single reactor by copolymerizing propylene and small amounts of ethylene (usually 7% and lower). The copolymerized ethylene changes the properties of the polymer chains significantly and results in thermoplastic products that are sold into markets in which slightly better impact properties, improved clarity, decreased haze, decreased melting point, or enhanced flexibility are required. The ethylene monomer in the PP chain manifests itself as a defect in the chain regularity, thus inhibiting the chain's crystallizability. As the ethylene content increases, the crystallite thickness gradually decreases, and this manifests itself in a lower melting point. The amount of ethylene incorporated into the chain is usually dictated by the balance between thermal, optical, and mechanical properties.

### 2.1.3 Impact Copolymers

Impact copolymers are physical mixtures of HPP and RCP, with the overall mixture having ethylene contents on the order of (6–15% wt%. These are sold into markets where enhanced impact resistance is needed at low temperatures, especially freezer temperature and below.

The RCP part of the mixture is designed to have ethylene contents on the order of 40–65% ethylene and is termed the rubber phase. The rubber phase can be mechanically blended into the ICP by mixing rubber and HPP in an extruder, or it can be polymerized in situ in a two-reactor system. The HPP is made in the first reactor and the HPP with active catalyst still in it is conveyed to a second reactor where a mixture of ethylene and propylene monomer is polymerized in the voids and interstices of the HPP polymer powder particle. The amount of rubber phase that is blended into the HPP by mechanical or reactor methods is determined by the level of impact resistance needed. The impact resistance of the ICP product is determined not only by its rubber content but also by the size, shape, and distribution of the rubber particles throughout the ICP product. Reactor products usually give better impact resistance at a given rubber level for this reason.

As the rubber content of the ICP product is increased, so is the impact resistance, but this is at the expense of the stiffness (flexural modulus) of the product. Consequently, polymer scientists often describe a product as having a certain impact–stiffness balance. The stiffness of the ICP product is dictated by the stiffness of the HPP phase and the volume of rubber at a given rubber size distribution in the product. The impact resistance is dictated by the amount and distribution of the rubber phase in the ICP product.

## 2.2 TACTICITY

The solid-state characteristics of PP occur because the propylene monomer is asymmetrical in shape. It differs from the ethylene monomer in that it has a methyl group attached to one of the olefinic carbons. This asymmetrical nature of the propylene monomer thus creates several possibilities for linking them together into polymer chains that are not possible with the symmetrical ethylene monomer, and gives rise to what are known as structural isomers and stereochemical isomers in the PP chain.

In structural isomerism, polymer scientists refer to the olefinic carbon with the methyl group on it as the "head" (h) and the other olefinic carbon as the "tail" (t) of the monomer. The most common method of polymerization uses catalysts that link the monomers together in the "head-to-tail" fashion. Occasionally there is a "mistake" made and the monomers form a "head-to-head" or a "tail-to-tail" linkage, but these tend to be rare.

Stereochemical isomerism is possible in PP because propylene monomers can link together such that the methyl groups can be situated in one spatial arrangement or another in the polymer. If the methyl groups are all on one side of the chain, they are referred to as being in the "isotactic" arrangement, and if they are on alternate sides of the chain, they are referred to as being in the "syndiotactic" arrangement. Each chain has a regular and repeating symmetrical arrangement of methyl groups that form different unit cell crystal types in the solid state. A random arrangement of methyl groups along the chain provides little or no symmetry, and a polymer with this type of arrangement is known as "atactic" polypropylene.

When polymer scientists discuss the stereochemical features of PP, they usually discuss it in terms of "tacticity" or "percent tacticity" of polypropylene, and in the marketplace the term "polypropylene" is generally used to refer to a material that has high tacticity, meaning high isotactic content. The high-tacticity PP materials have desirable physical, mechanical and thermal properties in the solid state. Atactic material is a soft, sticky, gummy material that is mainly used in sealants, caulks, and other applications where stickiness is desirable. Syndiotactic PP, not a large-volume commercial material, is far less crystalline than isotactic PP.

## 2.3 MOLECULAR WEIGHT AND MOLECULAR WEIGHT DISTRIBUTION

Unlike pure simple compounds, whose molecules are all of the same molecular weight, polymer samples consist of molecules of different molecular weights. This is a reflection of the fact that a polymer sample is a collection of molecules of differing chain lengths. Therefore, an average molecular weight concept was adopted for polymers. No single average, however, can completely describe a polymer sample, and a number of different averages are used. Ratios of some of these averages can be used to calculate a molecular weight distribution (MWD), which describes the breadth of the molecular weights represented.

Molecular weight averages of PP are measured by the technique of gel permeation chromatography (GPC), a chromatographic technique that sorts out the polymer chains by chain length after the PP is dissolved in a solvent. When dissolved, the PP is no longer a thermoplastic but instead is a bunch of long molecules dispersed in a solvent. From the GPC data, one can calculate the number-average ($M_n$), weight-average ($M_w$), and z-average ($M_z$) molecular weights. In PP, the $M_n$ relates to physical properties of the solid, the $M_w$ relates to viscosity properties of the melt, and the $M_z$ to elastic properties of the melt. Because the GPC chromatogram contains a lot of data that are not easy to tabulate and communicate, it is convenient to use a ratio, especially $M_w/M_n$, because it gives a good estimate of the MWD and is a simple number to tabulate and store. It is a good estimate because the $M_n$ is very sensitive to short chains and the $M_w$ is very sensitive to long chains in the products.

The GPC equipment is fairly expensive and prone to failure, and the actual experiment is slow, labor intensive, and requires dissolution of the PP at high temperatures in solvents like xylene and trichlorobenzene. Thus, other methods to estimate the molecular weight have been developed. The most popular one is termed the MFR test, and it gives a number that is easily correlatable to the $M_w$ average. Most HPP products are sold with MFR numbers ranging from 0.2 to 45, and these correspond to $M_w$ values from 1 million down to 100,000. Note that the molecular weight averages are inversely proportional to MFR numbers.

## 2.4 MECHANICAL PROPERTIES

The mechanical properties of most interest to the PP product design engineer are its stiffness, strength, and impact resistance. Stiffness is measured as the flexural modulus, determined in a flexural test, and impact resistance by a number of different impact tests, with the historical favorite being the Izod impact at ambient and at subambient temperatures. These mechanical properties are mostly used to predict the properties of molded articles. Strength is usually defined by the stress at the yield point rather than by the strength at break, but breaking strength is usually specified for fiber or film materials under tensile stress.

To understand the use and comparison of mechanical property data, one must remember that mechanical properties are not measured on the resins themselves but instead on specimens fabricated from the resin, and it is from the physics governing the fabrication and mechanical testing procedures that the mechanical properties are derived. Because there are so many variables that can affect mechanical properties, consensus testing organizations like ASTM and ISO were formed to bring some uniformity and consistency to specimen preparation and mechanical testing. Because the ASTM and ISO fabrication and testing methods allow some freedom within their guidelines, when one is asked what the mechanical properties of a material are, the first answer should be to ask by what tests, what specimens, and under what conditions. The latter includes such factors as the exact specimen type, age of specimen, how the specimen was conditioned, testing speed, testing temperature, data acquisition procedure, and method of calculation.

Flexural modulus or stiffness typically increases as the level of crystallinity increases in a PP product, but it also depends on the type of crystal morphology. Thus, stiffness generally decreases as the crystallizability (tacticity) decreases or, in random copolymers, as the ethylene content increases because this tends to decrease crystallizability.

## 2.5 RHEOLOGY

Rheology is the science that studies the deformation and flow of matter, and in PP there is interest in both viscosity and elasticity of the melt state and the solid state. The rheological properties of PP are important because of the broad range of processing techniques to which PP is subjected, including fiber and film extrusion, thermoforming, and injection molding. The viscosity of PP is of most importance in the melt state because it relates to how easily a PP product can be extruded or injection molded. In fiber extrusion, melt elasticity is important to processability of a PP product because it relates to how easily a material can be drawn into a fiber. In contrast to PP, most engineering resins are used mainly in injection molding operations.

The viscosity of a PP product is related to its $M_w$, and a good estimation of it at low shear rates can be obtained from the MFR test. This is only a single point test, and more information about the viscosity at different strain rates is needed to completely understand and characterize the processability of a product. The strain rate dependence of melt viscosity in PP is related to its molecular weight distribution, which is commonly described by the ratio of the $M_w$ to $M_n$ averages. As the MWD of PP gets broader, it shear thins (becomes less viscous) more than a narrower MWD PP at the same strain rate.

As indicated above, the rheological properties in the melt are related to the MWD. In PP. these are controlled mainly by the process used, although with Ziegler-Natta catalysts there is a small effect due the catalyst. Typical MWDs are in the 5–6 range. The MWD can be made more narrow by using postreactor polymer chain shortening. This may be accomplished by adding a peroxide in the extrusion compounding manufacturing step, in which stabilizers and other additives are normally incorporated into the PP reactor product before pelletization. These controlled rheology (CR) resins have higher MFR and reduced MWD than the unmodified reactor product. In the CR process, also known as visbreaking (for viscosity breaking), the longer higher molecular weight molecules are preferentially (statistics) broken.

The MWD can be made broader by using a two-reactor configuration that produces different melt flow rates in each reactor. Recently, metallocene PP catalysts have shown the ability to produce PPs with very narrow molecular weight distributions, on the order of 2–3. These resins have a great deal of value in fiber extrusion applications where less shear sensitivity of the viscosity is important.

## 2.6 MORPHOLOGY

Homopolymer PP exists as a two- and possibly a three-phase system of crystalline and amorphous phases with the amorphous phase comprising a crystallizable isotactic portion and a noncrystallizable atactic portion. The noncrystallizable, gummy, atactic PP phase has small amounts of a low molecular weight oily material at a level of 1% and lower. The latter has been characterized in some products as having some structural inversions of propylene monomers and some branches other than methyl. Typical levels of crystallinity in extruded PP pellets are in the 60–70% range. One way to describe the morphology of PP is to consider it an assemblage of crystallites that act as physical cross-links in an amorphous matrix.

In the crystalline phase, the alpha or monoclinic phase is the dominant crystal form of PP with a melting point of about 160°C. The beta or hexagonal phase is less common and less stable. The latter has a melting point of about

145°C. Typical levels of beta crystallites are less than 5% in injection-molded parts.

## 2.7 THERMAL ANALYSIS

A number of techniques fall under the thermal analysis heading. For PP characterization, one of the most useful is differential scanning calorimetry (DSC). A technique giving essentially the same information, although data are developed based on a somewhat different principle, is differential thermal analysis (DTA). In DSC, thermal transitions are recorded as a function of temperature, which is either increased or decreased at a defined heating or cooling rate.

Some of the useful information derived from DSC heating scans includes the melting temperature, which is taken as the maximum of the endothermic peak, and the heat of fusion, determined by integrating the area under the endothermic peak. The melting temperature of PP homopolymer is about 160°C, whereas that of usual PP random copolymers is about 145°C. Polypropylene impact copolymers exhibit the same melting temperatures as homopolymers, the rubber constituent not affecting the melting temperature. Impact copolymers do, however, have lower heats of fusion than homopolymers because the heat of fusion is related to the proportion of crystalline polymer present. The rubber portion is essentially noncrystalline and therefore does not melt.

In the DSC cooling of PP from the melt, crystallization occurs. The minimum of the exothermic peak defines the crystallization temperature. This temperature is an indication of how rapidly the PP crystallizes. The higher the temperature, the more rapid the crystallization. Nucleating agents added to PP increase the crystallization rate of PP, resulting in a higher crystallization temperature. PP crystallizes such that crystalline structures called spherulites are formed. Nucleation results in the formation of smaller spherulites than would otherwise have been formed. This, importantly, results in increased clarity and stiffness but also imparts some possibly undesirable features, such as warpage or brittleness.

Another important transition detected by DSC is the glass transition. This is the transition that amorphous (noncrystalline) materials undergo in changing from the liquid to rubbery state. In i-PP this is difficult to detect by DSC because the concentration of amorphous PP is small, but detection is easy in a-PP, the glass transition temperature being in the vicinity of $-15°C$.

There are several other thermal analysis techniques. In thermomechanical analysis (TMA), mechanical changes are monitored versus temperature. Expansion and penetration characteristics or stress–strain behavior can be studied. In dynamic mechanical analysis (DMA), the variations with temperature of various moduli are determined, and this information is further used to obtain fundamental information such as transition temperatures. In thermogravimetric analysis (TGA), weight changes as a function of temperature or time (at some elevated temperature) are followed. This information is used to assess thermal stability and decomposition behavior.

## 2.8 MANUFACTURING PROCESSES

The process technology for PP manufacture has kept pace with catalyst advances and the development of new product applications and markets. In particular, the relationship between process and catalyst technology was clearly symbiotic and that of a partnership. Advances in one technology had always exerted a strong push–pull effect on the other to improve its performance. The progress in process technology has resulted in process simplification, investment cost and manufacturing cost reductions, improvement in plant constuctability, operability, and broader process capabilities to produce a wider product mix.

The simplified block diagrams in Figs. 2.1–2.3 serve to illustrate the advances in PP process technology from a complex process in Fig. 2.1 to one that is simpler in Fig. 2.3. The slurry process technology as illustrated in Fig. 2.1 is typical of manufacturing units built in the 1960s and 1970s. This technology was designed for catalysts of the first and second generations. It required a solvent such as butane, heptane, hexane, or even heavier isoparaffins. The solvent served as the medium for dispersion of the polymer produced in the reactors and for dissolving the high level of atactic byproducts for removal downstream. The use of a solvent also facilitated the catalyst deactivation and extraction (or deashing) step, which required contacting the reactor product with alcohol and caustic solutions. Plants based on this technology required a large amount of equipment, a great deal of space, and complicated plot plans. They were high in both capital and operating costs, labor intensive, and energy inefficient. Moreover, there were environmental and safety issues associated with the handling of a large volume of solvent and the disposal of the amorphous atactic byproducts, and a large wastewater stream containing residual catalyst components. With the advent of third- and fourth-generation catalysts, many of these older slurry plants stayed viable by cost reduction aided by the higher catalyst activities and lower atactic production. They also benefitted from plant capacity creeps and de-bottlenecking.

The slurry process technology evolved into the more advanced slurry process (Fig. 2.2) in the late 1970s to take advantage of the higher performing third-generation catalysts initially and later the even better fourth-generation catalysts. The improved slurry processes were commonly referred to as the bulk (slurry) process. One major change from the older slurry technology was the substitution of liquid propylene in place of the solvent system. This became possible because catalyst de-ashing and atactic removal were no longer needed to

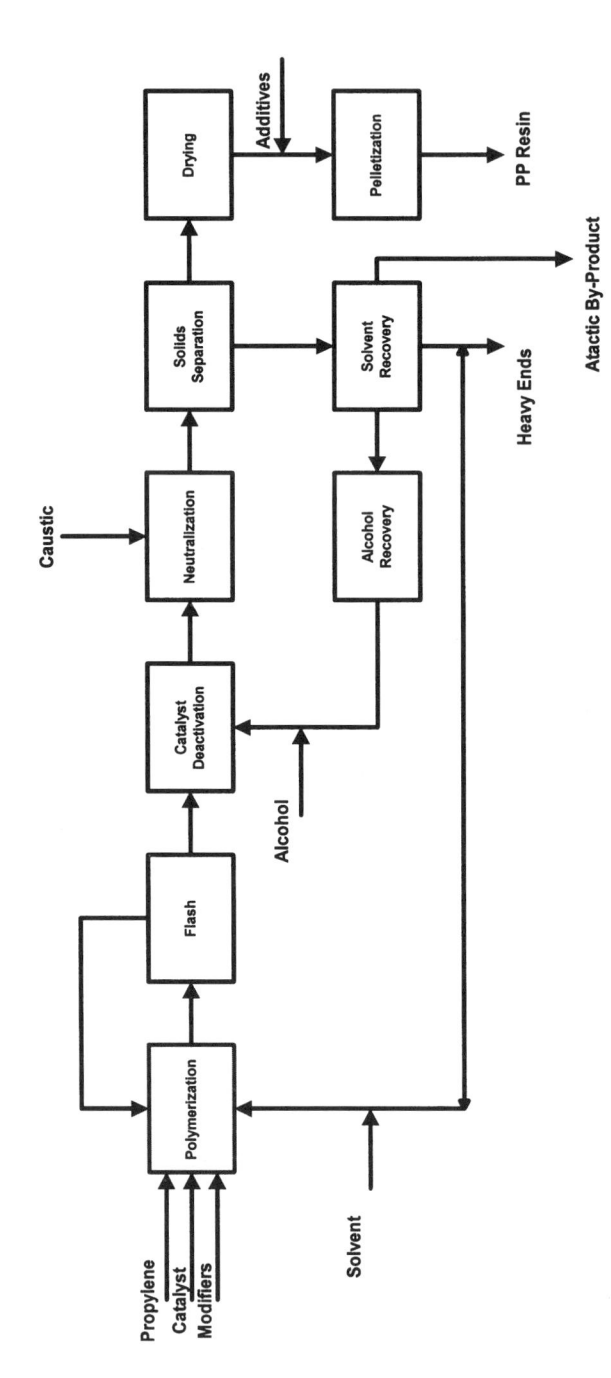

**FIGURE 2.1** Early slurry process technology.

# Structure and Properties of Polypropylene 21

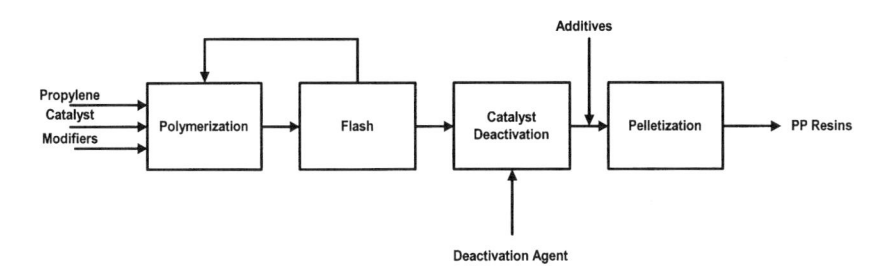

**FIGURE 2.2** Bulk (slurry) process technology.

produce acceptable PP resins. With very few exceptions, virtually all slurry plants built over the last two decades were based on bulk process technology. Montell's Spheripol process represents technology of this type, using pipe loop reactors operated liquid full, with a PP slurry in liquid propylene. Additionally, a fluidized bed reactor is used by Spheripol downstream of the bulk pipe loop reactors when impact copolymers are in the product slate.

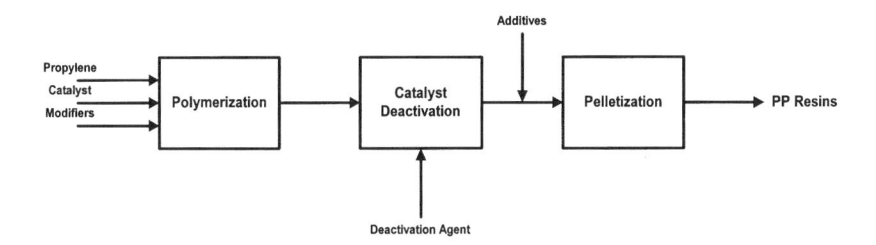

The emergence of gas-phase process technology for PP occurred about the same time as the bulk processes. Gas-phase technology was revolutionary in that it completely avoided the need for a solvent or liquid medium to disperse the reactants and reactor product. This process eliminates the separation and recovery of large quantities of solvents or liquid propylene required in slurry or bulk reactors. The PP products from the gas-phase reactors are essentially dry, requiring only deactivation of the very low level of catalyst residues before the incorporation of additives and pelletization. Thus, this process technology reduced the manufacturing of PP to the bare essential steps. Representatives of commercial gas-phase process technology include Amoco, Union Carbide (Unipol), and BASF (Novolen).

Amoco's technology features a horizontal stirred bed reactor system that uses mild mechanical agitation for reactor mixing and temperature control. The

**FIGURE 2.3** Gas-phase process technology.

heat of polymerization is removed by the use of quench cooling or evaporative cooling using a spray of liquid propylene. The Unipol process is based on a gas fluidization principle that relies on a large volume of fluidizing gas for reactor mixing, polymerization heat removal, and temperature control. According to trade literature, Unipol has claimed that the gas cooling can now be supplemented by some amount of liquid evaporation in the fluidized bed, referred to as the "condensing" mode cooling. The BASF gas-phase reactor is a vertical stirred bed reactor in which the polymerization heat is removed by vaporization of liquid propylene in the bed. In the above three gas-phase processes, a second reactor of a similar design as the first reactor is added for the production of impact copolymers. A. sketch of the reactor systems associated with the four types of commercial PP process technology described above—Amoco, Spheripol, BASF, Unipol—is shown in Fig. 2.4. The Amoco gas-phase process technology is more completely depicted in Fig. 2.5.

In summary, over four decades, PP process technology has never stopped creating value for resin customers through both incremental and generational changes. The changes came about through a partnership with advancements in catalysts to result in better manufacturing economics and simpler plants, making them easier to operate and at higher efficiencies. At the same time, the improved process technology has also added enhancements to many product properties and expanded the product applications.

### 2.8.1 World-Scale Technology

The PP industry is exciting and will continue to grow globally at a rate attractive to making new investments. Obviously, it is also highly competitive, and the resin customers have high expectations. To favorably compete and to satisfy customers, PP producers must have access to world-scale technology when new investment is being considered. The criteria for world-scale technology are the following:

1. Simple and efficient process
2. Attractive economics for resin manufacture: low plant investment and operating cost
3. Efficient and high performance with fourth-generation catalysts
4. Capability for a wide range of products, with the ability to allow easy product transitions in manufacturing
5. Environmentally clean and safe operations
6. Capability of plant design for high single-line capacity
7. Commitment of technology provider to continuous improvements and innovations

To improve capital utilization and remain competitive, we believe that new plants should have production capacity no less than 150,000 metric tons/yr. A

# Structure and Properties of Polypropylene

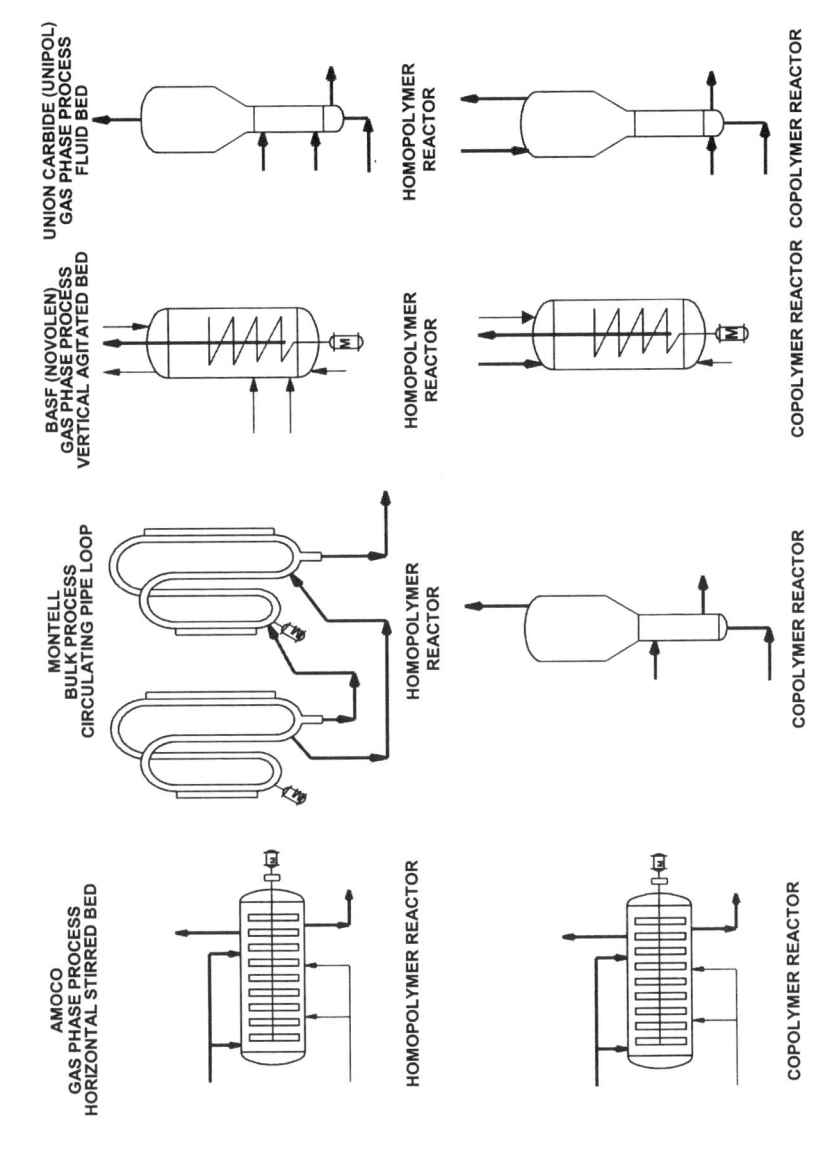

FIGURE 2.4  Reactor systems in polypropylene technologies.

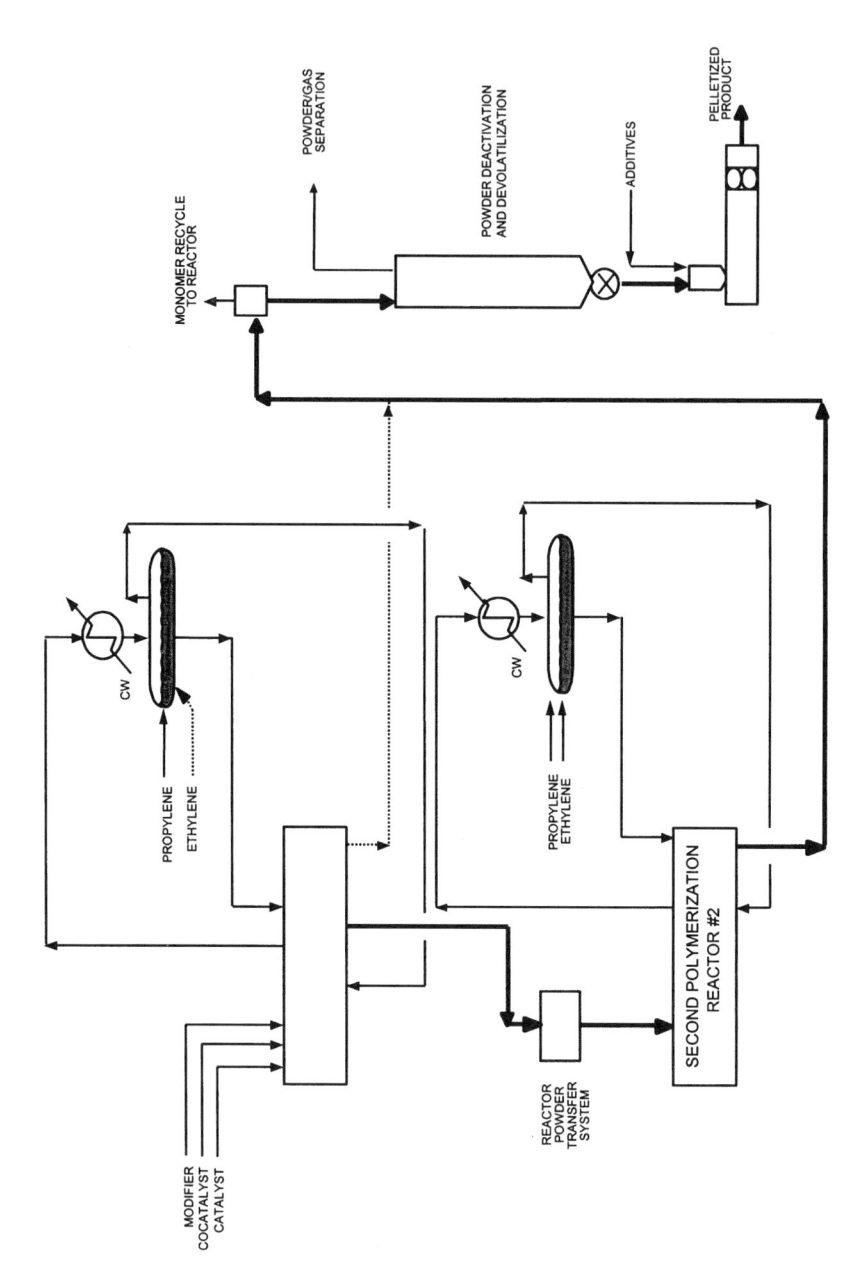

**FIGURE 2.5** Amoco gas-phase process technology.

new trend is to build larger units with production capacity over 200,000 metric tons/yr.

## 2.9 POLYPROPYLENE APPLICATIONS

As is obvious from the preceding discussion in this chapter, PP should really be considered a group of polymers, not just a single polymer. Because the properties of PPs cover a substantial range, the applications of PP are quite diverse. This, of course, belies the usual classification of PP as a commodity resin. The most important applications of PP are discussed in this section.

Organizing a discussion on applications is challenging because the question arises as to whether similarity of uses or similarity of the fabricated products or similarity of the fabrication techniques should be used as the criterion for arranging information. None of the methods is perfect. Here the material is organized in a fashion that intertwines these, but it seems logical to the authors based on our experience.

### 2.9.1 Fibers and Fabrics

A great volume of PP finds its way into an area that may be classified as fibers and fabrics. Fibers, which broadly speaking includes slit film or slit tape, are produced in various kinds of extrusion processes. The advantages offered by PP include low specific gravity, which means greater bulk per given weight, strength, chemical resistance, and stain resistance.

*Slit Film.* In slit-film production, wide web extruded film, which is oriented in the machine direction by virtue of the take-up system, is slit into narrow tapes. These tapes are woven into fabrics for various end uses.

In general, non-CR homopolymer of about 2–4 g/10 min is used in this application. Higher flow rate resins permit higher extrusion speeds, but lower MFR resins result in a higher tenacity at a given draw ratio.

A major application of slit film is in carpet backings, both the primary and secondary types. The primary carpet backing is not the one that is seen on the back of a carpet; that is the secondary backing. The primary backing is the one that is between the secondary backing and the face yarns and is the one to which the face yarns are tufted. The secondary backing protects the tufted fibers and adds substance to the carpet. Today, more carpet backing is produced from PP than from the natural jute fibers, which at one time were dominant. Jute suffers from its unsteady supply situation, being affected by weather and producing-country politics. Moreover, PP is not subject to damaging moisture absorption and mold attack in high humidity.

Slit film also finds its way into many other applications. These include twine, woven fabrics for feed and fertilizer sacks, sand bags and bulk container bags, tarpaulins, mats, screens for erosion prevention, and geotextiles to stabilize soil beds. Fibrillated slit film is used as a face yarn in outdoor carpets and mats.

*Continuous Filament Fibers.* Continuous filament (CF) fibers are more conventional fibers than slit-film fibers in that each strand results from extrusion through its own die hole. Polypropylene homopolymer is extruded through rather small holes in a die called a spinneret, each spinneret containing somewhere around 150 holes each, and spinning speeds are often high. The resulting filaments are very fine, being on the order of 5 deniers per filament (dpf; a denier being defined as 1 g/9000 m). Therefore, to reduce viscosity, relatively high melt flow rate PP (about 35) is used, and high process temperatures (about 250°C) are used. Usually, a CR resin is the choice because a narrow molecular weight distribution is desired.

Extrusion takes place through several spinnerets at the same time. The filaments are quenched with air, and the group from each spinneret is then taken up to produce yarn, the draw ratio and degree of orientation depending on the take-up equipment, on whether the process is single or two stage, and on the end use. Continuous filament fibers are not crimped or texturized as produced. Such bulking can be imparted in a secondary process or the CF fiber yarns can be used as is, often in combination with other types of yarns.

*Bulked Continuous Filament.* Bulked continuous filament (BCF) processes are similar to CF processes, but one main difference is that in BCF a texturizer is an integral part of the process, imparting bulk to the fibers through crimps or kinks. Commonly, BCF fibers are about 20 dpf, with yarns being in the vicinity of 2000 denier. Polypropylene homopolymer with MFRs in the range of 12–20 MFR and normal MWD is typically used. BCF yarns are mainly used as carpet face yarns and in fabrics for upholstery.

*Carpets.* The carpets constructed with PP face yarns are currently largely of the commercial type. For residential carpeting, dyability and deep pile construction have been desirable. Polypropylene cannot readily be dyed, and therefore PP fibers are colored via the addition of pigment during extrusion. Pigmented fibers have a somewhat different appearance from that of dyed fibers, such as those from wool, nylon, and polyester. Polypropylene fibers have also been less resilient in deep pile than those from wool or nylon. However, with advances in technology in PP resins, fiber, and carpet construction, PP usage in residential carpets is steadily growing.

Commercial carpets are of short pile, dense construction, and therefore resiliency of PP fibers is not an issue. Furthermore, for the same reason that PP cannot be dyed, it resists staining and soiling. Because muted tones are desirable for commercial carpeting, the colors available from pigmentation are ideal. Also, pigmented fibers are more color fast and fade less in sunlight. Another plus is that

## Structure and Properties of Polypropylene

in pigmented fibers, the color does not just reside on the surface but rather is distributed uniformly throughout; therefore, fiber breakage in rough treatment does not result in loss of color.

*Staple Fibers.* Staple fibers are short fibers, ranging from less than an inch to a little less than a foot in length, depending on the application. Staple fibers are spun fibers that are produced in either of two somewhat different processes. In the long-spin traditional process, fiber is spun similarly to CF fibers, wound undrawn in a tow (bundle) in one step and then drawn (if desired), crimped (if desired), and cut in a second step.

The other staple fiber process is known as the short-spin or compact spinning process. This is a one-step process with relatively slow spinning speeds of about 300 ft/min, but one in which the spinnerets have an extremely large number of holes (50,000–100,000 or more). Overall, the process is more economical than the traditional process.

Staple fibers can be carded and drawn into spun yarns in the same way as natural fibers, or they can be used in nonwoven fabrics. The PP resins used are usually homopolymers with MFRs from 10–30 g/10 min, depending on the application.

*Nonwoven Fabrics.* Nonwoven fabrics account for more PP usage than any other single fiber application. There are three types of nonwoven fabrics: thermobonded, from staple fibers; spun-bonded; and melt-blown. The spun-bonded and melt-blown processes are discussed below. The fabrics from each process differ from each other in properties and appearance, and often combinations of two types are used together. Spun-bonded fabrics are strong, whereas melt-blown fabrics are soft and have high bulk.

Nonwoven fabrics are used in several areas, probably the most well known being for the liners in disposable diapers. Similar fabric is also used in feminine hygiene products. At the other extreme, civil engineering fabric and tarpaulins are also produced from nonwoven fabrics.

SPUN-BONDED FABRIC. In this process, molten filaments are air quenched and then drawn by air at high pressure. To form a fabric, the filaments are then deposited on a moving porous belt to which a vacuum is applied underneath. Bonding of the fibers is accomplished by passing the fabric through heated calender rolls. The type of resin usually used is a CR homopolymer of about 35 g/10 min MFR.

MELT-BLOWN FIBER/FABRIC. In the melt-blowing process, polymer is extruded through a special melt-blowing die. The die feeds high-temperature air to the exiting filaments at a high velocity, and, in addition, the exiting filaments are quenched with cool air. In the process, drawn solidified filaments are formed very close to the die. Formation of the fabric is accomplished by blowing the filaments onto a moving screen. The melt-blowing process requires resins that have very high melt flow rates, on the order of 500 g/10 min or higher. A resin of narrow MWD is also desirable.

The fibers formed in the melt-blowing process are very fine and allow for the production of lightweight uniform fabrics that are soft but not strong. Fabrics from fine melt-blown fibers can be used in medical applications because they allow the passage of water vapor but prevent the penetration of liquid water and aqueous solutions.

MONOFILAMENT. As the name implies, single filaments are extruded; the molten filaments are cooled and solidified in a water bath and then drawn. Typically, monofilaments are fairly large, being on the order of 250 dpf. Rope and twine are produced by twisting bundles of monofilament together. Polypropylene rope and twine are strong and moisture resistant, making them very useful in marine applications.

Low MFR (usually between 1 and 4 g/10 min) homopolymer is most often used for monofilament. This provides the high tensile strength (tenacity) required in this application.

### 2.9.2 Strapping

Strapping is similar to slit film but thicker, being on the order of 20 mils. As the name implies, strapping is used to secure large packages or boxes or to hold stacks together. It takes the place of steel strapping, and its most important property is strength, although the moisture resistance of PP is also an important attribute. It is produced from either direct extrusion or from sheet that is slit. Uniaxial orientation is applied by drawing. Homopolymer resins of low MFR (between 1.0 and 1.5 g/10 min) are used for this application.

### 2.9.3 Film

By definition, film is less than 10 mils thick. There are two broad classes of film: cast and oriented.

*Cast Film.* In cast film processes, polypropylene is extruded through a die onto a chill roll and the resulting film is eventually taken up on winding equipment. Cast film is essentially unoriented but is still fairly clear because of the quench cooling that occurs. Film thickness usually ranges between 1 and 4 mils. An important feature of cast film is its softness and lack of cellophane-like crispness. Both homopolymers and random copolymers are used in cast film, the MFR most commonly being around 8 g/10 min. Random copolymers give slightly clearer, softer, and more impact-resistant film.

Cast film is converted to such products as bags for clothing articles such as mens shirts, pocket-pages for photographs, sheet protectors, and as the outer

# Structure and Properties of Polypropylene

nonporous sheet on disposable diapers. Cast film also is found in some tapes and pressure-sensitive labels.

*Biaxially Oriented Polypropylene Film.* Two methods are widely used for producing biaxially oriented PP (BOPP) film. One is the tenter process, and the other is the tubular or bubble process. In both, homopolymer of about 3 g/10 min MFR is most widely used, although random copolymer is used for better heat sealability.

In the tenter process, extruded sheet is drawn sequentially, first in the machine direction and then in the transverse or cross direction. Draw ratios of $4 \times 7$ to $6 \times 10$ are common. Film thickness ranges from 0.5 to 2.5 mils.

In the tubular process, a tube is extruded downward through an annular die over a mandrel that cools the tube. The tube is taken up through a water bath for further cooling. The flat tube is moved by rolls through an oven where it is reheated to a temperature close to the melting temperature. The tube is then inflated while being stretched in the machine direction, imparting orientation through roughly sixfold biaxial stretching. Films typically range in thickness from 0.25 to 2 mils.

Biaxially oriented PP film has excellent clarity and gloss and is printable through the use of some additional surface treatment technology. The main applications for BOPP film are in flexible packaging. A major use is in snack-food packaging, where the BOPP film is used in one or more layers in a multilayer bag construction. The BOPP film provides resistance to moisture vapor to keep snacks crisp and fresh tasting and provides a heat-sealable layer. Biaxially oriented PP film by itself is used for packaging of bakery products (e.g., bags for freshly baked bread). Many adhesive tapes are also produced from BOPP film.

A special kind of BOPP film, known as opaque film, is used for packaging products such as candy, chocolate bars, and soaps, and for labels such as those now used on most soft drink bottles. Opaque film is produced in the tenter process from PP to which a fine filler (e.g., calcium carbonate) or an incompatible polymer has been added.

## 2.9.4 Sheet/Thermoforming

Sheet is an extruded product that is greater than 10 mils in thickness (below this the product is identified as film), 40 mils being typical. Resin is extruded through a die and passes through a cooling roll stack and conveyed to nip rolls, after which sheet is wound on rolls or cut and stacked or conveyed directly to a thermoforming machine. Sheet width is usually between 2 and 7 ft.

Although there are other applications for sheet, the production of thermo-formed containers for rigid packaging applications predominates. However, PP, being a semicrystalline polymer, is not as ideal a material for the conventional thermoforming process as is polystyrene, an amorphous polymer. Polystyrene has been used with great success in the thermoforming of rigid packages for food for certain dairy and deli applications. Usually polystyrene is first extruded into a sheet that is reheated as it is sent to the thermoformer. The heated sheet is forced into the cavities of a multicavity mold by a combination of vacuum and pushing by a plug. The cooled containers are cut from the web and stacked. With polystyrene, the sheet possesses excellent melt strength, which is a problem with PP.

Notwithstanding what has been said above, thermoformed PP containers are making major inroads into the rigid packaging area. Thermoforming technology has been modified to overcome the deficiencies of PP so that its advantages can be exploited. For one thing, narrower sheet webs are sometimes used. Also, sag bands to prevent sheet from sagging are often used. Furthermore, a modified thermoforming process known as solid-phase pressure forming (SPPF) has been developed to minimize sag. In SPPF, fabrication takes place just below the melting point (about 160°C) and at applied pressures that are higher than encountered during normal forming processes.

One of the main advantages of PP resin is that it resists the stress cracking shown by polystyrene in the presence of fatty products, such as margarine. Also, it has a lower specific gravity than polystyrene, allowing for lighter weight containers at the same thickness. Furthermore, there are no health-related questions about the monomers from which PP and its copolymers are produced. Compared with high-density polyethylene (HDPE), PP has a stiffness advantage. PP impact copolymers are needed to compete with the low-temperature impact properties of HDPE, however.

A number of different PP resins, including homopolymers, random copolymers, and impact copolymers, are used in the production of thermoformed containers. In most cases, an MFR of about 2.5 g/10 min fits the need. The choice of polymer type depends on the desired end use. Where cost is the most important criterion, "plain vanilla" homopolymer is used. For added stiffness, a nucleated homopolymer is used. If high clarity and stiffness are required, a clarified homopolymer may be desirable (although more costly). If very high clarity is desired, a clarified random copolymer could be used. A random copolymer also offers a slight edge in flexibility, which is an advantage when toughness at somewhat lower than ambient temperature (e.g., refrigerator temperature) is desired. For impact resistance at low temperature, such as freezer temperature, an impact copolymer is required; high clarity is not a feature in such applications.

### 2.9.5 Injection Molding

In injection molding, polymer is softened and conveyed with a screw in a manner similar to extrusion, and the polymer is pushed through a runner system into a cavity or multiple cavities of a mold. The mold is cooled, and eventually the mold separates and solid parts are ejected and stacked. In thin-wall injection molding (TWIM) for rigid packaging container applications, the thickness does not usually exceed 25 mils and is often considerably less than that.

*Rigid Packaging Containers and Housewares.* Many of the same types of dairy and deli containers as produced by thermoforming are produced by injection molding. For PP, injection molding came first. Which technique is better can be argued at length. Arguments are related to cost, speed, efficiency, part appearance, and thickness uniformity. The latter is important not only in terms of top-to-bottom variation for rigidity and stackability considerations but in terms of the circumference, where minor variations can cause printability problems. Thickness uniformity has been reputed to be one of the virtues of injection molded containers, and another is part appearance.

For consumer items, such as food storage containers and sport/exercise water bottles, which take many shapes and can be round, square, flat, or tall, injection molding seems to predominate. In items like stadium cups and hotel and bathroom cups, both the injection molding and thermoforming techniques are well represented.

A number of different PP resins are used for injection molded rigid packaging and houseware applications. For the most part, the resin will have an MFR of 5 g/10 min or above. In TWIM, the MFR may be in the neighborhood of 75 g/10 min for ease of mold filling. Homopolymers and random copolymers, often nucleated or clarified, are commonly used, but impact copolymers are the resins of choice if low-temperature impact is needed and if high-MFR resins are required (as in TWIM). Homopolymers of such high MFR are brittle. Both CR and reactor resins are used, although reactor resins are often considered more desirable, for two reasons. A reactor resin's broader molecular weight distribution results in better processing in injection molding and a reactor resin is better for good organoleptics (taste and odor transfer characteristics) because of the absence of byproducts of peroxide decomposition.

*Caps and Closures.* Screw caps for bottles and jars are just one kind of closure produced from PP. Most tamper-evident closures, trigger sprayers, and dispensing closures that are resealable with, for example, a hinged flip-top lid are also produced from PP. One very desirable feature of PP is its ability to form a "living hinge" that can be used repeatedly. Generally, a homopolymer is used; the MFR can be anywhere from about 2.5 to about 20 g/10 min. It should be mentioned that some PP closures for soft-drink bottles are produced in a proprietary molding process that is different from injection molding.

*Appliances and Hand Tools.* Polypropylene is widely used in both small and large appliances. In small appliances, like electric drip coffee makers, can openers, blenders, and mixers, and in tools, like electric drills, PPs ease of molding, light weight, stiffness, durability, electrical properties, appearance, and cost make it a very attractive resin choice. Homopolymers and impact copolymers are both used, with an MFR of around 12 g/10 min being the norm.

In the large-appliance category, properly stabilized and often filled PP is used in parts such as dishwasher liners, which take the place of porcelain interiors, and in washing machine agitators, soap and bleach dispensers, and other parts.

*Consumer/Industrial.* Items from luggage to outdoor tables and chairs to stadium seats to portable coolers and ice chests to garden sprayer tanks to lawn sprinklers and many other items not mentioned either are produced from PP entirely or have PP components.

*Medical.* A large medical application for PP is disposable syringes, which are injection molded from homopolymer or sometimes random copolymer. The MFR is generally about 25 g/10 min and CR resins prevail. In this application, the ability to sterilize with gamma rays (from $^{60}$Co) is important, and part of the technology involves proper stabilization.

Medical vials, the amber-colored small containers that hold pills and capsules, from the pharmacy are almost exclusively injection molded from PP, homopolymer being most prevalent. An MFR within the range of 12–20 g/10 min is most common.

### 2.9.6 Blow Molding

In blow molding, bottles or jars are produced by blowing, with air, a molten previously formed part. There are three types of blow molding: extrusion blow molding (EBM), injection blow molding (IBM), and injection stretch blow molding (ISBM).

*Extrusion Blow Molding.* In EBM, a hollow molten plastic cylinder, called a parison, is clamped in a mold and blown into a bottle that is finally trimmed. In the blowing process, threads for screw caps and handles are produced. The most common bottles using this technique are milk bottles, which are produced from HDPE. An advantage of HDPE in this process is its melt strength, which is important in preventing parison sagging.

Polypropylene, however, is used for many extrusion blow-molded bottles, such as those that contain pancake syrup. Polypropylene bottles offer an advantage in hot-filling capability. Also, PP, even homopolymer, provides good contact clarity. The clarity of random copolymers is even better. The MFR of choice for both homopolymers and random copolymers is generally about 2 g/10 min.

*Injection Blow Molding.* In IBM, a mostly still-molten injection-molded preform, with solidified neck finish, is blown within a mold. The preform looks somewhat like a test tube with threads at the open end. The technique is used to produce relatively small (compared with EBM) bottles and wide-mouth jars, with neck finishes that are far better in appearance than those produced in EBM.

*Injection Stretch Blow Molding.* As in IBM, an injection-molded preform is blown within a mold in the ISBM process. However, the temperature is kept just below the melting point, and blowing is performed in combination with stretching in the long direction; the stretching is provided by a moving stretch rod. Biaxially oriented jars and bottles are produced with greater clarity, strength and barrier properties than those from IBM.

Polypropylene is slowly but surely finding its way into ISBM applications. Polyethylene terephthalate (PET) now dominates, being helped by its widespread use in carbonated soft-drink bottles up to 3 L in volume. With the incorporation of a good clarifier, the clarity of PP bottles rivals that of PET bottles, but the oxygen and carbon dioxide barrier properties of PET are significantly better than those of PP. The water vapor barrier properties of PP, however, are superior, and new technology promises to improve the oxygen and carbon dioxide barrier properties of PP bottles.

### 2.9.7 Automotive

Polypropylene has a large presence in cars and other vehicles. For the most part, impact copolymers predominate. One of the original uses was in battery cases; in this application, which goes back more than 25 years, injection-molded impact copolymer, colored black, replaced black hard rubber. Now, cases of other colors and of natural translucent material are the norm. Another long-standing use of PP in a car has been for heat and air conditioning ducts, which are mostly unseen. Fan blades of various types are produced from filled (usually with talc) PP.

With the need to reduce the weight of cars, many components went from other heavier materials, such as metals and wood flour, to plastics, and now much of the plastic in cars is PP. Interior trim and several exterior components are made completely of PP or PP compounds. Polypropylene has an advantage over most other thermoplastics in density, which is only about 0.9 g/mL. Another bonus is the fact that PP parts are not prone to squeak and rattle.

In interior trim, doors; A, B, and C pillars; quarter panels; and consoles are all molded of PP. Different types of resin are used depending on the requirements, but the choice is almost always an impact copolymer. Most often, the parts are produced with molded-in color, even though sometimes the parts are painted for perfect color matching and appearance. For interior trim, resins yielding low-gloss parts are desirable because reflections are reduced, the visibility of knitlines and mold imperfections is minimized, and low-gloss parts have less of a "plastic" appearance.

In interior trim applications, impact copolymers with melt flow rates in the range of 20–35 g/10 min are most often seen, and they can be CR or reactor resins. The exact choice of resin can be complex because of the many different and sometimes conflicting requirements. Impact requirements, often mandated by government crash test results, need to be balanced against stiffness and scratch and mar resistance desires and requirements for heat resistance (the interior of a closed car in the hot summer sun can get extremely hot). In some door applications, appearance is not important because the molded part is covered with fabric. In a low-pressure molding operation known as in-mold decorating, the polymer is essentially molded along with the fabric coating. Very high MFR resins, in the neighborhood of 70 g/10 min, are required.

Polypropylene is also quickly becoming a major material for automotive exterior parts. Again, weight reduction has been an important factor. In newly designed car bumpers, it is becoming more and more common that the bumper fascia, which is the visible part, is molded from a thermoplastic olefin (TPO), of which PP is a major constituent. The early TPOs were produced by incorporating a rubber into PP. More advanced TPOs are now produced right in the reactor, and these are called reactor TPOs. Thermoplastic olefins are also used for air dams, body side claddings, rocker panels, and even grills in some vehicles.

# 3

## Chemical Coupling Agents for Filled and Grafted Polypropylene Composites

**Darilyn Roberts**
Crompton Corporation, Middlebury, Connecticut, U.S.A.

**Robert C. Constable**
BRG Townsend, Mt. Olive, New Jersey, U.S.A.

### 3.1 INTRODUCTION

Polypropylene is a very versatile polymer. It has many qualities, such as excellent chemical resistance, good mechanical properties, and low cost, that make it the polymer of choice for various applications. There are many ways in which the mechanical properties of polypropylene can be modified to suit a wide variety of end-use applications. Various fillers and reinforcements, such as glass fiber, mica, talc, and calcium carbonate, are typical ingredients that are added to polypropylene resin to attain cost-effective mechanical properties. Fibrous materials tend to increase both mechanical and thermal properties, such as tensile and flexural strength, flexural modulus, heat deflection temperature, creep resistance, and even impact strength. Fillers, such as talc and calcium carbonate, are often used as extenders, thus reducing the final material cost. However, some improvement in stiffness and impact can be obtained with these materials.

    Other types of fillers are often used to enhance properties, such as flammability resistance. Nonhalogen flame-retardant materials, such as magnesium hydroxide, are often added to polypropylene to produce low-smoke, nonhalogen, flame-retardant compounds.

Most of the fillers and reinforcements used are polar in nature. Polypropylene, on the other hand, is nonpolar. Poor adhesion between the filler surface and the polymer matrix prevents necessary wet-out by molten polymer to help break up aggregates of filler particles, resulting in poor dispersion, insufficient reinforcement, and poor mechanical properties.

Several steps can be taken to overcome these problems. One frequently used method is to make the surface more hydrophilic by treating the filler with a surface treatment, such as stearic acid. Other additives, such as silanes, zirconates, and titanates, are also used. These materials will react with both the filler surface and the polymer to increase the adhesion between the two. Another approach is to modify the chemistry of the polypropylene by attaching polar groups, such as acrylic acid or maleic anhydride, onto the polymer backbone. This chapter is devoted to the review of polymeric coupling agents, mainly acrylic acid grafted polypropylene (PPgAA) and maleated polypropylene (PPgMAH), their manufacture, and use in filled and reinforced polypropylene composites.

According to BRG Townsend, approximately 40 million pounds of chemical coupling agents was used in the plastics industry in 2000. This number includes silane, titanate, and functional polyolefins. Silanes make up the largest use, mainly as part of the fiberglass sizing and surface treatment for other fillers such as mica, flame retardant, and wollastonite.

It is estimated that around 90% of short fiberglass reinforced polypropylene composites contain maleated polypropylene. Maleated polypropylenes are also used in long-fiber polypropylene composites and glass mat thermoplastic (GMT) materials. The consumption of these coupling agents was between 4 million and 5 million pounds in North America in 2000.

Mineral-filled polypropylene composites use very little functionalized polypropylene. However, there are some niche applications where these coupling agents provide some benefit other than mechanical property increases. Typically, PPgAA polypropylene is preferred over PPgMAH in mineral-filled systems.

## 3.2 REVIEW OF COUPLING AGENTS

There are many types of chemical coupling agents used in the production of polypropylene composites. A coupling agent by definition is a substance that couples or bonds the filler to the polymer matrix [1]. To do this effectively, the coupling agent must have a unique structure. On one hand, it must be able to interact with the filler, which is polar in nature, while on the other hand, it must be compatible with the nonpolar molecular chains of the polypropylene resin. Some of the oldest types of coupling agents are organofunctional silanes. Silane coupling agents are molecules that have the following unique structure. One end has a hydrolyzable group, which is an intermediate in the formation of silanol groups that chemically bond to the surface of the filler. On the other end, silanes have organofunctional groups that entangle with polymer molecular chains by physical-type interactions. The type of functional group needed depends on the chemical structure of the polymer. Many types of organofunctional groups are available: epoxy, methacrylate, amine, mercapto, and vinyl. For polymers that have functional groups present in their backbone, it is easy to select a silane with a specific organofunctional group with which it will react. For example, in polyamides, silanes that have amino groups are commonly used. The abundance of carboxylic acid end groups, present along the polyamide molecular chain, provides reaction sites for the silane amine groups.

However, polypropylene has no functional groups present for the silane to react with. Therefore, vinyl silanes are used. In this case, the vinyl group must react with the polymer molecule via a free-radical reaction mechanism. Free radicals can be generated through heat, shear, or with a catalyst, such as an organic peroxide. There are various examples of two-component silane-peroxide systems. In 1986, Union Carbide introduced a product called UCARSIL PC-1A/PC-1B. This system was a complex mixture of organosilane and peroxide. The 1A and 1B components were used at a 3 : 1 ratio. Typical usage levels were 0.6 wt%. Later, an improved version was introduced that was designated PC-2A/PC-1B [2]. Other companies have published papers describing the performance of their silanes in filled and reinforced polypropylene [3,4]. These systems give excellent mechanical properties, but handling them in a production environment can be difficult. Compounders must be able to accurately feed low levels of two liquids into their extruder or the fillers must be pretreated with the silane before compounding. Also, there are some health and safety issues with handling certain vinyl silanes that can lead to increased compounding costs.

There are several theories on how silanes work as coupling agents [5]. One is the chemical reaction theory. This mechanism is the oldest and most widely accepted. It simply states that a covalent bond is formed between the filler and polymer. The silanol group reacts with the surface of the filler, whereas the organofunctional group reacts with the polymer. Another mechanism is formation of an interpenetrating network. Instead, the silane molecules diffuse into the polymer matrix, forming an interphase network of polymer and silane. These two explanations are the most widely accepted, although neither one alone completely explains chemical coupling. There are a host of other theories, including wetting and surface energy effects, morphology effect, acid base reactions, and ionic bonding.

## 3.3 ACRYLIC ACID GRAFTED POLYPROPYLENE

The functionalization of polypropylene with acrylic acid was reported as early as the mid-1960s. Some of the most detailed work is reported in various patents issued to Exxon. Steinkamp and Grail [6] reported the use of a single-screw extruder with a specially designed reaction zone that produces a very thin film and high surface area so that the monomer and initiator are thoroughly mixed prior to reacting with the polymer. The grafting of acrylic acid onto polypropylene results in various side reactions. The molecular weight of the polypropylene is reduced and the distribution narrowed. Along with grafted polypropylene, homopolymerized acrylic acid is produced as well. Graft efficiencies are relatively high up to about 6 wt% acrylic acid. At this level of functionalization, grafting efficiencies of more than 90% are obtained. Increasing the level of functionalization results in a reduction of graft efficiency.

## 3.4 GRAFTING REACTIONS OF ACRYLIC ACID ONTO POLYPROPYLENE

Possible reaction mechanisms are shown in Fig. 3.1. For polypropylene, it is accepted that the free radical produced by peroxide decomposition attacks the tertiary hydrogen atom. The resulting polymer radical causes beta scission and polymer molecular weight reduction, as shown in pathway A. If there is an unsaturated monomer present to react with the polymer, radical grafting may occur prior to chain scission, pathway B. This results in attachment of a functional group pendant to the polymer backbone. Since acrylic acid is readily polymerized, it does so until terminated via recombination or transfer. If grafting does not occur, there are several reactions that can occur. These are depolymerization (pathway D), transfer (pathway E), and recombination (pathway F). Pathway G shows that the resulting polymer radical can still result in grafting. In this case, the monomer will be grafted onto the end of the polymer. After chain end grafting, various termination reactions can occur (pathways H, I, and J), with generation of homopolymer being the predominant reaction. A side reaction that occurs is polymerization of the acrylic acid (pathway K), without grafting onto the polymer. Analysis of extracted polyacrylic acid shows that the molecular weight of this material is between 2500 and 3000 g/mol.

Various methods have been tried to reduce or limit the homopolymerization that occurs. Some basic methods that can be used are process related. One method is to vary the order of addition of the monomer and initiator. Addition of the initiator to the polymer melt just before the monomer addition allows the initiator to mix thoroughly with the polymer. This allows free-radical groups from initiator decomposition to promote reactive sites along the polymer backbone for preferential monomer addition, instead of homopolymerization by monomer molecules. Multiple addition sites for initiator and monomer can also be used, but this requires multiple pumps, injectors, mixing zones, and a long extruder.

Proper screw design selection is another way of increasing mixing and distribution of monomer and initiator in the polymer. Twin-screw extruders, with

# Chemical Coupling Agents for Composites

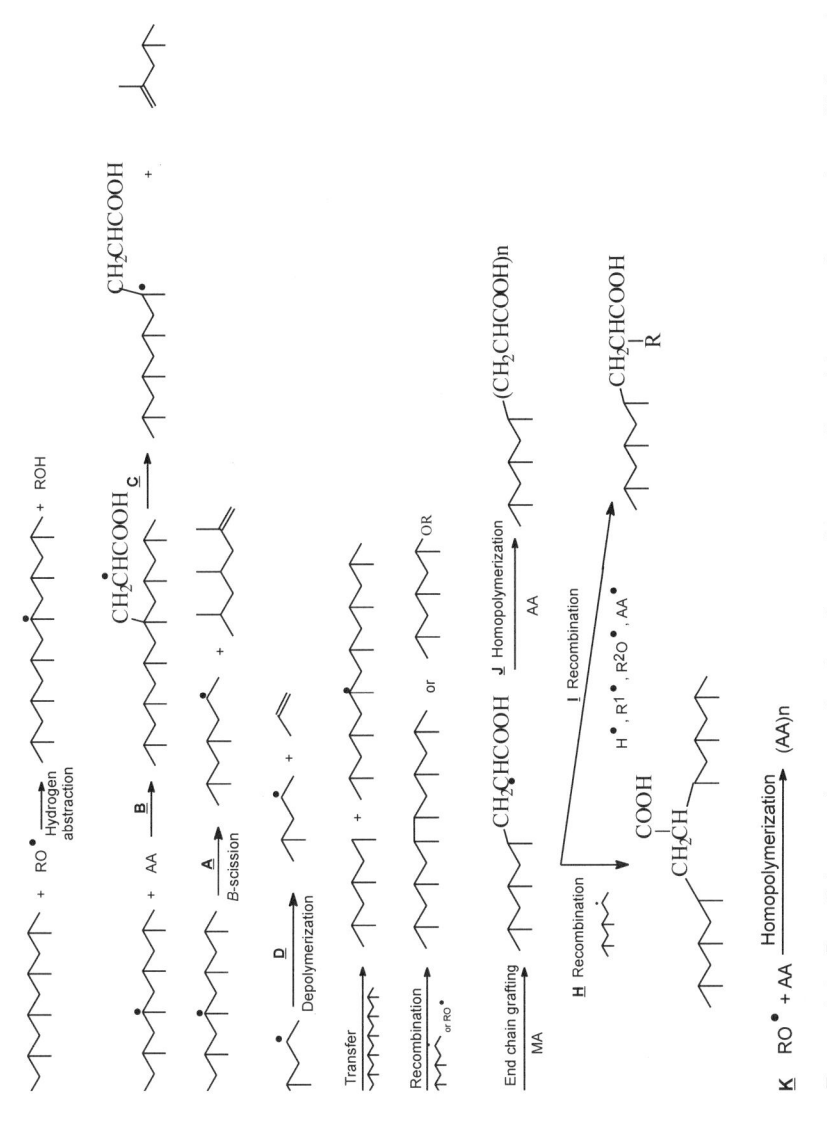

**FIGURE 3.1** Possible reaction mechanism for grafting acrylic acid onto polypropylene. (Adapted from Ref. 15.)

their modular screws, allow for greater flexibility in this area versus single screw extruders. There are many different types of screw elements that can be used to facilitate the reaction. Extruder manufacturers are always introducing new types of mixing elements that are designed for better distributive or dispersive mixing. The key to this type of reactive extrusion is to have a screw design that can disperse a low-viscosity material (i.e., monomer and peroxide, with a high-viscosity polymer). This must be done in a way that does not generate too much heat and that keeps the screw channels full. This can be achieved by using a combination of kneading blocks: narrow and wide, right- and left-handed pitch, and various degrees of stagger. There are specialty elements that can be used as well. Chapter 10 provides information concerning the underlying fundamentals of polypropylene functionalization via twin-screw reactive extrusion.

Clementini and Spagnoli [7] reported a method in which polypropylene was first functionalized with a hydroperoxide group before extrusion with acrylic acid. Other reactive sites could also be incorporated into polypropylene during its production. Chung [8] has several patents that discuss the incorporation of borane or *para*-methylstyrene into polypropylene. These functional sites can then be reacted with unsaturated monomers. Both of these methods supposedly limit both the amount of homopolymerization and the level of molecular weight reduction that occurs compared to other grafting methods.

Nongrafted poly(acrylic acid) is undesirable, due to the poor thermal stability. When processing PPgAA products, it is recommended that the process temperatures be kept below 220°C. If the processing temperature is higher than 220°C for any length of time, degradation of the poly(acrylic acid) will result. This degradation causes foaming of the extrudate, polymer degradation, and equipment corrosion. It is also recommended that whenever PPgAA products are used the processing equipment be purged with virgin polymer prior to shutdown to prevent degradation and possible metal corrosion.

## 3.5 MALEIC ANHYDRIDE GRAFTED POLYPROPYLENE

There is considerable information available on the grafting of polypropylene with maleic anhydride or its analogues. The grafting can be carried out using a variety of techniques, including thermal, solution, and extrusion. Eastman Kodak Company discusses one thermal grafting process in patent 3,433,777. In this patent, low-viscosity amorphous polypropylene is combined with maleic anhydride in an autoclave reactor. The material is heated to 325°C for 30 min. The resulting functional material has a high level of functionality, but it also has a drastically reduced molecular weight due to the thermal degradation of the polymer.

Another method developed by Hercules Inc. [9] consists of reacting crystalline polypropylene with maleic anhydride in the presence of organic peroxide in an inert liquid organic solvent. This process is expensive because it requires the separation of solvent and excess maleic anhydride from the modified polypropylene. A similar process, detailed in patent 3,414,551, uses a fluidized bed of polymer particles instead of a solvent.

## 3.6 GRAFTING REACTIONS OF MALEIC ANHYDRIDE ONTO POLYPROPYLENE

The use of organic peroxides in reactive extrusion to manufacture grafted polypropylene is widely documented. However, controversy remains over the exact grafting mechanism and the resulting graft structure. A summary of possible reaction mechanisms is shown in Fig. 3.2. For polypropylene, it is accepted that the free radical produced by peroxide decomposition attacks the tertiary hydrogen atom. The resulting polymer radical causes beta scission and polymer molecular weight reduction, as shown in pathway A. If there is an unsaturated monomer present to react with the polymer radical, grafting may occur prior to chain scission, as shown in pathway B. This results in a functional group attached pendant to the polymer backbone, shown in pathway C. There are several reactions that can occur if grafting does not occur. These are depolymerization (pathway D), transfer (pathway E), and recombination (pathway F). Pathway G shows the resulting polymer radical can still result in grafting. In this case, the monomer with be grafted onto the end of the polymer. After chain end grafting, various termination reactions can occur (pathways H, I, and J). Note that in pathway C the final graft structure is similar to chain end grafting. The polymer radical is deactivated through the various termination reaction steps.

There have been many studies conducted on the grafting of maleic anhydride (MAH) onto polypropylene. These studies have produced many different theories on the grafting mechanism. When comparing results from different studies, it is important to consider the methods used to conduct the grafting reaction. Different compounding methods, a variety of initiator types, and monomer and peroxide ratios have been studied. Hogt [10,11] has published several papers on this subject. His findings indicate that MAH grafts onto the polymer after beta scission. The addition of MAH did not contribute to molecular weight reduction of the polypropylene. At high MAH concentrations, there was evidence of suppression of the chain scission due to reaction of MAH with the polymer radical prior to beta scission. However, several researchers have found that at low levels of MAH addition, polymer degradation is enhanced [12–14], which does not substantiate the findings of Hogt. Several studies have reported the suppression of molecular weight reduction at high MAH concentrations, which was attributed to the formation of poly(maleic anhydride). In fact, at high MAH concentrations, oligimers of MAH have been observed by several researchers [10,15,16].

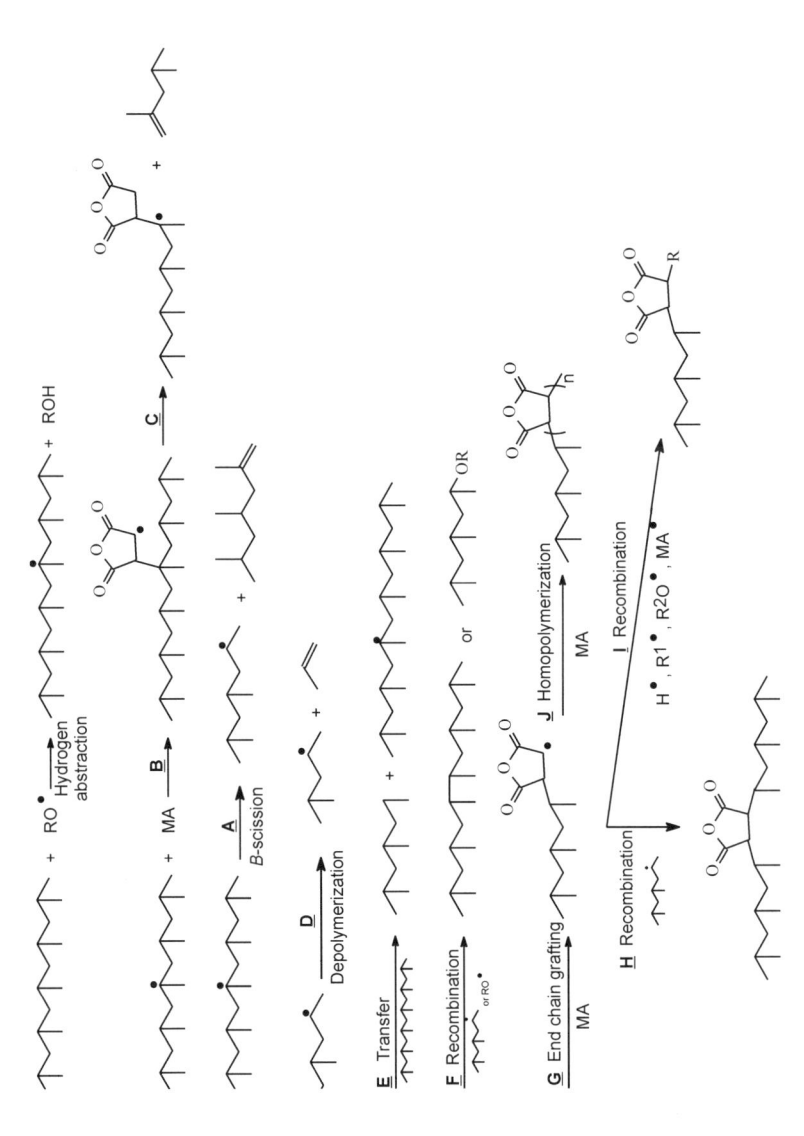

**FIGURE 3.2** Possible reaction mechanisms for grafting maleic anhydride onto polypropylene according to De Roover et al. (From Ref. 15. Reprinted with permission of John Wiley & Sons.)

## 3.7 MALEATED POLYPROPYLENE GRAFT STRUCTURE

There is a considerable amount of recent literature regarding the actual grafted structure of maleated polypropylene. Two of the most contested areas of debate center around the issue of where the maleic anhydride is attached to the polypropylene molecule (i.e., at the end of the chain or pendant along the chain) and whether or not the maleic anhydride is polymerized. Gaylord and Mishra [14] report that the grafting of maleic anhydride onto polypropylene occurs through the excitation of maleic anhydride caused by the rapid decomposition of the catalyst. The excimer is responsible for abstraction of the tertiary hydrogen from the polypropylene. The excimer then grafts onto the polypropylene and adds maleic anhydride by repetitive ionic coupling and electron transfer to maleic anhydride or through repetitive coupling with the maleic excimer, followed by ionic coupling. The net effect is the grafting and homopolymerization of maleic anhydride onto polypropylene. Their work also reports that this homopolymerization can be prevented by the addition of $N,N$-dialkylamides, such as dimethylformamide (DMF) or dimethylacetamide (DMAC). Both of these compounds presumably donate electrons to the maleic anhydride cation, followed by regeneration of the radical and maleic anhydride and amide.

De Roover et al. [15] report the presence of grafted poly(maleic anhydride) in polypropylene. The mechanism proposed for maleic anhydride grafting is shown in Fig. 3.3. They identified multiple grafted species through the use of Fourier transform infrared (FTIR) analysis. Using a curve fitting technique, absorption bands at $1784 \, \text{cm}^{-1}$ and $1792 \, \text{cm}^{-1}$ were identified as belonging to two different anhydride species. The absorption at $1784 \, \text{cm}^{-1}$ is assigned to the presence of poly(maleic anhydride) and the absorption band at $1792 \, \text{cm}^{-1}$ is assigned to succinic anhydride end groups. Additional studies by De Roover et al. [17] further substantiated the presence of poly(maleic anhydride) using $^{13}$C NMR and mass spectroscopy techniques.

Heinen et al. [18] conducted a study using $^{13}$C-labeled maleic anhydride to graft onto polyethylene, PP, and ethylene-polypropylene rubber. They found that in polypropylene, maleic anhydride grafting occurred prior to beta scission, to form an end-substituted unsaturated graft structure. Their proposed grafting mechanism and graft structure are shown in Fig. 3.4. They did not observe any homopolymerization of maleic anhydride occurring in polypropylene.

## 3.8 PEROXIDE SELECTION

Several researchers have looked at the effect of using various types of organic peroxides on the grafting of polyolefins. Callais and Kazmierczak [13] looked at a number of peroxides with different half-life temperatures. Their findings indicated that the peroxides with the highest grafting efficiency also had the greatest

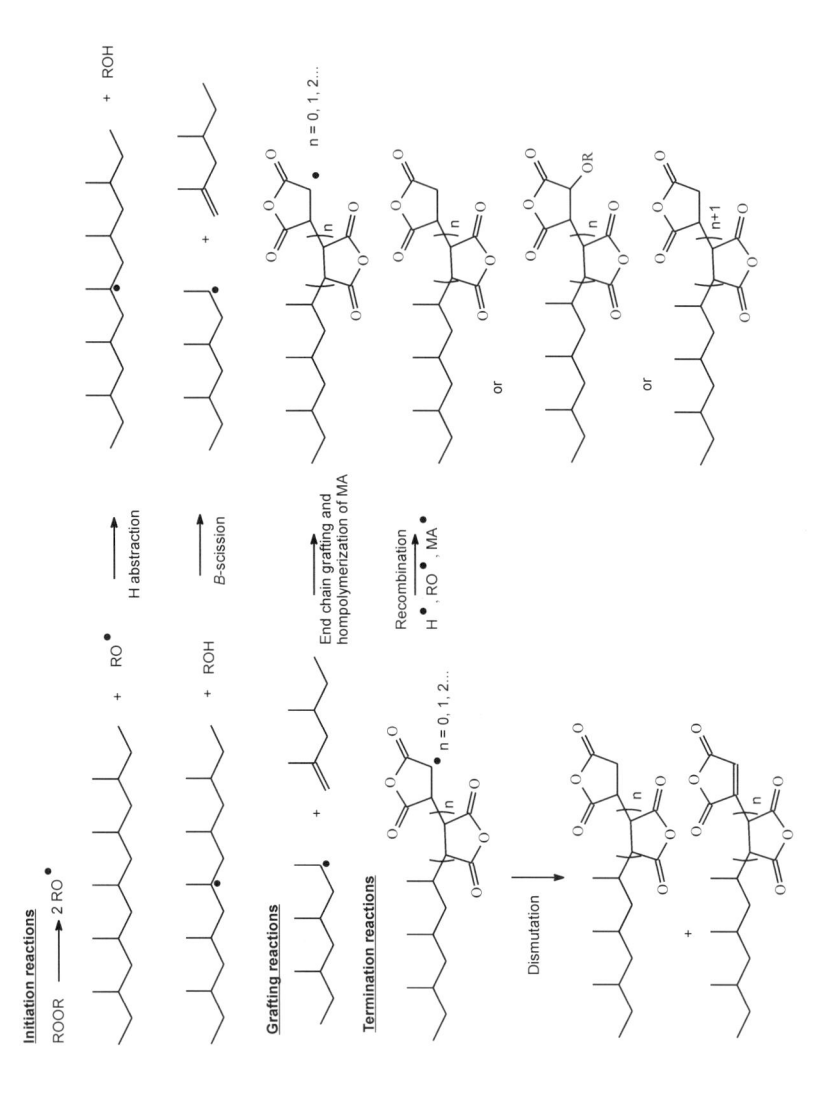

**FIGURE 3.3** Proposed PPgMAH grafting reaction. (From Ref. 15. Reprinted with permission of John Wiley & Sons.)

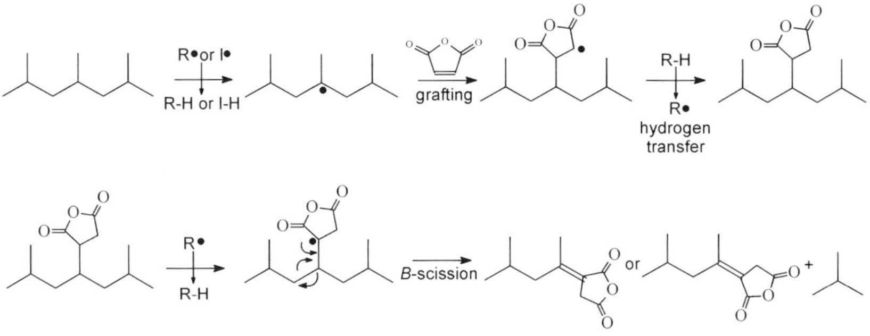

**FIGURE 3.4** Possible reaction mechanism for grafting maleic anhydride onto polypropylene according to Heinen et al. (From Ref. 21. Reprinted with permission from *Macromolecules*. 1996 American Chemical Society.)

effect on molecular weight characterized by melt flow rate measurement. They found no particular peroxide that had good grafting efficiency without causing significant molecular weight reduction. They did find that higher levels of peroxide increased the level of grafting. They also found that residence time and extrusion temperature had no effect on graft efficiency.

Hogt [10] also looked at various peroxide types and their effect on grafting efficiency. He found that the molecular structure of the peroxide, as it related to its solubility in both the polyolefin phase and maleic anhydride phase, played a role in grafting. His findings, although conducted in low-density polyethylene (LDPE), should apply to polypropylene as well since the solubility parameters are similar for LDPE and PP. The same correlation existed between graft efficiency and cross-link/visbreaking ability. Therefore, to increase the level of grafting, the molecular weight of the polypropylene must be reduced. Put another way, the corresponding measured melt flow rate by a melt indexer will necessarily increase with higher levels of grafted maleic anhydride adduct.

It is generally accepted that to increase the amount of grafted maleic anhydride, higher levels of catalyst are required. Consequently, higher levels of catalyst addition will be required for grafted polypropylene having lower molecular weight and higher incorporation of maleic anhydride monomer with greater grafted efficiency [10,13]. Table 3.1 shows four commercially available maleated polypropylenes and the effect of increasing maleic anhydride content on the melt flow rate, which is an indication of the grafted polypropylene molecular weight. As the level of maleic anhydride increases, so does the melt flow rate.

The inherent reduction in molecular weight accompanying the grafting process is an undesirable side reaction. If the molecular weight of the polypropylene becomes too low, the functionalized polypropylene becomes quite

TABLE 3.1 Effect of MAH Level on Melt Flow Rate

| Product | Wt% MAH | MFR (g/10 min) |
|---|---|---|
| PB 3001 | 0.1 | 5 |
| PB 3002 | 0.2 | 10 |
| PB 3150 | 0.5 | 50 |
| PB 3200 | 1.2 | 250 |

*Source:* Crompton Corporation Literature.

brittle with reduced strength and stiffness. Also, low molecular weight polypropylene may not cocrystallize with the polypropylene matrix, resulting in poor chemical coupling and low mechanical properties in the final composite.

There are a number of reported methods that can be used to limit the amount of polymer degradation while maintaining or increasing the graft efficiency. One method is the use of comonomers and other additives, which would limit polymerization or polymer degradation. U.S. patent 4,753,997 discusses the manufacture of maleic anhydride grafted polypropylene with the use of styrene as a comonomer and the use of a catalyst, such as $N,N$-dialkylethanolamine or $N,N$-dialkylaminoethyl acrylates. In this patent, the combined use of maleic anhydride, catalyst, peroxide, and styrene results in a grafted polypropylene with a higher graft level and less polymer degradation than those that did not contain the catalyst.

Another method reports the use of styrene without a catalyst. The styrene and maleic anhydride form a charge-transfer complex. This allows for higher graft efficiencies and higher molecular weight polypropylene [19]. However, the resulting graft structure is an alternating styrene and maleic anhydride block copolymer.

Other methods include reacting maleic anhydride with diene prior to grafting and modification of maleic anhydride by substituting one of its hydrogen atoms with a polar atom like bromine [19]. In fact, patents 3,882,194 and 3,873,643 both relate to grafting of a cyclic or polycyclic ethylenically unsaturated acid or anhydride onto polyolefins. The benefit in using these types of monomers is that they do not homopolymerize.

## 3.9 TEST METHODS

### 3.9.1 Fourier Transform Infrared Methodology

The testing for the presence of maleic anhydride in polypropylene at first appears to be quite simple. The carbonyl groups absorption peaks for both maleic anhydride and maleic acid show up at 1790 cm$^{-1}$ and 1712 cm$^{-1}$, respectively. In unmodified polypropylene, this area of the spectra is relatively devoid of other peaks. Therefore, one of the most common test methods used to determine the level of maleic anhydride in polypropylene is by pressing out a thin film and conducting FTIR analysis. Generally the peak height at 1790 cm$^{-1}$ is measured and a second peak, such as the CH$_3$ peak at 1165 cm$^{-1}$ or 810 cm$^{-1}$, is used to determine the pathlength. Dividing the peak intensity at 1790 cm$^{-1}$ by the peak intensity at 1165 cm$^{-1}$ yields a carbonyl index. The thickness of the film can also be measured manually and used in place of the 1165 cm$^{-1}$ absorption peak. Use of this type of method does not yield a quantitative measurement of the level of maleic anhydride present but can yield a relative measure of the amount present.

To quantify the amount of maleic anhydride present, a calibration curve is needed using levels of a standard material with known anhydride content. One method involves making solutions of known quantities of either hexadecylsuccinic anhydride (HDSA) or succinic acid (SA) in 1,2-methoxyethane. The absorbance of these solutions can be measured in a KBr cell of known pathlength. From measurements of absorbance per unit thickness as a function of concentration, a calibration curve can be generated [20]. Other researchers have used similar methods to develop calibration curves using different standards, e.g., *n*-octadecyl succinic anhydride or poly(maleic anhydride) [21].

Even with calibration curves, the exact measurement of maleic anhydride levels is difficult. There are several factors that will affect the test results that need to be considered. One is the level of unreacted MAH or free MAH. The most common method used to remove free MAH is to dry the material in a vacuum oven. Various temperatures and times are reported, but 120°C for a day or more seems to be the standard [30–32]. This drying also helps to convert any maleic acid back to the anhydride form.

Another factor is the presence of low molecular weight PPgMAH or oligimers of MAH. These products can be removed by dissolving the PPgMAH in a hot solvent, such as xylene or toluene, and precipitating it out of solution with acetone or methanol. This method will also remove any free MAH [21].

### 3.9.2 Titration

Titration is another method that is commonly used to measure the level of maleic anhydride in polypropylene. This method, too, has a number of problems. Again, there is the issue of removing the free MAH, and low molecular weight species must be dealt with. Also, care must be taken to prevent precipitation during titration, and of course there are the safety issues with handling hot solvents. The ratio of acid and anhydride present is also a concern. Several different test methods have been reported in the literature. Some involve converting any acid into anhydride via vacuum drying and then conducting nonaqueous titration in a hot solvent [10]. Another method is to convert all of the anhydride to acid by adding water [21,22].

## 3.10 THERMAL PROPERTIES OF PPgMAH AND PPgAA

Thermal gas chromatography–mass spectroscopy (GC/MS) analysis of PPgAA shows that various volatile components are generated during high-temperature exposure. There is evidence for peroxide degradation byproducts such as acetone and butyl alcohol. Other volatiles include acrylic acid monomer, carbon dioxide, and water. Carbon dioxide and water are formed when the acid dehydrolizes into anhydride and degradation of the anhydride occurs. The TGA curve shown in Fig. 3.5 is for PPgAA. The curve was generated in an $N_2$ atmosphere. The onset temperature is 320°C, compared to virgin polypropylene which is around 450°C.

TGA-FTIR and headspace GC/MS evaluation of PPgMAH shows that these materials are much more thermally stable than PPgAA. Figure 3.6 shows a TGA curve for PPgMAH. The onset temperature for degradation at 451°C is similar to that obtained for virgin polypropylene. Analysis of effluent gas composition shows evidence for carbon dioxide, carbon monoxide, maleic anhydride and water as degradation by-products.

## 3.11 OTHER FUNCTIONALIZATION TECHNIQUES

The incorporation of polar functional groups into polypropylene molecular structure is difficult. Unlike polyethylene where various polar monomers are copolymerized with ethylene, the addition of polar monomers to the polypropylene polymerization process cannot be done. These monomers react with the Ziegler-Natta catalysts and render them inactive. Therefore, postreactor functionalization has been the most traditional way of functionalizing polypropylene. In several recent patents, Chung et al. [8] disclosed the production of polypropylene, which contains copolymerized organoborane functional groups. Borane-containing groups can be incorporated into molecules of polypropylene during the copolymerizaton process without poisoning the catalyst. The borane group can then be reacted with unsaturated monomers, in the presence of an oxidative reagent, to produce grafted polypropylene without significant molecular weight reduction. The patent discloses the production of polypropylene grafted with polymethyl methacrylate (PMMA) with graft levels of 9–67% PMMA.

## 3.12 COUPLING MECHANISM

Maleated polypropylene and acrylic acid grafted polypropylene are used as chemical coupling agents in filled and reinforced polypropylene. These chemical coupling agents provide enhanced interfacial adhesion between the filler and

# Chemical Coupling Agents for Composites

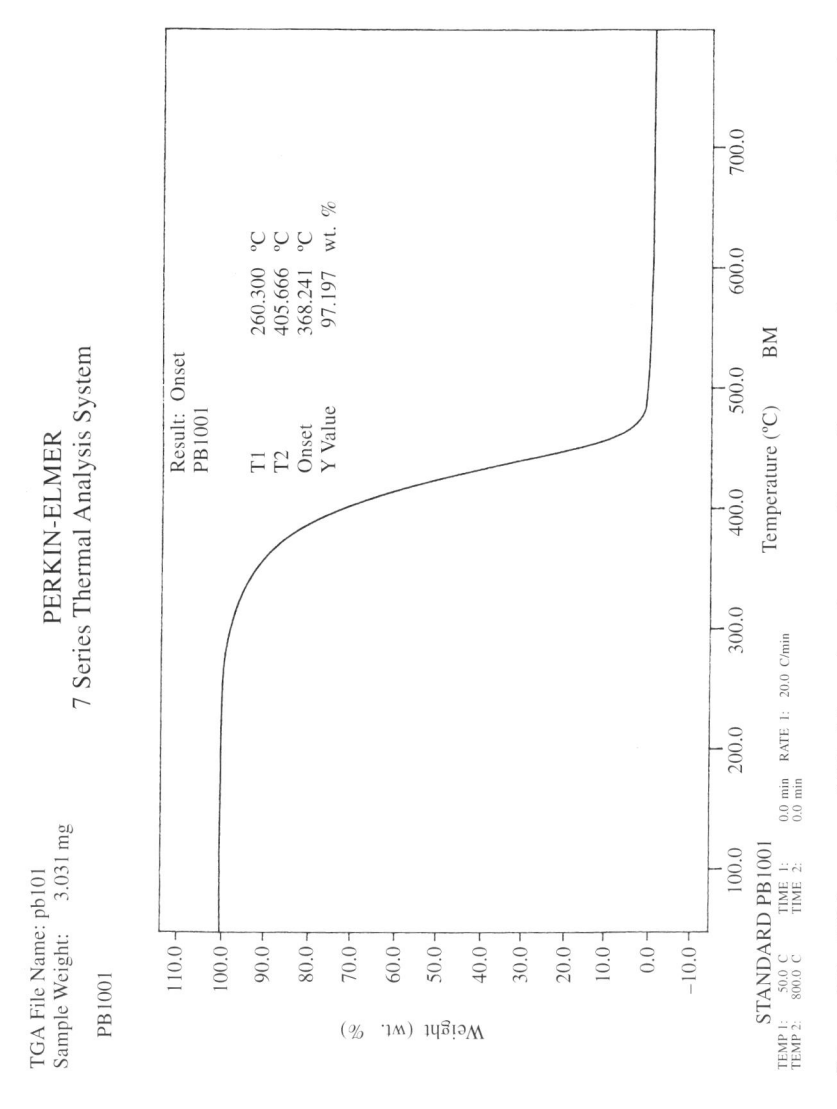

**FIGURE 3.5** TGA curve of PPgAA under nitrogen. (From Crompton Corporation Literature.)

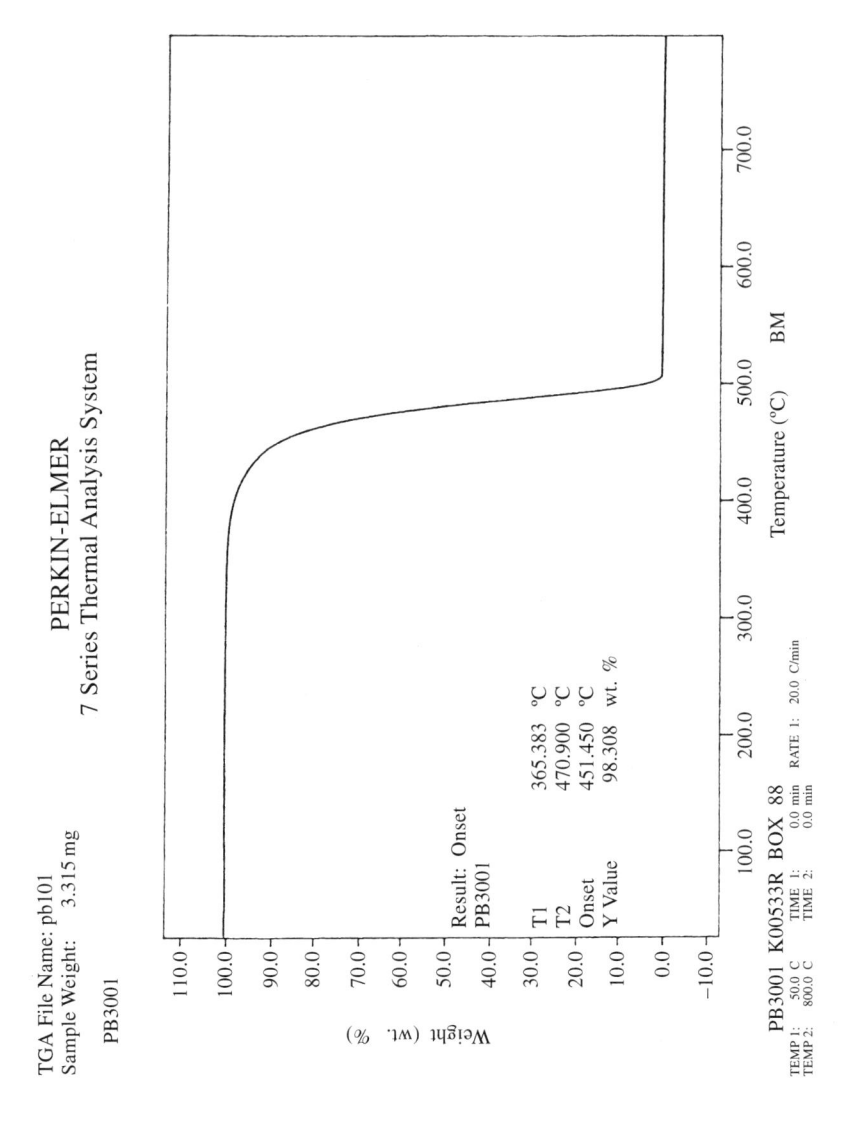

**FIGURE 3.6** TGA curve of PPgMAH under nitrogen. (From Crompton Corporation Literature.)

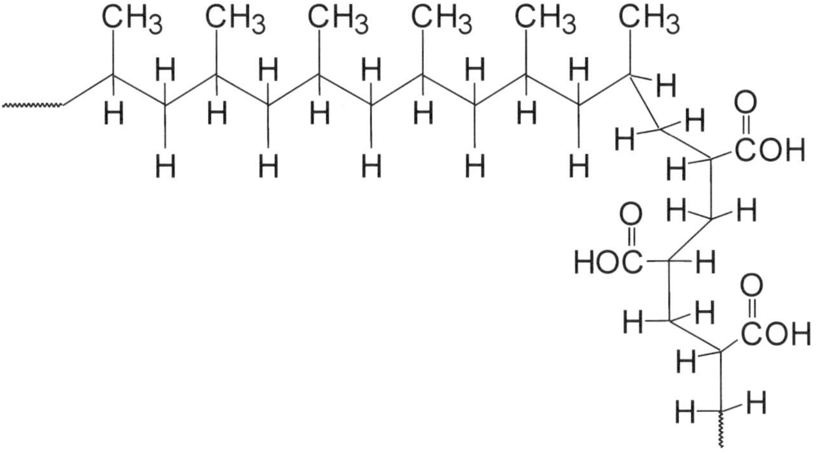

# Chemical Coupling Agents for Composites

**FIGURE 3.7** Chemical structure of acrylic acid grafted PP [AagPP]. (From Crompton Corporation Literature.)

polymer matrix. Chemical coupling involves two types of interactions that provide effective stress transfer from the relatively weak polymer matrix to filler reinforcement to give improved composite mechanical and thermal properties. Figure 3.7 shows a simplified chemical structure for PPgAA and Fig. 3.8 shows the structure for PPgMAH. The maleic anhydride group or the acrylic acid groups react with functional groups present on the surface of the filler to form chemical bonds as a primary interaction mechanism. Depending on the filler type, various surface functionalities are available for the acid or anhydride to react with. A second type of interaction consists of cocrystallization of the high molecular weight tail with the molecular chains of the polymer matrix giving physical entanglement.

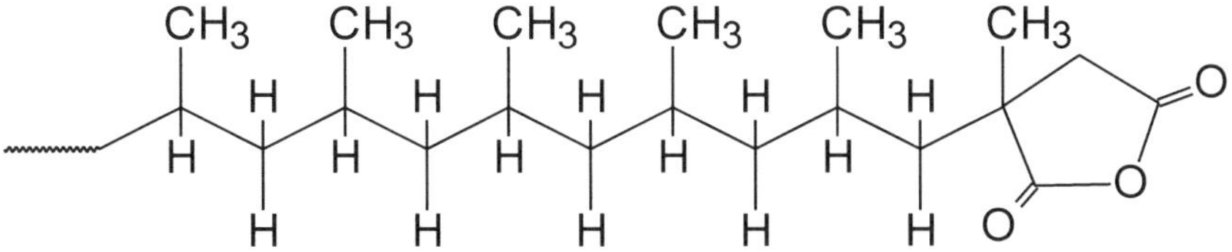

**FIGURE 3.8** Chemical structure of maleic anhydride grafted PP [PPgMAH]. (From Crompton Corporation Literature.)

The existence of interfacial adhesion due to chemical coupling can also be seen visually. Figures 3.9 and 3.10 are scanning electron microscopy (SEM) photomicrographs of fracture surfaces of glass fiber–reinforced polypropylene. Figure 3.9 is of an uncoupled composite and Fig. 3.10 is chemically coupled. The surface of the glass fiber in Fig. 3.9 is smooth, indicating that no polymer is adhered to the surface. Also, there is evidence at the base of the glass fiber that debonding has occurred between the glass fiber and the polymer. Figure 3.10 on the other hand, shows that the surface of the glass fiber is rough. This latter photomicrograph shows the presence of a layer polymer adhering to the fiber surface with the absence of debonding at the fiber–polymer interface.

The chemical structure of the grafted polypropylene plays an important role in its performance as a coupling agent. As discussed previously, the acid or anhydride functionality will react with the functional groups present on the surface. Therefore, the graft structure plays an important role. However, the presence of grafted homopolymerized acid or anhydride attached to a single polypropylene molecule is probably not the ideal structure. Numerous acid or anhydride groups would allow for multiple sites for bonding to the surface of the filler. However, because they are grafted onto one polypropylene molecule, there would be limited cocrystallization with the polypropylene matrix. The homopolymerized acid or anhydride groups may also reduce the number of available reactive sites present on the surface to react with other grafted polypropylene molecules.

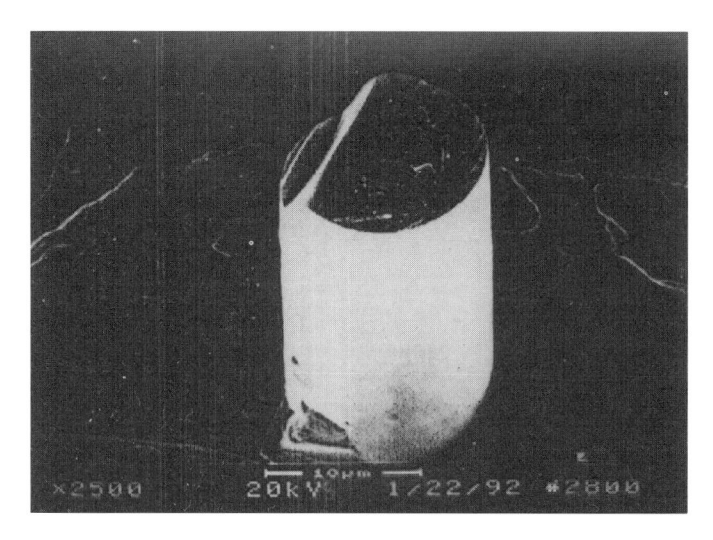

FIGURE 3.9 Scanning electron micrograph (SEM) of uncoupled GF-PP fracture surface. (From Crompton Corporation Literature.)

# Chemical Coupling Agents for Composites

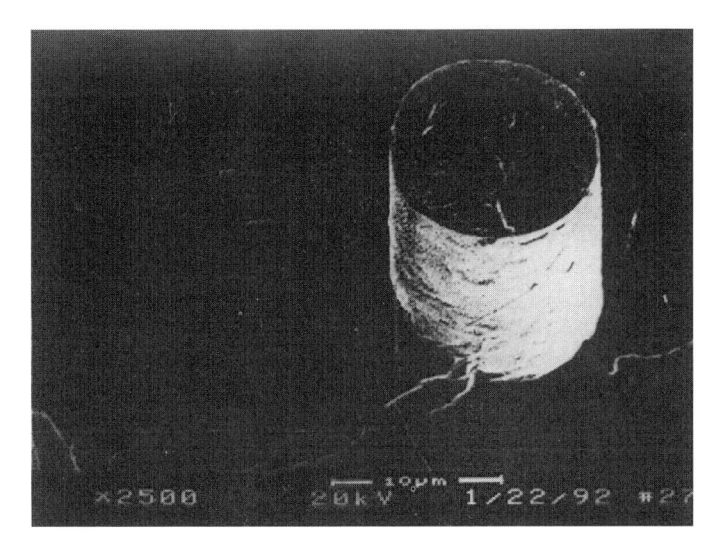

**FIGURE 3.10** Scanning electron micrograph (SEM) of coupled GF-PP fracture surface. (From Crompton Corporation Literature.)

Grafting that occurs at multiple sites along the polypropylene molecule may also not be desirable. If all of the maleic anhydride groups react with the surface of the filler, then again this may limit the cocrystallization of the grafted polypropylene backbone with the polypropylene matrix.

Evidence for the effect of graft location was reported in a study reported by Duvall et al. [23,24] using two different maleated polypropylenes as compatibilizers in polypropylene-polyamide blends. In their study, they evaluated both a high-maleated and a low-maleated polypropylene. The low maleic content PPgMAH was primarily end grafted, whereas the high maleic content PPgMAH had multiple maleic groups grafted along the polypropylene molecule. They found that the low anhydride content PPgMAH imparted higher maximum fracture strain versus the high maleic content PPgMAH. This was attributed to the difference in the grafted structure and melting point differences. The low maleic anhydride PPgMAH exhibited one melting peak and a single spherulite type. The PPgMAH phase with high graft content separated from the polypropylene and exhibited two spherulite types. These investigators also reported a 10-fold higher adhesive strength between the polypropylene and the low maleic content PPgMAH versus the high maleic content PPgMAH.

As discussed in the preceding paragraph, the ability of the PPgMAH to cocrystallize with the polypropylene matrix is very important. The molecular weight of the PPgMAH determines how well the grafted polypropylene molecules will diffuse into the polypropylene matrix and cocrystallize. If the molecular weight is too low, little or no cocrystallization occurs and poor mechanical properties are obtained. Several studies have been conducted in looking at the crystallization kinetics and spherulite structure of glass fiber–reinforced polypropylene [25,26]. Evidence for transcrystalline regions next to the glass fiber has been observed. Increases in transcrystallinity and decreased spherulite size have resulted in the enhancement of tensile strength [26].

Yin et al. [28] found that the addition of maleated polypropylene to wood fiber–filled polypropylene increased the nucleation capacity. The resulting interphase structure consisted of a transcrystalline region surrounding each wood fiber. They also reported that maleated polypropylene crystallized faster in the presence of wood fiber.

There are a few studies that investigated the effects of PPgMAH or PPgAA on the interphase thickness of polypropylene composites. In a study by Felix et al. [29], the interphase thickness was determined for a group of cellulose-polypropylene composites containing a set of three different coupling agents to promote interfacial adhesion between the polypropylene matrix and cellulose fibers. The coupling agents were maleic anhydride functionalized polymers encompassing a range of molecular weights. These investigators found that as the molecular weight of the coupling agent increased, so did the thickness of the interphase layer along with the mechanical properties.

Karian [30] reported similar findings in glass fiber–reinforced polypropylene (GFRP) composites. In this study he uses heat capacity measurements to estimate the interfacial thickness. He reported an excellent correlation between interfacial thickness and tensile strength. As the interfacial thickness increased so did the tensile strength. Chapter 12 describes development of megacoupled GFRP composites having mechanical strength at the perfect coupling limit.

A similar study by Karian and Wagner [31] used tensile creep measurements to determine the interfacial shear strength of GFRP composites when PPgAA is used as a coupling agent. They determined that there was a critical level of coupling agent addition where interfacial adhesion was greater than the matrix yield stress. Above this critical concentration of PPgAA, the failure mode is attributed to flaws in the matrix rather than debonding of the fibers.

## 3.13 USE OF PPgMAH AND PPgAA AS COUPLING AGENTS IN FILLED AND REINFORCED POLYPROPYLENE

### 3.13.1 Introduction

The following data will demonstrate the use of both maleated and acrylic acid grafted polypropylene as chemical coupling agents in filled and reinforced polypropylene. Various factors, such as the effect of molecular weight, functionalization level, and functionalization type will be examined.

### 3.13.2 Glass Fiber Reinforced Polypropylene

*Glass Fiber Sizing Chemistry.* The composition of the sizing or coating used in glass fiber manufacture is very important to the final mechanical properties of GFRP. There are a substantial number of patents related to this subject. The size usually comprises an oxidized polyolefin, a polymeric film former, an organofunctional silane, and an organic acid. The oxidized polyolefin commonly used is Epolene E-43 made by Eastman. This material is a low molecular weight maleated polypropylene. The type of silane used is often an aminosilane such as A1100, A1120, and A1130 available from Crompton Corporation. Patent 4,489,131 (Dec. 18, 1984), granted to Owens-Corning Fiberglas Corporation, demonstrates the need for an organic acid, preferably terephthalic acid as part of the sizing formulation. When the coated glass fiber surface contains terephthalic acid as part of the patented size formulation, the resulting GFRP has higher tensile strength, flexural strength, Izod impact, and heat deflection temperature.

Sizing is used on glass fiber for a number of reasons. It is used to hold the fibers in a bundle to prevent "fuzzing" of the glass fiber during handling. It also helps protect the glass fiber from damage and provides compatibility with the resin matrix, which improves the mechanical properties of the composite. A study by Daemen and den Besten [32] demonstrates this very well. Their study compares the use of both sized and unsized glass fibers in polypropylene as well as the use of PPgAA as a coupling agent. Figure 3.11 shows the tensile strength properties of glass-filled polypropylene with composites containing various levels of glass fiber. Both sized and unsized glasses were used. Data

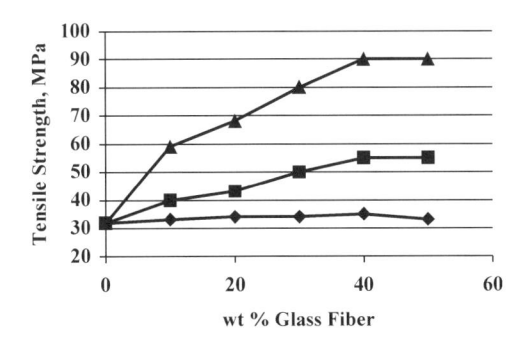

**FIGURE 3.11** Influence of compatibility and adhesion tensile strength. (From Ref. 32.) ▲ 10% PPgAA, ◆ noncompatible, ■ compatible.

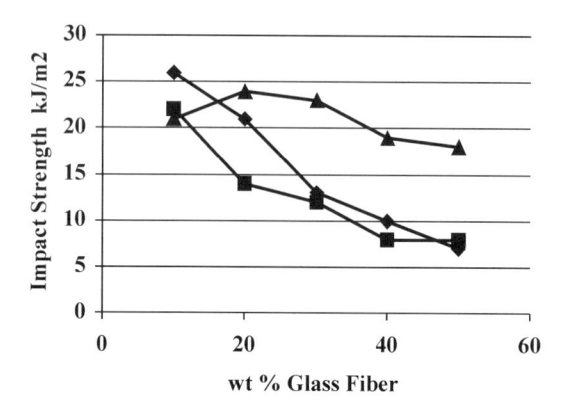

**FIGURE 3.12** Compatibility-adhesion unnotched Charpy impact strength. (From Ref. 32.) ▲ 10% PPgAA, ◆ noncompatible, ■ compatible.

on the use of 10 wt% PPgAA are also shown. When unsized glass fiber was used, no improvement in tensile strength was observed at any level of glass addition. The use of sized glass showed an increase in tensile strength as the percent glass was increased. With the addition of 10 wt% PPgAA, significant improvements in tensile strength were observed at each level of glass loading. In fact, 50–60% increases are seen. Impact strength, depicted in Fig. 3.12, shows a slightly different trend. Generally, the impact strength is higher with the unsized glass versus the sized glass. The addition of PPgAA improves the impact strength to yield the highest impact values, except at 10 wt% glass fiber loading. With the use of this coupling agent, up to 100% increases in impact strength are realized.

*Comparison of PPgAA versus PPgMAH Coupling Agents in GFRP.* Figures 3.13–3.15 show data on 30 wt% GFRP that contains either PPgAA or PPgMAH [33]. The PPgAA used in the study was Polybond 1001. This product is a homopolymer polypropylene grafted with 6 wt% acrylic acid. The PPgMAH used was also a homopolymer polypropylene grafted with 1.5 wt% maleic anhydride. Figure 3.13 shows the effect of coupling agent addition on the tensile strength properties. Note that different levels of coupling agent were used, depending on the type of functionality. The PPgAA was added at 5%, 10% and 20% by weight. The PPgMAH was added at 0.5%, 1%, 2%, and 5% by weight. The addition of both coupling agents shows significant increases in tensile strength. At the highest loading of each coupling agent, a 30% increase in tensile strength is obtained. The PPgMAH coupling agent is more efficient, showing mechanical property improvement with as little as 1% addition and reaching its maximal tensile strength at 5 wt% loading. At 5 wt% loading of

# Chemical Coupling Agents for Composites

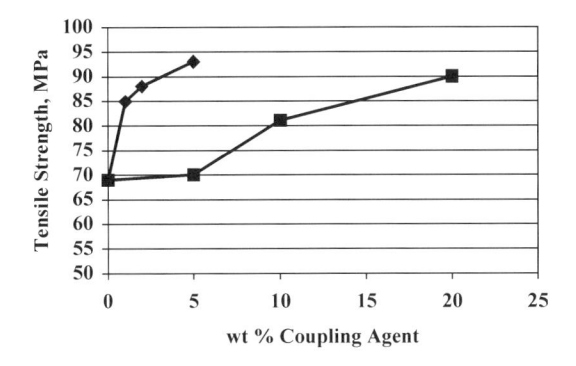

**FIGURE 3.13** Tensile strength comparison: PPgAA versus PPgMAH. (From Ref. 33.) ■ PPgAA, ♦ PPgMAH.

PPgAA, very little increase in tensile strength is seen. To attain the highest tensile strength, 20 wt% PPgAA is required. Figure 3.14 shows the reverse-notched Izod impact strength properties. This property shows the same trend as tensile strength, except that a 100% increase in impact strength is obtained at the highest coupling agent loading. Notched Izod impact strength is shown in Fig. 3.15. This property shows a slightly different trend in comparison with the other properties. At low levels of addition, 0.5 wt% PPgMAH, or 5 wt% PPgAA, a slight decrease in impact strength is seen. However, there is improvement in impact strength at higher loading levels. The reason for the decrease in impact strength at low loadings is attributed to suppression of microcrazing [31].

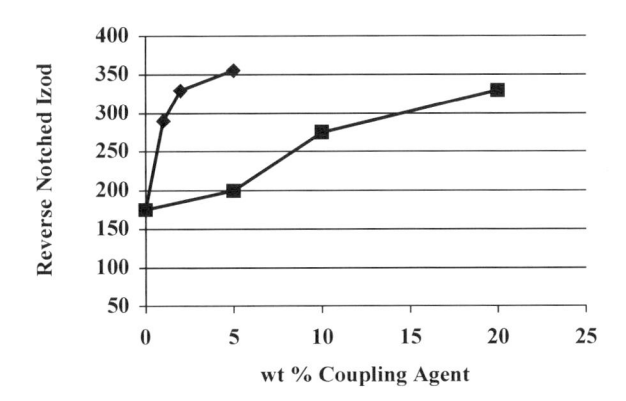

**FIGURE 3.14** Reverse notched Izod comparison: PPgAA versus PPgMAH. (From Ref. 33.) ■ PPgAA, ♦ PPgMAH.

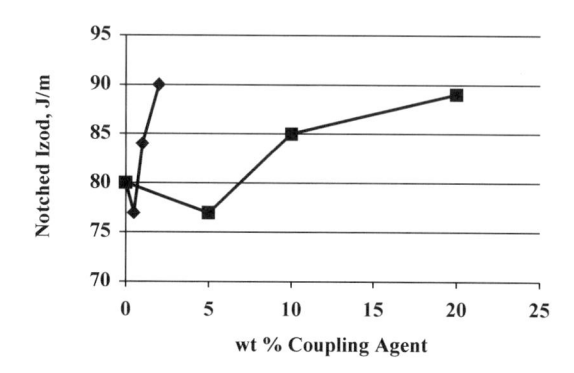

**FIGURE 3.15** Notched Izod comparison: PPgAA versus PPgMAH. (From Ref. 33.) ■ PPgAA, ◆ PPgMAH.

*Different PPgMAH Types in GFRP.* There are a number of commercially available coupling agents, most of which differ in the level of maleic anhydride content. The effect of MAH content on the efficiency of PPgMAH coupling agents is shown in Figs. 3.16, 3.17, and 3.18. The data in these figures show the mechanical properties of 30 wt% GFPP, containing various grades of PPgMAH. PB 3001 contains 0.1 wt% MAH, PB 3150 contains 0.5 wt%, and PB 3200 contains 1.2 wt%. The addition level of coupling agent varies depending on their concentration of MAH. The lowest functionalized material, PB 3001, was added at 2%, 5%, and 10% by total weight of the composite. PB 3150 was added at 1%,

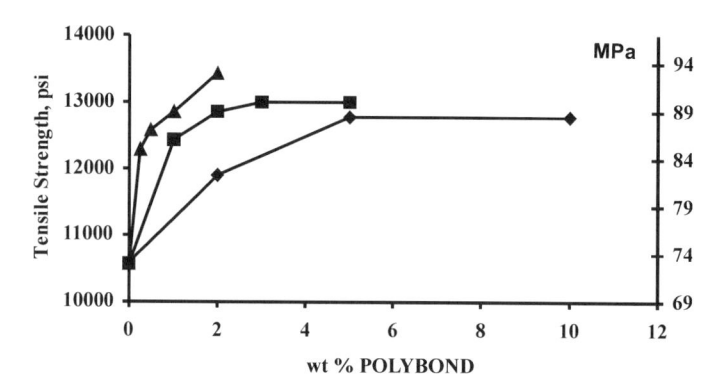

**FIGURE 3.16** Effect of PPgMAH type on tensile strength of 30 wt% GFRP composite. (From Cromptonn Corporation Literature.) ▲ PB 3200, ◆ PB 3001, ■ PB 3150.

# Chemical Coupling Agents for Composites

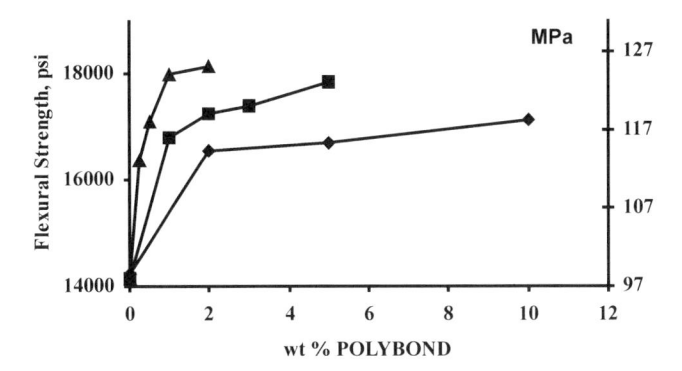

**FIGURE 3.17** Effect of PPgMAH type on flexural strength of 30 wt% GFRP composite. (From Crompton Corporation Literature.) ▲ PB 3200, ◆ PB 3001, ■ PB 3150.

2%, 3%, and 5%. The highest functionalized product, PB 3200, was added at 0.5%, 1%, 1.5%, and 2% by weight.

Figure 3.16 shows the effect of coupling agent addition on the tensile strength properties. With the addition of any of the three PPgMAH products, a significant increase in tensile strength is seen. Each exhibits a sharp increase in tensile strength followed by a gradual leveling off. For PB 3001, the tensile strength begins to level off at the 5 wt% addition level. An increase in addition level up to 10 wt% results in no further improvement. For PB 3150, the tensile strength begins to level off at the 2 wt% addition level. The PB 3200 shows a similar trend, but the plateau region has not yet been reached.

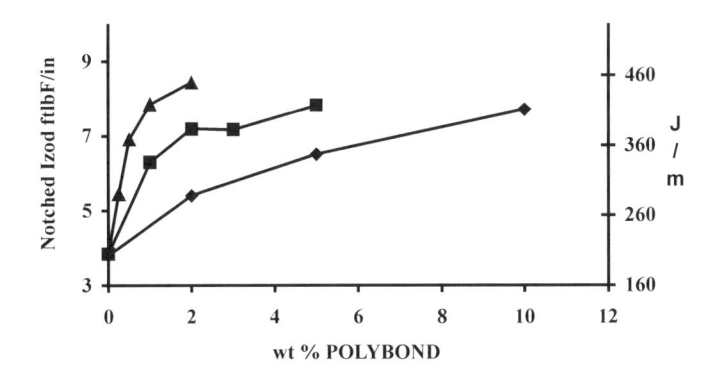

**FIGURE 3.18** Effect of PPgMAH type on reverse notched Izod of 30 wt% GF-PP composite. (From Crompton Corporation Literature.) ▲ PB 3200, ◆ PB 3001, ■ PB 3150.

Greater chemical coupling efficiency for higher maleated PPgMAH (PB 3200) is demonstrated by similar tensile strength obtained at 1 wt% addition level compared to 2 wt% and 5 wt% addition levels of PB 3150 and PB 3001, respectively. The flexural strength properties are shown in Fig. 3.17. Trends similar to those of the tensile strength properties are seen. However, impact properties show a different trend. Figure 3.18 shows that no plateau region is seen. Impact strength continues to increase with higher coupling agent loadings. The amount of increase gets smaller at the higher addition levels but improvement is still seen. Therefore, if high impact properties are desired, then high loadings of coupling agent should be used.

These data show the effect of using maleated polypropylenes with different levels of maleic anhydride. As the level of maleic anhydride in the coupling agent increases, the greater is its efficiency, resulting in lower required dosage levels.

Besides the level of functionality, molecular weight also plays an important role in coupling agent performance. Hyche et al. [34] conducted a study using various PPgMAH products with different levels of MAH and different molecular weights. Figures 3.19 and 3.20 show data on 30 wt% GFRP using two different PPgMAH types. "Hmah/Mmw" stands for high maleic anhydride content and medium molecular weight. "Lmah/Hmw" stands for low maleic anhydride content and high molecular weight. The material was considered to have a low maleic anhydride content if it contained less than 1.5%. The molecular differences between the two materials were as follows: Hmw material had an $M_n$ of 60,000, whereas the Mmw material had an $M_n$ between 15,000 and 25,000.

Figure 3.19 shows the tensile strength properties using both of these PPgMAH materials. Although the addition of either PPgMAH results in an improvement in the tensile strength, the Lmah/Hmw material gave the highest properties. The same is true for the flexural strength and impact properties as

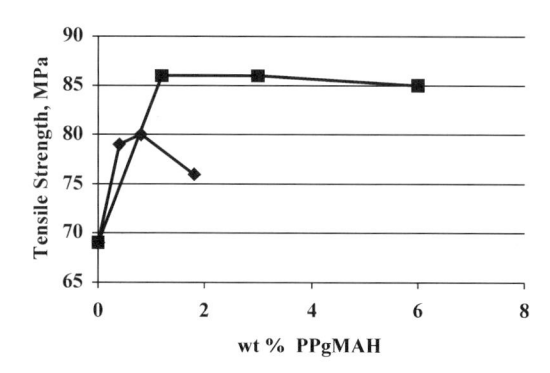

FIGURE 3.19 Effect of PPgMAH type-level on tensile strength. (From Ref. 34.) ■ Lmah/Hmw, ♦ Hmah/Mmw.

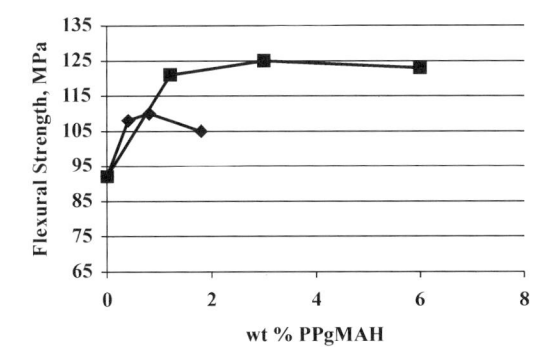

**FIGURE 3.20** Effect of PPgMAH type-level on flexural strength. (From Ref. 34.) ■ Lmah/Hmw, ◆ Hmah/Mmw.

shown in Figs. 3.20 and 3.21. Data indicates that for GFRP, a high molecular weight, low maleic content PPgMAH is the coupling agent of choice.

*Different Glass Fiber Types.* There are several different manufacturers of glass fibers. Each supplier applies its own proprietary size to the glass fiber. The performance of various glass fibers is shown in Figs. 3.22–3.24. The data shown are on 30 wt% GFRP that contains various levels of maleated polypropylene. The following glass fiber types were evaluated: PPG 3298, OCF 147, Star Strand 740, and Vetrotex 968. Figure 3.22 shows the tensile strength properties. Composites containing no coupling agent have tensile strengths between 60 and 80 MPa. The addition of coupling agent, even as low as 1 wt%, results in a significant increase in tensile strength with three of the glasses. Increasing the coupling agent level

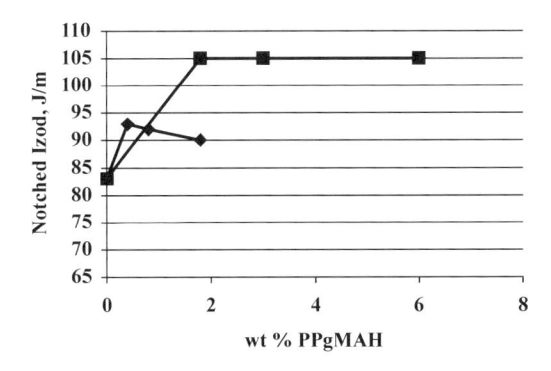

**FIGURE 3.21** Effect of PPgMAH type-level on notched Izod. (From Ref. 34.) ■ Lmah/Hmw, ◆ Hmah/Mmw.

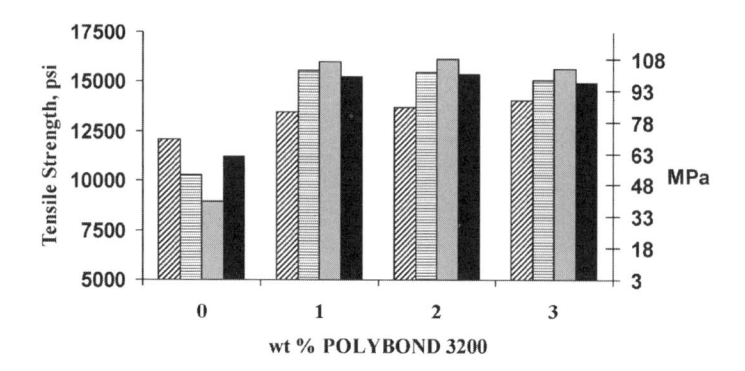

FIGURE 3.22 Effect of glass fiber type on tensile strength versus wt% Polybond 3150. (From Crompton Corporation Literature.) ▨ PPG 3298, ☰ OCF 147, ☐ Star Strand 740, ■ Vertrotex 968.

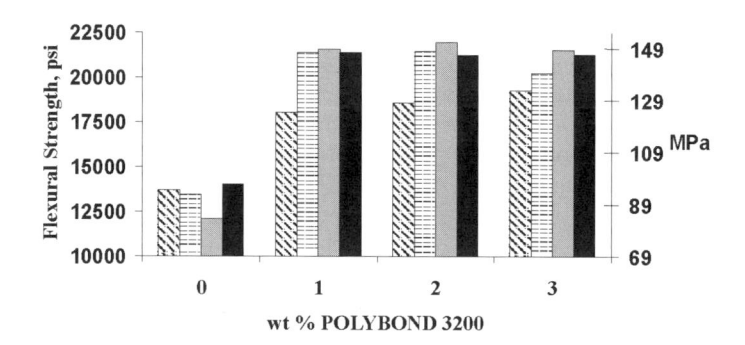

FIGURE 3.23 Effect of glass fiber type on flexural strength at various addition levels of Polybond 3150. (From Crompton Corporation Literature.) ▨ PPG 3298, ☰ OCF 147, ☐ Star Strand 740, ■ Vertrotex 968.

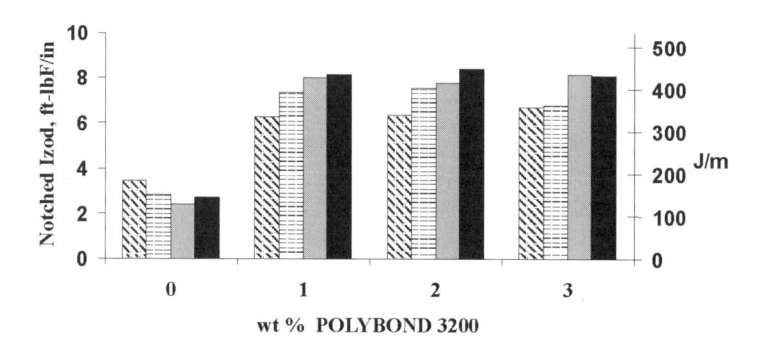

FIGURE 3.24 Effect of glass fiber type on reverse notched Izod impact with various addition levels of Polybond 3150. (From Crompton Corporation Literature.) ▨ PPG 3298, ☰ OCF 147, ☐ Star Strand 740, ■ Vertrotex 968.

## Chemical Coupling Agents for Composites 63

results in little additional increase in tensile strength for any of the glass fibers. Flexural strength properties are shown in Fig. 3.23 with similar trends as tensile strength. With the addition of 1 wt% coupling agent, up to a 50% increase in flexural strength is seen with some of the glass fibers. Chapter 9 describes glass fiber manufacturing and reinforcement technology. Chapter 13 deals with long-term aspects of tensile behavior characterized by creep measurements.

Reverse notched Izod impact property data are shown in Fig. 3.24. In uncoupled composites, there is little difference in glass performance. The PPG 3298 yields the highest properties with no coupling agent. By addition of coupling agent, impact performance increases significantly for all glass fibers.

*Polyamide Glass Fiber.* A good example of glass fiber sizing chemistry and mechanical property relationship is shown in Fig. 3.25. These data were generated using glass fibers designed for use in polyamide, not polypropylene. The particular glass used was Silenka 8045. Use of this fiber in polypropylene without coupling agent results in tensile strength properties lower than traditional uncoupled glass-filled polypropylene. However, with the addition of maleated polypropylene, a 300% increase in tensile strength is obtained. This shows that the sizing chemistry is not very compatible with the base polypropylene but that it is very compatible with the coupling agent. Most polyamide glass fiber contains a size that has a high level of amine functionality.

However, not all polyamide glass fiber types work in chemically coupled PP composites. Figures 3.26, 3.27, and 3.28 compare the use of two different glass fibers from PPG. PPG 3242 is a 14-μm glass fiber sized for polypropylene. PPG 3540 is a 10-μm glass fiber sized for polyamide (PA). When compounded into PP the two types of glass fibers perform differently. In composites containing no PPgMAH, the 3540 glass fiber yields very low tensile and flexural strength

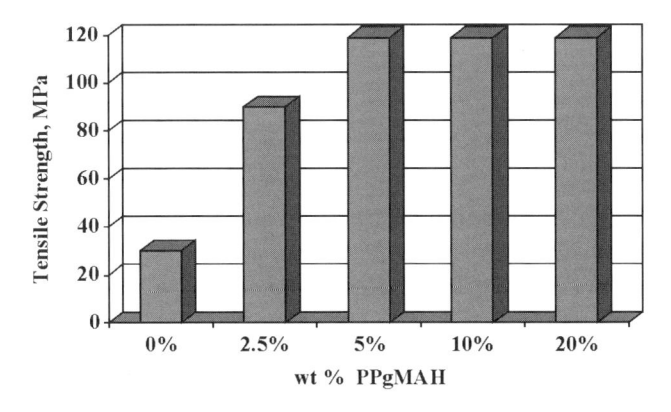

**FIGURE 3.25** Effect of polyamide glass fibers in polypropylene. (From Crompton Corporation Literature.)

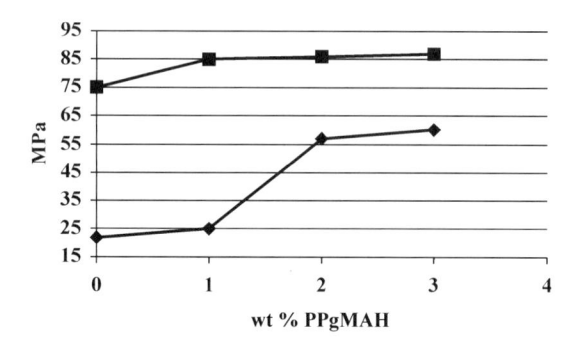

**FIGURE 3.26** Comparison of tensile strength values for polyamide and polypropylene glass fiber types. (From Crompton Corporation Literature.) ■ PPG 3242,

properties in comparison with the 3242. This shows very poor compatibility between the PP and the PA glass fiber. Although the addition of PPgMAH increases the adhesion and mechanical properties of the composites containing the 3540 glass fiber, they never reach those of the composites containing the 3242 glass.

*Detergent Resistance.* Another area where glass fiber size chemistry shows performance differences is in detergent resistance, often referred to in industry as "suds" resistance. Suds resistance is measured as mechanical property change after conditioning in hot water containing detergent. The data given in Fig. 3.29 shows reverse Izod impact performance of 30 wt% GFRP composites

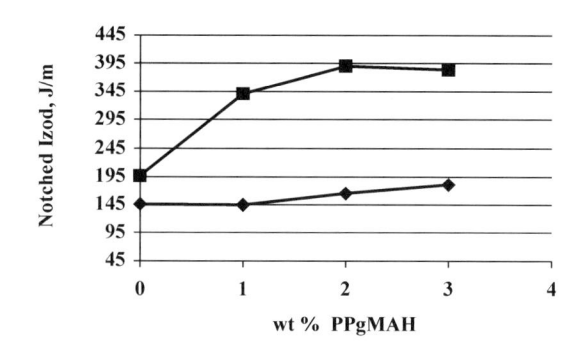

**FIGURE 3.27** Comparison of reverse notched Izod for polyamide and polypropylene glass fiber types. (From Crompton Corporation Literature.) ■ PPG 3242,

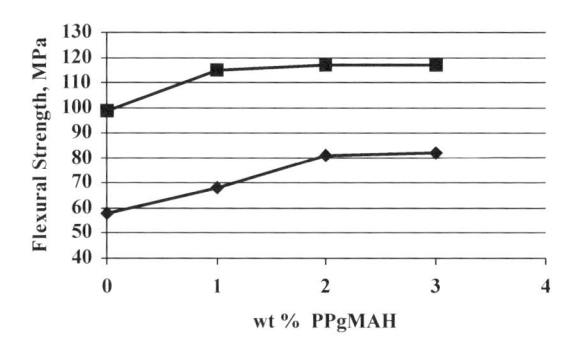

**FIGURE 3.28** Comparison of flexural strength values for polyamide and polypropylene glass fiber types. (From Crompton Corporation Literature.) ■ PPG 3242, ◆ PPG 3540.

with four different glass fiber types. Conditioning was done at 95°C in a 0.5 wt% phosphorus-free detergent solution. Testing was done after 20- and 40-day exposure. All of the composites contain PPgMAH coupling agent. The Vetrotex, Schuller, and OCF glass all had similar impact strength for the unaged samples. The PPG glass yielded higher impact properties. After aging, impact properties decreased for all glass fiber types. Vetrotex glass showed the lowest percent decrease, and no difference between 20- and 40-day aging. The PPG glass showed the highest percent decrease, but the final impact properties were similar to those of the Vetrotex glass. Both Schuller glass and OCF glass showed

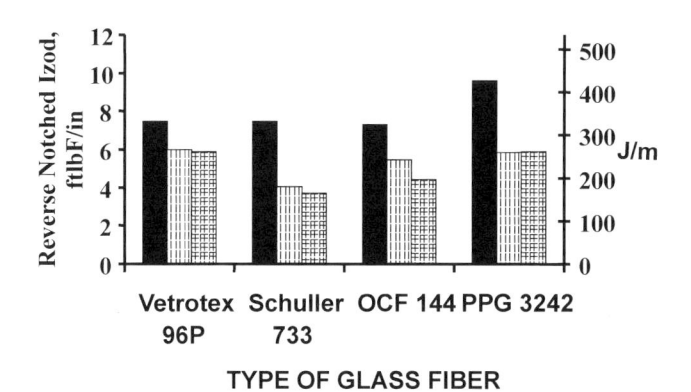

**FIGURE 3.29** Detergent resistance of 30 wt% GFPP: effect of glass fiber type on reverse notched Izod impact. (From Crompton Corporation Literature.) ■ Unaged, ▥ 20 days, ▦ 40 days.

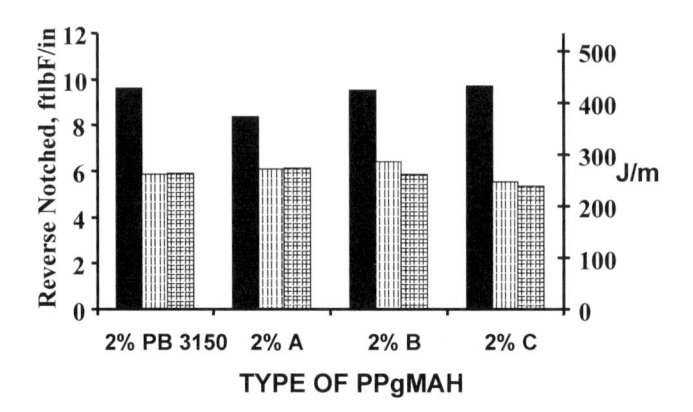

**FIGURE 3.30** Detergent resistance in 30 wt% GFPP: effect of PPgMAH type on reverse notched Izod. (From Crompton Corporation Literature.) ■ Unaged, ▥, 20 days, ▦ 40 days.

significant decreases after aging and with additional decrease seen after 40-day aging.

The type of coupling agent has little effect on suds resistance. Figure 3.30 shows the performance of various maleated polypropylenes in composites containing 30 wt% PPG 3242 fiberglass. All of the composites contain 2 wt% maleated polypropylene from different suppliers. All of the unaged composites have similar performance, about 500 J/m. After aging, impact performance drops to around 300 J/m for all composites. These data show that the sizing chemistry, rather than the presence or type of coupling agent, influences the suds resistance.

*Glass Fiber and Mineral Fiber Combinations.* One area of recent interest is in the combination of glass fiber with other fillers. Glass fiber with mica is commonly used to reduce part warpage. Another reason for glass and mineral fiber combinations is cost reduction. One type of mineral fiber that can be used is a wool fiber from Lapinus. This fiber is composed of silicon dioxide, aluminum trioxide and calcium oxide. It has a diameter of 5 μm and a fiber length of 200 μm. The mechanical properties of composites containing 15 wt% glass fiber and 15 wt% Lapinus fiber are shown in Figs. 3.31–3.33. Two levels of maleated polypropylene addition were studied. The composite containing no coupling agent yields mechanical properties that are significantly lower than those of uncoupled 30 wt% GFRP. However, with the addition of 1.5 wt% maleated polypropylene, the mechanical properties are improved so that they compare favorably with those of uncoupled 30 wt% GFRP.

# Chemical Coupling Agents for Composites

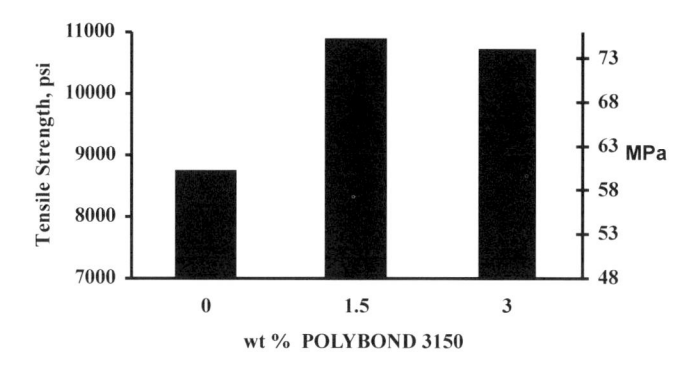

**FIGURE 3.31** Effect of PB 3150 level on tensile strength of glass fiber mineral-filled PP. (From Crompton Corporation Literature.)

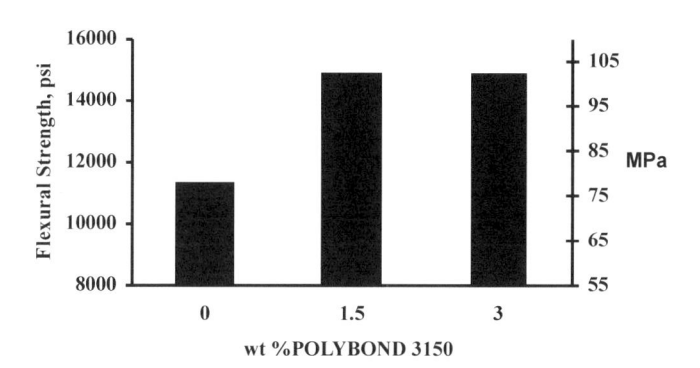

**FIGURE 3.32** Effect of PB 3150 level on tensile strength of glass fiber mineral-filled PP. (From Crompton Corporation Literature.)

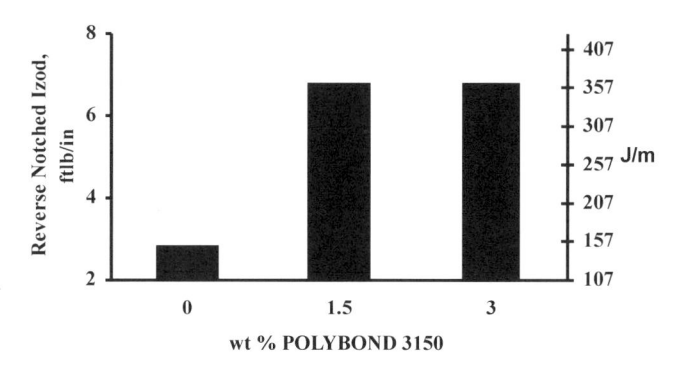

**FIGURE 3.33** Effect of PB 3150 level on reverse notched Izod impact of glass fiber/mineral filled PP composite. (From Crompton Corporation Literature.)

## 3.14 CHEMICAL COUPLING OF MICA POLYPROPYLENE

Mica is a high-aspect-ratio mineral that is often used in reinforced polypropylene. There are several types of micas, including muscovite, phlogopite, and biotite. These micas are basically complex hydrous aluminum silicate minerals. They differ slightly in chemical composition, especially with respect to iron content. The more iron that is present, the darker the color. Biotite is black, phlogopite is golden brown, and muscovite is white.

Chapter 14 provides a detailed discussion of mica reinforcement of polypropylene resins. When used in polypropylene, mica increases mechanical and thermal properties such as flexural strength, flexural modulus, tensile strength, and heat deflection. To improve the mechanical properties, mica is frequently coated with silanes, which are expensive and often difficult to apply. The use of acrylic acid or maleic anhydride grafted polypropylene as a chemical coupling agent in mica-filled polypropylene results in composites with enhanced mechanical properties. In fact, studies have shown that the use of these coupling agents with untreated mica results in mechanical properties that are similar to those using silane-treated mica [35].

Figure 3.34 shows the effect of PPgAA addition to composites containing 40 wt% phlogopite mica. Two different micas were evaluated, including 60 S, an untreated mica, and 60 NP, an aminosilane-treated mica. Addition levels of PPgAA are 10%, 15%, and 20% by weight of the total composite. With the addition of coupling agent, there is a significant increase in the tensile strength. With no coupling agent, the composite containing the treated mica has higher tensile strength than the composite with untreated mica. When PPgAA is added,

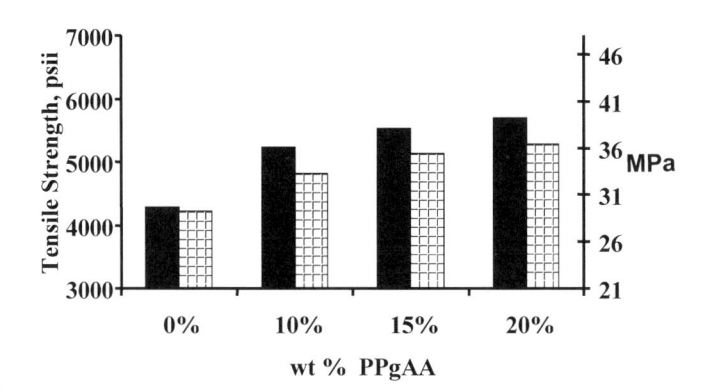

FIGURE 3.34 Tensile strength of 40 wt% mica-filled PP: effect of PPgAA addition. (From Ref. 35.) ■ 60 NP mica treated with aminosilane, ▦ 60 S untreated mica.

# Chemical Coupling Agents for Composites

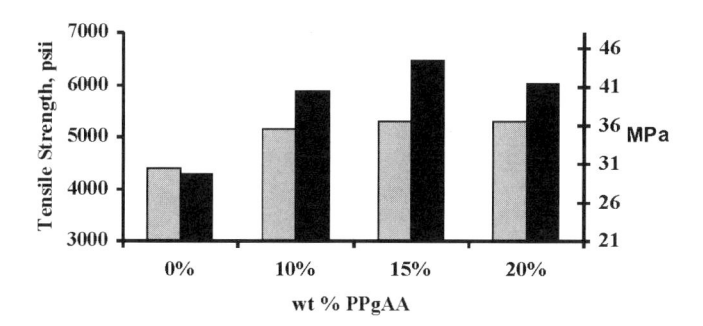

**FIGURE 3.35** Tensile strength of 40 wt% mica-filled PP: effect of PPgAA addition. (From Ref. 35.) ☐ 200 HK untreated mica, ■ 200 NP mica treated with aminosilane.

the tensile strength of the composite with untreated mica is equivalent to the composite with treated mica and no coupling agent. All of the composites containing both treated mica and PPgAA have higher tensile strength than the composites containing untreated mica and PPgAA. This indicates that there is a synergistic effect between the aminosilane surface treatment and the PPgAA coupling agent.

Similar results are shown in Fig. 3.35, which shows the tensile strength data for composites containing smaller particle size micas. Both a treated, 200 NP and untreated, 200 HK, micas were evaluated. Again, the NP grade is treated with aminosilane.

Not all surface treatments show this synergistic effect. Figure 3.36 compares an untreated, 200 P, mica with one that contains azidosilane, 200 PT. Without PPgAA, the mica with the azidosilane yields a composite with higher tensile strength than the untreated mica. However, when PPgAA is used as a coupling agent, no difference in tensile strength improvement is observed between the composites containing either treated or untreated mica. This lack of improvement suggests that there is no synergistic effect between the azidosilane and PPgAA.

Other researchers have reported the use of PPgMAH in mica-reinforced polypropylene composites. In a paper by Borden et al. [36], mechanical property improvements of 9–29% are shown with the addition of 10 wt% maleated polypropylene. In another paper, Hyche et al. [34] indicated that the use of medium molecular weight, highly functionalized PPgMAH works better than high-molecular weight, low functionalized material in mica-filled polypropylene.

Chapter 15 describes a new SC coupling system that enhances combinations of mica and glass fibers to make cost-effective hybrid composites of polypropylene.

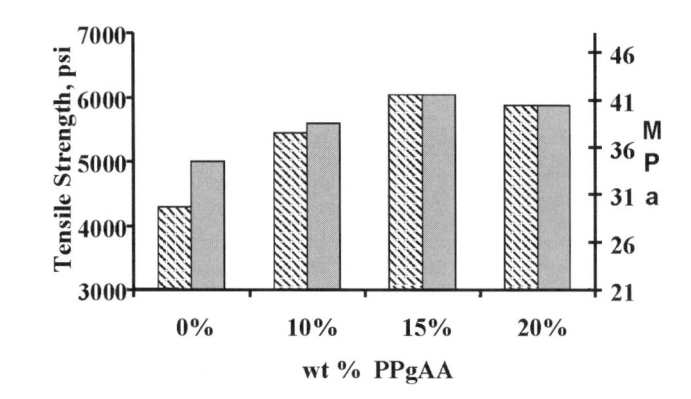

**FIGURE 3.36** Tensile strength of 40 wt% mica-filled PP: effect of PPgAA addition. (From Ref. 35.) ▨ 200 P untreated mica, ▢ 200 PT mica treated with azido silane.

## 3.15 TALC FILLED POLYPROPYLENE

The use of talc as a filler in polypropylene results in composites that exhibit excellent properties such as tensile strength, flexural modulus, low shrinkage, heat deflection, and scratch or mar resistance. Talc is a hydrated magnesium silicate, with a surface that is either hydrophilic or hydrophobic, depending on its source. Many talcs are supplied with various surface treatments designed to modify the surface to improve adhesion of the polypropylene to the talc. These surface treatments are often stearates or silanes. The use of acrylic acid grafted polypropylene as a chemical coupling agent has been shown to increase the mechanical properties of talc-filled polypropylene [37]. Figure 3.37 shows the effect of PPgAA addition to 40 wt% talc-filled polypropylene. Various types of talcs were evaluated. The untreated talcs were the Microtalc and Emtal grades, whereas the treated grades were Microtuff and Cyprubond. With the untreated talcs, 11–12% improvement in tensile strength is realized when 20 wt% PPgAA is added. For the treated talcs, 11–19% improvement is seen. The surface treatment for the Microtuff talc was aminosilane. Composites that contained this talc in combination with 20 wt% PPgAA had higher tensile strength than composites containing the untreated talcs.

Maleated polypropylenes were also shown to act as coupling agents for talc filled polypropylene. Hyche et al. [34] reported that medium molecular weight, highly functionalized PPgMAH is the preferred coupling agent for talc-filled polypropylene.

Two grades of 40 wt% talc (uncoated and silane treated)–filled polypropylene were evaluated with 5 wt% PB 3150 and without coupling agent. Figure 3.38 makes a comparison between impact properties for the given material design. The

# Chemical Coupling Agents for Composites

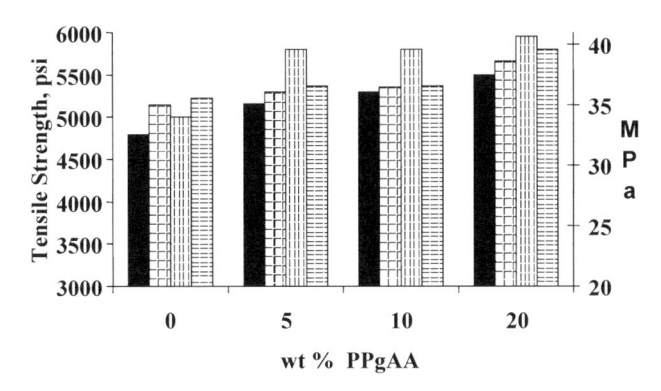

**FIGURE 3.37** Tensile strength of 40 wt% talc-filled polypropylene. (From Ref. 37.) ■ Microtalc (untreated), ▦ Emtal (untreated), ▥ Microtuff (treated), ▨ Cyprubond (treated).

Microtalc 12-50 (made by Specialty Minerals) has a particle size of 1.2 µm and is uncoated. The second talc, Microtalc 1200, has the same particle size and a silane surface treatment. The data indicate that silane treated talc is not effective in increasing impact strength. Addition of a coupling agent is most beneficial for increase in reverse notched Izod, particularly for the uncoated grade of talc. The effect of coupling agent addition on tensile and flexural strength properties is not shown because the amount of improvement is small at approximately 10–15%.

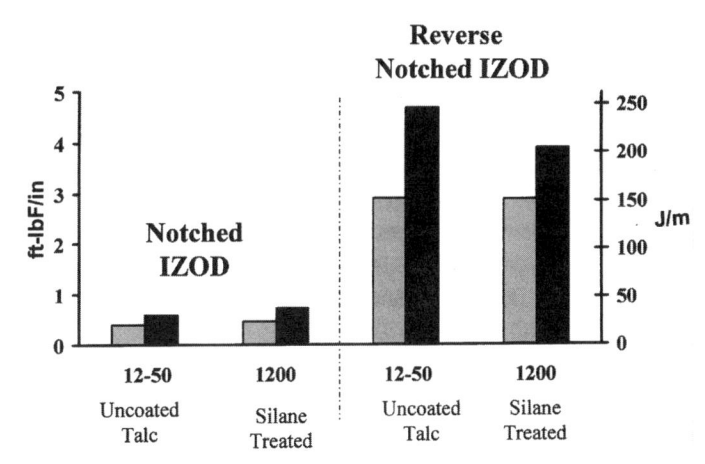

**FIGURE 3.38** Notched Izod–reverse notched Izod of 40 wt% filled PP for uncoated-coated talc fillers and two levels of PB 3150. (From Crompton Corporation Literature.) ▨ 0 wt% PB 3150, ■ 5 wt% PB 3150.

## 3.16 MAGNESIUM HYDROXIDE FILLED POLYPROPYLENE

There are many ways to impart flammability resistance to polymers. One of the oldest methods is to use halogenated materials. However, there has been a shift away from the use of these halogenated materials. The two most common nonhalogen additives used are alumia trihydrate (ATH) and magnesium hydroxide [Mg(OH)$_2$]. Magnesium hydroxide is used in polypropylene compounds due the high processing temperature. If ATH is used in polypropylene, it can thermally decompose during processing. Magnesium hydroxide is available from a variety of suppliers. As with other fillers, it is available in many different particle sizes and with a variety of surface treatments. To get the desired flame-retardant properties, high filler content at about 50–60 wt% is required. With this high level of filler, the mechanical properties of the final composite are low. Also, it is very difficult to disperse this level of filler into polypropylene. The flame-retardant properties require good wet-out by the polymer matrix and dispersion of the filler particles. To increase dispersion, the surface of the Mg(OH)$_2$ particles is coated with a thin layer of stearic acid or some other compatibilizer. The carboxyl group on the stearic acid reacts with the functional groups present on the surface. However, the length of the aliphatic tail of the stearic acid is too short to entangle and cocrystallize with the molecules of the polymer matrix.

Maleated or acrylic acid grafted polypropylene can also be used as a coupling agent for these types of compounds. The use of maleated polyethylene and polypropylene resins in this application is widely documented. Jancar [38] conducted a study of maleated polypropylene in Mg(OH)$_2$-filled polypropylene. His findings indicated that there was a critical concentration of carboxylic acid groups. At low levels of addition, the composite yield stress increased sharply. However, at a certain concentration based on filler content, very little increase in yield stress was observed with increased addition of PPgMAH. A possible explanation is that 100% of the filler surface is covered by PPgMAH or that the interfacial adhesion was stronger that the yield stress of the matrix.

Table 3.2 shows data on several different composites made with three types of magnesium hydroxides. The composites were evaluated with and without PPgMAH. Table 3.3 lists the properties of the various grades of magnesium hydroxides used.

The composites containing the untreated Mg(OH)$_2$, Magnifin H7, exhibit a significant increase in both tensile and flexural strength with 5 wt% PPgMAH addition. However, the impact properties are only slightly improved.

The use of the proprietary coated Mg(OH)$_2$, Magnifin H 5K V, also showed a significant increase in both tensile and flexural strength in the presence of chemical coupling agent. However, impact properties decreased. The surface coating is a type of fatty acid that apparently does not compatibilize well with the nonpolar polymer matrix, even in the presence of coupling agent. This would

# Chemical Coupling Agents for Composites

TABLE 3.2 Mechanical Properties of Magnesium Hydroxide–Filled Polypropylene

| Property | Type Mg(OH)$_2$ filler |||||||
|---|---|---|---|---|---|---|
| | H7 || H51 V || H5K V ||
| Wt% PPgMAH | 0 | 5 | 0 | 5 | 0 | 5 |
| Tensile strength, psi | 3200 | 5000 | 4200 | 4900 | 2550 | 4000 |
| Flexural strength, psi | 5900 | 8900 | 7500 | 9500 | 5550 | 8300 |
| Notched Izod, ft lb/in. | 0.5 | 0.5 | 0.5 | 0.6 | 0.9 | 0.7 |
| Reverse notched Izod, ft lb/in. | 1.1 | 2.5 | 1.8 | 3.8 | 12 | 3 |

*Source:* Crompton Corporation Literature.

explain the much lower composite strength with and without chemical coupling compared to composite properties for untreated grade of filler (Magnifin H7).

Magnifin H 51 V contains an aminosilane surface treatment. The combination of this surface treatment and maleated polypropylene resulted in the highest mechanical properties for all of the composites studied, again showing a synergistic effect between aminosilane and maleated polypropylene.

## 3.17 CELLULOSE FILLED POLYPROPYLENE

Cellulosic materials are good candidates for use as fillers in polymers, especially with the recent importance in recycling. There are many different types of cellulosic materials available, such as wood flour, recycled newspaper, nut shells, and starch. These materials have several advantages including low cost, low density, low abrasiveness, and are a renewable resource. However, they do have a number of disadvantages, such as low thermal stability, high moisture absorption, and poor interfacial adhesion. Despite these problems, there are a number of products available in the marketplace. One of the most successful applications is plastic lumber using polyethylene as the base resin with wood filler content up to

TABLE 3.3 Magnesium Hydroxide Properties

| Type | Particle size (μm) | Surface treatment |
|---|---|---|
| Magnifin H7 | 1 | Untreated |
| Magnifin H51 V | 1.5 | Aminosilane |
| Magnifin H5K V | 1.5 | Proprietary |

*Source:* Crompton Corporation Literature.

50 wt%. To reduce raw material costs and conserve natural resources, both resin and wood feedstocks come from recycled sources. For example, TREX is a plastic lumber that uses polyethylene scrap from the manufacture of plastic trash bags and sawdust from ground-up wood pallets.

A number of studies have looked at the use of acrylic or maleic anhydride grafted polypropylenes as coupling agents for composites of polypropylene with cellulose-type materials. Sanadi et al. [39] published results on using recycled newspaper fibers in polypropylene. In their study, they evaluated several different coupling agents consisting of maleated polypropylene with different molecular weights and different acid contents. Their findings indicated that all of the coupling agents improved the tensile and impact strengths. However, the coupling agents with both high molecular weight and high acid content produced the highest mechanical properties.

Olsen [40] reported similar findings on the use of maleated polypropylenes as coupling agents for wood flour–polypropylene composites. He reported that the addition of maleated polypropylene to these composites resulted in improvements in tensile strength, flexural strength, hardness, and heat deflection temperature. Also, molecular weight and acid or anhydride content played an important role in determining the effectiveness of the maleated polypropylene. He reported that PPgMAH with high anhydride content and high molecular weight yielded the highest mechanical properties.

Table 3.4 shows data on 40 wt% wood flour filled polypropylene with and without maleated polypropylene [41]. The data show improvements of 30% in tensile and flexural strength and a 16% improvement in reverse notched Izod impact with the addition of maleated polypropylene.

Jacoby et al. [42] have reported work on wood-filled, highly crystalline polypropylene. They compared polypropylene with wood filled polyethylene (HDPE and LLDPE)–based composites, which are typically used in such applications as plastic lumber. The results of their investigation are summarized in Table 3.5 and Figs. 3.38 and 3.39. Compared with other wood-filled polyolefins, highly crystalline PP offers the highest stiffness, strength, creep,

TABLE 3.4　40 wt% Wood Flour–Filled Polypropylene

| Property | 0 wt% PPgMAH | 5 wt% PPgMAH |
| --- | --- | --- |
| Tensile strength, MPa | 30 | 38 |
| Flexural strength, MPa | 51 | 67 |
| Notched Izod, J/m | 44 | 38 |
| Reverse notched Izod, J/m | 90 | 104 |

Source: Crompton Corporation Literature.

TABLE 3.5 Properties of 60 wt% Wood-Filled Polyolefin Composites

| Property | ACCPRO 9433 | Homopolymer PP | Impact modified PP copolymer | HDPE | LLDPE |
|---|---|---|---|---|---|
| Flexural modulus, MPa | 6000 | 4800 | 4400 | 3500 | 3000 |
| Tensile strength, MPa | 35 | 22 | 20 | 18 | 24 |
| Notched IZOD, ft-lb F/in. | 20 | 20 | 24 | 19 | 53 |
| Creep, % strain at 1000 hr | 0.2 | 0.34 | Failure 91 hr | Failure 24–48 hr | 0.50 |
| HDT-1.82 MPa, °C | 127 | 99 | 94 | 80 | 85 |

*Source*: BP Amoco Polymers [42].

[Bar chart showing Tensile Strength MPa for ACCPRO 9433 and Homopolymer PP at different maleation levels]

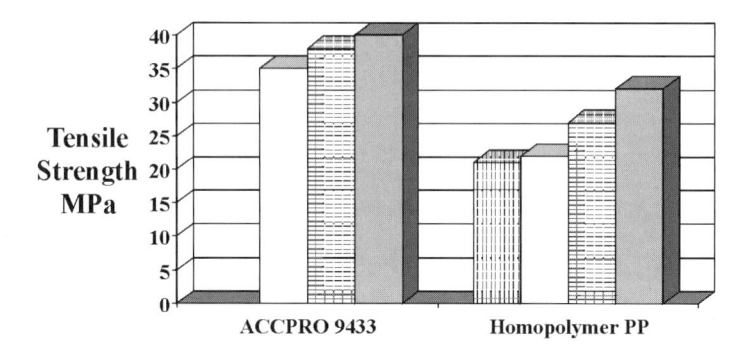

**FIGURE 3.39** Effect of maleation levels on tensile strength of wood-filled PP. (From Ref. 42 and BP Amoco Polymers.) ▨ 0 wt%, ☐ 1.0 wt%, ▨ 2.5 wt%, ☐ 4.0 wt%.

high-temperature resistance, and lowest moisture pickup. The addition of maleated polypropylene as a coupling agent increases the tensile and flexural strength. Maleated polypropylene works in both standard homopolymer polypropylene and the highly crystalline material.

Figure 3.39 shows the effect of maleated polypropylene addition on tensile strength. The high-crystalline polypropylene resin (ACCPRO 9433) already contains some level of maleation, so no control, with 0 wt% coupling agent, is shown. However, with the higher addition levels of maleated polypropylene, increases in tensile strength are observed. With the addition of 4 wt% maleated PP, a 14% increase is observed in the high-crystallinity PP, and in the standard homopolymer PP a 52% increase is seen.

The addition of maleated polypropylene to wood-filled polypropylene also reduces the moisture pickup as shown in Fig. 3.40. With as little as 1 wt% of maleated polypropylene moisture pickup is reduced by 1 wt%. This is attributed to the better fiber wet-out and coupling to the hydroxyl groups. The composite based on high-crystalline PP exhibits lower water absorption than the corresponding wood-filled homopolymer PP. The higher crystallinity of the ACCPRO 9433 reduces moisture absorption due to its higher crystalline structure.

## 3.18 NANOCOMPOSITE POLYPROPYLENE

Nanocomposite technology is one of the hottest areas in polymer science today. This technology consists of dispersing clays into polymers to create reinforced composites featuring valuable or, in some cases, unique properties, with low filler loading. The most common clay used in commercial applications is montmorillonite, a member of the smectite clay family. These clays have a unique platy

# Chemical Coupling Agents for Composites

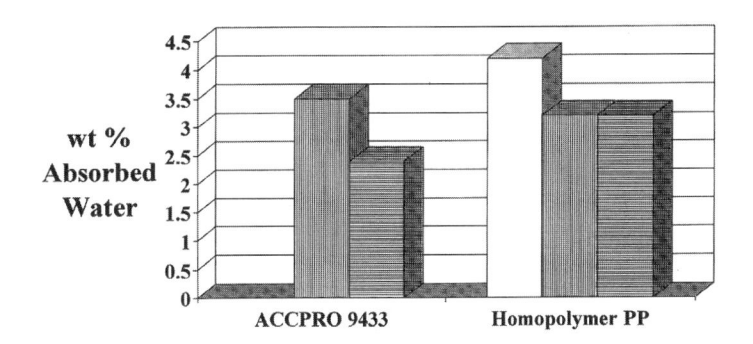

**FIGURE 3.40** Effect of maleation level on water absorption of 40 wt% wood-filled PP after 30 days immersion. (From Ref. 42 and BP Amoco Polymers.) ☐ 0 wt%, ▨ 1.0 wt%, ▨ 2.5 wt%.

structure and a unit thickness of less than 1 nm ($1 \times 10^{-7}$ cm), while the other dimensions are in the micrometer ($1 \times 10^{-4}$ cm) range. When totally dispersed, the average surface area is on the order of 700 m$^2$/gram. With a high aspect ratio, these clays are highly effective reinforcements, provided they are properly dispersed.

The surface of these clays is highly hydrophilic and requires modification to make it compatible with most polyolefins. Surface modification can be done using either of two methods: ion exchange and ion–dipole interaction. These surface treatments are used to intercalate the clay. Intercalation is defined as the separation of the individual layers of clay. Once intercalation is achieved, molten polymer can get between the clay platelets and exfoliate them. Exfoliation is the dispersion of the clay in the polymer. Only composites with good exfoliation exhibit increased mechanical properties.

Polyamide nanocomposites based on nylon 6 polymer produced in the reactor [43] have been around for several years. Toyota Central Research and Development laboratories pioneered this work. Toyota has licensed its patented in-reactor technology to Nanocor and Ube Industries. Nanocor then sublicenses the technology to companies who want to make nylon nanocomposites. Several companies have commercial nylon-based nanocomposite products, including Bayer, RTP, Ube, Unitika, Honeywell, and Showa Denko.

The manufacture of olefin-based nanocomposites has proved to be much more challenging than the pioneering nylon based nanocomposites. The major problem has been the low polarity and inherent incompatibility of the clay with polyolefins. This lack of compatibility makes intercalation and exfoliation of the clay very difficult.

Many different approaches have been taken to the production of olefin-based nanocomposites. Several researchers have reported good results using maleated polypropylenes as a compatibilizer. Toyota Central labs have reported the successful production of PP-based nanocomposites that use maleated PP as a compatibilizer [44].

Nanocor has also reported the use of maleated PP as a compatibilizer in a masterbatch of clay and PP [45]. The masterbatch contains 50–60% clay which can be let down to level of 2–7% by weight. Improvements of 13% in tensile strength and 50% in flexural modulus were realized with loadings of 6% of Nanomer clay. Improvements in heat resistance (indicated by elevated HDT values) were also reported. See Chapter 20 for more details.

Patent 5,910,523 by Hudson discusses the treatment of clay with both aminosilane and maleated polypropylene. When dispersed in isotactic polypropylene, good dispersion was observed using transmission microscopy. With a 1 wt% loading of this treated clay, a remarkable eightfold increase in tensile modulus was observed. At high loadings, increases in tensile strength and modulus were also seen.

General Motors recently announced the first commercial application for nanocomposite TPO material, for use in stepboards for the 2002 GMC Safari and Chevrolet Astro minivan. The TPO was developed in conjunction with Basell Polyolefins, Blackhawk Automotive Plastics, and Southern Clay. The composition of the formulation is proprietary but most likely involves a compatibilizer additive with adhesive properties like PPgMAH.

## REFERENCES

1. ED Plueddeman. Silane Coupling Agents. New York: Plenum Press, 1982, p. 4.
2. R Godlewski. Efficient reinforcement promoters for reinforced polypropylene. SPE (Society of Plastics Engineers) ANTEC (Annual Technical Exhibition and Conference) Preprints, Atlanta, 1988, pp. 1481–1484.
3. AD Ulrich, WG Joslyn. Improved hydrolytic resistance of filled organic polymer composites using modified organofunctional silanes. SPE ANTEC Preprints, New York, 1989, pp. 1318–1322.
4. CH Johnston. A new silane coupling agent for glass reinforced polypropylene. SPE ANTEC Preprints, New York, 1989, pp. 1891–1893.
5. ED Plueddeman. Silane Coupling Agents. New York: Plenum Press, 1982, pp. 17–20.
6. U.S. Patent 3,886,227.
7. U.S. Patent 4,578,428.
8. U.S. Patent 5,286,800 and U.S. Patent 5,543,484.
9. U.S. Patent 3,414,551.
10. AH Hogt. Modification of Polyolefins with Maleic Anhydride. 1990 Compalloy, pp. 181–193.

11. AH Hogt. Modification of polypropylene with maleic anhydride. SPE ANTEC Preprints, Atlanta, 1988, pp. 1478–1480.
12. RM Ho, AC Su, CH Wu. Functionalization of polypropylene via melt mixing. ACS Polym Prepr, 33:941–943, 1992.
13. PA Callais, RT Kazmierczak. The maleic anhydride grafting of polypropylene with organic peroxides. SPE ANTEC Preprints, Dallas, 1990, pp. 1921–1923.
14. NG Gaylord, MK Mishra. Nondegradative reaction of maleic anhydride and molten molypropylene in the presence of peroxides. J Poly Sci Polym Lett 21:22–30, 1983.
15. B De Roover, M Sclavons, V Carlier, J Devaux, R Legras, A Momtaz. Molecular characterization of maleic anhydride-functionalized polypropylene. J Polym Sci A Polym Chem 34:1195–1202, 1996.
16. NG Gaylord, S Maiti. J Polym Sci Polym Lett 11:253, 1973.
17. B De Roover, J Devaux, R Legras. Maleic anhydride homopolymerization during melt functionalization of isotatic polypropylene. J Polym Sci Polym Chem 34:1195–1202, 1996.
18. W Heinen, M van Duin, CH Rosenmoller, CB Wenzel, HJM de Groot, J Lugtenburg. $^{13}$C NMR Study of the grafting of maleic anhydride on polyethylene, polypropylene and ethene-propene copolymers. Macromolecules 29:1151–1157, 1996.
19. GH Hu, JJ Flat, M Lambla. Free radical grafting of chemically activated maleic anhydride onto polypropylene by reactive extrusion. SPE ANTEC Preprints, San Francisco, 1994, pp. 2775–2778.
20. TH Kozel, RT Kazmierczak. A rapid Fourier transform method for the determination of grafted maleate on polyolefins. SPE ANTEC Preprints, Montreal, 1991, pp. 1570–1575.
21. M Scavons, V Carlier, B De Roover, P Franquinet, J Devaux, R Legras. The anhydride content of some commercial PPgMA: FTIR and titration. J Appl Polym Sci 62:1205–1210, 1996.
22. S. Hacker. Not all maleated polyolefins are created equal. SPE ANTEC Preprints, Dallas, 2001, pp. 2673–2676.
23. J Duvall, C Sellitti, C Myers, A Hiltner, E Baer. Interfacial effects produced by crystallization of polypropylene with polypropylene-g-maleic anhydride compatibilizers. J Appl Polym Sci 52:207–216, 1994.
24. J Duvall, C Sellitti, C Myers, A Hiltner, E Baer. Effect of compatibilization on the properties of polypropylene/polyamide-66 (75/25 Wt/Wt) blends. J Appl Polym Sci 52:195–206, 1994.
25. A Misra, BL Deopura, SF Xavier, FD Hartley, RH Petters. Transcrystallinity in injection molded polypropylene glass fibre composites. Die Angewandte Makromole Chem 113:113–120, 1983.
26. HJ Tai, WY Chiu, LW Chen, LH Chu. Study on the crystallization kinetics of PP/GF Composites. J Appl Polym Sci 42:3111–3122, 1991.
27. A Misra, D Tyagi, SF Xavier. Proceedings of Nuclear Physics and Solid State Physics Symposium, Indian Institute of Technology, Bombay, India, 1980.
28. S Yin, T Rials, M Wolcott. Crystallization behavior of polypropylene and its effect on woodfiber composite properties. The Fifth International Conference on Woodfiber-Plastic Composites, USDA Forest Service, Madison, Wisconsin, 1999.

29. JM Felex, P Gatenholm. Interphase design in cellulose fiber/PP composites. Polym Mater Sci Eng. ACS Fall Meeting Preprints, 1990, pp. 315–316.
30. HK Karian. Thermodynamic probe of inter-molecular coupling in the interphase region of glass fiber reinforced polymer composites. SPE ANTEC Preprints, Boston, 1995, pp. 1665–1669.
31. HG Karian, HR Wagner. Assessment of interfacial adhesion in chemically coupled glass fiber reinforced polypropylene. SPE ANTEC Preprints, New Orleans, 1993, pp. 3449–3453.
32. JMH Daemen, J den Besten. The effects of glass fibre size and coupling additives on the properties of glass fibre reinforced polypropylene. PPG Silenka Glass Fibre Technical Literature.
33. AM Adur, RC Constable. Chemical coupling of glass filled polypropylene using acid or anhydride modified polypropylenes. SPE ANTEC Preprints, Montreal, 1991, pp. 1892–1896.
34. KW Hyche, DA Jervis, RD Hollis, I Seppa. Improved polyolefin composites through the use of maleated polyolefin coupling additives. Functional fillers for thermoplastics, thermosets, elastomers; Intertech Conferences (Proc.), Amsterdam, The Netherlands, 1996.
35. AM Adur, RC Constable, J Humenik. Performance enhancement in mica filled polypropylene obtained by addition of chemically modified polyolefins. SPE ANTEC Preprints, Atlanta, 1988, pp. 1474–1477.
36. KA Borden, RC Weil, CR Manganaro. The effect of polymeric coupling agent on mica-filled polypropylene: optimization of properties through statistical experimental design. SPE ANTEC Preprints, New Orleans, 1993, pp. 2167–2170.
37. AM Adur, S Flynn. Performance enhancement in talc-filled polypropylene obtained by addition of acrylic acid modified polypropylene. SPE ANTEC Preprints, Los Angeles, 1987, pp. 508–513.
38. J Jancar. Influence of the filler particle shape on the elastic moduli of PP/$CaCO_3$ and PP/$Mg(OH)_2$ composites. J Mater Sci 24:4268–4274, 1989.
39. AR Sanadi, RA Young, C Clemons, RM Rowell. Recycled newspaper fibers as reinforcing fillers in thermoplastics: Part I. Analysis of tensile and impact properties in polypropylene. J Reinforced Plastics Composites 13:54–67, 1994.
40. DJ Olsen. Effectiveness of maleated polypropylene as coupling agents for wood flour/polypropylene composites. SPE ANTEC Preprints, Montreal, 1991, pp. 1886–1891.
41. DH Roberts, RC Constable. The Use of Functionalized Polymers in Recycle Applications. Annual Recycling Conference, 1997.
42. P Jacoby, R Sullivan, W Crostic. Wood filled high crystallinity polypropylene. The 6th International Conference on Wood-Fiber Plastic Composites, USDA Forest Service, Madison, Wisconsin, 2001.
43. A Okada, A Usuki. Mater Sci Eng, C3:109–115, 1995.
44. M Kato, A Usuki, A Okada. Synthesis of polypropylene oligomer-clay intercalation compounds. J Appl Polym Sci 66:1781–1785, 1997.
45. T Lan, G Qian. Preparation of high performance polypropylene nanocomposites. Additives 2000. Clearwater Beach, Florida, 2000.

# 4

## Stabilization of Flame-Retarded Polypropylene

**Robert E. Lee, Donald Hallenbeck, and Jane Likens**
Great Lakes Chemical Corporation, West Lafayette, Indiana, U.S.A.

### 4.1 BACKGROUND

Ultraviolet (UV) stabilization of polypropylene (PP) systems containing flame retardants proves to be a difficult technical challenge. The reason for this difficulty is expanded on later. However, generation of acidic products from bromine-based flame retardants during processing or exposure can cause a catastrophic deactivation of the hindered amine light stabilizer (HALS). An understanding of the mechanism for generation of acidic products from aliphatic- and aromatic-based flame retardants has led to formulation approaches based on flame-retardant structure.

As PP fiber continues to expand its share of the textile market [1], new application areas are identified. One such critical area is flame-retardant fiber for wall coverings, upholstery, commercial carpeting, and automotive uses. For flame-retardant PP fiber, the key issues are processing, UV stability, and economics. Recent advances in flame-retardant technology significantly improved processing, but UV stability and formulation of economically acceptable additive packages are difficult to achieve.

In addition to fiber, molded applications require UV-durable flame-retarded features from polypropylene. Perhaps better put, if these features were available, the polypropylene market share could expand rapidly. The following mechanisms for degradation, stabilization, and flame retardation present information required to develop improved systems. Methods of performance evaluation are key to proper formulation and are covered in some detail. It will be of value to see the commercial formulations included and the insights into next-generation systems.

## 4.2 MECHANISMS

Those close to the end-use applications of UV/flame-retardant (FR) PP generally "formulate." Formulate means the use of components as ingredients as when cooking. As such, additives and their interactions are not considered in a chemical or mechanistic way. The combination of additives for UV/FR PP requires a greater level of scrutiny because the interaction of additives and reaction byproducts can lead to catastrophic results. Specifically, the wrong combination of independently effective additives can hurt performance based on one or more criteria. The mechanisms of degradation, stabilization, and flame retardation indicate which interactions are possible.

### 4.2.1 Degradation

As outlined in a simplified mechanism in Fig. 4.1, degradation proceeds through a radical chain mechanism [2,3]. Initiation typically occurs through exposure to heat generated during production. Presence of trace metal impurities, such as copper or iron, accelerates radical formation. Reactive hydroperoxides are formed after reaction of the carbon-centered radical with oxygen. Thermally induced homolytic cleavage of hydroperoxides leads to additional reactive radical formation and subsequent polymer chain scission.

Degradation of polymeric material with heat and oxygen initially involves breaking a bond between carbon and hydrogen atoms to make uncharged species called free radicals, as shown (Fig. 4.2) in the chain initiation reaction (1). These react quickly with oxygen to form peroxide radicals in a chain propagation reaction (2). These peroxy radicals in turn lead to many other reactive species, including peroxides as shown in reaction (3). The problem becomes the exponential increase in these reactive species through branching reactions (4).

### 4.2.2 Stabilization

An effective approach to thermal stabilization is through the use of a radical-terminating primary antioxidant. The most common class of antioxidant for radical termination is a hindered phenol [4]. Phenolic antioxidants are highly effective at relatively low concentrations (i.e., < 0.5 wt%) in inhibiting decom-

# Stabilization of Flame-Retarded Polypropylene

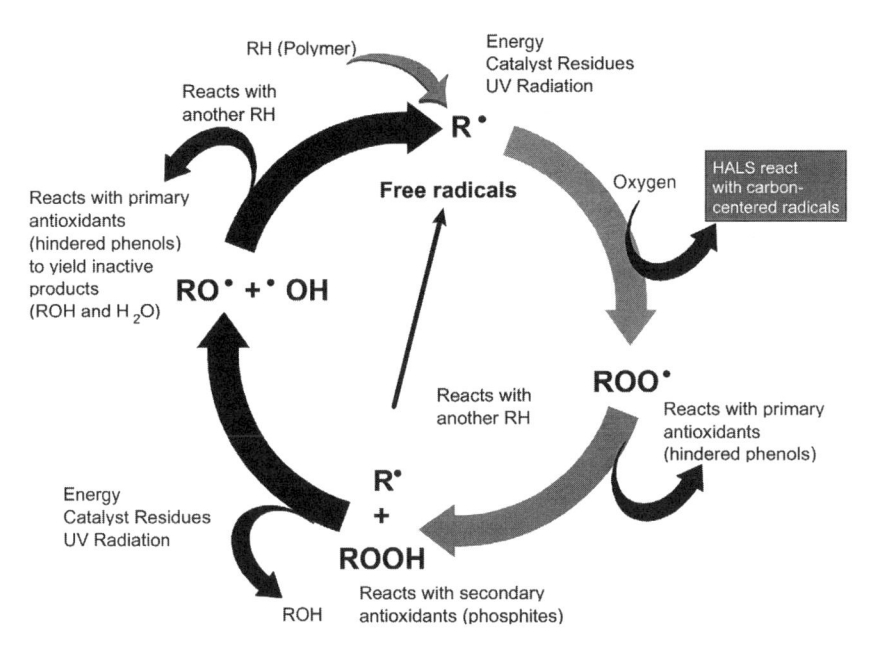

**FIGURE 4.1** Auto-oxidation of polymers.

CHAIN INITIATING STEP

RH + Δ(energy) → R• + H•         polymer cleaved into 2 radicals         Reaction (1)

CHAIN PROPAGATION STEP

R• + O2 → RO2•         radicals react with oxygen         Reaction (2)

RO2• + RH → ROOH + R•         peroxides formed         Reaction (3)

CHAIN BRANCHING STEP

ROOH + Δ→ RO• + •OH         peroxides produce more radicals         Reaction (4)

**FIGURE 4.2** Radical reactions (1) through (4).

position. The mechanism involves a chain-breaking donation of a hydrogen atom from the antioxidant to the reactive peroxy radical (Fig. 4.3). This produces a less reactive resonance-stabilized phenolic radical. Peroxycyclohexadienones can then be formed after reaction with a second peroxy radical [5]. Each phenolic moiety is capable of trapping a total of two radicals before it is completely consumed.

Preventative or secondary antioxidants act at the initiation stage of the radical chain mechanism to prevent the formation of radical products. Their mechanism involves the decomposition of hydroperoxides to form stable nonradical products. In the absence of peroxide scavengers, hydroperoxides thermally or photolytically decompose to radical products and accelerate decomposition. The most common secondary antioxidants are sulfur-based "thiosynergist" or phosphorus-based "phosphites."

Thiosynergists have been shown to decompose several moles of hydroperoxide per mole of stabilizer [6]. The hydroperoxide is typically reduced to an alcohol, and the thiosynergist is transformed into a variety of oxidized sulfur products, including sulfenic and sulfonic acids. Synergistic combinations with phenolic antioxidants are often used to enhance thermal stability in polyolefins at elevated temperatures ($> 100°C$).

Phosphites are also commonly used in combination with phenolic antioxidants to inhibit polymer degradation and to improve color [7]. As in the case of thiosynergists, phosphites reduce hydroperoxides to the corresponding alcohols and are transformed into phosphates at temperatures above 180°C. The temperature limits for secondary antioxidants can be linked to both reaction kinetics and diffusion control characteristics.

Figure 4.4 shows thermogravimetric analysis (TGA) data for selected phenolic antioxidants. Volatility is an important criterion because of the potential for loss during manufacturing. For example, butylated hydroxytoluene (BHT) (Fig. 4.5) is the most volatile of the group, with a 5 wt% loss at about 90°C and a 90 wt% loss at about 142°C. These temperatures are below normal processing temperatures, so BHT can be seen fuming from the extruder. However, dispersion and solubility of additives in the polypropylene prevent total loss. Dibutylnonyl phenol (DBNP) (Fig. 4.6) is structurally similar to BHT except for the nine-

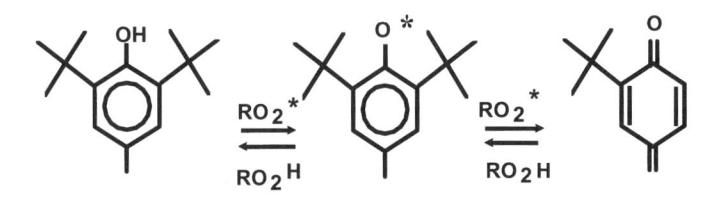

FIGURE 4.3  Radical trapping mechanism for BHT, a phenolic antioxidant.

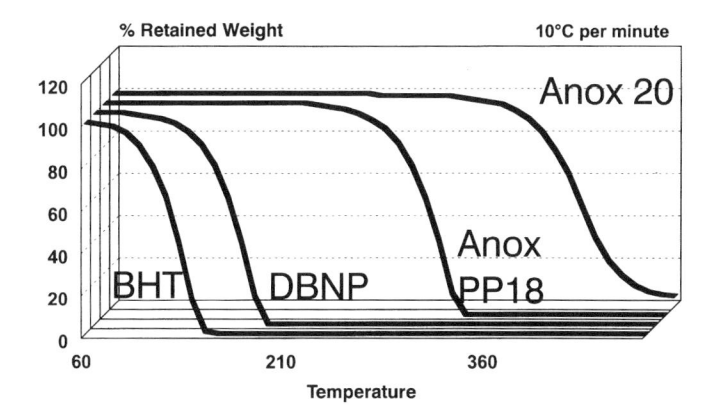

# Stabilization of Flame-Retarded Polypropylene

**FIGURE 4.4** TGA of four specific phenolic antioxidants: BHT, Lowinox DBNP, Anox PP18, and Anox 20.

**FIGURE 4.5** Lowinox BHT, CAS no. 128-37-6, 2,6-*di-tert*-butyl-4-methylphenol butylated hydroxytoluene.

**FIGURE 4.6** Lowinox DBNP, CAS no. 4306-88-1, 2,6-*di-tert*-butyl-4-nonylphenol.

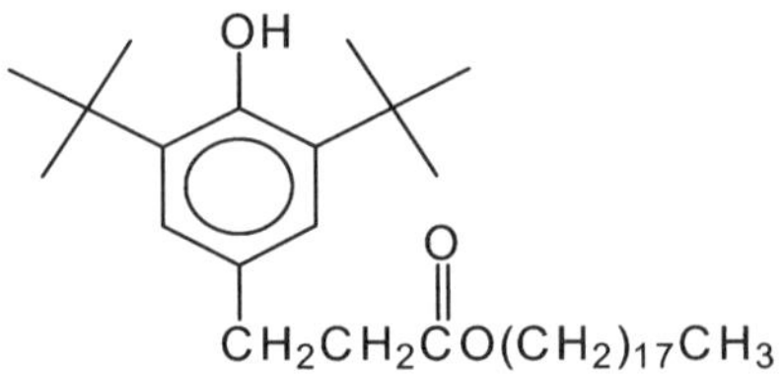

**FIGURE 4.7** Anox PP18, CAS no. 2082-79-3, octadyl-3-(3′,5′-*di-tert*-butyl-4′-hydroxyphenyl)propionate.

carbon group in the 4-position. Its resistance to volatility is improved in that it has a range for the 5 and 90 wt% losses of 120–182°C, respectively. Anox PP18 (Fig. 4.7) and Anox 20 (Fig. 4.8) have TGA volatility values that are predominantly above the peak processing temperature of polypropylene. They would be expected to evaporate significantly less during process if incorporated than BHT.

The decreased volatility is achieved by increasing the molecular weight of the antioxidant. This is accomplished through the addition of long hydrocarbon chains in the case of Anox PP18 and DBNP. Anox 20 achieves a high molecular weight without dilution of the active content by coupling four phenolic moieties. DBNP is higher in molecular weight than BHT and also a liquid. Liquid antioxidants are generally easier to handle and meter into liquid color concentrate production than solids like BHT, Anox PP18, and Anox 20.

Volatility and migration are also two issues affecting performance. Although some degree of volatility is generally considered desirable, excessive migration and volatility can have a deleterious impact. It has been demonstrated that large-scale migration of volatile BHT can occur during normal production [8]. This migration results in a dramatic decrease in antioxidant concentration. Migration of stabilizers to the surface leads to loss during rain, laundering, or other wash and clean events. Thus, higher molecular weight additives having a greatly reduced rate of migration and loss are an advantage.

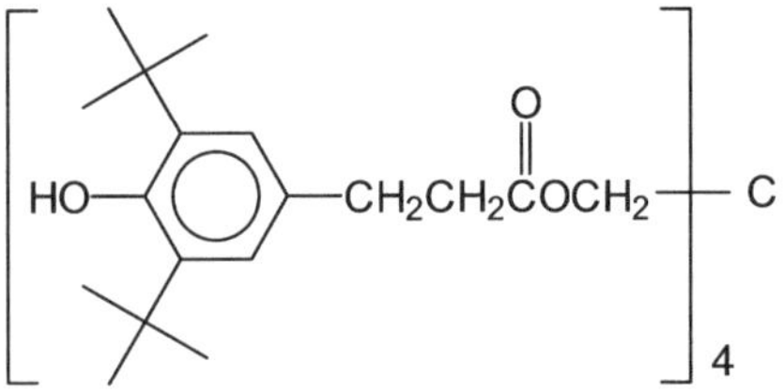

**FIGURE 4.8** Anox 20, CAS no. 6683-19-8, tetrakismethylene (3,5-*di-tert*-butyl-4-hydroxyhydrocinnamate)methane.

## Stabilization of Flame-Retarded Polypropylene

Primary antioxidants, like phenolics, scavenge radical species to prevent further reaction. Reactions (5) and (6), shown in Fig. 4.9, illustrate the fact that a phenolic antioxidant can scavenge two radical species. Therefore, the efficiency of a phenolic antioxidant will be related to its ratio of active phenolic weight to its total weight. For example BHT scavenges two radicals and has a molecular weight of 220 g/mol. However, 2,6-*di-tert*-butyl-4-nonylphenol scavenges two radicals and has a molecular weight of 332 g/mol. Therefore, the DBNP is more than 50% greater in molecular weight without an increase in activity. The tradeoff is that DBNP is a liquid with low volatility compared with BHT, which is a volatile solid.

The advantage of effective polyphenolic antioxidants like Anox 20 (Fig. 4.8) is that they contain more than one active group. Here the molecule traps up to eight radicals and has a molecular weight of 1178 g/mol. This is 147 molecular weight units per radical, which is between BHT and DBNP, but it has a much greater advantage in terms of volatility and other attributes.

Unlike primary antioxidants, all secondary antioxidants work by decomposing reactive species like peroxides as shown in Fig. 4.9 for reactions (7) and (8). They do not have the ability to trap radicals initially [9]. However, one of the most popular phosphite secondary antioxidants is made from three phenolic groups (Alkanox 240, Fig. 4.10). As those groups are made available, they have primary activity. Unfortunately, most phosphites typically do not become effective until 180–200°C. Therefore, they cannot work well by themselves. Primary and secondary antioxidants work by different mechanisms and often are synergistic.

The stabilization methods above are predominately involved during processing temperatures or molten conditions. However, they cannot be neglected when considering long-term thermal stabilization or light stabilization. This is because

CHAIN TERMINATION STEP

R• + AH → RH + A•     antioxidant scavenges 1st radical     Reaction (5)

R• + A• → RA          antioxidant scavenges 2nd radical     Reaction (6)

DECOMPOSITION OF REACTIVE INTERMEDIATES

ROOH + SR'2 → ROH + O=SR'2     1st peroxide decomposed     Reaction (7)

ROOH + O=SR'2 → ROH + O2SR'2   2nd peroxide decomposed     Reaction (8)

FIGURE 4.9 Radical reactions (5) through (8) involved in stabilization.

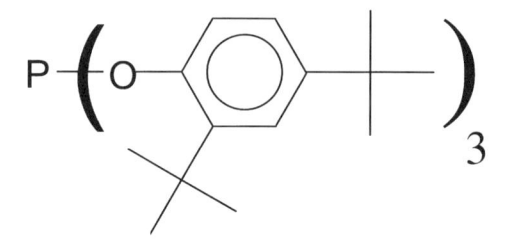

**FIGURE 4.10** Alkanox 240, CAS no. 31570-04-4, *tris*(2,4-*di-tert*-butylphenyl)-phosphite, an effective phosphite secondary antioxidant.

of the residual stabilizers and reaction byproducts carried into the next phase of the product life. A clear example of this is the loss of UV durability experienced as a function of multiple extrusion of the same material. Multiple extrusion creates color in PP and increases in melt flow. Figure 4.11 shows the reduction in UV durability of a single system processed at 230°C and 300°C.

The material formulated for Fig. 4.11 had a melt flow of 14 g/10 min at 230°C with a mass of 2160 g as pelletized. The fiber processed at 230°C had a subsequent melt flow of 20 (g/10 min at 230°C with a mass of 2160 g). However, the fiber processed at 300°C had a melt flow of 39 (g/10 min at 230°C with a mass of 2160 g). The change from 14 to 20 is typical because of a second processing step that causes degradation and reduction of residual stabilizers that

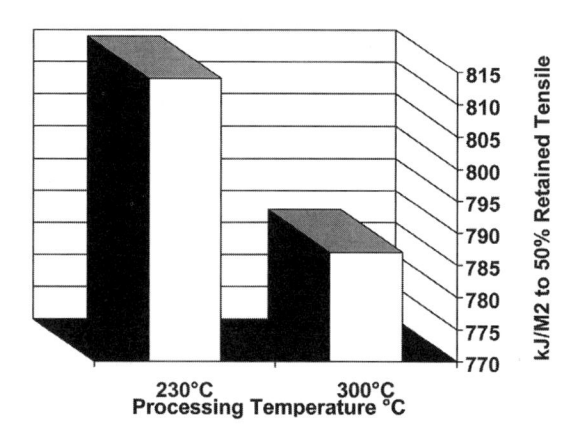

**FIGURE 4.11** Loss of UV durability (kJ/M exposure to reach 50% tensile strength) with changes in processing temperatures for polypropylene homopolymer, 1296 denier, 72 filament partially oriented yarn for carpet fiber. Fiber was stabilized with 0.2 wt% Chimassorb 944 and 0.15 wt% Alkanox P24.

## Stabilization of Flame-Retarded Polypropylene

protect the polymer during the measurement of melt flow (third heat history). However, the increase to a melt flow of 39 in the material processed at 300°C clearly indicates that additional degradation has occurred in the polymer that also had the shortest lifetime for UV exposure. Thus, the degradation from processing contributes to a reduction of UV durability.

To stop degradation of polypropylene by light, there are several common and effective methods. The simplest may be coating the polymer surface with either a clear coat or a color coat of paint that can provide UV durability. This is done for better color matching in automotive applications. However, lack of UV absorbers in these coatings may allow damaging light penetration to the polymer surface where coating adhesion is lost. Additives to the polypropylene to prevent light damage directly may work by competing for the light or interfering with degradation chemistry like the antioxidants previously discussed.

A loading of 2.5 wt% carbon black will allow many systems to retain their physical integrity during exposure to UV light. In this system, the light is absorbed at the surface, and damage is reduced and contained to the surface layers. For flame-retarded polypropylene, this is seldom a viable option because it limits the color options and highlights a common problem with blooming of white coadditives.

A UV light absorber for polypropylene can be of a number of different chemical classes. Benzophenones and benzotriazoles are the most commercial. Their mechanism involves absorption of the light and subsequent dissipation of the energy as heat. Figure 4.12 indicates the mechanism for Lowilite 22, a common benzophenone-type absorber. The benzotriazoles work by a similar mechanism. Figure 4.13 shows Lowilite 28, a common benzotriazole for polypropylene.

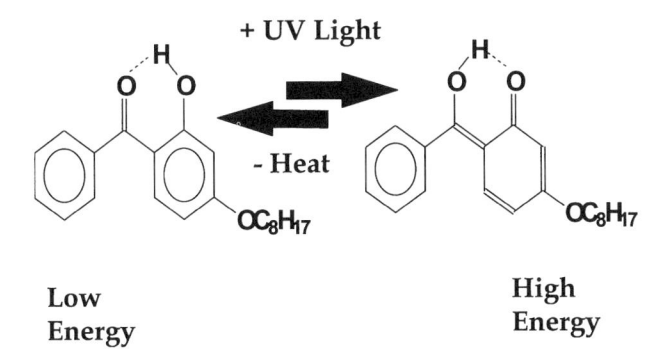

**FIGURE 4.12** Nonconsuming mechanism for a benzophenone-type UV light absorber, Lowilite 22.

FIGURE 4.13 Lowilite 28, CAS no. 25973-55-1, 2-(2'-hydroxy-3',5'-di-tert-amylphenyl)benzotriazole, a common benzotriazole-type UV light absorber.

The particular advantage of absorbers is the protection they afford coadditives and the polypropylene. Pigments and brominated flame retardants are coadditives that undergo light damage independently of the polypropylene. Figure 4.14 shows the UV reflectance spectra for two brominated flame retardants. The lack of reflectance for DE83R, decabromodiphenyl oxide, at damaging UV light wavelengths shows that damage from light can occur directly with the flame retardant. Thus, a costabilizing UV absorber would be specifically helpful with aromatic flame retardants like DE83R.

Most polypropylene uses a radical trapping stabilizer of the hindered amine type. These hindered amine light stabilizers are often referred to as HALS. Figure

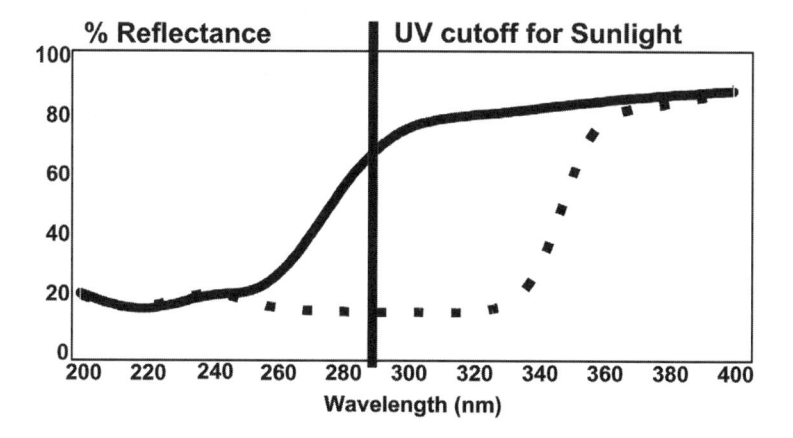

FIGURE 4.14 UV spectra of flame-retardant particles and the reflectance of damaging light for CD 75 (—), hexabromocyclododecane and absorption for DR83 (---), decabromodiphenyloxide.

# Stabilization of Flame-Retarded Polypropylene

**FIGURE 4.15** Lowilite 77, CAS no. 52829-07-9, *bis*-(2,2,6,6-tetramethyl-4-piperinyl)-sebacate, a common HALS for thick polypropylene applications.

4.15 shows a common example sold as Lowilite 77. Figure 4.16 shows the cyclic mechanism of HALS that attributes to their good performance-to-weight ratio.

The fact is that HALS are so effective in polypropylene they often preclude the uses of absorbers in many applications. However, a significant limitation is the alkaline nature of the common secondary HALS. If coadditives are acidic or produce acidic degradation byproducts, these can coordinate with the HALS to both interfere with stabilization and contribute to color variations. The alkylation of secondary HALS to form tertiary HALS like Lowilite 76, shown in Fig. 4.17, greatly reduces these harmful coordinations.

Another way to reduce the alkalinity of a HALS is to form a polymeric molecule through the active amine groups as Lowilite 62, shown in Fig. 4.18. With flame-retarded polypropylene, a combination of stabilizers is often found to be synergistic. Later, common commercial systems and some next-generation systems are discussed.

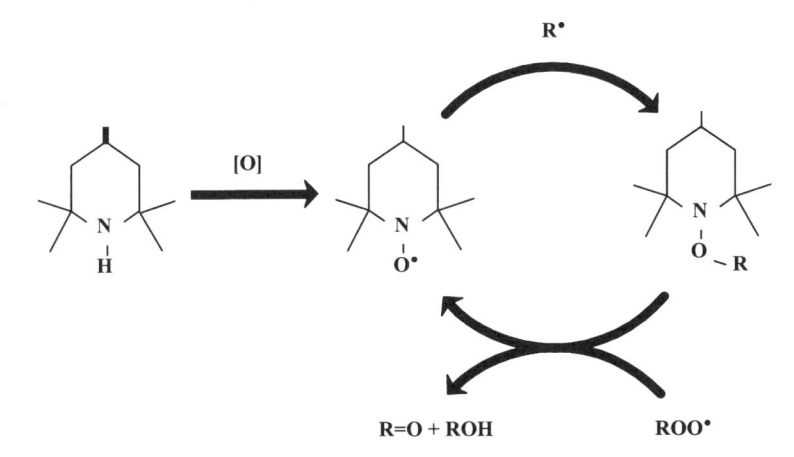

**FIGURE 4.16** Cyclic mechanism for HALS stabilization through radical trapping. Alcohol and carbonyl species are also produced, which can have degrading effects.

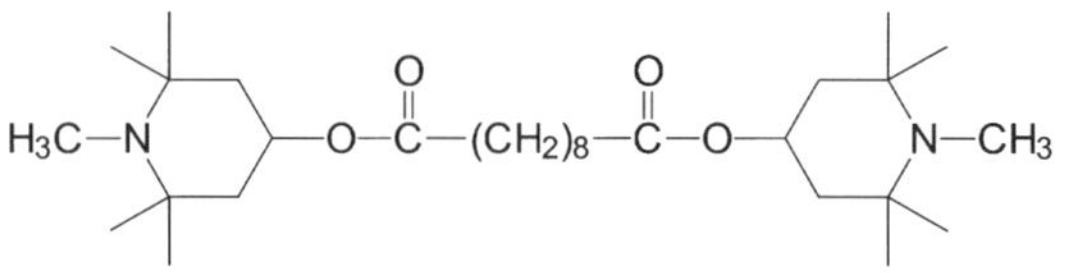

**FIGURE 4.17** Lowilite 76, CAS no. 41556-26-7, *bis*(1,2,2,6,6-pentamethyl-4-piperidinyl)sebacate, a common tertiary HALS having reduced alkalinity.

### 4.2.3 Flame Mechanisms

Combustion is a very complex combination of physical and chemical phenomena that must interact in balance for combustion to occur. The three components necessary to support combustion are fuel, heat, and oxygen, which form the classic fire triangle (Fig. 4.19).

Fire suppression is accomplished by means that affect one or more of the legs of the fire triangle. Additives that liberate water of hydration upon heating cool the substrate and dilute the combustible gases. Although these additives are inexpensive, this is a relatively inefficient approach to flame retardation because high load levels of about 30 wt% are required.

Intumescent additives functioning in the condensed phase form a thermal barrier that protects the substrate and limits diffusion of combustible gases out of the substrate and oxygen into the substrate. Significant effort in developing these systems is underway, as they are characterized by relatively low heat release.

Additives functioning in the vapor phase exhibit the highest efficiency because they interfere with the combustion chemistry. Total load levels as low as 3 wt% are sufficient for certain flammability performance requirements. The halogen-containing systems that function in the vapor phase are characterized by higher heat releases than systems operating in the condensed phase. The steps

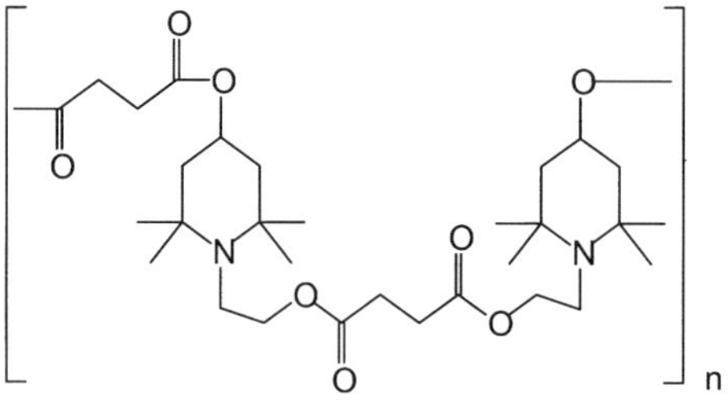

**FIGURE 4.18** Lowilite 62, CAS no. 65447-77-0, dimethyl succinate polymer of 4-hydroxy-2,2,6,6-tetramethyl-1-piperidine ethanol, a tertiary HALS that is polymeric in nature.

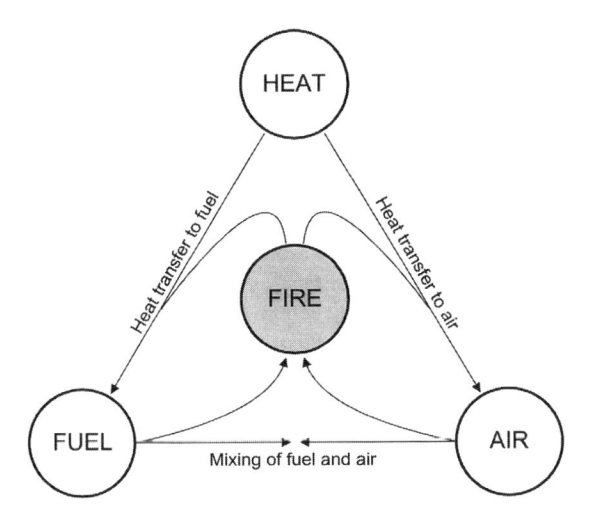

# Stabilization of Flame-Retarded Polypropylene

**FIGURE 4.19** Classic fire triangle shows the three components necessary to support combustion.

involved in the combustion of polymers as described by Troitzsch [10], shown in Fig. 4.20, are summarized below.

The condensed phase is heated by an ignition source or by thermal feedback of radiant heat from the gas phase oxidation reactions. Thermolytic cleavage of the polymer supplies combustible and noncombustible gaseous products to the gas phase combustion zone. These products react with oxygen

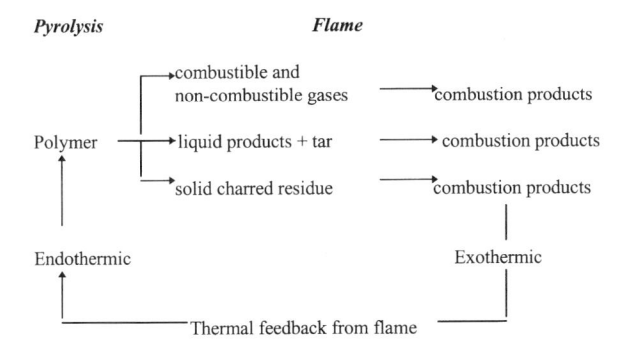

**FIGURE 4.20** General steps of polymer combustion. The flame processes differ from light and low-temperature thermal degradation processes in that temperatures are hundreds of degrees centigrade and vapor phase.

```
Start (polyolefin)    RH           ⟶   R• + H•

Growth                R• + O2      ⟶   ROO•
                      ROO• + RH    ⟶   ROOH + R•

Branching             ROOH         ⟶   RO• + •OH
```

**FIGURE 4.21** Thermal oxidation radical products of combustion.

and release heat during the production of carbon dioxide, carbon monoxide, water, and soot.

Oxidative degradation of polymers (Fig. 4.21) leads to the formation of highly reactive H and OH radicals as described by Thiery [11]. Thermal feedback reinforces pyrolysis to further fuel the flame. A simple model of growth and branching is shown in Fig. 4.22.

The free radicals of H and OH, which are proliferated by the chain-branching reactions, confer a high velocity to the flame front. They attack the hydrocarbon species and participate in reactions that yield the various terminal combustion products.

### 4.2.4 Mechanisms of Flame Retardancy

Combustion is prevented or stopped by affecting one or more of the three components necessary to support combustion (heat, fuel, and oxygen). Flame-retardant mechanisms cluster into four general classes.

*Heat Sink Mechanisms.* Endothermic processes inherent to additives such as aluminum trihydrate ($Al_2O_3 \cdot 3H_2O$, also known as aluminum hydroxide, $Al(OH)_3$) and magnesium hydroxide, $Mg(OH)_2$, cool the substrate to tempera-

```
Growth        CH4 + OH•       ⟶   CH3• + H2O
              CH4 + H•        ⟶   CH3• + H2
              CH3• + O        ⟶   CH2O + H•
              CH2O + CH3•     ⟶   CHO• + CH4
              CH2O + H•       ⟶   CHO• + H2
              CH2O + OH•      ⟶   CHO• + H2O
              CH2O + O        ⟶   CHO• + OH•
              CHO•            ⟶   CO + H•
              CO + OH•        ⟶   CO2 + H•

Branching     H• + O2         ⟶   OH• + O
              O + H2          ⟶   OH• + H•
```

**FIGURE 4.22** Growth and branching model for pyrolysis that further fuels the flame.

## Stabilization of Flame-Retarded Polypropylene

$$2Al(OH)_3 \rightarrow Al_2O_3 + 3H_2O$$

$$Mg(OH)_2 \rightarrow MgO + H_2O$$

**FIGURE 4.23** Flame retardancy mechanisms occurring by heat sink.

tures below those required to sustain combustion. Water vapor evolved in their endothermic decomposition dilutes the combustible fuel in the gaseous phase, whereas the alkaline inorganic residues afford a level of thermal barrier protection (Fig. 4.23).

Thermogravimetric analysis shows that aluminum trihydrate (ATH) yields its waters of hydration in the range of 205–225°C. Because this temperature is often achieved during the compounding of thermoplastics, ATH is not generally used in these systems. Magnesium hydroxide dehydrates in the range of 300–320°C and can be safely processed in polypropylene, with load levels on the order of 60 wt% being used to achieve desired flammability performance. Figure 4.24 shows the TGA thermograms of polypropylene, $Mg(OH)_2$, and a formulation containing polypropylene and about 30 wt% $Mg(OH)_2$. Observe the degradation profile of the formulation shifting to lower temperatures and its increased residues content due to the presence of $Mg(OH)_2$.

Flame retardants that inhibit combustion by the physical actions of cooling, diluting, and insulating are less effective on a weight basis than those that chemically inhibit the flame chemistry.

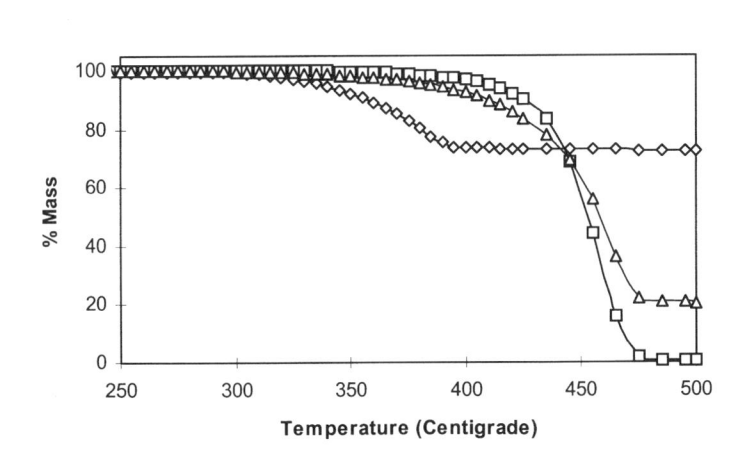

**FIGURE 4.24** TGA thermograms, % mass retention vs. temperature, of (□) polypropylene, (◇) $Mg(OH)_2$, and (△) polypropylene with 30% $Mg(OH)_2$, under nitrogen.

*Condensed-Phase Mechanisms.* Condensed-phase systems decompose upon heating to form a large amount of thermally stable residue, or char. This char acts as a thermal shield for radiant heat transfer from the flame to the polymer and as a physical barrier to limit diffusion of flammable gases from the polymer to the combustion zone.

Intumescent chars result from a combination of charring and foaming of the polymer surface, resulting in a thick protective barrier. Because of their low environmental impact and their relatively low heat release, there is considerable interest in the development of intumescent systems. These systems typically contain an acid source, such as ammonium polyphosphate (APP), and a polyhydroxy compound, such as pentaerythritol, that dehydrates and chars due to acid attack. APP is a linear polymer that decomposes (Fig. 4.25) upon heating in three successive steps as reported by Camino et al. [12]. Effective intumescent systems require that gases are evolved between the gelation and vitrification stages of the material and that the char has sufficient strength to maintain its structure.

*Vapor-Phase Mechanisms.* Halogen flame retardants are very effective because they interfere with the radical chain mechanism in the gas phase of the combustion process. We have previously seen that during the combustion of polymers, radical chain reactions involving the OH and H radicals occur, as shown in Fig. 4.26.

Because this system is regenerative, halogenated flame retardants are very effective at relatively low load levels. Bromine is more effective than chlorine because of the lower bond energy of the carbon to bromine bond. This allows the liberation of halide at the more favorable point in the combustion process. It is also believed that HCl is formed over a wider temperature range and therefore is present at lower concentrations in the flame front and less effective than HBr.

*Antimony Halogen Synergism.* Combinations of antimony trioxide (ATO, $Sb_2O_3$) with halogenated flame retardants exhibit superior performance to those without $Sb_2O_3$. Antimony trioxide reacts as shown in Fig. 4.27 with hydrogen halide to provide a controlled release of $SbX_3$ over a combustion range of 245–565°C as shown [13].

The $SbX_3$ formed in these reactions provides a supply of HX that interferes with the radical chain mechanisms as previously described and illustrated in Fig. 4.28. Synergism likely arises from the fact that the antimony halogen chemistry increases the lifetime of reactive species in the combustion zone and therefore increases the probability that the flame-branching reactions will be interrupted as described above. Further mechanistic details are provided by Read and Heighway-Bury [14], Kirk-Othmer [15], Katz and Milewski [16], Hastie and McBee [17], and Brauman [18].

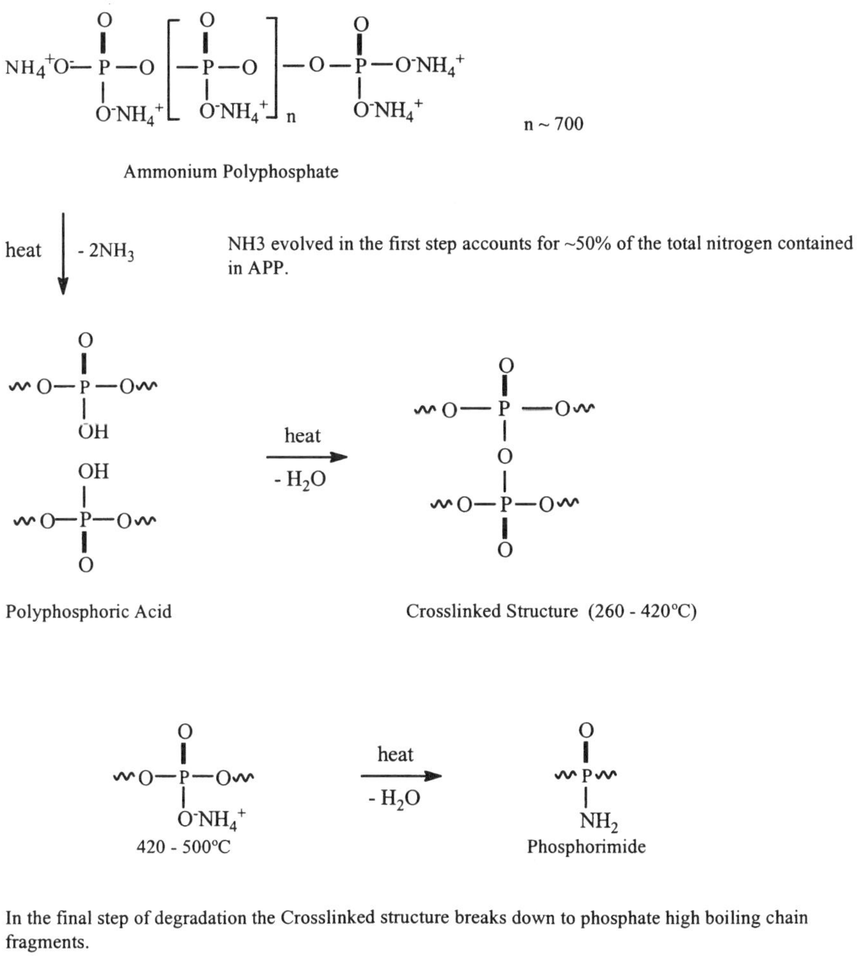

# Stabilization of Flame-Retarded Polypropylene

$$NH_4^+O-\overset{\overset{O}{\|}}{\underset{O^-NH_4^+}{P}}-O\left[-\overset{\overset{O}{\|}}{\underset{O^-NH_4^+}{P}}-O\right]_n-O-\overset{\overset{O}{\|}}{\underset{O^-NH_4^+}{P}}-O^-NH_4^+$$

$n \sim 700$

Ammonium Polyphosphate

heat ↓ - 2NH$_3$    NH3 evolved in the first step accounts for ~50% of the total nitrogen contained in APP.

Polyphosphoric Acid →(heat, -H$_2$O)→ Crosslinked Structure (260 - 420°C)

420 - 500°C →(heat, -H$_2$O)→ Phosphorimide

In the final step of degradation the Crosslinked structure breaks down to phosphate high boiling chain fragments.

**FIGURE 4.25** Three stages of ammonium polyphosphate (APP) decomposition.

## 4.3 METHODS OF EVALUATION

Methods used for evaluation of stabilization and flame retardancy are key for formulation development. Polypropylene process stability might be evaluated by either multiple processing at nominal conditions or significantly increasing the temperature and shear during single processing. As mentioned previously, the multiple extrusion followed by measurement of the melt flow and color variations

$$H\bullet + O_2 \longrightarrow OH\bullet + O$$
$$O + H_2 \longrightarrow OH\bullet + H\bullet$$

Halogenated flame-retardants decompose to form halide radicals and hydrogen halide as follows:

$$RX \longrightarrow R\bullet + X\bullet$$
$$X\bullet + RH \longrightarrow R\bullet + HX$$
$$\text{where } X = Br, Cl$$

Hydrogen halide then interferes with the radical chain mechanisms:

$$HX + H\bullet \longrightarrow H_2 + X\bullet$$
$$HX + OH\bullet \longrightarrow H_2O + X\bullet$$
$$X\bullet + H\bullet \longrightarrow HX$$

**FIGURE 4.26** Halogen mechanism for interfering with the flame process.

is particularly suited for polypropylene because there are seldom cross-linking or other processes to confuse the results as happens with polyethylene.

Evaluation of UV durability is specific to the application. For example, office business equipment will use an accelerated aging method like ASTM D4459 to mimic sunlight through office window glass in a dry and cool environment. This xenon arc exposure method accelerates the evaluation such that a three-unit composite color shift after 300 hr is a standard failure criteria. At this point the physical integrity has not significantly changed but the color variation would be objectionable to a customer. Figure 4.29 shows the performance for a number of stabilizers in polypropylene plaques. The criterion of interest is the retention of impact strength.

For polypropylene fibers, UV durability is usually measured in terms of retention of tensile strength or elongation with accelerated aging. Figure 4.30 shows data for aging in Florida. Natural and accelerated aging methods are useful screening tools for UV durability. However, the prediction of service life from aging results requires extensive correlation studies.

$$Sb_2O_3 + 2HX \longrightarrow 2SbOX + H_2O$$
$$5SbOX \longrightarrow Sb_4O_5X_2 + SbX_3 \quad 245\text{-}280°C$$
$$4Sb_4O_5X_2 \longrightarrow 5Sb_3O_4X + SbX_3 \quad 410\text{-}475°C$$
$$3Sb_3O_4X \longrightarrow Sb_2O_3 + SbX_3 \quad 475\text{-}565°C$$

**FIGURE 4.27** Mechanism for endothermic antimony trioxide (ATO) synergy with halogen flame retardants.

# Stabilization of Flame-Retarded Polypropylene 99

$$SbX_3 + H\bullet \longrightarrow HX + SbX_2$$

$$SbX_2 + H\bullet \longrightarrow HX + SbX$$

$$SbX + H\bullet \longrightarrow HX + Sb$$

**FIGURE 4.28** Mechanisms for antimony liberation of HX.

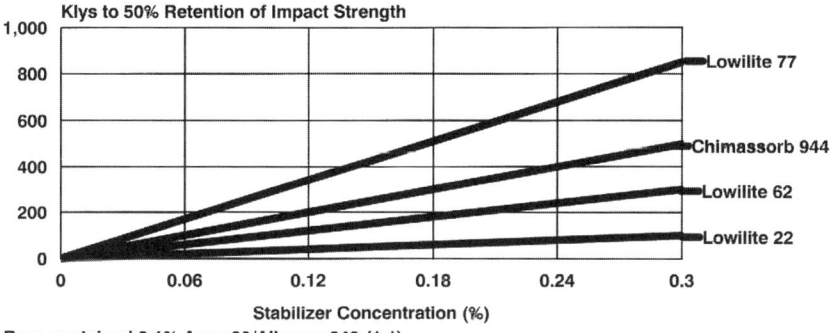

**FIGURE 4.29** Relevant light stabilizer performance for actual outdoor aging.

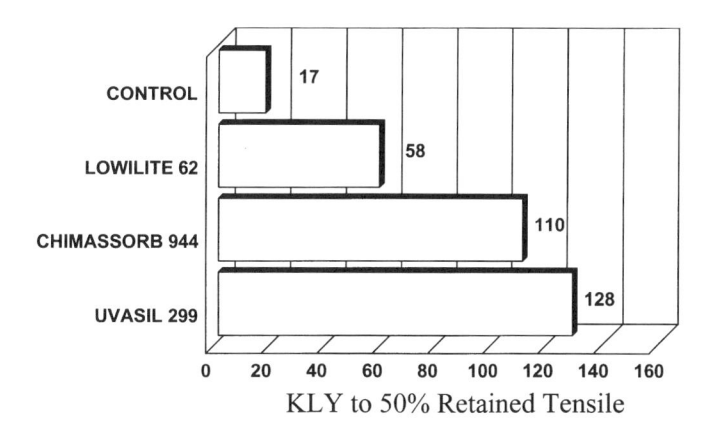

Base contained 0.1% Anox 20/Alkanox 240 (1:1)
100 mil Plaques, 45°S Florida exposure, Dynstat Impact

| Stabilizer | Klys to 50% Retention of Impact Strength |
|---|---|
| Lowilite 77 | ~850 |
| Chimassorb 944 | ~550 |
| Lowilite 62 | ~300 |
| Lowilite 22 | ~150 |

(at 0.3% stabilizer concentration)

**FIGURE 4.30** Tensile strength retention of 1296 denier/72 strand polypropylene fiber having 0.2 wt% HALS Lowilite 62, Chimassorb 944, or Uvasil 299 under Florida exposure conditions (45° South). Base stabilization includes 0.16 wt% Anox TB311 and 0.1 wt% calcium stearate. The control has no HALS. kLy is a unit for outdoor exposure for which 1 year equals 110–140 kLys.

KLY to 50% Retained Tensile

| Sample | KLY |
|---|---|
| CONTROL | 17 |
| LOWILITE 62 | 58 |
| CHIMASSORB 944 | 110 |
| UVASIL 299 | 128 |

TABLE 4.1 Flammability Assessment of Flame-Retarded Polypropylene Systems

| Test type | ASTM method | ASTM title | Test description |
|---|---|---|---|
| Horizontal | D635 | Rate of burning or extent and time of burning of self-supporting plastics in a horizontal position | The scope of this test is to determine the maximal burn rate for a bar specimen. To get an HB rating, the sample must meet the following criteria:<br>a. If the test specimen has a thickness <3 mm, the burn rate cannot exceed 40 mm/min.<br>b. If the test specimen has a thickness >3 mm, the burn rate cannot exceed 75 mm/min. |
| Vertical | D3801 | Measuring the comparative burning characteristics of solid plastics in a vertical position | The scope of this test is to determine the ability of materials to self-extinguish (i.e., upon removal of an ignition source, the material ceases to burn). The flame is applied to the sample twice for 10 sec. There are three possible ratings for this test based on the duration of burning after the ignition source has been removed and the dripping behavior of the material. The criteria for each rating are shown below: |

# Stabilization of Flame-Retarded Polypropylene

| | | | |
|---|---|---|---|
| | | V-0: | Individual burn times < 10 sec; total burn time for all samples < 50 sec; no flaming drips that ignite cotton located below sample. |
| | | V-1: | Individual burn times < 30 sec; total burn time for all samples < 250 sec; no flaming drips that ignite cotton. |
| | | V-2: | Individual burn times < 30 sec; total burn time for all samples < 250 sec; flaming drips that ignite cotton. |
| Vertical | D5048-90 | Measuring the comparative burning characteristics and resistance to burn-through of solid plastics using a 125-mm flame | The scope of this test is to determine the ability of a material to resist burn-through after the ignition source has been removed. The flame is applied to the material 5 times for 5 sec (total ignition time 25 sec). To achieve a 5V rating, the total flame and glow time for each sample after the last ignition cannot exceed 60 sec. (5VA indicates the flame burned through the plaque and 5VB has no burn-through after the ignition source has been removed.) |

To design a flame-retardant polypropylene system, the following key factors need to be determined:

- What is the end-use application (i.e., appliance, automotive, textile)?
- What are the flammability ratings the product must achieve?
- Will the product be exposed to excessive temperature fluctuations?
- What type of UV stability is required for the product?
- Will the product be recycled?

The next phase of the assessment is divided into flammability performance and effect on overall physical properties.

For flammability performance, three flame tests established by Underwriters Laboratories (UL) protocol UL94 are used for polypropylene. The specimen orientation, flame size, and duration have been modified to reflect possible exposure situations. Table 4.1 shows the type of flame, the ASTM number, and a brief description of the scope of the tests. The goal of the horizontal test is to determine the burn rate of a sample and not its ability to self-extinguish. Conversely, the vertical burn test is designed to gauge a sample's ability to withstand flame propagation and dripping once an ignition source has been removed. An alternative test to the vertical burn test is the 125-mm test in which the samples are mounted horizontally and monitored for burn-through. Once the flammability performance has been established, the effect of the flame retardant on physical properties must be determined.

Physical property assessment of polypropylene is divided into short- and long-term evaluations. The UL746 A and B standards describe the protocols for the short- and long-term evaluations, respectively. A list of the tests measured in the short term is shown in Table 4.2. The tests designated by UL as "Performance Level Categories," or PLCs, are used to assist the end user to select materials with equivalent performance and to identify an alternative, if original system becomes unavailable.

The short-term evaluation is conducted on all systems to determine the effect of the flame retardant on physical properties and to establish a baseline for the long-term evaluation. If the composite has a UL94 V2 or better rating, then long-term evaluations (UL746B) are conducted. For the UL746B, the same properties are assessed as in the short term, but now the samples are aged at four different temperatures. When the samples begin to visibly degrade or exhibit more than a 50% reduction in physical property performance, the samples are removed from the oven. Physical properties and flammability performances are measured on the aged samples. The maximum temperature and time at which physical properties are maintained at greater than or equal to 50% unaged values is designated as the relative thermal index, or RTI. Polypropylene systems with stabilizer packages geared toward long-term heat aging will typically have RTIs in the 100–125°C range. Addition of organohalogen flame retardants increases

# Stabilization of Flame-Retarded Polypropylene

TABLE 4.2  Typical Short-Term Properties Assessed in UL-746A for Polypropylene

| Property | Test method |
| --- | --- |
| Tensile strength | ASTM D638 |
| Tensile or Izod impact | ASTM D1822 or D256 |
| Dielectric strength | ASTM D149 |
| Volume resistivity | ASTM D257 |
| High voltage, low current, high arc resistance | ASTM D495 |
| Comparative tracking rate | ASTM D3638 |
| High voltage arc tracking rate | UL |
| Hot wire ignition | ASTM D3874 |
| High current arc ignition | UL |
| High voltage arc resistance to ignition | UL |
| Glow wire resistance to ignition | UL |

the thermal properties of polypropylene composites and consequently can result in higher RTIs.

## 4.4 COMMON COMMERCIAL SYSTEMS

Polypropylene has begun to infiltrate applications previously deemed too demanding. New catalysts, modified fillers, and compounding procedures have elicited the introduction of polypropylene grades with improved stiffness, toughness, temperature resistance, scratch resistance, better gloss, and dramatically improved load-bearing ability at elevated temperatures. Experts in the polyolefin industry speculate that the full impact of metallocene is still at least 3 years and perhaps as many as 10 years away. End-use applications have now expanded to include automotive side body panels, structural ducts, and heat-generating appliances, which, historically, have been made from thermoplastic urethanes (TPUs), acrylonitrile-butadiene-styrene (ABS) copolymer, and glass-filled systems.

Consequently, the need for flame-retardant polypropylene has grown. As with other resin systems, there are two approaches that are taken to achieve acceptable combustion resistance. The first and most predominantly used method works in the vapor phase and relies on the synergism of organohalogen and metal oxide additives, such as PE68 (Fig. 4.31); DE83R, CD75P, or GPP36 (Fig. 4.32); or Dechlorane Plus (Fig. 4.33) used with antimony oxides. DE83R, PE68, and CD75P are products of Great Lakes Chemical Corporation. Dechlorane Plus is an Occidental Chemical product. The second method focuses on intumescence and replaces halogen additives with organophosphorus, phosphoric acid, and melamine salts or inorganic complexes. Examples of typical flame-retardant polypropylene systems are shown in Table 4.3 [19,20]. In addition to DE83R, PE68,

TABLE 4.3  Flame-Retarded Propylene Systems

| Flame retardant | DE-83R | | PE-68 | FR-1025 | | BA-59 | FR-1808 | PB-370 | Nonnen 52 | Dechlorane Plus | Mg(OH)$_2$ |
|---|---|---|---|---|---|---|---|---|---|---|---|
| Homopolymer | 93 | | 96 | 90.0 | | 91 | 90.9 | 94 | 100 | 100 | 35 |
| Block copolymer | | 64.3 | 100 | | 62.5 | | | | | | |
| Flame retardant | 4.8 | 23.8 | 11 | 5.8 | 25 | 5.2 | 8.5 | 4 | 10 | 52.6 | 65 |
| Antimony trioxide | 4.6 | 11.9 | 3 | 2.6 | 12 | 1.7 | 2.8 | 2 | 5.5 | 15.8 | |
| Misc. additives | 0.6 | | 15.15 | 0.7 | | 2.1 | 0.6 | | 14.8 | 42.6 | |
| % Bromine | 4 | 19.8 | 5.8 | 6 | 17.5 | 3 | 6 | 2.8 | 5.1 | | |
| % Chlorine | | | | | | | | | | 5.9 | |
| % Phosphorus | | | | | | | | 0.12 | | | |
| % ATO | 1.6 | 11.9 | 1 | 2.8 | 12.1 | 1.7 | 2.8 | 2 | 4.2 | 7.5 | |
| UL-94 rating | V-2 | V-0 | V-2 | V-2 | V-0 | V-2 | V-2 | V-2 | V-0 | V-0 | V-0 |

Melt flow ranges of homopolymer, 2–12 g/10 min; melt flow index of block copolymer, 17 g/10 min; DE-83R; decabromodiphenyl oxide (Great Lakes Chemical Corporation); PE-68; bis(2,3-dibromopropylether) of tetrabromobisphenol A (Great Lakes Chemical Corporation); FR-1025; poly(pentabromobenzyl)acrylate (Dead Sea Bromine, Ltd.); BA-59P; tetrabromobisphenol A (Great Lakes Chemical Corporation); FR-1808; octabromotrimethylphenyl indan (Dead Sea Bromine, Ltd.); Reoflam PB-370; *tris*(tribromoneopentyl)phosphate (FMC); Nonnen 52; *bis*(3,5-dibromo-4-dibromopropyloxyphenyl)sulfone (Marubishi Oil Chemical Corporation); Dechlorane Plus, Diels–Alder diadduct of hexachlorocyclopentadiene and 1,5-cyclooctadiene (Occidental Chemical).
*Source:* Data from Refs. 19 and 20.

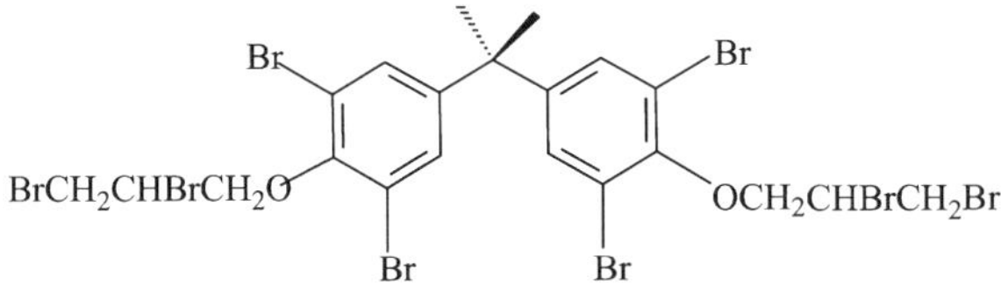

# Stabilization of Flame-Retarded Polypropylene 105

**FIGURE 4.31** Great Lakes Chemical Corporation PE-68, CAS no. 21850-44-2, tetrabromobisphenol A *bis*(2,3-*di*-bromopropylether), which contains both aliphatic and aromatic bromine accounting for 68% of the molecular weight.

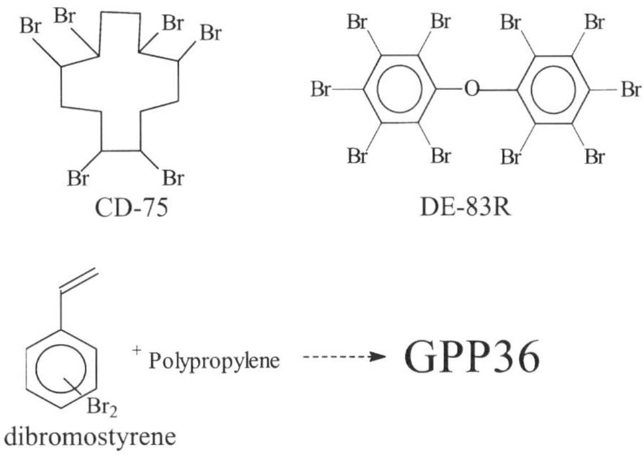

**FIGURE 4.32** Popular flame retardants CD-75, CAS no. 3194-55-6; DE-83R, CAS no. 1163-19-5; GPP-36, CAS no. 148993-99-1, which differ greatly in their bromine type and processing characteristics.

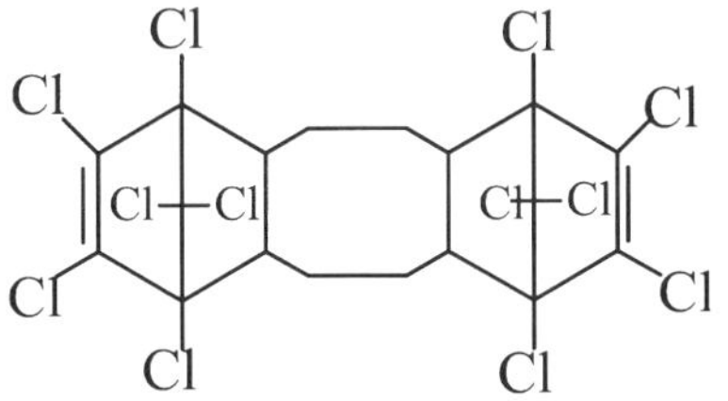

**FIGURE 4.33** Dechlorane Plus.

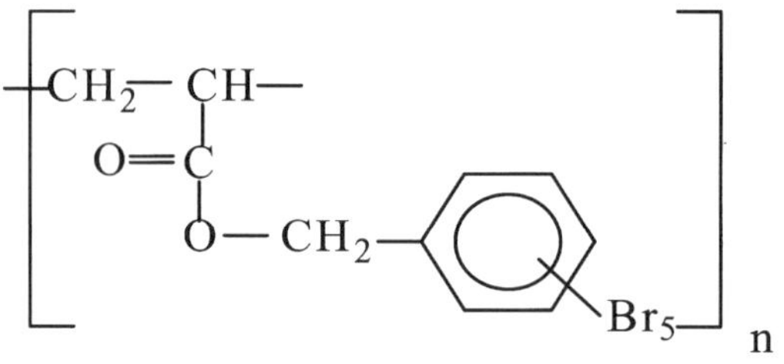

**FIGURE 4.34** FR-1025.

and Dechlorane Plus shown earlier, these examples include FR1025 (Fig. 4.34), BA59 (Fig. 4.35), FR1808 (Fig. 4.36), PB370 (Fig. 4.37), Nonnen 52 (Fig. 4.38), and $Mg(OH)_2$.

For the first approach, alkyl organohalogen additives (such as CD75P) tend to be more efficient because of the lower temperature required to release the halogen. However, these products consequently are not as thermally stable and limit processing temperatures. Also, current research indicates that the alkyl organohalogen additives suffer severe degradation during regrind studies, which in turn prohibits recycling. The aromatic organohalogen additives are more thermally stable and can be processed at higher temperatures and shear rates, but loading must be increased. Products with both aromatic and aliphatic halogens, such as PE68, can be processed at higher temperatures and at lower loading without significantly reducing activation temperature. Additional synergism or higher halogen content can be achieved by incorporating phosphorus into the molecule as an FR or as a carrier. FMC Corporation has an alkyl organohalogen phosphate, Reoflam PB370, that can be used in polypropylene. In this case, the apparent primary function of the phosphorus is as a carrier to increase the brominated alkyl substitution. The processing parameters of PB370 are similar to PE68 but require slightly higher loading.

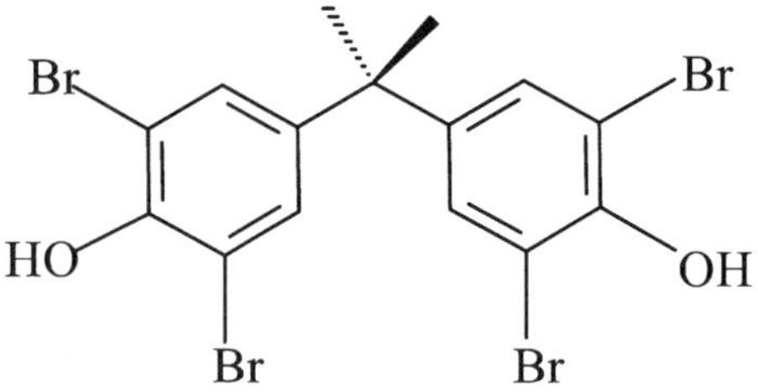

**FIGURE 4.35** Great Lakes Chemical Corporation BA-59, CAS no. 79-94-7, tetrabromobisphenol A.

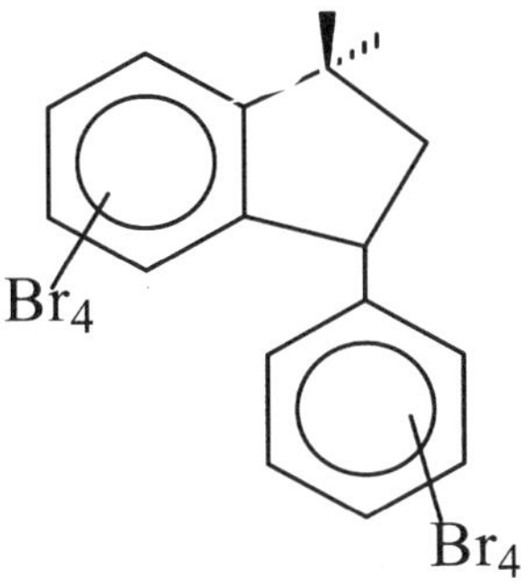

## Stabilization of Flame-Retarded Polypropylene 107

**FIGURE 4.36** FR-1808.

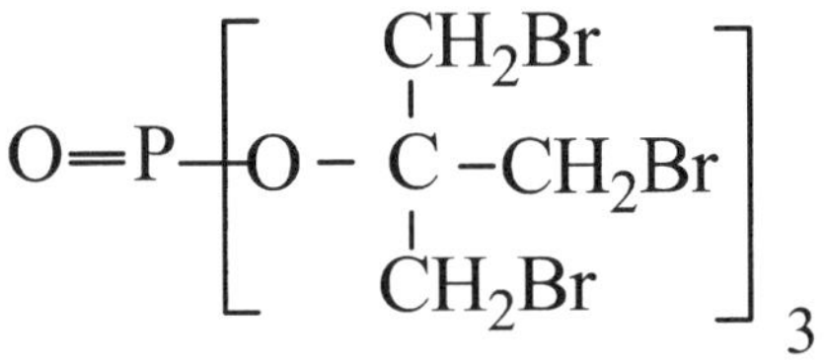

$$O=P\left[O-\underset{\underset{CH_2Br}{|}}{\overset{\overset{CH_2Br}{|}}{C}}-CH_2Br\right]_3$$

**FIGURE 4.37** Reoflam PB-370.

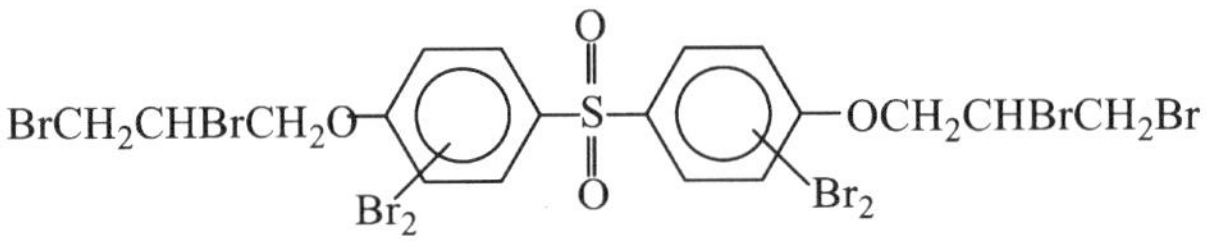

**FIGURE 4.38** Nonnen 52.

The theory behind the second approach is that incorporation of nonhalogen or intumescent flame retardants will produce lower smoke and less potentially harmful byproducts. The most commonly used inorganic additives are magnesium hydroxide [$Mg(OH)_2$] and ATH. Both must be used at extremely high loadings, detrimentally effect the physical property of the composite, and have disassociation temperatures within the processing temperature profiles for polypropylene. Commercial phosphorus-based alternatives are the ammonium poly-

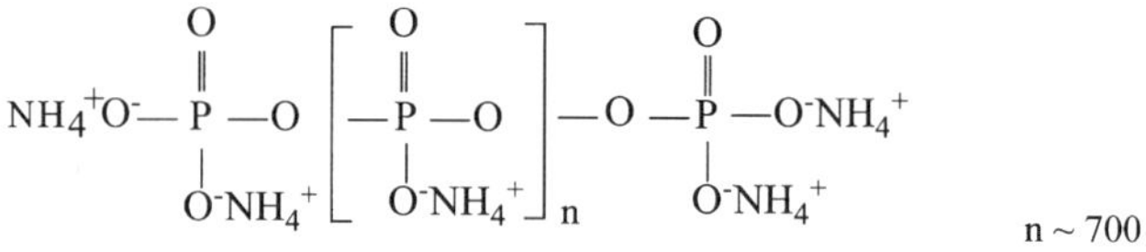

$$NH_4^+O^-\!\!-\overset{\overset{O}{\|}}{\underset{\underset{O^-NH_4^+}{|}}{P}}\!\!-O\left[-\overset{\overset{O}{\|}}{\underset{\underset{O^-NH_4^+}{|}}{P}}\!\!-O\right]_n\!\!-O-\overset{\overset{O}{\|}}{\underset{\underset{O^-NH_4^+}{|}}{P}}\!\!-O^-NH_4^+$$

$n \sim 700$

**FIGURE 4.39** Commercial ammonium polyphosphate salts.

Proprietary Intumescent Flame Retardant
15% phosphorous

**FIGURE 4.40** Great Lakes Chemical Corporation NH-1511.

phosphate salts (Fig. 4.39; Monsanto) or NH-1511 (Fig. 4.40; Great Lakes Chemical Corporation) and NH-1197 (Fig. 4.41; Great Lakes Chemical Corporation). Like their inorganic counterparts, effective loadings are higher than the organohalogen additives. Also, these products are hydroscopic, have lower thermal stability, and tend to be less economically feasible. Increased research emphasis on this approach has been observed.

## 4.5 NEXT-GENERATION SYSTEMS

Past- and present-generation systems struggled with problems of meeting physical requirements while maintaining performance and cost constraints. Next-generation systems must leap beyond these issues and address higher processing temperatures and higher use temperatures. For example, the interior automotive trim applications have pressed the long-term aging requirements up from 110°C to 150°C.

Reduction of worker chemical exposure is also driving next-generation systems to include low-dusting and no-dusting physical forms of single products and product blends. Combination of all additives into a single pellet without a binder improves dosing while lowering dust exposure. No-dust blends and next-

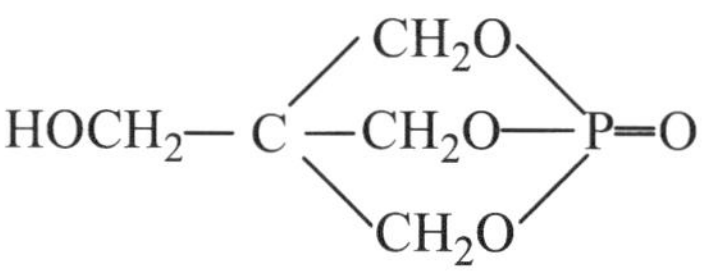

$$HOCH_2-C\underset{CH_2O}{\overset{CH_2O}{\diagup}}\!\!\!\diagdown\!\!\!P\!=\!O$$

**FIGURE 4.41** Great Lakes Chemical Corporation intumescent FR NH-1197, CAS no. 501-78-0, which has 17 wt% phosphorus.

## Stabilization of Flame-Retarded Polypropylene

generation systems have been the subject of recent polyolefin and polypropylene conferences [20–23].

Another specific requirement for next-generation systems is better component compatibility [23,24]. The use of a silicone backbone for a stabilizer, as shown in Fig. 4.42, addresses the need to have more compatible chemical species for light stabilizers. With increased compatibility between polypropylene and the HALS, increased efficiency was noted along with a substantial increase in extraction resistance to most common solvents. For filled polypropylene as mentioned above, the stabilizer had a reduced migration rate into fillers. This increased polypropylene-to-light-stabilizer compatibility maintained efficiency by avoiding coadditive interactions.

Figure 4.43 shows a less alkaline silicone-based stabilizer that is particularly beneficial in polypropylene with brominated flame retardants. By reducing acid–base interaction, coadditive interactions are also reduced. Fiber tensile retention with UV aging shown in Fig. 4.44 illustrates this point. Next-generation systems will modify these additives to add a graftable function [23]. Unlike grafted monomeric stabilizers, polymeric stabilizers with graftable functions only anchor the stabilizer to the polymer. This prevents the complete loss of migration that reduced the monomeric grafted stabilizer's performance.

In some applications, it is actually possible to segregate antagonistic coadditives such as the flame retardant from the stabilizer by coextrusion. For example, bicomponent fiber extrusion is one such application. Figure 4.45 shows a filament cross-section where the flame retardant is restricted to the core whereas the sheath is UV stabilized. Figure 4.46 demonstrates the advantage in UV durability from bicomponent construction.

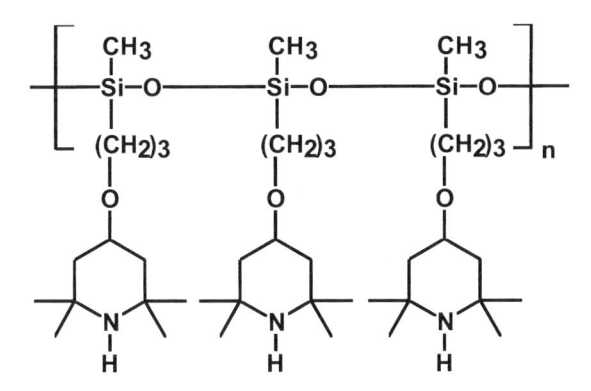

FIGURE 4.42 Great Lakes Chemical Corporation high molecular weight silicone-based hindered amine light stabilizer, Uvasil 299, CAS no. 182935-99-0.

FIGURE 4.43 Great Lakes Chemical Corporation's less alkaline, high molecular weight, silicone-based amine light stabilizer, Uvasil 816.

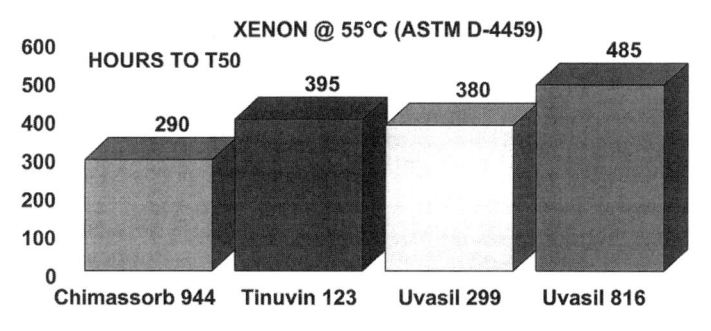

FIGURE 4.44 Advantage of improved stabilizer compatibility in flame-retarded polypropylene UV durability. Values are hours of xenon exposure under ASTM D-4459 required to reach 50% of initial tensile strength. Base formulations include 6 wt% bromine from GPP-39, 0.5 wt% of the specific HALS, and 1.5 wt% Tinuvin 234. Amoco 7556 was the polypropylene used.

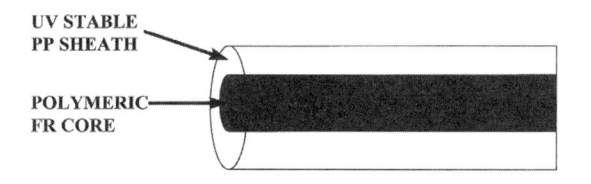

FIGURE 4.45 Cross-section of a single-fiber filament with coextrusion of bicomponent construction.

**% RETAINED STRENGTH vs ASTM D4459 EXPOSURE**

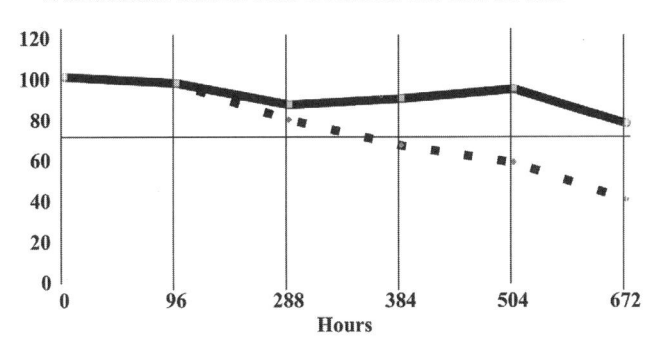

FIGURE 4.46 Advantage of bicomponent construction (3.9 wt% Br, —) on UV durability compared with monocomponents (3.0 wt% Br, ---). Formulations include GPP 39 as the bromine source, 0.5 wt% Uvasil 299, 1.5 wt% Tinuvin 234, and 0.15 wt% Anox BB011.

Super submicrometer technology is now available to produce antimony synergist with particle sizes below 1 µm. The advantages include applications in flame-retarded fiber without screen-packing problems and reduced UV light absorbency. This will allow next-generation systems to enhance flame-retardant performance and UV durability at the same time.

## 4.6 SUMMARY

Stabilization of flame-retarded polypropylene requires attention to detail, including performance requirements, cost constraints, and coadditive interactions. Improvements to traditional additive performances often fall short of requirements. Therefore, the focus is shifting to the full formulation components, often with new chemistry types. System compatibility is a key concept. Next-generation systems will be more worker and environment friendly as the polypropylene industry displaces both commodity and engineering resins at ever harsher requirements.

## REFERENCES

1. NF Rainey. Polypropylene and Polyethylene Fibers: Innovative Building Blocks for the Textile Industry. Symposium Proceedings, Atlanta, GA, March 1990.
2. JL Bolland, G Gee. Trans Faraday Soc 42:236, 1946.
3. JL Bolland, Trans Faraday Soc 44:669, 1948.
4. PK Das, MV Encinas, JC Scaiano, S Steenken. J Am Chem Soc 103:4162, 1981.

5. J Pospisil. Adv Polym Sci 36:69, 1980.
6. C Armstrong, MJ Husbands, G Scott. Eur Polym J 15:241, 1979.
7. K Schwetlick. Pure Appl Chem 55:1629, 1983.
8. G Combs, TH Austin, TA Craig, RD Duffy. Polyurethanes World Congress, Vancouver, BC, Canada, 1993.
9. C Neri, S Costanzi, RM Riva, R Farris, R Colombo. Mechanism of action of phosphites in polyolefin stabilization. Polymer degradation and stability. 49:65–69, 1995.
10. J Troitzsch. International Plastics Flammability Handbook, 2nd ed. New York: Hanser Publishers, 1990, p. 16.
11. P Thiery. Fireproofing. Amsterdam: Elsevier, 1970.
12. G Camino, MP Luda, L Costa. ACS Symp Ser 76:599, 1995.
13. J Edenbaum, ed. Plastics Additives and Modifiers Handbook. New York: Van Nostrand Reinhold, 1992, p. 1042.
14. NJ Read, EG Heighway-Bury. J Soc Dyers Colour 74:823–829, 1958.
15. Kirk-Othmer Encyclopedia of Chemical Technology, Vol. 10, 3rd ed. New York: John Wiley & Sons, 1980, p. 355.
16. HS Katz, JV Milewski, eds. Handbook of Fillers and Reinforcements for Plastics. New York: Van Nostrand Reinhold, 1978.
17. JW Hastie, CL McBee. Halogenated Fire Suppressants. ACS Symposium Series, 1975, p. 16.
18. SK Bauman, J Fire Retard Chem 3:117–137, 1976.
19. U.S. patents 5,409,980 (PQ Corporation) and 5,420,183 (HPG International, Inc.), 1996 BCC proceedings, FMC product literature for Reoflam PB370.
20. RL Gray, C Neri. New UV Stabilization System for Polypropylene. Houston: SPE PO-RETEC, 1998.
21. RE Lee, C Neri. No Dust Blend Technology. Houston: SPE PO-RETEC, 1998.
22. RE Lee, C Neri, BM Sanders. Stabilizing Effects of Hindered Amines. Atlanta, GA: HALS History SPE ANTEC, 1998.
23. RL Gray, RE Lee. Polymer bound stabilizers—a review past and present. Additives Conference, Orlando, FL, Feb. 16–18, 1998.
24. RE Lee, OI Kuvshinnikova, JM Zenner. Interactions of pigments and hindered amine light stabilizers in polyolefin systems. Hilton Head, SC, March 2–4, 1998.

# 5

## Recycling of Polypropylene and Its Blends: Economic and Technology Aspects

**Akin A. Adewole and Michael D. Wolkowicz**
Basell USA, Inc., Elkton, Maryland, U.S.A.

In the late 1980s, many communities in the United States were struggling with the relatively new problem of how to manage waste disposal. Landfills were filling up, and public opposition to new landfills and incineration plants, with the feeling that "nobody wants it in their neighborhood," was strong. As landfills closed, the shock of increasing disposal costs was tremendous, and the federal government responded through the Environmental Protection Agency (EPA) to help prioritize the problems concerning landfills and solid waste disposal.

In a proactive response, major resin-producing companies formed the Council for Solid Waste Solution in 1988 [1]. Its mission is to ensure the reuse or safe disposal of plastics in accordance with the four-part waste management hierarchy put forth by the EPA, with recycling as the greatest imperative. In EPA criteria, source reduction is the number one priority, followed by recycling, incineration, and landfill only as a last resort [1]. States enacted legislation to encourage recycling, and thousands of communities nationwide established local waste management programs. Meanwhile, the Council for Solid Waste Solutions developed a blueprint that is a source for mapping the elements of a successful infrastructure for all aspects of the plastics recycling process. The Council's

Municipalities Program supports community pilot projects. Recycling of plastics continues to grow as a result of legislation and efforts by environmental and trade-based organizations. The public image of plastics has climbed dramatically in the last 5 years. The industry now has a favorability rating on par with materials that compete with it in many applications, such as steel and paper [2]. Even though much of the focus on plastic recycling in the United States has been of a postindustrial nature [3,4], household recycling is growing in volume. Curbside recycling collection programs in municipalities increased from 1000 to 8800 during the period 1988–1996. Such programs are not available to 51% of the population [5].

Postconsumer recycling of packaging materials has grown approximately 10% per year since 1992. In 1993, the United States recycled 6.9% of all packaging materials produced. In the years 1989–1991 the growth rates were 40–50%, resulting mainly from newly enacted legislation that varied state by state in mandate, thus affecting the amount and type of packaging recycled. Polypropylene (PP) postconsumer packaging materials recycled at a rate of 0.8% of pounds sold in 1993. This was an increase of 25% from the previous year. This compares with a rate of 1.9% for low-density polyethylene (LDPE) and linear low-density polyethylene (LLDPE) packaging materials, 16% for high-density polyethylene (HDPE) packaging materials, and 28% for polyethylene terephthalate (PET) packaging materials. In the years 1996–1997 there was a decrease in the rate of recycling worldwide, reflecting the cyclibility of demand. Free market economics is now the dominating factor in postconsumer recycling, forcing institutions to reevaluate how to manage waste [6–8]. The viability of a particular nonintegrated reclamation infrastructure is severely tested by the impact of new virgin resin capacity.

Postconsumer recycling in Europe has taken a broader technological approach (as opposed to primarily mechanical separation in the United States) and has realized higher rates of recycling primarily due to increasing legislative pressures. Europe uses a mix of technologies to recycle plastics. Most waste is incinerated (thermal recovery) followed by mechanical and chemical recovery [9,10]. Legislation mandating postconsumer recycling in Europe is typically more aggressive than that in the United States; thus, the amount of plastics being recycled in Europe is growing at about 20% per year [3,11]. For example, Germany's federally mandated corporate-funded green dot recovery system required 64% recovery of plastics packaging in 1996 [7]. In comparison, the state of Oregon mandated in 1996 that rigid-container manufacturers must source-reduce 10% or incorporate 25% postconsumer recycle content [6]. The Association of Plastics Manufacturers in Europe estimated that more than 25% of waste plastics (8.83 billion pounds) were reused in 1995 through mechanical recycling, feedstock (chemical) recovery, and incineration for energy recovery. This compares with the 1.61 billion pounds of postconsumer plastics packaging recycled in the United States in 1996.

## Recycling of Polypropylene and Its Blends 115

In Europe, recycling is beginning to change the way plastic suppliers think. Industry is looking beyond its normal supply chain and taking into consideration the material after its normal use (lifetime) in both durable and nondurable goods [9,12]. To this end, resin producers are teaming up with customers to form partnerships in recycling. One such partnership involved Montell Polyolefins working in Fiat's auto recycling scheme. The project involved disassembling old vehicles to yield recyclable materials and designing future vehicles and versatile forms of polypropylene to facilitate future recycling. Twenty-five thousand vehicles processed in 1993 generated about 1 million pounds of recycled materials [13–15]. VAG collects plastic polypropylene bumper fascia through its dealer network in Germany. Bumpers are regranulated and used within the group in the production of new items. German automotive manufacturers have formed a group to work with materials suppliers and materials reprocessors for the purpose of achieving performance specifications for recycled materials in order to use them in other automotive parts [12,16]. Ford established a vehicle dismantling center in Cologne, Germany to study recycling economics. Ford also introduced a listing of recommended materials for use in auto recycling, including polypropylene, acrylonitrile-butadiene-styrene (ABS), polyethylene, polyamide, acrylic, polyvinyl chloride (PVC), and polycarbonate [16]. The current emphasis to design durable goods for future recycling applies to almost any type of manufactured plastic product.

The continuing efforts of researchers to successfully expand the property and/or processing envelope of polypropylene has led to its becoming an industry favorite. This is especially true in the automotive industry where applications require different physical and/or mechanical properties. The versatility of polypropylene suits many of the applications, thus simplifying recycling. Most of Europe's car manufacturers recognize polypropylene as the preferred plastic material when designing for the entire life cycle [17]. In 1989, Hoechst collaborated with Opel to show that materials from used polypropylene bumpers and battery casings could produce quality parts for "out of sight" applications. To augment recycling applications, Hoechst brought on stream a 10 million pound per annum polypropylene recycling plant in 1992. Waste polypropylenes from bumpers, appliance housings, or packaging materials are collected, separated from nonviable PP, pulverized, cleaned, modified with property-enhancing polymers and additives, and pelletized to give new recycled grades [18].

The advent of metallocene catalyst–polymerized polypropylene technology will expand the opportunities to make new PP-based polymers significantly different from current technology. Not only will these polymers have new properties but, because of their tailored structure, they can be inherently compatible with nonolefinic polymers [19]. This would make the technology of mechanical recovery and separation much simpler and bring new recycled products to the marketplace. Presently, postreactor technologies are available that make polypropylene compatible with other nonolefinic polymers. These include reactive extrusion and the Montell Hivalloy technology PP alloy resins.

## 5.1 ECONOMICS OF POLYOLEFIN RECYCLING

The recyclability of a material depends on the existence of a market for that material. Postconsumer recycling has always raised issues of quality and cost. Quality issues are a factor when the recycled material is to reenter the marketplace. This is the case whether recycled material moves directly to fabricators or enters the value chain as depolymerized feedstock or monomers [3]. The general scheme to recycle materials requires that they first be segregated from a waste stream, then processed into a usable form, and finally sold to a user. Recycling technologies typically fall into three categories: mechanical recovery, chemical recovery, and incineration. The infrastructure can be complex, requiring private institutions and municipalities practicing specific collection tasks, a variety of materials processors, and brokers all coordinated to send new, sustainable, quality products to a stable receptive marketplace [20].

The cost of collecting, sorting, and reclaiming materials, whether they are plastic or aluminum, is about the same. Thus, the potential price for the reclaimed material is a key factor in the equation. On this basis, in 1991, aluminum was most attractive followed by PET and HDPE. Adding other materials to the stream stresses the economics of the system. In the last several years, the technology to collect, sort, and reclaim has improved, thus reducing these costs [21–27].

Reko, the plastics recycling subsidiary of DSM in the Netherlands, was established in 1981. The primary motive for this was the generation of profit from recycling. Its current capacity is 50 million pounds per year, and it handles a wide variety of materials in both film and solid form. Reportedly the market will pay approximately 80% of the cost of virgin material for a similar recycle product. Taking LDPE as an example, Reko needs to charge 1.1 DM/kg to make a profit. This means that the price of virgin material must be above 1.3 DM/kg [12]. The main factor limiting opportunities for profit from plastic recycling is the cyclic nature of the virgin polyolefin market. The financial backing of a large firm is necessary to survive in times when virgin polymer prices are depressed. Recycling is still a young industry, and many improvements in infrastructure and production costs are pending.

In the United States, as the public consciously chooses to purchase recycled products or products that contain recycle content, the value of recycled materials changes [21]. The consciousness of the consumer public translates this into law in Oregon, California, Wisconsin, and Florida, with other states to follow. This plus the realization that recycling is a part of waste management, like landfills or incinerators, and is a cost society must bear in effect separates the supply–demand curve of the recycled material from the supply–demand curve of the virgin material. Also, as recyclers obtain Food and Drug Administration (FDA) status and International Standards Organization (ISO) certification, new market opportunities will open the possibility for higher prices for materials derived from the waste stream [28].

In the United States, Montell Polyolefins (now Basell USA, Inc.) is the only worldwide PP producer to set up a 100% owned recycling facility dedicated to postconsumer PP packaging. The conversion of PP bottles and other packaging materials into recyclable-containing resins under the trade name Refax are done through the company's wholly owned Polymer Resource Group, Inc. located in Baltimore, Maryland. This product is a compounded material containing up to 25% postconsumer PP mixed with 75% Montell virgin or wide-spec resin. The design of each Refax product targets the specific needs of the customer. Current markets utilizing Refax are packaging, automotive, and sheet extrusion. According to industry consultants [29], achieving low or no-cost recycling of any commodity is a complex task. To accomplish this, some universal criteria must be met. A processor producing recycle materials must have:

1. A market
2. A collection infrastructure that provides an economical bridge between generators and end users
3. A jurisdiction with a long-term recycling commitment willing to make changes
4. A trade association that will support the development of a sound approach to recycling its commodity

In the best of times, recycled PET can bring the reclaimer an average profit of 8 cents per pound. Recycled polyolefins typically bring the reclaimer an average profit margin of 4 cents per pound [1]. Many other manufacturers have formed partnerships or ventures with materials recovery facilities [1,30,31].

## 5.2 TECHNICAL INVESTIGATIONS

### 5.2.1 Alternative Technologies for Recycling of Polypropylene and Its Blends

A variety of plastics recycling technologies have been proposed and developed in the last two decades, partly in response to legislative and consumer demands. These technologies are based on processing, raw materials formulation, and product properties requirements. Plastics recycling technologies cover a broad range, from depolymerization of single polymers to recover the corresponding monomers to novel processing of commingled, postconsumer, and industrial plastic waste streams. These technologies are relevant for the recycling of polypropylene and its blends and are briefly described in this section.

*Depolymerization to Monomer.* Concern over the environmental impact of waste plastic and the sensitivity of polypropylene production costs to monomer cost have led to an interest in the conversion of waste polymer to monomer. The major method for depolymerization is *pyrolysis*, although other methods, such as oxidative, catalytic, and solution degradation, are also known. Polymer degradation proceeds by several mechanisms, one of which favors the production of monomer. Pyrolysis in conjunction with silica-alumina catalysts was used by Smith [32], Vasile et al. [33], and Yamamoto and Takamiya [34]. These catalysts were found to increase the yields of alkanes and butenes at the expense of propylene and ethylene. Schutze [35] used free-radical initiators to produce branched paraffin solvents and lube oils. Coenan and Hagen [36] used pulverization, heating, and solvents to produce $C_5$–$C_{30}$ alkanes, cycloalkanes, and aromatics. Several authors have reported varying propylene yields (18–26%) from the pyrolysis of polypropylene [37–39]. Kiang et al. [40] reported a propylene yield of 28% from pyrolysis of atactic polypropylene.

Several authors [39–41] have studied the mechanism of polypropylene pyrolysis with similar results. The primary and secondary alkyl radicals formed during pyrolysis react further either with hydrogen transfer to produce nonmonomeric hydrocarbons or without hydrogen transfer to produce monomer. The relative rates of reactions are dependent on reaction conditions. Currently, production of fuels tends to drive depolymerization technology. The Procedyne process [42,43] produces either no. 2 or no. 6 fuel oil. It appears that the most cost-effective technology for production of monomer from polymer is the production of fuel from polymer, followed by production of monomer from fuel using standard technology. Overall, existing technologies suggest that high conversion of polymer to monomer will require a technological breakthrough, such as a catalyst that preferentially drives the degradation reaction that produces monomer.

### 5.2.2 Selective Dissolution Process

The Rensselaer's Selective Dissolution Process pioneered by Nauman and colleagues [44–48] is aimed to recover functionally pure polymers by dissolving one polymer at a time from a complex mixture such as the household packaging waste stream. Such a waste stream consists of six major polymers: HDPE, PET, LDPE, PP, polystyrene (PS), and PVC. The various polymers are incompatible, and direct recycling of such a mixture would result in poor physical properties and little commercial value. The selective dissolution process also eliminates the rather expensive mechanical sortation process, thus making the recycling economics more attractive. The same thermodynamic differences that cause incompatibility provides an opportunity to separate the components of a mixed plastic waste stream. Pure components can be obtained by selective dissolution using a controlled sequence of solvents [xylene, tetrahydrofuran (THF)] and solvation temperatures, followed by low solids flash devolatilization and compositional quenching. Dissolution is fast at low (about 10 wt%) polymer concentrations and is usually followed by a filtration step to remove insoluble contaminants, such as metals, glass, and pigments. Stabilizers and impact modifiers can be added to the filtered mass to obtain a value-added product. The dissolution solvent is separated from the plastics and returned to the process for reuse. A typical polymer recovery using xylene as the solvent is shown below:

| Temp. (°C) | Polymer | Measured recovery (%) |
| --- | --- | --- |
| 25 | PS | 99.6 |
| 75 | LDPE | 98.3 |
| 105 | HDPE | 99.3 |
| 120 | PP | 99.6 |
| 138 | PVC | 99.4 |
| ND | PET | 100 |

### 5.2.3 Extrusion/Compounding-Based Technologies

One of the most challenging areas of plastics recycling, from both technical and economic standpoints, is the upgrading of commingled plastics to marketable properties. Commingled plastics of polyolefin, PS, PET, and PVC are typical of the postconsumer curbside collection waste stream. As in virgin polymer blends, the processing (mixing) of such a mixture suffers from both interdiffusional limitations, due to the presence of large macromolecules having low diffusion coefficients, and unfavorable thermodynamics of mixing, due to their incompatibility. To increase mixing by reducing the diffusion time (i.e., $t_{\text{diff}} = S^2/D$, where $S$ is the striation thickness and $D_{\text{AB}}$ is the diffusion coefficient between polymers A and B), one selects a process equipment that would reduce the striation thickness as much as possible. For instance, twin-screw extrusion can generate both shear and elongational stresses and is therefore more efficient than single-screw extrusion, where there are mainly shear stresses. The kneading disks in the screw profile of a twin-screw extruder are responsible for the elongational flows achievable in this mixer because of the periodic expansion and compression of material in three-dimensional flow [49]. Other process factors, such as feed rate, screw speed, and screw design (type and arrangement of screw elements), can affect dispersion and breakup of minor components. Dispersion and the mixing quality of the incompatible multicomponent polymer system are also influenced by rheological considerations, which depend on material-related parameters, such as the ratio of viscosity of the minor to major components and the concentration of the dispersed minor phase. Taylor dispersion, resulting from both shear and extensional flows generated in the kneading screw elements, is responsible for minor phase breakup in typical two-phase polymer blends.

From theoretical and experimental studies, a generalized equation proposed by Wu [50] gives the domain size in polymer blends in a shear field as follows:

$$d_n = (\eta_d/\eta_m)^\alpha [\gamma/(G \times \eta_m)] \tag{5.1}$$

when

$\eta_d/\eta_m > 1$, $\alpha = 0.84$
$\eta_d/\eta_m < 1$, $\alpha = -0.84$

where $d_n$ is number-average particle diameter, $\eta_d$ is melt viscosity of dispersed phase, $\eta_m$ is melt viscosity of matrix, $\gamma$ is interfacial tension between matrix and dispersed phase, and $G$ is shear rate.

It is clear from Eq. (5.1) that higher matrix viscosity, higher shear rate, and lower interfacial tension are necessary for smaller particle size. One way to reduce interfacial tension in polymer blends is to incorporate a small amount (2–3 wt%) of a compatibilizer during the compounding operation. Several studies using the twin-screw extrusion/compounding for recycled polyolefins and other mixtures of postconsumer plastic wastes have been reported in the literature [51–53]. Twin-screw extruders vary in design such as corotating, counterrotating, intermeshing, and nonintermeshing varieties. Therefore, equipment selection may depend on other factors such as the nature of the materials feed stream (powder, pellets, liquid), the viscoelasticity and concentration of the various components, and the protocol of addition of the components in the product formulation. In a cold-treatment extrusion of recycled commingled plastics, Sangani and Estepp [54] used a counterrotating nonintermeshing twin-screw extruder to process unwashed and unsorted granulated plastic mixture at ambient temperature directly into its finished profile. In recycling of nylon carpet that consists of nylon, styrene-butadiene-rubber latex adhesive with calcium carbonate filler, and the PP backing, Hagberg and Dickerson [55] used a corotating nonintermeshing twin-screw extruder for the compounding operation.

Apart from the diffusional limitation in polymer mixing, a second limitation is the inherently unfavorable thermodynamics. For a polymer blend to be a single phase at equilibrium, the requirement that $\Delta G_m < 0$ must be fulfilled, where $\Delta G_m = \Delta H_m - T \Delta S_m$. The combinatorial entropy term is small for polymers due to the large molar volumes; therefore, the enthalpic contributions that are usually positive often dominate the free energy of mixing in polymeric systems. This ultimately leads to positive contributions to $\Delta G_m$, which are unfavorable for mixing. Introduction of interacting groups by chemical modification of a polymer or by copolymerization can accelerate the degree of interaction between polymers, resulting in a favorable (exothermic) heat of mixing. Such interacting groups of compounds are known as compatibilizers and have been shown to improve the energetics between otherwise immiscible polymer pairs. Compatibility of immiscible blends may be improved through one of the following general routes [56,57]:

1. By adding a nonreactive or physical compatibilizer, separately synthesized, such as block or graft copolymers with segments capable of specific interactions and/or chemical reactions with the blend constituents.
2. By blending suitably functionalized polymers or chemical compatibilizers capable of enhanced specific interactions, such as coulombic and/or chemical reactions, such as amide formation. In reactive systems, a copolymer may be formed in situ during mixing, acting in a similar manner as a chemical compatibilizer.
3. By using in-reactor grafting polymerization to produce a copolymer compatibilizer, such as polypropylene-graft-polystyrene (PP-g-PS) and the corresponding nonolefinic polymer component polystyrene (PS) simultaneously from the monomer styrene. This approach is used for the Hivalloy-PET system, discussed in the next section.

A properly chosen compatibilizer will preferentially locate at the interface of the immiscible homopolymers, acting to reduce interfacial energy, enhance finer dispersion during mixing, improve phase stability against gross segregation or motion-induced particle coalescence, and promote interfacial adhesion.

Some studies have been reported [58–60] that explored the effectiveness of different compatibilizers for the enhancement of recycled plastics properties in a variety of applications. Xanthos et al. [60] studied compatibilized blends of refined postconsumer commingled plastics, typical of tailings from a community curbside collection. Similarly, Dagli et al. [59] studied value-added blends and composites from recycled plastic fishing gear. These authors [61] also reported significant property improvement for a recycled polyolefinic blend of PP-HDPE using the Kraton rubber as compatibilizer. Similarly, Hagberg and Dickerson [55] also showed improved Izod impact strength for the Kraton-compatibilized postconsumer carpet.

## 5.2.4 Chemically Modified Polypropylene in Recycling Application

A polymerized reactor copolymer, obtained by in-reactor grafting polymerization technology, is a two-phase rigid copolymer that combines the best attributes of semicrystalline polypropylene with those of amorphous polystyrene. In the process, the compatibilizer PP-g-PS and the nonolefinic polymer component PS are simultaneously generated from the monomer styrene. This styrenic Hivalloy (Hivalloy is a trademark of Basell USA, Inc.) system can be regarded as "self-compatibilized," with a reactor product composition of about 70% PP, 20% free PS, and 10% g-PS at a styrene monomer addition level of 45 pph of the precursor homopolymer, at a grafting efficiency of about 30%. The presence of the graft copolymer PP-g-PS makes the reactor product a potentially compatible material for polyolefin, polystyrene, and PET recycled streams.

## 5.3 PP-BASED BLENDS WITH RECYCLED PET

### 5.3.1 Introduction

There has been a major push in the last decade or so to find new uses for recycled plastic materials. The 2-L PET beverage container, which carries polypropylene labels and caps, has been successfully recycled in the United States. However, a readily available material that could easily compatibilize PET and PP would be attractive to the recycling industry by simplifying the recycling process and generating value-added products. The low cost and excellent mechanical properties and availability of recycled PET make this material a logical choice for alloying with the styrenic Hivalloy reactor product. The following experiment was designed to compare the differences between blends of polypropylene, Hivalloy reactor product, functionalized Hivalloy reactor product, and recycled PET, respectively.

### 5.3.2 Experimental

Polypropylene (homopolymer precursor), Hivalloy reactor product (PP-g-PS), and functionalized Hivalloy [PP-g-PS(S-Co-GMA)] glycidyl methacrylate graft were blended with a recycled PET material obtained from Star Plastics in pelletized form. The materials were bag blended (Table 5.1) and stabilized with 0.1% Irganox 1010 and PEPQ based on total batch weight.

Each blend was starve fed and compounded on a Haake twin-screw extruder at 150 rpm, using a temperature profile of 475 °F, 525 °F, 550 °F, and 550 °F. The compounded materials were dried at 100 °C for 2 hr. Each blend was injection molded into flex bars, single-gated and dual-gated tensile bars, and Izod impact bars using a 1.5-ounce Battenfeld injection molding machine. Specimens

## Recycling of Polypropylene and Its Blends

**TABLE 5.1** Recycled PET Blends with Polypropylene-Based Materials

| Factor | A | B | C | D | E | F | G | H | I | J | K | L | M |
|---|---|---|---|---|---|---|---|---|---|---|---|---|---|
| **Material Description** | | | | | | | | | | | | | |
| Crystalline PET Regrind (g) | 998 | | | | 249.5 | 499 | 748.5 | 249.5 | 499 | 748.5 | 249.5 | 499 | 748.5 |
| RA-063 Precursor (g) | | 998 | | | 748.5 | 499 | 249.5 | | | | | | |
| Batch 21 PP-g-PS (g) | | | 998 | | | | | 748.5 | 499 | 249.5 | | | |
| PP-g-P(S-co-GMA)5%GMA (g) | | | | 998 | | | | | | | 748.5 | 499 | 249.5 |
| PEPQ (g) | 1 | 1 | 1 | 1 | 1 | 1 | 1 | 1 | 1 | 1 | 1 | 1 | 1 |
| Irganox 1010 (g) | 1 | 1 | 1 | 1 | 1 | 1 | 1 | 1 | 1 | 1 | 1 | 1 | 1 |
| **Mechanical Properties** | | | | | | | | | | | | | |
| Flexural modulus (kpsi) | 408.4 | 188.7 | 343.4 | 334.5 | 259.9 | 267 | 314.4 | 354.4 | 352 | 364.6 | 350.4 | 356.3 | 361.4 |
| *Flexural modulus (MPa)* | 2814 | 1300 | 2366 | 2305 | 1791 | 1840 | 2166 | 2442 | 2425 | 2512 | 2414 | 2455 | 2490 |
| Flexural strength (psi) | 15020 | 6409 | 10360 | 10700 | 8703 | 9255 | 10730 | 11760 | 12350 | 13240 | 11060 | 12440 | 13640 |
| *Flexural strength (MPa)* | 103.5 | 44.2 | 71.4 | 73.7 | 60 | 63.8 | 73.9 | 81 | 85.1 | 91.2 | 76.2 | 85.7 | 94 |
| Notched Izod (ft-lb/in.) | 0.48 | 0.38 | 0.31 | 0.2 | 0.36 | 0.59 | 0.47 | 0.29 | 0.29 | 0.43 | 0.33 | 0.19 | 0.50 |
| *Notched Izod (J/m)* | 26 | 20 | 17 | 11 | 19 | 32 | 25 | 15 | 15 | 23 | 18 | 10 | 27 |
| **Tensile Strength** | | | | | | | | | | | | | |
| Single gated (psi) | 7011 | 4740 | 5795 | 6201 | 5046 | 6097 | 6893 | 6943 | 7481 | 8273 | 6829 | 7649 | 8727 |
| *Single gated (MPa)* | 48.3 | 32.7 | 39.9 | 42.7 | 34.8 | 42 | 47.5 | 47.8 | 51.5 | 57 | 47.1 | 52.7 | 60.1 |
| Dual gated (psi) | 9782 | 4527 | 3813 | 3638 | 1818 | 1087 | 1519 | 3050 | 2553 | 3669 | 2343 | 2669 | 5087 |
| *Dual gated (MPa)* | 67.4 | 31.2 | 26.3 | 25.1 | 12.5 | 7.5 | 10.5 | 21 | 17.6 | 25.3 | 16.1 | 18.4 | 35 |
| Weldline retention (%) | 140% | 96% | 66% | 59% | 36% | 18% | 22% | 44% | 34% | 44% | 34% | 35% | 58% |
| Heat distortion 264 psi (F) | 153 | 127 | 172 | 160 | 145 | 154 | 156 | 160 | 156 | 154 | 156 | 154 | 151 |
| *Heat distortion 264 psi (C)* | 67 | 53 | 78 | 71 | 63 | 68 | 69 | 71 | 69 | 68 | 69 | 68 | 66 |

from each blend were tested for tensile properties at 2 in./min (50 mm/min), flexural properties at 0.5 in./min (12.5 mm/min), and impact properties using a 2-pound hammer on an Izod bar.

### 5.3.3 Results/Discussion

The compounding and molding operations confirmed good processability for each blend. However, a difference, noticeable in processing, was that the functionalized blends had a tendency to break during extrusion and required slightly higher injection pressures. An examination of the blend specimen's fractured surface for delamination was used as a qualitative way to compare the compatibility of the blends. The polypropylene blends showed a characteristic peeling effect, indicating delamination. On the other hand, the Hivalloy-PET blends broke cleanly, indicating greater compatibility compared with the PP-PET blends. Furthermore, the glycidyl methacrylate graft blends showed the smoothest fractured surface.

The Hivalloy, PPgPS-PET blends showed superior tensile and flexural properties compared with the PP-PET blends, whereas the functionalized Hivalloy-PET blends showed slightly improved physical properties compared with the straight Hivalloy, PPgPS-PET blends. This suggests that the glycidyl methacrylate may be reacting with the PET to improve the adhesion between the two phases. The results of the mechanical properties versus blend compositions are shown in Fig. 5.1a–e.

### 5.3.4 Conclusions

A qualitative examination of the fractured and external surfaces of injection-molded specimens shows PET to be compatible with Hivalloy materials. Hivalloy-PET blends showed superior flexural and tensile properties compared with PP-PET blends. Glycidyl methacrylate–functionalized Hivalloy-PET blends showed slightly higher mechanical properties than straight Hivalloy-PET blends. On the whole, recycled PET appears to be a good candidate for use as a cost-effective blend component in Hivalloy materials. Other innovative Hivalloy-PET blend formulations, which include the addition of impact modifiers and compatibilizers, can be studied to optimize the stiffness–impact balances of these blends. The effectiveness of functionalized Hivalloy as a potential blend component for the postconsumer commingled plastics should be explored.

# Recycling of Polypropylene and Its Blends

(a)

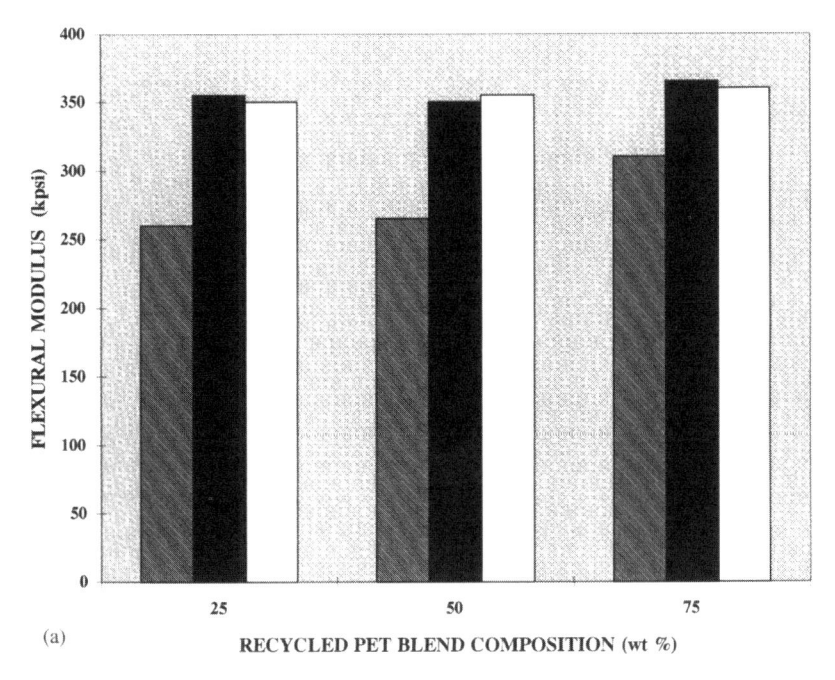

**FIGURE 5.1** Mechanical properties vs. blend compositions for recycled PET blends. ▨, RA-063; ■, PP-g-PS; □, PP-g-P(S-co-GMA). (a) Flexural modulus vs. recycled PET blend composition. (b) Flexural strength vs. recycled PET blend composition. (c) Notched Izod vs. recycled PET blend composition. (d) Tensile strength vs. recycled PET blend composition. (e) Weldline strength vs. recycled PET blend composition.

## 5.4 COMPATIBILIZATION, PROCESS, AND ECONOMIC ANALYSES OF RECYCLED PP-HDPE BLENDS

In this section, we discuss a specific example of a compatibilized recycled PP-HDPE system. An extrusion-compounding process that is representative of a typical commercial operation was used for blending. Detailed process and economic analyses are presented.

### 5.4.1 Compatibilized PP-PE Blend Systems

The technology for polyolefinic plastics recycling has evolved considerably in the past few years. To be a viable business in an entrepreneurial American economic system, there must be continued demonstration that polyolefinic recycling is technically credible and cost competitive.

(b)

(c)

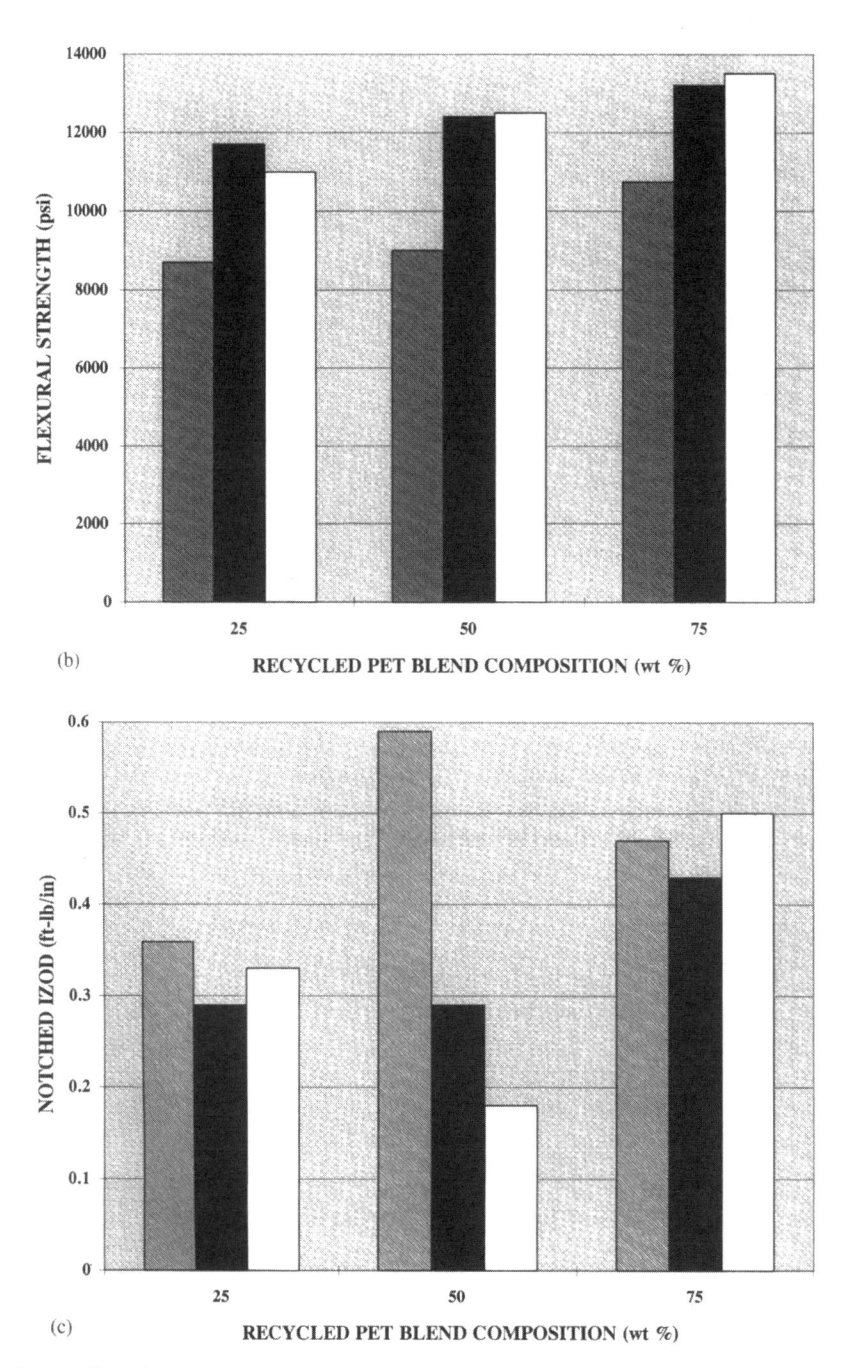

FIGURE 5.1 (continued)

# Recycling of Polypropylene and Its Blends

(d)

(e)

FIGURE 5.1 (*continued*)

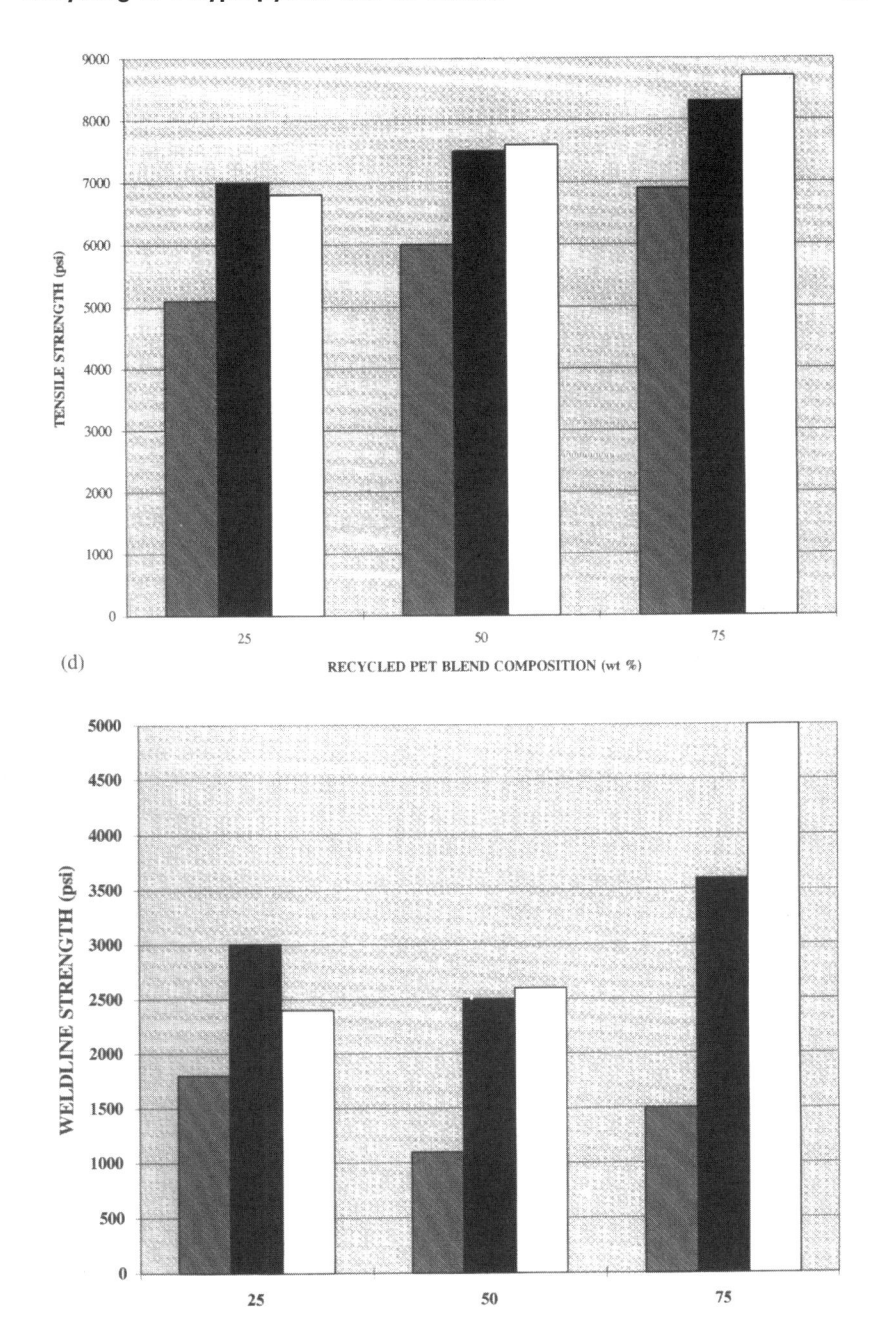

Consistent sourcing of polyolefin feedstock for large-scale recycling activities is in itself a major challenge. The so-called commingled plastics from the postconsumer rigid packaging waste stream is largely composed of polyolefins, especially HDPE. Commingled plastics are the tailings left after HDPE milk, water, or juice bottles and PET soda bottles have been sorted out of a pile of postconsumer rigid plastic containers, typical of a community curbside collection.

No infrastructure on a national scale exists for commingled plastics collection. As such, commingled plastics composition varies with each community, and the attainment of one "model" feedstock remains elusive. Nevertheless, improved properties of commingled plastics via blend modification by reactive functionalization and compatibilization have been reported by several workers [58–61].

However, a polyolefinic substream sourced from HDPE PET soda bottle, base cups, and rigid bottle packaging for household detergents and other cleaners are being reclaimed cleanly and consistently by many recyclers across the country. This substream provides a consistent feedstock for recycling along with other polymer types such as industrial waste PP, thus diminishing the overall amount of postconsumer plastics wastes that could otherwise be landfilled or incinerated.

*Multiphase Polymer System.* Complex morphologies have been shown to result from modification of immiscible commingled plastics with significant (up to 20%) PET component by reactive functionalization or compatibilization [58]. Factors such as composition, interfacial tension, viscoelastic properties, and processing are known to affect blend morphology of immiscible polymer systems [62–64].

Improved properties of commingled plastics via blend modification by reactive functionalization and compatibilization have been reported by several workers [58,65]. This work is confined to a two-phase PE-PP morphology. In the two-phase immiscible PE-PP system, poor interfacial adhesion results in poor blend mechanical properties. The lack of stability in the morphology causes gross separation or stratification during later processing or use. Block and graft copolymers of the form A-B have been used as compatibilizers to improve interfacial adhesion and reduce interfacial tension between A-rich and B-rich phases to provide A-B alloys with improved and unique balances of properties.

Kraton FG1901X is used as the compatibilizer. This maleated SEBS (styrene-ethylene-butene-styrene) triblock presumably reduces the PE-PP interfacial tension to promote compatibilization by improving interfacial adhesion.

### 5.4.2 Processing and Material Factors

It is now clear that state-of-the-art technologies in polymer processing are required to convert recycled plastics to useful blends with marketable properties. As with virgin materials, blend formulation, compounding, and mold fabrication are used to achieve specific properties for targeted applications.

Therefore, an understanding of the relationship between recycled polyolefinic blend formulation and processing on one hand and blend morphology and properties on the other should facilitate a more cost-effective approach both to product formulation and processing and to product design for applications using recycled polyolefins.

In this work, we attempt to show the influences of PE/PP ratio, processing, and compatibilizer concentration on blend morphology and properties before and after glass fiber reinforcement for blends of recycled HDPE and an industrial waste PP.

*Blend Formulation.* Compounded blends consist of varying fractions of HDPE (20–50%), PP (43–79%), and Kraton rubber FG1901X (1–7%). This is the compatibilized blend. The reinforced composite blend is a 20% glass fiber–OCF/BA457 loading of the compatibilized blend. Hercoprime (proprietary tradename of a Montell coupling agent) HG201 (1%) was used in all glass fiber–reinforced (GFR) formulations to promote glass fiber adhesion to the polymer matrix.

Recycled HDPE sourced from PET soda bottle basecups and postconsumer household cleaner rigid bottle was supplied by Wheaton Plastics, Polymer Resource Group, or the now defunct Plastics Recycling Alliance. The feedstock appeared consistent from supplier to supplier. The PP component is a mixture of homopolymer and copolymer industrial off-grade materials. The mixture was peroxide visbroken to a consistent melt flow rate (MFR) of 20 before compounding.

*Compounding/Injection Molding.* HDPE and PP comprise an immiscible pair. It has been reported that dissipative mix melting of multicomponent polymer systems with compatibilizers exhibit super- and submicrometer morphologies that are important to final properties. The HDPE-PP-Kraton rubber FG1901X system is a case in point. For such a system, the resulting morphology can be affected by the protocol of addition of participating components, and mixing. For the HDPE-PP system, we generated both the rheology curves (Fig. 5.2) and the plot of viscosity ratio versus shear rate (Fig. 5.3). The mismatch in the rheology curves suggests possible difficulties in dispersive melt mixing. However, mixing during dissipative mix melting can be substantially improved via "viscosity matching," by adjusting the temperatures in the mix-melting region (first two to

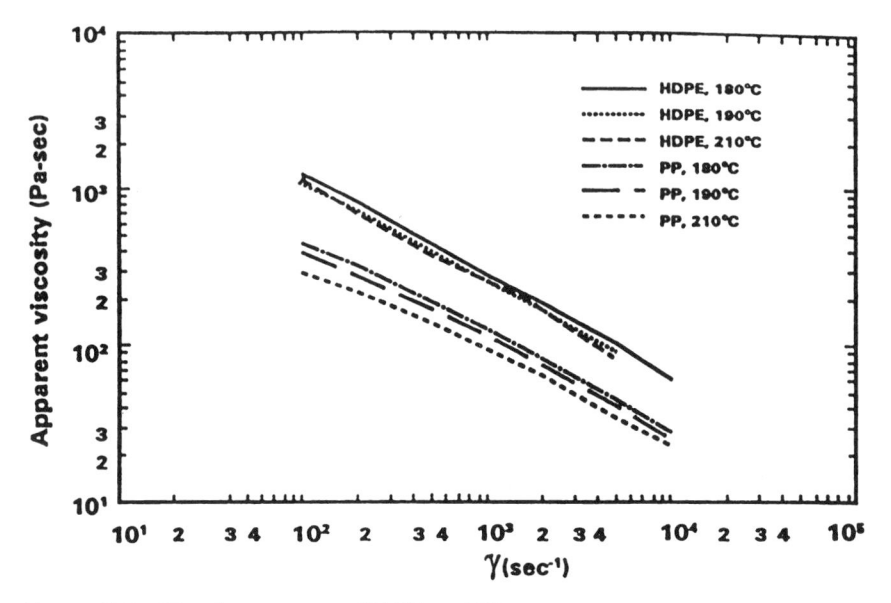

FIGURE 5.2　Rheology curves: HDPE and PP.

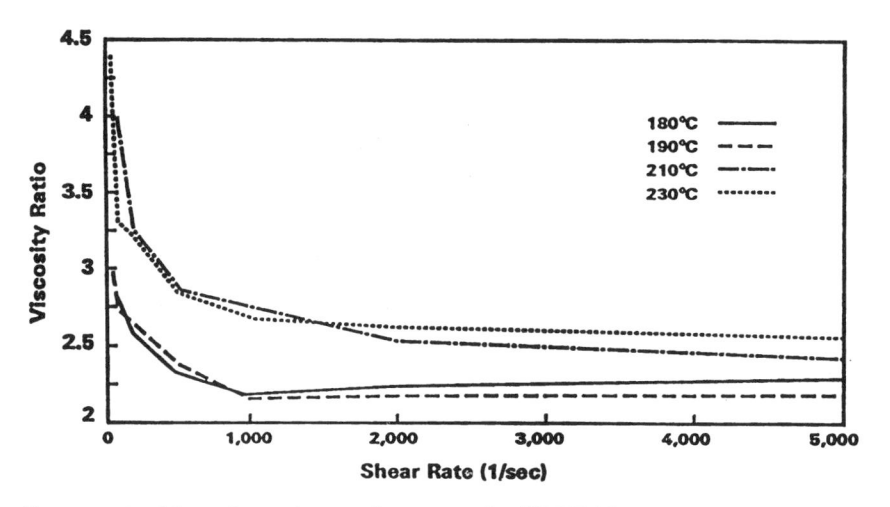

FIGURE 5.3　Viscosity ratio vs. shear rate for HDPE-PP system.

# Recycling of Polypropylene and Its Blends

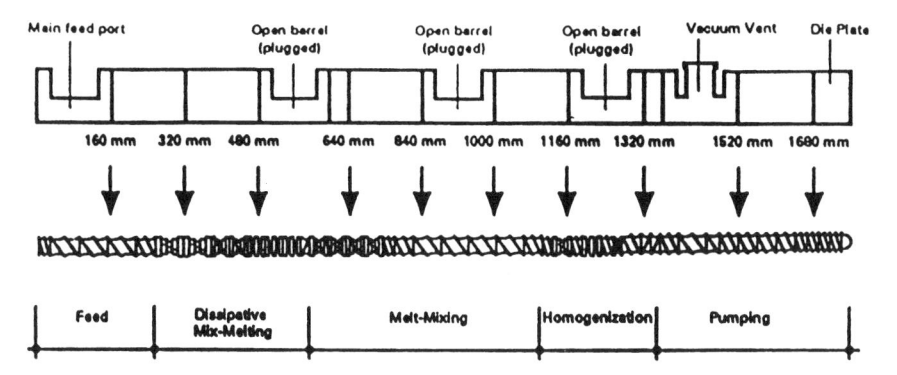

FIGURE 5.4 Compounding screw profile.

three barrels) of the extruder using high-shear kneading blocks. This was successfully accomplished as shown by the micrometer to submicrometer morphologies of the compounded blends. Extruder screw speed ranged from 300 to 500 rpm. An inclining temperature profile from feed to die ranged from 160°C to 210°C. The screw profile used for the compounding process is shown in Fig. 5.4. Blend formulations were compounded on a 40-mm, corotating, intermeshing, 42 (L/D), Werner Pfleiderer twin-screw extruder, equipped with a loss-in-weight K-Tron feed system. The extruder, with 10 barrel sections, has down-

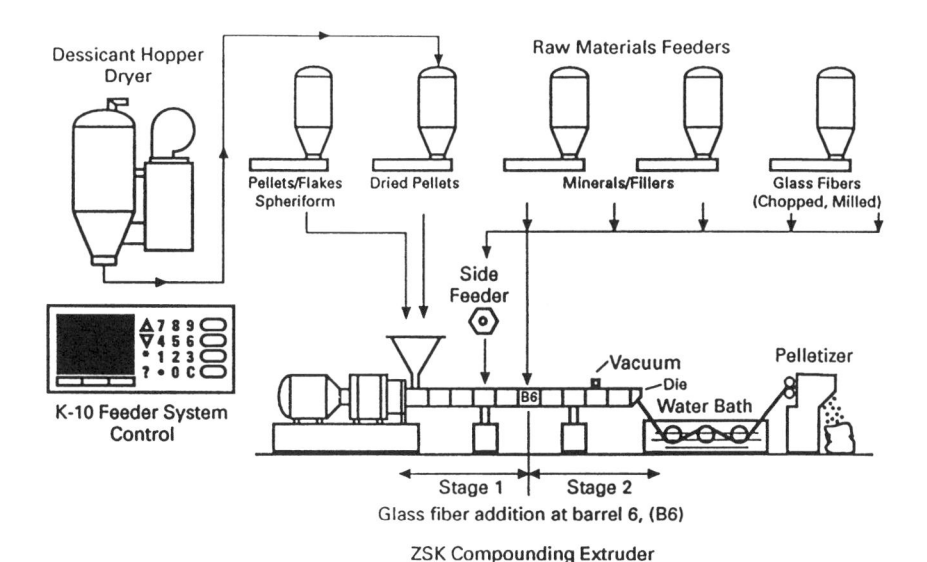

FIGURE 5.5 Compounding system.

stream feeding and venting capabilities. Glass fiber was added at Kombi barrel 6 (Fig. 5.5). The $\frac{3}{16}$-in. OCF/BA 457 glass fiber was fed into the melt to preserve aspect ratio and reduce equipment wear. Accurate feeding of the chopped glass fibers was ensured using a loss-in-weight feed system and a properly configured screw design. A mass balance closure, using ash analysis, was determined to verify loading accuracy. Standard water bath stranding and pelletizing were used. Strand samples were collected for morphology testing and characterization. Pelletized samples were also collected for MFR determination and injection molding of specimens to test for mechanical properties.

Injection-molded specimens were prepared according to ASTM D4101 methods. The specimens were tested for flexural modulus (MPa), tensile yield strength (MPa), tensile elongation at break (%), notched Izod impact (J/m), and heat distortion temperature (HDT) (°C).

*Microscopy.* Morphological characterization of selected samples was accomplished using phase contrast optical microscopy (PCOM) with a Leitz Aristomet and scanning electron microscopy (SEM) using a Hitachi S-800 instrument. Samples for PCOM were prepared by cutting semithin sections from extruded strands and flex bars in both the parallel and cross-flow directions using a Ruchart Ultracut E microtome. SEM specimens were prepared by fracture in liquid nitrogen. The optics of the PCOM system are such that the PE phase appears as the darker phase in the ensuing optical photomicrographs.

### 5.4.3 Compounding Using Statistical Design of Experiments

*Design of Experiments.* We selected an orthogonal rotatable response surface design known as the Box-Behnken design in three variables. The design used a subset of the three points in the corresponding full three-level factorial (Fig. 5.6). All 12 "edge" points (solid dots) lie on a single sphere about the center

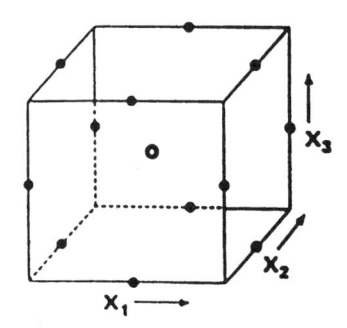

FIGURE 5.6 Response surface DOE. Geometry of Box-Behnken design for three variables.

# Recycling of Polypropylene and Its Blends

TABLE 5.2  Three-Variable Box-Behnken Design

| $X_1$ | $X_2$ | $X_3$ |
|---|---|---|
| +1 | +1 | 0 |
| +1 | −1 | 0 |
| −1 | +1 | 0 |
| −1 | −1 | 0 |
| +1 | 0 | +1 |
| +1 | 0 | −1 |
| −1 | 0 | +1 |
| −1 | 0 | −1 |
| 0 | +1 | +1 |
| 0 | +1 | −1 |
| 0 | −1 | +1 |
| 0 | −1 | −1 |
| 0 | 0 | 0 ⎫ |
| 0 | 0 | 0 ⎬ Center Point |
| 0 | 0 | 0 ⎭ |

Where: $X_1$ is screw speed, $X_2$ is HDPE fraction (%), and $X_3$ is Kraton fraction (%).

of the experimental region, making them equally distant from the center—a geometric property associated with rotatability. The three replicated center points (shaded circles) have two additional functions. First, they provide a measure of inherent experimental error. Second, they give relatively constant prediction variance as a function of distance from the center within the design region [66].

The Box-Behnken statistical design is shown in Table 5.2. Material-related input variables (HDPE fraction, Kraton rubber concentration) and process-mixing-related input factors (screw speed, in revolutions per minute) and their design ranges are shown in Table 5.3 and 5.4, respectively. Responses of interest are compounded blend mechanical properties and overall conversion costs.

TABLE 5.3  Materials and Compounding

| Material | Composition (%) | Variable codes − | 0 | + |
|---|---|---|---|---|
| HDPE (postconsumer) | 20–50 | 20 | 35 | 50 |
| Kraton (FG1901X) | 1–7 | 1 | 4 | 7 |
| PP (industrial waste) | 43–79 | | | |

TABLE 5.4 Process and Compounding: Dispersive and Distributive Mixing via ZSK 40 Extruder

| Parameter | Range | Variable codes − | 0 | + |
|---|---|---|---|---|
| Screw speed | 300–500 rpm | 300 | 400 | 500 |
| Feed rate-fixed | 45.5 kg/hr | | | |
| Temperature | 160°C (feed)–210°C (die) | | | |

The design is capable of full polynomial models up to order 2. Therefore, a response model ($y = a_0 + a_1x_1 + a_2x_2 + a_3x_3 + a_{11}x_1^2 + a_{22}x_2^2 + a_{33}x_3^2 + a_{12}x_1x_2 + a_{13}x_1x_3$) includes linear, interaction, and curvature effects for each input variable. Detailed statistical analysis and predictive models are presented for each response as a function of the input factors using multiple linear regression.

### 5.4.4 Process Analysis

*Response Surface Analysis.* The input–response data set for both the compatibilized blend and reinforced composite are shown in Tables 5.5(a) and 5.5(b), respectively. All data analyses were performed using Design Expert 3.0 software. Stepwise multiple linear regression was used with the general criterion that statistical significance at the 95% level of confidence was required to retain variables in the final regression equations. Responses were analyzed individually for flexural modulus, Izod impact strength, elongation at break, tensile strength, and heat distortion temperature. As a reference point, Table 5.6 shows results for starting materials and uncompatibilized blends.

Response surfaces showing the effects of composition on mechanical properties are compared with the compatibilized blend and the glass-fiber-reinforced composite in Fig. 5.7 and 5.8. Regression models for the compatibilized blends are shown below the response surface graphs (Fig. 5.7, a–e) versus reinforced (Fig. 5.8, a–e) blends shows a marked difference in the nature of the responses. Most notably, the curvature in the response observed in the compatibilized blends has vanished, and the response is a function of Kraton rubber only for the flexural modulus, notched Izod impact, and tensile strength. Similarly, the heat distortion temperature is now only a linear function of Kraton and HDPE levels. Finally, elongation at break has been reduced to a single value (3.43 ± 0.45%), as more than 90% of the variability in the data was explained by the mean value. Thus, the interaction between the fiber and the matrix is stronger than the more subtle PP–PE interaction. The Kraton rubber is an impact

# Recycling of Polypropylene and Its Blends

**TABLE 5.5(a)** DOE Input Response Data Set: Compatibilized Blend

| Experiment | | Input variables | | | Response variable | | | | |
|---|---|---|---|---|---|---|---|---|---|
| Random order | Run # | $X_1$ rpm | $X_2$ % HDPE | $X_3$ Tensile % Kraton | $Y_1$ strength (MPa) | $Y_2$ Elong. notched flex mod (MPa) | $Y_3$ @ impact (J/m) | $Y_4$ Brk. (%) | $Y_5$ HDT (°C) |
| 9  | 1  | + 500 | + 50 | 0 4 | 28 | 1157 | NBF[a] | 83  | 85  |
| 15 | 2  | + 500 | − 20 | 0 4 | 30 | 1316 | 77.4   | 113 | 97  |
| 12 | 3  | − 300 | + 50 | 0 4 | 27 | 1102 | NBF    | 112 | 94  |
| 10 | 4  | − 300 | − 20 | 0 4 | 30 | 1337 | 84.4   | 130 | 109 |
| 5  | 5  | + 500 | 0 35 | + 7 | 27 | 1109 | 149.5  | 171 | 89  |
| 13 | 6  | + 500 | 0 35 | − 1 | 30 | 1316 | 60.3   | 36  | 100 |
| 1  | 7  | − 300 | 0 35 | + 7 | 27 | 1116 | 145.8  | 139 | 87  |
| 11 | 8  | − 300 | 0 35 | − 1 | 30 | 1268 | 65.7   | 52  | 97  |
| 4  | 9  | 0 400 | + 50 | + 7 | 26 | 985  | NBF    | 189 | 78  |
| 6  | 10 | 0 400 | + 50 | − 1 | 29 | 1346 | 127.6  | 150 | 91  |
| 8  | 11 | 0 400 | − 20 | + 7 | 29 | 1388 | 105.0  | 283 | 90  |
| 14 | 12 | 0 400 | − 20 | − 1 | 31 | 1344 | 71.6   | 48  | 111 |
| 3  | 13 | 0 400 | 0 35 | 0 4 | 28 | 1151 | 83.8   | 98  | 96  |
| 7  | 14 | 0 400 | 0 35 | 0 4 | 29 | 1178 | 70.5   | 86  | 102 |
| 2  | 15 | 0 400 | 0 35 | 0 4 | 29 | 1199 | 80.1   | 104 | 96  |

[a] NBF, No break at flex.

**TABLE 5.5(b)** DOE Input Response Data Set: 20% GFR Composite

| Experiment | | Input variables | | | Response variable | | | | |
|---|---|---|---|---|---|---|---|---|---|
| Random order | Run # | $X_1$ rpm | $X_2$ % HDPE | $X_3$ % Kraton | $Y_4$ tensile strength (MPa) | $Y_1$ flex mod (MPa) | $Y_2$ notched impact (J/m) | $Y_3$ Elong. @ brk. (%) | $Y_5$ HDT (°C) |
| 9 | 1 | + 500 | + 50 | 0 4 | 51 | 3114 | 87 | 3.9 | 129 |
| 15 | 2 | + 500 | − 20 | 0 4 | 56 | 3417 | 68 | 3.4 | 150 |
| 12 | 3 | − 300 | + 50 | 0 4 | 51 | 3011 | 79 | 3.8 | 128 |
| 10 | 4 | − 300 | − 20 | 0 4 | 60 | 3652 | 55 | 3.2 | 151 |
| 5 | 5 | + 500 | 0 35 | + 7 | 47 | 2949 | 73 | 3.6 | 140 |
| 13 | 6 | + 500 | 0 35 | − 1 | 53 | 3190 | 42 | 3.5 | 139 |
| 1 | 7 | − 300 | 0 35 | + 7 | 47 | 2983 | 80 | 3.5 | 135 |
| 11 | 8 | − 300 | 0 35 | − 1 | 55 | 3328 | 53 | 3.1 | 144 |
| 4 | 9 | 0 400 | + 50 | + 7 | 45 | 2956 | 84 | 3.1 | 131 |
| 6 | 10 | 0 400 | + 50 | − 1 | 60 | 2645 | 41 | 3.1 | 150 |
| 8 | 11 | 0 400 | − 20 | + 7 | 44 | 2501 | 91 | 3.1 | 128 |
| 14 | 12 | 0 400 | − 20 | − 1 | 58 | 3459 | 56 | 4.8 | 147 |
| 3 | 13 | 0 400 | 0 35 | 0 4 | 51 | 3114 | 67 | 2.9 | 139 |
| 7 | 14 | 0 400 | 0 35 | 0 4 | 51 | 3225 | 68 | 3.2 | 140 |
| 2 | 15 | 0 400 | 0 35 | 0 4 | 53 | 3183 | 73 | 3.8 | 139 |

[a] NBF, no break at flex.

# Recycling of Polypropylene and Its Blends

TABLE 5.6  Uncompatibilized Blend

| Properties | PP (only) | HDPE (only) | PP (80%) + HDPE (20%) | PP (65%) + HDPE (35%) | PP (50%) + HDPE (50%) |
|---|---|---|---|---|---|
| Flexural modulus, MPa | 1729 | 1036 | 1619 | 1489 | 1341 |
| Notched Izod impact, J/M | 35 | NBF | 40 | 43 | 61 |
| Tensile elong. & break, % | 214 | 167 | 187 | 87 | 109 |

modifier and a compatibilizer, thus the effect on Izod impact. Modulus decreased due to matrix softening caused by the Kraton rubber. The singularity in the elongation at break is due to surface and volume imperfections caused by the introduction of the glass fiber, which generates deformations under tension that end catastrophically in break. Details of the major differences between the compatibilized and reinforced blends follow.

As noted above, flexural modulus for the reinforced blend is described by

$$\text{FlexMod} = 3554 - 93 \times \text{Kraton} \tag{5.2}$$

with FlexMod in MPa and Kraton level in percent. Equation (5.2) is statistically significant at the 99.7% level of confidence with $R^2 = 0.504$. The relatively low value of $R^2$ coupled with the high level of confidence suggests that Kraton level is a significant factor, but the variability of the test is too high to allow the use of Eq. (5.2) for predictive purposes. There is no curvature as evident in the compatibilized blend, and the flexural modulus has approximately tripled ($\sim 1200$ MPa for compatibilized versus 3500 MPa for reinforced).

Similarly, notched Izod impact strength for the reinforced blend is described by

$$\text{Izod} = 45.13 + 5.57 \, (\text{Kraton}) \tag{5.3}$$

with Izod in J/m and Kraton level in percent. Equation (5.3) is statistically significant at the 99.9% level of confidence with $R^2 = 0.679$. As with flexural modulus, the relatively low value of $R^2$ coupled with the high level of confidence suggests that Kraton is a significant factor, but the variability of the test is too high to allow the use of Eq. (5.3) for predictive purposes. Again, there is no curvature as evident in the compatibilized blend; however, the notched Izod values for the reinforced blend occupy the low end of the range ($\sim 40$–$100$ J/m for reinforced versus 60–160 J/m for compatibilized blends).

**Flexural Modulus**

(a)

**Model:**
FlexMod = 1577 - 15.73 HDPE + 21.75 Kraton
+ 0.2586 HDPE$^2$ + 3.605 Kraton$^2$
-2.25 HDPE x Kraton
**Significance:**
Confidence Level > 99%
$R^2$ = 0.978

**Notched Izod Impact**

(b)

**Model:**
Izod = 256.4 - 12.42 HDPE - 23.51 Kraton
+ 0.1904 HDPE$^2$ + 2.221 Kraton$^2$
+ 0.5694 HDPE x Kraton
**Significance:**
Confidence Level > 99%
$R^2$ = 0.964

**Elongation at Break (300 RPM)**

(c)

**Model:**
Elongation = 369.2 + 2.184 RPM - 8.338 HDPE
+ 31.4 Kraton - 0.0027 RPM$^2$ + 0.1813 HDPE$^2$
+ 3.41 Kraton$^2$ - 1.089 HDPE x Kraton
**Significance:**
Confidence Level > 99%
$R^2$ = 0.959

**Tensile Strength**

(d)

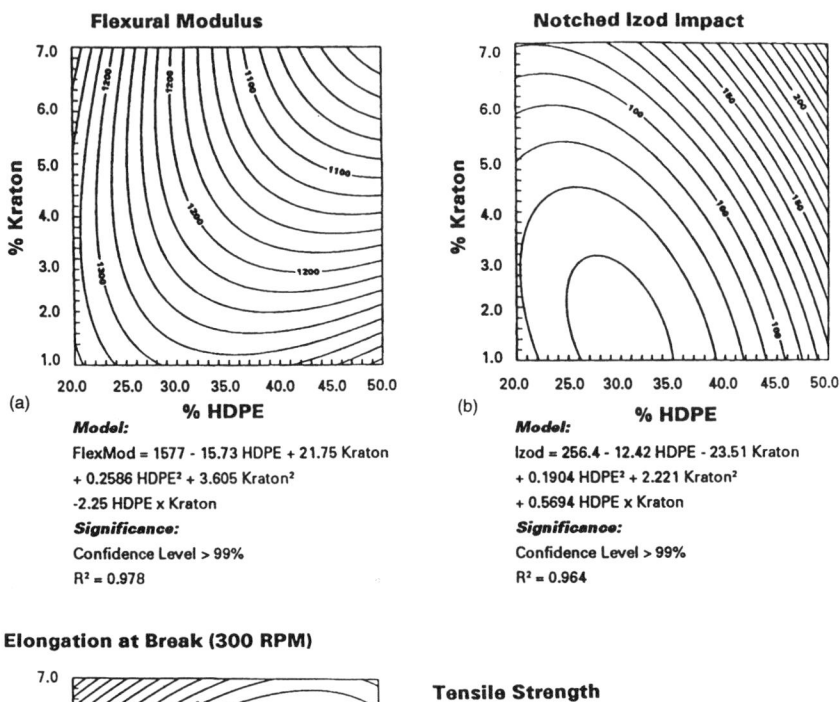

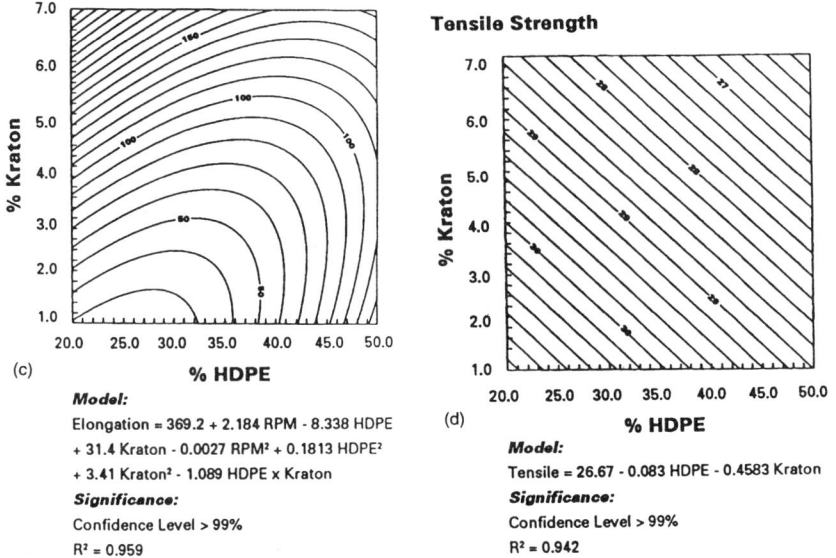

**Model:**
Tensile = 26.67 - 0.083 HDPE - 0.4583 Kraton
**Significance:**
Confidence Level > 99%
$R^2$ = 0.942

FIGURE 5.7  Overview of response surface plots for compatibilized blend. (a) Flexural modulus; (b) notched Izod impact; (c) elongation at break (300 rpm screw speed); (d) tensile strength; (e) heat distortion temperature.

### Heat Distortion Temperature

(e)

**Model:**
HDT = 116.0 - 0.4917 HDPE + 1.375 Kraton
-0.4583 Kraton$^2$

**Significance:**
Confidence Level > 99%
$R^2 = 0.822$

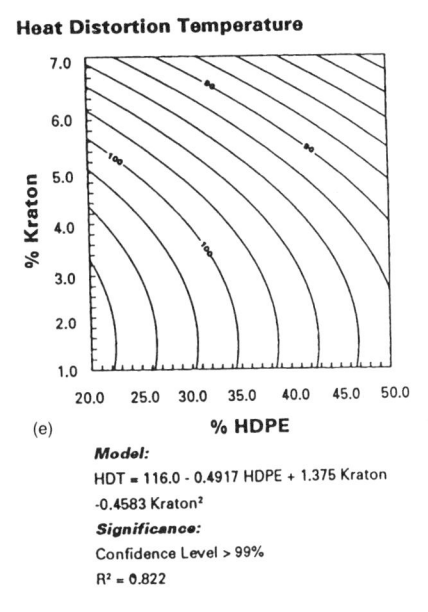

FIGURE 5.7 (*continued*)

As discussed above, elongation at break was severely compromised by the addition of glass fibers; elongation values for the composite blends are adequately represented by a mean value ± standard deviation. This result is based on the ANOVA, which shows that approximately 98% of the data variability is due to the mean value; hence, elongation at break (expressed in %) is stated as 3.43 ± 0.45. However, for the compatibilized blend, Fig. 5.7c shows the response surface at the lowest shear rate (corresponding to a screw speed of 300 rpm). All of these values were in excess of 30% elongation. As reported in a previous publication [61], elongation increased with increased shear rate for the compatibilized blend.

Tensile strength for the reinforced blend is also a linear function of Kraton level only:

$$\text{Tensile} = 59.3 - 1.79 \times \text{Kraton} \tag{5.4}$$

with tensile strength in MPa and Kraton level in percent. Equation (5.4) is statistically significant at the 99.9% level of confidence with $R^2 = 0.646$. As with flexural modulus, the relatively low value of $R^2$ coupled with the high level of confidence suggests that Kraton is a significant factor, but the variability of the test is too high to allow the use of Eq. (5.4) for predictive purposes. Although there was no curvature present for the compatibilized blend (Fig. 5.7d), increasing HDPE resulted in decreasing tensile strength at a given level of Kraton. However,

## Flexural Modulus

(a) % Kraton vs % HDPE contour plot
Model: FlexMod = 3554 − 93 × Kraton

## Notched Izod Impact

(b) % Kraton vs % HDPE contour plot
Model: Izod = 45.13 + 5.57 × Kraton

## Elongation at Break (300 RPM)

(c) Observed Singularity
Value: 3.43 ± 0.45

Value: 3.43 ± 0.45

## Tensile Strength

(d) % Kraton vs % HDPE contour plot
Model: Tensile = 59.3 − 1.79 × Kraton

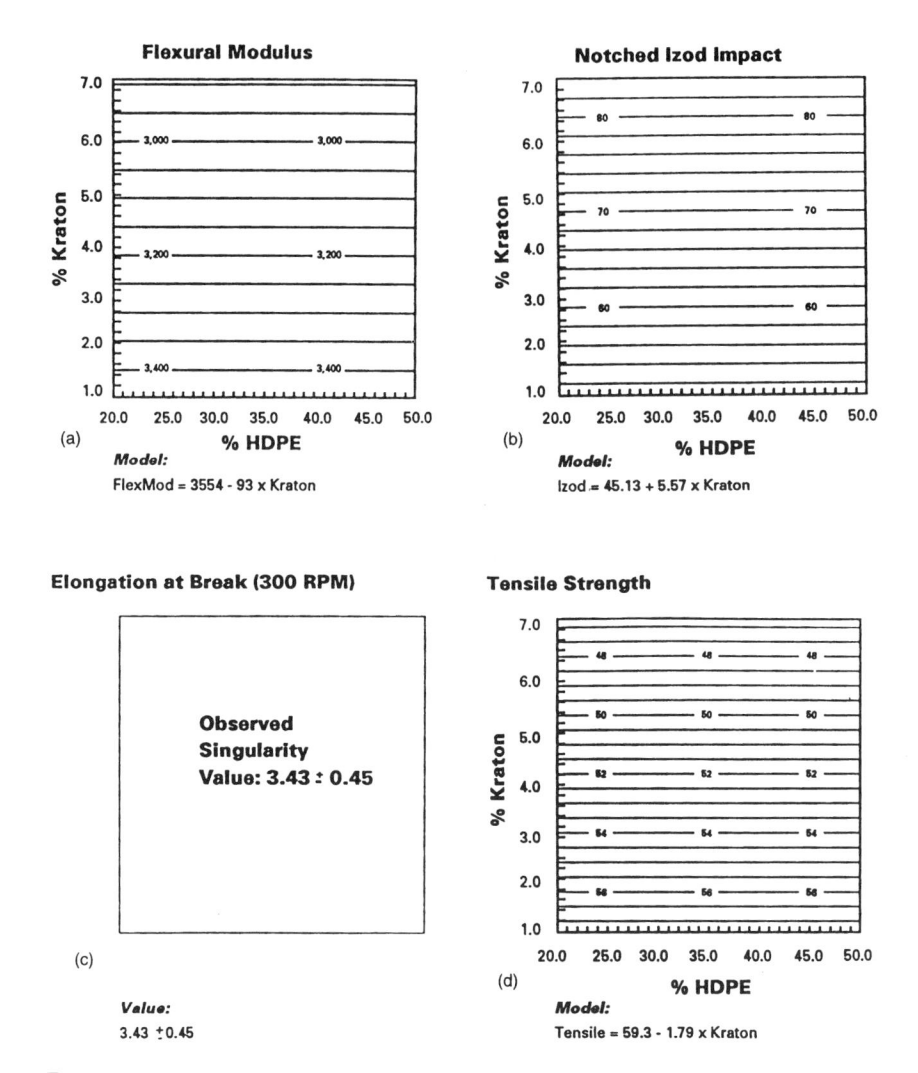

**FIGURE 5.8** Overview of response surface plots for reinforced composites. (a) Flexural modulus; (b) notched Izod impact; (c) elongation at break (300 rpm screw speed); (d) tensile strength; (e) heat distortion temperature.

in the reinforced blend, there is no effect from increasing HDPE, most probably due to the greater effect of the glass fibers. This is supported by the much wider range in tensile strength seen in the reinforced blends (∼45–60 MPa) versus the compatibilized blends (∼27–32 MPa).

### Recycling of Polypropylene and Its Blends

**Heat Distortion Temperature**

[Contour plot: y-axis "% Kraton" from 1.0 to 7.0, x-axis "% HDPE" from 20.0 to 50.0, with diagonal contour lines labeled 135, 136, etc.]

(e)

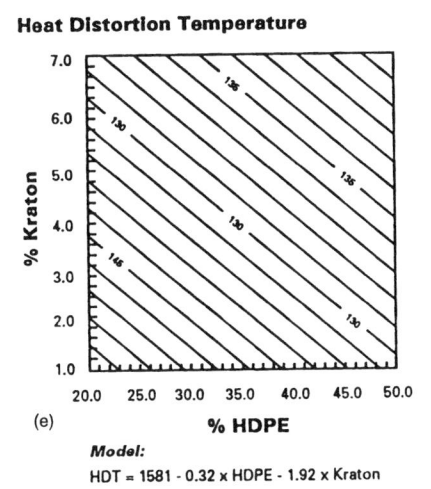

*Model:*
HDT = 1581 - 0.32 × HDPE - 1.92 × Kraton

**FIGURE 5.8** (*continued*)

The HDT for the reinforced blend is a linear function of both Kraton and HDPE.

$$\text{HDT} = 1581 - (0.032 \times \text{HDPE}) - (1.92 \times \text{Kraton}) \tag{5.5}$$

with HDT in °C, Kraton level in percent, and HDPE level in percent. Equation (5.5) is statistically significant at the 98.3% level of confidence with $R^2 = 0.496$. As with previous responses, the relatively low value of $R^2$ coupled with the high level of confidence suggests that both HDPE and Kraton are significant factors, but the variability of the test is too high to allow the use of Eq. (5.5) for predictive purposes. Again, there is no curvature as observed in the compatibilized blend (Fig. 5.7e); however, the values of HDT for the reinforced blends are much higher (~130–150°C) than those of the compatibilized blends (~85–110°C). A much higher value of HDT for the composite blend is consistent with the high value for flexural modulus.

#### 5.4.5 Morphology Analysis

Initial morphologies of extruded strands were examined as a function of extruder rpm and blend composition. Phase contrast optical microscopy shows a 35% HDPE dispersed phase domain size to be comparable (1 µm < size ≤ 5 µm) at both 300 and 500 rpm (Fig. 5.9). The effect of compatibilizer on domain size of the dispersed phase is shown in Fig. 5.10.

142                                                                    Adewole and Wolkowicz

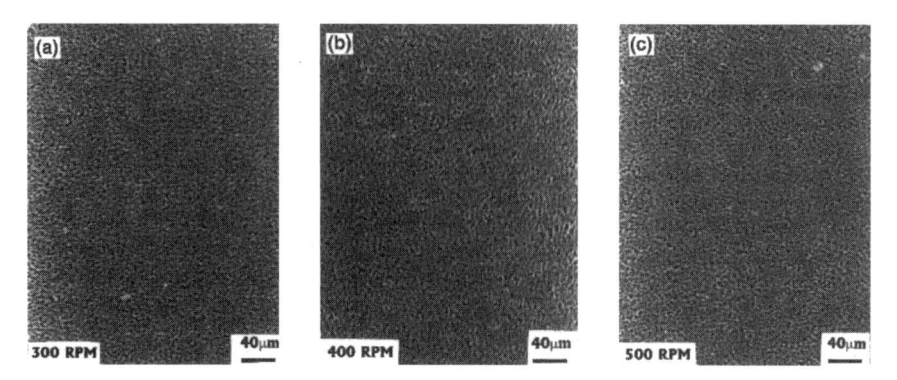

FIGURE 5.9  HDPE dispersed phase domain size as a function of extruder speed: (a) 300 rpm; (b) 400 rpm; (c) 500 rpm.

By comparing the morphology of stranded pellets before and after glass fiber addition, we show that the glass fiber has no effect on the polymer micromorphological phase structure. Figure 5.11 shows cross-sections of tensile bars fractured in liquid nitrogen to expose the phase morphology. The photomicrographs show the dispersed phase morphology of HDPE and Kraton rubber to exhibit similar distribution with or without glass fiber.

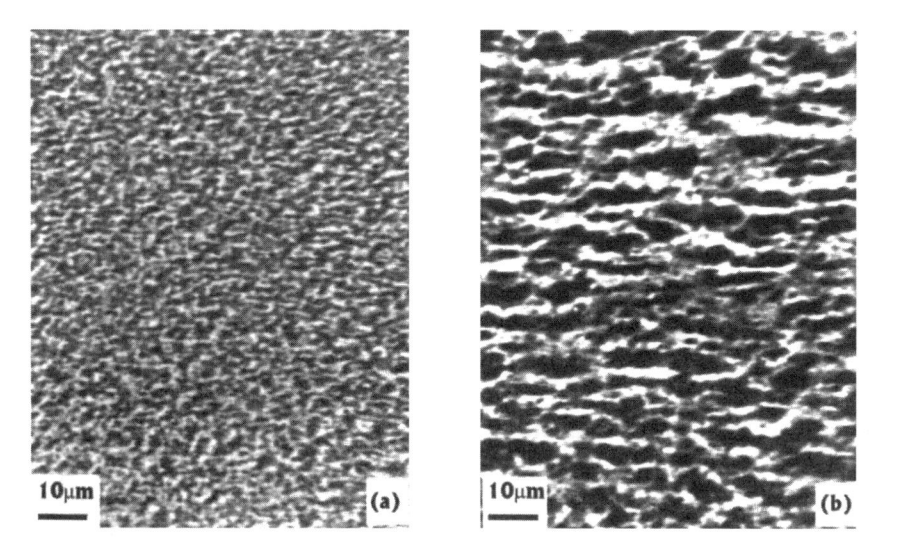

FIGURE 5.10  Effect of compatibilizer at 4 wt% level on domain size of dispersed phase at 35 wt%. (a) With compatibilizer; (b) without compatibilizer.

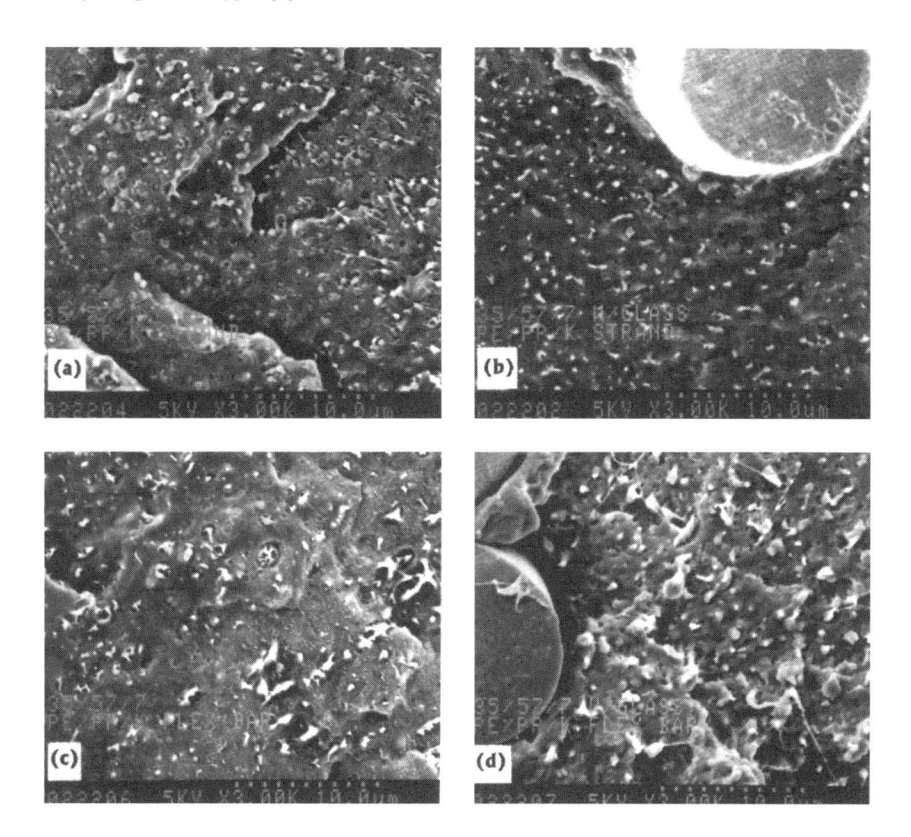

**FIGURE 5.11** Blend morphology of extruded strand and molded bar. (a) Extruded strand without glass fiber; (b) extruded strand with glass fiber; (c) molded bar without glass fiber; (d) molded bar with glass fiber.

We successfully produced micrometer-size HDPE dispersed-phase morphology at 300, 400, and 500 rpm in PP, due to the shear intensive screw design. Improved stiffness and HDT properties due to glass fiber reinforcement are balanced with corresponding reduction in impact and elongational properties.

## 5.5 ECONOMIC ANALYSIS

Because recycled plastics have had a first use and have been environmentally exposed, required conversion processes are often more stringent. For example, compounding may have to be done on a twin- rather than a single-screw extruder for dispersive mixing. In addition, compounding purification steps such as devolatilization and melt filtration may be required for some recycled feedstock.

Consequently, the compounding process can have a significant impact on the overall conversion cost. Another contributor to conversion cost is the compatibilizer additive. The most effective of these additives cost about $2.00 per pound. This adds 10 cents per pound of product at a 5% level of compatibilizer additive.

Statistical design of experiments (DOE) is often used in the early stages of process optimization. This is followed by a validation of the predictive model using actual plant production data. The response surface model described below captures the process performance window and shows the effect of changing composition and extruder screw speed on blend properties. A validation step can be easily implemented.

### 5.5.1 Blend Properties/Cost Optimization

Individual response maps for each important property are useful for determining general trends as a function of composition. However, they may be used to determine operating windows for a desired balance between mechanical properties and cost of the blended resin.

We have the following assumptions in the current analysis.

1. The cost of compounding (processing) either virgin material or the compatibilized blend is the same.
2. PP and HDPE reclaim is $0.15/lb.
3. The price for Kraton rubber is $2.00/lb.
4. The maximal acceptable price for the blend is $0.25/lb.

These assumptions were based on cost estimates when preparing the manuscript; however, the method presented here may be easily adjusted for different assumptions. The cost constraint implies that any blend with more than 5.4% Kraton rubber would be too expensive to satisfy assumption 4. We require the blend to have the following properties:

1. Flexural modulus greater than 1100 MPa
2. Izod impact greater than 65 J/m
3. Elongation at break not less than 125% and not greater than 175%
4. Tensile strength greater than 28 MPa
5. HDT greater than 90°C

Determination of blends that satisfy the above criteria is performed by overlaying the appropriate response curves from Figs. 5.12–5.16, as shown in Fig. 5.17a (300 rpm), 5.17b (400 rpm), and 5.17c (500 rpm).

At 300 rpm (Fig. 5.17a), the desired balance of properties and cost appears as a narrow triangular region along the left axis with vertices at (20,4.4), (20,5.4), and (27,5.4). At 400 rpm (Fig. 5.17b), the operating window shows two distinct

# Recycling of Polypropylene and Its Blends

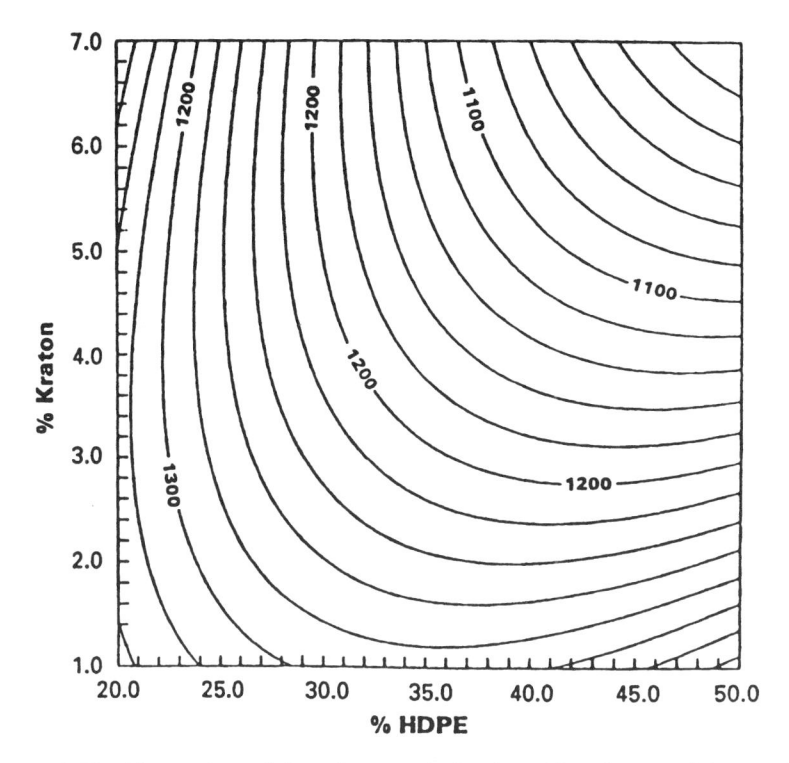

**FIGURE 5.12** Flexural modulus characteristic plots. Fitted to model equation: FlexMod = 1577 − 15.73 HDPE + 21.75 Kraton + 0.2586 HDPE$^2$ + 3.605 Kraton$^2$ − 2.25 HDPE × Kraton.

regions: one for HDPE levels below 36% and one for HDPE levels above 46%. The region with lower HDPE levels requires higher levels of Kraton rubber, whereas the region with higher HDPE levels requires lower levels of Kraton rubber to make blends with the desired properties. At 500 rpm (Fig. 5.17c), the operating window expands to include the entire range of HDPE composition, generally with lower levels of Kraton rubber. However, Kraton rubber level does span the entire range below 5.4%, the cost constraint. Note that not every combination of Kraton rubber and HDPE is sufficient to satisfy the requirements for the blend performance.

The shift in allowable compositions is the direct result of the elongation at break response shifting with changing rpm. As described above, as rpm increased, elongation at break increased for a given composition. Mechanical properties for a given blend composition (% HDPE, % Kraton rubber) and

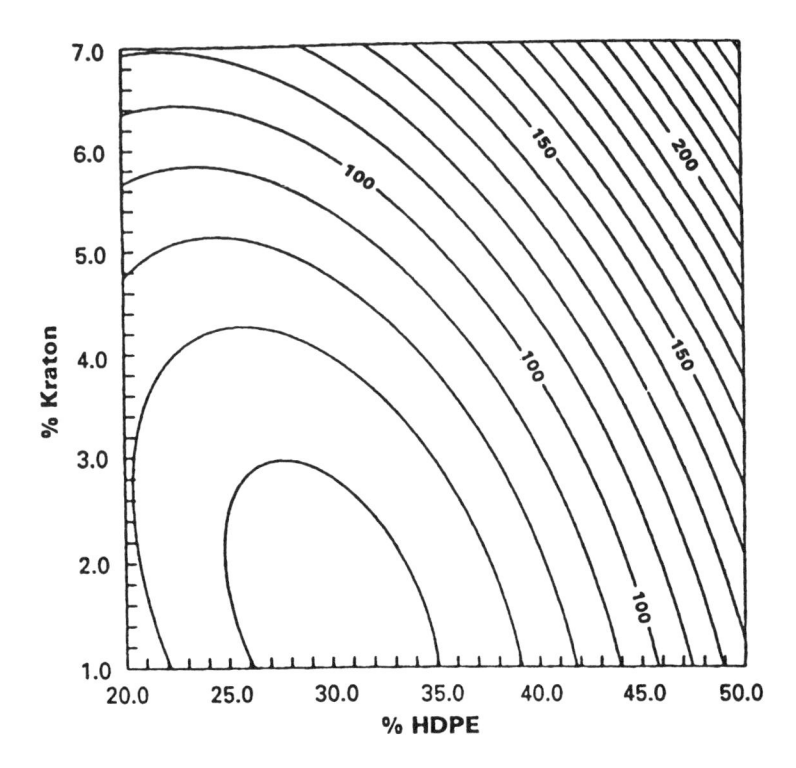

**FIGURE 5.13** Notched Izod impact characteristic plots. Fitted to model equation: Izod = 256.4 − 12.42 HDPE − 23.51 Kraton + 0.1904 HDPE$^2$ + 2.221 Kraton$^2$ + 0.5694 HDPE × Kraton.

operating rpm selected from Table 5.5(a) can be located in the corresponding Figs. 5.12–5.16. The overall stiffness–impact balance for the compatibilized and reinforced blends is shown in Fig. 5.18.

### 5.5.2 Conclusions

Response surface statistical DOE is an efficient approach to determine an operating window for attaining a balance of mechanical properties and cost for the compatibilized recycled PP-HDPE blend.

The dependence on composition and processing of the micromorphological phase structure and mechanical properties of both unfilled and glass-filled recycled PP-HDPE blends is demonstrated in this study.

# Recycling of Polypropylene and Its Blends

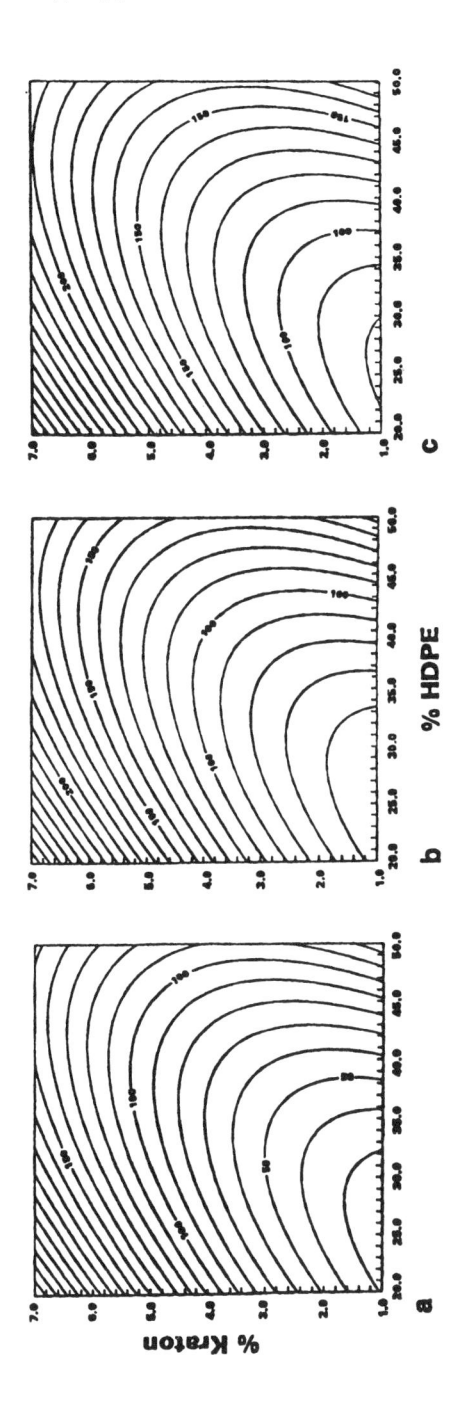

**FIGURE 5.14** Elongation at break characteristic plots vs. extruder screw speed. Fitted to model equation: Elongation = 369.2 + 2.184 rpm − 8.338 HDPE + 31.4 Kraton − 0.0027 rpm$^2$ + 0.1813 HDPE$^2$ + 3.41 Kraton$^2$ − 1089 HDPE × Kraton. (a) 300 rpm; (b) 400 rpm; (c) 500 rpm.

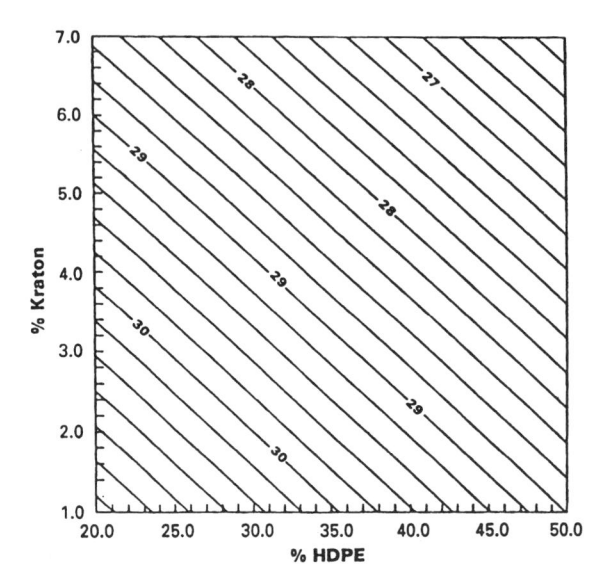

**FIGURE 5.15** Tensile strength characteristic plots. Fitted to model equation: Tensile= 26.67 − 0.083 HDPE − 0.4583 Kraton.

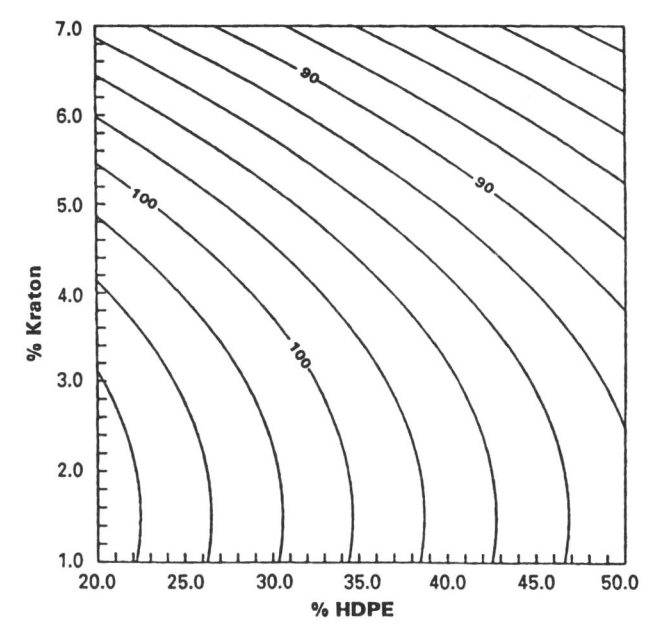

**FIGURE 5.16** Heat distortion characteristic plots. Fitted to model equation: HDT= 116.0 − 0.4917 HDPE + 1.375 Kraton − 0.4583 Kraton$^2$.

# Recycling of Polypropylene and Its Blends

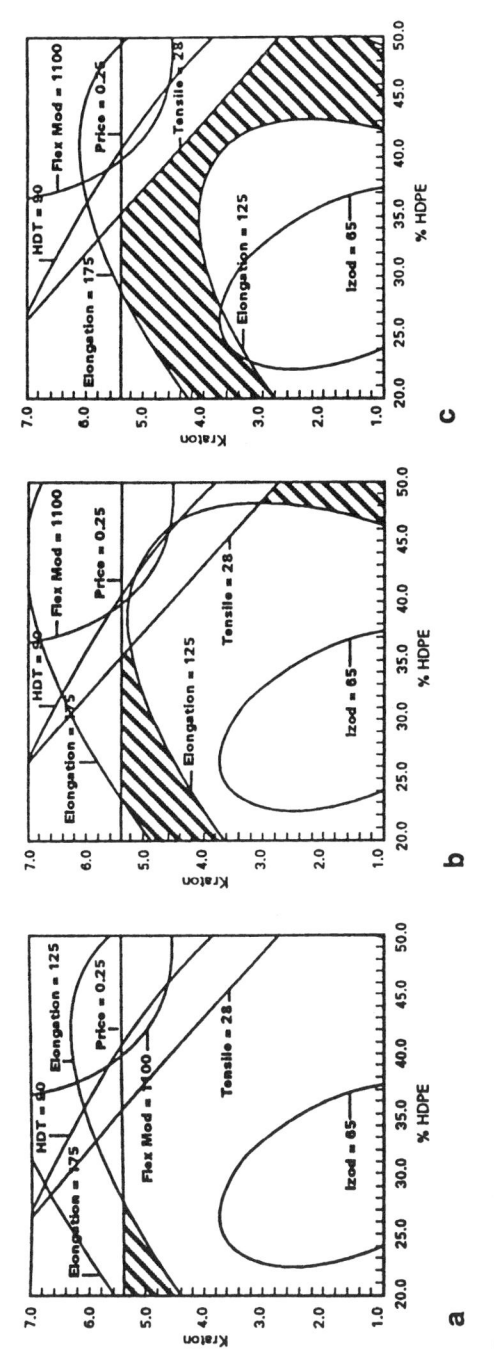

**FIGURE 5.17** Property–cost balance vs. extruder screw speed: (a) 300 rpm; (b) 400 rpm; (c) 500 rpm.

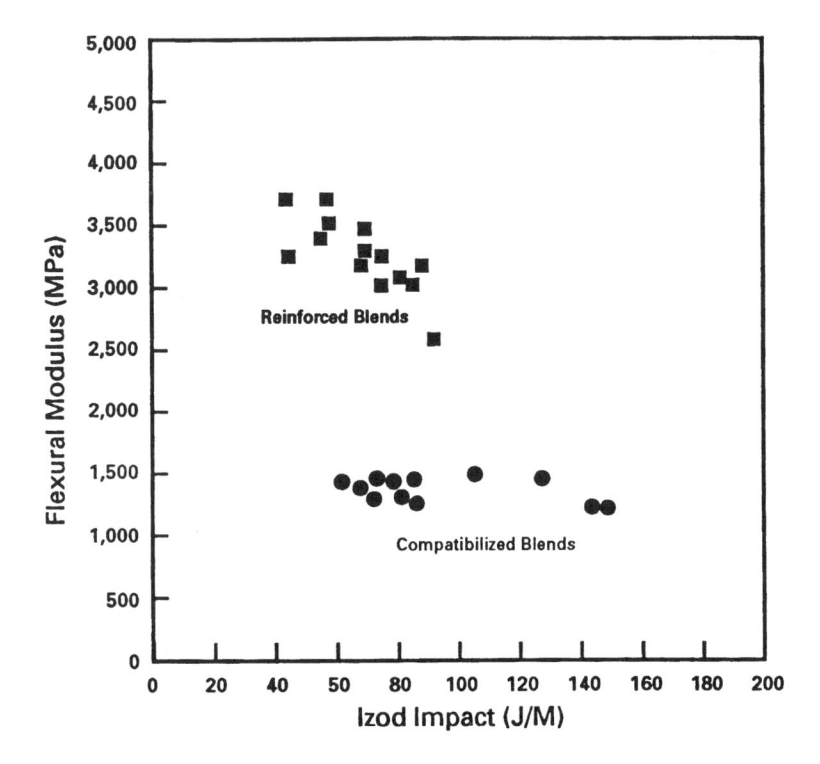

FIGURE 5.18 Stiffness–impact balance. ■, Reinforced blends; ●, compatibilized blends.

Well-dispersed micrometer to submicrometer HDPE domain size morphology is created during extrusion and compounding. For the blend, the compatibilizer concentration has a strong influence on impact and elongation properties. In addition, the polyolefin blend ratio and compatibilizer concentration exhibits interaction effects in determining the balance of properties for the blend system.

It is known that polypropylene, when melt blended with polyethylene, increases its impact resistance, whereas other mechanical properties, like Young's modulus, decrease [8]. A similar trend is shown in Table 5.6. As HDPE concentration increases, flexural modulus decreases and notched Izod impact increases. These blends were prepared at 500 rpm using all other compounding conditions as in the Response Surface Design Program.

By using a compatibilizer such as Kraton rubber, the stiffness–impact balance can be tailored, depending on the blend composition. For the compatibilized blend, we have shown in this work that response surface analysis demonstrates the effect of changing composition for each property and that elongation depends on shear rate, whereas other properties are not affected by shear rate.

Figure 5.18 shows the stiffness–impact balance for the reinforced versus the compatibilized blends. The compatibilized blends exhibit a lower flexural modulus and a wider range of Izod impact strengths than the reinforced blends. Compatibilized blend properties depend strongly on composition, whereas the composite properties depend more on Kraton level than on polyolefin composition. This offers a wide operating window to meet stiffness–impact requirements with respect to daily feedstock supplies.

The addition of glass fiber reinforcement to the compatibilized blends has a strong effect in altering the resulting mechanical properties of the material. Most notably, the glass fibers tend to remove the effect of HDPE on the properties and, for this series of experiments, to eliminate the two-factor interactions observed in compatibilized blends. In general, glass fiber reinforcement increases the flexural modulus, tensile strength, and heat distortion temperature while reducing the elongation at break and narrowing the Izod impact range, in contrast to compatibilized blends of equivalent composition.

These effects strongly suggest that the critical factor in determining cost–property balances for the reinforced blends lies in the physical property most critical for the chosen application. For example, if HDT is not a critical property, HDPE level is not an issue; cost is driven by the amount of Kraton necessary to achieve the desired property. This is in contrast to the compatibilized blends, where operating windows were a strong function of composition and shear rate in determining the cost–property balance.

## REFERENCES

1. Council for Solid Waste Solutions. The Blueprint for Plastics Recycling. Washington, DC: Society of the Plastics Industry, 1991.
2. PA Toensmeier, ed. Modern Plastics. New York: McGraw-Hill, 1997, p. 12.
3. FN Aronhalt. Integrated waste recovery approach allows maximum use of resources, In: PA Toensmeier, ed. Modern Plastics Encyclopedia '95. New York: McGraw-Hill, 1994, pp. A38–A39.
4. JH Schut, Sr. ed. Plastics World. Melville, NY: PTN Publishing Co., 1997, p. 61.
5. RA Denison, JF Ruston. Recycling is not garbage. In: S Hackman, ed. M.I.T.'s Technology Review. Burlington, VT: Lane Press, 1997, pp. 55–60.
6. WA Kaplan, ed. Modern Plastics Encyclopedia '96. New York: McGraw-Hill, 1995, pp. A39–A55.
7. F Aronhalt, R Perkins. The wave of recycling bumps into the seawall of economic reality. In: WA Kaplan, ed. Modern Plastics Encyclopedia '98. New York: McGraw-Hill, 1997, pp. A34–A35.

8. L Erwin, LH Healy, Jr. Packaging and Solid Waste Management Strategies. New York: AMA Membership Publications Division, 1990, pp. 11–94.
9. A Beevers, ed. European Plastics News. London: EMAP Business Publishing, 1992, pp. 19–29.
10. A Beevers, ed. European Plastics News. London: EMAP Business Publishing, 1995, pp. 28–29.
11. A Beevers, ed. European Plastics News. London: EMAP Business Publishing, 1994, p. 17.
12. A Beevers, ed. European Plastics News. London: EMAP Business Publishing, 1993, pp. 20–21, 37.
13. A Beevers, ed. European Plastics News. London: EMAP Business Publishing, 1993, pp. 3–28.
14. A Beevers, ed. European Plastics News. London: EMAP Business Publishing, 1994, pp. 3–22.
15. A Beevers, ed. European Plastics News. London: EMAP Business Publishing, 1991, pp. 75–76.
16. A Beevers, ed. European Plastics News. London: EMAP Business Publishing, June 1994, pp. 18–20.
17. A Beevers, ed. European Plastics News. London: EMAP Business Publishing, 1994, pp. 20–21.
18. A Beevers, ed. European Plastics News. London: EMAP Business Publishing, 1993, p. 37.
19. A Beevers, ed. European Plastics News. London: EMAP Business Publishing, 1995, p. 24.
20. The Society of the Plastics Industry, Inc. Plastics. Perspective. Washington, DC, 1992.
21. WE Pearson. Plastics Recycling Foundation, Annual Report 1991, Washington, DC, 1991.
22. A Beevers, ed. European Plastics News. London: EMAP Business Publishing, 1991, p. 14.
23. A Beevers, ed. European Plastics News. London: EMAP Business Publishing, 1995, p. 14.
24. JH Schut, Sr. ed. Plastics World. Melville, NY: PTN Publishing Co., 1997, pp. 36–40.
25. MH Naitove, ed. Plastics Technology. New York: Bill Communications, Inc., 1997, p. 64.
26. JJ Callari, ed. Plastics World. Melville, NY: PTN Publishing Co., 1996, p. 4.
27. A Beevers, ed. European Plastics News. London: EMAP Business Publishing, 1996, p. 9.
28. JJ Callari, ed. Plastics World. Melville, NY: PTN Publishing Co., 1997, p. 1.
29. E Taylor, B Burroughs. World Wastes, Argus Business Communications, November 1996.
30. PA Toensmeier, ed. Modern Plastics International. New York: McGraw-Hill, 1997, p. 16.
31. A Beevers, ed. European Plastics News. London: EMAP Business Publishing, 1993, p. 21.

32. VC Smith. Catalytic cracking of byproduct polypropylene. U.S. Patent 4151216 4/24/79.
33. C Vasile, P Onu, V Barboiu, M Sabliovschi, G Moroi. Acta Polym 36:543–550, 1985.
34. M Yamamoto, N Takamiya. J Rikogaku Kenkyusho 111:8–14, 1985.
35. D Schutze. Degradation of polypropylene wastes. Patent: Germany Offen., DE 2164073, 7/6/72.
36. H Coenen, R Hagen. Liquid hydrocarbons from wastes. Patent: Germany Offen., DE 3326284A1 02/21/85.
37. T Sawaguchi, T Kuroki, T Ikemura. J Bull Jpn Pet Inst 19:124–130, 1997.
38. T Sawaguchi, S Niikuni, T Kuroki. J Polym 22:1403–1406, 1981.
39. P de Amoyim, MT Sousa, C Comel, P Vermande. J Anal Appl Pyrol 4:73–81, 1982.
40. JKY Kiang, PC Uden, JCW Chien. J Polym Degrad Stab 2:113–127, 1980.
41. M Ishiwatari. J Polym Sci Polym Lett Ed 22:83–88, 1984.
42. HK Staffin, R Staffin, RB Roaper. Proc Intersoc Energy Convers Eng Conf 14:1656–1657, 1997.
43. HK Staffin, RB Roaper. U.S. Patent No. 4613713, 9/23/86.
44. EB Nauman, MV Ariyapadi, NP Balsara, TA Grocela, JS Furno, SH Lui, R Mallikarjun, R Chem Eng Commun 66:29–55, 1988.
45. JS Furno, EB Nauman. Polymer 32:87–94, 1991.
46. EB Nauman. Ency Polym Sci Eng Suppl, New York: John Wiley & Sons, 1989, pp. 317–323.
47. JC Lynch, EB Nauman. Proc SPE-RETEC, Charlotte, NC, October 30, 1989.
48. JC Lynch, EB Nauman. Proc 10th Intl Coextrusion Conf, 1989, pp. 99–110.
49. A Kiani, HJ Samman. SPE Antec Proceedings, 2758–2762, 1993.
50. S Wu. Polym Eng Sci 27:343, 1987.
51. J Grenci, S Dey, C Jacob, A Patel, SS Dagli, SH Patel, M Xanthos. Proceedings, SPE ANTEC Annu Conf 39, 488, May 1993.
52. T Vivier, M Xanthos. Proc 51st SPE Antec Annu Conf 39, 1997, May 1993.
53. KE Van Ness, WT Fielder, LW Strickler. Proc 51st SPE Antec Annu Conf 39, 2851, May 1993.
54. HA Sangani, G Estepp. J Plast Eng, pp. 27–29, February 1997.
55. C Hagberg, J Dickerson. J Plast Eng, pp. 41–43, April 1997.
56. M Xanthos. Polym Eng Sci 28:1392, 1988.
57. AA Adewole. Doctoral thesis, Loughborough University, Leicestershire, UK, 1998.
58. M Xanthos, A Patel, SJ Dey, SS Dagli, TJ Nosker. SPE ANTEC Annu Conf 1, 1992, pp. 596–601.
59. SS Dagli, M Xanthos, JA Biesenberger. SPE ANTEC Annu Conf 1994, pp. 3017–3020.
60. M Xanthos, A Patel, S Dey, SS Dagli, C Jacob. Adv Polym Technol 13:231–239, 1994.
61. AA Adewole, K Dackson, M Wolkowicz. Adv Polym Technol 13:219–230, 1994.
62. BD Favis, D Therrien. Processing/Morphology Relationships in Polymer Blends During Twin-Screw Extrusion. 2nd Int Congress on Compatibilizer and Reactive Polymer Alloying, Compalloy '90, 1990, p. 59.

63. KD Webber, SA Swint. Processing of Post-Consumer Commingled Plastics Waste for Consistent and Good Properties. ANTEC '91, 1991, p. 2175.
64. AA Adewole, MD Wolkowicz. Processability-Morphology-Properties Relationship in Compounding Polyolefinic Recycled Plastics. Proc 51st SPE Antec Annu Conf 1993, p. 39.
65. M Xanthos, TJ Nosker, K Van Ness. Compatibilization for Reuse of Commingled Post-Consumer Plastics. 6th Intern Congress on Compatibilizers and Reactive Polymer Alloying, Short Hills, NJ, Compalloy '92, 1992, p. 59.
66. G Box, D Behnken. Some new three level designs for the study of quantitative variables. Technometrics 2:455–475, 1960.

# 6

# Impact Behavior of Polypropylene, Its Blends and Composites

**Josef Jancar**
Technical University Brno, Purkynova, Brno, Czech Republic

## 6.1 INTRODUCTION

Significant product growth in commercially successful polypropylene (PP) composites has spurred increased activity in both academic and industrial communities to better understand structure–property relationships. It has been demonstrated that PP filled with calcium carbonate, talc, mica, clay, alumina trihydrate, magnesium hydroxide, carbon black, and other fillers can successfully bridge the gap in performance between neat commodity plastics, engineering resins and short-fiber reinforced thermoplastics. Moreover, commercial success of compounded PP has been indisputable in recent years with substantial growth in diverse markets such as household items, durable goods, appliances, passenger cars, mass transport vehicles, aircraft, sporting and leisure goods, electrical and construction industry, electronics, packaging, and medical devices. In one way or another, perhaps more than 50% of all manufactured polymer resins are filled with inorganic fillers to achieve desired properties.

Plastic products are exposed to many impact encounters during their service life. Recently, Perkins [1] wrote a comprehensive review of the many factors that influence impact resistance and determine toughness of a fabricated or molded part subjected to an end-use application. Toughness is defined as a measure of the ability of material structure or molded part to endure the application of a sudden applied load without experiencing "failure."

The toughness of composites, based on semicrystalline resins for engineering applications, is of major concern in meeting finished product requirements necessary for good performance. High strain rates, low temperatures, and the presence of stress risers often lead to brittle failure of materials even though they behave in a ductile manner at low strain rates or higher temperatures. During early plastic product development, the commercial success of high-impact polystyrene (HIPS) and acrylonitrile-butadiene-styrene (ABS) led to the development of a whole new group of rubber toughened plastics [2–4]. Since then, about 80% of blended or filled thermoplastics are compounded with some type of modifier to give products having improved impact resistance during their service lifetime. Indeed, toughness enhancement of the polymer matrix has become a major new field of polymer science and is very often the decisive characteristic used in material selection for a large variety of applications (e.g., automotive, home appliances, construction, utilities, sporting goods).

Chapter 1 provides trend data for the use of PP resins as cost-effective replacement of engineering polymers. With the development of tailored PP copolymers (see Chapter 2) and plastomers for impact modification (see Chapter 7), composites of fillers and fibrous reinforcement can be effectively toughened to compete with polyamide-type products.

Because impact modification of PP blends and composites represents an important area of commercial interest, materials scientists seek a fundamental understanding of the mechanisms underlying fracture failure processes. Most of these mechanisms also operate in the neat polymers; however, the incorporation of a secondary phase or component alters their modus operandi or introduces impact behavior that does not occur in the neat polymer. Combinations of PP with fillers or thermoplastic blends affect the balance of stiffness and impact resistance. The challenge to product design is how to attain a favorable balance of properties that suit the particular end-use application. Rather than depending on guesswork, the development of cost-effective formulations requires guidelines based on proven hypotheses of impact fracture mechanisms.

The following extremes of material properties exemplify the difficult task that is involved in new product design. Increased filler addition enhances stiffness and raises heat distortion temperature (HDT). Composites with high filler loadings above 60 wt% are used to reduce flammability. However, increased filler loadings correspond to a tendency for brittle fracture at low temperatures or part failure during high-strain impact events. Polypropylene composites can be toughened by blending with a rubbery thermoplastic or elastomer. However, PP-elastomer blends have reduced HDT or lower Vicat softening temperature. Consequently, the impact-modified PP composite may not withstand automotive

# Impact Behavior of Polypropylene

underhood applications with elevated service temperatures of 150–170°C. Any limitation at either extreme of the temperature range would partially restrict wider use of PP-based materials in the construction industry, cars, mass transportation vehicles, and outdoor electrical appliances.

In particulate-filled PP, the matrix is the load-bearing component since all deformation processes take place in the matrix. Particulate fillers with low aspect ratios are not capable of carrying any substantial portion of the load because there is an absence of interfacial friction as the means for stress transfer. This shortcoming is indicated by a lack of broken particles on the surfaces of fractured filled thermoplastics. It seems appropriate to begin this chapter with a brief overview of the fundamental concepts of structure–property relationships that govern the mechanical properties of semicrystalline thermoplastics used in compounding.

With addition of inorganic filler, various changes occur in the molecular and supermolecular structure of a thermoplastic resin. Composite properties depend on a variety of material-process variables (e.g., polymer matrix structure, filler content, chemical composition, surface activity, particle size and shape, compounding extruder design, mold design, and extruder-molding process conditions). Some changes may be latent or delayed (i.e., occurring later in the service life of the plastic part as a result of surrounding conditions). Reduction of molecular weight, crystal and spherulite size, and molecular mobility are among the most profound effects that solid filler has on the polymer matrix structure. The microstructure of the polymer–filler interphase is mirrored by the mechanical integrity of the molded part and long-term durability to extremes of surrounding temperatures and applied stresses.

## 6.2 DISSIPATION OF MECHANICAL ENERGY DURING FRACTURE PROCESS

Brittle fracture and ductile fracture are two general modes of failure. Brittle fracture is characterized by a linear relationship between impact load and sample deflection up to an abrupt failure during impact testing. Ductile fracture is manifested as plastic yielding with a gradual decrease in slope of load displacement curves, prior to eventual failure.

Brittle fracture frequently results from highly delocalized crazing. Crazes are pictured as pseudocracks that are oriented in a direction normal to applied stress load. These crazes are spanned by cold-drawn elongated fibrils of material that are confined to a very small volume of the polymer matrix microstructure. Ductile failure is due to either multiple crazing (crazes initiated in a comparatively large volume of the polymer matrix by a multiple deformation mechanism) or by shear yielding (plastic flow without crazing).

Depending on the type of failure mode, major contributors to the dissipation of mechanical energy at the cross-sectional area of possible fractures are crazing and shear yielding mechanisms. Crazing is most important for glassy thermoplastics [polycarbonate (PC), HIPS, polymethyl methacrylate (PMMA), and ABS]. Shear yielding is a major deformation mechanism in thermosets (epoxy, unsaturated polyesters) and semicrystalline thermoplastics with sufficiently high degree of crystallinity [PP, polyamides, polyethylene (PE)]. Shear yielding and crazing are not mutually exclusive to the given groups of materials. Very frequently, both modes of fracture can operate in the same specimen at the same time. Furthermore, relative contributions will depend on testing conditions (temperature, strain rate, crack tip radius) and on the material structural variables (crystalline morphology, tie molecules, molecular structure).

Frequently, a transition in major deformation mechanism from shear yielding to crazing or vice versa is accompanied by a sudden change in crack resistance. This is often referred to as a ductile-brittle transition (DBT). Hence, low-temperature impact resistance of PP-based materials is a key factor in material selection for molded parts subjected to ambient temperatures as low as −40°C. In this regard, plastomers described in Chapter 7 provide reduction of DBT well below end-use requirements.

Additional dissipative processes arise from the presence of a secondary component in the polymer matrix or from interactions between the host polymer and the secondary particles. For a given polymer composite, there may exist inclusion cavitation, interfacial cavitation, particle deformation, and crack pinning by constituent particles.

## 6.3 MECHANISMS OF TOUGHNESS ENHANCEMENT IN POLYMERS

It is generally accepted that the most effective dissipative processes are those involving large plastic deformations before crack initiation. These large plastic deformations take place in shear banding and crazing. However, extreme localization of plastic deformations into small volumes leads to macroscopic failure initiated from these areas of large plastic flow. This is even more pronounced in the vicinity of preexisting cracks or notches loaded at high strain rates or at low temperatures. An obvious method of increasing the amount of dissipated energy is extension of the volume of a polymer involved in shear yielding or crazing [2,5,6]. This is most effectively achieved by incorporating a secondary component of suitable elastic properties, inclusion size, and interfacial adhesion to the matrix. Most frequently used secondary components, or toughening agents, are elastomers, followed by thermoplastic inclusions and, in some instances, by rigid inorganic particles and reinforcements. This type of secondary inclusion controls the primary deformation mechanism on

# Impact Behavior of Polypropylene

the microscopic scale. In the case of brittle behavior of rigid thermosets, extension of shear yielding is achieved by reduction of the degree of cross-linking by increasing the molecular weight and flexibility of the chains between the cross-links. Among the toughening processes, nonexistent in neat polymers, the most important are crack pinning by rigid particles in thermosets, secondary component tearing, elastomer particle cavitation, and dewetting in rubber-modified polymers.

## 6.3.1 Delocalization of Shear Banding

Extensive shear yielding at the crack tip is a major mechanical energy dissipation mechanism in many tough polymers. On the other hand, localized shear yielding (i.e., shear banding) is believed to be a precursor of brittle fracture in many polymers. It is now well established that a delocalization of shear banding can be achieved in many otherwise brittle polymers by incorporating a secondary discontinuous component. This phenomenon is also described as the spreading of shear yielding over a larger volume of a material at the crack tip. This especially occurs when the elastic modulus of the secondary polymer is substantially lower than that of the matrix [5]. Newman and Strella [7] observed typical features of shear yielding in heterogeneous ABS terpolymer subjected to uniaxial tensile loading. The macroscopic features are necking, drawing, and orientation hardening. On the microscopic or structural level of individual rubber particles, it was found that the polymer matrix had undergone localized plastic deformation around virtually every single rubber particle. In the case of preexisting cracks or defects, the same effect was observed for material within the crack tip plastic zone, which was, additionally, substantially greater than in the case of unmodified polymer.

The explanation of the effect of secondary component on the spreading of shear yielding (i.e., delocalization of shear banding) is based on a concept of local stress fields and stress concentrations in the matrix due to presence of inclusions. This leads to reduction of the external load needed to plastically deform the material. The original Goodier's solution [8] for an isolated particle in an isotropic matrix resulted in a maximal stress concentration of about 1.9 at the equator of the inclusion [9]. It should be borne in mind that this solution belongs to a class of analytical "single-particle solutions." These do not take into account possible stress field interactions or overlaps in multiparticle systems [10]. The solution for interacting particles (i.e., for volume fractions above 0.09) was obtained using numerical finite element analysis [11–13]. The results indicate that due to overlap of the stress fields for neighboring particles, stress concentration up to 6 can be achieved. The location of the maximal stress concentration moves away from the particle surface and its actual position depends on the spatial packing of the inclusions. It was also shown that there is not much of a difference in stress fields between elastomer inclusion and a void [14]. Additional consideration should be given to the fact that a morphology of a semicrystalline matrix can be affected substantially when the interinclusion distance becomes of the order of lamella size.

The stresses around particles have the character of triaxial tension only in the case of perfect interfacial adhesion. This is exemplified by rubber-toughened epoxies, ABS, and toughened nylons due to a comparable bulk modulus between matrix and elastomer. In such a case, elastomer particle cavitation occurs. Microshear bands were observed in rubber-toughened PVC initiated at an angle of 55–64° with respect to the direction of the applied stress [15]. In addition, the authors have observed cavities in the rubber particles aligned in the planes of shear microbands. This explains a cause for the generation of stress whitening in this material. For ABS, it has been shown that the mechanism of delocalized shear yielding can lead to similar cavitation without crazing. Donald and Kramer [16] studied the effect of particle size on the deformation mechanism in ABS and found that in the case of 0.1-µm-diameter inclusions, crazing is suppressed and shear deformation promoted by particle cavitation. In the system with 1.5-µm-diameter inclusions, crazing was the major deformation mechanism for energy dissipation.

It is now believed that the process of particle cavitation relieves the local buildup of hydrostatic tension caused by constant volume shear banding. This allows additional enhancement of shear yielding in both thermosets and thermoplastics [17]. Hence, soon after development of some initial shear yielding, the local triaxial constraints are relieved by cavitation, even in relatively thick specimens. Similar explanations were proposed for PC/PE, PC/MBS [18], and rubber-modified epoxies [19–24]. In polymers or under test conditions favoring crazing as a major deformation mechanism, cavitation and voiding of the rubber particles leads to a premature craze breakdown that is damaging to the polymer.

### 6.3.2 Crazing in Semicrystalline Polymers Subjected to Impact

Jang et al. [25–28] studied extensively craze formation in semicrystalline polymers, namely, in virgin and rubber-modified polypropylene. They determined the effect of injection molding conditions on PP morphology and its relation to crazing at low temperatures and high strain rates. Their design of experiments (DOE) characterized the effects of various thermoplastic elastomer impact modifiers on crystallization, fusion, and crystalline morphology of molded PP specimens. Jang and coworkers found that crazes are formed preferentially in the core of the injection-molded test specimen, where a regular spherulitic structure existed. The surface "skin" and "shear" zones, which are characterized by oriented molecules and oriented spherulitic growth, respectively, contained few or no crazes. This effect was attributed to the molecular orientation effect, which is known to prevent craze formation and growth [29]. However, no consideration was given to the possibility that, in addition to the orientation of molecules or spherulites, the surface layers are in the plane stress state, which favors shear yielding rather than dilatational crazing. Even though there is some experimental evidences supporting the idea of crazing in semicrystalline polymers, the presence of crazes in these materials has not yet been unambiguously accepted. One of the major reasons is the fact that crystalline morphology represents material with translational symmetry, and none of the current craze initiation theories is capable of predicting formation of a craze in such a continuum.

Despite some experimental evidence of craze-like deformation, shear yielding [26] was proposed as the main deformation mechanism for fracture failure initiation in molded parts of semicrystalline polymers at moderate to high temperatures and low strain rates. Since brittle fracture only occurs at relatively low temperatures and/or high strain rates in amorphous regions, crazing is considered to be a minor component of plastic deformation in these polymers. For a given temperature, a critical strain rate exists above which craze-like deformation dominates and below which shear yielding is dominant [30–31]. This hypothesis is supported experimentally by an observation of appreciable increase in shear yield strength of polymers with decreasing temperature and/or increasing strain rate. On the other hand, the triaxial crazing stress is relatively unaffected by these test conditions. It is believed [32] that the specimen either shear yields if a shear component of the stress field exceeds the shear yield strength or crazes if the triaxial tension exceeds the craze formation stress. In most of these interpretations, formation of microcracks rather than crazes is neglected, even though brittle strength is far less temperature and strain rate sensitive than shear yielding strength. Wu [33] measured the energy dissipation contributions from crack formation, crazing, and shear yielding in rubber-modified nylon-66. The energy consumed by formation of new surfaces during impact loading was five orders of magnitude smaller than that dissipated by crazing and shear yielding at 23°C. Crazing dissipates about 25% of the total energy, and shear yielding dissipates the remaining 75% mostly in the form of heat, causing an increase in temperature within the stress-whitened zone by about 10°C. The crazing energy contribution is due to energy stored in craze fibrils (60%) and energy associated with fibril formation (40%).

Crazes in PP were found to be morphologically similar to those occurring in glassy polymers. They exhibit high reflectivity, large area-to-thickness ratio, and substantial planarity. Craze planes were usually perpendicular to the direction of principal tensile stress, although local deviations of up to 15° existed. It is assumed that these deviations are due to local order (i.e., crystalline structure and superstructure). Moreover, crazes in PP have a larger tendency to bifurcate or branch off than crazes in glassy polymers. They propagate through spherulites, and their length is not restricted to one spherulite diameter nor do they grow preferentially in the radial direction. As was shown by Friedrich [34], crazing can be affected substantially by alterations of the crystalline structure. If most crazes develop at the interfaces between large spherulites, the measured fracture toughness is low. Distinct fracture resistance values exist for crazes developed within larger spherulites or at the boundaries of the smaller ones. This phenomenon could be one of the possible mechanisms of how elastomers can affect crack resistance of semicrystalline polymers when added to modify crystalline morphology.

As in the crazing in rubber-modified glassy polymers, it appears reasonable to assume that a critical size of rubber particle size domain also exists for rubber-modified semicrystalline polymers. Jang et al. [26] observed that for elastomer particles of diameter, $D$, smaller than about 0.5–1.0 µm, no crazes were initiated. It was suggested [27] that the loss of craze nucleation efficiency of small particles was due to the small size of the stress concentration region, which cannot achieve the critical porosity necessary for craze nucleation [18]. The critical particle size for no craze formation will vary for different blends of PP modified with EPR, ethylene-propylene-diene rubber (EPDM) or styrene-butadiene rubber (SBR) depending on the particular matrix/rubber pair. One reason for such a variation can be the change in the rubber-to-matrix modulus ratio that affects the stress concentration ability [35,36].

The actual morphology of crazes in PP consists of fibrils spanning the craze surfaces and, in addition, interconnecting fibrils randomly oriented in respect to the principal stress direction [27]. Apparently, a large number of voids exist in these crazes, distinguishing them from the crazes observed in glassy polymers. One can speculate about the reasons for the observed differences. For example, from concepts concerning the proposed role of weakly bonded rubber particles as the only stress raisers, experimental observations suggest that these particles do not act as craze terminators in any case. On the other hand, due to the inherent anisotropy of semicrystalline domains, preferential directions for large deformation to occur will exist locally. As a result, conditions for microcavitation/ microcracking/craze initiation may deviate from the maximal dilatational stresses calculated for homogeneous isotropic solid. At present, there is no conclusive explanation for this observation.

### 6.3.3 Crack Bridging and Particle Tearing

Crack bridging by elongated elastomer inclusions was the first model attempting to explain rubber toughening in HIPS [37]. It was also suggested as a possible toughening mechanism in thermoplastic-modified epoxies [38]. The contribution of crack bridging and particle tearing to the total energy dissipated during fracture of rubber-modified polymers was shown to be negligible because of the very low shear elastic modulus of the elastomer [7]. Several authors [39,40] presented experimental data suggesting that crack bridging and particle tearing operates in thermoplastic-modified epoxies; however, the understanding of this phenomenon is far from complete. Current research in the area of ionomer compatibilized PP/TLCP (thermotropic liquid crystalline polymer) blends [41] provided first experimental proves that it is possible to design PP-TLCP interphase capable of transferring stresses from PP into TLCP domains in such a way that these domains exhibited drawing. During this process, substantial increase in elastic modulus of the TLCP domain has been observed resulting in a large energy dissipation even during the impact loading. This approach seems very promising in introduction of additional deformation process into failure mechanisms existing in neat and rubber-modified PP.

In the case of thermoplastic-modified epoxies, the proposed role of elongated rigid plastic particles is to span the two fracture surfaces and apply tractions that effectively reduce stress intensity factor applied at the crack tip [42]. Additional contribution to the energy dissipation is assumed from the plastic deformation of the inclusions [43]. Pearson and Yee [42] suggested that the contribution to the overall toughness from this mechanism in thermoplastic-modified epoxies is between limits given by the case of rubber-modified [38] and glass-filled epoxy resin [44].

The model of Ahmad et al. [38] predicts that the size of the particle affects the total tear energy consumed. In particular, toughness improvements should be greater for larger particles, which contradicts the rubber-toughening concept in PP. In addition, increases in particle stiffness and tear energy should result in enhanced crack resistance. Two to three orders of magnitude larger elastic moduli of thermoplastics compared with common elastomers can conceivably increase the importance of this toughening mechanism. Moreover, when the secondary polymer possesses large deformability, the increase in tear energy also enhances the contribution from this mechanism. As was shown by Cerere and McGrath [39] and Raghava [45], increase in molecular weight of the secondary thermoplastic polymer led to improvement in toughness in agreement with expected increase in deformation to break with increasing molecular weight. Rose's model [44] explains the observed maximum on toughness versus filler volume fraction by considering the ratio between the ease of circumventing the particle and matrix cracking.

## 6.4 CHARACTERIZATION OF IMPACT BEHAVIOR OF POLYPROPYLENE

The presence of a supermolecular structure, namely, crystalline regions and their aggregates, substantially complicates the failure mechanisms in semicrystalline thermoplastics (e.g., polypropylene under impact loading compared to glassy polymers). As described in Chapter 2, degree of crystallinity, spherulite size and shape, lamellae thickness and size, and crystallographic structure of the basic unit are primary structural variables that affect impact behavior of polypropylene resins. Chemical structure of the chain, average molecular weight and molecular weight distribution, molecular relaxations, and related molecular properties are defined as secondary variables. This group of molecular characteristics influences toughness through their effect on crystalline structure and mechanical response of amorphous regions in the semicrystalline matrix. The occurrence of an isothermal transition under adiabatic conditions in these materials, during impact loading and crack initiation, also plays a role in determining the amount of material fracture toughness. In addition, all standard impact tests have the specimen and loading geometry as variables embodied in the measured values via control over the state of stress.

In Section 6.9, a detailed analysis of the most frequently utilized impact tests (i.e., Charpy and Izod impact tests) is used to characterize fracture toughness. Temperature, strain rate, crack tip curvature, specimen thickness, annealing, aging, irradiation, and environmental effects are discussed as test variables within the framework of fracture mechanics.

A simple spring mass method proposed by Williams [46] can be used to interpret the dynamic effects observed during the uniaxial test. However, even if one can eliminate artifacts related to dynamic effects during impact and other test-related artifacts, impact strength remains dependent on both material composition and specimen geometry. In order to separate these two sets of variables, full-fracture toughness measurement has to be performed under conditions of small-scale yielding and assuming that the major deformation mechanism does not change with the addition of secondary component. In the next section, analytical procedures are utilized to separate the effects of the test geometry from geometry-independent material properties. The mathematical tools in the lumped weight model are based on linear elastic fracture mechanics. This will provide a fundamental understanding for the significance of experimental data obtained using common impact tests.

To extend the applicability of this method from brittle to quasi-brittle materials, a concept of small-scale yielding will be applied. This approach allows incorporation of general views on both structural and test variables affecting resulting impact behavior.

## 6.5 LINEAR ELASTIC FRACTURE MECHANICS

A major impulse for great scientific effort was determining the root cause for seemingly inexplicable failures of the Liberty ships during World War II. This resulted in the creation of a new scientific discipline—fracture mechanics. The

U.S. Navy was the major research sponsor in this area during the 1940s and 1950s. This effort was built on a previous discovery by Griffith [47] that the strength of materials is a stochastic parameter depending on the distribution of defects that are always present in any material. The inherent material strength is controlled by the size of these defects, mode of loading, and material properties. Most of the initial effort was devoted to studies of metals [48–51] using the concepts of linear elastic behavior prior to brittle fracture. As a result, the term *linear elastic fracture mechanics* (LEFM) is used by material scientists in the polymer field. The use of LEFM for analysis of fracture in polymers started in the early 1960s. The development of early concepts of fracture mechanics for this young discipline is described in Williams's book [46] published in 1984.

Theoretical estimates of tensile stress necessary to break a covalent bond result in a required force of about 5 nN/bond. Assuming an ideal structure and uniform stress distribution, theoretical strength of a polyethylene in the direction of macromolecules would be about 20 GPa. Measured strengths in oriented polymers are at least one but more frequently two orders of magnitude smaller [52]. Structural imperfections in the form of voids, chain ends, and local entanglements, causing sequential rather than simultaneous and uniform loading of individual bonds, are the reasons for the observed discrepancy. One can expect considerably smaller strengths in the direction perpendicular to the chain direction because the bonding energies of van der Waals or hydrogen intermolecular interactions are substantially smaller than those of primary chemical bonds. Cooperative rupture of van der Waals bonds should require a stress of 100 MPa, and similar rupture of densely hydrogen-bonded structures, such as nylon, should require about 500 MPa [53].

### 6.5.1 Modified Griffith Criterion

To determine the relation between the material properties, defect size, and the measured strength of a test specimen, Griffith [47] used phenomenological thermodynamics to describe the energy balance in an infinite body containing a sharp crack. Assuming an ideal material with linear elastic behavior, he treated crack growth as a process of creating new surfaces. This process increases the total energy content of the body. This energy must be supplied by the work performed by the external applied forces and the elastic energy stored in the body. A crack can grow only when enough strain energy is available.

This approach was later modified by Irwin [48,49], Orowan and Weld [50], and Orowan [51] by including a strain energy release rate term $G_c$ to account for changes caused by small-scale plastic deformations at the crack tip and surfaces. The Griffith-Orowan fracture criterion can be demonstrated for a very large sheet of brittle material with a stress field existing around an opening of 2 ka shown in

Fig. 6.1. The change in total energy $U_T$ is shown to be a sum of energy components $G_d$, $U_s$, and $E_0$ that are functions of surface crack depth $a$.

The Griffith-Orowan criterion for spontaneous fracture is defined at a critical crack depth $a_c$ where the slope $dU_T/da$ changes abruptly to negative values as depicted in Fig. 6.1. The modified Griffith-Orowan criterion predicts a critical stress limit $\sigma_c$ that corresponds to a critical energy release rate $G_c$.

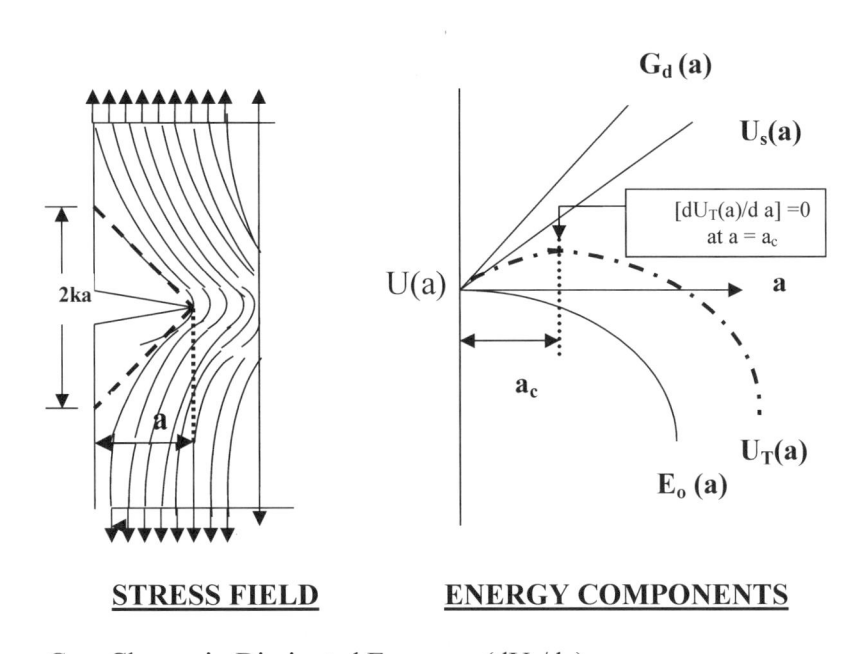

**STRESS FIELD**  **ENERGY COMPONENTS**

$G_d$ = Change in Dissipated Energy = $(dU_d/da)$
$U_s$ = Surface Free Energy $(\gamma)$ Gradient = $2\gamma a$
$E_o$ = Change in Elastic Strain Energy = $-2 (ka^2/2)(\sigma \varepsilon/2)$
   = $-[\sigma^2 ka^2]/2E$

$\varepsilon$ = strain = stress $\sigma$/ tensile modulus of elasticity E

$U_T$ = Change in Total Energy = $U_s + G_d + E_o$

$[dU_T/da]_{a=a_c} = [2\gamma + G_c - \sigma_c^2 ka_c/E] \leq 0 \Rightarrow \sigma_c = [E(2\gamma+G_c)/ka_c]^{1/2}$

FIGURE 6.1 Griffith-Orowan fracture criterion.

## 6.6 PROBLEMS ASSOCIATED WITH UNINSTRUMENTED STANDARD IMPACT TESTS

Impact fracture of polymers depends heavily on stress distribution at a notch tip and how the material responds to the mechanical stress state. Table 6.1 indicates stress states that correspond to the type of loading for different test methods. Polymer fracture under impact loading is very complicated and is affected by many testing parameters (e.g., temperature, strain rate, specimen thickness, notch length and blunting, specimen/impact geometry, and type of loading). The initial concept of fracture mechanics, implicit in the development of standardized impact tests, was to determine conditions for initiation of a brittle fracture leading to a catastrophic failure of the material. Figure 6.2 shows the specimen geometries for notched Izod, reverse notched Izod, and Charpy notched beam impact tests that are most commonly used in the plastics industry. Unfortunately, despite practical importance of Izod and reverse Izod tests, only the simple Charpy test geometry provides conditions for more or less rigorous mechanical analysis and, thus, allows interpretation of the experimental data in terms of structural and geometric variables.

The Ceast Resil 5.5 tester shown in Fig. 6.3a consists of free-swinging pendulum and Izod-type hammer assembly of known mass $m$. The impact test begins with the pendulum latched at height $h_0$. It is manually released, striking and breaking the notched sample at the bottom of the swing. The pendulum attains final height $h_f$ after impact. The absorbed energy of the specimen is assumed to be the loss of potential energy $mg(h_0 - h_f)$. The measured total energy to break is the only information obtained, since the impact hammer is not instrumented with a transducer. Without provision to measure impact force, no information about magnitude of load or time of contact is obtained. The stochastic nature of fracture phenomena and the inevitable effects of test geometry make it very difficult to find quantitative relations between the performance of a polymer in the laboratory test on simplified test specimen and performance of a final part. Hence, the fundamental defect in these types of

TABLE 6.1 Stress States Generated By Test Method

| Test Method | Stress State |
| --- | --- |
| Tensile Testing | Uniaxial |
| Tensile Impact | Uniaxial |
| Charpy Impact Notched | Triaxial |
| Izod Impact Notched | Triaxial |
| Puncture Test | Biaxial |

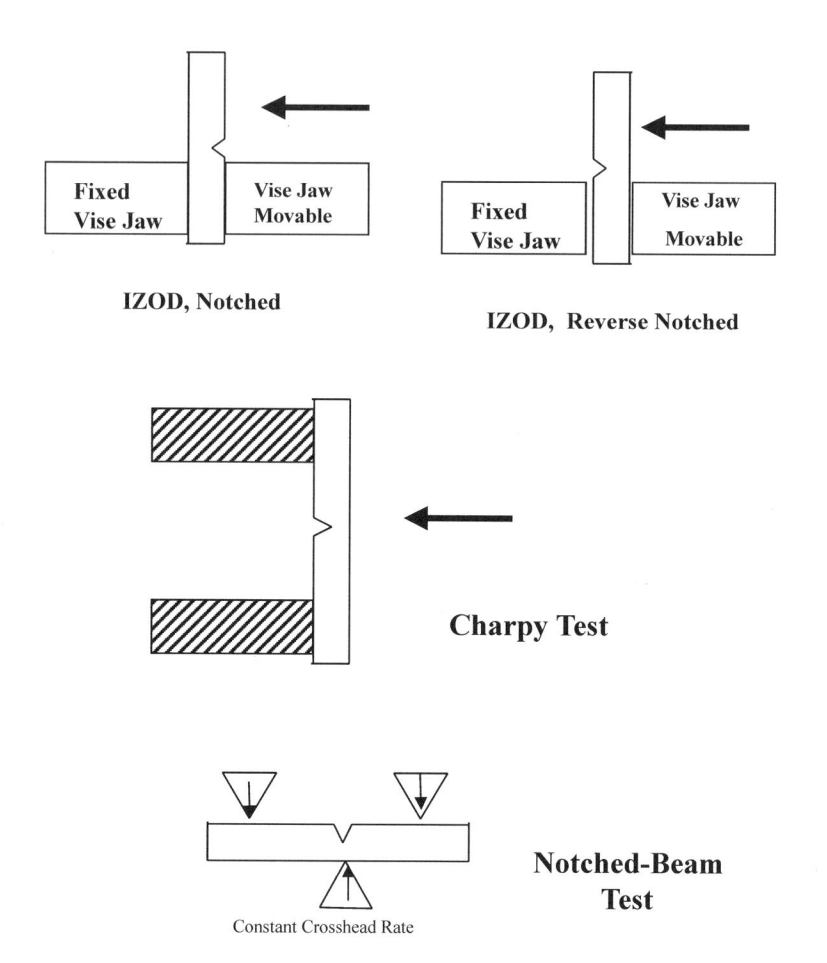

**FIGURE 6.2** Impact test geometries.

impact testing is that measured total fracture energy values only provide global information concerning impact behavior of materials.

Impact tests are intended to simulate the most severe abuse to which molded plastic parts can be subjected. In comparison, impact tests on the final parts are the ultimate tests of material performance, part design, and quality of processing. There is an apparent lack of correlation between measured impact fracture energy and end-use impact resistance due to the extreme complexity of the microscopic fracture processes. This deficiency in methodology restricts attempts to relate impact strength to material structure. In particular, the influence of specimen geometry is sometimes poorly matched with the type of failure

# Impact Behavior of Polypropylene

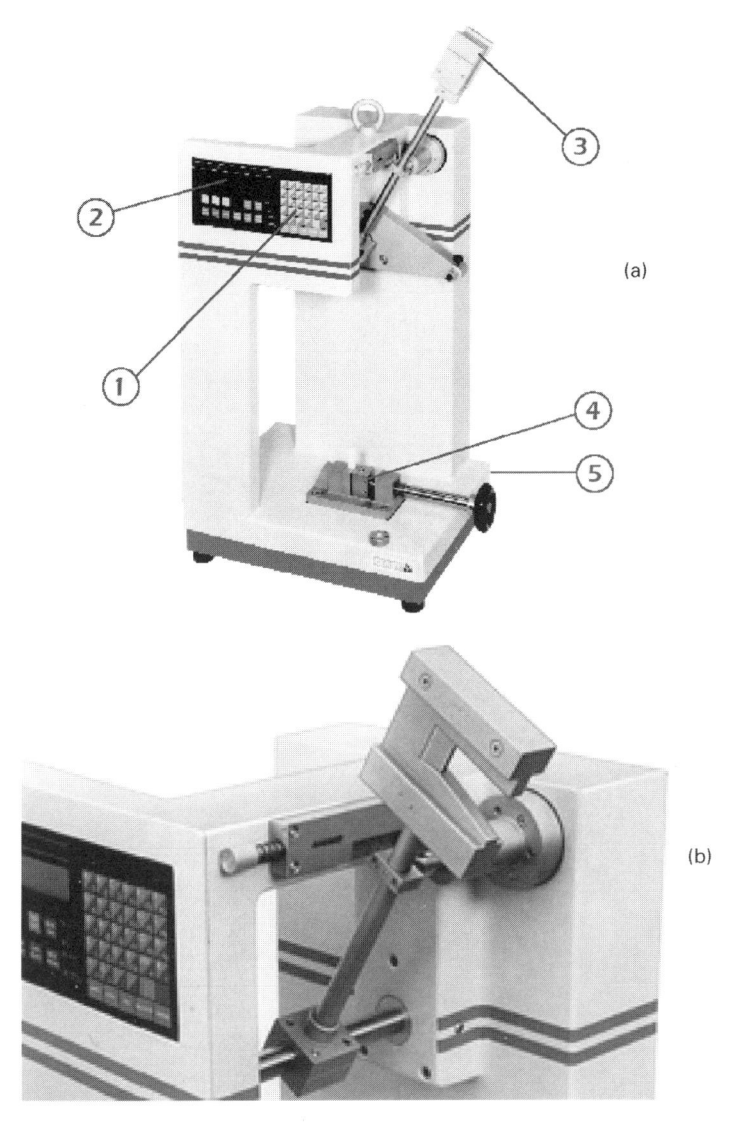

**FIGURE 6.3** Photographs of impact test equipment. 6.3a Resil 5.5 Impact tester-Izod hammer (Courtesy CEAST S.p.A.) 1. Keypad 2. LCD display 3. Izod hammer 4. Vice for Izod specimen 5. PC connection port. 6.3b Resil 5.5 impact tester-instrumented Charpy hammer (Courtesy CEAST S.p.A.). 6.3c Fractovis falling weight tester-system (Courtesy CEAST S.p.A.) 1. DAS 4000 data acquisition 6.3d Fractovis falling weight tester-instrumented tup (Courtesy CEAST S.p.A.) 1. Weights 2. Tup with transducer 3. Specimen support.

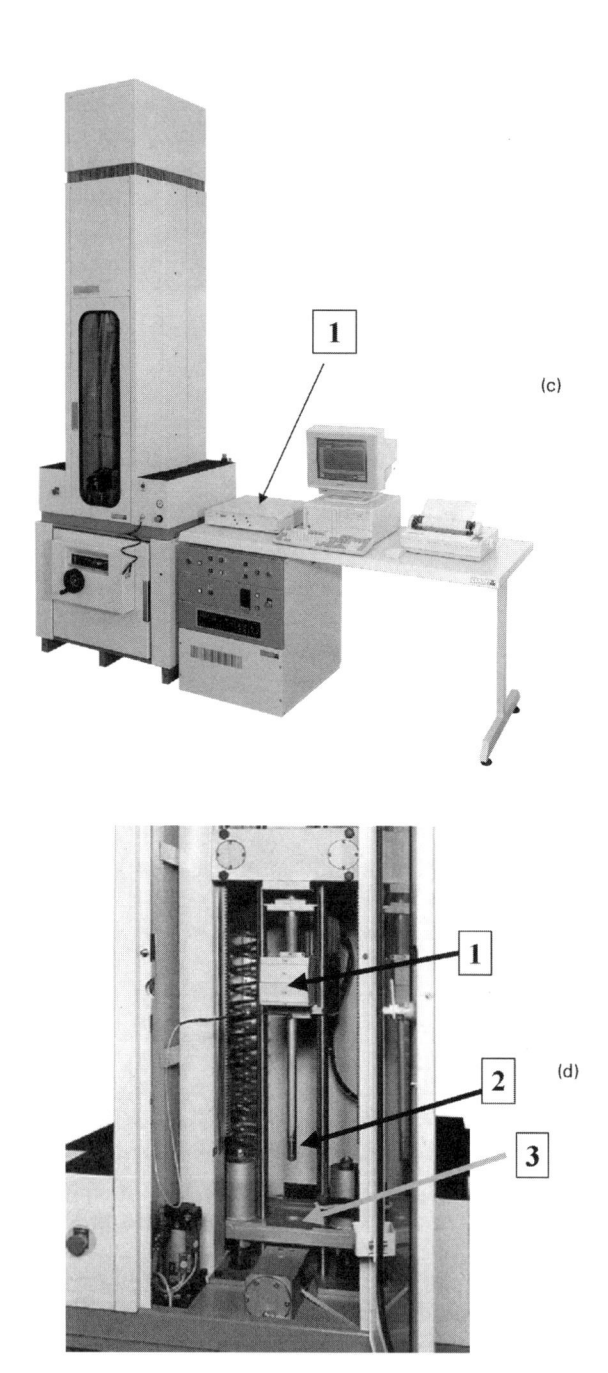

FIGURE 6.3 (continued)

mechanism of defects present in the sections of the actual molded part subjected to impact loads. "Testing the material for a car bumper is not the same as testing an actual finished car bumper" [54].

Sherman [54] wrote a review article that summarized interviews with a number of individuals representing various companies to assess traditional types of impact testing equipment that have become the standards of the industry but are now being challenged by recent trends toward "real-life testing."

Despite experimental evidence of changes in the mode of fracture with composition, only a limited number of authors have taken this into account and modified test procedures accordingly [55–57]. As a result, a large volume of published material has been generated that relates geometry independent impact strength to a wide variety of structural variables (e.g., filler volume fraction, rubber properties, particle size and size distribution, matrix-inclusion adhesion, molecular structure of the polymer). Likewise, the effects of processing cannot be properly determined for a realistic measure of material toughness in a given application. While there is an abundance of material data, there is serious doubt about the fundamental basis for making even relative comparisons between toughness values for a set of different materials.

Over the past few years, instrumented impact testing has emerged from major industrial laboratories into a broad range of research/development projects to characterize toughness as a function of material structure subjected to real-life impact load conditions. Recording and analyzing the load deformation history provides valuable information regarding the failure event. The shape of different types of force–time plots provides a fingerprint of the actual failure modes for material under impact load [54]:

1. Brittle—The specimen shatters.
2. Brittle/ductile—There is some yielding in the specimen with many propagating radial and circumferential cracks present.
3. Ductile/brittle—The plot has a well-developed yield peak with some radial cracking, but most radial cracks are intact just prior to penetration.
4. Ductile—The plot exhibits fully ductile failure with no indication of radial cracks.

It is not surprising that instrumented impact methods have become the preferred route for obtaining data for quantitative analysis of fracture processes and development of performance criteria for effective material design. Very often, however, data from instrumented impact tests are misinterpreted. One has to be very careful when analyzing force–time curves from instrumented strikers taking into account effects such as sample acceleration, impactor natural vibrations, bouncing, and so forth. These effects can greatly obscure experimental data especially for brittle materials and in tests with small specimen mass compared to the impactor mass.

## 6.7 USE OF LEFM FOR IMPACT TEST DATA REDUCTION

After proper data extraction procedures are applied, impact tests provide fracture energy $U$ for test specimen of a given geometry, i.e., width $B$, thickness $D$, span length $L$, and notch-to-thickness ratio $a/D$. The fracture energy must be corrected for the kinetic energy of the flying pieces and instrument corrections. When performing impact tests using specimens of systematically varying geometry, one can determine the fracture energy as a function of specimen geometry.

Williams [46] suggested that under simplified conditions of static loading, the measured fracture energy of a linearly elastic solid of finite dimensions is directly proportional to the ligament area (BD) and the geometry factor $\Phi$. The measured fracture energy $U$ is corrected for the kinetic energy of the flying pieces and for the instrument corrections. The proportionality constant is the apparent critical strain energy release rate, $G'_c$:

$$U = G'_c (BD) \Phi \left( \frac{a}{D} \right) \tag{6.1}$$

For class I and class II fractures of linear elastic solids, fracture energy may be considered directly proportional to $[BD\Phi(a/D)]$ with $G'_c$ as the proportionality constant. This concept is based on the assumption that the fracture is completed immediately after initiation. Since needed energy is supplied by strain energy stored in the specimen, no further energy is necessary to propagate the crack. This also means $G'_c$ is significantly greater than that required for propagation of a running crack.

The geometric factor $\Phi$ is dependent on the compliance $C$ and the ratio between the crack length "$a$" and thickness $D$ of the specimen:

$$\Phi = \frac{\dfrac{C}{dC}}{d\left(\dfrac{a}{D}\right)} \tag{6.2}$$

The compliance $C$ is the quotient of displacement/load for a cracked specimen with no crack growth. Based on Eq. (6.1), $G'_c$ can be determined from measurements of $U$ for several $a/D$ ratios as a slope of the plot $U$ versus $BD\Phi$, which should be linear.

To use LEFM in data reduction from impact tests, one should make sure that the mode of fracture does not change to a quasi-ductile one by changing test temperature, notch radius, or material composition. If large deviations from linearity exist, a plot of $U$ versus $BD\Phi$ will not be linear and another approach to determine $G'_c$ is needed. One can assume that in the case of a quasi-ductile

## Impact Behavior of Polypropylene

fracture that measured fracture energy is proportional to the ligament area $[B(D - a)]$. Plotting $U$ versus $[B(D - a)]$ will give an intersect with the vertical axis equal to $G'_c$ (critical energy at zero crack length).

It appears useful to define conditions for instability and derive the requirements necessary for completing the fracture without additional external work, i.e., at a constant displacement or, in other words, conditions for applicability of Eq. (6.1). From definition of $G$ we have:

$$G = \frac{1}{2B}\left(\frac{u}{C}\right)^2 \frac{dC}{da} \tag{6.3}$$

The crack becomes unstable when:

$$\frac{dG}{da} > 0 \tag{6.4}$$

under an assumption of constant $G'_c$ and independence of the kinetic energy on the crack length. Using the geometry calibration factor $\Phi$, one can define the instability conditions in the form:

$$-\frac{d\Phi}{d\left(\frac{a}{D}\right)} > 1 \tag{6.5}$$

From the dependence of $\Phi$ on $a/D$ for a given ratio between the span length $L$ and specimen thickness $D$ for Charpy test geometry, it can be seen that for $L/D$ equals 6, the fracture initiation is always unstable when $a/D < 0.36$. It was shown that for Charpy test with $L/D$ equal to 6, the fracture process is always unstable for $a/D < 0.13$. The fracture will initiate in a brittle, unstable manner for $0.13 < a/D < 0.36$ and after initiation it will arrest (when the crack length reaches $a/D = 0.36$). For $a/D > 0.36$ the fracture is always stable, requiring additional external work to propagate the crack after it initiates.

The LEFM concept is based on the assumption that critical strain energy release rate $G_c$ in propagation ($G_{cp}$) is the same as for initiation ($G_{ci}$). In the majority of routine impact tests, the notches are machined less than perfectly sharp. This results in blunted rather than sharp initial cracks. As a result, one can expect that $G_{cp} < G_{ci}$. After initiation, the crack loses its blunted character and becomes a natural, sharp crack. One can transform this idea into a mathematical form of an instability criterion using the ratio between propagation and initiation values of $G_c$:

$$-\frac{d\Phi}{d\left(\frac{a}{D}\right)} > \frac{G_{cp}}{G_{ci}} \tag{6.6}$$

This criterion can be interpreted for a given $L/D$ that there will be a critical ratio $G_{cp}/G_{ci}$ less than 1, below which all tests will be unstable with no arrests.

## 6.8 EXPERIMENTAL TECHNIQUES IN IMPACT TESTING

A considerable number of impact tests (Table 6.2) have been devised to characterize the impact resistance of plastics. To provide a fundamental understanding of the relationship between laboratory-measured uniaxial and biaxial impact strength and the performance of the final plastic part, fracture mechanical analysis of impact tests must be discussed in terms of the specific setup of various types of impact equipment [58].

The widespread use of standardized impact tests by the plastics industry for material specification and part design requires a careful analysis of the test methods and test data extracted from them. The first and easier task is connected with establishing conditions under which the test data reflect true fracture processes occurring during impact. Test values should be corrected for a variety of dynamic effects, such as bouncing. The second task is much more complicated and involves evaluation of the physical meaning of test data and their possible correlation to external test conditions and material structure.

Tests performed on unnotched test pieces can, to some extent, simulate behavior of in-service molded part but have no direct relation to material structural variables. A combination of some sort of stress concentrator together with high strain rates and bending loading geometry provides a severe test of material tendency to exhibit brittle fracture. The presence of natural defects, flaws, and design features (e.g., sharp edges, ribs) is simulated by impact testing using precracked or manually notched specimens. Ultimately, the tests should provide the most conservative values of fracture toughness for design purposes. In addition, these tests are frequently used for material selection and development. Rather poor reproducibility of impact tests reflects the test-dependent character of the most frequently used form of impact data to quantify impact strength.

In order to understand the true meaning of impact strength, one should include concepts of fracture mechanics to obtain well-defined measures of toughness in terms of test-independent material parameters. The development and wider application of new engineered plastics requires that true toughness values be a function of both microscopic and macroscopic variables. This desire for a fundamental meaning in toughness determination is the underlying purpose of the following subsections.

### 6.8.1 Dynamic Effects in Impact Testing

Most specimen geometries recommended by the test standards require somewhat small specimen dimensions. Hence, many dynamic effects observed in impact testing are not due to stress-induced crack propagation in the solid, so that stresses within the test piece are assumed to be static. This condition allows external behavior of the specimen and the loading system to be considered separately as some type of correction factor to a static model. Then conceptual models of fracture processes can be expressed by mathematical relationships based on the LEFM approximation.

In order to evaluate the nature and extent of dynamic effects in impact tests, Glover et al. [59] have proposed a simple elastic model for interaction between the striker and a specimen prior to fracture. They have also shown that this simple model can predict the load fluctuations seen in instrumented impact tests. Williams and Birch [60] modeled the specimen–striker interaction by an imperfectly elastic rigid body with prediction of very similar system response. The following summary of this approach is described in Williams's book [46]:

1. The impact event is described by a lumped spring-mass model.
2. Assuming mass of striker is much greater than that of specimen, there is no reduction in striker speed during impact.
3. The bouncing effect results from repeated loss of striker contact with specimen during the impact test.

Most commonly, the actual contact stiffness is considerably greater than that of the specimen, resulting in large load oscillations over the brief instant of the impact test. The problem of bouncing is more serious in brittle materials where load increases to low levels and there is only a brief onset of oscillations. Since a majority of standardized impact tests involve measurement of energy lost as due to strain energy for crack initiation and propagation, the onset of bouncing can produce large deviations of the measured energy-loss from the strain energy stored in the impacted body.

The bouncing effect in instrumented impact tests is usually detected by oscillations in the force response signal. Smoothing such a signal with all sorts of electronic and software filters can produce smooth force–time plots. On the other hand, these curves may not reflect the true time dependence of force in the impacted specimen during the impact event. This can lead to incorrect interpretations of measured fracture energy from peak force measurements as being energy transfer from the striker.

The main objective of the impact test is to determine the strain energy within a specimen, since the rate of its release controls the fracture initiation. On the other hand, the data measured by impact pendulums are the energies lost by the striker. It is clear that a reliable impact test procedure requires a direct proportionality between energy loss by the striker and strain energy dissipation.

In practical tests, the recorded energy will always be in discrete steps. This represents an inherent limitation to the ability of impact testing to discriminate between cumulative sources of measured energy other than strain energy dissipation for reliable routine impact testing. If the bouncing occurs during an impact test, force–time plots would consist of rows of points along the horizontal portions of each discrete step. These rows may also arise from varying the notch depth and radius. The use of instrumented impact strikers would allow recording

TABLE 6.2 Impact Test Methods

| Test Method | Standards | Description |
|---|---|---|
| FLEXURAL IMPACT TESTING | DIN 51222-85 | Pendulum impact testing machine. |
| | ISO CD-1993 | Pendulum impact testing machine. |
| | ISO 179-1993 | Plastics—Determination of Charpy impact strength. |
| | ISO 180-1993 | Plastics—Determination of Izod impact strength. |
| | ASTM D256-2000 | Standard test methods for impact resistance of plastics and electrical insulating materials. |
| | JIS K711 | Method of Charpy impact test for rigid plastics 1977. |
| | JIS K7110 | Method of Izod impact test for rigid plastics 1977. |
| | DIN 53453-1982 | Impact bending test—notched and unnotched. |
| | DIN 53753-1981 | Impact bending test. |
| | DIN 53435-1983 | Bending test and impact bending test—Dynstat |
| TENSILE IMPACT TESTING | ISO 8256-1990 | Plastics—Determination of tensile impact strength. |
| | ASTM D 1822-1989 | Standard test method for tensile impact energy to break plastics and electrical insulating materials. |
| | DIN 53448-1977 | Shock rupture test. |
| FALLING DART IMPACT TESTING | ISO 7765-1-1988 | Plastics film and sheeting—Determination of impact resistance by the free-falling dart method, Part 1: Staircase methods. |
| | ISO 6603-1-1985 | Plastics—Determination of multi-axial impact behavior of rigid plastics, Part 1: Falling dart method. |
| | ASTM D 3029-1990 | Standard test methods for impact resistance of flat, rigid plastic specimen by means of a tup (falling weight). |
| | ASTM D 2444-1992 | Standard test methods for determination of impact resistance of thermoplastic pipe and fittings by means of a tup. |
| | ASTM D 1709 | Standard test method for impact resistance of plastic films by the free-falling dart method. |

# Impact Behavior of Polypropylene

|  | | |
|---|---|---|
| | ASTM D 4277-1990 | Standard test method for total energy impact of plastic films by dart drop. |
| | JIS K7211-1976 | General rules for testing impact strength of rigid plastics by the falling weight method. |
| | DIN 53443-Teil 1, 1984 | Impact test. |
| | ISO 7765-2-1993 | Plastics film and sheeting—Determination of impact resistance by the free-falling dart method, Part 2: Instrumented puncture test. |
| | ISO 6603-2-1989 | Plastics—Determination of multiaxial impact behavior of rigid plastics—Part 2: Instrumented puncture test. |
| | ISO 6603-3 | New work proposal, N574 Plastics—Determination of multiaxial impact behavior of rigid plastics—Part 2: Instrumented puncture test. |
| INSTRUMENTED IMPACT FOR BOTH PENDULUM AND FALLING WEIGHT MODES | ASTM D 3763-1992 | Standard test method for high speed puncture properties of plastics using load and displacement sensor. |
| | DIN 53443-Teil 2, 1984 | Impact test. |
| | ISO 179-Part 2 | New work proposal N 573 1993 Plastics—Determination of Charpy impact strength, Part 2: Instrumented impact test. |
| | JIS B 7756-1993 | Impact testing machines for plastic materials—Instrumentation. |
| | ISO W.I. 572-1993 | A linear elastic fracture mechanics standard for determining $K_c$ and $G_c$ for plastics. |

of the entire force–time plots to permit detection of the bouncing effect, however, it cannot eliminate it.

### 6.8.2 Specimen Inertia Effects

During impact loading, a complex signal will be delivered from the specimen and all of the components involved in the impact event through the data acquisition system. This signal is a complex combination of contributions from mechanical response of the specimen, inertial loading of the striker as a result of acceleration of the specimen by the tup and various mechanical oscillations caused by the elastic energy stored in the striker, and other dynamic effects.

The inertial load may be significant for instrumented impact testing massive (large and heavy) specimens of brittle materials at high loading velocities. The inertial load is generally the load required to accelerate the specimen from zero velocity to the velocity of the impacting striker. The inertial loads usually cause sharp spike on the recorded force–time curve. The force spike is commonly followed by a decaying oscillations superimposed over the actual loading curve. For materials requiring small loads to initiate fracture, the presence of inertial force spike will result in inaccurate data collection. Both fracture energy and maximal peak load can be grossly overestimated when not accounting for inertia effects and can lead to a complete obscuring of the desired data. Hence, a correction for inertia effect is required.

One of the means to diagnose the test suspected of containing the inertia effect is to perform the test at different impact velocities, both lower and greater than the setting chosen to conduct the test. The magnitude of an inertial load is proportional to the impact velocity [61]. If a peak on the force–time curve is caused by inertia effect for a setting of lower impact velocity, a corresponding lower force spike should be observed. Mechanical response of plastics may be very sensitive to strain rate but not as sensitive as the inertia effect described above.

At least qualitatively, the inertia load can be determined by using a sticky tape to join the broken specimen together and retest it. The fracture energy will be negligible; however, the inertia load should not change much in comparison with the test on unbroken specimen.

### 6.8.3 Thermal Blunting under Impact Loading

Another effect that can contribute to the fracture toughness, determined under impact loading, is thermal blunting. This effect occurs as a result of rapid loading that results in localized plastic deformation in notched specimens. This process can be considered adiabatic in terms of phenomenological thermodynamics. Thermal blunting is manifested by an increase of material temperature at the crack tip. Due to poor thermal conductivity of polymers, localized melting or at

## Impact Behavior of Polypropylene

least substantial softening may be caused by this temperature increase. The adiabatic temperature increase, $\Delta T$, can be estimated from the critical stain energy release rate $G_b$ for a blunted crack:

$$\Delta T = \frac{G_b}{\pi \rho c t} \tag{6.7}$$

where $\rho$ is the density, $c$ is the specific heat, and $k$ represents thermal conductivity of the material. The loading rate is controlled by the loading time $t$.

Williams and Hodgkinson [62] derived an empirical expression for a parameter $N$ as a function of yield strength $\sigma_y$, yield strain $\varepsilon_y$ and fracture stress $\sigma_c$:

$$N = \left(\frac{\varepsilon_y}{2}\right)^{1/2} \frac{\sigma_{tc}}{\sigma_y} \tag{6.8}$$

Computed values of $N$ determine whether blunting occurs by thermal blunting mechanism or by self-blunting due to plastic flow caused by shear. For $N \geq 0.77$, self-blunting due to plastic flow occurs; for $N < 0.71$ crack tip blunting by thermal heating at the crack tip is the main blunting mechanism.

Critical strain energy release rate of a blunted crack, $G_b$, can be calculated using the following expression:

$$G_b = \frac{\rho \sigma_y^2}{\pi E} + \frac{G_{1c}}{2} \tag{6.9}$$

where $G_{1c}$ is the plain strain for the crack tip region.

Localized melting of the polymer at the crack tip leads to effective blunting of the crack tip and sharp increase in the fracture toughness expressed in terms of $G_c$ [62]. Experimentally, increase in strain rate was achieved by increasing the $a/D$ ratio or by increasing impact velocity using servohydraulic, pneumatic, or charge-fired strikers.

At high $a/D$, the loading time is greatly reduced causing heat generated at the running crack front to be contained within a small volume of polymer. This, along with poor thermal conductivity of polymer, results in an adiabatic heating of the material causing effective blunting of the crack tip. Therefore, additional energy is required to break the specimen. Specimens with shorter cracks have a longer loading time associated with their fracture. Any heat generated during fracture is dissipated isothermally. The effective blunting was expressed in terms of two parameters $N$ and $t_1$:

$$\left(\frac{G_b}{G_c} - \frac{1}{2}\right)^{1/2} = N^{-1}\left[1 - \left(\frac{t_1}{t}\right)^{1/2}\right] \tag{6.10}$$

using $N$ value computed according to Eq. (6.8). The parameter $t_1$ is the time scale for the rapid increase in toughness. The parameter $t_1$ can be estimated by using the following expression proposed by Williams and Hodgkinson [62]:

$$t_1 = \frac{G_c^2}{(T_s - T_0)} \pi \rho c k \tag{6.11}$$

where $T_s$ is the softening temperature and $T_0$ is the test temperature. The material parameters $\rho$, $c$, and $k$ are density, specific heat, and thermal conductivity, respectively.

### 6.8.4 Effect of Notch Radius and Blunt Cracks

Derivations of crack tip stress fields and other fracture mechanics terms are based on an assumption of ideally sharp natural cracks. However, it is not a common practice to prepare these sharp cracks in routine impact testing because of the time-consuming procedures that have to be employed. Hence, standardized tests are carried out with cracks or notches of finite radius. Commonly, standardized tests prescribe notch tip radius of 0.25 mm. This type of manually cut notch or crack can be viewed as a good approximation of the defects present in final products or damage caused during the service life.

Unlike sharp cracks, notches with a finite tip radii do not have a stress singularity at the very tip of the notch since the singularity occurs as a consequence of crack tip radius $\rho \to 0$. By utilizing the mathematical concepts based on LEFM approximation, a model was proposed to use critical stress acting at a critical distance $l_0$ from the crack tip to deduce an equivalent fracture parameter for these so-called blunt cracks. From Williams's book [46], the final equation relating to the ratio of a critical strain energy release rate for the blunted crack, $G_b$, and the sharp crack, $G_c$, to the crack tip radius $r$ and the critical distance $l_0$:

$$\frac{G_b}{G_c} = \frac{\left(1 + \frac{r}{2l_0}\right)^3}{\left(1 + \frac{r}{l_0}\right)^2} \tag{6.12}$$

Typically, $l_0$ ranges from 10 to 100 μm; whereas $r = 250$ μm for the Charpy notched test, i.e., $r \gg l_0$. Expansion of Eq. (6.12) with elimination of insignificant $l_0^2$ and $l_0^3$ terms and cancellation of like terms gives:

$$\frac{G_b}{G_c} = \frac{(4l_0 r + r^2)(2l_0 + r)}{(2l_0 r + r^2)(2^3 l_0)} = \frac{4l_0 + r}{8l_0} \tag{6.13}$$

Finally, a much simpler linear expression results:

$$\frac{G_b}{G_c} = 0.5 + \frac{r}{8l_0} \tag{6.14}$$

Plotting the measured $G_b$ versus $r$, one should obtain a straight line with a slope of $(G_c/8l_0)$ and an intercept of $G_c/2$. Blunt notches tend to induce ductile failure because of the higher $G_b$ values.

Equation (6.14) can be used to estimate the effect of crack blunting for different types of plastics. Polymers having low $l_0$ exhibit rapid increase in $G_b$ for increased notch radius. Since materials with low $\sigma_y$ tend to melt on the surface of machined notch, generation of blunt cracks is more likely to occur with an apparent increase in fracture toughness. Also, blunt cracks can substantially alter the temperature dependence of $G_b$ and lead to the onset of an apparent DBT in materials like PC and PMMA. This effect is of particular interest for PMMA, which exhibits a transition peak in the $G_b$ versus $T$ plot at $-60°C$. This peak rises significantly when increasing the crack tip radius.

### 6.8.5 Loss Peak Effects

The effects of cooperative viscoelastic segmental motions (released within a narrow temperature interval for fracture toughness of polymers) can sometimes be observed as an increase in fracture toughness at temperatures corresponding to the maxima of the loss tangent, $\tan\delta$ [46]. Several authors have used these experimental observations to explain how molecular relaxations can contribute to fracture resistance of a polymer. However, the phenomena described in the previous paragraphs should make the reader aware of the conceptual limits of such a hypothesis. One of the main objections to these attempts is based on the stochastic nature of failure. In other words, brittle fracture occurs in the weakest point of the fracturing solid and, thus, fracture is a highly localized phenomenon. This means that only a very limited number of molecules are involved in failure process. On the contrary, each molecule of the polymeric solid contributes to the viscoelastic response such as the loss tangent. Therefore, it is difficult to believe that there might be a correlation between these two phenomena.

One possible explanation for experimentally observed "correlation" between $\tan\delta$ and fracture toughness is based on the molecular mechanism of localized yielding involving cooperative rearrangements of a large number of segments. These rearrangements can be easier at temperatures where segmental mobility greatly increases, i.e., at maxima in $\tan\delta(T)$ versus $T$ plot. Blunting of the sharp crack tip can result, leading, in accordance with Eq. (6.14), an increase of measured fracture toughness.

One should expect that the critical length $l_0$ will be constant over a wide temperature range with an apparent drop at elevated temperatures with ductile failures. An increase in $G_b$ is not due solely to the $l_0$ effect. For example, PVC, PMMA, and PC exhibit drops in slope for $l_0$ versus $T$ curves at temperatures corresponding to β transitions on tan δ curves. In other cases, such as polytetrafluoroethylene (PTFE), peaks in $G_c$ measured for sharp cracks are observed at the positions of secondary relaxation transitions.

Initially, it was proposed that these peaks in fracture toughness originate directly from increased energy absorption by the released molecular or segmental motions. However, this explanation is not satisfactory for the following reasons. It doesn't seem likely that a substantial increase in critical strain energy release rate can be caused by a release of some sort of segmental motion if only a very small, localized volume of polymer is involved in fracturing. The apparent increase in fracture toughness is now believed to be caused by promotion of additional deformation processes, e.g., multiple crazing and microvoiding. These processes are stabilized by the presence of viscoelastic material, even if they cause a brittle fracture under less favorable conditions. Supporting evidence for this hypothesis is provided by the data obtained for PTFE. Due to the molecular structure of PTFE, the fracture process involves diffuse energy absorption (mode IV fracture), even with sharp notches. Other glassy polymers have their dissipative processes fairly localized in the small plastic zones when sharp notches are used. Only polymers with blunt cracks exhibit wide-scale energy absorption to allow for a tan δ effect to become apparent. An increase in fracture toughness is then a consequence of additional deformation processes being available.

### 6.8.6 Standardized Charpy Impact Tests

The geometry of the standard Charpy impact test is based on a three-point bending of a rectangular beam with freely supported ends and the loading in the midway between the supports (Fig. 6.2). The striker to exert impact force can be a free-swing pendulum, falling-weight, or servohydraulic driven tup.

The relevant standards applied to this test are ASTM D256, ISO R-180, and DIN 53453. However, these standards are not sufficiently meaningful because they restrict the geometry of test specimens, especially the notch radius (except for ASTM D256, Method D) and notch length to specimen width ratio. Moreover, these methods do not provide any guidance for fabrication of test specimens, especially in the case of injection molding. The shortcomings of standardized impact tests were recognized long ago. However, relatively little change has occurred in methodology [63]. The square-ended notch prescribed by the German DIN 53453 standard is ill defined and unsatisfactory. The significant role of specimen width or, in other words, the state of stress at the notch tip is only partially recognized. The lack of a clear definition of the reference state is one of the major shortcomings of all standards on Charpy impact testing of plastics. A well-defined reference state is crucial, especially in the case when the impact strength is used to compare materials of different composition. The current practice in using impact data is much like using melt indexers to obtain melt flow rate as a quality control tool.

The excess energy pendulum impact test is an indication of the energy to break standard specimens using a standard pendulum under stipulated conditions of specimen mounting, notching, and pendulum velocity at impact. According to the ASTM standard, the energy lost by the pendulum is the sum of energies required to (1) initiate fracture, (2) propagate crack, (3) toss the free ends of the specimen, (4) bend the specimen to the point of fracture, (5) produce vibrations in the device, (6) overcome friction in the pendulum bearings, (7) indent specimen in the point of impact, and (8) overcome the friction caused by the rubbing of the striker over the face of the bent specimen. Terms (1) and (3) are the most significant for relatively brittle materials, with (3) being a large portion of the overall energy for brittle and dense materials. The suggested correction for kinetic energy of the flying pieces is only approximate because it does not account for the rotational and rectilinear velocity satisfactorily. For ductile materials, term (2) may be large compared to (1), and factors (5) and (8) can become quite significant.

The other questionable area in the standards is the definition of a break. ASTM recognizes four types of break: complete break, hinge break, partial break, and nonbreak. To properly use data in a given set of standard results, breaks other than complete break should not be included in reports. Measured impact strengths cannot be directly compared for any two materials that undergo different types of failure as defined above. According to the ASTM standard, the reported average must be derived from specimens contained within a single failure category. In addition, if two failure modes are observed, the average must be done for each mode separately.

Classical tests such as Charpy, Izod, or falling-drop methods give only a single parameter for characterization of the material impact behavior. This single number can be energy absorbed by a specimen in the case of pendulum impact modes or energy (height × mass) required to crack or break specimens. For the mean failure energy obtained by Bruceton staircase procedure, evaluation of the results must be carried out manually.

Charpy impact testing can be used to determine critical strain energy release rates $G'_c$ and to estimate the values of fracture toughness in states of plane strain and plane stress, $G_{1c}$ and $G_{2c}$, respectively [46]. Hence, experimental data can be obtained from standardized Charpy notched impact strength (CNIS) as to their relevance for material selection and design. In Section 6.9, CNIS data are compared with measurements of strain energy release rates $G'_c$ obtained under the well-defined criteria of linear elastic fracture mechanics of materials under the conditions of small-scale yielding. A variety of polypropylenes as well as their blends with elastomers, rigid fillers, or both are included in this study.

*Uniaxial Loading.* To better simulate "real life" testing [54], there is increased interest in understanding the factors affecting fracture failure of molded parts exposed to a sudden blow by an emphasis on evaluating crack resistance at strain rates well above those used in standardized tensile tests. Impact loading, with striker speeds of the order of 1–5 m/sec, can successfully simulate the most severe abuse to which material can be exposed in an end-use application. Numerous tests to evaluate fracture resistance in plastics have been devised over the years. However, since the simple standardized tests do not measure test-independent fracture toughness but rather complex geometry–dependent fracture energy, the results do not often correlate well with the service performance. The most common impact tests used, such as the Charpy and Izod impact tests, utilize bending of a test bar resulting in load being applied *uniaxially*. The more complicated stress distribution at the point of impact and near the specimen supports is generally neglected. In order to control the position of crack initiation, notch or starter crack, representing the largest defect in the solid, is introduced. In this chapter, only notched impact tests are discussed, since the unnotched tests cannot be reasonably interpreted because the position and size of the largest defect are unknown. This can result in nonsymmetrical loading and in tremendous effects of specimen preparation procedure on the measured fracture energy. The standard Charpy and Izod tests (ASTM D256, ISO R-180, ISO R-179, DIN 53453) are commonly performed with little awareness that nothing more than simple quality control can be expected. The ranking of materials according to impact strength is possible only under very restricted conditions, when the impact strength represents true fracture toughness of the material and materials of different composition are compared at equivalent states of stress [64,65].

### 6.8.7 Standardized Izod Impact Tests

The Izod impact test geometry consists of a cantilever beam mounted vertically in a vise up to the middle of the notch (Fig. 6.2). Figure 6.3a shows a Ceast Resil 5.5 pendulum impact tester equipped with an Izod hammer. A special jig should be used to position the test specimen in the vise. The top plane of the vise shall bisect the angle of the notch. The top edge of the vise-fixed jaw shall have a radius of 0.25 mm. Variations in clamping pressure can alter the results substantially. Thus it is necessary to standardize this pressure because it affects the specimen at the most critical place, i.e., notch tip and/or crack plane. It should be noted that standardization of an external clamping force does not ensure equal stresses inside the solid body.

Although a large portion of the mass must be concentrated in the head, the arm should be sufficiently rigid to minimize vibrational energy losses. The suggested radius of the striker nose is 0.79 mm. The vertical height of fall of the striker is 610 mm, which produces the velocity at the point of impact of about

# Impact Behavior of Polypropylene

3.46 m/sec. The point of impact is on the side of the machined notch at a distance of 22 mm above the top surface of the vise. It is suggested that the working range of the hammer should not exceed 85% of the total energy it can deliver. However, in order to use any reasonable mathematical analysis of the test by assuming constant striker velocity during the impact event, the energy interval should be substantially reduced. The processes contributing to loss of the striker's energy during impact are the same as discussed in the previous section on the Charpy impact test.

According to ASTM D256, the molded specimens should have a width between 3.17 and 12.7 mm. In ASTM E399-82, which is used to determine static fracture toughness, the specimen thickness is related to the material properties based on the results of LEFM analysis, which implicitly contains a requirement of a defined reference state of stress. In the case of ASTM E399, the reference state is a plane strain state of stress, allowing for a meaningful comparison of relative fracture toughness of materials of different compositions or for a use of the data for design purposes. Arbitrary choice of specimen width within the recommended interval can cause large data scatter and change of failure mode and, ultimately, lead to results that are unique with little physical relevance for materials comparison or for design purposes. Based on the analysis of Jancar and DiBenedetto [65,66], one should use LEFM to determine appropriate specimen thickness.

Another serious problem area is a machining of the notch. There are several parameters that can affect the quality and reproducibility of the notching procedure. Among the most important are (1) cutter and feed speed, (2) choice of a cutter, and (3) cooling. Kreiter and Knoedel [67] have recently investigated the effect of the machining technique on the standardized impact strength for PP.

### 6.8.8 Instrumented Impact Tests

Numerous empirical tests, discussed in previous sections, have been devised to measure energy absorption in plastics during impact. Their greatest virtue is ease and quickness in assessing impact strength [68]. However, these uninstrumented impact tests are not intrinsic measures of toughness and lack any quantitative correlation between toughness and structure–performance parameters. Variation in test geometry alters relative contributions from energy-absorbing processes. These deficiencies in standardized impact tests can be overcome by use of instrumented impact testers instead. Instrumented pendulums and falling-weight testers can be used in any of the test geometries [69–76]. In particular, the geometry of the Charpy test allows for the most comprehensive description of the fracture processes during an impact event. A major step to fully characterizing an impact event is introduction of transducers in the impact strikers or hammers for pendulum testers and tups for falling-weight impact instruments. Instrumented impact testing involves a variety of different types of impact and test specimen geometries. Figure 6.3b shows the instrumented hammer on the pendulum impact tester. A transducer with an implanted strain gauge to either a swinging-pendulum hammer or falling-weight tup is used to measure force on impact contact with the test specimen. The basic transducer instrumentation is essentially the same for each type of test. The main response signal detected by instrumented equipment is in the form of force–time curves.

*Theoretical Considerations of Falling-Weight Testing.* Analysis of the force–time or force–deformation curves is based on Newton's second law relationship connecting the impulse to the change of momentum, and the first principle of thermodynamics of energy conservation.

The basic calculation of velocity, deformation, and energy is derived from the following equations. A servohydraulic drive used for the Ceast Fractovis falling-weight tester (Fig. 6.3c and 6.3d) increases the initial velocity $v_0$ to speeds up to 20 m/sec. Therefore, the following relationships need only be slightly modified for the influence of gravity:

$$F(t) = -a(t) \cdot m \tag{6.15}$$

$$a(t) = -\frac{F(t)}{m} = \frac{dv}{dt} = \frac{d^2x}{dt^2} \tag{6.16}$$

$$-\int_0^t \frac{F(t)}{m} = \int_0^t a(t) \cdot dt = \int_0^t \frac{dv}{dt} dt = v(t) - v_0 \tag{6.17}$$

$$v(t) = v_0 - \frac{1}{m}\int_0^t F(t)dt \tag{6.18}$$

$$E(t) = \int F(x)dx = E_0 + \int_0^t F(t')v(t')dt' \tag{6.19}$$

where $a$ = deceleration, in m/sec
$F$ = total force on the instrumented hammer or tup, in N
$v_0$ = impact velocity just before impact, in m/sec
$E_0$ = energy of the tup just before the impact, in J
$m$ = mass of the pendulum-hammer or tup just before impact, in kg
$x$ = deformation (vertical displacement of tup), in m

Figure 6.4 provides a family of curves of $v_0$ versus potential energy plots for given sets of dropping masses ranging from 1.8 to 50 kg. The upper limit of velocity is 20 m/sec. At about 50 kg falling weight, the attainable initial energy of impact is 1300 J.

Displacement, when not a derived parameter, is determined by noncontacting measurement of the displacement of the tup relative to the anvil by using optical, inductive, capacitive, or other methods. The signal transfer characteristics

# Impact Behavior of Polypropylene

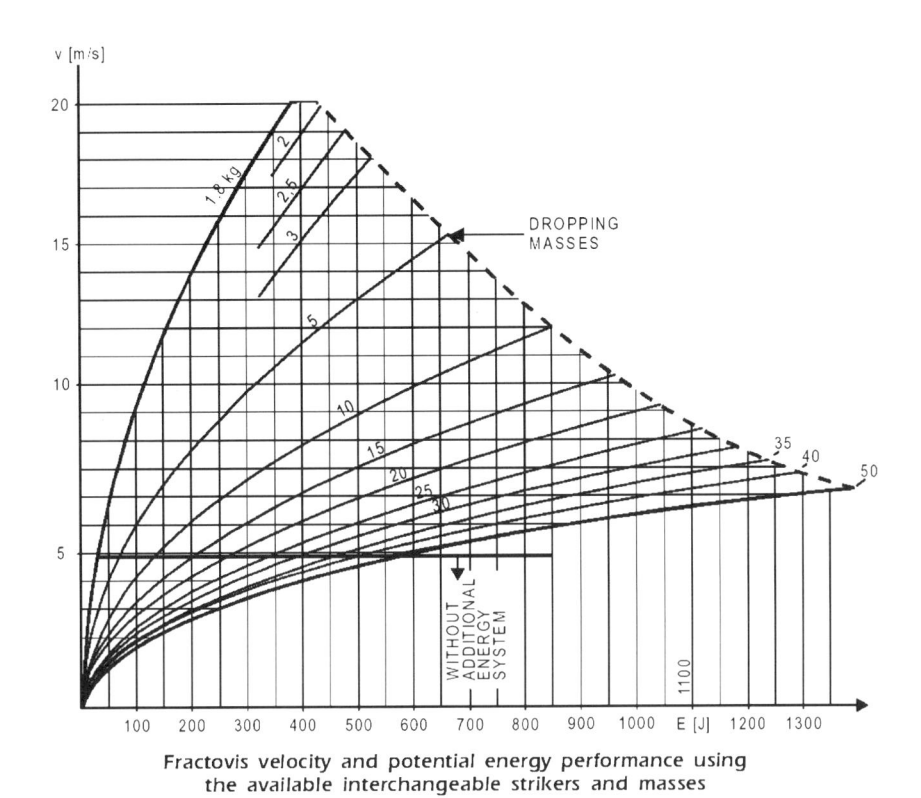

Fractovis velocity and potential energy performance using
the available interchangeable strikers and masses

**FIGURE 6.4** Fractovis velocity versus potential energy plots-dropping mass (Courtesy CEAST S.p.A.).

of the displacement measurement system must correspond to that of the force measuring system in order to make the two recordings comparable. For displacement measurement less than 1 mm, it is recommended to determine the displacement only from force–time measurement and hammer or tup impact velocity. In conclusion, by instrumented impact testing with only a force transducer, curves $F(t)$, $v(t)$, $E(t)$, $l(t)$, and $F(x)$, $v(x)$, $E(x)$ can be obtained. The parameters characterizing an impact event as curves, peak, and total values are reliable only when the instrument setup is chosen correctly. The main problems associated with instrumented impact testing are:

Transducer type and position
Signal reliability
Frequency response

A transducer can be located at the tup near the load application point, on the specimen in the crack area, or on the anvil or clamps. Each position has some advantages and disadvantages. For a variety of reasons, the preferred transducer position is the tup location.

Strain gauges or piezoelectric quartz transducers are commonly used. In regard to frequency response, the equipment records all possible phenomena during an impact event. Therefore, the natural frequency of the striker (hammer or tup) must be high. Piezoelectric transducers have natural frequency for about twice that of strain gauges but built into the striker this frequency decreases significantly (e.g., titanium striker tup of mass 30 g inserted in front of the transducer reduces the frequency from 50 kHz to 12 kHz). Strain gauges that are very small can be placed very near to the striking edge and, if properly designed and backed, can reach frequencies up to 25 kHz.

Since all parameters characterizing an impact event are based on the force measurement, it is critical to receive reliable signals from the transducer. Maximal care must be taken to the load cell calibration. The instrumented striker is a dynamic load cell, and therefore the most applicable calibration procedures should be dynamic loading techniques. Dynamic calibration can be done with the low-blow elastic impact test by striking the tup with a known elastic impulse or by equating a secondary determination of specimen fracture energy to the area under force–deformation record. The latter is the most commonly employed technique for Charpy impact instruments. Another possibility is to compare measured displacement during the impact by testing a special sandwich specimen of precisely known distance between impacted layers. A static calibration is generally in very good agreement with the dynamic case. Care should be taken to ensure that the loading geometry is exactly the same as that for impact test. When a specimen is loaded in impact, the force sensors of the instrument give a signal that is a complex combination of the following components:

- True mechanical response of the specimen
- Inertial loading of the tup as a result of acceleration of the specimen from the tup
- Low-frequency fluctuation caused by stored elastic energy and reflected stress waves

Instrumented impact testing can be carried out on each type of Ceast impact apparatus, including both swinging-pendulum and falling-weight equipment. The basic instrumentation is the same; the data acquisition system can be DAS 2000 or DAS 4000, according to the instrument.

Figures 6.3c and 6.3d illustrate the major components of the Fractovis instrumented falling-weight impact tester. Figure 6.5 provides typical data output for an impact test. The cursor can be moved to any position on the CRT screen to obtain the corresponding values of loading force, integrated energy, striker

## Impact Behavior of Polypropylene

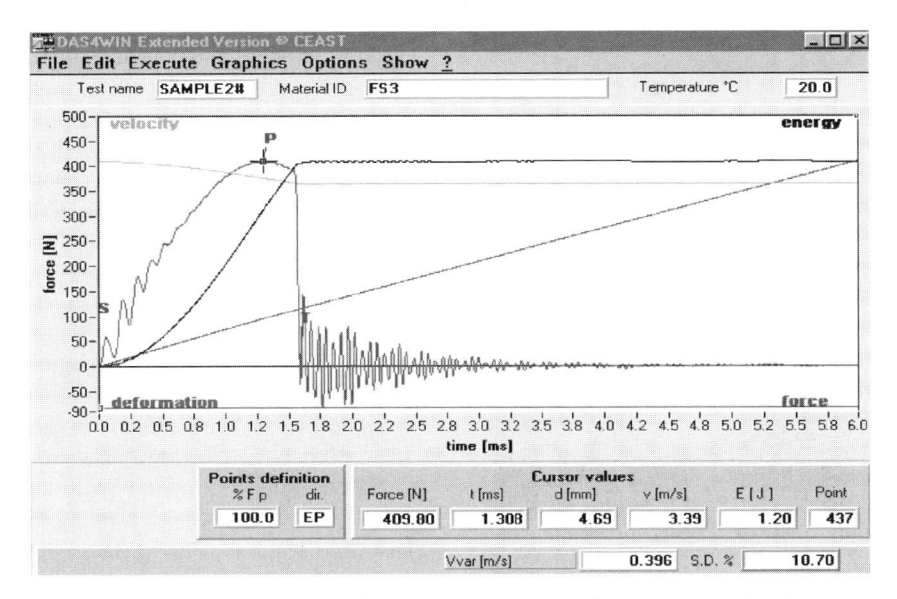

**FIGURE 6.5** DAS 4000 WIN-CR display measured data analysis (Courtesy CEAST S.p.A.).

velocity, and deformation plots. The given point at maximal load gives force load of 409.8 N after 1.308 msec of contact with the integrated energy of initiation at 1.2 J for the displacement of 4.69 mm.

### 6.8.9 Fundamentals of Ductile–Brittle Transition

All commercially significant polymers exhibit, in a more or less pronounced manner, abrupt changes in fracture behavior from ductile to brittle or vice versa over relatively narrow intervals of test or structural variables. Most frequently, this phenomenon is encountered in measuring temperature dependence of a fracture parameter such as $K'_c$, $G'_c$ or $J'_c$ or impact strength. This usually large and sudden change in failure mode is often termed ductile–brittle transition (DBT). DBT can occur when varying temperature, strain rate, notch tip radius, and specimen width. Similar effects have been observed for polymers with varying structural parameters (e.g., elastomer type, filler content, rubber domain size/shape, interparticle distance between rubber domains, and interfacial adhesion).

Since many plastic parts (e.g., automotive, appliance, sporting goods) are designed to perform over a wide temperature range, any onset of brittle behavior may negatively affect the service performance. Consequently, substantial effort is devoted to the investigation of DBT with emphasis on interpretation of this phenomenon in terms of structural and test variables. In spite of this effort, there is no consistent fundamental model on which to base interpretation of empirical data regarding this transition.

*Effect of Material Properties.* It is of paramount importance for the plastics engineer to select materials and design parts that do not undergo DBT or that can endure any onset of brittle behavior within the required temperature range. Due to the complex nature of fracture processes, use of material property as a variable is more appropriate than use of any specific test or structural variable. For example, fracture toughness is indirectly proportional to the square of yield strength. Hence, as long as fracture is satisfied by LEFM requirements, dependence of fracture toughness on yield strength will remain the same for a given polymeric material, regardless of the way in which yield strength was varied (temperature, strain rate, elastomer volume fraction).

The ultimate goal is to separate the effects of the state of stress at the crack tip from the true effects of structural variables on DBT. In other words, we want to determine when the DBT manifests only a mere change in the state of stress and under what conditions. True change in material properties underlies this transition. Since impact strength determination does not permit separation of effects of geometry from those of materials properties, a simplified approach involving determination of $G_c$ and $K_c$ cannot be utilized.

A single value of impact strength can only be used to compare relative toughness of materials tested at different temperatures under highly restricted conditions described by fracture mode types I–IV (Fig. 6.6). Transition from mode I or II to mode III fracture (plastic hinging) can occur within the range of temperatures investigated, causing a substantial increase in measured fracture energies. However, data obtained from mode III fractures are not comparable with those obtained in modes I and II. Moreover, application of LEFM method to determine toughness parameters for mode III fracture is very limited, so that extreme care must be taken in interpreting analytical results.

To understand and analyze processes relating to DBT, algorithms based on fracture mechanics should be employed. Extensive surveys by the author and published data on various aspects of DBT suggest that mode I and mode II fracture classes should be included in such an analysis. Most often, however, authors of published papers on DBT have observed the occurrence of mode III fracture failure in large plastic zones comparable in size with the specimen dimensions to develop prior to fracture initiation.

Mode III fracture with plastic hinging initiates and propagates into a fully plastic deformation of the polymeric material. This condition hinders any attempt to characterize mechanical response in the elastic region of the material. Obviously, any process leading to such an extension of the plastic zone prior to fracture initiation will eventually lead to DBT. The plastic zone size ($R_p$) for

# Impact Behavior of Polypropylene

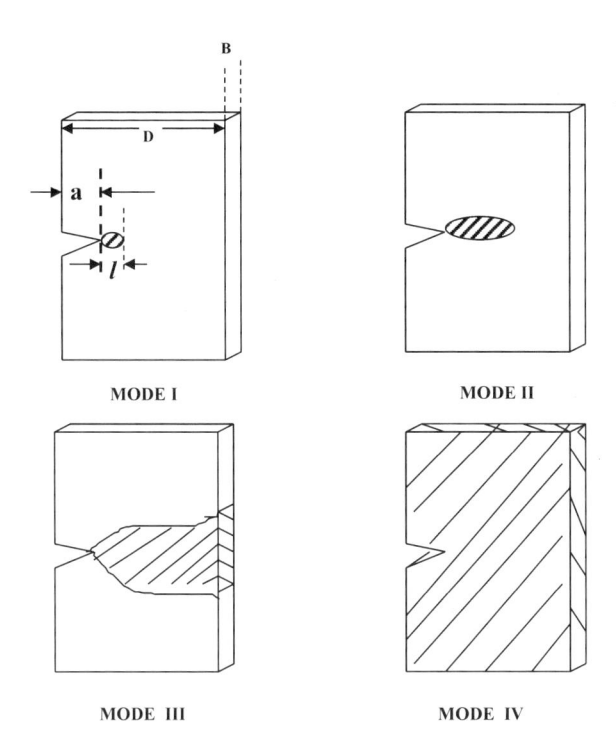

**FIGURE 6.6** Class types of fracture modes [46]: Mode I: $l \ll a, D, B$ for elastic fracture mechanics. Mode II; $l < D-a$ for contained yielding. Mode III: $l > D-a$ for fully yielded. Mode IV: Diffuse dissipation.

crack tip growth depends on $v_y$, $E$, $\sigma_y^{-2}$, and $G_{2c}$. At temperatures well below polymer $T_g$, there is only very weak temperature dependence of $E$ and $v$. The thermal behavior of critical strain energy release rate exhibits only gradual increase of $G_{2c}$ with temperature. Therefore, the temperature dependence of $\sigma_y$ is assumed to be the main source for propagation of the crack tip size $R_p$ in the plastic zone with increase in temperature $T$.

Another way to look at the DBT phenomenologically is based on the assumption of greatly different temperature dependence of the brittle strength and tensile yield strength [77]. Generally, yield strength exhibits much stronger temperature dependence than brittle strength, which is almost temperature independent. For a given strain rate, there will always be a temperature at which the two curves of brittle strength and yield strength cross. For temperatures below this $T_{DBT}$, brittle strength occurs at lower external loading compared with yield strength. Hence, the solid fails in a brittle manner prior to development of

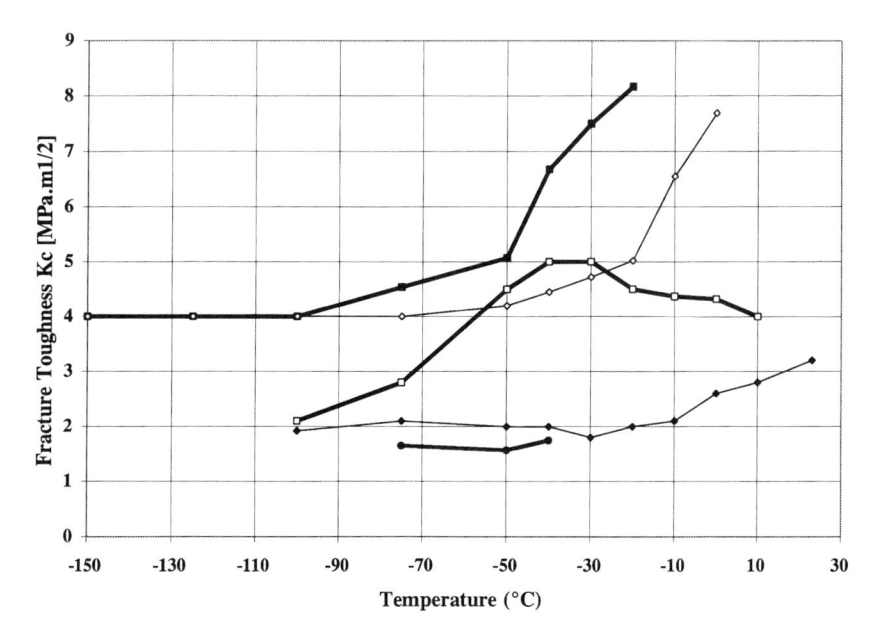

**FIGURE 6.7** Temperature dependence of fracture toughness for various polymers. □ ABS bending. ■ PP/EPR thickness 10 mm. ◇ PP/EPR thickness 50 mm. ◆ PP/CaCO$_3$. ● Epoxy/rubber.

any substantial plastic deformation. At temperatures above $T_{DBT}$, macroscopic yielding occurs at stress lower than brittle strength and, thus, large-scale plastic deformation results instead of brittle failure.

*Temperature Dependence of $G'_c$.* Usually temperature and thickness effects are treated together because both are assumed to arise from variations in the yield strength. Here they are considered separately to provide deeper insight into the problem. The temperature dependence of fracture toughness parameters ($G_c$ and $K_c$) for several polymers have been determined over a range of temperatures. In Fig. 6.7, all curves are sigmoidal with substantial change in fracture toughness over a narrow temperature interval. This threshold is often referred to as a ductile–brittle transition (DBT). The temperature at which DBT occurs suggests that structure-dependent processes are responsible for this primarily mechanical transition [78].

Several theories exist correlating the DBT obtained from impact tests to structural variables expressed in terms of thermodynamic parameters such as $T_g$ of the polymer [79] using impact strength as a measure of toughness. It is doubtful, however, that any viscoelastic process can play such an important role

## Impact Behavior of Polypropylene

because fracture is a highly localized process and the contributions to $T_g$ come from the whole volume of the test specimen. In comparison, the contribution to fracture energy comes only from a small volume represented by the crack tip plastic zone. In addition, the observed sharp change in fracture toughness can arise from a change in the failure mode.

The above-mentioned thermodynamic approach can probably be used in the case of mode IV diffuse fracture, when the whole specimen contributes to the energy dissipation during fracture. More realistic interpretation of DBT, based on an idea of mechanical nature of this transition, was proposed for structures of PC, PMMA, PE, rubber-toughened PP, and nylon [14].

The concept of mixed mode of fracture has been used to analyze experimental data to make an estimate of temperature dependence of $G'_c$ [46]. In this case, the most important material parameter is the yield strength ($\sigma_y$) that controls the size of the crack tip plastic zone ($R_p$), which is proportional to $\sigma_y^{-2}$. The temperature range considered here covers the range where LEFM is applicable and it assumes that the main deformation mechanism remains unchanged. One can attempt to determine the conditions for DBT from a simple mechanistic model proposing that DBT occurs when $G'_c = G_{2c}$ or $B = B_0 = 2R_p$. If $\sigma_y$ versus $T$ is assumed to be linear in its temperature dependence and goes to zero at temperature $T_s$, with the reference value $\sigma_0$ at $T_0$, one can write an empirical expression for temperature $T_{DBT}$ at which DBT occurs:

$$T_{DBT} = T_s - (T_s - T_0)\sqrt{\frac{2R_0}{B}} \tag{6.20}$$

where:

$$R_0 = \frac{1}{2\pi}\left(\frac{EG_{2c}}{\sigma_{y0}^2}\right) \tag{6.21}$$

The condition for the yield strength going to zero at $T_s$ can be satisfied by choosing $T_s = T_g$ for glassy or $T_s = T_m$ for semicrystalline polymers. One can show that the temperature dependence of the apparent critical strain energy release rate, $G'_c$, can be expressed as:

$$G'_c = G_{1c} + \frac{EG_{2c}}{\pi B \sigma_{y0}^2}(G_{2c} - G_{1c})\frac{1}{(T_s - T)^2} \tag{6.22}$$

There are three main regions on the temperature dependence of $G'_c$:

1. At very low temperatures:

$$\frac{1}{(T_s - T)^2} \to 0 \to G'_c = G_{1c} \tag{6.23}$$

2. In the transition region, where $G'_c$ corresponds to the condition:

$$\frac{2R_p}{B} = 1 \rightarrow G'_c = G_{2c} \tag{6.24}$$

3. Eq. (6.22) can be simplified by setting $T = T_{DBT}$ and $G'_c = G_{2c}$ to give:

$$T_{DBT} = T_s = \sqrt{\frac{EG_{2c}}{\pi B \sigma_{y0}^2}} \tag{6.25}$$

All of the effects described above are taken for constant specimen thickness, $B$, and strain rate, $d\varepsilon/dt$.

### 6.8.10 Strain Rate Effects

It is a well-established fact that deformation behavior of polymeric material is controlled by changes in conformational entropy of long molecular chains. Hence, this feature of polymer molecular structure makes mechanical response very sensitive to variations in the rate at which the polymeric solids are deformed.

The strain rates considered in this chapter are well above those for standard tensile or flexural tests and well below the ballistic deformation rates at which bullet-proof vests and other ballistic polymer applications are tested [4]. Impact speeds vary from 1 to 10 m/sec. The actual strain rates in the solid body depend on the loading and specimen geometries. Depending on the specimen and loading geometry, strain rates ranging from $10^4$ to $10^6$ sec$^{-1}$ are generated in impact test specimen. Generally, the strain rates during an impact event are four to six orders of magnitude greater than those used in standardized tensile tests. This range is compared to strain rates from $10^{-1}$ to $10^2$ sec$^{-1}$ existing in static tests such as tensile testing.

The strain rate has a major effect on the yielding of plastics [77]. However, it is very difficult to accurately determine the strain rates during the brief period of time for an impact test. Bucknall estimated the strain rates ($d\varepsilon/dt$) for HIPS, PC, and toughened nylon [80]. From experimental observations indicating that stress–strain curve for HIPS was relatively linear and the plastic zone developed in about 0.1 msec, Bucknall estimated strain rate at the notch tip during the loading portion of the impact test to be about $10^2$ sec$^{-1}$. Strain rates ahead of a propagating crack are even more difficult to estimate. By making very simplified assumptions for quasi-static analysis, the strain rate expression can be written as:

$$\frac{d\varepsilon}{dt} = \frac{\pi E^2 v \varepsilon_y^3}{K_c^2} \tag{6.26}$$

where $v$ is the speed of the crack front. For crack speeds at about 500 m/sec, Eq. (6.26) gives strain rates on the order of $2.3 \times 10^4$ sec$^{-1}$ for HIPS. High strain rates ensure that plane strain conditions occur at the crack tip. High strain rates also suppress the viscoelastic nature of the mechanical response of polymers during very short relaxation times. This condition can locally increase $T_g$ of glassy materials by 30–45°C, since the corresponding frequencies are of the order of 1–100 kHz. Fracture can change from isothermal to adiabatic condition during an increase in the test speed. This thermal change can have adverse consequences due to onset of localized melting of the polymer at the crack tip [81].

Jancar and DiBenedetto [82] estimated the strain rate at the crack tip, assuming elastoplastic behavior with the deformation confined to the region of the size of the crack tip plastic zone. Calculated strain rates of about $10^5$ sec$^{-1}$ are two to three orders of magnitude greater than the maximal strain rates attainable with universal tensile testing devices ($10^2$ sec$^{-1}$). Observed shifts in glass transition temperature are in agreement with interpretation of the semiempirical Williams-Landel-Ferry equation. The $T_g$ of polymer deformed at calculated strain rate should drop by some 30–40°C compared with the $T_g$ measured using differential scanning colorimetry or low-frequency dynamic mechanical thermal analysis. Performing a notched impact test on PP at room temperature should provide results similar to the static tensile testing at −30°C to −40°C.

When the rate at which a material is externally loaded corresponds to a time scale of a specific chain mobility relaxation, ductile failure may result. The relaxation processes are inelastic events in which energy is absorbed in the system. As long as the loading time is shorter than the relaxation time, brittle failure occurs when no energy can be absorbed via resonance with local chain motion. Since a time–temperature superposition operates in viscoelastic polymers, a relaxation process occurring at low temperatures for slow loading rates may shift to higher temperatures when loaded at higher rates [83].

Usually, sub-$T_g$ segmental motions do not absorb enough energy to affect impact fracture substantially. This is due to the fact that low values of tan δ associated with these relaxations indicate very small, predominantly elastic deformations. Bulk ductility is not guaranteed by the presence of these maxima. The volume of material strained at large deformations and involved in failure process is not substantially affected by these maxima. In order to achieve measurable bulk ductility, large-scale main-chain cooperative motions are required. However, even in this case, extreme localization of the fracture process zone leads to marginal effects of this chain mobility [84].

From empirical evidence, it is known that yield strength and elastic modulus are both rate dependent. The question is whether $G_c$ also varies with the strain rate and how such a variation affects the use of this parameter for material selection and design. Since the majority of impact tests involve bending load geometry, we will show that it is possible to use a simple beam theory to estimate the nominal strain rate within an uncracked elastic body. In three-point bending geometry of the Charpy impact test, strain rate can be related to the striker speed $v$, the span length $L$, and specimen thickness $D$ in the form:

$$\varepsilon = 6\frac{v}{D}\left(\frac{D}{L}\right)^2 \tag{6.27}$$

Equation (6.27) provides an estimate of strain rate of $75\,\text{sec}^{-1}$ for Charpy test with span length of $L$ value of 41 mm, thickness $D$ equal to 6 mm, and impact velocity $v$ value of 3.5 m/sec. This strain rate is about 30–50 times greater than strain rates encountered in common tensile tests. In the case of notched specimen, the deformation is concentrated in a small plastic zone at the crack tip and, thus, strain rates at the crack tip can be considerably higher than elastic deformation rates in the test specimen. Using Eq. (6.25) and assuming that the deformation in the elastic portion of the specimen can be neglected (i.e., all of the deformation processes take place in the crack tip plastic zone), an estimated strain rate of $2.1 \times 10^5\,\text{sec}^{-1}$ is calculated for the same test geometry with notched specimen. It is worthwhile to note that with the change of the $R_p$, the same test conditions will result in different strain rates.

## 6.9 FRACTURE TOUGHNESS–IMPACT STRENGTH OF PARTICULATE FILLED POLYPROPYLENE

The objective of this section is to provide a basis for predicting the limits of fracture toughness and impact strength of particulate-filled polypropylene by using the combined concepts of LEFM and small-scale yielding. To describe a concentration dependence of the composite yield strength, a microscopic model based on concepts of localized yielding and percolation theory is used [82].

### 6.9.1 Analysis

The toughness of a material is related to the amount of energy dissipated during a fracture and can be described in terms of either a stress intensity factor $K$ or a strain energy release rate $G$ [55]. Critical values of these terms (i.e., $K_c$ and $G_c$) define the fracture criterion for a given material. An opening mode I of fracture is assumed solely in this analysis, and $K_c$ and $G_c$ are the critical values of $K$ and $G$ in mode I fracture. Within the region of validity of LEFM, there is a simple relation between $K$ and $G$:

$$K^2 = EG \tag{6.28}$$

where $E$ is the material Young modulus.

## Impact Behavior of Polypropylene

Yielding at the crack tip occurs under shear controlled conditions and is independent of hydrostatic stress. Hence, the stress level at which yielding occurs can be expressed in terms of multiaxial stresses by the von Mises yield criterion:

$$(\sigma_1 - \sigma_2)^2 + (\sigma_2 - \sigma_3)^2 + (\sigma_3 - \sigma_1)^2 = 2\sigma_y^2 \qquad (6.29)$$

where $\sigma_y$ is the material yield strength and $\sigma_i$ ($i = 1, 2, 3$) are the principal stresses. Under plane strain conditions, for example:

$$\sigma_3 = \nu(\sigma_1 + \sigma_2) \qquad (6.30)$$

and the constraint factor $M$ describing the elevation of the applied stress at yield due to constraints imposed by the surrounding material can be defined as:

$$M = \frac{\sigma_1}{\sigma_0} = \left[\left(1 + \frac{\sigma_2}{\sigma_1}\right)^2 (1 - \nu + \nu^2) - \frac{3\sigma_2}{\sigma_1}\right]^{-1/2} \qquad (6.31)$$

At the crack tip $\sigma_1 = \sigma_2$ and for small-scale yielding with the stress field controlled by elastic stresses, we have

$$M = (1 - 2\nu)^{-1/2} \qquad (6.32)$$

The size scale for crack tip yielding, $\ell$, in the crack plane can be written using Irwin's approach [48,49] as

$$\ell = \frac{K_c^2}{2\pi M^2 \sigma_y^2} \qquad (6.33)$$

It is commonly assumed that the size of the plastic zone in the crack plane and in the direction perpendicular to crack length is of the same size $\ell$. The size scale $\ell$ of the crack tip plastic zone relative to specimen dimensions ($B$, $D$) and crack length ($2a$) is used to rank failure modes [55] and can be used to describe the displacement of the state of stress from the reference state of plain strain.

The energy dissipated during a localized yielding around a critical defect is roughly proportional to the extension of a plastic zone in front of a natural crack or a notch. Irwin approximated the length of the plastic zone $R_y$ in front of a crack tip in the direction of crack propagation by Eq. (6.33) where $\sigma_y$ is the material yield strength of the material and $M$ is a plastic constraint factor. When the material around the crack tip is in a state of plane stress, $M$ is equal to unity. It assumes a higher value in a state of plane strain. A state of plane stress exists at the edges of the specimen while a state of plane strain exists in the center

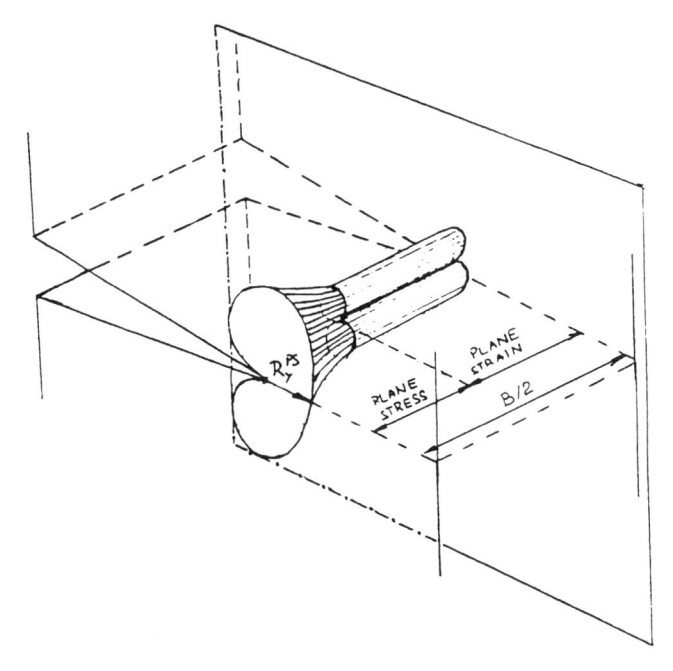

**FIGURE 6.8** Schematic representation of plane stress-strain regions of specimen cross-section. (Ref. 66 courtesy of SPE.)

(Fig. 6.8). The respective sizes of plastic zones in plane strain, $R_y^{PN}$, and in plane stress $R_y^{PS}$, can be expressed as:

$$R_y^{PN} = (1/2\pi)[K_{1c}/\sigma_y]^2$$
and  (6.34)
$$R_y^{PS} = (1/2\pi)[K_{2c}/\sigma_y]^2$$

where $K_{1c}$ and $K_{2c}$ are critical stress intensity factors in plane strain and plane stress, respectively. The measured fracture toughness, $G_c$, is expected to be minimal when the plastic zone is smallest. This is achieved with maximal constraints under full plane strain conditions (i.e., $G_{1c}$). As the constraints decrease, $G_c$ may be expected to increase to an upper limit, which, for a small-scale yielding condition, is the plane stress value $G_{2c}$. This phenomenon is of great importance in edge-notched specimens, where the center region gives a plane strain value and the surfaces yield plane stress value. This results in redistribution of $G_c$ values across the specimen cross-section, and any measured value of strain energy release rate will be an average one ($G_c'$). For small-scale

## Impact Behavior of Polypropylene

yielding in the crack tip region along with crack line an approximate relation between plane strain, $G_{1c}$, and plane stress, $G_{2c}$, fracture toughness is given by:

$$G_{1c} = G_{2c}(1 - 2\nu)^2 \qquad (6.35)$$

where $\nu$ is the material Poisson's ratio. Thus, for a typical plastic with $\nu = 0.3$, $G_{1c}$ in plane strain is approximately 0.16 of that in plane stress $G_{2c}$. In reality, fracture occurs first in the center, and a complete fracture occurs when the $G_{2c}$ condition is achieved at the surface by redistribution of the load [86].

It was shown by Williams [46] and Fernando and Williams [87] that the measured stress intensity factor $K'_c$ for a mixed mode of failure can be approximated as a combination of contributions from the plane strain ($K_{1c}$) and plane stress ($K_{2c}$) regions using a ratio between specimen width $B$ and the size of the plastic zone $R_y^{PS}$ and assuming a bimodal distribution of $K_c$. For $B >= 2R_y^{PS}$:

$$K'_c = K_{1c} + [K_{2c}/(\pi B)^{1/2}\sigma_y]^2(K_{2c} - K_{1c}) \qquad (6.36)$$

and for $B < 2R_y^{PS}$ $K'_c$ is given by Eq. (6.33) and is proportional to the square root of the specimen width:

$$EG'_c = K'^2_c = \sigma_y^2(\pi B) \qquad (6.37)$$

Equation (6.36) may also be written in terms of the corresponding strain energy release rates [14]:

$$(G'_c)^{1/2} = [G_{1c}/(1 - \nu)]^{1/2} + [EG_{2c}/B\sigma_y^2]\{G_{2c}^{1/2} + [G_{1c}/(1 - \nu)^{1/2}]\} \qquad (6.38)$$

Equations (6.36) and (6.38) are valid only until the whole cross-section of the specimen is in the state of plane stress (i.e., $B = 2R_y$). Further reduction of the specimen thickness leads to a decrease in the measured fracture toughness, as expressed by Eq. (6.37). Although truly valid for homogeneous and isotropic solids, Eqs. (6.36) to (6.38) can be applied to randomly filled particulate composites as long as the plastic zone is significantly larger than the particle diameter. Then one may assume that the crack is propagating through a quasi-homogeneous medium and one may use the macroscopic properties of the composite to calculate $K$ and $G$. Within the limits of LEFM, assuming small-scale yielding and an unchanged mechanism of crack propagation, one can write the fracture toughness and strain energy release rate relative to those of the unfilled matrix:

$$(K_c)^2/(K_m)^2 = (\sigma_c)^2/(\sigma_m)^2 = (E_c G_c)/(E_m G_m) \qquad (6.39)$$

Critical to this argument is the assumption that the minimal value of the plane strain fracture toughness $K_{1c}$ is a purely matrix property, independent of the presence of particulate filler. Nikpur and Williams [85] and Williams and

Fernando [87] found little change in $K_{1c}$ with the addition of a second phase in high-impact polystyrenes and polyethylene-modified polypropylenes. Thus, the variation in $K'_c$ and $G'_c$ were shown to be due entirely to changes in modulus and yield strength. It is our hypothesis that this will also be true for calcium carbonate–filled polypropylene, and we can therefore determine the effect of filler concentration on toughness from the changes in apparent moduli and yield strengths. From Eq. (6.35) one can express a strain energy release rate relative to that of the matrix:

$$(G'_c)^{rel} = (\sigma_c^{rel})^2 / E_c^{rel} \tag{6.40}$$

Using the Kerner-Nielsen model for expressing the effect of filler concentration on Young's modulus [85]:

$$E_c/E_m = (1 + ABv_f)/(1 - B\varphi v_f) \tag{6.41}$$

where

$$A = (7 - 5v_m)/(8 - 10v_m) \tag{6.42}$$
$$B = [(E_f/E_m) - 1]/[(E_f/E_m) + A] \tag{6.43}$$
$$\varphi = (1 + 0.89 v_f) \tag{6.44}$$

for random packed monodisperse spheres, and the model proposed earlier for yield strength of $CaCO_3$-filled polypropylene [64]:

$$\sigma_{yc}/\sigma_{ym} = 1 + 1.06 v_f^2 \tag{6.45}$$

for $0.56 > v_f > 0$ ("perfect" adhesion, brittle matrix, rigid interphase),

$$\sigma_{yc}/\sigma_{ym} = 1.33 \tag{6.46}$$

for $v_f > 0.56$ ("perfect" adhesion, brittle matrix, rigid interphase),

$$\sigma_{yc}/\sigma_{ym} = (1 - 1.21 v_f^{2/3}) \tag{6.47}$$

for $0.64 > v_f > 0$ ("no" adhesion, soft interphase), one can predict lower and upper bounds for the concentration dependence of the strain energy release rate for the composites in the case of "no" and "perfect" adhesion which equals to those for soft and rigid interphase.

## 6.9.2 Effect of Specimen Geometry

Williams [46], Ward et al. [77], and Jancar and DiBenedetto [88] have proposed an approximate model of mixed mode of fracture to account for the effect of finite specimen dimensions for $K'_c$ and $G'_c$ respectively. The basic idea in both theories is substitution of the actual distribution of fracture toughness across the cross-section by a simple bimodal distribution, assuming plane strain value in the center and plane stress value at the surface area of the specimen. Size of the plastic zone $2R_p$ relative to the specimen width $B$ gives the contribution of plane stress regions

## Impact Behavior of Polypropylene

and is a measure of the displacement of the state of stress at the crack tip from the plane strain conditions. Note that this approach can be used only if the mode of failure does not change with the test conditions or material composition (i.e., it attains its brittle character).

There is a critical specimen width $B_0$ above which this approximation can be used. This critical width is reached when the whole specimen cross-section is under the conditions of plane stress and thus the measured fracture toughness reaches its upper plane stress value $G_{2c}$. Mathematically, critical width is defined using critical stress intensity factor as:

$$B_0 = \frac{1}{\pi}\left(\frac{K_{2c}}{\sigma_y}\right)^2 = \frac{1}{\pi(1-2\nu)^2}\left(\frac{K_{1c}}{\sigma_y}\right)^2 \tag{6.48}$$

or in terms of $G_c$, using the relation between $K_c$ and $G_c$ [Eq. (6.28)], in the form

$$B_0 = \frac{E}{\pi}\frac{G_{2c}}{\sigma_y^2} = \frac{E}{\pi(1-2\nu)^2(1-\nu^2)}\frac{G_{1c}}{\sigma_y^2} \tag{6.49}$$

From the above equations, it is clear that the critical specimen thickness is a material parameter depending upon the elastic modulus $E$, yield strength $\sigma_y$, and Poisson's ratio $\nu$. Thus, changing the test conditions or material composition, one can vary $E$, $\sigma_y$, and $\nu$, resulting in substantial changes in $B_0$. The effects of changes in $E$ are small in comparison with the effects of changes in $\sigma_y$ and $\nu$.

In determining the temperature dependence of fracture toughness according to ASTM E399 standard, a requirement on the specimen width $B > 2.5\,(K_{1c}/\sigma_y)^2$ must be satisfied for the above-mentioned reasons. In general practice, the specimen thickness is determined at room temperature and then kept constant during the experiments. If, for example, the temperature is decreased or increased, the ratio between $B$ and $B_0$ can change. This results in a variation in the relative contributions from plane strain and plane stress regions. One can in this way vary the apparent fracture toughness of the material without changing its true material properties $G_{1c}$ or $G_{2c}$. Eventually, $B$ can become smaller than $B_0$ and the concept of mixed mode of failure can no longer be used. In conclusion, the fracture toughness tests should be performed using a constant $B/B_0$ ratio. In other words, one should vary the specimen thickness when varying test conditions or material composition.

Further decrease of $B$ below $B_0$ causes an experimentally observed decrease in measured fracture toughness due to the reduction in the cross-section of the material capable of resisting a crack. Specimen width becomes the controlling factor, and the dependence of $G'_c$ on $B$ can be expressed in the form:

$$G'_c = \frac{\sigma_y^2}{E}(\pi B) \tag{6.50}$$

since the whole cross-section is under the plane stress conditions.

TABLE 6.3 Particulate Filled and Rubber Modified Composites

| Group | Polymer Matrix | Filler | Structural Variable Ingredient | Vol% | Particulate Interphase Morphology |
|---|---|---|---|---|---|
| I | iPP homopolymer | CaCO3+ phlogopite Mica | Hybrid Blend Combinations | Total = 20 | No Adhesion |
|  | MPP | CaCO3+ phlogopite Mica | Hybrid Blend Combinations | Total = 20 | Strong Adhesion |
| II | iPP+MEPR | Mg(OH)2 filler type 1 @ 30 vol% | MEPR | 0–20 | Complete Encapsulation of Filler by MEPR |
|  | cPP+MEPR | Mg(OH)2 filler type 1 @ 30 vol% | MEPR | 0–20 | Complete Encapsulation of Filler by MEPR |
|  | MPP | CaCO3+ muscovite mica | Hybrid Blend Combinations | Total = 20 | Strong Adhesion |
|  | PP | CaCO3+ muscovite mica | Hybrid Blend Combinations | Total = 20 | No Adhesion |
|  | PP | CaCO3 | CaCO3 | 0–20 | No Adhesion |
|  | MPP | CaCO3 | CaCO3 | 0–20 | Strong Adhesion |
|  | MPP+EPR | Mg(OH)2 filler type 1 @ 30 vol% | EPR | 0–20 |  |
|  | iPP+MEPR | Mg(OH)2 filler type 2 @ 30 vol% | MEPR | 0–20 | Complete Encapsulation |

# Impact Behavior of Polypropylene

| | | | | |
|---|---|---|---|---|
| III | MPP+EPR | Mg(OH)2 filler type 1 @ 30 vol% | EPR | 0–20 | Complete Separation No Adhesion |
| | PP+MEPR | | MEPR | 0–20 | |
| | PP+EPR | | EPR | 0–20<c> No Adhesion | |
| | MPP+EPR | Mg(OH)2 filler type 1 @ 30 vol% | EPR | 0–20 | Complete Separation |
| | PP+MEPR | Mg(OH)2 filler type 2 @ 30 vol% | EPR | 0–20 | Complete Encapsulation |
| | PP+MEPR | CaCO3 | MEPR | 0–20 | Complete Encapsulation |
| | PP+MEPR | Mg(OH)2 filler type 1 @ 30 vol% | MEPR | 0–20 | Complete Encapsulation |
| | PP+MEPR | Mg(OH)2 filler type 2 @ 30 vol% | MEPR | 0–20 | Complete Encapsulation |

iPP = isotactic PP homopolymer
PP = low tacticity PP
cPP = copolymer
MPP = MA-g-PP
EPR = random ethylene-propylene
MEPR = maleic anhydride grafted EPR
Mg(OH)2: filler type 1 [specific surface area 7 m$^2$/gm] and filler type 2 [specific surface area = 18 m$^2$/gm]

It is now believed that the energy redistribution across the cross-section, rather than stress redistribution, reflects the physical processes connected with the mixed mode of failure more accurately. Hence, the use of an energy term $G_c$ rather than a stress term $K_c$ in describing the mixed mode of failure appears to be physically more correct. The simple derivation of the relation between apparent critical strain energy release rate $G'_c$ and the two limiting values $G_{1c}$ and $G_{2c}$ can be found elsewhere [89]. It will be discussed in the following section in more detail.

### 6.9.3 The Reference State

Using simplified assumptions and neglecting the term $(1 - v^2)$, one can derive a relation between the measured, apparent fracture toughness $G'_c$ and its plane strain value $G_{1c}$ in the form [64]:

$$G'_c = G_{1c}(1 + f)^2 \qquad (6.51)$$

where the scale factor $f$ is a function of the specimen width $B$, material Young's modulus $E$, Poisson's ratio $v$, and yield strength $\sigma_y$:

$$f = \left[\frac{2v}{(1 - 2v)^3}\right]\left[\frac{G_{1c}}{\pi B}\right]\left[\frac{E}{\sigma_y^2}\right] \qquad (6.52)$$

The scale factor can be interpreted as a measure of the displacement of the state of stress during a particular test from the state of plane strain. To calculate $f$, values of material properties should be determined at the strain rate and temperature of the experiment. In other words, one can use the plane strain value of fracture toughness as a reference state and then determine the displacement of the actual state of stress at the crack tip from this reference state. Because the measured apparent fracture toughness depends on both the material parameters and the state of stress at the crack tip, materials of different compositions or at different test conditions should be compared at an equivalent displacement from the same reference state or, in other words, at the same scale factor. The limitations of this approach are given by restrictions of LEFM and the fact that the type of fracture should remain in mode I or mode II (Fig. 6.7).

### 6.9.4 Computation of Optimal Filler Level for CaCO₃-Filled Polypropylene

By analyzing the $G'_c$ data and comparing it to measurements of CNIS, one can determine the appropriateness of using the CNIS for comparison of the toughness of materials of different composition. This type of methodology is demonstrated for calcium carbonate–filled polypropylene composite. Strain energy release rates

## Impact Behavior of Polypropylene

were calculated from experimental CNIS data using Eq. (6.35). The limiting values of $G_{1c}$ and $G_{2c}$ were then calculated using Eqs. (6.35) to (6.38). The solution of Eq. (6.35) provided values of $G_{1c}$ equal to $1.0\,\text{kJ/m}^2$ and $0.9\,\text{kJ/m}^2$ for PP and maleated PP (MPP), respectively, and $G_{2c}$ equal to $5.9\,\text{kJ/m}^2$ and $5.4\,\text{kJ/m}^2$ for PP and MPP, respectively (Table 6.3). The size of the plastic zone at the crack tip, $2R_y^{PS}$, was calculated from Eq. (6.34) using yield strength values for PP and MPP reported previously (Table 6.3) and Eqs. (6.45) to (6.47) for the relative yield strength of the composites. Figure 6.9 shows that $2R_y^{PS}$ increases with increasing $v_f$ for soft interphase (no adhesion) and remains nearly constant with $v_f$ for rigid interphase (perfect adhesion). The relative strain energy release rates corresponding to upper and lower bounds were calculated using Eqs. (6.40) and (6.41) to (6.44). The results are shown in Fig. 6.10.

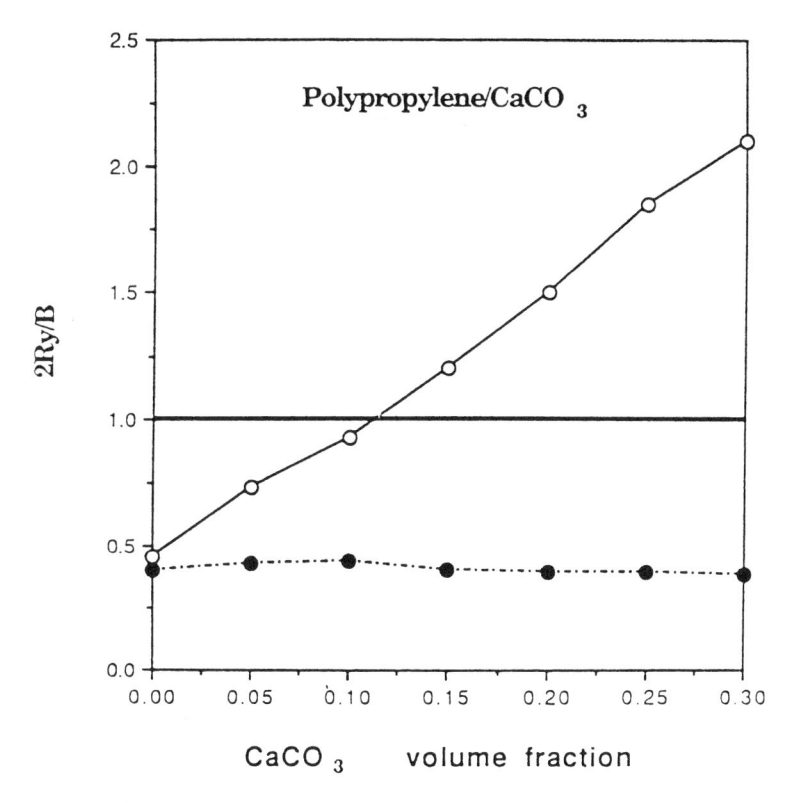

FIGURE 6.9 Concentration dependence of plane stress plastic zone size. (Ref. 66 courtesy of SPE.) ○, Upper limit: "no" adhesion; ●, lower limit: "perfect" adhesion.

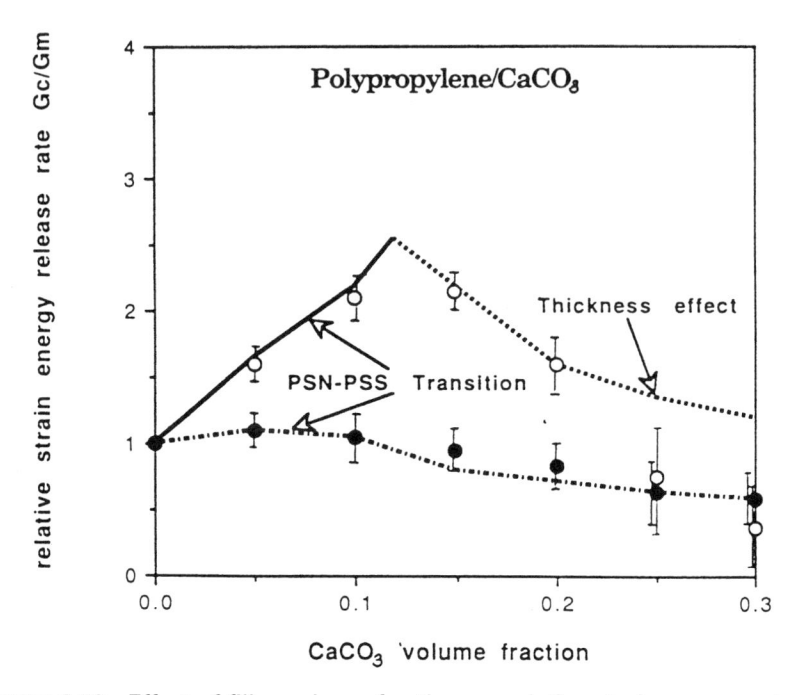

**FIGURE 6.10** Effect of filler volume fraction on relative strain energy release rate. (Ref. 66 courtesy of SPE.) ○, upper limit: "no" adhesion; ●, lower limit: "perfect" adhesion.

In the case of soft interphase, the contribution of the plane stress region to the measured strain energy release rate $G'_c$ increases with $v_f$ due to the increase in the size of plastic zone in accordance with Eq. (6.38). The measured value of $G'_c$ reaches maximal value at the filler volume fraction of 0.12 where the whole specimen width at the crack tip is in the state of plane stress. Above $v_f = 0.12$, the effect of a reduction in matrix effective cross-section due to the presence of the filler becomes the controlling parameter, causing a reduction in fracture toughness. In the interval $0.12 < v_f < 0.20$, Eq. (6.47) describes the experimental data well. Above $v_f = 0.2$, the effects of agglomeration and increasing local constraints near filler particles due to the reduction of interparticle distances become the controlling factors, causing a steeper drop than expected in strain energy release rate.

In the case of rigid interphase, the size of the crack tip plastic zone does not change within the range of filler volume fraction investigated. At the same time, the matrix is constrained to a greater degree due to interaction with the filler

# Impact Behavior of Polypropylene

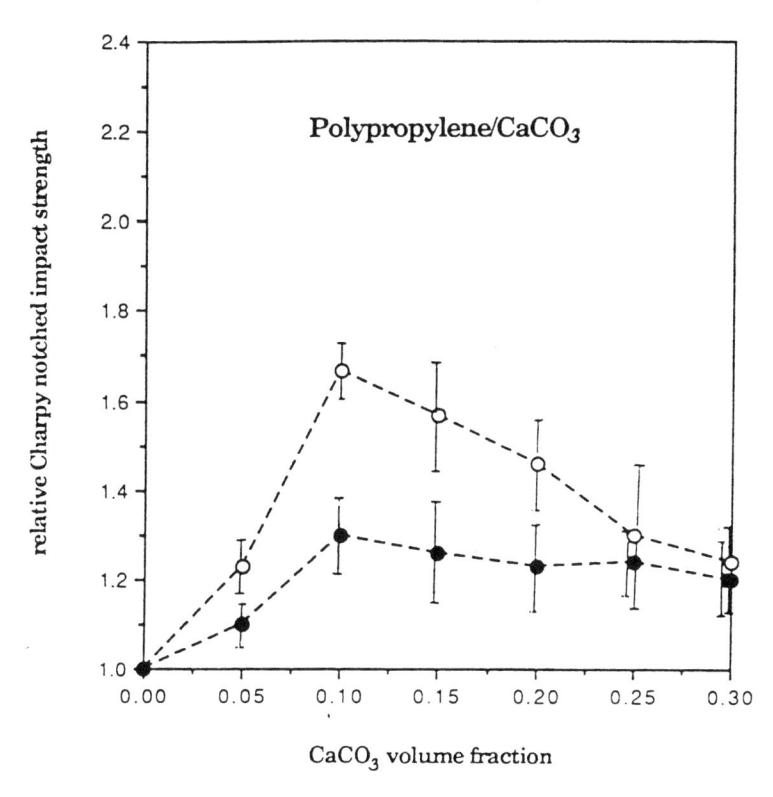

**FIGURE 6.11** Dependence of Charpy notched impact strength on filler content. (Ref. 66 courtesy of SPE.) ○, upper limit: "no" adhesion; ●, lower limit: "perfect" adhesion.

surface, leading to additional embitterment of the material. The relative values of $G'_c$ are in good agreement with prediction based on a macroscopic plane stress–plane strain transition. The notched impact strength (NIS) values follow the same type of concentration dependence as those of $G'_c$ in all of the cases studied (Fig. 6.11). Thus, one can conclude that the same explanation as described above for $G'_c$ holds also for NIS.

## 6.9.5 Effect of Inclusion Volume Fraction

Experimental data for standard Charpy impact strength was analyzed for array of particulate-filled and rubber-modified polypropylene composites [64]. Table 6.3 summarizes interphase design characteristics.

The functional dependence of the ratio (CNIS/$G_c'$) and the scale factor $f$ on the volume fraction of secondary phase, $v_f$, in a variety of polypropylene matrices are plotted versus $v_f$ in Figs. 6.12–6.14. One may assume that the CNIS/$G_c'$ ratio is constant in the case when both CNIS and $G_c'$ have the same functional dependence on $v_f$. On the other hand, substantial variation in the CNIS/$G_c'$ ratio indicates different functional dependences of the two terms.

It is clear from Eq. (6.51) that to compare the toughness of two materials of different composition at an equivalent state of stress, the comparison must be done at the same value of the scale factor. It has been shown that the addition of a secondary phase into these polypropylene matrices does not affect the most conservative, plane strain value of $G_{1c}$ significantly [87]. The observed changes in the measured toughness were ascribed to the changes in the scale factor $f$

**mica volume fraction**

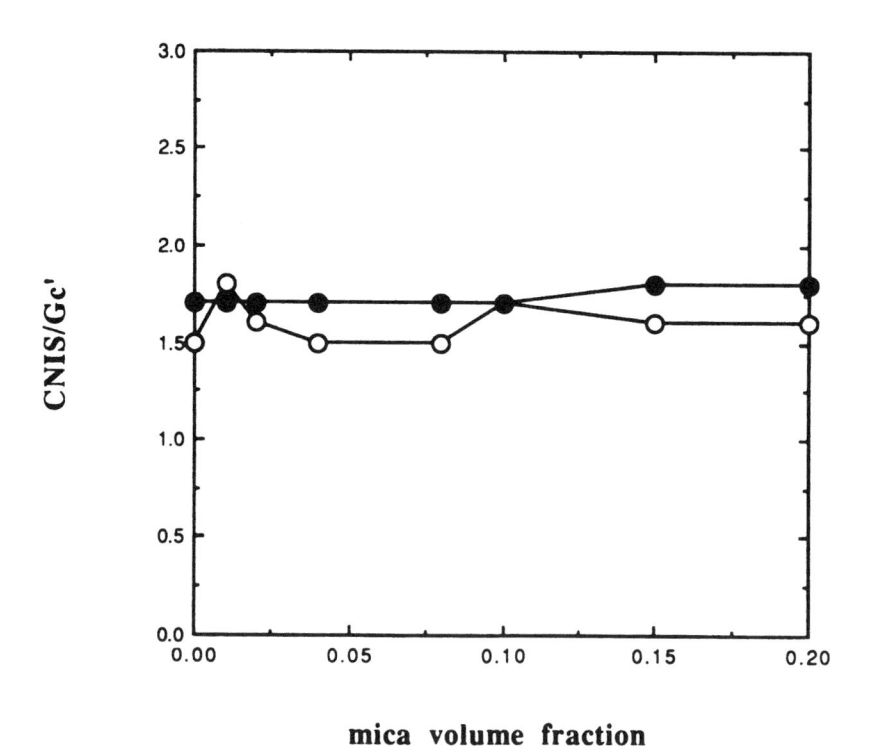

FIGURE 6.12 Dependence of impact parameters on mica concentration for Group I filled PP Composites. (Ref. 64 courtesy of SPE.) 6.12a. CNIS/Gc'. 6.12b. Scale Factor. ○ CaCO$_3$/phlogopite mica: strong adhesion ● CaCO$_3$/phlogopite mica: no adhesion.

# Impact Behavior of Polypropylene

[Figure: plot of $(1+f)^2$ vs mica volume fraction, showing two curves—filled circles near 3.2 and open circles near 2.2—across volume fractions 0.00 to 0.20]

**mica volume fraction**

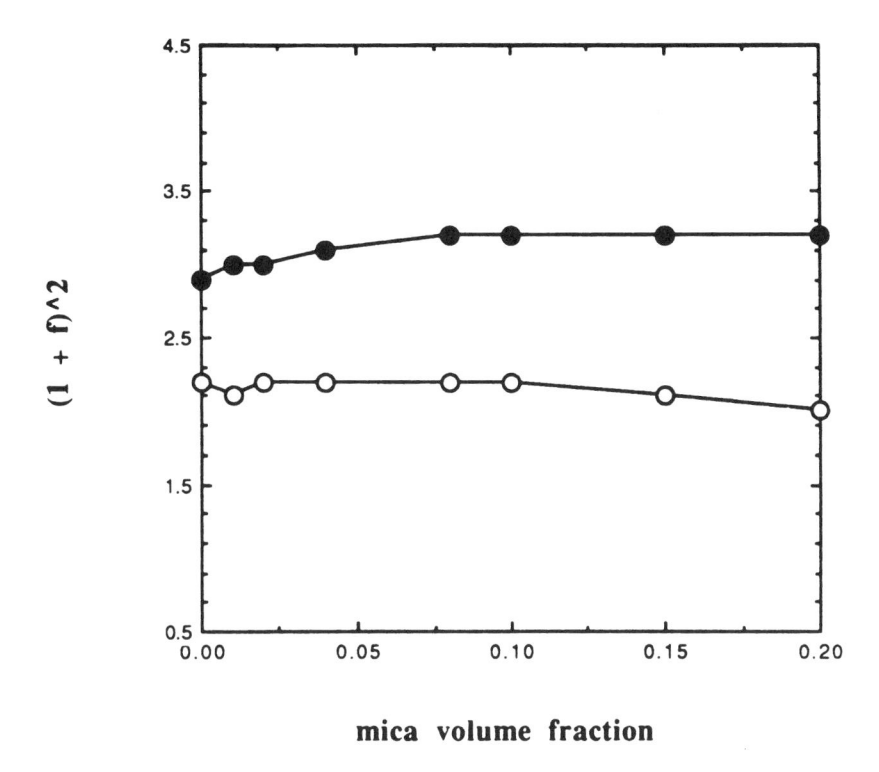

FIGURE 6.12 (*continued*)

representing the increase of the crack tip plastic zone size, controlled by a reduction in the material yield strength, $\sigma_y$. This appears to occur in other filled thermoplastic materials as well [14].

For group 1, consisting of a small number of materials, both the ratio $\text{CNIS}/G'_c$ (Fig 6.12a) and the scale factor $f$ (Fig. 6.12b) are independent of the composition. This group of materials is characterized by a constant Poisson ratio. A constant $\text{CNIS}/G'_c$ ratio indicates that CNIS has the same functional dependence on $v_f$ as $G'_c$. To use CNIS for a meaningful comparison of the toughness of materials of a different composition, the scale factor, $f$, should be independent of composition. Under these conditions a single value of CNIS is a reasonable measure of relative toughness because the materials of different composition are compared at the same state of stress. Thus, the observed differences in CNIS with composition reflect true changes in the fracture toughness independent of the state of stress on the test specimen.

**inclusion volume fraction**

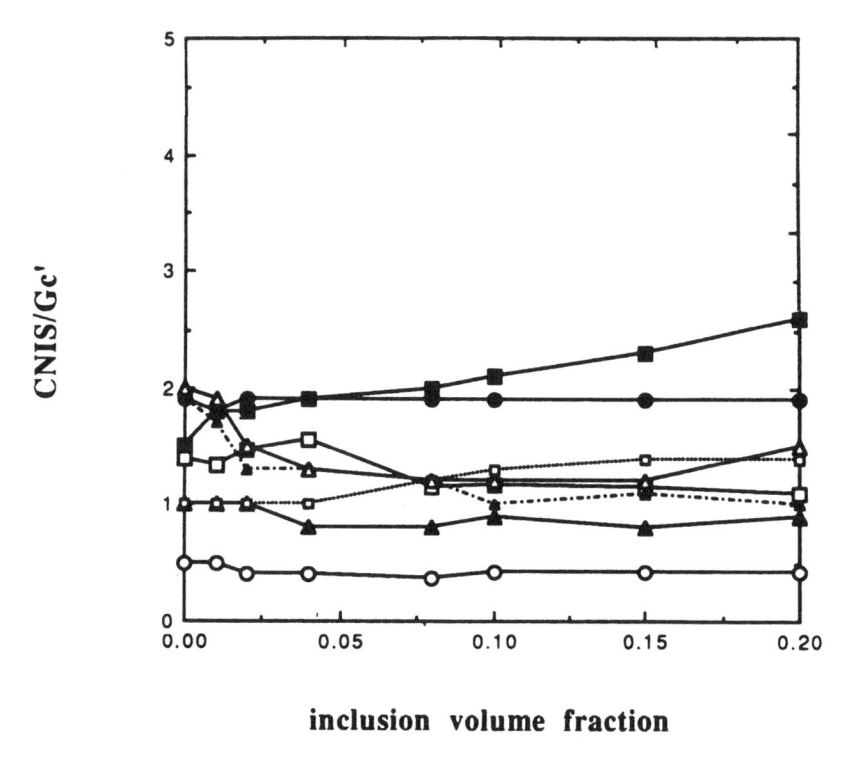

FIGURE 6.13 Dependence of impact parameters on material composition for Group II filled PP composites. (Ref. 64 Courtesy of SPE.) 6.13a. CNIS/Gc? 6.13b. Scale Factor. ○ iPP/MEPR/Mg(OH)$_2$-filler 1: Complete encapsulation. □ cPP/MEPR/Mg(OH)$_2$-filler 1: Complete encapsulation. ■ MPP/[CaCO$_3$/musco-vite mica): Strong adhesion. ● PP/[CaCO$_3$/muscovite mica): No adhesion. ▲ PP/CaCO$_3$: No adhesion. □ MPP/CaCO$_3$: Strong adhesion. ▲ MPP/EPR/ Mg(OH)$_2$-filler 1: Complete separation. △ iPP/MEPR/Mg(OH)$_2$-filler 2: Com-plete encapsulation.

For group 2 materials (Fig. 6.13a), the CNIS/$G'_c$ ratio does not change significantly with composition; however, the scale factor $f$ (Fig. 6.13b) varies substantially. For this group of materials, a single value of CNIS is not sufficient to separate the effect of geometry from the change in material properties or mechanism of failure. A complete fracture mechanical study providing the fracture toughness in terms of $G'_c$ or $K'_c$ is needed to separate the effects of geometry from those of material properties. A single value of CNIS cannot be used as a meaningful comparison of the relative toughness of two materials of different composition.

# Impact Behavior of Polypropylene

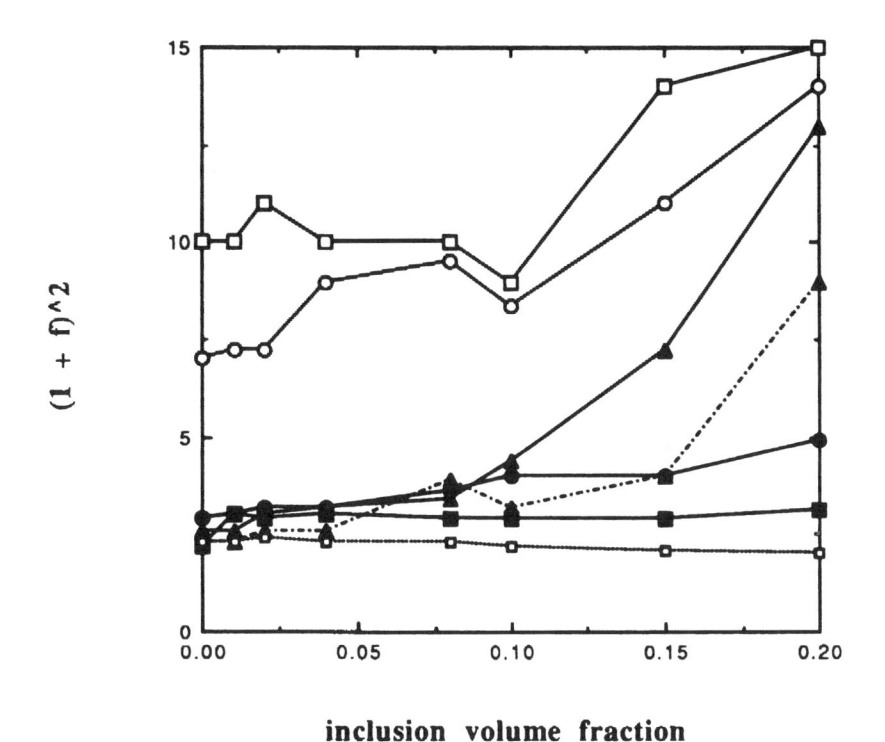

**FIGURE 6.13** (*continued*)

The largest group 3 material consists mostly of rubber-modified blends. The corresponding Fig. 6.14a exhibits large nonsystematic variations in the CNIS/$G'_c$ ratio with composition. This suggests that the functional dependences of the CNIS and $G'_c$ on the volume fraction of the secondary phase are different. Moreover, the scale factor increases substantially with increasing concentration of dispersed phase (Fig. 6.14b). Hence, the state of stress at the crack tip changes with composition under fixed test conditions. For this class of materials, a comparison of CNIS values for materials of different compositions is not a reliable indication of the relative toughness and is of little value as a parameter for material selection and design.

## 6.9.6 Effect of Interfacial Adhesion

The CNIS and $G'_c$ of a group of ternary composites of polypropylene filled with 30 vol% rigid Mg(OH)$_2$ and 10 vol% elastomeric EPR inclusions were studied as

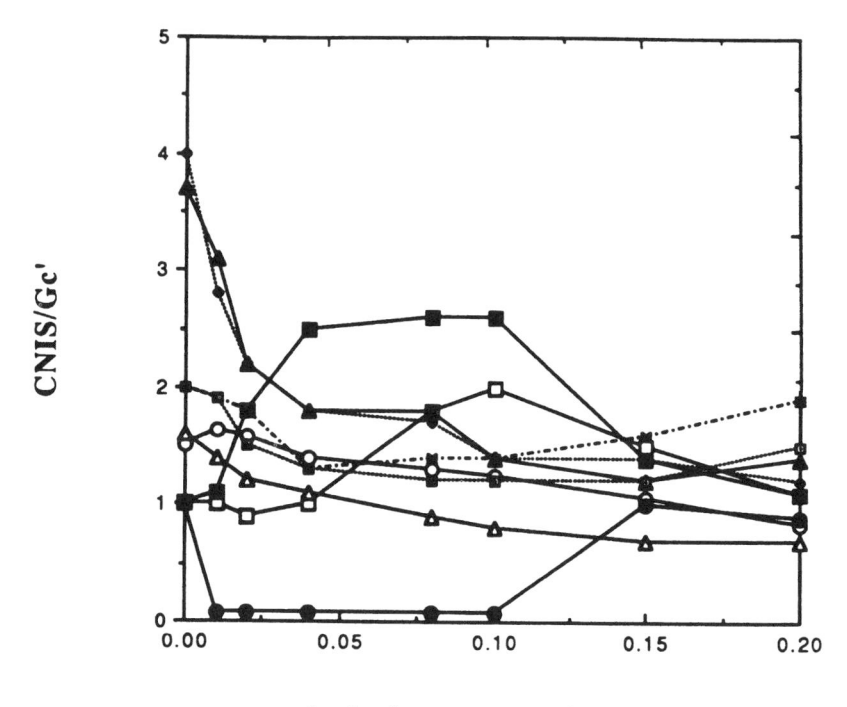

**FIGURE 6.14** Dependence of impact parameters on material composition for Group III filled PP composites. (Ref. 64 courtesy of SPE.) 6.14a. CNIS/Gc' 6.14b. Scale Factor ○ MPP/EPR/Mg(OH)$_2$-filler 1: Complete separation. ● PP/MEPR: No adhesion. □ MPP/EPR: No adhesion. ■ PP/EPR: No adhesion. ◆ MPP/EPR/Mg(OH)$_2$: Complete separation-copolymer matrix. △ MPP/EPR/Mg(OH)$_2$-filler 2: Complete separation. ▲ PP/MEPR/CaCO$_3$: Complete encapsulation. ⊡ PP/MEPR/Mg(OH)$_2$-filler 1: Complete encapsulation. X PP/MEPR/Mg(OH)$_2$-filler 2: Complete encapsulation.

a function of a concentration of maleic anhydride (MAH). In all of the systems studied, the variation in adhesion was achieved by increasing the concentration of MAH in either the matrix (complete separation of the elastomer and filler) or the elastomer (complete encapsulation of the filler by the elastomer). The CNIS/$G'_c$ ratios and scale factor are plotted versus MAH concentration in Figs. 6.15a and 6.15b, respectively.

Composites with a complete encapsulation of Mg(OH)$_2$ of submicrometer size by the elastomer exhibited relatively small variation in CNIS/$G'_c$ ratio over the range of MAH concentration investigated (Fig. 6.15a). Only one system with

# Impact Behavior of Polypropylene

*Y-axis:* $(1 + f)^2$
*X-axis:* inclusion volume fraction

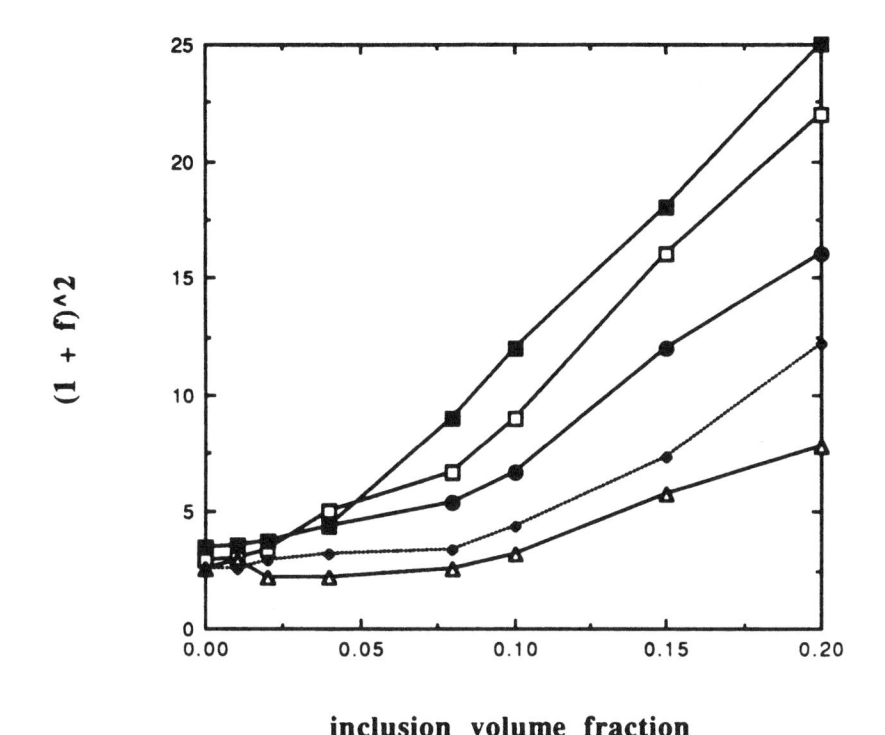

**FIGURE 6.14** (*continued*)

a complete encapsulation of the high-surface-area filler has exhibited CNIS/$G_c'$ independent of MAH concentration. However, the scale factor, $f$, increases substantially with MAH concentration for all the materials investigated (Fig. 6.15b). This leads to a conclusion that a single value of CNIS is not sufficient to separate the effect of test geometry from that of change in the material properties. Thus, single values of CNIS cannot be used to determine how adhesion will affect fracture toughness under equivalent states of stress. This is in agreement with previously published results [14] showing that enhancement of adhesion results in a change of the state of stress at the crack tip resulting in a change of fracture mechanism. A complete fracture mechanical study has to be performed to separate the effect of test geometry from that of material properties.

Both the CNIS/$G_c'$ ratio (Fig. 6.16a) and the scale factor $f$ (Fig. 6.16b) decrease substantially with the MAH concentration for the composites with complete separation. This behavior suggests that single values of CNIS are of little value for a comparison of relative toughness of materials of different composition or as a parameter for materials selection and product design. In

**MAH concentration [wt%]**

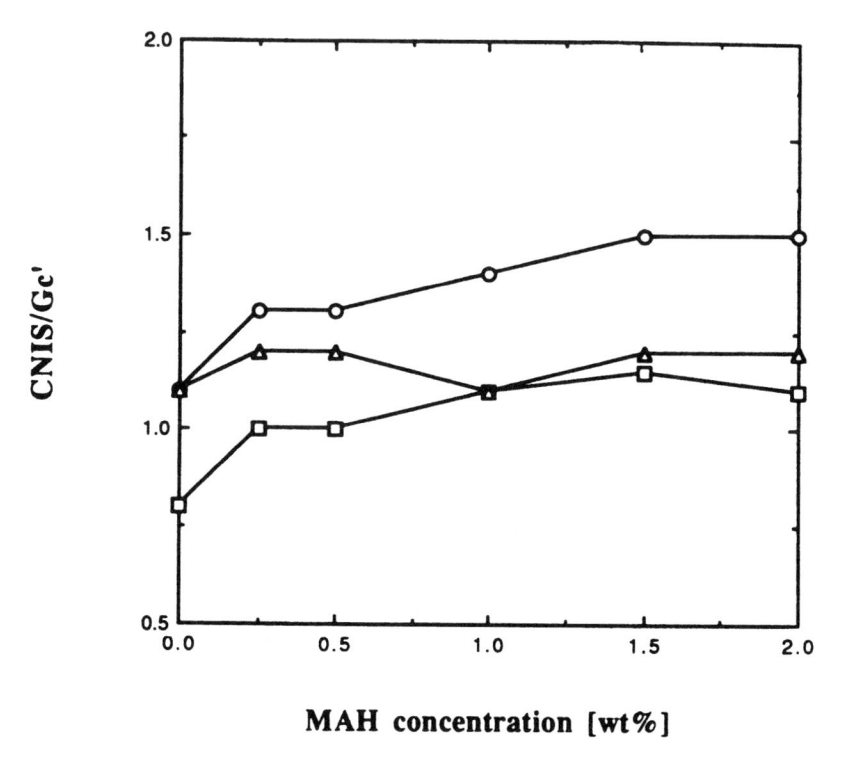

FIGURE 6.15 Dependence of impact parameters on maleic anhydride level for Group II filled PP composites. 6.15a. CNIS/Gc' 6.15b. Scale Factor (Ref. 64 courtesy SPE.) ○ PP/MEPR/CaCO$_3$; Complete encapsulation. □ PP/MEPR/Mg(OH)$_2$-filler 1: Complete encapsulation. △ PP/MEPR/Mg(OH)$_2$-filler 2: Complete encapsulation.

these cases, a fracture mechanics approach should be carried out so as to make comparisons of materials with different degree of adhesion relative to an equivalent state of stress.

## 6.10 CONCLUSIONS

A large volume of experimental CNIS data was analyzed for a wide range of polypropylene-based materials to determine their usefulness as a comparative measure of fracture toughness and as a parameter for materials selection and

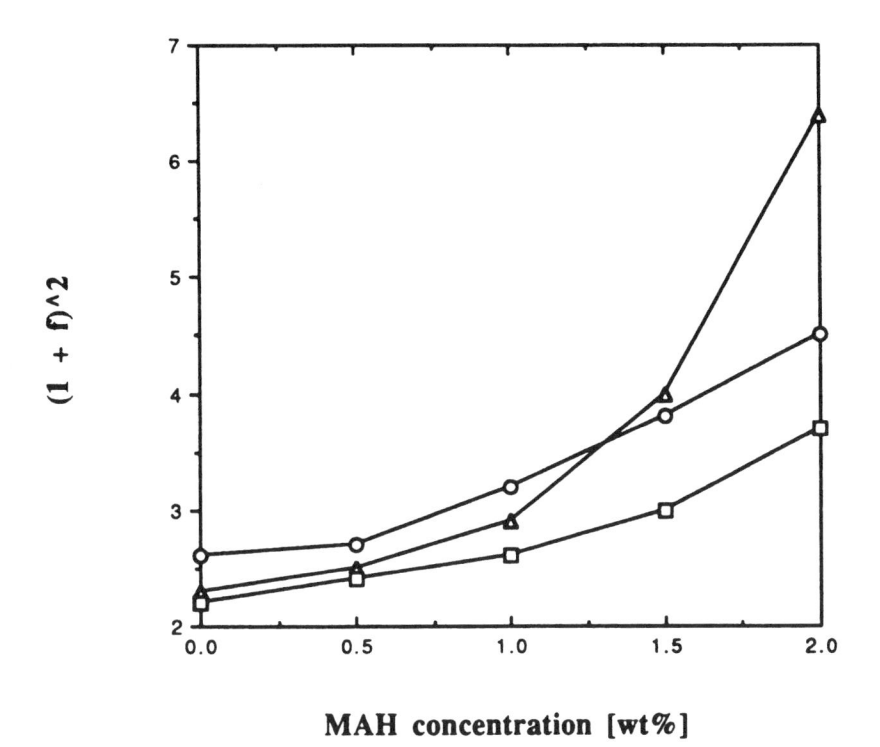

**FIGURE 6.15** (continued)

design. LEFM was used as a tool for comparison of the CNIS with the fracture toughness expressed in terms of $G'_c$.

Three major groups of materials were identified according the $\text{CNIS}/G'_c$ ratio and the scale factor dependence on material composition:

1. The first group (hybrid-filled polypropylene composites) exhibited $\text{CNIS}/G'_c$ ratio *independent of composition*, suggesting that the same functional dependence on the composition existed for both CNIS and $G'_c$. CNIS is in this case a good measure of *geometry-dependent material fracture toughness*. Moreover, the scale factor expressing the state of stress at the crack tip was independent of composition. Hence, CNIS can be used to compare toughness of materials of different composition. This group was characterized by Poisson's ratio independent of composition.

**MAH concentration [wt%]**

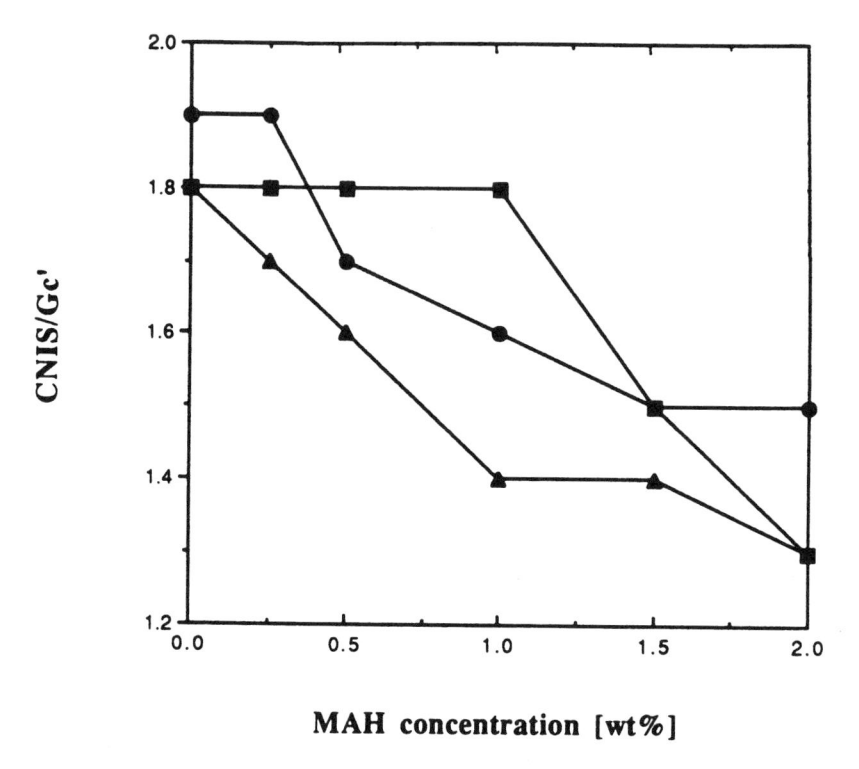

FIGURE 6.16 Dependence of impact parameters on maleic anhydride level for Group III filled PP composites. 6.16a. CNIS/Gc' 6.16b. (Ref 64. courtesy SPE.) Scale Factor ● MPP/EPR/CaCO$_3$: Complete separation. ■ MPP/EPR/Mg(OH)$_2$-filler 1: Complete separation. ▲ MPP/EPR/Mg(OH)$_2$-filler 2: Complete separation.

2. The second group *also* exhibited CNIS/$G'_c$ ratio independent of composition. However, they showed *substantial variation in the scale factor*, prohibiting CNIS from being used to compare materials of different composition. A complete fracture mechanical study providing $G'_c$ is needed to separate the effect of test geometry from that of change in material properties.
3. Materials from the third group were characterized by substantial variations in both the CNIS/$G'_c$ ratios and the scale factors, suggesting different functional dependence of each term on the composition. In this case, CNIS is not a meaningful measure for comparing toughness of materials of different composition. Moreover, large variation in the scale factor exists for this class of materials.

# Impact Behavior of Polypropylene

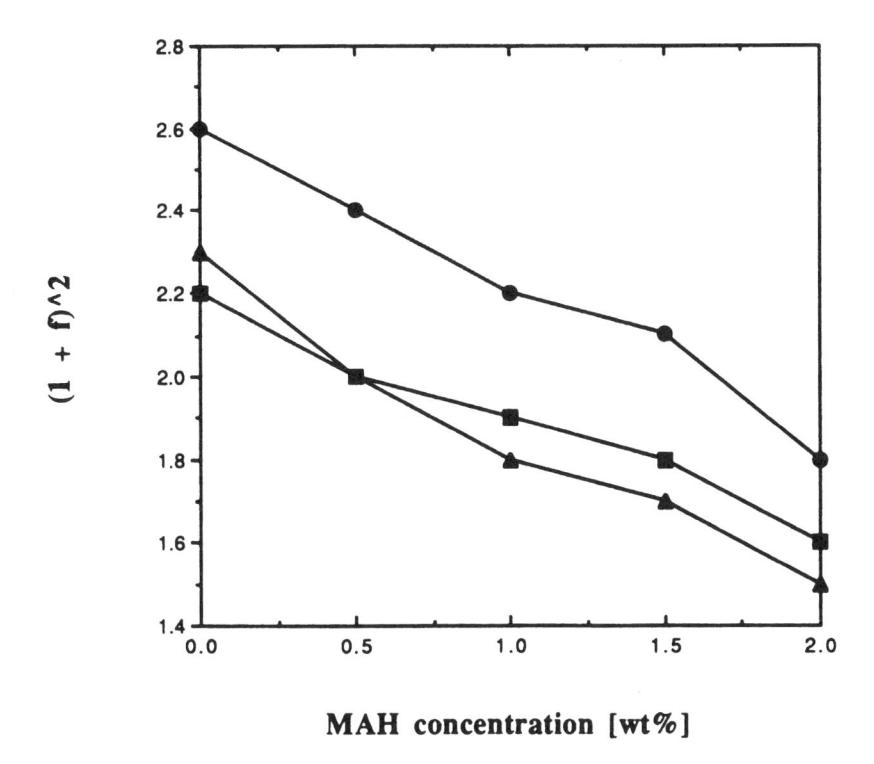

y-axis: $(1 + f)^2$
x-axis: MAH concentration [wt%]

FIGURE 6.16 (*continued*)

Similar classification of materials can be used to evaluate the significance of CNIS for investigations of effects of adhesion on material toughness. In most cases, single values of CNIS cannot be used to compare relative toughness of materials of different adhesion and phase morphology because they are not compared under equivalent states of stress.

## REFERENCES

1. WG Perkins. Polymer toughness and impact resistance. Polym Eng Sci 39:2445–2460, 1999.
2. CB Bucknall. Toughened Plastics. London: Applied Science, 1977, p. 188.
3. CK Riew, JK Gillham, eds. Rubber Modified Thermoset Resins. Washington, DC: American Chemical Society, ACS Adv. in Chem #208, 1984.
4. SK Gagger. In: SL Kessler, GC Adams, SB Driscoll, DR Ireland, eds. Instrumented Impact Testing of Plastics and Composite Materials. Baltimore: ASTM STP 936, Philadelphia: American Society for Testing and Materials, 1987.

5. AJ Kinloch. Polymer Fracture. London: Elsevier, 1983, pp. 421–425.
6. Y Huang, DL Hunston, AJ Kinloch, CK Riew. In: CK Riew, CK Kinloch, eds. Toughened Plastics I. Washington, DC: ACS233, 1993, p. 1.
7. S Newman, S Strella. J Appl Polym Sci 9:2297, 1965.
8. JN Goodier. Trans Am Soc Mech Engs 55:39, 1933.
9. RJ Oxborough, PB Bowden. Phil Mag 30:171, 1974.
10. J Jancar, A DiAnselmo, AT DiBenedetto. Polym Eng Sci 32:1394–1399, 1992.
11. LJ Broutman, G Panizza. Int J Polym Mater 1:95, 1971.
12. FJ Guild, RJ Young. J Mater Sci 24:298, 1989.
13. FJ Guild, RJ Young. J Mater Sci 24:2454, 1989.
14. J Jancar, A Di Anselmo, AT DiBenedetto, J Kucera. Polymer 34:1684–1694, 1993.
15. HB Fhaaf, H Breuer, J Stabenow. J Macromol Sci Phys B14:387–, 1977.
16. AM Donald, EJ Kramer. J Mater Sci 17:1765, 1982.
17. HJ Sue, AF Yee. J Mater Sci 29:3456 (1994).
18. AF Yee. J Mater Sci 12:757, 1977.
19. S Kunz, PWR Beaumont. J Mater Sci 16:3141, 1981.
20. WD Bascom, RY Ting, RJ Moulton, CK Riew, AR Siebert. J Mater Sci 16:2657, 1981.
21. JR Bitner, JL Rushford, WS Rose, DL Hunston, CK Riew. J Adhesion 13:3, 1982.
22. AJ Kinloch, JG Williams. J Mater Sci 15:987, 1980.
23. SC Kim, HR Brown. J Mater Sci 22:2589, 1987.
24. HJ Sue, RA Pearson, AF Yee. Polym Eng Sci 31:793, 1993.
25. RZ Jang, DR Uhlmann, JB Vander Sande. J Appl Polym Sci 29:4377, 1984.
26. BZ Jang, DR Uhlmann, JB Vander Sande. J Appl Polym Sci 30:2485, 1985.
27. BZ Jang, DR Uhlmann, JB Vander Sande. Proc SPE Antec 1984, p. 523.
28. BZ Jang, DR Uhlmann, JB Vander Sande. Proc SPE Antec 1984, p. 549.
29. P Beardmore, S Rabinowitz. J Mater Sci 10:1763, 1975.
30. BZ Jang, DR Uhlmann, JB Vander Sande. Polym Eng Sci 25:98, 1985.
31. BZ Jang, DR Uhlman, JB Vander Sande. ACS Org Coat Appl Polym Sci Proc 49:129, 1983.
32. PI Vincent. Polymer 1:425, 1960.
33. S Wu. J Appl Polym Sci 21:699, 1983.
34. K Friedrich. Fracture 1977, Vol. 3. ICF4, Waterloo, Canada, 1119, 1977.
35. RH Beck, S Gratch, S Newman, KC Rush. Polym Lett 6:707, 1968.
36. A Pavan, T Ricco. J Mater Sci 11:1180, 1976.
37. EH Merz, GC Claver. J Polym Sci 22:325, 1956.
38. ZB Ahmad, MF Ashby, PWR Beaumont. Scr Metall 20:843, 1986.
39. JA Cerere, JE McGrath. Polym Preprints 27:299, 1986.
40. CB Bucknall, AH Gilbert. Polymer 30:213, 1989.
41. J Jancar. J Proc 5th International Meeting on Polymer Blends, Bratislava, SK, October 4–7, 2000.
42. RA Pearson, AF Yee. Polymer 34:3658, 1993.
43. V Flaris, J Stachurski. J Appl Polym Sci 44:2354, 1992.
44. LRF Rose. Mech Mater 8:11, 1987.
45. RS Raghava. J Polym Sci Polym Phys B26:65, 1986.

46. JG Williams. Fracture Mechanics of Polymers. Chichester: Ellis Horwood, 1984.
47. AA Griffith. Phil Trans R Soc A221:163, 1920.
48. GR Irwin. Fracturing of Metals. Cleveland: Am Soc Metals, 1984, p. 147.
49. GR Irwin. In: S Fluge, ed. Encyclopedia of Physics, Vol. 6. Berlin: Springer-Verlag, 1958, p. 551.
50. JL Orowan, JL Weld. J Res Suppl 34:157, 1955.
51. JL Orowan. Fatigue and Fracture of Metals. New York: John Wiley & Sons, 1952, p. 139.
52. HD Keith. In: MB Bever, ed. Encyclopedia of Materials Science and Engineering. New York: Pergamon Press, 1986, p. 3110.
53. HF Mark. In: P Weiss, ed. Adhesion and Cohesion. New York: Elsevier, 1962, p. 240.
54. LM Sherman. Outfiting your lab: Impact testers—Which test to use? Which instrument to buy? Plastics Technol October 2001, p. 56.
55. E Plati, JG Williams. Polymer 16:915, 1975.
56. R Jakvsik, F Jamarani, AJ Kinloch. J Adhesion 32:245, 1990.
57. AJ Kinloch, GAK Kodokian. J Adhesion 24:109, 1987.
58. RE Evans. A rationale for testing to establish suitability for an end-use application. In: RE Evans, ed. Physical Testing of Plastics—Correlation with End-Use Performance. Baltimore: ASTM STP 736. Philadelphia: American Society for Testing and Materials, 1981, pp. 3–14.
59. AP Glover, FA Johnson, JC Radon, CE Turner. Proc Int Conf on Dynamic Fracture Toughness. Welding Inst./ASM, London, 1976.
60. JG Williams, MW Birch. Proc. 4th Int Conf on Fracture. 1. Part IV, University of Waterloo, 1977, p. 501.
61. MC Cheresh, S McMichael. Impact testing and fracture toughness impact testing. In: S Kessler, GC Adams, SB Driscoll, DR Ireland, eds. Instrumented impact testing of plastics and composite materials, Philadelphia: American Society for Testing and Materials, 1987, p. 15.
62. JG Williams, JM Hodgkinson. Proc Roy Soc London A375:231, 1981.
63. PI Vincent. In: RM Ogarkiewicz, ed. Thermoplastics: Properties and Design. London: John Wiley & Sons, 1974, p. 68.
64. J Jancar, AT DiBenedetto. Polym Eng Sci 34:1799–1807, 1994.
65. J Jancar, AT DiBenedetto. Proc ANTEC 1994, San Francisco, vol. 2, p. 1710.
66. J Jancar, AT DiBenedetto. Polym Eng Sci 33:559–563, 1993.
67. J Kreiter, R Knoedel. Kunstoffe 83:889, 1993.
68. GC Adams, TK Wu. Materials characterization by instrumented impact testing. In: W Brostow, RD Corneliussen, eds. Fracture of Plastics, New York: Hanser, 1986, pp. 144–168.
69. WE Wolstenholme, SE Pregun, CF Stark. J Appl Polym Sci 8:119, 1964.
70. CB Arends. J Appl Polym Sci 9:3531, 1965.
71. LC Cessna, JP Lehane, RH Ralston, T Prindle. Polym Eng Sci 16:419, 1976.
72. H Gonzales, WJ Stowell. J Appl Polym Sci 20:1389, 1976.
73. T Casiraghi, G Castiglioni, G Ajroldi. Plast Rubber Proc Appl 2:353, 1982.
74. PE Reed. Composite Polym 1:157, 1988.
75. GC Adams, RG Bender, BA Crouch, JC Williams. Polym Eng Sci 30:241, 1990.

76. JG Williams, GC Adams. Int J Fract 33:209, 1987.
77. IM Ward. Mechanical Properties of Solid Polymers, 2nd ed. New York: John Wiley & Sons, 1983.
78. AJ Wruk, TC Ward, JE McGrath. Polym Eng Sci 21:313, 1981.
79. W Brostow. Impact strength: determination and prediction. In: W Brostow, RD Corneliussen, eds. Failure of Plastics. New York: Hanser, 1986, pp. 196–207.
80. CB Bucknall. Makromol Chem, Macromol Symp 16:209, 1988.
81. K Dijkstra. Deformation and fracture of nylon 6/rubber blends. PhD thesis dissertation, University of Twente, Netherlands, 1993.
82. J Jancar, AT DiBenedetto. J Mater Sci 30:1601, 1995.
83. MT Takemori. Rate, temperature, ductile-brittle-transition. In: Impact Fracture of Polymers. Kyushu University Press, Kyushu, 1992, p.1.
84. G Menges, HE Boden. Deformation and failure of thermoplastics on impact. In: W Brostow, RD Corneliussen, eds. Failure of Plastics. Munich: Hanser, 1986, pp. 169–193.
85. K Nikpur, JC Williams. J Mater Sci 14:467, 1979.
86. J Jancar, AT DiBenedetto. Sci Technol Composite Mater 3:217, 1994.
87. PL Fernando, JG Williams. Polym Eng Sci 21:1003, 1981.
88. J Jancar, A Di Anselmo, AT DiBenedetto, J Kucera. Polymer 34:1684–1694, 1993.
89. J Jancar, AT DiBenedetto. J Mater Sci 30:2438, 1995.

# 7

## Metallocene Plastomers as Polypropylene Impact Modifiers

**Thomas C. Yu**
ExxonMobil Chemical Company, Baytown, Texas, U.S.A.

**Donald K. Metzler**
ExxonMobil Chemical Company, Houston, Texas, U.S.A.

### 7.1 INTRODUCTION

Polypropylene (PP) is characterized by high tensile strength, high stiffness, and high heat deflection temperature under load. However, one major deficiency of PP is its low impact resistance, particularly at low temperatures. Blending PP with an elastomeric modifier provides a simple way to significantly improve impact strength of the base resin. At present, ethylene-propylene rubber (EPR) and ethylene-propylene diene rubber (EPDM) are the most common types of elastomer used to modify the impact properties of PP resins [1]. Other elastomeric modifiers include natural rubber [2], styrene-butadiene-styrene block copolymer (SBS) and its hydrogenated analogue SEBS [3], polyisobutylene (PIB) [4], and very low density polyethylene (VLDPE).

In general, any elastomer-modified PP blend is known as a thermoplastic olefin (TPO). By manipulating the ratio of materials, as well as judicious selection of PP resin and elastomeric modifier, TPO formulators can achieve wide ranges of stiffness, flexural modulus, and impact strength. The use of TPO in automotive interior, exterior, and under-the-hood applications increased considerably during the 1980s and early 1990s. In North America, TPO is used in more than 70% of bumper fascia. The next major growth will be in interior parts such as air bag covers, door skins, and instrument panel skins. The annual growth rate of TPO compounds designed for automotive applications is estimated to be 10.7% in the years 1997–2002 [5]. For interior parts, the growth rate is predicted to exceed 30%.

As compounders developed and refined TPO blend technology, PP manufacturers pursued an alternate approach to PP impact modification by developing different grades of impact copolymer (ICP) resins produced in a series of reactors (see Chapter 2). The first reactor produces PP homopolymer, followed by one or two gas-phase reactors in which ethylene is introduced to produce EPR. The gas-phase reactor can be either a vertical fluidized bed or a horizontal stirred bed design [6]. Because of different reactivity of propylene and ethylene inside the gas phase reactor, complex mixtures of PP with ethylene-propylene copolymers and linear low-density polyethylene (LLDPE) are produced.

Compounders responded by developing new TPO formulations based on elastomer-modified ICP. Recently, resin suppliers have introduced even higher elastomer content reactor products called reactor-grade TPOs (RxTPOs) [7]. The result has been an ever broadening array of product types and continually improving performance in a variety of applications. The advent of metallocene plastomers [8] for PP modification represents the latest step in this continuous improvement process.

## 7.2 METALLOCENE PLASTOMERS

### 7.2.1 Definition

Plastomers are ethylene-$\alpha$-olefin copolymers with compositions and properties spanning the polyolefin spectrum between *plast*ics such as LLDPE and elas*tomers* such as EPR or EPDM. Comonomer content typically ranges from about 10 wt% to about 30 wt%. Density ranges from about 0.860 to about 0.910. Figure 7.1 illustrates the relationship between plastomers and other polyolefins. Some writers distinguish between "polyolefin plastomers" (POPs) and "polyolefin elastomers" (POEs) by dividing the groups at about 0.89 density, with POPs above and POEs below this density (9). For the purposes of this chapter, the definition of plastomers includes both of these categories.

Metallocene plastomers are plastomers made using metallocene single-site catalysts. Other polyolefins exist in this composition range made with conventional catalysts: Dow Attane [10], Union Carbide Flexomer (11); and Mitsui Tafmer products [12]. These are generally called "very low density polyethy-

# Metallocene Plastomers as PP Impact Modifiers

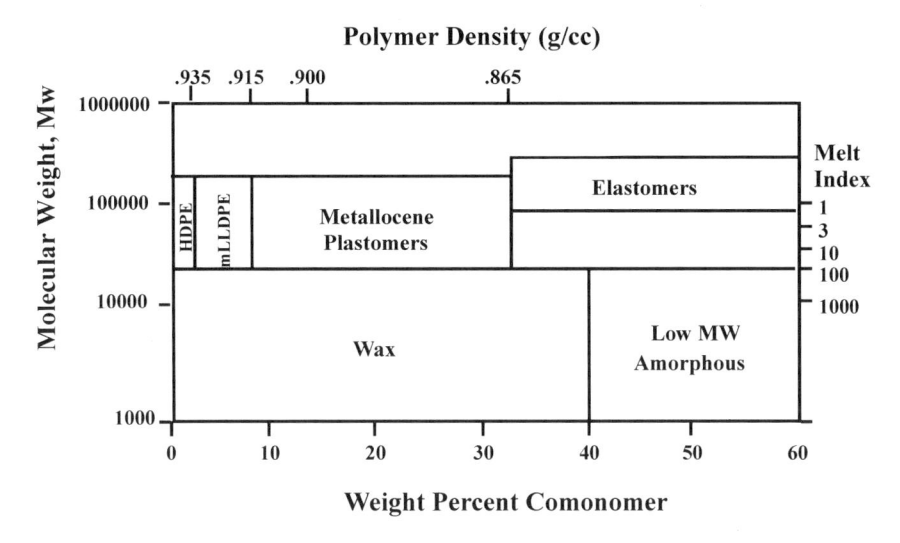

**FIGURE 7.1** Polyolefin product regions.

lenes" (VLDPEs) and "ultralow density polyethylenes" (ULDPEs), and are not discussed in this chapter.

### 7.2.2 Metallocene Catalysts

A metallocene is a coordination compound consisting of a transition metal ion, such as zirconium or titanium, with one or two cyclopentadienyl ligands [13]. The ligands are frequently joined by a short "bridge," which constrains the shape of the complex to a "clam shell" geometry. The discovery that opened the door to commercial success was that substitution of the cyclopentadienyl rings allows the metallocene to produce high molecular weight polymers [14,15]. The length and structure of the bridge and the nature of cyclopentyl ring substitution are critical factors that control the activity and selectivity of the catalyst, the structure, and sometimes the stereochemistry of the polymer product. The structures of two typical metallocene catalysts are shown in Fig. 7.2. It is well known that the critical success of Ziegler-Natta catalysis was based on the discovery of metal alkyls as activators. The discovery of methyl alumoxane [16] and noncoordinating anions has contributed similarly to the success of metallocene catalysts in polyolefin production [17,18].

### 7.2.3 Advantages of Metallocene Catalysts in Plastomer Production

Metallocenes are generally 'single-site' catalysts, meaning that all catalytic sites are identical. This feature yields products that are extremely uniform in composi-

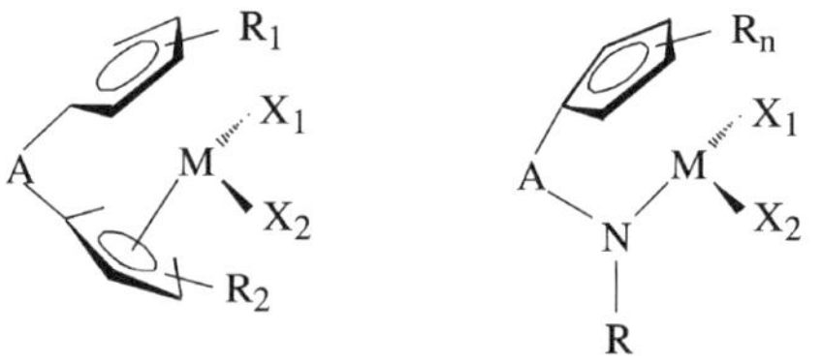

Bis-Cp(Cyclopentadienyl ring)

M = Transition metal, usually group 4b (Zr, Ti, Hf)
A = Optional bridge atom, generally Si or C atom
R = H, alkyl, or other hydrocarbon groups
X = Halogen atom (generally Cl) or alkyl group

**FIGURE 7.2** Structures of typical metallocene catalysts.

tion and molecular weight, approaching statistical limits. In contrast, most conventional Ziegler-Natta catalysts have multiple active sites exhibiting a range of reactivity and producing a spectrum of composition and molecular weight. (Some Ziegler-Natta catalysts, such as soluble vanadium catalysts, are single site). In addition, metallocene catalysts are generally more effective than conventional catalysts for incorporating high levels of a wide variety of olefin comonomers. This characteristic allows the large-scale production of ethylene-$\alpha$-olefin copolymers over the entire composition range. Commercial metallocene catalysts demonstrate high catalyst efficiency, yielding products that are low in catalyst residue and, importantly, free of acidic residues such as the chlorides commonly left by Ziegler-Natta catalysts. Efficacy of metallocene catalysts has been proven in a variety of commercial processes, including solution, slurry, gas-phase, and high-pressure bulk polymerizations.

### 7.2.4 Commercial Plastomer Suppliers

Six companies currently market metallocene plastomers. Exxon Chemical manufactures and markets ethylene-butene and ethylene-hexene copolymers under the Exact tradename. Dex Plastomers, a joint venture of Exxon Chemical and DSM, manufactures ethylene/octene copolymers under the Exact tradename, which are sold by both partners. Dow Chemical and DuPont Dow Elastomers offer ethylene-octene copolymers under the trade names Affinity and Engage, respectively. In the near future DuPont Dow will also offer ethylene-butene copolymers. Japan Polychem manufactures and markets ethylene-hexene copolymers under the Kernel tradename. Mitsui Petrochemical manufactures propylene, butene, and octene copolymers, which are sold under the Tafmer trade name, along with nonmetallocene plastomers.

## 7.2.5 Physical Properties

Physical properties of metallocene plastomers span the range between plastics and elastomers. Compared with LLDPE, plastomers are lower in density, tensile strength, flexural modulus, hardness, and melting point. They exhibit higher elongation and toughness. They are exceptionally clear, with very low haze values at lower densities. Physical properties of some typical plastomer resins are shown in Table 7.1. For all of the mechanical property measurements, compression-molded 1/8-in.-thick test plaques were prepared and conditioned for 40 h before test specimens were die cut for the conditioned sample plaques.

Compared with polar-ethylene copolymers such as ethylene vinyl acetate (EVA), ethylene methyl acrylate (EMA), and ethylene ethyl acrylate (EEA) copolymers, plastomers are significantly more compatible with PP. Compared with EVA, plastomers have exceptional thermal stability. Densities are significantly lower than those of polar-ethylene copolymers.

Because of their narrow composition distribution, differential scanning calorimetry (DSC) of metallocene plastomers exhibits sharp endotherm (melting) and exotherm (crystallization) peaks. Melting and crystallization temperatures decrease sharply with increasing comonomer content and are significantly lower than the melting points ($T_m$) of Ziegler-Natta products of the same composition. Figure 7.3 illustrates the relationship between composition and $T_m$ for ethylene-butene copolymers, compared with melting points of Ziegler-Natta products. Melting points of the latter are depressed only slightly by increasing comonomer because of the presence of high molecular weight (MW), low comonomer fractions that dominate the melting behavior of the material. With their sharp melting points, plastomers are readily pelletized and remain as discrete free-flowing pellets in spite of unusually low densities.

## 7.2.6 Thermal Transitions

Figure 7.4 shows the dynamic mechanical spectrum of an ethylene-butene plastomer [0.8 dg/min melt index (MI) and 0.880 g/cm$^3$ density] used for PP modification. Data were generated at 3 Hz, 2°C/min heat rate, and in uniaxial tensile mode. Several tan δ peaks indicate two distinct transitions designated β (−30°C) and γ (−110°C). The α-transition represents the local mobility in the crystalline region of polyethylene lamella. Because softening of the test specimen leads to excessive extension, the α-transition is not detected by this test.

The β-transition represents the glass transition temperature ($T_g$) of the polymer and represents motion or displacement of polymer segments. The γ-transition results from the localized relaxation of polymer chains and is believed to provide the major mechanism for low-temperature impact absorption. In well-dispersed PP-plastomer blends, the β-transition temperature is generally not a

TABLE 7.1 Physical Properties of Ethylene-α-Olefin Plastomers

| Property | Test method | Butene copolymers Exact 3035 | Exact 4011 | Exact 4033 | Exact 4049 | Hexene copolymers Exact 3132 | Exact 4150 | Octene copolymers Exact 0201 | Exact 8201 |
|---|---|---|---|---|---|---|---|---|---|
| Density, g/cm$^3$ | ASTM D1505 | 0.900 | 0.888 | 0.880 | 0.873 | 0.900 | 0.895 | 0.902 | 0.882 |
| Melt index, dg/min | ASTM D1238 | 3.5 | 2.2 | 0.8 | 4.5 | 1.2 | 3.5 | 1.1 | 1.1 |
| Melt flow rate, dg/min | ASTM D1238 | 4.5 | 2.8 | 1.2 | 7.0 | 2.0 | 6.7 | 2.3 | 2.2 |
| Mooney viscosity, ML(1 + 4) @125°C | ASTM D1646 | 8 | 12 | 26 | 7 | 21 | 8 | 17 | 19 |
| DSC peak melting point, °C | ASTM D3417 | 91 | 71 | 61 | 54 | 99 | 89 | 96 | 73 |
| Hardness, Shore A, 15 sec | ASTM D2240 | 90 | 84 | 77 | 71 | 87 | 87 | >90 | 81 |
| Hardness, Shore D, 15 sec | | 43 | 34 | 29 | 21 | 47 | 37 | 43 | 31 |
| Vicat softening point, 1000 g, °C | ASTM D1525 | 79 | 61 | 54 | 40 | 91 | 73 | 85 | 56 |
| Ultimate tensile stress, Kpsi | ASTM D412 | 2.6 | 3.7 | 1.0 | 0.9 | 4.7 | 4.6 | 4.1 | 3.5 |
| Elongation @ break, % | ASTM D412 | 710 | 710 | 590 | 860 | 650 | 700 | 790 | 730 |
| Tensile modulus, Kpsi | ASTM D412 | | | | | | | | |
| @ 100% | | 0.9 | 0.6 | 0.5 | 0.4 | 1.1 | 0.9 | 1.0 | 0.6 |
| @ 200% | | 0.9 | 0.7 | 0.6 | 0.5 | 1.1 | 0.9 | 1.0 | 0.7 |
| @ 300% | | 1.0 | 0.8 | 0.7 | 0.5 | 1.3 | 1.0 | 1.1 | 0.8 |
| Flexural modulus, 1% Secant, Kpsi | ASTM D790 | 12.0 | 5.4 | 3.5 | 1.8 | 14.0 | 9.5 | 12.0 | 5.0 |

# Metallocene Plastomers as PP Impact Modifiers

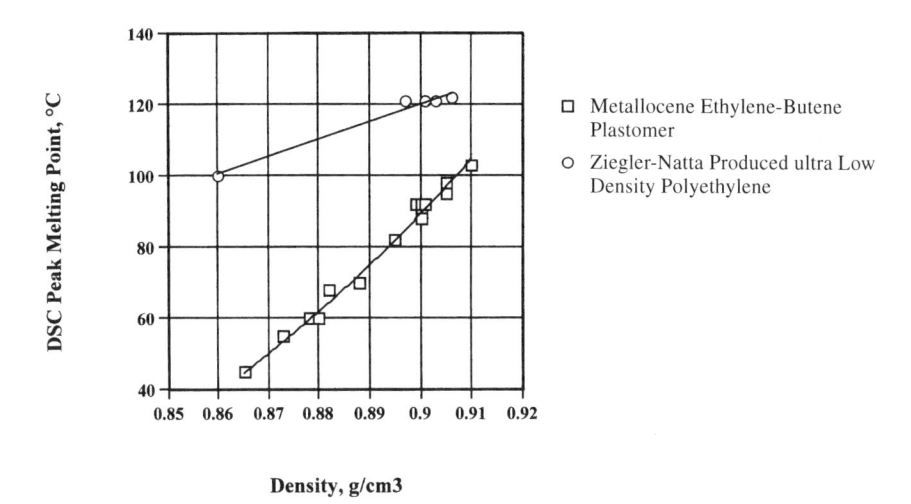

**FIGURE 7.3** Relationship of melting point to composition of ethylene-butene plastomers. □, Metallocene ethylene-butene plastomer; ○, Ziegler-Natta-produced ULDPE.

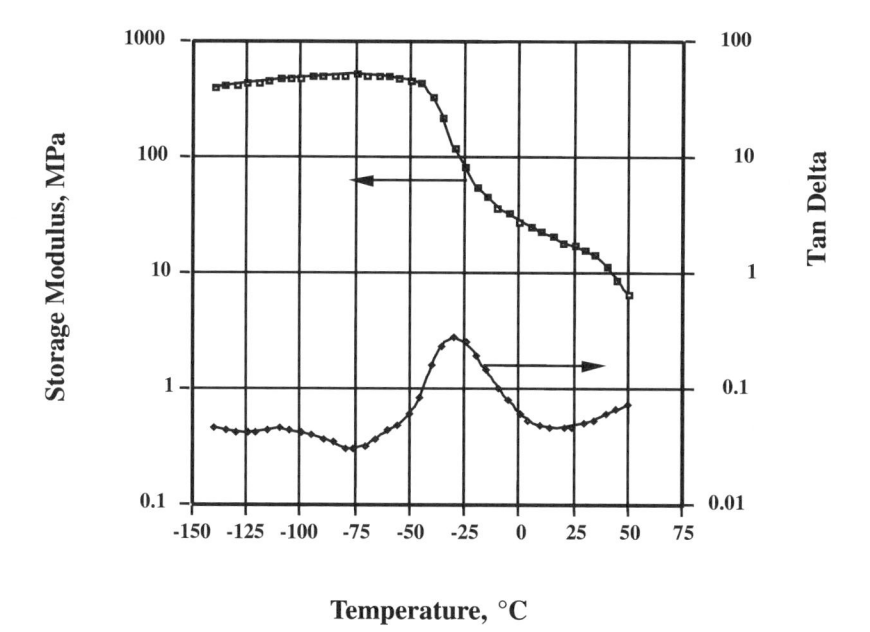

**FIGURE 7.4** Dynamic mechanical properties of ethylene-butene plastomer.

limitation for low-temperature impact performance. Figure 7.5 shows that $T_g$ is dependent on the density or comonomer content.

### 7.2.7 Molecular Weight and Composition Distribution

The unique features of plastomer can be illustrated by its molecular architecture. Figure 7.6 shows the molecular weight distribution for both a metallocene catalyst–produced ethylene-butene copolymer (Exact 4033 with 0.8 dg/min MI and 0.880 g/cm$^3$ density) and a gas-phase-polymerized VLDPE based on Ziegler-Natta catalyst (Flexomer 1085 with 0.8 dg/min MI and 0.884 g/cm$^3$ density). The measurement was carried out using gel permeation chromatography coupled with differential refractive index detection. It is clearly shown that the VLDPE possesses more of both comonomer-poor high molecules and comonomer-rich low-end molecules than plastomer. The presence of high percentages of high ends in VLDPE makes it opaque. The large amount of low ends makes the product tacky and difficult to handle. The common fractionation techniques used for the determination of composition distribution of polyethylenes are temperature-rising elution-fractionation (TREF) and crystallization analysis fractionation (CRYSTAF). Both analytical techniques are inadequate when the crystallinity of the sample is very low, i.e., <0.900 density plastomer. A third technique based on the stepwise isothermal segregation technique (SITS) [19] can be employed to obtain approximate composition distribution of plastomers.

Using conventional DSC, Fig. 7.7 shows differences in the melting behavior for Exact 4033, Exact 3035, and Flexomer 1085:

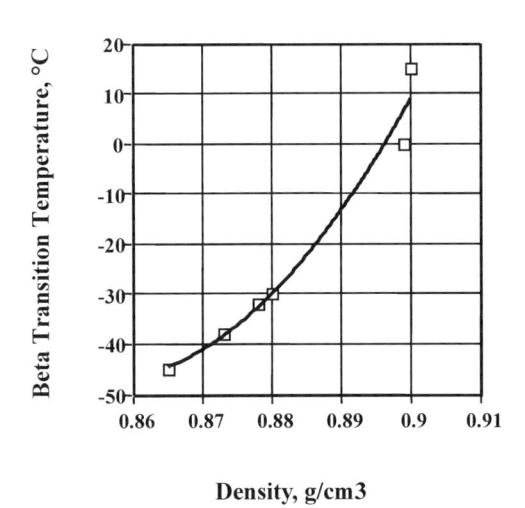

FIGURE 7.5 Effect of copolymer composition on β-transition temperature.

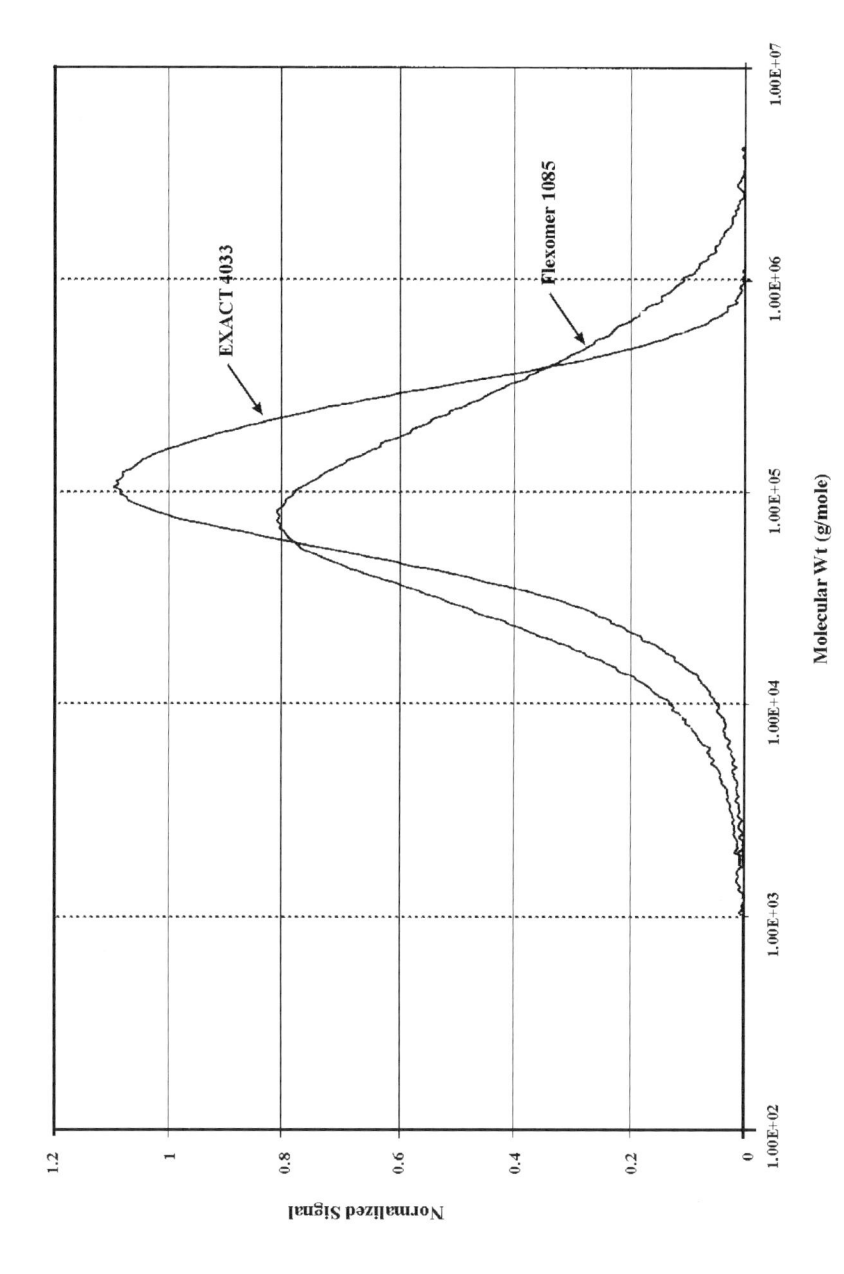

# Metallocene Plastomers as PP Impact Modifiers

**FIGURE 7.6** Molecular weight distribution of Exact 4033 and Flexomer 1085.

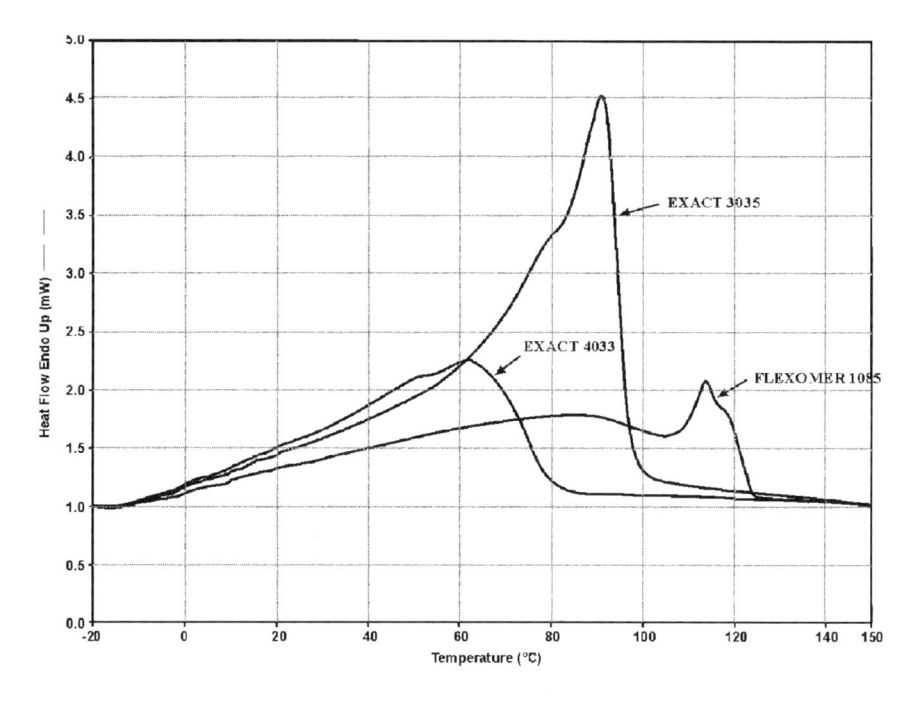

**FIGURE 7.7** DSC peak melting point of Exact 4033, Exact 3035, and Flexomer 1085.

1. Exact 4033 exhibits a broad melting endotherm with peak temperature at about 60°C.
2. Exact 3035 (3 MI and 0.900 density) has low butane comonomer content with reduced intramolecular heterogeneity. Consequently, the endotherm peak is quite narrow and has a melting point located at about 90°C.
3. In comparison, Flexomer 1085 has a much higher melting point at 114°C.

Using the SITS methodology, the number of methylene units in Exact 4033 is determined to range from 8 to 29 compared with 9 to 125 for Flexomer 1085. Flexomer 1085 therefore shows a much broader composition heterogeneity than Exact 4033. The SITS procedure consisted of the following steps:

1. The weighed sample is heated to 210°C and held at that temperature for 5 min to destroy the effect of thermal history.

# Metallocene Plastomers as PP Impact Modifiers

2. Afterward, the sample is fast cooled at a rate of 20°C/min to approximately the sample's peak melting points determined by the initial DSC measurement.
3. The sample is then successively cooled by increments of 10°C at the rate of 1°C/min and maintained at each temperature for 30 min for crystallization to reach steady-state equilibrium.
4. The fractionated sample was next analyzed for determination of its multiple melting points as shown in Fig. 7.8.
5. The phenomenon of fractionation during crystallization is well described in the literature. For each melting point, one can determine a sequence length of crystallizable methylene (CH2) units based on a monodispersed model for polymers [20] as tabulated in Table 7.2.

## 7.3 MATERIALS

### 7.3.1 Modifiers

Unless otherwise noted, the modifiers used in the experiments described below were the following. All are commonly used in PP modification. MI values given

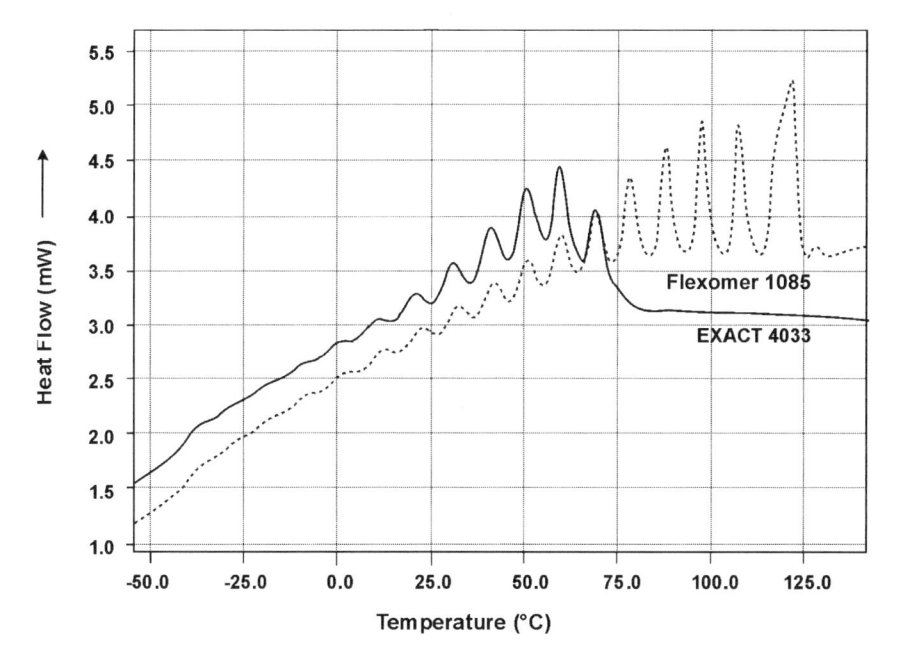

FIGURE 7.8 Thermal fractionation experiment for Exact 4033 versus Flexomer 1085.

TABLE 7.2 Methylene Distribution of Exact 4033 and Flexomer 1085

| Temp.°C | Heat of fusion, J/g Exact 4033 | Number of methylene CH₂ units Exact 4033 | Temp.°C | Heat of fusion, J/g Flexomer 1085 NT | Number of methylene CH₂ units Flexomer 1085 NT |
|---|---|---|---|---|---|
| −28.8 | 0.7 | 8 | −26.4 | 0.7 | 10 |
| −19.0 | 2.3 | 9 | −17.4 | 1.4 | 10 |
| −8.8 | 3.2 | 10 | −7.5 | 2.3 | 11 |
| 0.6 | 4.3 | 11 | 2.1 | 2.9 | 12 |
| 10.5 | 6.5 | 13 | 12.1 | 4.0 | 13 |
| 20.4 | 7.9 | 14 | 22.0 | 4.2 | 14 |
| 30.1 | 9.3 | 16 | 32.0 | 5.0 | 15 |
| 40.9 | 11.7 | 19 | 41.7 | 5.1 | 17 |
| 50.5 | 14.5 | 21 | 50.9 | 5.6 | 19 |
| 59.6 | 13.5 | 25 | 59.6 | 6.3 | 21 |
| 68.9 | 8.5 | 29 | 68.8 | 6.7 | 24 |
|  |  |  | 78.4 | 7.2 | 28 |
|  |  |  | 87.6 | 6.8 | 33 |
|  |  |  | 97.2 | 6.6 | 41 |
|  |  |  | 106.8 | 5.7 | 55 |
|  |  |  | 121.1 | 9.7 | 107 |
|  |  |  | 127.6 | 0.4 | 190 |

as degrees per minute are measured at 190°C and 2.16 kg load. The modifier density values are given as grams per cubic centimeter.

*Ethylene-butene (EB) plastomer:* Exact 4033 (0.80 dg/min, 0.880 g/cm$^3$ ), Exact 3035 (1.0°/min, 0.900 g/cm$^3$), Exact 4049 (4.5°/min, 0.873 g/cm$^3$), Exact 3035 (3.5°/min, 0.900 g/cm$^3$) are all metallocene-catalyzed products made by ExxonMobil Chemical Company. Flexomer 1085 (0.8°/cm$^3$, 0.884 g/cm$^3$) is an ethylene-butene copolymer that is produced via a gas-phase reactor process using Ziegler-Natta catalyst originally at Union Carbide and now at Dow Chemical. The tacky surfaces of Flexomer pellets are dusted with talc powder to reduce the tendency for sticking together.

*Ethylene-octene (EO) plastomer:* Engage 8100 (1.0°/min, 0.870 g/cm$^3$) is manufactured by Dupont Dow Elastomers. Exact 0201 (3.0°/min, 0.920 g/cm$^3$) is an ExxonMobil product.

*Ethylene-hexene (EH) plastomer:* EXACT 9106 (2.0°/min, 0.900 g/cm$^3$) is made by ExxonMobil.

*Ethylene-propylene rubber (EPR):* JSR07P ethylene-propylene rubber (0.3°/min, Mooney viscosity is 47, 0.864 g/cm$^3$) comes from Japan Synthetic Rubber Company.

### 7.3.2 Polypropylenes

All PP resins used were commercial grades produced by ExxonMobil. Homopolymers were general-purpose or injection molding grades. Random copolymers were high-clarity grades containing 2% ethylene and a clarifying agent. ICPs were injection molding grades containing 9% or more ethylene. Melt flow rates of PP resins are measured as degrees per minute at 230°C and 2.16 kg load, and are specified in the individual experiments.

### 7.3.3 Mineral Filler

The grade of talc used in talc-filled PP experiments was Cimpact 710, a non-surface-treated micronized talc, with an average particle size of 1 μm, produced by Luzenac America. Many other types of particulate and layered inorganic fillers have been evaluated, but the layered talc gained wide acceptance due to its nonabrasive nature and its ability to serve as a nucleating agent for PP.

Recently, composites containing montmorillonite (a special grade of treated clay platelets) have generated a lot of interest in polypropylene modification. Montmorillonite is a naturally occurring clay phyllosilicate that has the same layered and crystalline structure as talc but has different layers of ionic charges. Maleic anhydride–grafted PP compatibilizer is often blended with octadecyl-ammonium surface-treated montmorillonite to make a master batch [21]. The master batch is next compounded under high shear with PP resin to exfoliate multiple layers of clay into separate layers. This new type of mineral reinforced PP composite contains a relatively low loading of nanometer ($1 \times 10^{-9}$ m) thick clay platelets with nominal 10- to 30-µm cross-sections. Therefore, the resulting nanocomposites have high aspect ratios (>1000) to give marked improvement in mechanical (high stiffness) and thermal (elevated heat distortion temperatures) properties. Chapter 20 provides a detailed description of nanocomposite technology.

The inherent thermal stability of nanocomposites and tendency to form char upon ignition has led to increased activity in the development of flame-retardant materials based on PP and other thermoplastics [22]. The observed synergetic effect on flame retardancy caused by adding small amounts of exfoliated clay has led to reduced loading levels of conventional halogenated flame retardants, as discussed in Chapter 20.

In comparison, the commercialization of nanocomposites based on TPO to utilize the enhancement of physical properties has been limited thus far to just a few cases. Nanocomposites having low loading levels of 6% by weight of exfoliated montmorillonite layers are beginning to show up in automotive exterior claddings such as a TPO step-assist for General Motors vans [23]. These polyolefin materials are considered potential replacements for engineering resin applications with molded parts having low density and relatively low cost.

## 7.4 TEST METHODS

### 7.4.1 Impact Testing

Two types of impact testing are generally used to characterize TPO compounds: the notched Izod test (ASTM D256) [24] and the instrumented impact test (ASTM D3763) [25]. The notched Izod test is familiar to most workers in the plastics field. A standard notch is cut in a rectangular specimen. The specimen is broken by a dropping pendulum. The impact strength of the material is computed from loss of momentum of the pendulum. Please refer to Chapter 6 for a complete description of impact testing equipment design and methods.

The instrumented impact test is an automated upgrade of the Gardener impact test. The conventional Gardner test uses a falling dart to pierce a disk specimen. The drop height of the dart is used to calculate the impact energy of the material at failure. The instrumented test uses a similar falling dart, with a strain gauge attached to the tip of the dart. The instrument records force as a function of distance as the dart penetrates the specimen. Test speed can be adjusted to provide testing at 5–25 miles/hr test speed by varying dart height. Tests are frequently carried out in a controlled-temperature chamber to provide access to the low temperatures required for certain automotive fascia specifications. A typical instrument uses a dart with a 0.5 in hemispherical tip. The test specimen is a 4-in.-diameter by 1/8-in.-thick injection-molded disk, clamped with a 3-in.-diameter test area.

A variant of the instrumented impact test employs a spring-driven ram rather than a falling dart as the source of kinetic energy. Principles of the test are otherwise similar to those of the dart test.

A typical instrumented impact test output is shown in Fig. 7.9 [26]. The force versus distance data are integrated to provide the total energy needed to break the specimen. In addition, the shape of the curve provides information about the failure mode. If the force trace drops nearly vertically from its maximal value, brittle failure occurs. If the force trace is elongated, ductile failure occurs. A "ductility index" (DI) can be computed:

$$DI = \{(\text{total energy} - \text{yield energy})/\text{yield energy}\} \times 100$$

Total energy represents the total area under the force–distance curve. Yield energy is the area under the portion from the start of the experiment to the point of maximal force. As a rule of thumb, a ductility index less than 35% represents brittle failure. It is necessary to report both the total energy and ductility index to characterize impact strength in the instrumented impact test.

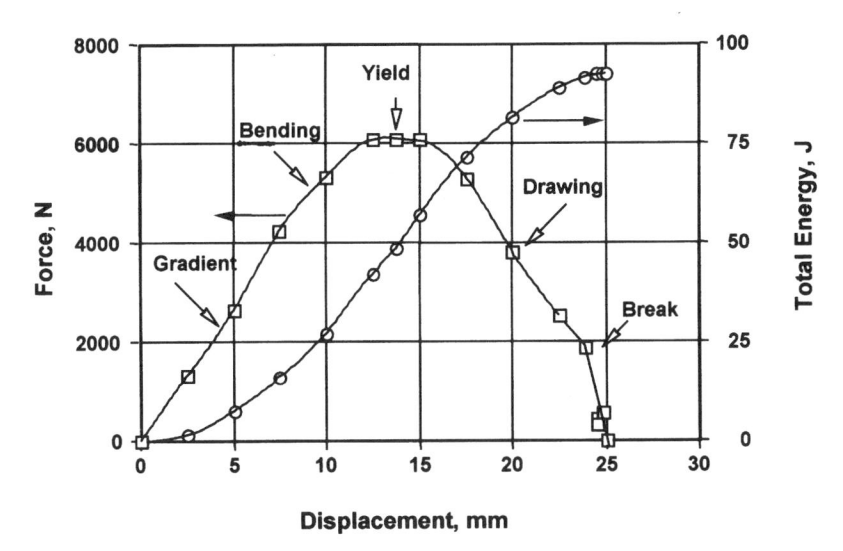

FIGURE 7.9 Falling-weight instrumented impact test. □, Force versus displacement data; ○, integrated total energy.

## 7.4.2 Analysis of Morphology

Because of low contrast between metallocene plastomers and PP, a new method [27] has been developed to characterize polyolefin blend morphology using heavy $RuO_4$ staining and low-voltage scanning electron microscopy (LVSEM). A glass knife, then stained in $RuO_4$ vapor for 2.5 h; first cryogenically sections the sample face. Sections about 100 nm thick are cut from the face to reach a depth of approximately 0.5 mm. The sample is then examined by LVSEM at 2.0 kV. Computer image analysis of the micrographs is used to determine average particle size and particle size distribution of dispersed plastomers in the PP matrix. The scanned images are analyzed using image analysis software [28,29].

The software calculates the area of the dispersed phase, the major and minor axes ($D_{major}$ and $D_{minor}$), and orientation of each modifier particle. Typically about 80 particles are analyzed from each micrograph and three to four images of every blend examined for statistically meaningful results. The following parameters calculated:

$$D_{avg1} = (D_{major} + D_{minor})/2$$
$$D_{avg2} = (4\ area/\pi)^{1/2}$$
$$D_n = \Sigma n_i d_i / \Sigma n_i$$
$$D_v = \Sigma n_i d_i^3 / \Sigma n_i d_i^2$$
$$AR = D_{major}/D_{minor}$$

$D_{avg1}$ and $D_{avg2}$ represent two ways to represent average particle diameter. $D_{avg1}$ is the average of the particle major and minor axes lengths. $D_{avg2}$ is the diameter of a circle with an area equal to the measured area of the dispersed particle. $D_n$ and $D_v$ are the number average and volume average diameter, respectively. AR is the aspect ratio.

Particle diameters measured or computed from cut sections generally underestimate actual particle diameters since some sections are cut at the top or bottom of a particle rather than from its midsection. Several methods have been proposed in the literature to estimate the actual diameter from two-dimensional projections. A recent study [30] concluded that applying these corrections provided only an 8% increase in $D_n$ values and produced a small shift in the overall distribution to lower diameters. These corrections were not applied in the analyses described below.

## 7.5 MODIFICATION OF PP HOMOPOLYMERS

Four types of polypropylene resins are available commercially: homopolymer, random copolymer, impact copolymer, and reactor-grade TPO. Homopolymer polypropylene (HPP) is the most difficult to impact enhance. All other types of

# Metallocene Plastomers as PP Impact Modifiers

PP are easier to modify because they contain varying amounts of ethylene linkages or ethylene-propylene bipolymers that reduce the stiffness of the base resin and increase its impact resistance. These structures also provide some degree of compatibility with ethylene-α-olefin plastomers. Basic studies conducted using HPP in polypropylene-plastomer blends illustrate principles that apply to other types of PP as well.

## 7.5.1 Impact Enhancement of a Plastomer Compared with an EP Elastomer

Formulation of an impact-enhanced PP typically requires a trade-off of other physical properties (particularly flexural modulus) to achieve improved impact strength. Flexural modulus values decrease continuously as modifier content is increased. However, impact strength generally increases to a maximal plateau beyond which no significant impact enhancement is realized. Often, the formulator's goal is to find the minimal modifier level needed to reach the maximal impact plateau. This task is complicated by the fact that this optimal level varies with the impact test used and the temperature of the test. Figure 7.10 illustrates this situation by showing the effects of increasing the loading level for two types of impact modifier (EB plastomer and EPR) on the Izod impact of homopolymer PP. Plots for flexural modulus decrease in a linear fashion. In comparison, the trend line for impact strength follows an 'S'-shaped curve. Initially, impact

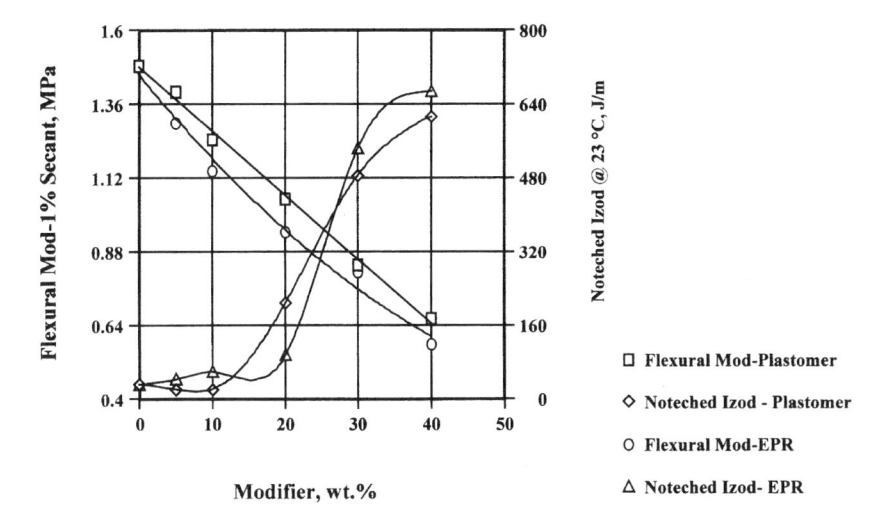

**FIGURE 7.10** Stiffness/Izod impact balance in HPP blends. □, Flexural modulus plastomer; ◇, notched Izod plastomer; ○, flexural modulus EPR; △, notched Izod EPR.

strength values change little between 0-15 wt% modifier, but from 15 wt% to about 40 wt% impact strength increases sharply. Eventually, impact strength values reach a constant plateau above about 40 wt% modifier.

The following experiments compare the impact modification effects of an EB plastomer with EPR in PP homopolymer. The design of experiment matrix for TPO formulations consisted of two PP homopolymer resins (MFR values of 5°/min and 35°/min) and two different modifiers (EB plastomer and EPR). Each PP resin was blended with 20 wt% and 30 wt% of a given modifier and compounded by a twin screw extruder. Injection molded parts were tested for flexural modulus, Izod impact, and instrumented impact. Impact tests were run at several temperatures.

Test data are summarized in Table 7.3 and Figs. 7.11 to 7.13. Moduli of the compounds decrease with added modifier, but appear independent of modifier type. In notched Izod tests in 5 MFR PP, good impact results were achieved with both modifiers at the 30 wt% modifier level. They also showed some impact enhancement at the 20 wt% level and at 0°C with 30% modifier. In 35 MFR PP only limited impact enhancement was realized, and then only at 23°C and 30 wt% modifier. In most instances where impact was above 1 ft-lb, the values were higher for the plastomer blends than the EPR blends. The data show that impact strength is greater at higher temperature, higher modifier loading, and for the PP resin with lower MFR (5°/min).

TABLE 7.3 Properties of Impact-Modified Homopolymers

|  | Base resin | 20% EB plastomer | 30% EB plastomer | 20% EPR | 30% EPR |
|---|---|---|---|---|---|
| *Base resin: 5 MFR HPP* | | | | | |
| Flexural modulus, 1% secant, MPa | 1622 | 1145 | 869 | 1083 | 883 |
| Notched Izod, J/m | | | | | |
| 23°C | 30 | 325 | 581 | 96 | 640 |
| 0°C | 13 | 53 | 277 | 27 | 69 |
| −40°C | 13 | 14 | 23 | 17 | 28 |
| *Base resin: 35 MFR HPP* | | | | | |
| Flexural modulus, 1% secant, MPa | 1249 | 925 | 731 | 835 | 704 |
| Notched Izod, J/m | | | | | |
| 23°C | 21 | 46 | 427 | 50 | 74 |
| 0°C | 11 | 24 | | 24 | |
| −40°C | 10 | 13 | | 12 | |

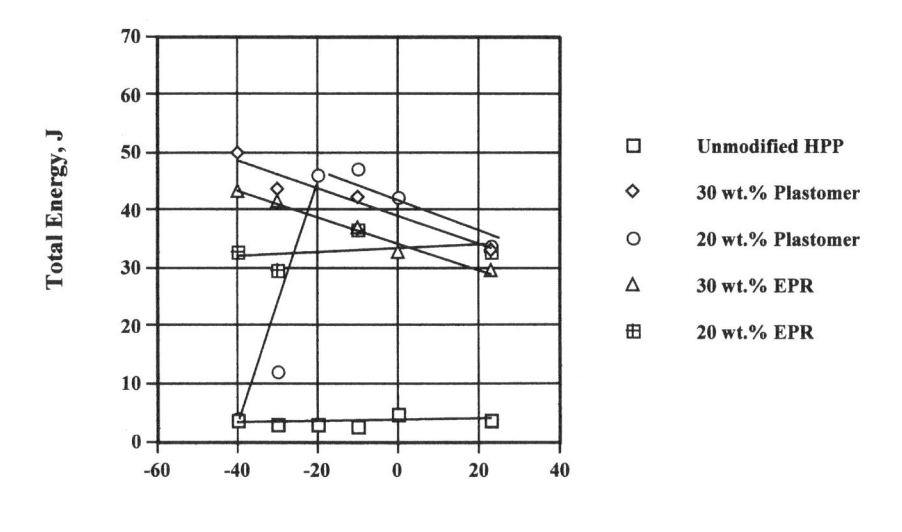

## Metallocene Plastomers as PP Impact Modifiers

**FIGURE 7.11** Modification of 5 MFR homopolymer: total energy. □, Unmodified HPP; ◇, 30 wt% plastomer; ○, 20 wt % plastomer; △, 30 wt% EPR; ⊞, 20 wt% EPR.

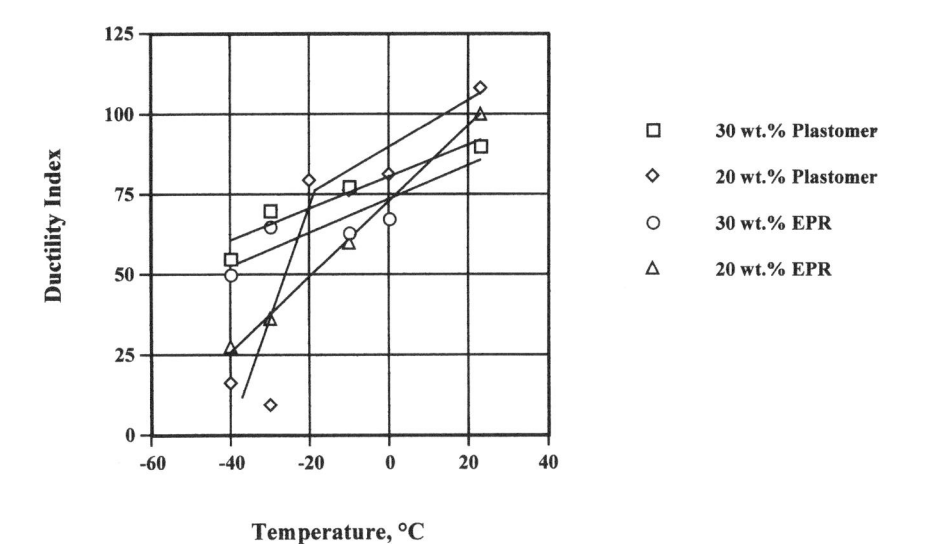

**FIGURE 7.12** Modification of 5 MFR homopolymer: ductility index. □, 30wt% plastomer; ◇, 20 wt% plastomer; ○, 30 wt% EPR; △, 20 wt% EPR.

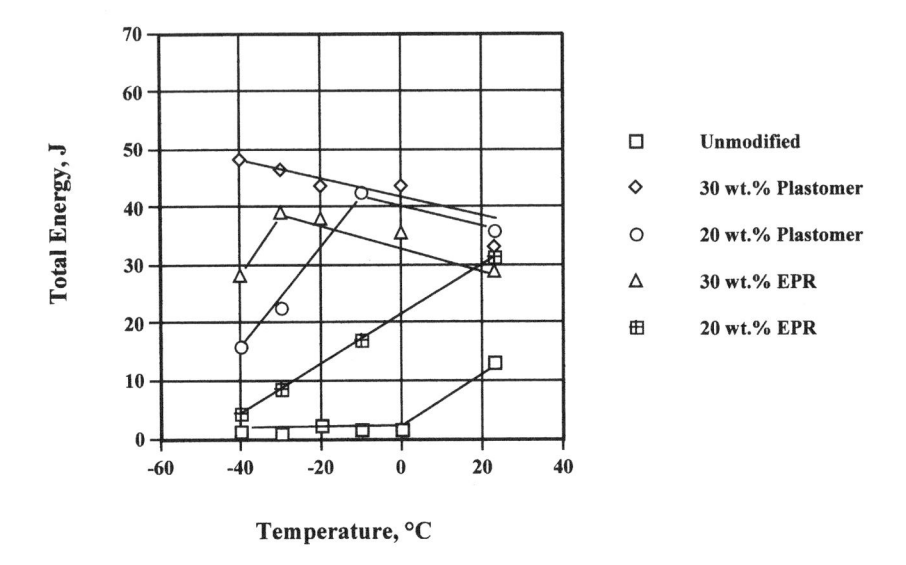

**FIGURE 7.13** Modification of 35 MFR homopolymer: total energy. □, Unmodified HPP; ◇, 30 wt% plastomer; ○, 20 wt% plastomer; △, 30 wt% EPR; ⊞, 20 wt% EPR.

The instrumented impact data are somewhat counterintuitive. Figure 7.11 shows for both modifiers at 30 wt % content, total impact energy *increases* as the test temperature *decreases*. The energy-absorbing components of these blends apparently become stiffer but not brittle at lower temperatures. At 20 wt% modifier, total impact energy increases to a maximum as test temperature decreases, then drops off sharply, in the case of the plastomer. It appears that 20 wt% modifier is not enough to fully reinforce the PP matrix at very low temperatures. At $-10°C$ and above, total energy varies little between 20 wt% and 30 wt % modifier. Raising the modifier level extends the effective low-temperature service range but adds little to impact performance at ambient temperature. At $-20°C$ and above, total impact energy was higher for the plastomer samples than for the analogous EPR samples. Figure 7.12 shows the computed DI data for these blends. Only the 20% plastomer blend showed a sharp transition between ductile and brittle failure, evidenced by a sharp drop in the DI.

Data for the 35 MFR homopolymer blends (Fig. 7.13) show total energy increasing with decreasing temperature for the 30 wt% EB plastomer blend, but passing through a maximum for the 20 wt% plastomer and 30 wt% EPR blends. The 20 wt% EPR blend had total energy decreasing continuously across the test temperature range. Total energy results for the plastomer blends were again greater for the plastomer blends than for the EPR blends across the temperature

# Metallocene Plastomers as PP Impact Modifiers

range. DI data for all four blends show a gradual loss of ductility with reduced temperature (Fig. 7.14).

The most effective modifier system across the entire temperature range was 30 wt% EB plastomer. Total energy values were about equal for this system in 5 MFR and 35 MFR PP, showing that effective impact modification could be achieved using PP resins with relatively high MFR and at low temperatures.

## 7.5.2 Effect of Mixing Intensity on Impact Enhancement

We hypothesize that much of the difference in impact modifier effectiveness among the four experiments described above is related to the effectiveness of modifier dispersion. The high-molecular-weight low-MI EPR dispersed more readily in the lower MI PP and gave better impact results there. The plastomer provided a closer viscosity match to both PP resins than the EPR and gave better impact results than the EPR.

In recent years, compounders have tended to replace batch-compounding equipment, such as Banbury mixers, with continuous processing equipment, such as twin-screw extruders. Plastomers are readily dispersed in PP using twin-screw compounding extruders. However, adequate dispersion for impact enhancement can also be achieved with less intensive mixing, including single-screw extruders and batch mixers. In some cases, impact modification can also be realized simply

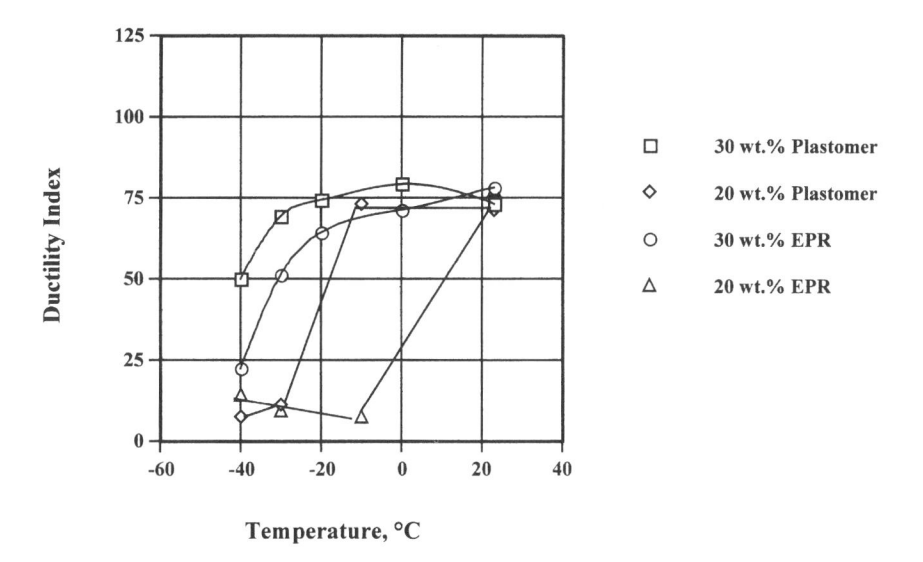

FIGURE 7.14 Modification of 35 MFR Homopolymer: ductility index. □, 30 wt% plastomer; ◇, 20 wt% plastomer; ○, 30 wt% EPR; △, 20 wt% EPR.

by using dry blends of plastomer and PP resin fed directly to an injection molding machine.

The experiment described below demonstrates the effect of varying degrees of mixing intensity associated with different compounding methods on the impact strength of a PP homopolymer-plastomer blend. The polymers used were the same 35 MFR HPP and EB plastomer used in the above series, blended at a 70:30 HPP plastomer ratio. Blends were compounded on several common commercial-size batch and continuous mixing systems. The continuous mixers were: 57-mm twin-screw extruder, $2\frac{1}{2}$-inch single-screw extruder, the same single-screw extruder with a Maddock mixing element and Farrel CP-500 continuous mixer. The batch machine was a Stewart Bolling no. 10 mixer. To complete the study, dry-blended ingredients were injection molded on the same machine as the compounds.

Figure 7.15 shows total impact energy as a function of temperature for all blends. The wide range of total impact energy at $-40°C$ among the experiments suggests a significant difference in degree of dispersion. The highest total impact energy and thus the best dispersion was achieved with the twin-screw extruder and the single-screw extruder with a Maddock mixing head. This was followed by Farrel continuous mixer, single-screw extruder without mixing head, and the Stewart Bolling intensive mixer. The least improvement in impact resistance was shown by the uncompounded dry blend. The DI data shown in Fig. 7.16 indicate that the only blends ductile at $-40°C$ were those compounded on the twin-screw and single-screw extruder fitted with the Maddock mixing segment.

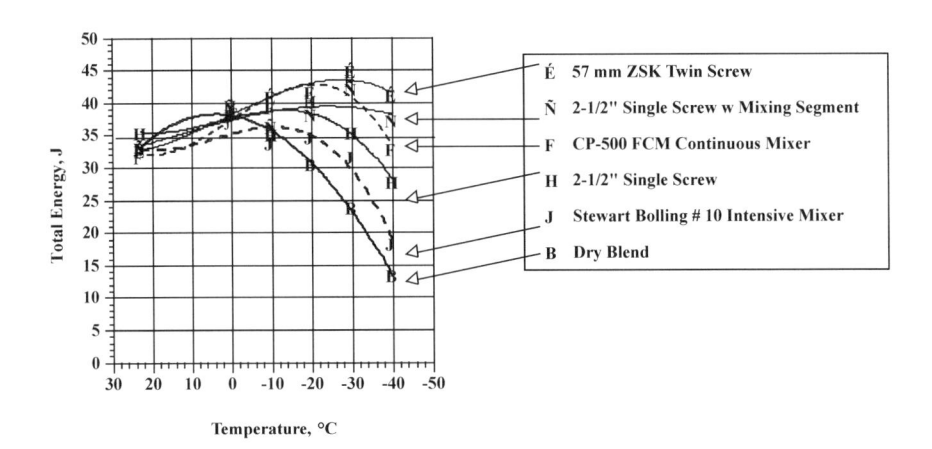

FIGURE 7.15 Effect of compounding method/mixing intensity on impact energy of PP-plastomer blends.

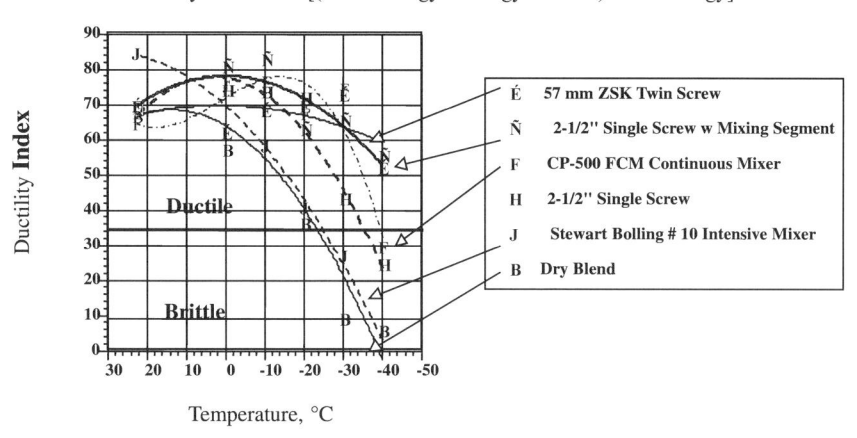

# Metallocene Plastomers as PP Impact Modifiers

Ductility Index DI = [( Total Energy - Energy at Yield)/Yield Energy] × 100

**FIGURE 7.16** Effect of compounding method/mixing intensity on ductility index of PP-plastomer blends.

It is interesting to note, and important for many practical applications, that at temperatures of $-10°C$ and above, the dispersion differences are not sufficient to produce a measurable difference in impact energy. For applications not requiring very-low-temperature service, even dry blending a modifier provides significant impact enhancement.

As an extension to the above study, several types of dispersive and distributive mixing elements [31] were evaluated using a single-screw extruder [32] compared with the Maddock mixing element used in the previous study. Elements evaluated included dispersive elements such as straight and tapered Maddock, pineapple and Twente mixers; and distributive elements such as pin and gear mixers. The results indicated that all of these elements would perform well for mixing PP and plastomers.

## 7.5.3 Morphology and Impact Resistance

The studies described in the previous section demonstrate the critical importance of mixing intensity on impact performance of PP modifiers. Plastomers and EPRs are not miscible with PP, but form highly dispersed two-phase polymer blends. Mixing is important because it determines the average domain size and particle size distribution of modifier domains dispersed in the PP matrix. In well-dispersed blends, average particle size can be less than 1 µm.

The two-phase microstructure is critical to impact modification. Chapters 6 and 11 describe various models for interpreting impact behavior. Impact energy is absorbed during fracture when a propagating crack meets a rubbery domain and fracture energy is dispersed by deformation of the rubbery material [33]. As a rule of thumb, an average particle size of about 1 µm must be achieved to provide suitable low-temperature impact.

Figure 7.17a shows a low-voltage scanning electron micrograph for a 35-MFR HPP-EB plastomer blend. Figure 7.17b is the corresponding particle size distribution determined by computer analysis of the image. Figure 7.18 characterizes a PP-EPR blend using the same PP base resin. The microstructure of dispersed elastomer particles is shown in Fig. 7.18a and the particle size distribution plot is given in Fig. 7.18b. The HPP-plastomer blend particle sizes ranged from 0.1 to 2.3 µm and averaged 0.78 µm. In the HPP-EPR blend, particle sizes were about double, ranging from 0.2 to 4.5 µm with an average value of 1.79 µm.

The cumulative particle size distributions for a 70:30 wt% blend of HPP-Exact 4033 were determined for five polypropylenes with MFR values of 1.7, 5.3, 13.0, 20.0, and 35.0 g/10 min. In Fig. 7.19, the particle size distribution is very nearly the same for the HPP resins ranging from 1.7 to 20 MFR. However, the particle size distribution plot for 35 MFR HPP tends to have increased domain

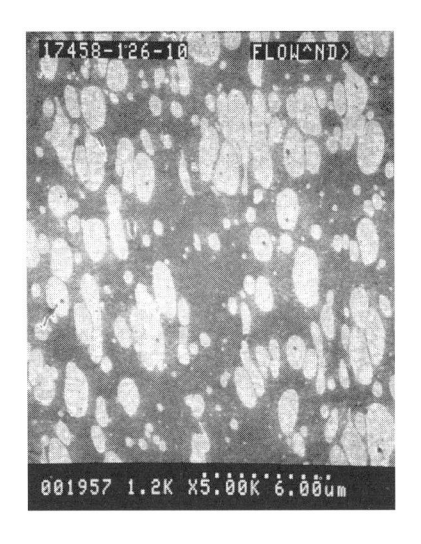

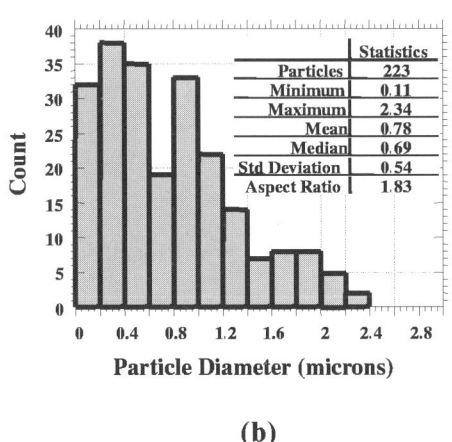

FIGURE 7.17 (a) SEM photomicrograph of morphology and (b) particle size distribution of HPP-plastomer blend.

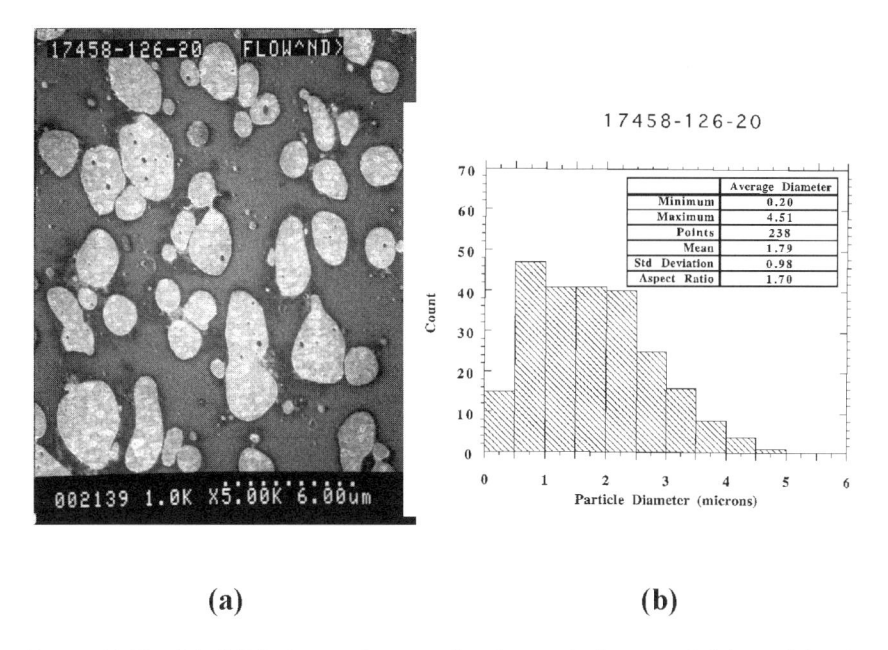

# Metallocene Plastomers as PP Impact Modifiers 245

**FIGURE 7.18** (a) SEM photomicrograph of morphology and (b) particle size distribution of HPP-EPR blend.

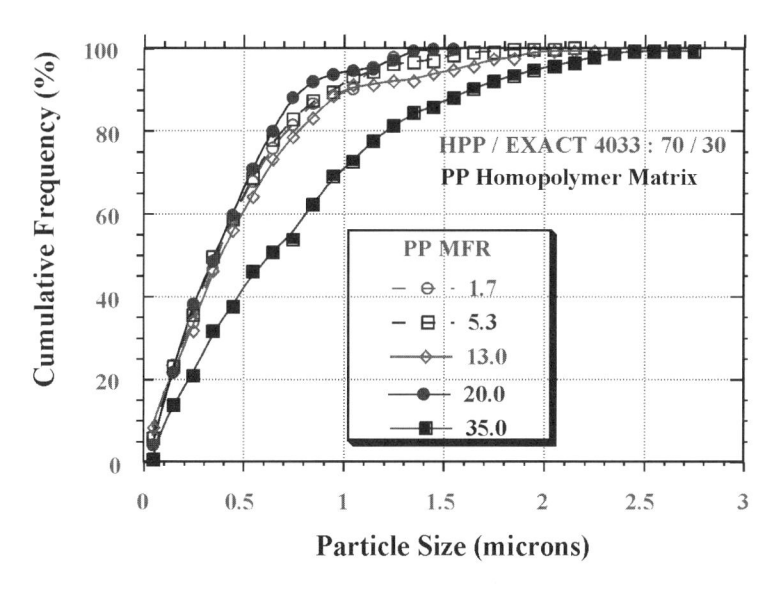

**FIGURE 7.19** Particle size distribution plots for given set of HPP MFR values.

size due to an increasingly large difference between the melt viscosities of the plastomer and the PP ingredients. For 20 MFR HPP, more than 85% of the plastomer particles have a diameter less than 1 µm whereas in the 35-MFR PP only 70% of the particles are smaller than 1 µm. Since submicrometer dispersion was achieved from low- to high-flow HPP, good retention of impact resistance from 1.7 MFR to 35 MFR is expected. Figure 7.20 shows a comparison of the room temperature notched Izod as a function of the MFR for plastomer and EPR. High notched Izod values were observed for all plastomer blends, but the EPR exhibited poor impact for blends consisting of medium- and high-MFR HPP resins.

### 7.5.4 Blush Resistance

When used to modify HPP, conventional EPR causes stress-induced whitening or "blush." Substituting metallocene plastomer for EPR drastically diminishes blush marks. Blush appears during instrumented impact testing with a white ring formed on the test plaque after penetration of the test dart. A blush index can be calculated by dividing the diameter of the whitened ring by the diameter of the dart. The extent of blush varies with time but usually stabilizes after overnight aging.

Figure 7.21 shows photographs of test plaques for an array of PP-elastomer modifier blends after instrumented impact testing. These specimens exhibit varying degrees of stress whitening primarily as a function of modifier type.

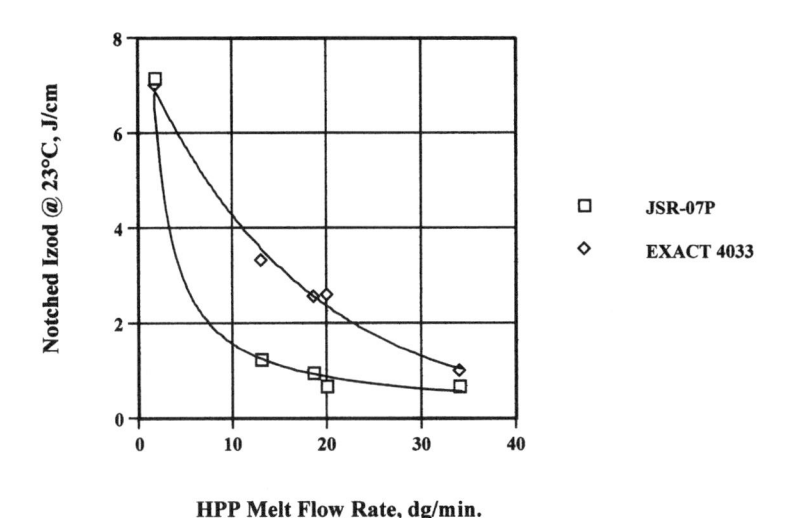

FIGURE 7.20 Effect of HPP MFR on Izod impact. □, JSR-07P; ◇, Exact 4033.

# Metallocene Plastomers as PP Impact Modifiers

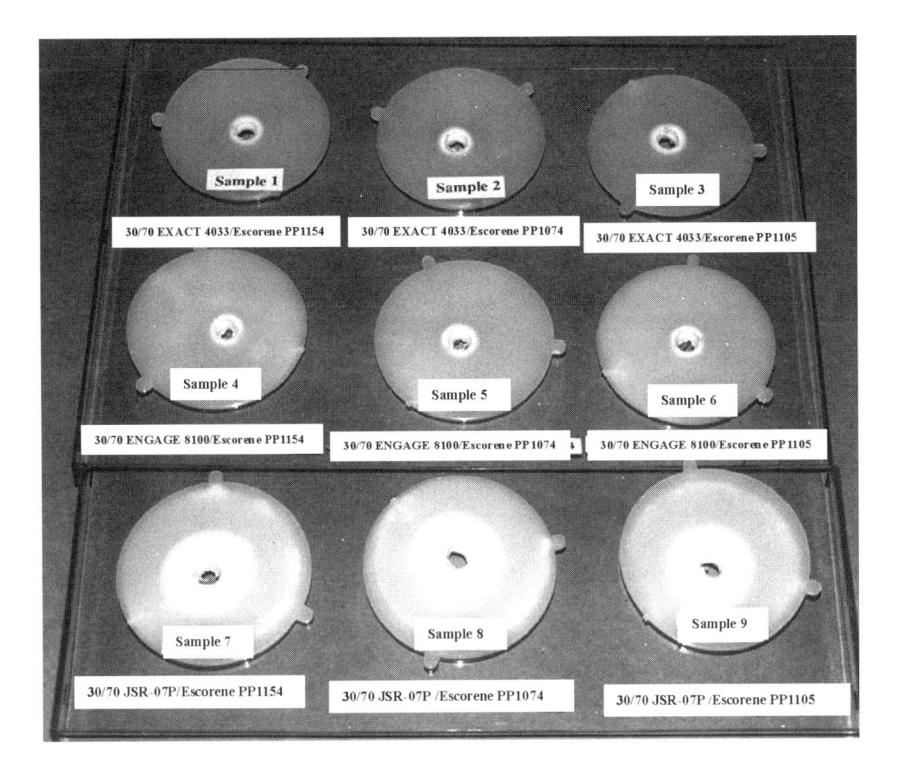

**FIGURE 7.21** Effect of modifier type on blush.

The elastomers were EB plastomer (Exact 4033), EO plastomer (Engage 8100), and EPR(JSR-07P). All three modifiers were blended in three homopolymer PPs with different flow rates: 13 (Escorene PP 1154), 20 (Escorene PP1074), and 35 MFR (Escorene PP1105). The blend ratio was 30% modifier in 70% PP.

Visual assessment of the test specimens in Fig. 7.21 indicates that the extent of stress whitening of the EPR-based blends is broader than that of both plastomers with little dependence on polypropylene resin used. Figure 7.22 shows the calculated blush index for the test results. The blush index for each EPR panel is more than twice that of the comparable plastomer blend.

## 7.6 MODIFICATION OF RANDOM COPOLYMERS

The development of sorbitol-based clarifying agents in the early 1980s led to the commercialization of clarified polypropylene for making clear rigid containers by injection stretch blow molding. The injection stretch blow molding process is similar to that used to make polyethylene terephthalate (PET) beverage bottles.

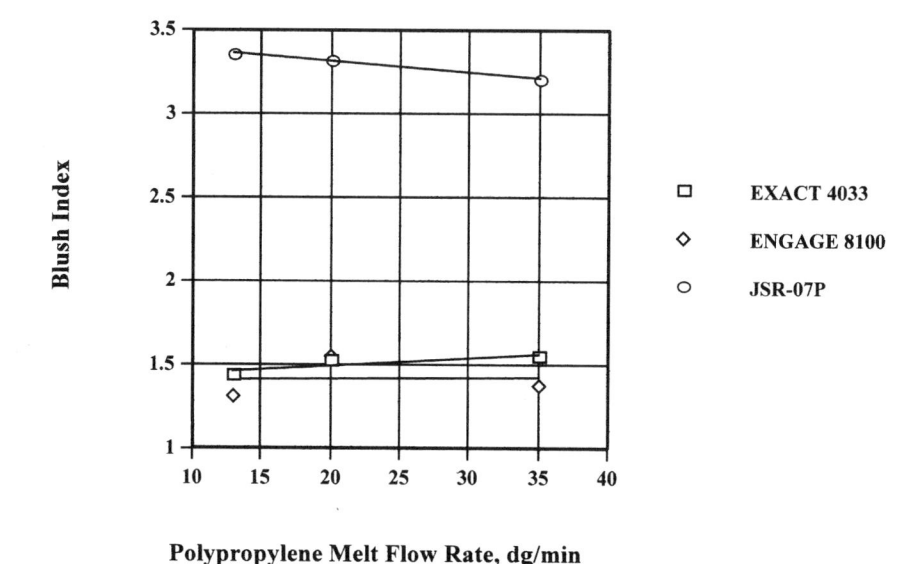

**FIGURE 7.22** Effect of modifier type on blush index. □, Exact 4033; ◇, Engage 8100; ○, JSR-07P.

Commercial injection stretch blow molding machines for PET resins can be used to mold clarified PP with only minor adjustments. Because of the difference in shrinkage, PET tooling may not perform properly when running PP. Applications for this type of container include medications (especially tablets), baby bottles, and baby food jars.

### 7.6.1 Nucleation of Polypropylene

The polypropylene molecular chain conformation is known to be a threefold helix. Three different crystalline forms arise because of the positioning of the pendant methyl groups. These are monoclinic $\alpha$ form, the hexagonal $\beta$ form and the triclinic $\gamma$ form (34). Compared to unnucleated PP, addition of an $\alpha$ nucleator to polypropylene makes PP crystallize faster and at higher temperatures. This crystallization behavior corresponds to faster injection molding or blow molding cycles. Nucleation ($\alpha$) also increases the PP resin mechanical properties such as higher stiffness and elevated heat distortion temperature. Typical $\alpha$-nucleating agents are insoluble materials such as talc or salts of carboxylic and phosphonic acids. Insoluble nucleating agent is difficult to disperse, and the nucleating agent itself scatters light and often increases the haze level of the finished part.

## Metallocene Plastomers as PP Impact Modifiers

Clarifying agents are a special subclass of nucleating agents that not only improve its cycle time and mechanical property but also enhance its clarity. For instance, the sorbitol-based clarifying agent dissolves in molten polypropylene [35] to give a homogeneous solution. As the polymer cools, a fibrous network of the clarifying agent forms, and the surface of this network becomes the nucleation site for crystallization. Because the diameter of the fibers was measured to be around 100 Å (smaller than the wavelength of visible light), the dispersed nucleating agent produces minimal haze in polypropylene. Figure 7.23 shows the effect of clarifier addition on polypropylene crystal sizes, when examined under a polarizing light microscope, the regular polypropylene (Fig. 7.23a) shows large and uneven crystals that refract light and increase opacity. The clarified polypropylene (Fig. 7.23b), on the other hand, generates smaller, highly dispersed crystals that allow light to pass through with less refraction.

### 7.6.2 Refractive Index Matching

One significant deficiency of clarified random copolymer (RCP) containers is poor low-temperature impact strength. This shortcoming limits the use of RCPs in applications such as containers for chilled fruit juices; since containers may fail if dropped when being taken out of the refrigerator. The addition of a metallocene plastomer has been shown to sufficiently toughen containers to minimize this problem while retaining the clarity of the RCP. Figure 7.20 illustrates the improvement in impact strength resulting from the addition of a metallocene plastomer to a clarified RCP resin (12 MFR, 3% ethylene). A different ethylene-butene plastomer (0.90 density, 3.5 MI) was used in this study. This higher density plastomer was selected to provide a close room temperature refractive index match with the RCP to minimize haze. Figure 7.24 shows the variation of

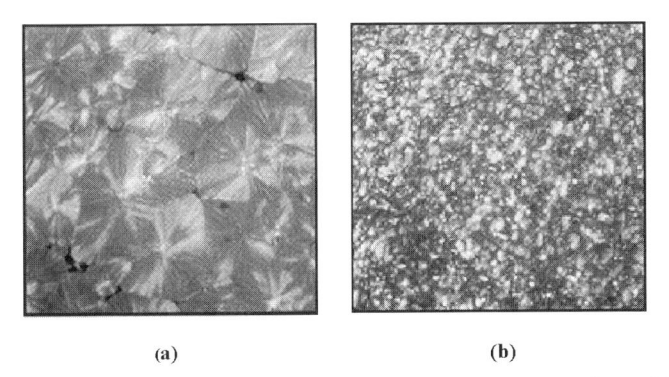

**FIGURE 7.23** Effect of clarifier addition on polypropylene clarity. (a) Polypropylene and (b) clarified polypropylene

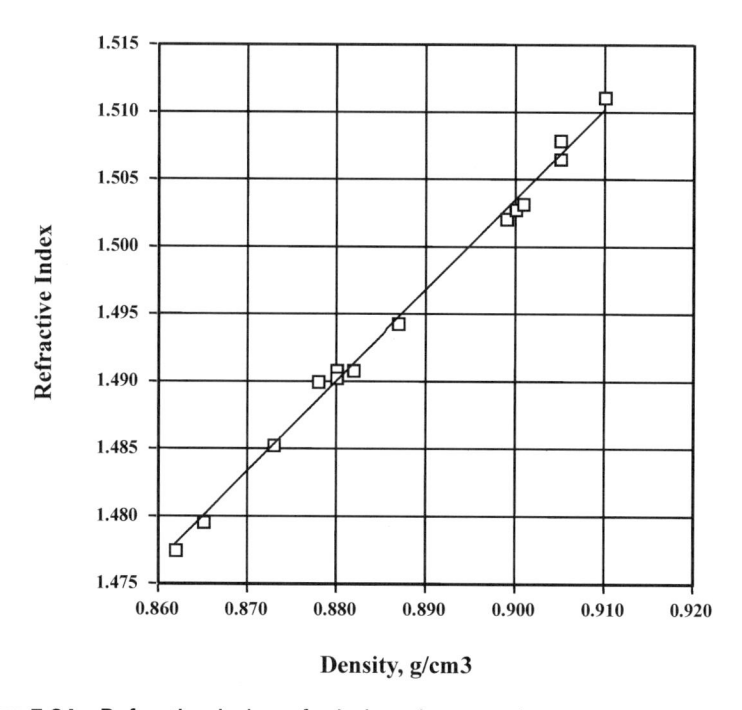

**FIGURE 7.24** Refractive index of ethylene-butene plastomers.

the room temperature refractive index of ethylene-butene plastomer with density. At room temperature, the refractive index of a clarified polypropylene RCP is measured to be 1.504 and Exact 3035 at 0.900 density have almost identical 1.503 index. Therefore, addition of EXACT 3035 to that particular polypropylene causes almost no haze increase in molded parts. In general, regardless of comonomer type, plastomers at 0.900 density were found to be suitable impact modifiers for clarified RCPs with minimal haze increase.

Total impact energy was measured at 23°C, 0°C, and −10°C using the instrumented impact method as shown in Fig. 7.25. The impact energy of the unmodified RCP is about 26 J at 23°C, enough to provide significant impact resistance in room temperature service. However, the impact energy drops off to nearly zero at 0°C. Addition of 10 wt% plastomer provided impact resistance at 0°C that is approximately the same as the room temperature impact value. The important feature of clarified RCP resin is that this improvement in low-temperature impact performance was attained while maintaining a low haze. Haze was measured on 125-mil compression-molded plaques. The haze of the compound containing 20 wt% plastomer was 8%, compared with 7.5% for the unmodified clarified RCP.

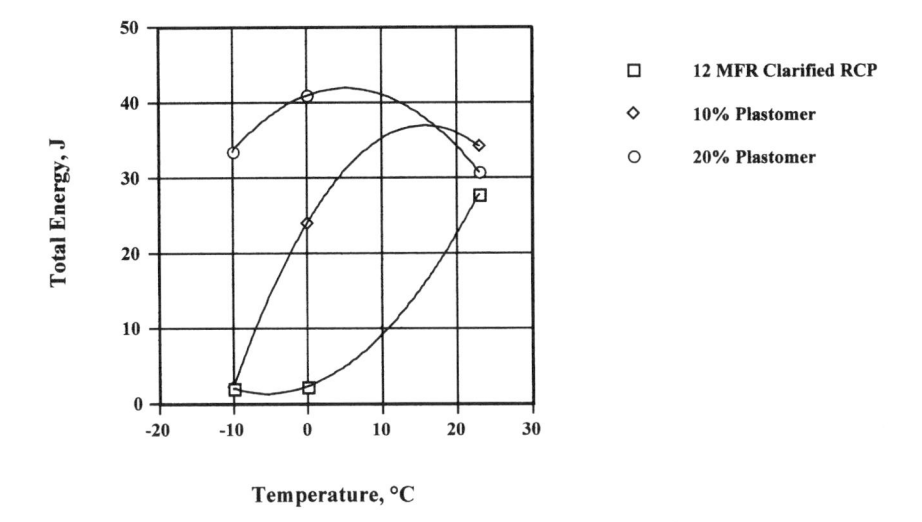

FIGURE 7.25 Impact modification of clarified polypropylene random copolymer. □, 12 MFR-clarified RCP; ◇, 10% plastomer; ○, 20% plastomer.

### 7.6.3 Fabrication of Clear Polypropylene Parts

Plastomers are supplied in free-flowing pellets and have similar molecular weight as clarified polypropylene. It is feasible to use a dry blend of polypropylene and plastomer to directly injection mold into finished parts without an upstream compounding step. In most instances, the check valve of the injection molding machine serves as the high-shear mixing zone to homogenize the molten plastomer into polypropylene base resin.

A large storage container was directly molded using a dry blend of 20% plastomer-modified 35 MFR clarified PP dry blend [36]. The dimensions of each container were 17.5 × 27 cm at the base and 22 cm in height with an average wall thickness of 2 mm. The mold had a single center gate at the bottom of the container. Figure 7.26 shows the LVSEM images from both bottom and side. Next both images were inverted and digitized and average particle size and average aspect ratio of particles determined. The desirable submicrometer dispersion of plastomer is evident even under these conditions. A similar dry blend was evaluated on a Bekum Model H121S single-station shuttle-type blow molder with a 2-L handle ware mold [37]. It was found that a clear bottle with enhanced drop impact resistance could be produced.

### 7.6.4 Effect of Plastomer Addition on Mold Cooling Time

The addition of a large amount of low-melting plastomer might require extended mold cooling during injection molding in order to avoid parts sticking. Isothermal

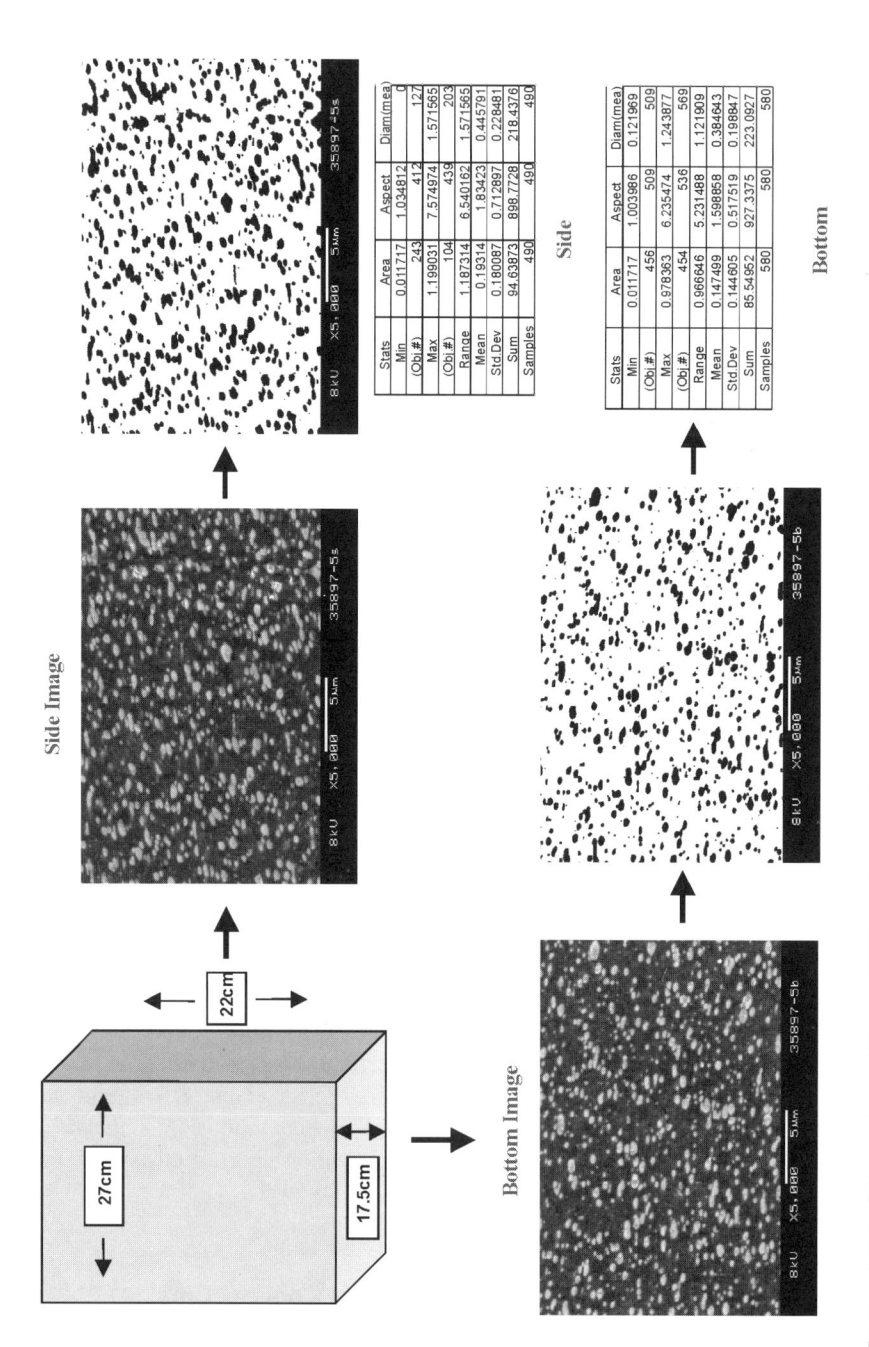

**FIGURE 7.26** Dispersion of Exact 0203 in dry-goods container.

# Metallocene Plastomers as PP Impact Modifiers

crystallization was carried out using DSC at different temperature settings to study mold cooling requirements. Three different types of plastomer were selected for the study, i.e., EB plastomer (Exact 3035), ethylene-hexene plastomer (Exact 9106), and EO plastomer (Exact 0201) at 15 wt% addition level in a 30 MFR clarified RCP (Escorene PP 9505). Each sample was cooled rapidly to the crystallization temperature and crystallized isothermally for 30 min. The half time ($t_{1/2}$) was determined when 50% crystallization had occurred. Figure 7.27 shows the plot of crystallization half-time at various temperatures. Almost no change was observed in $t_{1/2}$ for PP 9505 and its blends. In other words, there was no change in the crystallization rate of PP 9505 with addition of all three type of plastomers. The molding cooling time is unaffected by plastomer addition.

## 7.7 MODIFICATION OF IMPACT COPOLYMERS

### 7.7.1 Effect of Plastomer on Izod Impact

Figure 7.28 illustrates the effect on Izod impact and flexural modulus of modifying ICP with EB plastomer and EPR. Figure 7.28 should be compared

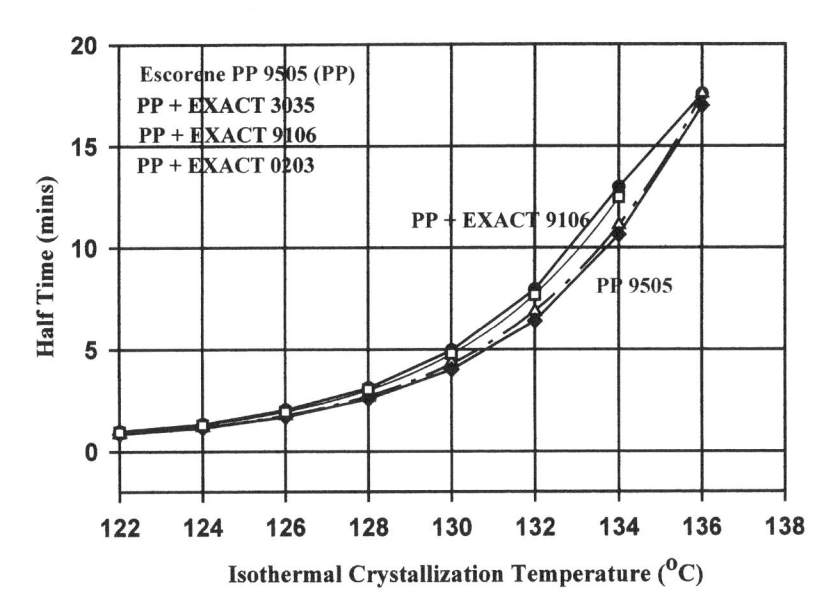

FIGURE 7.27 Effect of plastomer addition on crystallization rate of Escorene PP 3505. ♦, neat PP resin; △, PP + Exact 0203; ●, PP + Exact 9106; □, PP + Exact 3035.

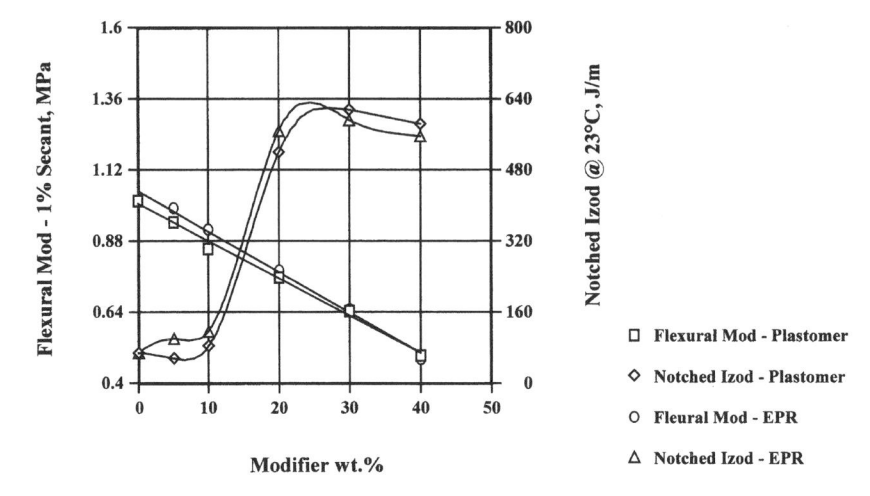

FIGURE 7.28 Stiffness/Izod impact balance in ICP blends. □, Flexural modulus plastomer; ◇, notched Izod plastomer; ○, flexural modulus EPR; △, Notched Izod EPR.

with Fig. 7.10, which shows an analogous experiment using HPP. In both cases, flexural modulus decreases roughly linearly with modifier addition, whereas an increase in Izod impact follows an 'S'-shaped curve. Using ICP, the impact curve rises at a lower modifier level (about 10% compared to 15% with HPP) and plateaus at a lower level (about 20% versus about 40% for HPP).

### 7.7.2 Effect of Plastomer on Low Temperature Instrumented Impact

In automotive TPO applications, impact strength is often required at −30°C to −40°C. Most medium-impact ICP resins, which contain around 9% ethylene, do not perform satisfactorily at these temperatures. In practice it is found that a minimal concentration of elastomeric phase of about 10–15 wt% is required for significant impact enhancement at these temperatures. Figure 7.29 shows data on plastomer-PP blends for total impact energy and maximal force as a function of modifier level at −40°C and 3.8 m/sec test speed. At 5 wt% plastomer concentration the sample exhibits brittle failure. At 10 wt% plastomer, there is a distinct yield point with some drawing of the sample after the yield. However, the energy absorbed from yield to break is far less than the energy to yield. Therefore, it is a brittle failure. At 20 wt% plastomer, a bell-shaped, ductile failure curve results. When more than 20 wt% plastomer is added, no further increase in force or displacement is observed. The only effect is a reduction in stiffness and yield force.

# Metallocene Plastomers as PP Impact Modifiers

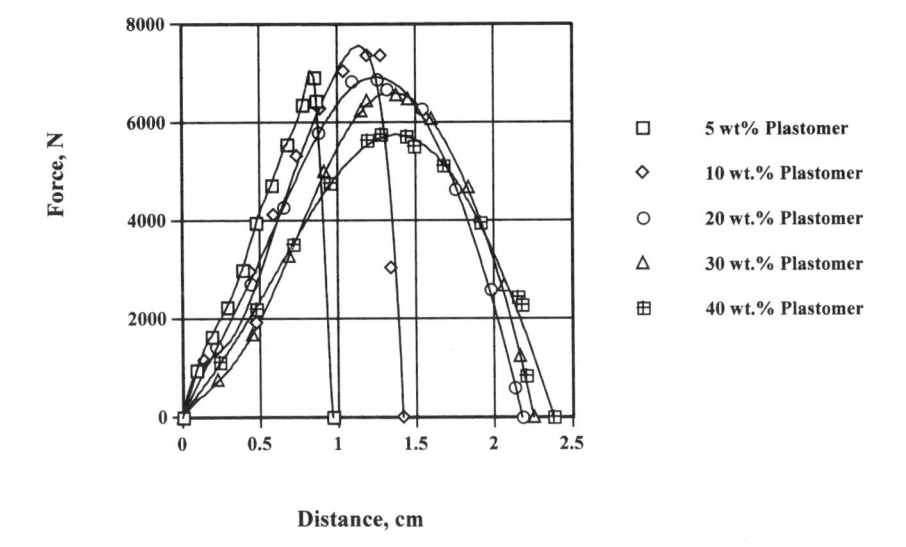

FIGURE 7.29 Effect of modifier content on force/distance profiles of plastomer/ICP compounds in instrumented impact test at −40°C Wt% plasticizer level: □, 5; ◇, 10; ○, 20; △, 30; ⊞, 40.

An interesting feature shown in Fig. 7.30 is that the sample plaque absorbed more impact energy at −20°C than at 23°C. Rather than becoming brittle at low temperature, the material actually becomes tougher as the modulus increases. At lower temperatures, the total energy is reduced as ductility is reduced.

## 7.7.3 Modification of Filled Impact Copolymers

One of the most important applications for impact-modified PP is automotive TPO, commonly used for automotive bumper fascia, external body cladding, and interior trim. Compounds are typically three-component blends of ICP with talc and an impact modifier such as a metallocene plastomer or EPR. The impact modifier provides low-temperature toughness but softens the matrix. The talc particles restore the stiffness of the compound but detract from impact strength. The objective of TPO compounding is to select the combination of the three components that provides an optimal balance of stiffness and impact strength. The experiment described below was carried out to illustrate the effects of metallocene plastomer and talc on the properties of a TPO composition. Experiments were designed around a typical automotive bumper composition of 60 wt% ICP, 30 wt% impact modifier, and 10 wt% talc (38).

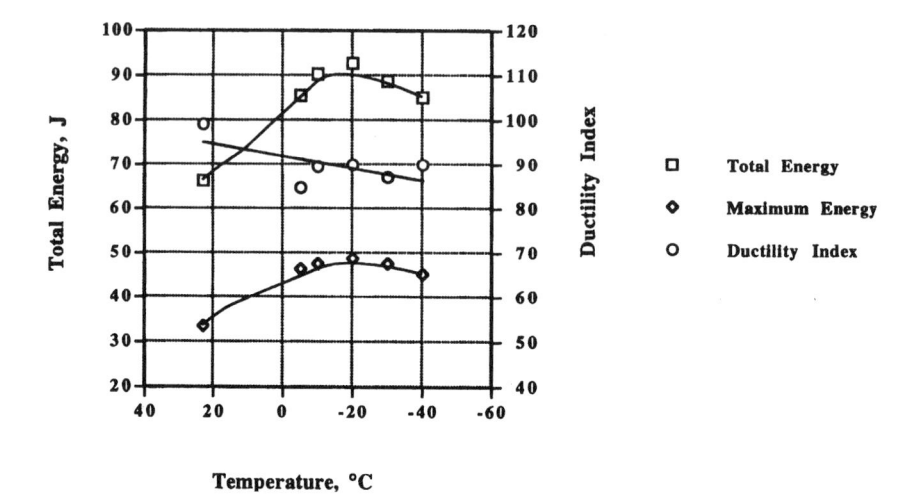

**FIGURE 7.30** Effect of temperature on instrumented impact of plastomer-ICP compounds.

*Experimental Design.* The PP base resin used was a 35 MFR ICP resin containing 9% ethylene. The filler was a non-surface-treated micronized talc, with an average particle size of 1 μm. The modifiers used were the standard EB plastomer and EPR. An extreme vertices mixture designed experiment was employed [39]. The three ingredients were subject to high- and low-percentage constraints:

ICP:      $55\% \leq X_1 \leq 70\%$
Modifier: $15\% \leq X_2 \leq 30\%$
Talc:     $5\% < X_3 < 15\%$

A mixture design is different from an ordinary factorial design in that the ingredients must total 100 parts. In an ordinary factorial experiment the level of each component is independent of the other components. The dimensionality of its factor surface is equal to the number of components. But for a mixture experiment the factor space will have a dimensionality equal to the number of components minus 1. For instance, for a three-component factorial experiment the factor space is a cubic. For a three-component mixture it is 2. In mathematics this type of space is called a simplex.

A full cubic, three-component experimental design [40] was used for this study. The factor space and design points are illustrated in Fig. 7.31.

*Results.* Although the experiment was designed on a full cubic model, analysis of variance of the experimental data showed that the system obeyed the simple linear model (i.e., no interactions were statistically significant). This

# Metallocene Plastomers as PP Impact Modifiers

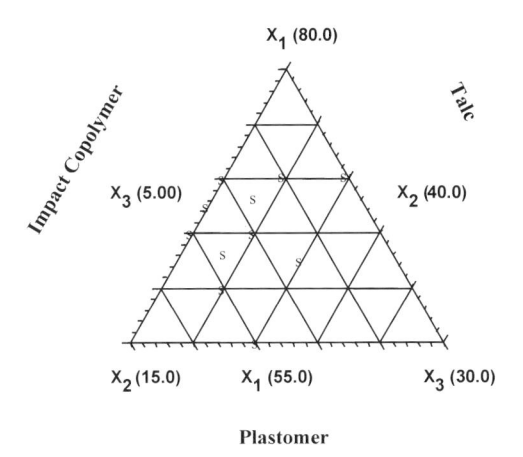

**FIGURE 7.31** TPO experimental design.

suggests that the PP-modifier-talc formulation behaves as a base material modified by two additives acting independently. Given the nature of the formulation (a large-volume fraction PP continuous phase, and two additives without coupling or chemical interactions with the PP), this result is not surprising.

The response contour surfaces for several key properties were generated using the mixture polynomial to highlight the performance differences of plastomer versus conventional EPR. The metallocene plastomer and EPR provided very similar impact enhancement in both notched Izod and instrumented impact tests (Figs. 7.32 and 7.33). In all four cases, impact strength depended primarily on the impact modifier and less on the talc content. Flexural moduli were similar for plastomer and EPR compounds (Fig. 7.34). The plastomer formulations showed significantly less shrinkage than the EPR formulations (Fig. 7.35). Shrinkage varied primarily with the PP volume fraction because the PP component has the largest degree of crystallinity and thus the largest potential to shrink. The plastomer formulations had higher melt flow rates than the elastomer formulations (Fig. 7.36) because of the higher melt index of the plastomer relative to the EPR. This difference would be expected to provide improved flow in injection molding.

## 7.7.4 Molecular Tailored Blends

Recently, high-flow TPO compounds have been introduced for injection molding complex automotive exterior parts. The performance of parts in the field requires TPO compounds to possess not only low-temperature impact resistance but also higher stiffness to reduce part weight and high surface quality to facilitate

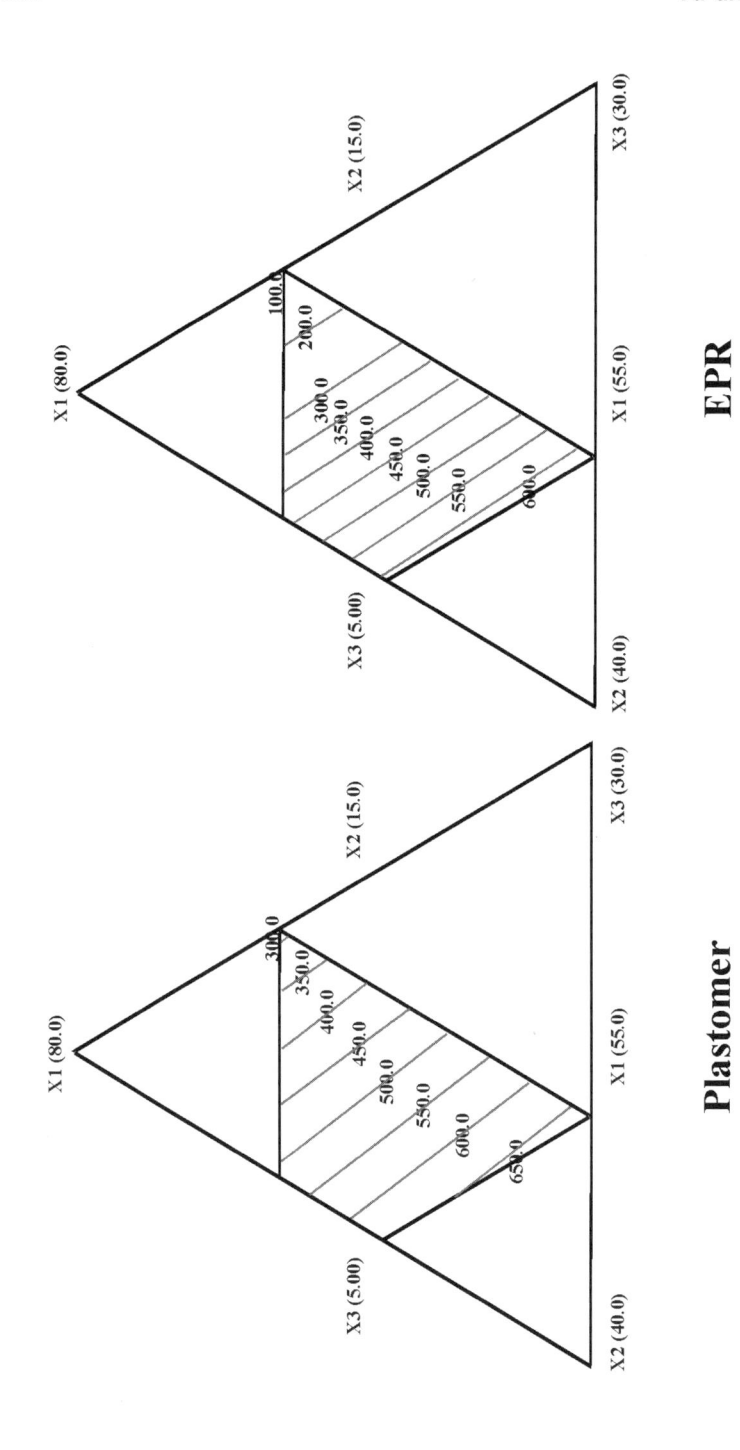

**FIGURE 7.32** Effect of TPO composition on Izod impact.

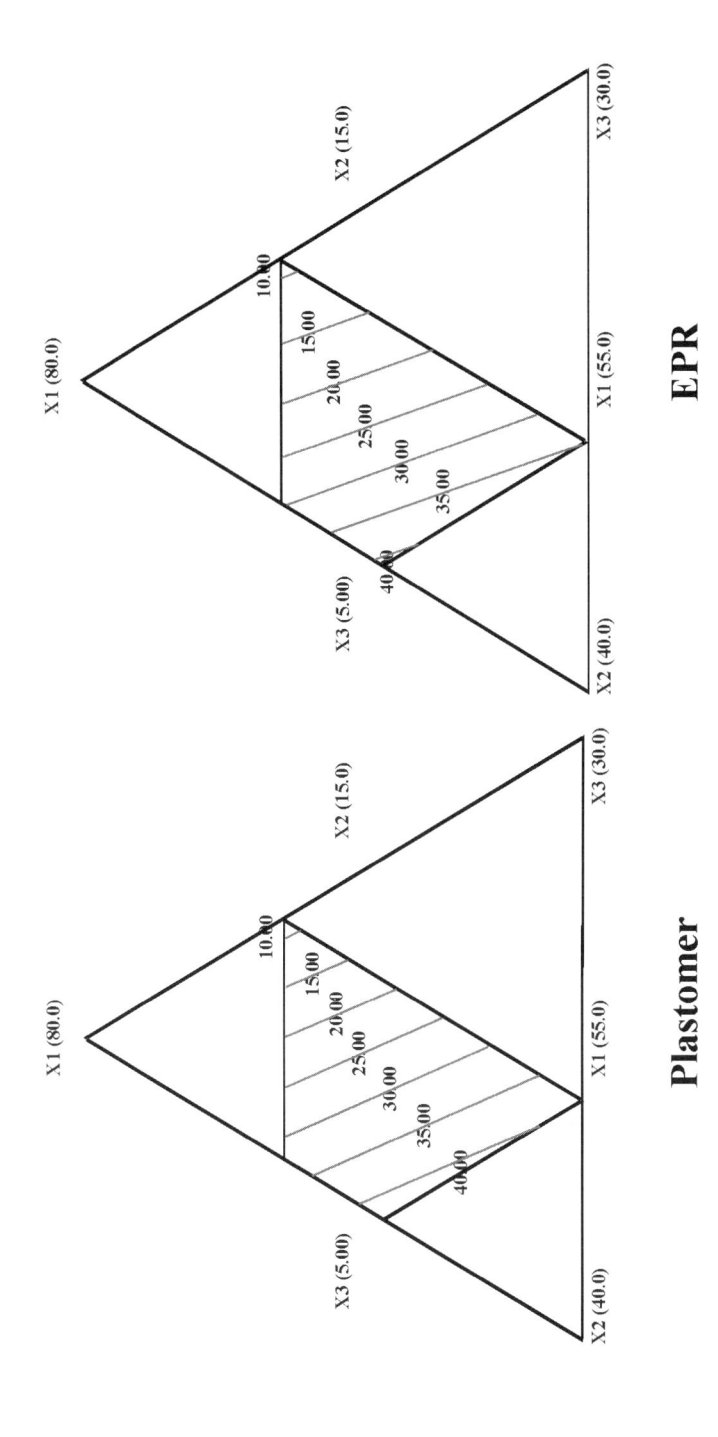

# Metallocene Plastomers as PP Impact Modifiers

**FIGURE 7.33** Effect of TPO composition on low-temperature impact.

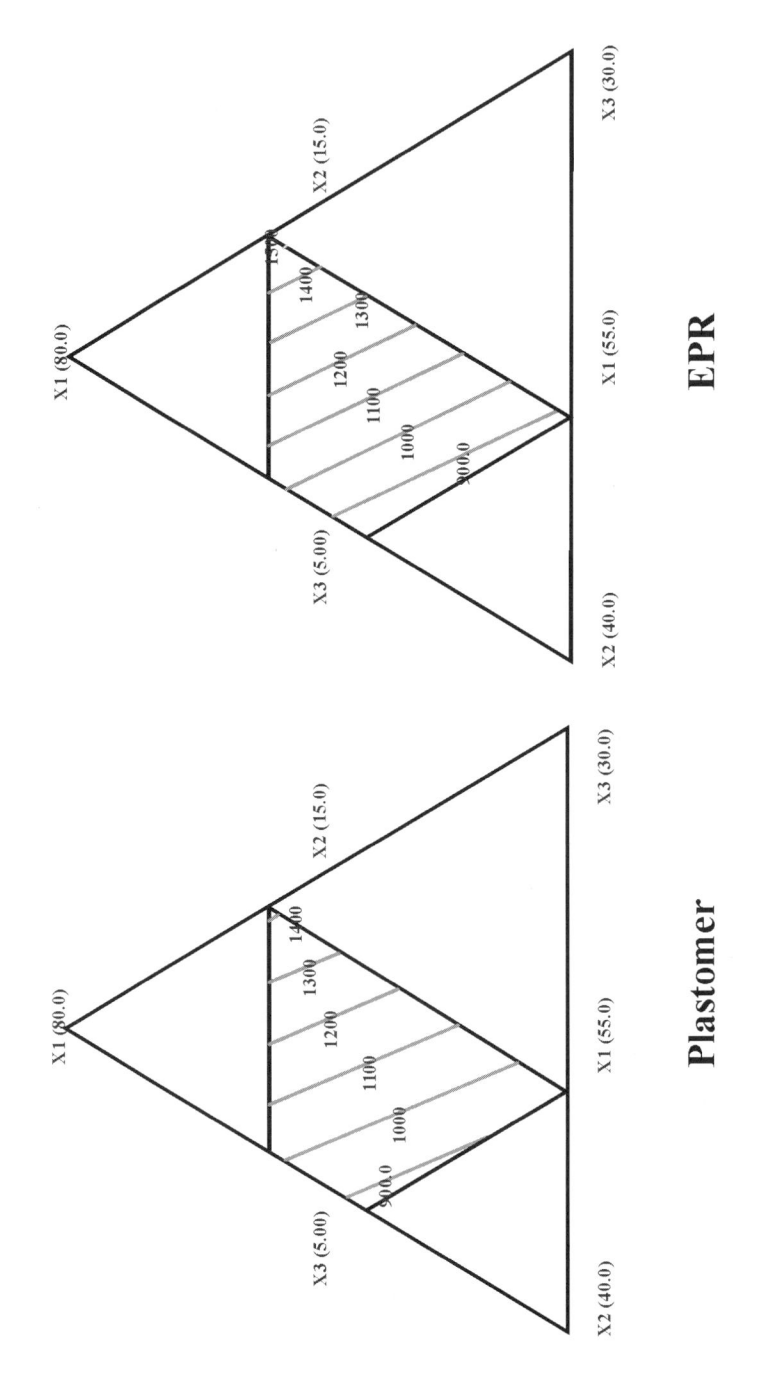

FIGURE 7.34  Effect of TPO composition on flexural modulus.

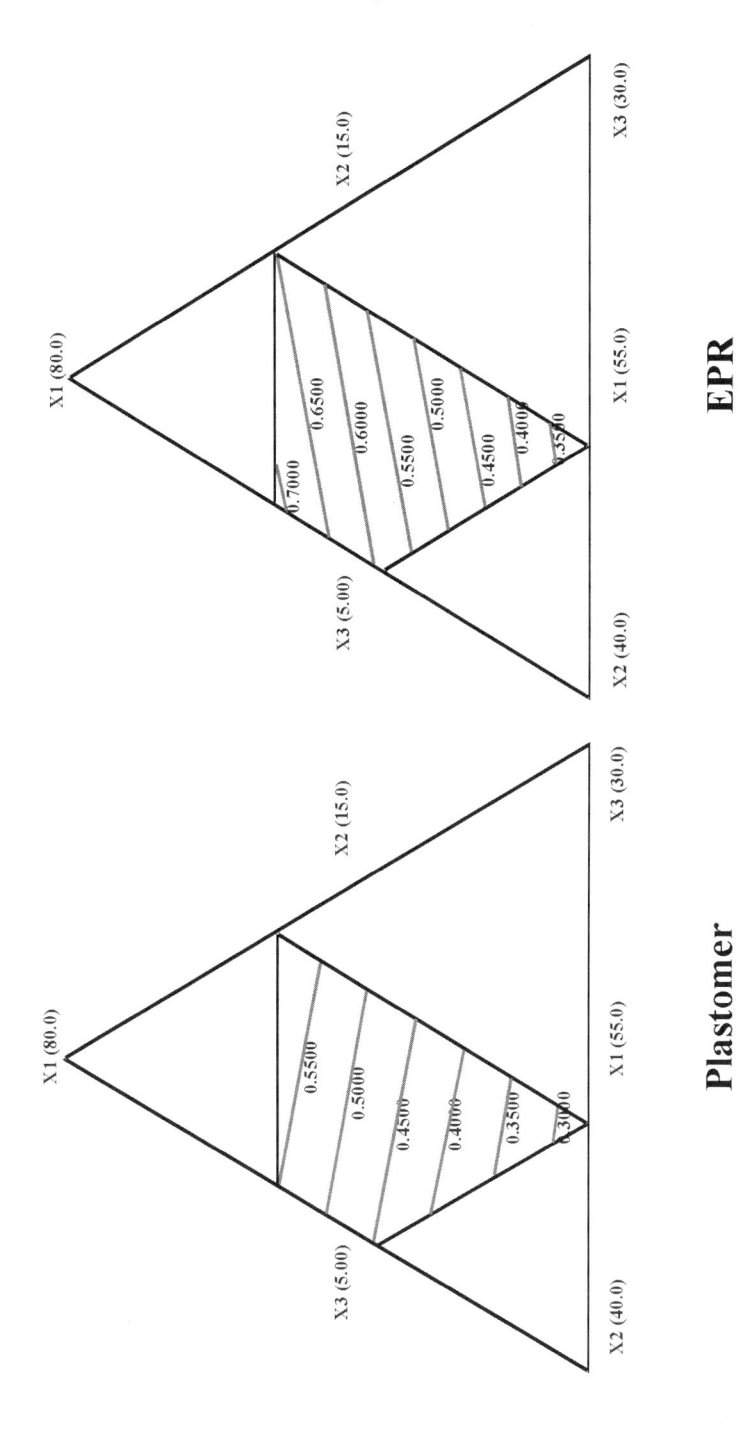

# Metallocene Plastomers as PP Impact Modifiers 261

**FIGURE 7.35** Effect of TPO composition on shrinkage.

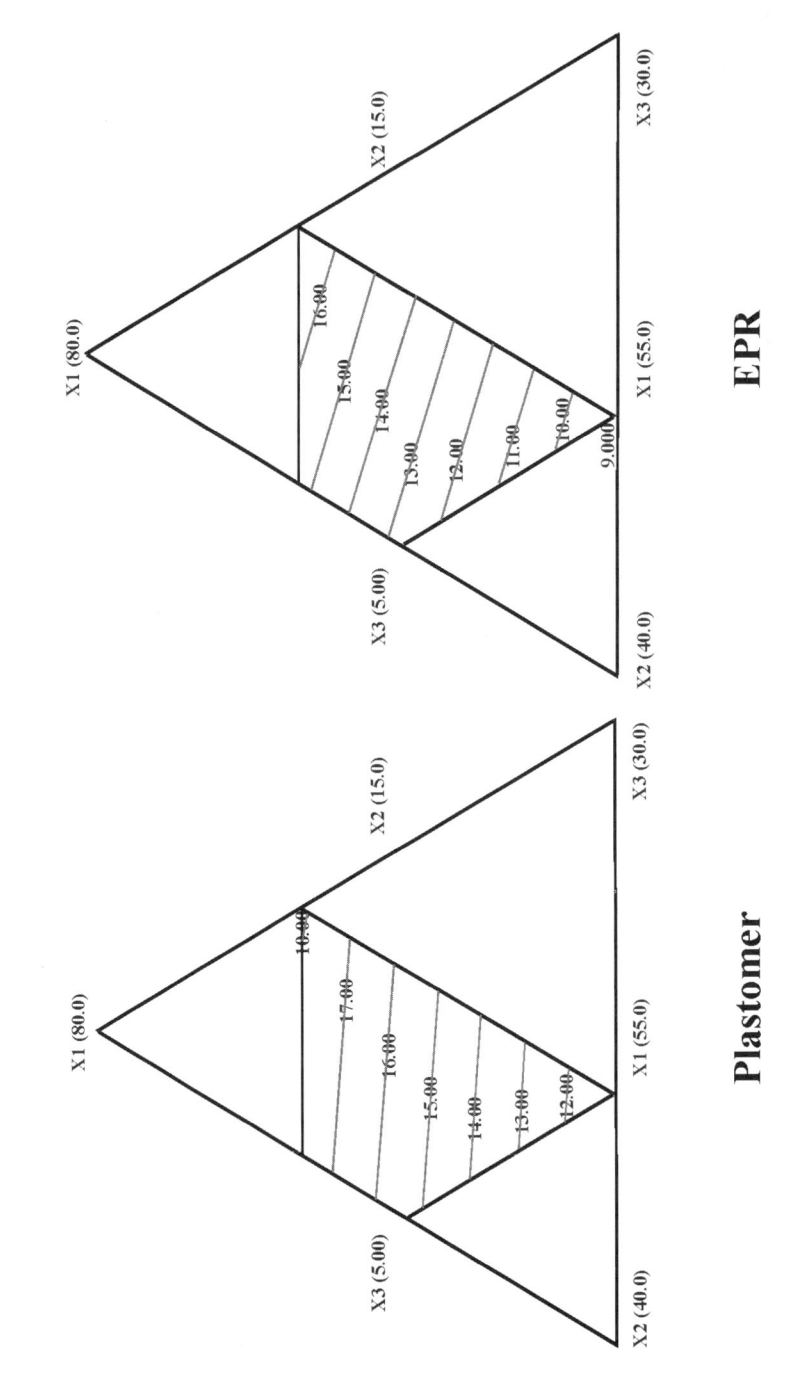

**FIGURE 7.36** Effect of TPO composition on melt flow rate.

painting. To meet this performance challenge, TPO compounds have been developed with controlled crystallinity in both the matrix resin and the dispersed rubber phase. For example, an ICP resin can be made that consists of a high-flow high-crystallinity PP and a high molecular weight EPR. This ICP can be modified with a lower molecular weight plastomer and mineral filler to achieve the desired stiffness–impact combination. An example of such a TPO compound is the Toyota Super Olefin Polymer (TSOP) developed by Toyota Motor [41,42].

An important characteristic of this type of compound is that it contains crystallizable segments dispersed in the rubbery domains. Figure 7.37 shows a transmission electron micrograph of the bulk phase of a TSOP-made bumper fascia under high magnification (30,000×). In this micrograph, the PP matrix resin is shown as a white background and the dispersed elastomer particles are shown as dark irregularly shaped objects. Polyethylene lamellae are visible inside the dispersed elastomer particles. These lamellae originate from both the ethylene segments of the EPR and the plastomer modifier dispersed in the rubber. Crystallinity in the rubber phase enhances both the stiffness and impact performance of the compound.

## 7.8 PAINTABILITY OF AUTOMOTIVE TPOS

TPO compounds are frequently used for automotive exterior parts, such as bumper fascia, that are painted to match the rest of the vehicle. Getting paint to adhere to a nonpolar polyolefin surface can be a difficult technical challenge.

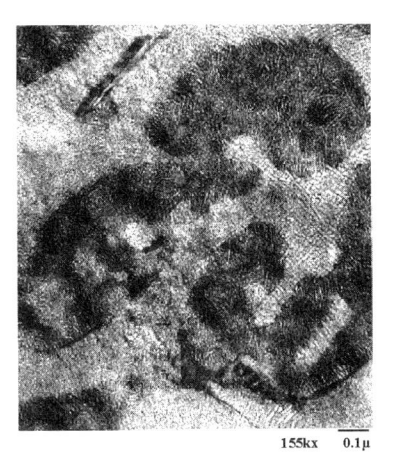

FIGURE 7.37 Transmission electron micrograph of tailored blend.

Studies have shown that the impact modifier in a TPO formulation can have a significant influence on paint adhesion.

### 7.8.1 TPO Painting

Robotic flame treatment of automotive bumper fascia surface to promote adhesion is quite popular in Europe. In North America, chlorinated polyolefin (CPO) adhesion promoter is the dominant choice. Figure 7.38 shows the steps for obtaining a TPO painted panel for qualification:

1. The injection-molded plaque was first washed in pressurized deionized water and air dried for a few minutes.
2. A solvent-borne or a water-borne adhesion promoter layer was applied to a specified dry film thickness and the solvent was allowed to flash off for several minutes.
3. After sufficient time, the color coating was applied and allowed to flash off one more time for a few minutes.
4. Application of the final protective clear coating was completed.
5. After the final flashing, the coated sample plaque was baked to a 121°C (250°F) surface temperature for 30 min to complete the cross-linking chemical reaction for both the color and clear coats.

* Dry Film Thickness

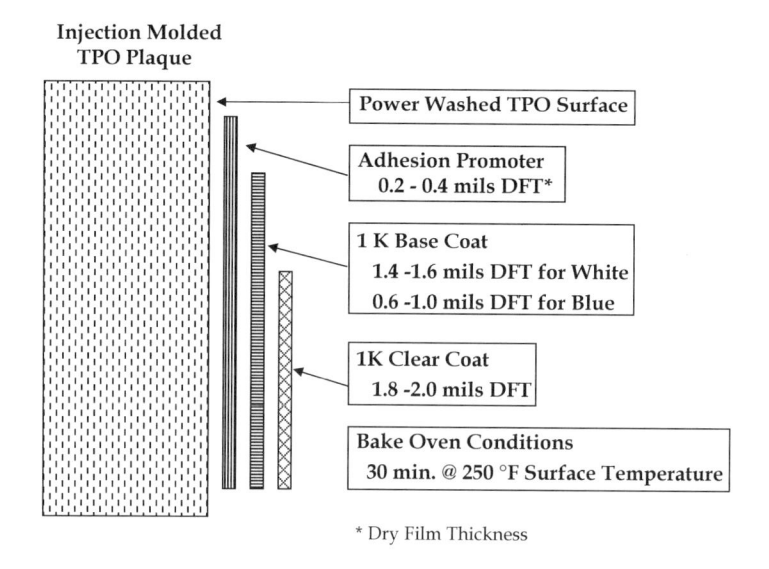

**FIGURE 7.38** Diagram of paint plaque preparation steps.

# Metallocene Plastomers as PP Impact Modifiers

Commercially available color coats and clear coats based on melamine and urethane chemistries are commonly know as 1K and 2K paint, respectively.

One of the common mechanisms proposed for use of adhesion promoter is that the CPO in the adhesion layer diffuses into the solvent-softened TPO substrate and anchors inside the dispersed plastomer particles. This mechanism generates polar surface bonding sites along the TPO surface. Figure 7.39 shows a comparison between the morphology of the painted surfaces of test plaques molded from TPO materials based on Exact 4033 (Fig. 7.39a) and JSR-07P (Fig. 7.39b). The composition of this compound was 60 parts of 35 MFR ICP, 30 parts impact modifier, and 10 parts talc. In each micrograph, the heavily stained phase is the plastomer phase and the continuous dark phase is the poorly stained PP phase. The large, irregularly shaped particles are talc. Both micrographs show the paint layer, the TPO surface, and about 10 µm in depth from the TPO surface. The fine spider web type of dispersion of plastomers near the paint plaque surface provided better anchors to the CPO than the less dispersed JSP-07P near the paint plaque surface.

One drawback to the utilization of the adhesion promoter is that the diffusion of CPO into TPO surface is analogous to a stress cracking phenomenon, i.e., penetration of the adhesion promoter into the polymer surface. This penetration could lead to reduced impact resistance of the painted parts. The effect of painting on impact resistance for Exact 4033–based TPO samples is shown in Fig. 7.40. Due to this so-called stress cracking effect, the ductility of the painted TPO was not as good as that of the freshly molded TPO test panel. Both

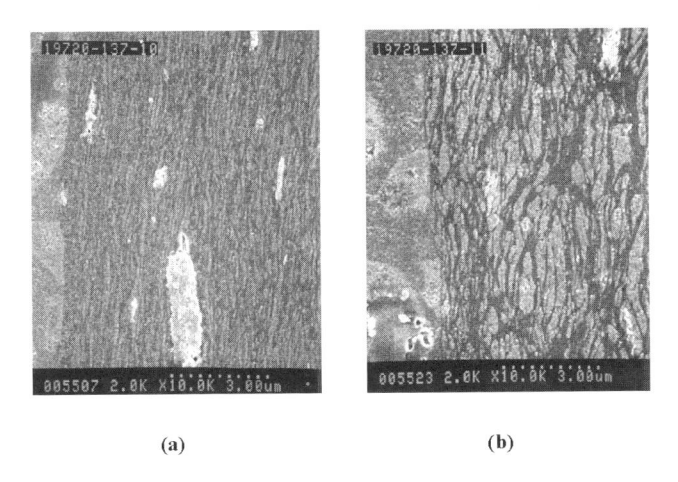

(a)          (b)

FIGURE 7.39 Comparison between morphology of SEM cross-sections for TPOs based on (a) Exact 4033 versus (b) JSR-07P plastomers.

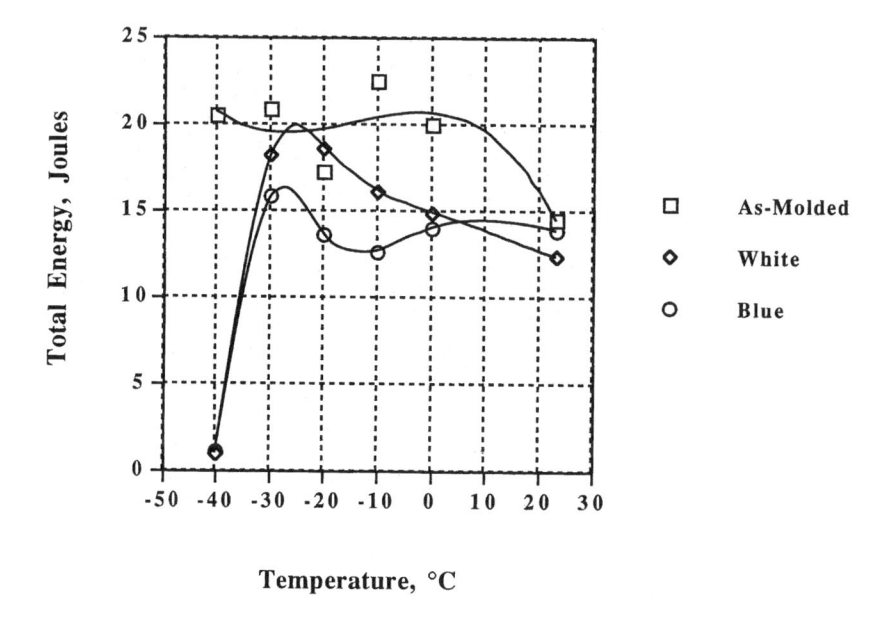

FIGURE 7.40 Instrument impact plots for unpainted versus painted panels. □, As-molded (unpainted); ◇, white; ○, blue.

white and blue painted panels showed ductility only at −30°C whereas the unpainted test panel was found to be ductile at −40°C.

### 7.8.2 Testing of Painted TPO Parts

A recent study [43] demonstrated that TPOs modified with plastomer resins provided significantly better adhesion than EPR-modified formulations. Improvements were demonstrated with both conventional paint and with an adhesion promoter free "olefinic" paint based on a hydrogenated polybutadiene diol-melamine coating resin. Painted panels based on both Exact 4033 and JSR-07P were tested with the Ford Statram test. This method, which was developed to simulate the rub of a car bumper against a stationary object such as a safety post, employs a ram with a hemispheric head that rubs the length of the test panel surface at a programmed speed and vertical force. The force and total energy of the abrasion are recorded during the test and paint removal measured by image analysis. A schematic of the test is shown in Fig. 7.41. Tests were conducted using three combinations of plaque temperature and ram forces to simulate a range of field conditions. Using both paint systems and under all three test conditions, the plastomer-modified TPOs demonstrated better paint adhesion than the EPR-based TPOs. Results were generally better with the olefinic paint than

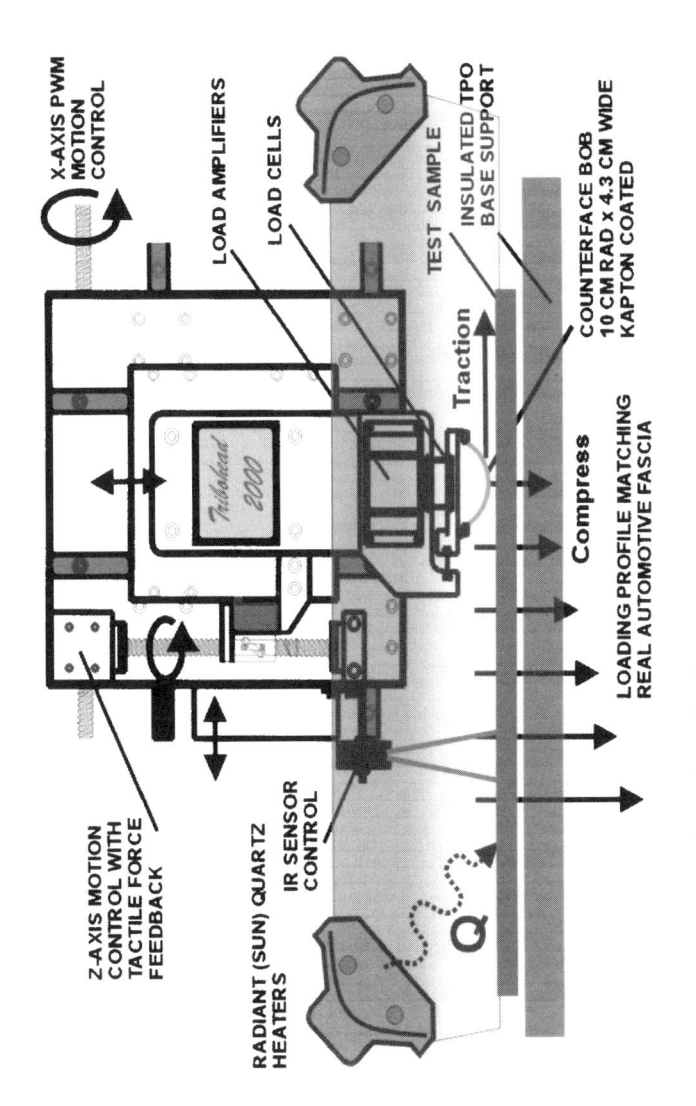

# Metallocene Plastomers as PP Impact Modifiers

**FIGURE 7.41** The STATRAM test for paint adhesion.

with the conventional paint. Both Exact 4033 and JSP-07P paint panels after testing are shown in Fig. 7.42. The Exact 4033 paint panel showed no paint peeling in comparison with complete paint peeling for the JSR-07P test panel.

## 7.9 FLEXIBLE THERMOPLASTIC OLEFIN

Flexible TPOs or flexible polyolefins (FPOs) have high heat resistance and steam sterilization resistance. Compounds with low stiffness are tailored for medical, construction, and automotive markets. Huntsman previously manufactured blends of reactor-grade high molecular weight atactic and isotactic PP resins, called Flexible polyolefins, under the trade name Rexflex. However, Rexflex is no longer available in the marketplace. Instead, the same type of product can be duplicated using a combination of polypropylene and plastomer. Figure 7.43 shows the effect of plastomer type and addition level on flexural modulus for materials based on impact-grade polypropylene (Escorene PD 8191, 1 MFR). At around 40 wt% addition, the flexural modulus of the TPO compound matches that of flexible PVC.

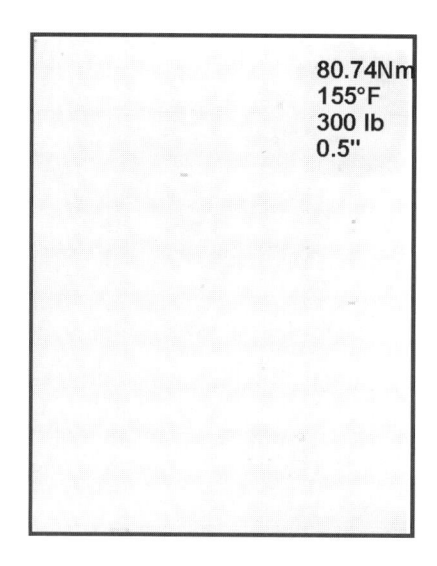

**Plastomer**  **EPR**

FIGURE 7.42 Paint panels after STATRAM test (68.3°C, 2.1 MPa load, 0.21 mm/sec test speed, and 10.2 cm travel).

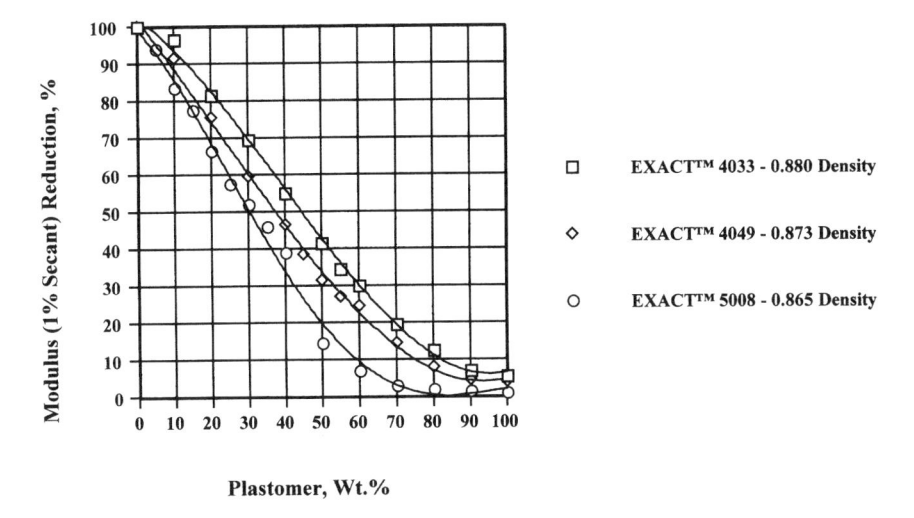

# Metallocene Plastomers as PP Impact Modifiers

**FIGURE 7.43** Effect of plastomer addition on flexural modulus. □, Exact 4033-0.888 g/cm$^3$; ◇, Exact 4049-0.873 g/cm$^3$; ○, Exact 5008-0.865 g/cm$^3$.

One recent commercial activity is the development of a white single-ply roof membrane. The current nonblack single-ply roof membrane is mostly on flexible PVC produced by the calendering process. Other flexible polymers include chlorosulfonated polyethylene and butyl rubber. The traditional black roof membrane consists of a heavily filled and cross-linked EPDM. In comparison with thermoset EPDM membrane, the light-colored TPO membrane reduces a substantial amount of the air conditioning load. Rapid installation is accomplished by heat seaming instead of solvent welding .The inherent flame retardancy of the PVC material can be matched by the addition of magnesium hydroxide to the TPO compound [44]. To take advantage of recent advances in compounding technology, in-line compounding for sheet extrusion is now feasible [45]. The sequence of process steps for making flame-retardant TPO formulation are as follows:

1. The individual ingredients in the TPO formulation are metered separately to feed ports located along the barrel of a twin-screw compounding extruder.
2. Polypropylene and plastomer pellets are fed to the main feed port.
3. Magnesium hydroxide masterbatch is fed via a side port downstream of the melting zone of the extruder.
4. The compounded melt discharge is dropped into a gear pump that pressurizes it sufficiently to be extruded through the sheeting die.
5. A three-stack roll converts the polymer melt flat sheeting.

The major hindrance to PVC replacement by flexible TPO lies in the failure of the TPO to use all current PVC welding techniques. For example, due to the polar nature of PVC resin, flexible PVC can be easily welded together using radiofrequency (RF) welders. In the case of the nonpolar flexible TPO, a substantial amount of polar-ethylene copolymer needs to be included in the compound to provide the necessary polarity for RF welding.

## 7.10 THERMOPLASTIC ELASTOMER

When subjected to flow-induced stress, the morphology of the TPO compound undergoes stratification as shown in Fig. 7.39 for two TPO formulations based on Exact 4033 (Fig. 7.39a) and JSR-07P (Fig. 7.39b) plastomers. By controlling stress or partially cross-linking the dispersed-phase morphology, TPO compounds are converted to commercially known thermoplastic elastomers (TPEs). When the degree of cross-linking of the dispersed rubber or plastomer particles is more than 90%, this particular type of TPE compound is sometimes referred to as a thermoplastic vulcanizate (TPV).

### 7.10.1 Dynamic Vulcanization

In traditional thermoset technology, static vulcanization is commonly practiced. This allows a peroxide-laden compound to be first extruded into a continuous profile and then profile vulcanized inside a long heat tunnel. The static mode of the vulcanization process, practiced by the rubber industry, is not well suited to thermoplastic blends due the absence of shearing forces to break down the domain size of the elastomer particles during curing.

A better alternative to making TPEs from blends of PP and plastomers is the dynamic vulcanization process [46]. Selective cross-linking of the plastomer occurs preferentially during melt extrusion under dynamic shear to give a final compound having dispersed cross-linked rubber particles.

### 7.10.2 Melt Rheology

When the dispersed plastomer is cross-linked, its molecular weight increases many fold with corresponding reduction in melt flow rate. The polymer blend also undergoes a phase separation process in which the high molecular weight rubber particles are dispersed in the un-cross-linked low molecular weight polypropylene phase. One may envision the TPE or TPV compounds as many plastomer boulders inside a polypropylene fishnet. During subsequent melt processing, the low molecular weight polypropylene will elongate into fine fibrils that serve as lubricants for the higher molecular plastomer boulders. Figure 7.44 makes a comparison between shear viscosity versus shear rate plots for lightly cross-linked and highly cross-linked thermoplastic elastomers. The measurements

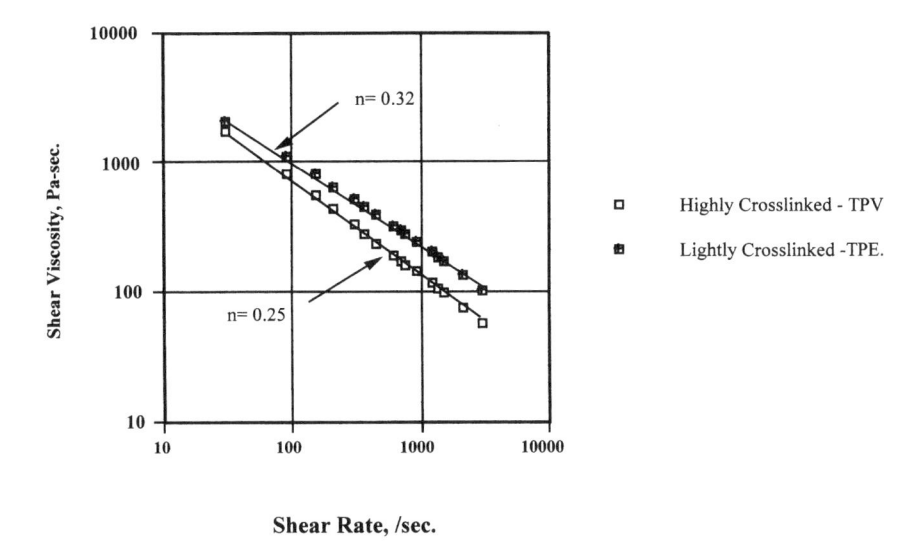

# Metallocene Plastomers as PP Impact Modifiers

**FIGURE 7.44** Shear viscosity of thermoplastic elastomers. ☐, Highly cross-linked TPV; ⊞, Lightly cross-linked TPE.

were conducted using a Rosand precision rheometer at 230°C test temperature and shear rates ranging from 30 to 3000 sec$^{-1}$. The shear-dependent viscosities for both materials fit the power law rheological model. In this model, a logarithmic plot of shear stress versus shear rate is linear with a slope "$n$" called the flow index or non-Newtonian parameter. Flow index values of 0.32 and 0.25 were determined for the highly cross-linked and the lightly cross-linked compounds, respectively. These materials are pseudoplastic fluids by virtue of exhibiting shear thinning behavior. The inherent decrease in melt viscosity with increase in shear rate makes it much easier to fabricate shaped articles by injection molding or extrusion processes.

### 7.10.3 Cross link Plastomer by Peroxide

The most common cross-linking technique involves the addition of a small amount of chemically reactive additives, such as a peroxide initiator alone or in combination with vinylalkoxysilane, to generate either carbon-carbon (C-C) and silicon-oxygen-silicon (Si-O-Si) bonds between neighboring polymer molecular chains. Both techniques are practiced by the wire and cable industry to produce cross-linked low-voltage and medium-voltage wire coatings.

The effect of peroxide-initiated cross-linking on phase morphology is shown in Fig. 7.45. A 50:50 dry blend of 35 MFR PP and Exact 4033 was melt blended on a 30-mm twin-screw extruder. The morphology of the melt-

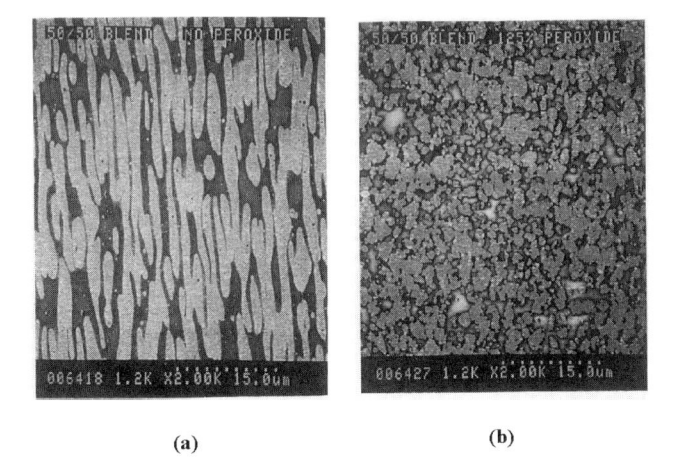

**FIGURE 7.45** Effect of peroxide addition on phase morphology. (a) No peroxide and (b) 0.125 wt% peroxide.

blended pellets with no peroxide (Fig. 7.45a) exhibits a significant change when peroxide is added. Although submicrometer dispersion was achieved, visible flow induced elongated sausage-like plastomer particles. When 0.125% of dicumyl peroxide was added to the same dry blend, the phase morphology (Fig. 7.45b) changed significantly to a distribution of agglomerated spherical particles.

It is generally known that addition of peroxide tends to degrade PP resin by ($\beta$) scission of the molecular chains in the molten state. In one recent application [47], a peroxide master batch (Vulcup 40KE) was added to a TPO compound during compounding to simultaneously partially cross-link the plastomer phase and degrade the PP phase to increase the melt flow of the product. In a second extrusion step, processing oil was added to reduce the hardness of the compound.

### 7.10.4 Cross-link Plastomer by Organosilane

In order to prevent degradation of polypropylene ingredient in making a partially cross-linked TPE or fully vulcanized TPV compound, it is better to react plastomer with mixtures of peroxide and any other reactive group before blending it with polypropylene resin.

Cross-linking processes to make TPEs and TPVs are based on existing wire and cable cross-linking technology. The initial step involves using a small amount of peroxide that can be used as a free-radical initiator to graft a reactive organosilane vinylalkoxysilane (vinyltrimethoxysilane or vinyltriethoxysilane) onto the plastomer molecule backbone. A small amount of tin catalyst (dibutyltin dilaurate) is also added at the end of the extrusion process to later accelerate subsequent cross-linking reactions by "static vulcanization" of the compounded product.

Typically, fabricated parts can be exposed to moisture in a steam chamber to hydrolyze the alkoxy groups attached to the silicon atoms (Si-O-C) so as to give hydroxyl end groups (-Si-OH). This leads to intermolecular cross-linking of dispersed plastomer to give Si-O-Si linkages between neighboring plastomer molecules. Since moisture diffusion into the parts is required to complete the cross-linking reaction under this static vulcanization, the degree of cure is therefore depth dependent. Because of this difficulty, most of the silane-cross-linkable compounds by static cure are not suitable for thick-wall injection molding or extrusion.

In a recent European patent, TPVs based on silane-grafted plastomer under *dynamic vulcanization* was disclosed [48]. A specially designed twin-screw extruder had a sufficient number of mixing zones to make finished product in a single pass:

1. Peroxide initiated grafting of organosilane unto the plastomer backbone.
2. Blending of polypropylene resin after grafting step is complete.
3. Instead of the tin catalyst to promote cross-linking reaction, an aqueous buffer solution having a pH in the range of 8–15 was found to be effective in accelerating hydrolysis of the siloxane end groups and promote faster cross-linking reactions. The solution is injected into the polymer molten blend of grafted plastomer and polypropylene resin.

Upon exiting the twin-screw extruder, the TPV product is completely vulcanized with no need for posttreatment. The blend with polypropylene allows the compound to be melted and fabricated into a variety of end-use products.

## 7.11 CONCLUSION AND FUTURE DIRECTIONS

Metallocene plastomers are already widely used as PP impact modifiers and are positioned to become the leading modifier type in the near future. The increased utilization of plastomers to modify PP resins continues to grow. Plastomers provide equivalent impact enhancement (including at low temperatures) with additional performance advantages, including easier dispersion, superior impact in high-flow PP resins, and low haze in clarified RCP resins.

Current work in many development labs focuses on modification of reactor-grade TPOs to make soft, flexible compounds for extruded or calendared sheets. In a growing number of similar applications, flexible PP sheet can replace plasticized PVC or PVC blends, chlorinated polyethylene, or EPDM.

Flexible PP sheet products are expected to find use in industrial applications such as geomembranes, pond liners, and single-ply roofing. Advantages over PVC include tensile and tear strength, puncture resistance, and recyclability. Polyolefin sheets are 30% lighter than PVC sheets at equivalent thickness to provide easier handling and installation. Polyolefin sheets can be thermally welded, unlike conventional elastomer sheets that must be bonded by vulcanization or adhesives.

One major emerging application is in the leather-like automotive interior skin application. Instrument panel surfaces and door panel skins based on a fixed-morphology technique are under development. Compared to PVC, flexible PP provides improved aging and reduced window fogging because no plasticizer is used. Recycling is enhanced as more parts are made from polypropylene.

The current instrument panel fabrication based on thermoforming PVC compound is a tedious process:

1. The flexible embossed skin needs to be first laminated with a cross-linked polypropylene foam layer.
2. The laminate sheet stock is next heated in an infrared chamber and draped onto a positive vacuum forming (male plug) process, as shown in Fig. 7.46a.
3. The TPO sheeting shows excessive sheet sagging when heated to the vacuum temperature and the embossed grain pattern tends to wash out after vacuum formation.

One modification to the TPO sheeting is the incorporation of a small amount of pre-cross-linked rubber concentrate [49]. Figure 7.47 shows the morphology change that occurred when cross-linked rubber concentrate was added to the TPO sheeting. In the TPO sheeting (Fig. 7.47a) the plastomer phase (light phase) is oriented with the machine direction of the sheeting. Due to its low melting point of plastomer, it cannot increase the melt strength of the PP matrix resin. With the addition of pre-cross-linked high melting rubber concentrates to give network-enhanced sheeting (Fig. 7.47b), the high-temperature stiffness (storage modulus) of the sheeting increases to prevent sagging of the hot sheet under its own weight. The increased cross-linked rubber also provides additional elastic memory that resists grain washout. In general, a storage modulus of at least 1 MPa (145 psi) is required to prevent excessive sagging during thermoforming. Figure 7.48 shows the DMTA scan of a cross-linked elastomer-enhanced TPO. At the thermoforming temperature of 125–140°C, a storage modulus of 5 MPa (725 psi) is maintained.

A recent development leading toward the simplification of the thermoforming process is the negative vacuum–forming process [50] shown in Fig. 46c. In this process, an unembossed TPO and foam-laminated sheet drapes over a female mold. The combination of heat and vacuum draws it evenly into the mold cavity, which has the etched grain pattern. Better grain retention around sharp contours can be obtained at the higher tooling cost.

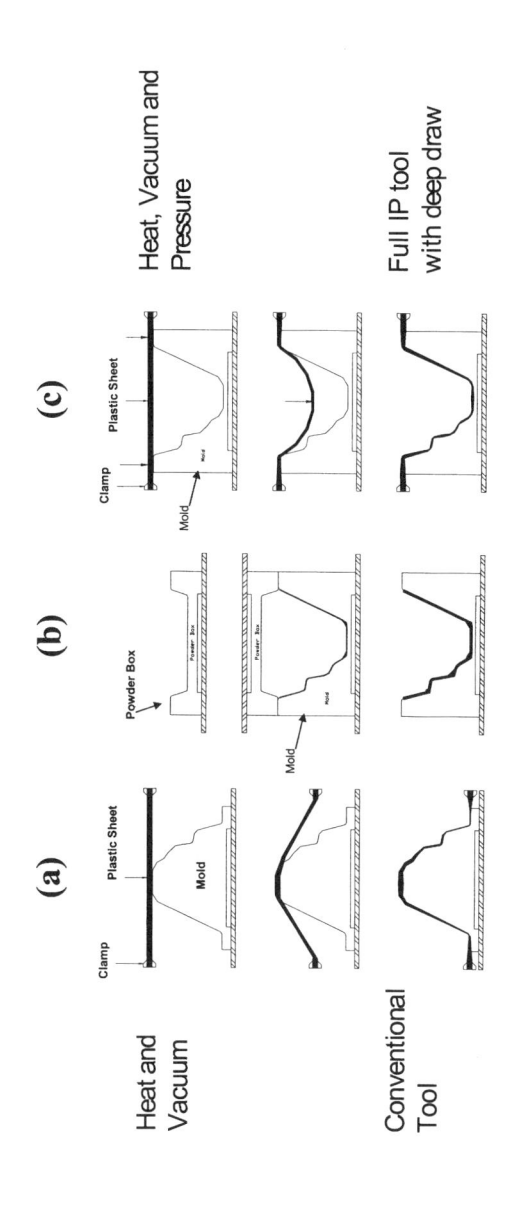

**FIGURE 7.46** Comparison of thermoforming processes to fabricate automotive interior skin for various compounds. (a) Positive vacuum forming PVC, (b) powder slush molding TPU and (c) negative vacuum forming PO.

9kx 0.5µ  9kx 0.5µ

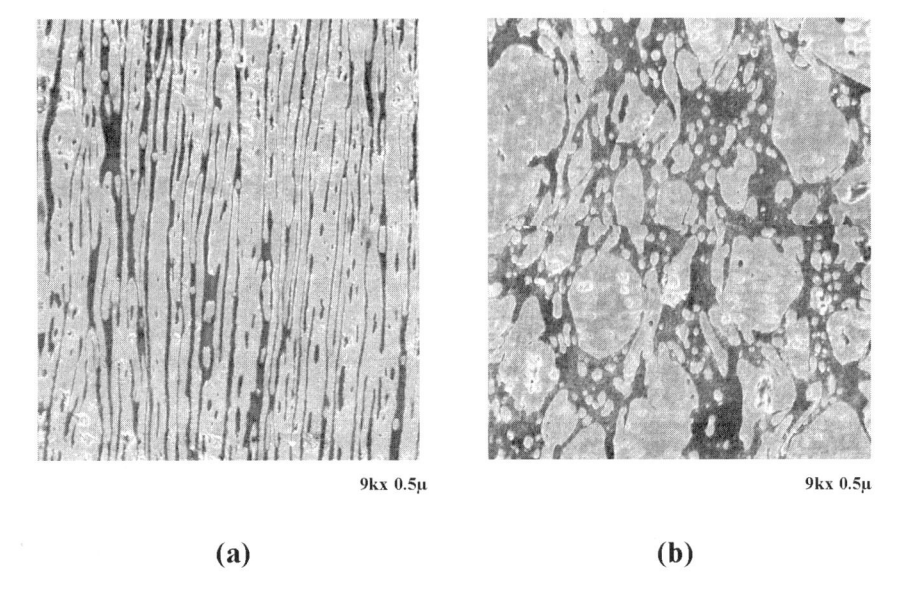

(a)  (b)

**FIGURE 7.47** Morphology comparison of TPO and network enhanced sheeting. (a) TPO sheeting and (b) network-enhanced sheeting.

Another development is the adaptation of the slushing molding process (Fig. 7.46b) currently used for thermoplastic polyurethane powder and plastisols for TPO powders or micropellets. The TPO powder was produced [51] in a three-step process:

1. In the first step polypropylene, rubber, and additives were compounded using a twin-screw extruder.
2. The TPO pellets were next cryogenically pulverized. The pulverized powders have complex shapes and tailings. They were also somewhat sticky due to their large rubber content.
3. An inorganic filler was next used to polish the pulverized powders to increase its flowability and bulk density.

Because the polished powder showed an average particle size of 0.224 mm, as well as a broad particle size distribution, both contribute pinhole-free molded parts.

In another development [51], the TPO compound was first converted to micropellets using a single-screw extruder and a specially designed die plate. A rotomolding study on a laboratory machine showed promising results, but the narrow particle size distribution and larger average particle size of the micropellets produced pinholes on mold surface.

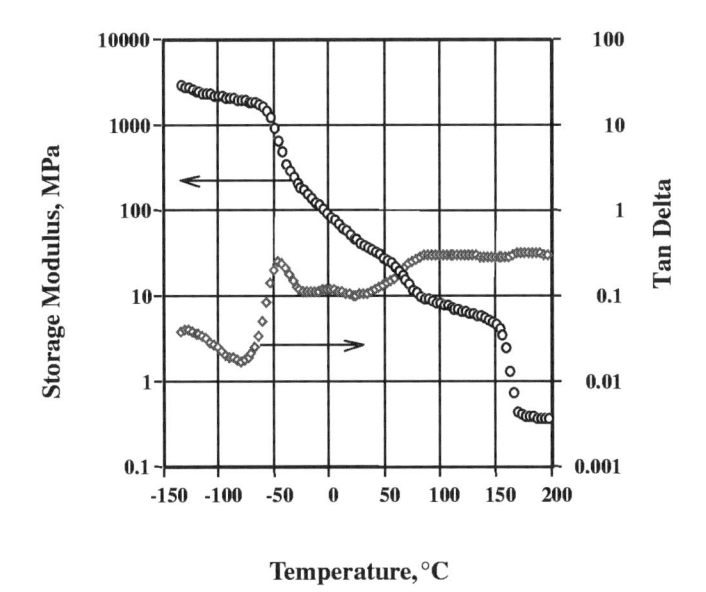

**FIGURE 7.48** Storage modulus of automotive interior skin.

## REFERENCES

1. DJ Synnott, DF Sheridan, EG Kontos. EPDM-polypropylene blends. In: SK De and AK Bhowmick, eds. Thermoplastic Elastomers from Rubber-Plastic Blends. Chichester: Ellis Horwood, 1990, pp. 130–158.
2. DJ Elliott. Natural rubber-polypropylene blends. In: SK De, AK Bhowmick, eds. Thermoplastic Elastomers from Rubber-Plastic Blends. Chichester: Ellis Horwood, 1990, pp. 102–129.
3. G Holden, NR Legge. Thermoplastic elastomers based on polystyrene-polydiene copolymers. In: NR Legge, G Holden, HE Schroeder, eds. Thermoplastic Elastomers. New York: Hanser, 1987, pp. 47–65.
4. Vistanex polyisobutylene properties and applications, Exxon Chemical Company Tech Bull SYN-74-1434, 1974.
5. RF Price. North America automotive TPO market outlook. Proceedings of TPOs in Automotive '97, Novi, MI, 1997.
6. AK Ford, HG Lollis, C Metaxas. Improved TPO durability using horizontal bed reactor design. Proceedings of TPOs in Automotive '97, Novi, MI, 1997.
7. Himont: Catalloy process stretches PP properties. Modern Plastics, December 1991, pp. 22–23.
8. CS Speed, BC Trudell, AK Mehta, FC Stehling. Structure/property relationships in EXXPOL polymers. SPE RETEC Polyolefins VII International Conference, Houston, 1991, pp. 45–66.

9. SP Chum, CI Kao, GW Knight. Structure and properties of polyolefin plastomers and elastomers produced from single site constrained geometry catalyst. Proceedings of Polyolefins IX International Conference, 1995, p. 471.
10. Attane ultra low density ethylene-octene copolymers, Form No. 305-1605-790X SMG, Dow Plastics, Midland, MI.
11. FG Stakem, HK Ficker, MA Corwin. Flexomer polyolefins: material of choice for enhanced product performance in polypropylene blends. Proceedings of TPE '92 Conference, Orlando, FL, 1992, pp. 287–305.
12. Mitsui Petrochemical Industries Ltd. Tafmer A/P ethylene/alpha-olefin copolymer product bulletin, 1986.
13. AM Thayer. Metallocene catalysts initiate new era in polymer synthesis. Chem Eng News, Sept 11, 1995.
14. HC Welborn, JA Ewen. Process and catalyst for polyolefin density and molecular weight control. US Patent 5,324,800, 1994.
15. AA Montagna, RM Burkhardt, AH Dekmezian. Single site catalysis: its evolution and impact on the polymer field. Proceedings of Metallocene Technology '97 Conference, Chicago, 1997, pp. 1–9.
16. H Sinn, W Kaminsky, HJ Wollmer, R Woldt. Living polymers on polymerization with extremely productive Ziegler catalysts. Angew Chem Int Ed Engl 19:390–392, 1980.
17. HW Turner, GG Hlatky. Catalysts, method of preparing these catalysts and polymerization processes wherein these catalysts are used. PCT Int Appl WO 88300698.3, 1988.
18. HW Turner. Catalysts, method of preparing these catalysts and method of using said catalysts. PCT Int Appl WO: 88300699.1, 1988.
19. B Wolf, S Kenig, J Klopstock, J Miltz. Thermal fractionation and identification of low density polyethylenes. J Appl Polym Sci, 62:1339–1345, 1996.
20. M Varma-Nair. Unpublished results, 2000.
21. E Manis, A Touny, L Wu, K Strawhecker, B Lu, TC Chung. Polypropylene/ montmorillonite nanocomposite: review of the synthetic routes and material properties. Chem Mater 13:3516–3523, 2001.
22. JW Gilman, CL Jackson, AB Morgan, R Harris Jr. Flammability properties of polymer-layered-sillicate nanocomposites. Chem Mater, 12: 1866–1873, 2000.
23. R Leaversuch. Nanocomposites. Plastics Technol pp. 64–69, October 2001.
24. ASTM D–256–90b. Standard test method for impact resistance of plastics and electrical insulating materials. American Society for Testing and Materials, 1990.
25. ASTM D-3763-86. Standard test method for high-speed puncture properties of plastics using load and displacement sensors. American Society for Testing and Materials, 1986.
26. TC Yu. Impact modification of polypropylenes. Proceedings of Society of Plastics Engineers Annual Technical Conference, San Francisco, 1994, pp. 2439–2445.
27. GM Brown, JH Butler. New method for the characterization of domain morphology of polymer blends using ruthenium tetroxide staining and low voltage scanning electron microscopy (LVSEM). Polymer 38:3937–3945, 1997.
28. W Rasband. NIH Image Shareware Version 1.2. 1990.
29. KaleidaGraph, Version 2.1. Abelbeck Software, 1988.

30. CE Scott, CW Macosko. Polymer 36:442, 1994.
31. C Rauwendaal. Mixing in single screw extruder. In: I Manas-Zloczower, Z Tadmor, eds. Mixing and Compounding of Polymers, New York: Hanser, 1994, pp. 251–329.
32. TC Yu. Plastomer-polypropylene blend mixology. Proceeding of SPE RETEC Polyolefins X Conference, Houston, 1997, pp. 227–239.
33. E. N. Kresge. Rubber thermoplastic blends. In: DR Paul, S Newma, eds. Polymer Blends. Vol. 2. New York: Academic Press, 1978, pp. 293–310.
34. PJ Phillis, K Mezghani. In: JC Salamone, ed. Polymeric Materials Encyclopedia, Boca Raton: CRC Press, Vol 9, 1996, p.6637.
35. RD Leaversuch. Advances enhance optical properties in PP. Modern Plastics 75(8): 8, January, 1998, pp. 50–53.
36. TC Yu, DK Metzler, M Varma-Nair. Impact enhancement of clarified polypropylene with selected metallocene plastomers. Proceedings of Society of Plastics Engineers Annual Technical Conference, Dallas, 2001, pp. 1688–1693.
37. TM Miller, TC Yu. Extrision blow molding of clarified polypropylene rigid containers: influence of EXACT plastomers. Proceedings of SPE RETEC Polyolefins Conference, CD Version, Houston, 2002.
38. TC Yu. EXACT$^{TM}$ plastomer optimization in high flow thermoplastic olefins. Proceedings of Society of Plastics Engineers Annual Technical Conference, Boston, 1995, pp. 2358–2368.
39. JA Cornell. Experiments with Mixtures, 2nd ed. New York: Wiley Interscience, 1990, pp. 139–227.
40. Design Expert software version 2.05, Minneapolis: Stat-Ease, 1989.
41. T Hishio, T Nomura, T Yokoi, H Iwai, N Kawamura. Development of super olefin bumper for automobiles. Toyota Tech Rev 42:11–22, 1992.
42. T Nomura, T Nishio, K Iwanami, K Yokomizo, K Kitano, S. Toki. Characterization of microstructure and fracture behavior of polypropylene/elastomer blends containing small crystal in elastomeric phase. J Appl Polym Sci 55:1307–1315, 1995.
43. TC Yu. Evaluation of polar conventional and non-polar olefinic paint technologies in thermoplastic olefins. Olefinic Paint Technology (SP-1334), Society of Automotive Engineers, Detroit, 1998, pp. 35–50.
44. NR Dharmarajan, TM Miller, TC Yu. Plastomer based compounds for TPO roof membrane applications. Proceedings of Polyolefins International Conference, Houston, 2002, pp.1694–1698.
45. C Martin. High speed twin screw extrusion technology for TPE pellets, sheet and profile. Proceedings of Thermoplastic Elastomers Topical Conference, Houston, 2002, pp.79–94
46. A Coran, RP Patel. Thermoplastic compositions of high unsaturation diene rubber and polyolefin rubber. U.S. Patent 4,104,210, Aug.1, 1978.
47. L Weaver, H Heck, D Parikh. Extending ethylene/1-octene compound performance through modification for extrusion and injection molding. Proceedings of Automotive Global Conference, CD Version, Troy, MI, 2001
48. C-P Kirchner, H-G Fritz, Q Cai. A thermoplastic composition and a process for making the same. EP 1050548A1, 1999.
49. TC Yu. Low Modulus Thermoplastic olefin compositions. U.S. Patent 6,207,754 B1, 2001.

50. R Colvin. Negative vacuum forming. Modern Plastics 79(4):38–39, 2002.
51. H Sugimoto, A Imai. Novel TPO compound for powder slush molded auto interior skins. Proceedings of SPE Automotive Global Conference, Troy, MI, 1999, pp. 71–79.
52. W Weng, M Kontopoulou. Rotomolding of polyolefin plastomer and TPOs. Proceedings of Society of Plastics Engineers Annual Technical Conference, CD Version, San Francisco, 2002.

# 8

## Talc in Polypropylene

**Richard J. Clark and William P. Steen**
Luzenac America, Englewood, Colorado, U.S.A.

### 8.1 INTRODUCTION

Talc is an industrial mineral [1–4] that is found in deposits throughout the world. Major commercial deposit areas are shown in Fig. 8.1. The major sources of talc used in polypropylene are available from mines in the United States, Canada, France, Austria, Italy, Australia, China, and India.

Also known as soapstone, talc has been used for thousands of years. It is characterized by its soft and slippery feel. It is the softest of the minerals, with a Mohs hardness of 1. This softness also makes it easy to carve and shape. Exquisite ornamental serving utensils made from carved talc have been found dating back to antiquity.

#### 8.1.1 Geology

Geologically, talc is typically formed by the alteration of a dolomite or serpentinite host rock. The talc formed from a dolomite host is typical of the type found in Montana, France, and China. These large deposits are characterized by a microcrystalline talc structure, with talc concentrations in the deposit ranging from 93% to 99% talc by weight. Talc from these deposits can be sorted manually, optically, or mechanically to enhance color and talc content.

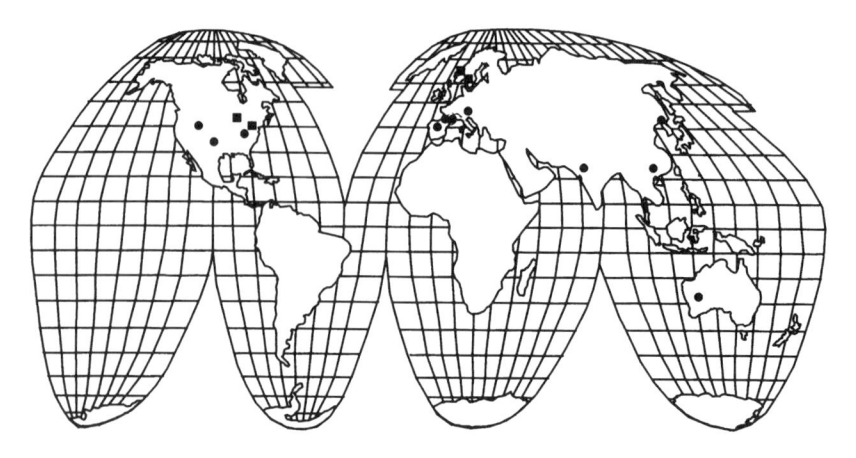

FIGURE 8.1 Worldwide locations of talc deposits. ● Dolomite, host talc deposits; ■ Serpentinite, host talc deposits.

The talc/magnesite ores are derived from serpentinite hosts and are typically found in Vermont, Quebec, and Finland. This type of ore is macrocrystalline in nature and composed of approximately 45–65% talc by weight, with the remainder of the ore made up of magnesite (magnesium carbonate). In some low-cost applications, such as automotive under-the-hood, these ores are coarsely ground and used as they are. These ores are also frequently beneficiated to 93–99 wt% talc by flotation. Since talc is hydrophobic, it is easy to float with surfactants. The talc floats on an aqueous bubble while the comineral impurities sink in the water phase.

### 8.1.2 Chemistry

Chemically, talc is hydrous magnesium silicate ($3MgO \cdot 4SiO_2 \cdot H_2O$). The typical chemical composition [2] of talc is 31.7% MgO, 63.5% $SiO_2$ and 4.8% $H_2O$, although this can vary depending on the source of the ore. Talc is a layered silicate (phyllosilicate) consisting of sheets of magnesium in octahedral coordination similar to brucite, $Mg(OH)_2$, sandwiched between sheets of silicon in tetrahedral coordination. Alternating sandwiched sheets, 19 Å thick, are held together by weak van der Waals forces giving rise to talc's low hardness and slippery feel. The molecular structure of pure talc is shown in Fig. 8.2. Cominerals typically include *dolomite* [$Ca(CO_3)_2 \cdot Mg(CO_3)_2$], *chlorite* [$(Mg, Al, Fe)_6 (Si, Al)_4 O_{10}(OH)_8$], or *magnesite* ($MgCO_3$). Upon heating, the carbonate minerals decompose at temperatures ranging from 500° to 600°C. This can lead to unexplained weight loss when ashing a polypropylene composite

## Talc in Polypropylene

**Talc Basic Structure: 3MgO-4SiO$_2$-H$_2$O**

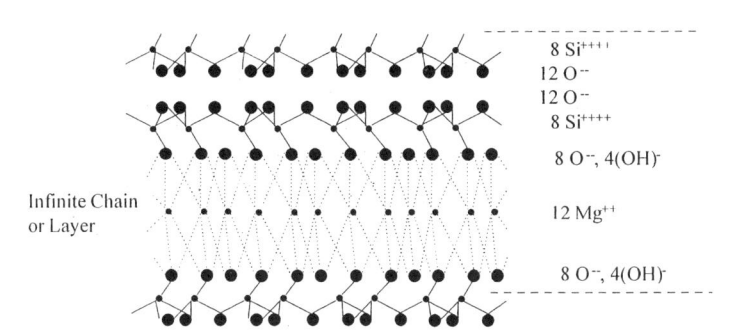

**Formulation per repeating Layer: (OH)$_8$Mg$_{12}$Si$_{16}$O$_{40}$**

Infinite Chain or Layer

8 Si$^{++++}$
12 O$^-$
12 O$^-$
8 Si$^{++++}$
8 O$^-$, 4(OH)$^-$
12 Mg$^{++}$
8 O$^-$, 4(OH)$^-$

**FIGURE 8.2** Molecular structure of pure talc mineral. (OH)$_8$Mg$_{12}$Si$_{16}$O$_{40}$ or (OH)$_2$Mg$_3$(Si$_2$O$_5$)$_2$ = talc.

containing impure talc. A weight loss versus temperature curve is shown for a typical talc in Fig. 8.3.

Talc loses its water of hydration at 1000°C. At this temperature, the talc crystal restructures itself to form the mineral enstatite (MgSiO$_3$). Enstatite is significantly harder than talc with a Mohs hardness of 5–6. Talc is inert to most chemicals. It normally will absorb an equilibrium 0.2–0.3% by weight of moisture under normal atmospheric conditions.

The surface chemistry of talc is not fully understood. It is theorized that when the talc crystal is fractured reactive groups are formed at the edge. These reactive groups can potentially react with the polymer or deactivate stabilizers. The theorized surface chemistry of a talc crystal is shown in Fig. 8.4. The planar surfaces of talc are hydrophobic and for the most part unreactive except for some Mg$^{2+}$ sites. The edges of talc platelets are much more polar and contain many reactive sites as listed in Fig. 8.4. The reactive sites of the edges are used when reacting surface treatments to talc.

When slurred with water, talc has a pH of 9–9.5 yet the majority of the proposed reactive sites are acidic in nature. The reversible nature of these surface reactions is probably what leads to the basic response in the water slurry.

### 8.1.3 Mining/Processing

Talc deposits are typically mined by open-pit mining. In a few cases talc is mined underground. The talc is removed from the mine, crushed, and ground to a specific size on a variety of processing equipment including roller mills, hammer mills, and/or jet mills. The end-use requirements in plastic determine the fineness

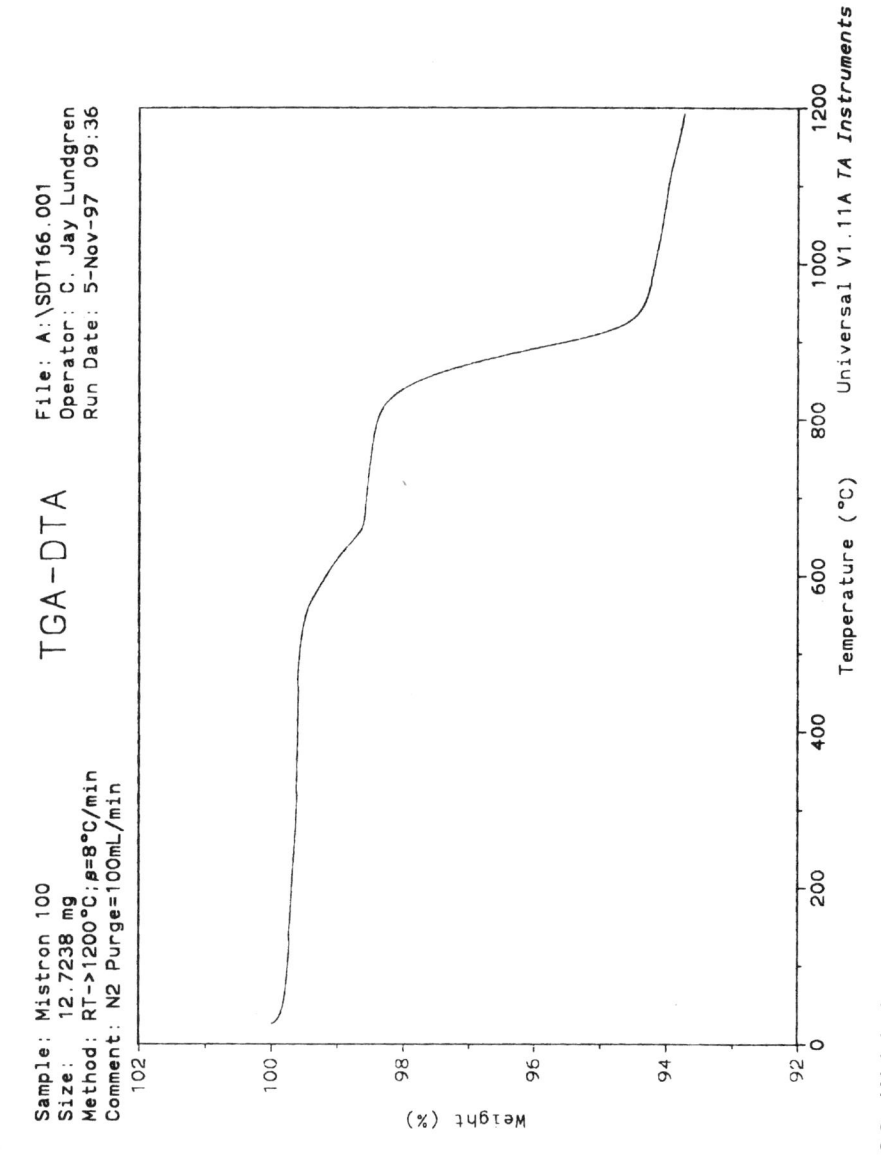

**FIGURE 8.3** Weight loss versus temperature.

# Talc in Polypropylene

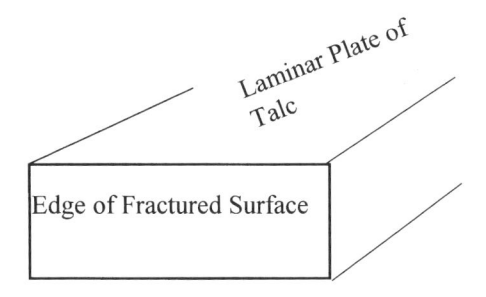

*Chemistry of Possible Reactive Sites located along cross-section of the fractured surface submerged in water:*

1.) **Weakly Acidic Terminal Hydroxyl**

   HO-Si-  into  $H^+$ + $[O\text{-}Si]^-$

2.) **Proton Release through Polarization of Water Molecules**

   $Mg^{+2}$ + $H_2O$  into  $2H^+$ + $MgO$

3.) **Lewis Acid Site**

   Metal ion reacts with paired electrons on water molecule.

4.) **Octahedral Iron Ox/Redox Site:**

   $Fe^{II}/Fe^{III}$ for any present in talc ore.

5.) **Strongly Acidic Bronsted Site**

   $Mg^{+2}$ + HO-Si-  into  $H^+[Si\text{-}O\text{-}Mg]^+$

6.) **Weakly Basic Site**

   Magnesium Hydroxide- $Mg(OH)_2$

**FIGURE 8.4** Possible surface sites on talc. 1, Weakly acidic terminal hydroxyl; 2, Proton release through polarization of water molecules; 3, Lewis acid site; 4, Octahedral iron ox/redox site; 5, Strongly acidic Bronsted site; 6, Weakly basic $Mg(OH)_2$.

of grind. Roller-milled products are used in low-impact applications, such as fans and fan shrouds in automotive underhood parts. Hammer- and jet-milled products are used where impact requirements are important in the finished product. Besides the plastics industry, talc is also used in ceramics, as a flux; in paper to control pitch or as a filler; in coatings to control gloss and sag; and in rubber to increase stiffness and processability. Talc's most widely recognized application is in the cosmetics industry although this market represents only a small volume of sales.

Since talc is a naturally occurring mineral, the color of the talc can vary from layer to layer of the same mine deposit. The color of talc is affected by very low levels of impurities such as iron oxides, chlorite, dolomite, and magnesite. The true color-controlled tabs must be mined from well-studied ore sources.

### 8.1.4 Grind Measurement

Talc has a platy structure that leads to its unique ability to reinforce polypropylene and other polymers. Grinding of talc is generally done by a shearing mechanism to increase the delamination of the platelets thereby increasing the aspect ratio. Although the aspect ratio of the talc is a valuable number, it is not widely reported because it is difficult to report a representative number for a large sample.

Within the talc processing plant, top size and loose bulk density are used to control the grind. The median particle size may also be used as the controlling parameter. Median particle size can be measured via laser light scattering techniques or via Stokes law settling rates. The latter is the most commonly reported value by talc companies. The Sedigraph measures the settling rates of the talc particles using an X-ray detector. A scan of particle size and number of particles is used to generate a typical particle size distribution curve, as shown in Fig. 8.5. The chart plots the cumulative mass finer than a given particle size

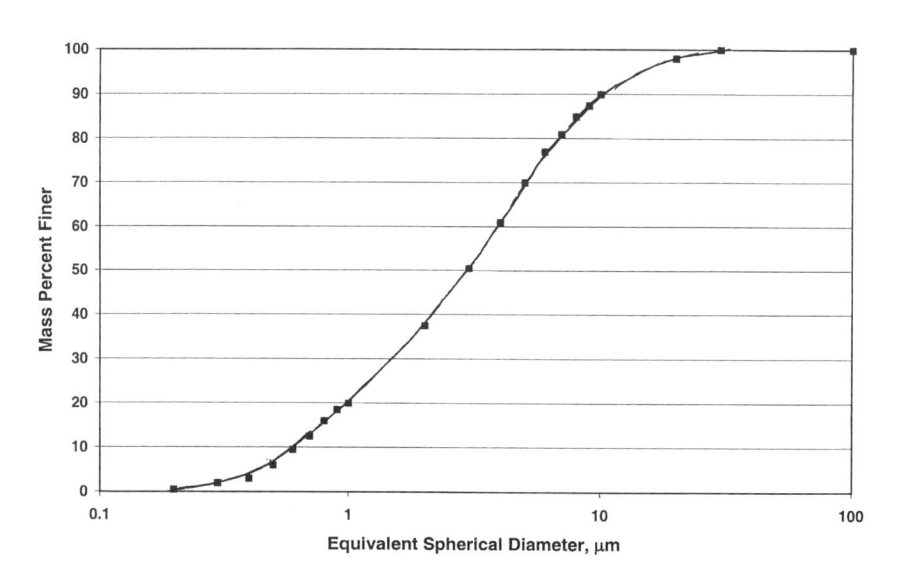

FIGURE 8.5 Cumulative mass percent finer versus diameter.

versus the log of particle size. A median particle size of 2.9 μm is determined at 50 wt% percent of the sample.

## 8.2 EFFECT ON REINFORCEMENT

A scanning electron photomicrograph of a typical ground talc sample is shown in Fig. 8.6. It features a platy morphology of the talc particles. These platy particles tend to orient along flow lines during molding of a filled polymer composite. As a result, talc reinforcement gives increased stiffness and heat deflection temperature (HDT) compared with the unfilled polymer resin. Tensile strength is less influenced by talc filler but does increase somewhat as particle size decreases.

### 8.2.1 Historical Perspective

Talc was first used in polypropylene homopolymer in the 1960s, for under-the-hood automotive parts. The first major applications were in fan shrouds and blades. In the mid-1970s, talc-reinforced copolymers replaced stamped metal parts in appliances. These parts included pump housings, washer tubs, and spin baskets. In the mid 1970s, ultrafine talc was used in thermoplastic olefin (TPO) compounds to replace polyurethane and acrylonitrile-butadiene-polystyrene (ABS) blends in fascias and kick plates due to lower costs and the ability to

FIGURE 8.6 SEM photomicrograph of typical ground talc.

meet the new 5 mile per hour automotive crash test. In the late 1980s, the Japanese began using ultrafine talc to improve the stiffness of high-impact copolymers and TPO blends for bumpers, dashboards, instrument panels, and other automotive components.

### 8.2.2 Effect of Talc Loading

The primary use for talc in polypropylene and its copolymers is to increase the stiffness of the compound. The amount of talc used in a compound has a direct effect on the level of stiffness obtained. The addition of 20% talc loading to a homopolymer with a flexural modulus of 200 kpsi (1382 MPa) will increase the stiffness from 200 kpsi (1382 MPa) to approximately 400 kpsi (2764 MPa). Forty percent loading will increase the stiffness to 600 kpsi (4146 MPa). Similar changes are noted in impact copolymers and high crystalline polypropylene (Fig. 8.7).

### 8.2.3 Macro-versus Microcrystalline

The platyness of the specific talc used in the compound contributes to the level of stiffness obtained. A talc that has a large aspect ratio (ratio of length to thickness) is macro-crystalline. Macrocrystalline talc is found in deposits located in Italy,

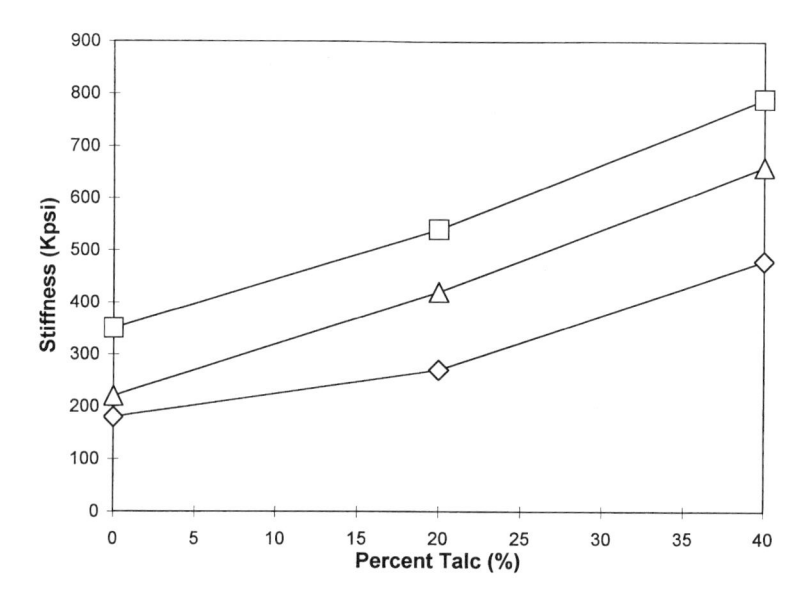

**FIGURE 8.7** Effect of talc on plastic stiffness. ◇ Copolymer; □ HCPP; △ homopolymer.

# Talc in Polypropylene

Vermont, and Canada. Macrocrystalline talc will give greater stiffness at a given loading than a microcrystalline product from mines in Montana, Australia, and China. Figures 8.8 to 8.12 compare the difference in stiffness, strength, color, long-term oven aging, and impact properties versus the particle size distribution of similar particle-sized microcrystalline and macrocrystalline talc samples in the same polymer at a 30 wt% loading. Particle size has a large effect on impact strength and a slight effect on flexural modulus.

### 6.2.4 Effect on HDT

Since HDT is closely related to the stiffness of the composite, it increases with increasing talc loading in polypropylene, as shown in Fig. 8.13 at 264 psi loading. The increased HDT means that the part can be used at a higher operating temperature without softening.

## 8.3 IMPACT/STIFFNESS BALANCE

### Effect of Talc Particle Size

Representing a discontinuous phase in a polymer matrix, talc, like other minerals, will decrease the impact strength of a polymer. A photomicrographic analysis of a

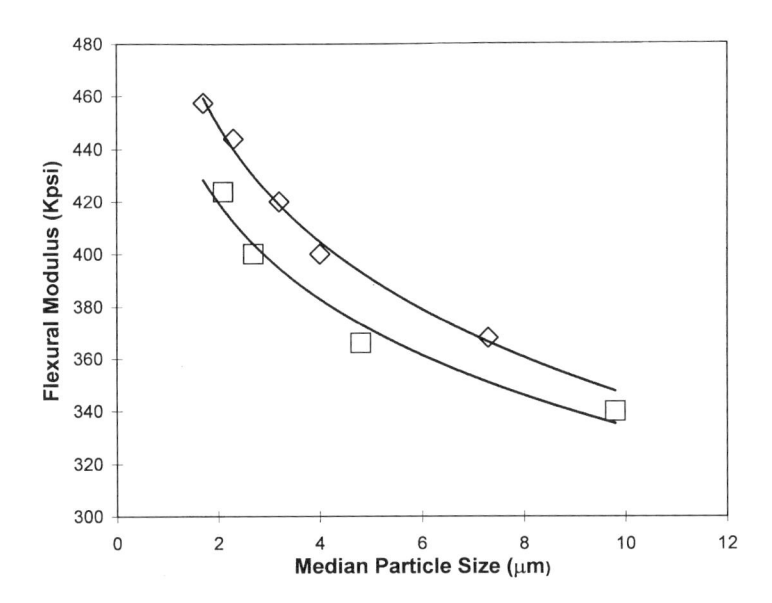

FIGURE 8.8 Particle size effect on flexural modulus. ◇ Jetfil (Canadian); ☐ Stellar/Cimpact (Chinese).

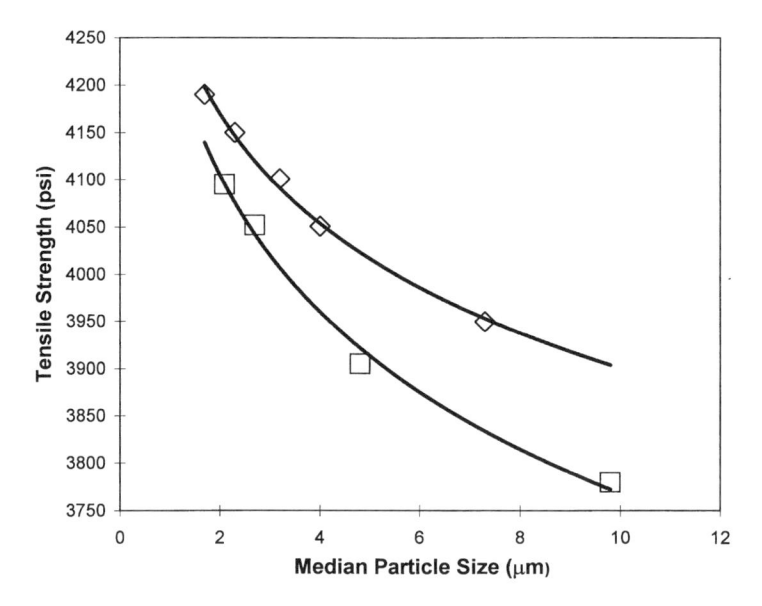

**FIGURE 8.9** Particle size effect on tensile strength. ◇ Jetfil (Canadian); ☐ Stellar/Cimpact (Chinese).

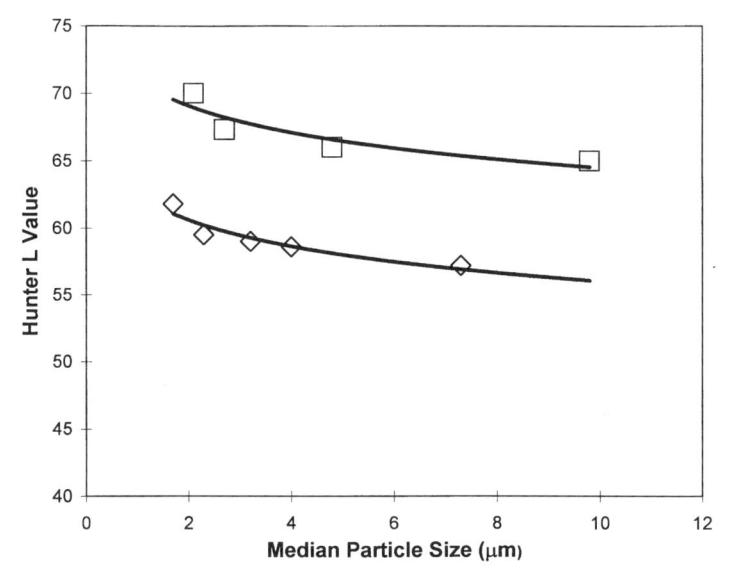

**FIGURE 8.10** Particle size effect on Hunter L values. ◇ Jetfil (Canadian); ☐ Stellar/Cimpact (Chinese).

# Talc in Polypropylene

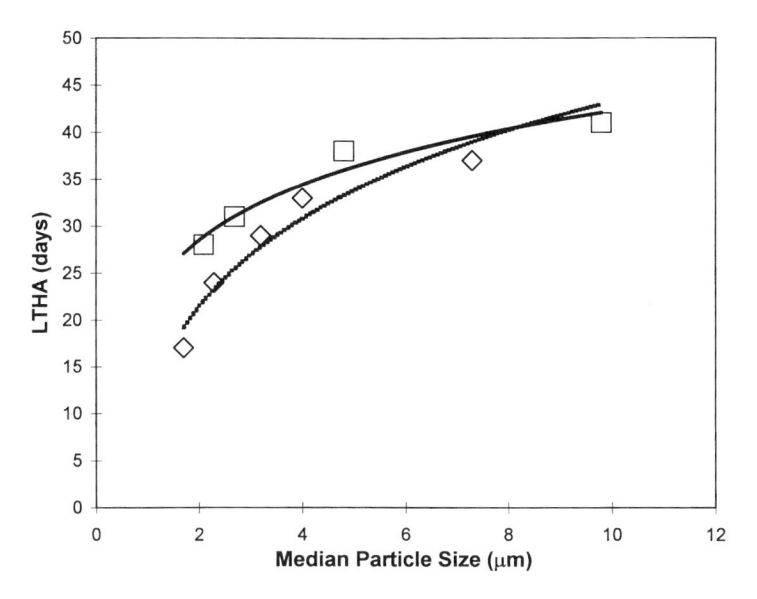

**FIGURE 8.11** Particle size effect on LTHA. ◇ Jetfil (Canadian); ☐ Stellar/Cimpact (Chinese).

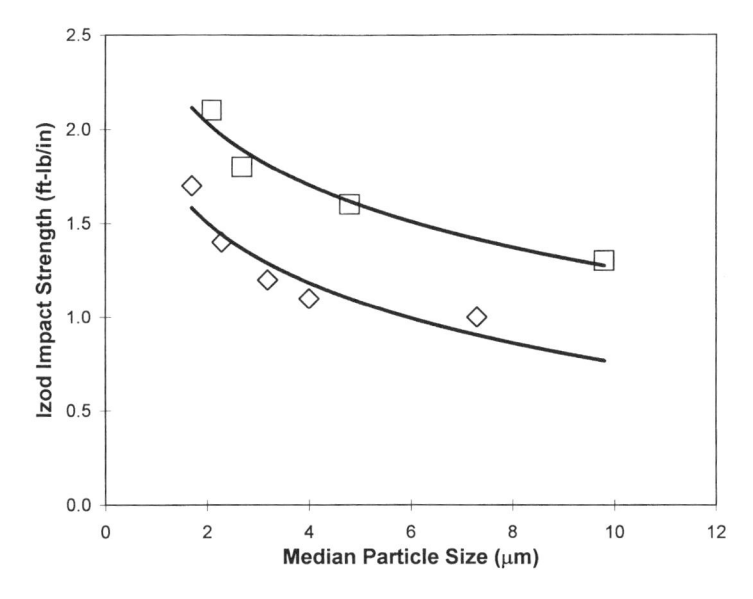

**FIGURE 8.12** Particle size effect on Izod impact strength. ◇ Jetfil (Canadian); ☐ Stellar/Cimpact (Chinese).

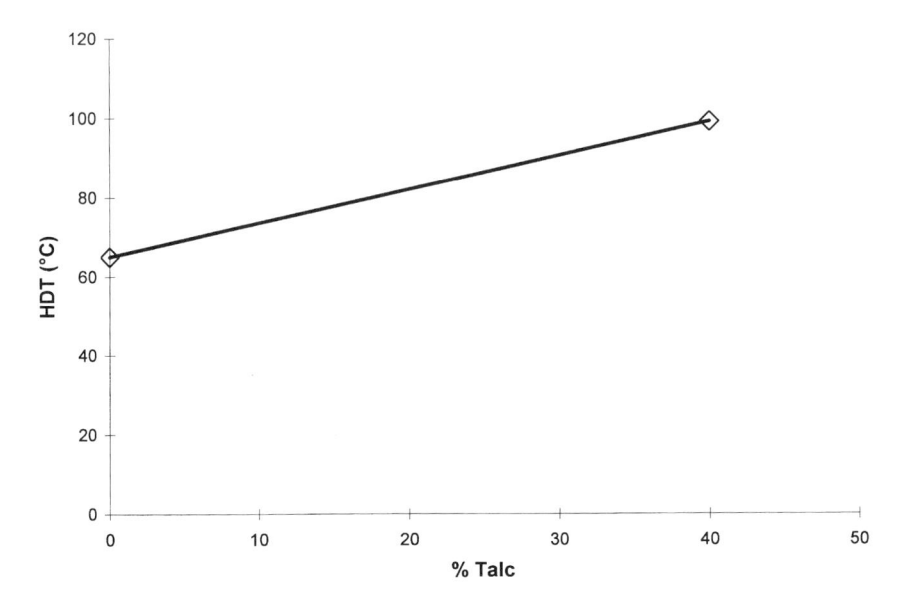

FIGURE 8.13 Effect of talc level on HDT.

falling-weight impact area is shown in Fig. 8.14–8.15. In Figure 8.14, a medium-impact copolymer with a 30% loading of a large particle talc fails in a brittle manner when compared with the same copolymer with a fine talc (Fig. 8.15) that gives ductile failure. Note the stretching of the polymer during failure in the ductile sample.

As shown in Fig. 8.16, the top size, or largest particle in the talc product size, directly influences the level of impact performance in a given polymer. In high-impact applications such as impact copolymers or TPOs, a talc with a very fine top size is required to minimize the adverse effect of talc on impact strength. The choice of talc fineness is particularly important in obtaining a ductile failure in impact at low temperature. The compounder often has the option of using a more flexible polymer and a coarser talc or a more rigid polymer and a finer talc to meet a given stiffness–impact balance. The relative costs of the talc and polymer dictate the final choice.

### 8.3.2 Designing Composite Performance

The addition of talc to a high-impact copolymer increases the stiffness. As increasingly finer talc is used, the stiffness–impact balance is improved. This is shown in Fig. 8.17, against a matrix of the typical balance of properties shown for several different engineering resins. The stiffness–impact balance moves from the

# Talc in Polypropylene

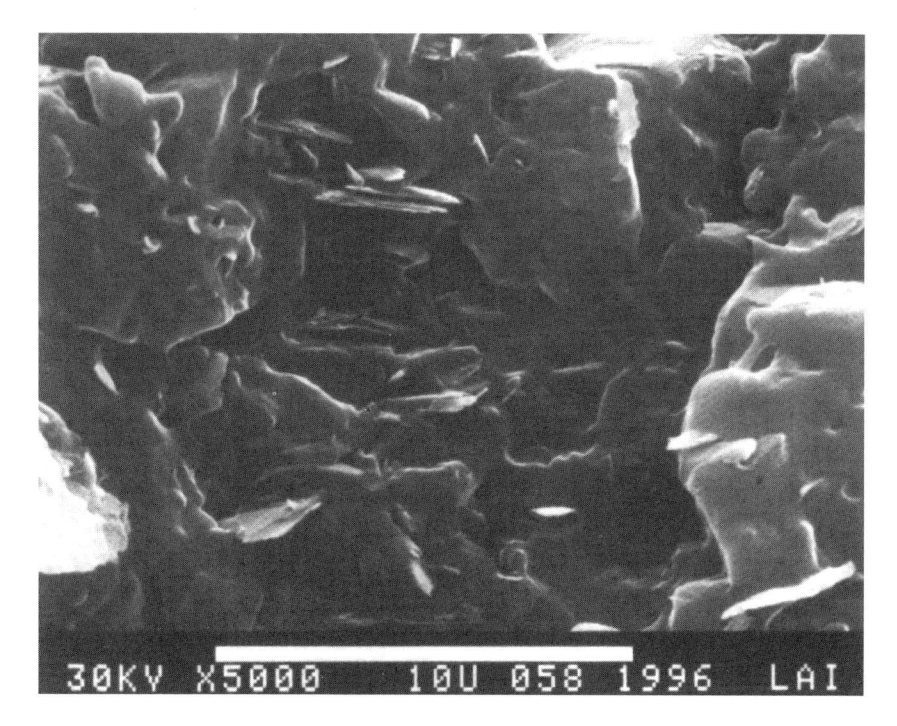

**FIGURE 8.14** Brittle failure due to large talc particles.

left, for the neat resin, to the right with the addition of equal loadings of different top-sized talc samples. The addition of finer and finer talc can take a polypropylene copolymer to the realm of properties previously obtained only with engineering resins.

## 8.4 EFFECTS ON OTHER PROPERTIES

### 8.4.1 Mold Shrinkage

Another benefit from the addition of talc to a polypropylene is reduced mold shrinkage. In Table 8.1, mold shrinkage rates are shown comparing impact copolymer and homopolymer with and without talc versus ABS. At a 30% loading, talc reduces the mold shrinkage by 39% in the impact copolymer and 57% in the homopolymer. This reduction in mold shrinkage can be used to minimize mold dimension changes caused by the polymer.

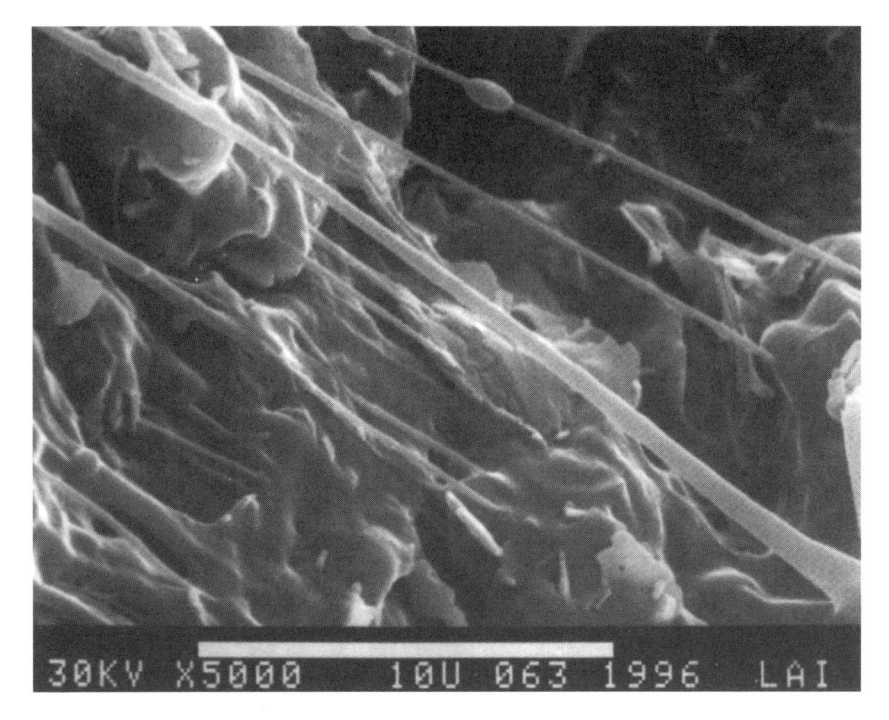

**FIGURE 8.15** Ductile failure due to small talc particles.

### 8.4.2 Thermal Expansion

The addition of 30% talc to polypropylene reduces the coefficient of thermal expansion by about 50% in the temperature range 50–150°C ($6 \times 10^{-5}$ in./in./°C for impact copolymer resin versus $3 \times 10^{-5}$ in./in.°C). The effect appears to be independent of the fineness of grind of the talc.

### 8.4.3 Melt Flow Rate

The addition of talc to polypropylene does not seem to have a significant effect on melt flow rate. Talc causes a slight increase in measured melt as shown in Table 8.2.

Recent trends in the industry have been toward the use of higher melt flow resins to improve mold filling rates. The dispersion of talc in high melt flow resins (50 g/10 min and above) can be accomplished with screw designs that increase shear mixing. Generally, single-screw extruders do not provide enough shear to disperse talc in high melt flow resins

# Talc in Polypropylene

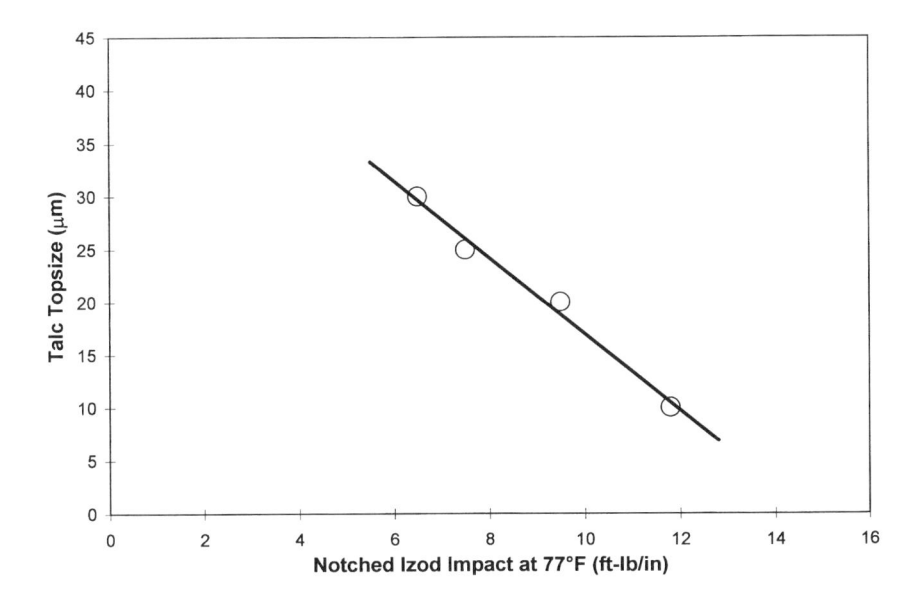

**FIGURE 8.16** Effect of talc top size on Izod impact at 77°F.

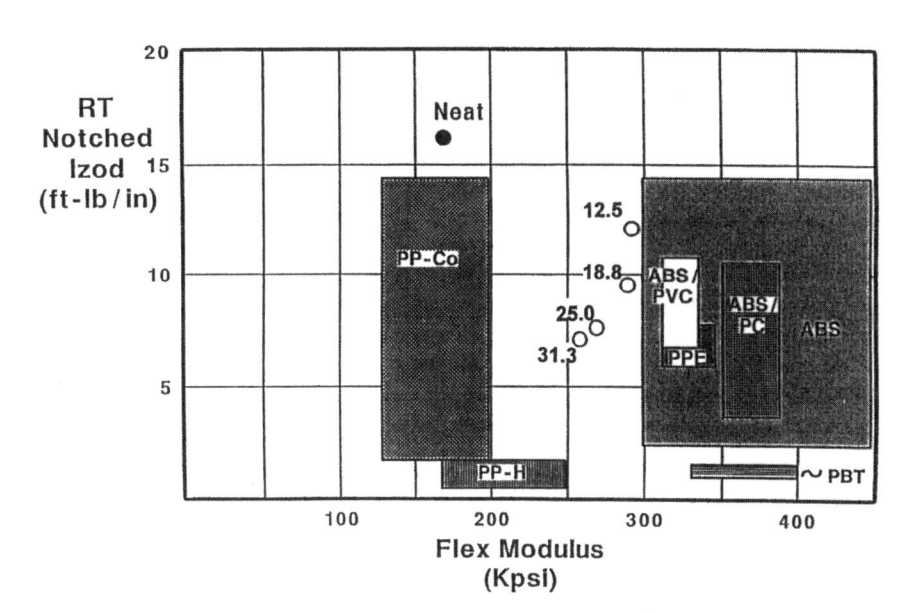

**FIGURE 8.17** Stiffness-impact balance for talc versus engineering resins. ● Neat high-impact copolymer; ○ Data for 30 wt% talc with indicated top size values.

TABLE 8.1  Effect of 30% Talc on Mold Shrinkage

| Polymer | Talc (wt%) | Mold shrinkage (in./in.) | Change due to talc (%) |
|---|---|---|---|
| Polypropylene Homopolymer | 0.0 | 0.018 | — |
|  | 30.0 | 0.011 | 39 |
| Polypropylene Copolymer | 0.0 | 0.021 | — |
|  | 30.0 | 0.009 | 59 |
| ABS (comparison) | 0.0 | 0.005 | — |

### 8.4.4 Crystallinity

Talc acts as a nucleating agent in crystalline polymers. It is not as effective as some chemical additives but it does increase the crystallization temperature as shown on the differential scanning calorimetry (DSC) chart (Fig. 8.18). The crystallization temperature of a polypropylene homopolymer is increased with talc concentration as shown in Table 8.3. A very fine, high surface area, microcrystalline talc is usually used to increase crystallinity. The addition of talc to polypropylene for any reason will increase the crystallinity of the polymer. See Chapter 16 for a more detailed discussion of nucleating effects due to submicrometer talc.

### 8.4.5 Long-Term Oven Aging

The origin of the talc ore used in polypropylene will have a significant influence on the color and long-term oven aging performance of the composite. Numerous tests have been performed to determine the variables in talc that contribute to this difference. It appears that talc surface chemistry can lead to talc–polymer and talc–stabilizer interactions, which can cause discoloration and depolymerization under long-term heat aging in a hot oven set at 150–160°C. High iron content in the talc can also adversely influence the level of performance.

TABLE 8.2  Effect of Talc on Melt Flow Rate

| Polymer | Talc (wt%) | Melt flow rate (g/10 min) |
|---|---|---|
| Impact Copolymer | 0.0 | 1.4 |
|  | 30.0 | 1.9 |

# Talc in Polypropylene

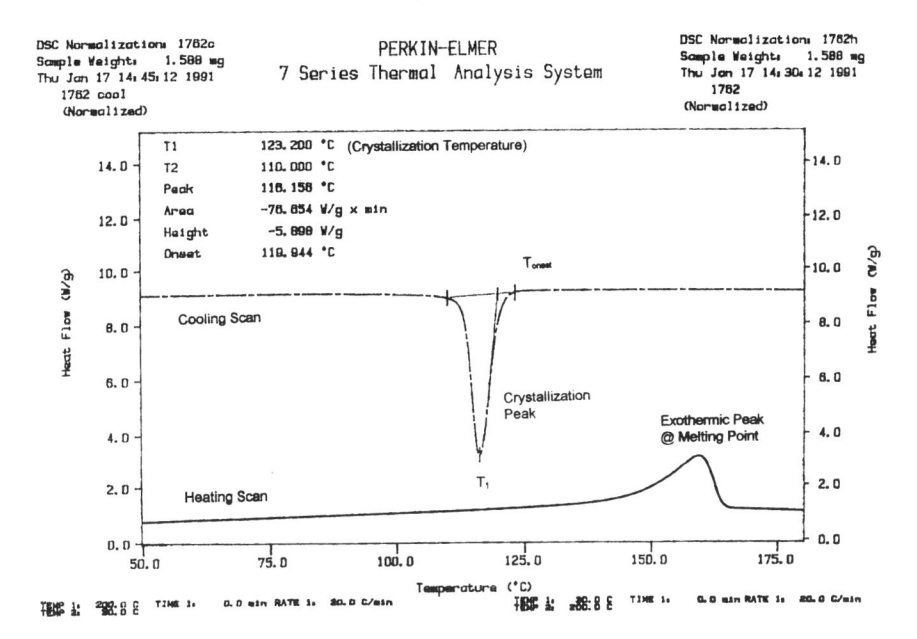

**FIGURE 8.18** DSC melting-crystallization peaks.

The long-term thermal stability of the talc-polymer composite can have an effect on the cost of the compound by dictating the amount and type of stabilizer used. Based on Fig. 8.19, the wt% of typical stabilizer blend with a 2 : 1 ratio of di stearyl thio dipropionate (DSTDP) to hindered phenol can be predicted for the talc ore used and time-to-failure requirement. Lower talc concentrations give better oven aging results. Coarser talc particles give better heat aging performance than fine particles.

**TABLE 8.3** Effect of Talc Loading on Crystallization Temperature

| Talc loading (%) | DSC crystallization temp. (°C) |
|---|---|
| 15 | 123.204 |
| 5 | 122.131 |
| 2 | 121.103 |
| .5 | 117.526 |
| .1 | 116.241 |
| 0 | 115.319 |

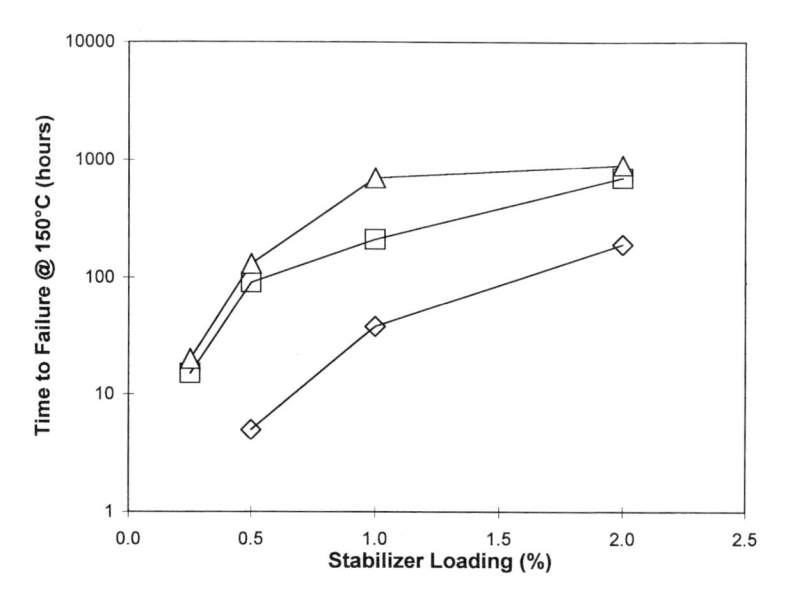

**FIGURE 8.19** Effect of talc filler type stabilizer loading combination on LTHA. ◇ Beaverwhite 325; ☐ Vertal 410; △ Stellar 410.

The absorption of antioxidants onto the surface of talc particles is responsible for deactivating a portion of the stabilizer package. Consequently, more stabilizer would be required to offset the amount deactivated. A filler deactivator [6] is added to put a coating on the talc particles to reduce absorbed stabilizer, e.g., 0.5% epoxy resin (Araldite 7072).

### 8.4.6 Effects of Iron on LTHA

Iron oxide is frequently a contaminant in a talc deposit. To study the effect of iron contamination on the long-term oven aging of talc, a talc product with good aging characteristics was intentionally doped with a purified grade of iron oxide ($Fe_2O_3$). The long-term oven aging dropped in direct proportion to the iron content, as shown in Fig. 8.20. A similar test with naturally occurring iron oxide gave similar results. Filler deactivator, such as epoxy resin, can be added to the talc-reinforced compound to minimize the acceleration of thermo-oxidation reactions catalyzed by iron and other transition metals. The epoxy resin coats the surfaces of talc particles, thereby isolating the iron impurity from polymer contact.

# Talc in Polypropylene

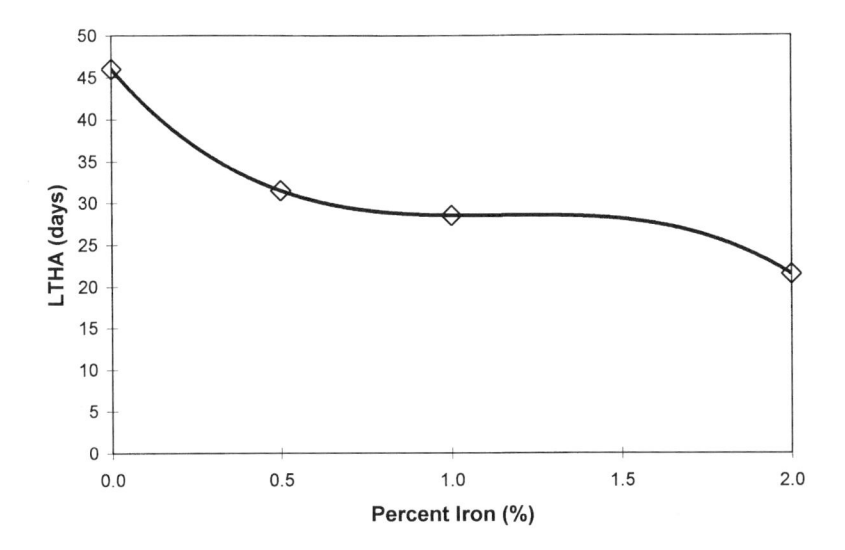

**FIGURE 8.20** Effect of iron level on LTHA. ◇ Stellar 410.

## 8.4.7 Response to TiO$_2$

The color of the talc-polypropylene composite can influence the amount of stabilizer required and the amount of prime pigment used for a given color match. The response of several talc-containing compounds to changing loadings of TiO$_2$ is shown in Fig. 8.21. This figure can be used to determine the amounts of TiO$_2$ that would be required to achieve a given composite color based on the color of the talc in the composite. Assuming a Hunter "L" value of 90 was required in a finished part, a compounder would use 1.5% of TiO$_2$ with Cimpact 710, over 3% with JetFil 700, or over 4.5% with Cimpact 699 assuming a 15% talc loading would meet the mechanical property requirements. Cost differences between the various talc choices balance against the total cost savings obtained with the TiO$_2$.

## 8.4.8 Effect on Ultraviolet Stability

The source of the talc ore also appears to be important in applications where resistance to ultraviolet (UV) light is required. As shown in Figs. 8.22 and 8.23, Cimpact 710 retains more of its stiffness than a Montana-based product after 2000 hr of UV exposure in a QUV unit. The tensile property reductions (Fig. 8.22) are about the same for both talc-filled products but less than the neat resin.

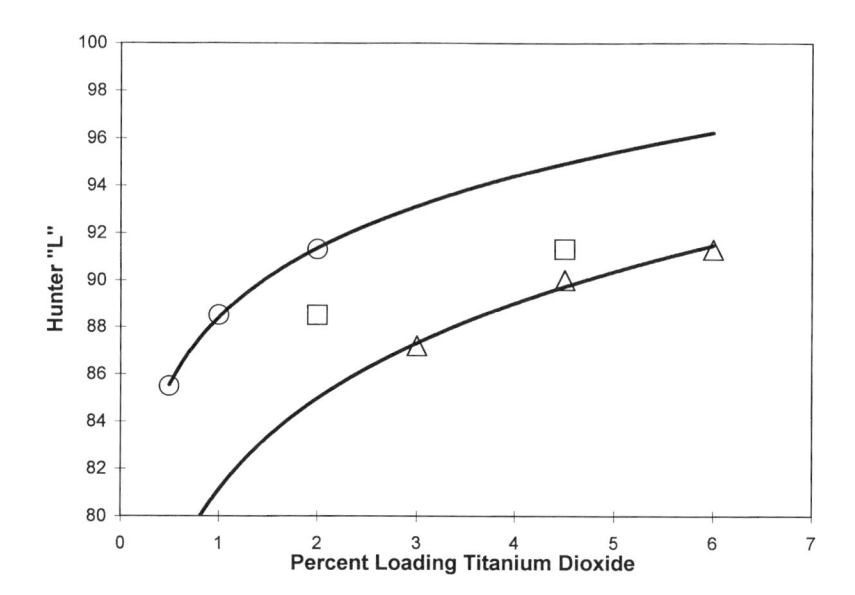

**FIGURE 8.21** Hunter "L" versus wt%TiO$_2$ with talc filler level of 15 wt%. △ Cimpact 710; □ Jetfil 700C; ○ Cimpact 699.

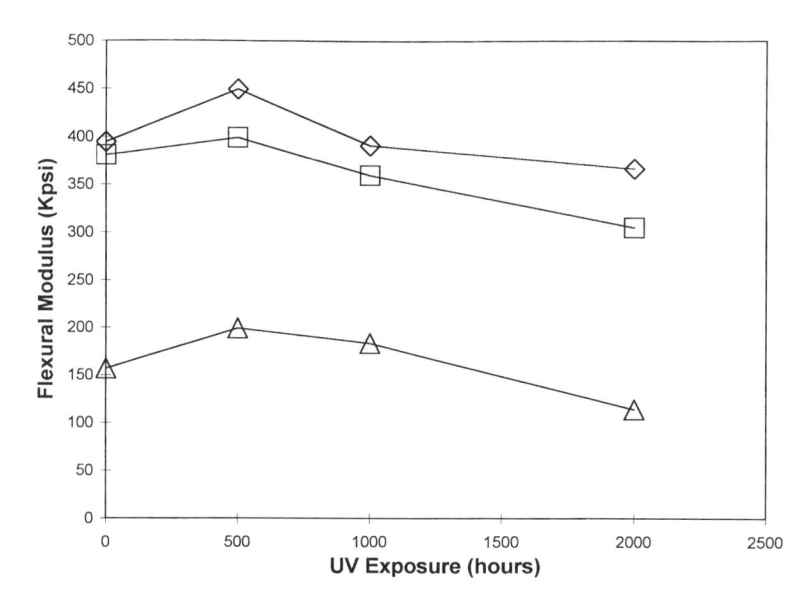

**FIGURE 8.22** Effect of UV exposure on flexural modulus. ◇ Cimpact 710; □ Cimpact 699; △ neat polypropylene resin.

# Talc in Polypropylene

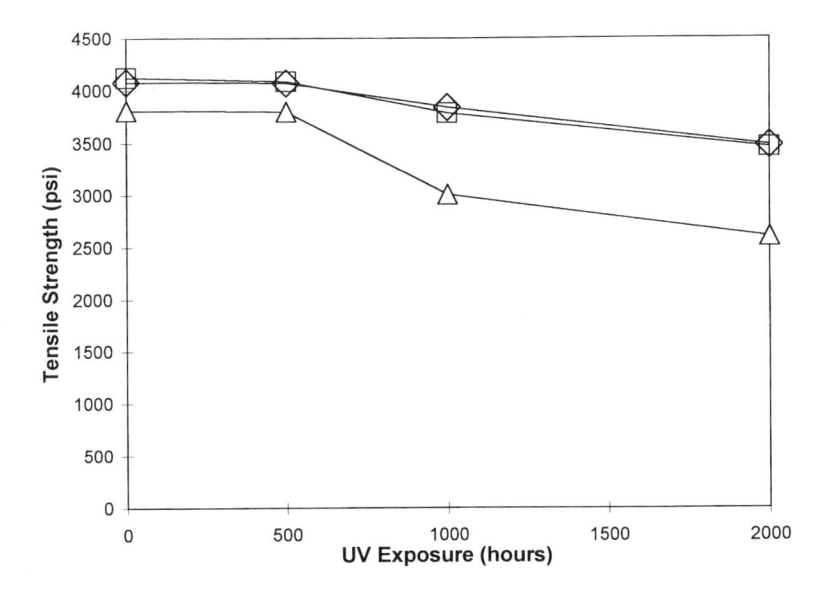

**FIGURE 8.23** Effect of UV exposure on tensile strength. ◇ Cimpact 710; □ Cimpact 699; △ neat polypropylene resin.

## 8.4.9 Moisture Barrier

Talc can be used to provide barrier properties in polypropylene film. The moisture vapor transmission rate is cut in half with a 20 wt% loading of talc in an 8- to 10-mil film. The results are shown in Table 8.4.

## 8.5 ECONOMICS OF TALC USE

Because talc has a specific gravity of 2.78, the addition of talc to polypropylene will increase the specific gravity of the compound. The lower cost of most talc

**TABLE 8.4** Effect of Filler Type on Moisture Barrier[a]

| Filler | Moisture vapor transmission (g./100 in.$^2$/24 hr) |
| --- | --- |
| None | 0.090 |
| Talc | 0.045 |
| Mica | 0.055 |

[a]10-mil polypropylene film.

**Compound Cost ($/lb.)**

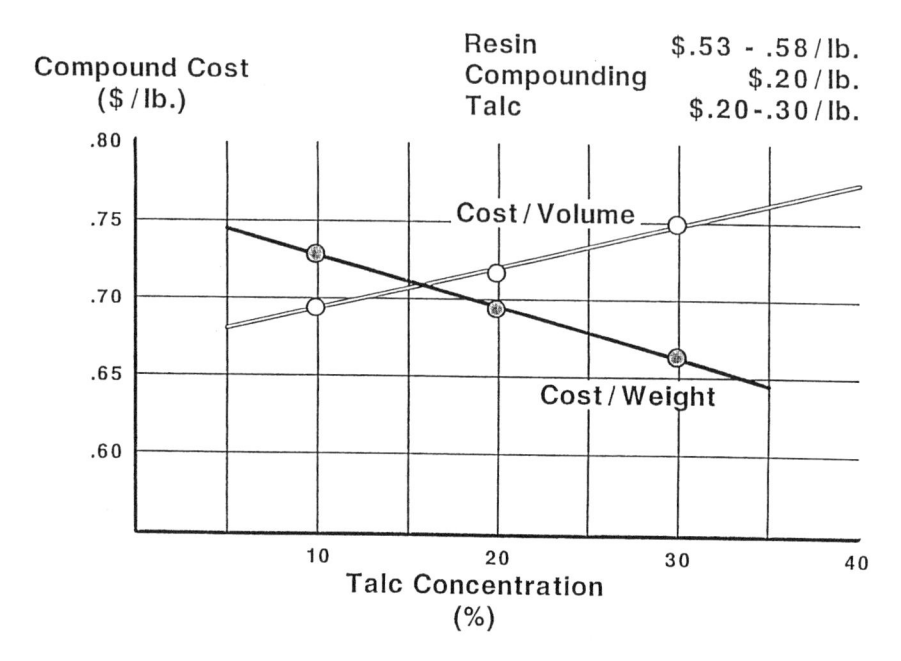

**Resin** $.53 - .58/lb.
**Compounding** $.20/lb.
**Talc** $.20 - .30/lb.

FIGURE 8.24 Effect of talc on unit volume/weight cost.

($0.10–0.30 per pound), relative to the polymer cost ($0.53–0.58 per pound, combined with $0.20 per pound compounding costs), will give a lower cost per pound for the compounded material. As shown in Fig. 8.24, the higher specific gravity of the compound can result in the cost/volume actually increasing.

From a practical standpoint, the cost–performance balance frequently dictates which talc is used in a given application. In automotive under-the-hood applications, where cost drives the equation, low-cost talcs are commonly used. Color is not important since most of the parts are black. Transportation costs considerations frequently play a part in determining the overall cost of the talc to the compounding plant particularly in low-cost, non-color-sensitive applications.

Talc is the softest and least abrasive of all minerals. To compare the abrasiveness of talc with other commonly used minerals, such as calcium carbonate and calcined silica, the wear of a brass plate in an Einlehner test is measured after a given number of cycles. A 4-$\mu$m talc is shown to be four times less abrasive than calcium carbonate in Table 8.5.

# Talc in Polypropylene

TABLE 8.5  Relative Abrasion of Minerals (Einlehner Method)

| Mineral | Relative abrasion (g/m$^2$) |
| --- | --- |
| Talc | 20 |
| Calcium carbonate | 80 |
| Calcined silica | 260 |

## 8.6 EFFECT ON COMPOUNDING EXTRUSION

Talc must be compounded into the polymer prior to molding to ensure good dispersion. Improperly dispersed talc will cause agglomerate formation. The aggregates of talc particles clumped together will lead to poor falling-weight impact strength and poor surface appearance. Undispersed talc is particularly critical in thin-walled molded parts. In the case of some ultrafine or submicrometer grades of talc, where the loose bulk density can drop to 6–10 lbs. per cubic foot, feeding talc into the processing equipment can be a problem. The low bulk density of ultrafine talcs generates slower feed rates and lower output. Split feeding talc into the feed throat and a downstream feed port can help overcome the feeding problem.

### 8.6.1 Compacted Talc

Another solution to the handling problems created by low loose bulk density talc is the wide spread use of compacted ultrafine or submicrometer talc products [6]. In the compaction process, talc is made into a slurry with water. The resulting slurry is forced through a pellet die to form a strand similar to the strand coming from an extruder. During the process the strand breaks into short pellets. These pellets are then dried to less than 0.5% moisture level. This process changes the loose bulk density from 6–10 pounds per cubic foot to 40–50 pounds. The pellet must be compact enough to withstand transportation stress but friable enough to be dispersed in compounding equipment. Sometimes single-screw extruders do not have sufficient shear to disperse the pellet adequately. Throughput rates can be significantly improved with the use of compacted talc. Instead of the output rate being determined by the rate at which the talc can be fed into the feed throat, the amount of compacted talc added is limited only by the torque of the processing equipment. In Chapter 16, new technology for compacted submicrometer talc is discussed in detail.

## 8.6.2 Splay on Molded Parts

The appearance of splay and related surface imperfections can lead to significant rejection of defective molded parts. The major root cause for this problem is the presence of residual moisture or some type of volatile trapped on the molded surface. Depending of the feedstock, splay can be caused by incompatible polymer contaminants, e.g., small amounts of polyamide resin in recycled polypropylene flake. Splay can also be caused by nonoptimal molding conditions.

Particular care should be taken during the compounding extrusion process to provide adequate venting by applied vacuum to remove air, moisture in the filler, and other volatiles. Without adequate vacuum, internal moisture is difficult to remove by a postextrusion drying step. Entrainment of moisture during water quench in the pelletization of polypropylene compounds is generally below about 0.2 wt%. External moisture present in hydrophobic polypropylene is easily vaporized by sensible heat remaining in quenched pellets from the extrusion process. However, if splay is encountered, postdrying is often the solution. This can be accomplished by allowing sufficient time for hot pellets in the gaylord to vent to the air before covering. Actual post-drying in a hot-air dryer can be utilized in the last resort.

Plugged venting and improper runners in the mold design can be causes for entrained volatiles that manifest as splay on the molded part. Splay due to trapped gases can be minimized by adjustments in injection speed, injection pressure, mold temperature, and barrel heat.

## 8.6.3 Surface Modification

Surface modification of talc is a major area for development of the next generation of value-added talc products. Surface modification of talc can alter the interaction between talc and the polymer matrix. Silanes, glycols, and stearates have been used commercially to improve dispersibility and processing as well to react with other components in a polymer blend.

The increasing use of polypropylene compounds by the automotive industry for highly engineered applications is discussed in Chapter 1. This development has driven the need for greater performance from mineral-filled polypropylene, e.g., calcium carbonate, mica (see Chapters 14 and 15), wollastonate (see Chapter 18), and talc-filled compounds.

*Replacement of PVC by Talc-Filled Polypropylene.* Due to polyolefin composite replacement of PVC in the interior of automobiles, a distinct change in odor has become an issue [7]. The "new-car smell" that people experience is attributed to phthalate plasticizers, e.g., dioctyl phthalate (DOP), that evolve from flexible PVC components. As PVC is replaced, the odor changes. Talc is known to generate a 'sulfur-like' odor when compounded at high levels (greater than

35 wt%) into polypropylene. The odor is enhanced by viscosity-broken polypropylene and oxidative degradation of secondary antioxidant used in polypropylene formulations. For example, use of DSTDP thioether in talc-filled materials has a tendency to have an unpleasant smell. Presumably, the sulfur bridging atom is oxidized and transformed into $SO_2$ and $SO_3$ byproducts (discussed in Chapter 4, Section 4.2.2).

Müller [6] suggests replacement of DSTDP by hindered amine stabilizer (Chimassorb 119) in an odor-free stabilizer package with a dual function as being both a thermal stabilizer and light stabilizer. Steen and Bloomfield have developed a surface modification technology [8] that significantly reduces or eliminates the odor generated when talc is compounded into polypropylene compounds. This new technology also significantly improves the thermal stability of talc-reinforced composites. Talc made with this novel surface modification technology has been commercialized on automotive platforms in North America and Europe.

*Replacement of Engineering Thermoplastics by TPO Compounds.* TPO compounds are replacing engineering thermoplastics in automotive instrument panels. While TPOs have a considerable cost advantage over engineering thermoplastics, talc-reinforced TPOs have much poorer scratch and mar resistance. Traditional talc-reinforced TPOs have a tendency to whiten when scratched (Chapter 7, see Fig. 7.18). Initially, the whitening was thought to be talc in the valley of the scratch, but electron microscopy reveals that whitening is due to crazing or light diffraction due to fibril formation in the scratch. While considerable work continues on improving the scratch and mar resistance of TPOs, a new generation of reactive talcs has proven successful in improving the scratch and mar resistance of specially formulated TPOs. These functionalized talcs react with functionalized polymers in the blend to increase surface toughness. TPOs containing reactive talcs have been commercialized in instrument panels on North American–built automobiles.

*Development of R Talc.* Improvements in surface modification technology have led to chemically coated talc platelets called R Talc [9]. Addition of this grade of talc to polypropylene homopolymer does not result in any noticeable increase in reinforcement compared with unmodified talc filler. However, R Talc interacts synergistically with rubber domains in the TPO matrix to significantly enhance flexural modulus of the talc-filled composite while maintaining impact strength.

Figure 8.25 shows the effects of talc filler type and polypropylene resin/rubber blend ratio on the stiffness–impact balance for TPOs with 25 wt% talc filler loading. For the purpose of comparison, talc filler types were R Talc and unmodified talc. In order to bracket a wide range of impact strength, TPO formulations A and B consisted of two levels of (ethylene-propylene-diene)

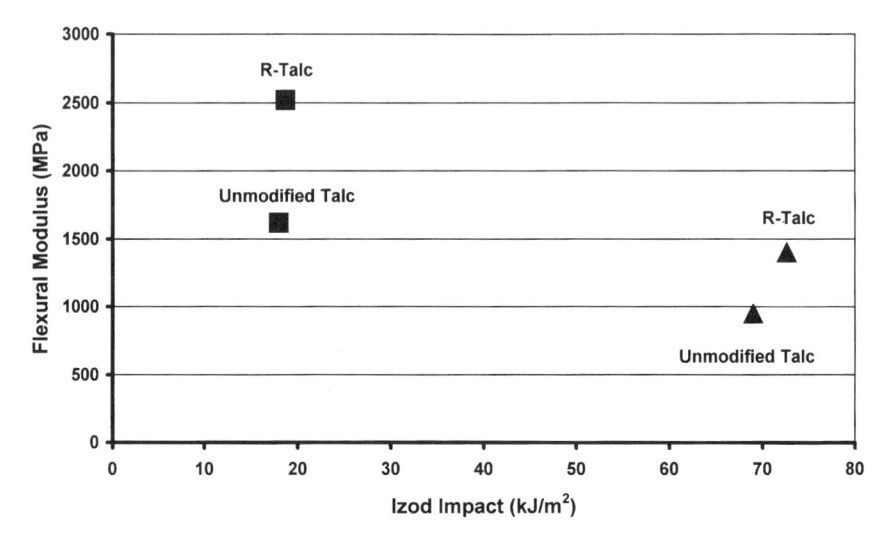

**FIGURE 8.25** Stiffness-impact balance for R talc versus unmodified talc [10]. ■ Formulation A, low rubber; ▲ formulation B, high rubber. Data values labeled by type of talc filler: R talc or unmodified talc.

monomer (EPDM) in rubber-modified polypropylene resin. Formulation A had an 80:20 blend ratio of polypropylene resin to EPDM that corresponded to a low rubber loading. The increased impact of formulation B corresponded to a higher rubber loading with 60:40 blend ratio. Incorporation of R Talc reinforcement gave significant increase in flexural modulus or stiffness with no appreciable loss in impact strength for the two levels of rubber loading.

R Talc works successfully with EPDM and EPR rubber but does not generate the same increased level of flexural modulus with ethylene-butene and ethylene-octene metallocene elastomers. Work continues on modifying the surface treatment to make R Talc compatible with metallocene elastomer containing TPOs.

## 8.7 SAFETY AND HANDLING

Talc is approved by the U.S. Food and Drug Administration for use in polymeric compounds in contact with food. Talc is listed in Title 21, Code of Federal Regulations, Part 178.3297, "Colorants for Polymers."

Talc can be handled and used safely provided reasonable workplace practices are observed. Skin contact is generally of minimal concern, except that continuous exposure can dry the skin because talc absorbs the skin's natural oils. The use of moisturizing lotion is recommended in this instance. Eye contact

# Talc in Polypropylene

with talc can cause a mild mechanical irritation of the eye. A thorough rinsing of the eye with water or an approved eye wash is recommended as first aid. Accidental ingestion of talc is of no concern. Inhalation of talc dust should be avoided and the use of National Institute of Occupational Safety and Health–approved dustmasks should be worn if the ACGIH TLV of 2 mg/m$^3$ (respirable dust) is exceeded. Long-term inhalation of talc dust can lead to a mild pneumoconiosis (Greek for "dusty lungs", which is characterized by shortness of breath, wheezing, and a chronic cough.

## 8.8 FUTURE TRENDS

New technology for surface modification of talc filler has maximized the performance of talc reinforced polyolefins [9,11,12]. Furthermore, current programs are underway to modify talc to improve the interaction with polar engineering thermoplastics. Hence, new formulations have been designed to directly influence the interphase microstructure between talc particles and surrounding polymer matrix. These developments will drive talc-reinforced polyolefin composites to even higher levels of performance.

### 8.8.1 Fundamental Understanding of Talc Microstructure

The need for higher performance materials will be the key driving force behind the development of new surface modification technology. The goals of current development programs are to attain a better fundamental understanding of the relationship between thermal-mechanical properties of composites and particle sized–surface characteristics of talc. For example, the effects of particle size and surface modification have been determined for surface treated and unmodified talc reinforcement of polypropylene compared with composites of wollastonite [11] and mica [12] mineral fillers. Particle size recommendations [12] for talc-filled polyolefin applications range from fine to coarse grades as shown in Fig. 8.26. Each application regime corresponds to a range of upper–lower limits, e.g., TPO (median diameter 1.0–1.5 μm, top size of 8 μm) and automotive under-the-hood for black color (median diameter 9–11 μm, top size 50 μm).

### 8.8.2 Control of Molded in Color

The automotive industry is driving the development of molded-in-color plastic components. Many of the plastics components used in a car are painted for continuity of color or texture. Painting adds anywhere between 20% to 50% to the cost of a plastic component. In an effort to cut costs, automobile manufacturers are pursuing the development of molded-in-color components. Molded-in-color components will require new levels of color consistency for polymer additives.

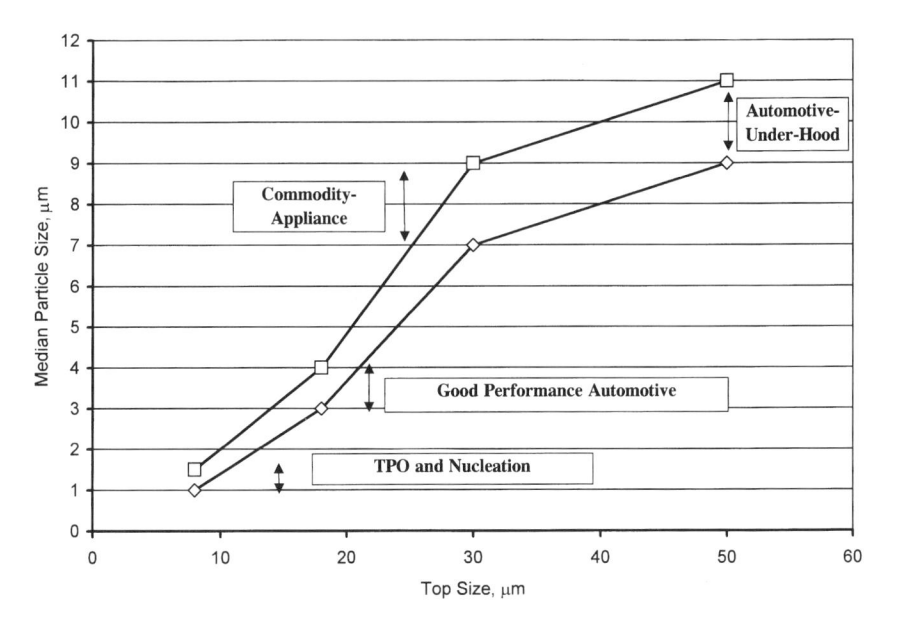

**FIGURE 8.26** Ranges of particle sizes for talc-filled polyolefin applications [12]. ◇ Lower limit plot of median particle size vs. top size plot; □ upper limit plot of median particle size vs. top size plot.

Talc color has traditionally been graded by color measurements via Photovolt brightness or GEB. This type of measurement has been geared to the paper industry which is the largest consumer of talc. Due to increased use of talc in paints and polymer applications, there is a need to make more accurate assessment of color by Cielab or L*a*b* color space. The tristimulus values of L* (white–black axis), a* (red–green axis), and b* yellow–blue axis) are measurements of the whiteness, hues, and undertones that talc imparts to plastics.

Control of molded-in color for TPO automotive applications is dependent on an understanding of the correlation [10] between the molded color cast of the talc-filled polypropylene composite and pasted color of ground talc filler. A detailed understanding of the ore source (see Section 8.1.1) is necessary for consistent color over a 2- to 5-year production period of a molded-in-color part. Talc will be blended after the mining process, during purification, and after the talc has been milled for consistency of color.

### 8.8.3 Compacted Talc–Zero Force Technology

The use of compacted talc will allow compounders to improve output rates. Enhanced compacting methods will allow the talc to be dispersed more completely in the polymer with an improvement in mechanical properties. ZF ("zero force") technology has been recently presented [13] as a improvement in talc compaction. The ZF technology densifies talc without using the highly compressive forces of pelletization. Talcs densified with ZF technology have been reported to develop better mechanical properties and higher throughput rates during processing. Other methods of talc densification are under development that promise even greater performance enhancement and production rates.

## REFERENCES

1. G Graff, ed. Modern Plastics Encyclopedia. New York: McGraw-Hill, 1994.
2. HS Katz, TV Milewski, eds. Handbook of Fillers for Plastic. New York: Van Nostrand Reinhold, 1987.
3. RJ Piniazkiewicz, EF McCarthy, NA Genco. Industrial Minerals and Rocks. 6th ed. Littleton, CO: AIME, 1994.
4. DH Solomon, DG Hawthorne. Chemistry of Pigments and Fillers. New York: John Wiley and Sons, 1991.
5. D Müller. Stabilization of talc reinforced polypropylene. Paper presented at Functional Fillers for Thermoplastics, Thermosets and Elastomers Conference, The Netherlands, Oct. 28–30, 1996.
6. WP Steen, WR Sevy. A new generation of ultrafine talc for use in high impact polyolefins. Proceedings of the SPE 38th Annual Technical Conference, Detroit, May 1992, pp. 1685–1688.
7. RI Clark, OF Noel. Talc filled polypropylene-modified for odor elimination and improved UV stability. Proceedings of 9th International Business Forum on Specialty Polyolefins, Houston, Oct. 1999, pp. 215–231.
8. WP Steen, DR Bloomfield. Compacted mineral filler pellet and method for making the same. U.S. Patent 5,773,503. June 30, 1998.
9. RJ Clark, OF Noel. New generation talc reinforcement for TPOs, Proceedings of SPE Automotive TPO Global Conference, Detroit, Oct. 2–4, 2000.
10. MJ Lorang, RJ Clark. Reinforcing pigments for molded in color TPO. Proceedings of SPE Automotive TPO Global Conference, Detroit, Oct. 1–3, 2001.
11. TL Wong, CMF Barry, SA Orroth. The effects of filler size on the properties of thermoplastic polyolefin blends. J Vinyl Add Technol 5: 235–240, 1999.
12. JM Zazyczny, JZ Keating. Reinforcing TPO compounds: an investigation of functional fillers. Proceedings of SPE Automotive TPO Global Conference, Detroit, Oct 1–3, 2001.
13. RJ Clark. Recent developments in talc technology. Proceedings of Exec. Conf Management' Additives 1999, San Francisco, March 23, 1999.

# 9

# Glass Fiber–Reinforced Polypropylene

**Philip F. Chu**
Saint-Gobain Vetrotex America, Wichita Falls, Texas, U.S.A.

## 9.1 THE MAKING OF GLASS FIBERS

### 9.1.1 History of Glass Fiber Industry

For centuries people dreamed of a new fiber to enhance the usage of natural fibers such as cotton and wool. Ever since Phoenicians discovered glass, it became evident that glass could be melted and drawn into filaments. The Venetians mastered the art of drawing fibers from molten glass for decorating glassware in the 16th century. The attempt to use glass fiber as an apparel material was highlighted at the Columbian Exposition in Chicago in 1893. Fiber glass neckties, lamp shades, and a woman's dress, all made in Toledo, generated tremendous interest in this new material.

Glass fiber was the second man-made fiber, next to rayon, to be of commercial importance. In 1938, the same year DuPont was making nylon fibers at a small pilot plant at Wilmington, Delaware, Owens Corning was formed for the sole purpose of producing glass fibers, initially as an insulation material. It was a joint venture between Owens-Illinois and Corning Glass Works, both of which had made significant progress in the glass fiber manufacturing process during the early 20th century.

Later, Pittsburgh Plate Glass (now PPG), Libby-Owens-Ford, Vetrotex (a division of St. Gobain) of France, and Pilkington of UK all acquired the licenses from Owens Corning. Interesting reviews can be found in the two books specifically written on this subject [1,2] and recent articles [3,4]. In the last decade, Owens Corning, PPG, and Vetrotex International have expanded their presence globally. In these days, glass fibers are produced in every corner of the world [5].

### 9.1.2 Manufacturing Process of Glass Fibers

*Glass Composition.* In the very beginning, glass fibers were produced in two stages. First, raw materials were melted in the furnace and glass marbles were formed. The marbles were remelted into liquid and filaments were drawn. This was described in an early book on glass-reinforced plastics first published in 1954 [6]. Although this process is still being used, the modern fiberglass plant is built for direct process.

All raw materials purchased by glass producers have to pass very stringent requirements for composition and particle size distribution. Silica is from sand and alumina from clay. Boric oxide is from colemanite and boric acid. Limestone or calcite is for calcium oxide. A summary is given in Table 9.1.

Each ingredient is stored separately in the silo and metered to a mixing tank by accurate weighing. Then the mixture is transferred to the batch silo above the batch charger. In a fully automated glass plant, the whole system is computer controlled and totally enclosed to prevent the dispersion of dust [7].

The physical and chemical properties of glass fiber depend on its composition: The E-glass, which has a low content of alkali ions, is known for its excellent electrical resistance, an important property for the textile glass yarns. From there it has become the most widely used fiberglass composition. The soda-lime-silica glass, known as A-glass, has also been used in small quantity as a plastic reinforcement. Many other glass compositions are commercially produced for special applications. C-glass is more chemically resistant than E-glass, especially against acidic solution. AR-glass improves the alkali resistance of the glass, as in a cement environment. AR-glass is rich in zirconium oxide. D-glass is rich in boric oxide and has lower dielectric strength. Two high-strength glasses, S-glass from Owens Corning and R-glass from Vetrotex International, have a 30% higher strength-to-weight ratio. Hollow S-glass filaments are also available to further increase the strength-to-weight ratio. The chemical composition of each type of glass is given in Table 9.2.

*Fiber Forming Process.* The preblended batch material is continuously charged into the furnace where melting and fusion take place. The furnace is normally rectangular with a longitudinal channel at the exit. The length-to-width ratio of the smelter depends on the daily tonnage required. Because of the

# Glass Fiber–Reinforced Polypropylene

TABLE 9.1 Requirements of Raw Material Quality for Glass Fibers

| Acceptable analysis (% min) | Compound | | | | | | |
|---|---|---|---|---|---|---|---|
| | $SiO_2$ | $Al_2O_3$ | $B_2O_3$ | MgO | CaO | $F_2$ | $Na_2SO_4$ |
| $SiO_2$ | 98.05 | 44 | 4–5 | 1 max | | 1 max | |
| $Al_2O_3$ | 0.1 | 37 | 0.5 | 0.6 max | 0.2 max | 1 max | 0.03 max |
| $Na_2O$ | 0.1 | 2 max | 0.25–0.3 | 0.4 max | | | |
| $KO_2$ | 0.1 | | | 0.2 max | | | |
| $Fe_2O_3$ | 0.1 | 1 max | 0.1 max | 0.4 max | 0.05 max | 0.25 max | 0.16 max |
| CaO | | 0.8 | 27–29 | 31 | 55.4 | 70.2 | |
| $B_2O_3$ | | | 40–42 | | | | |
| MgO | | | 2–3 | 20 | | | |
| $P_2O_5$ | | | | | 0.1 max | | |
| $MnO_2$ | | | | | 0.1 max | | |
| S | | | | | 0.1 max | | |
| $F_2$ | | | | | | 47.3 | |
| PbO | | | | | | 0.2 max | |
| $Na_2SO_4$ | | | | | | | 95 |
| NaCl | | | | | | | 3 max |
| $H_2SO_4$ | | | | | | | 2 max |
| $H_2O$ | 0.1 | 1 max | 22 | | 0.1 max | 0.4 max | 0.4 max |
| Particle size | | | | | | | |
| > 150 μm | 0.1 | 1 max | | 2 | 2 | | |
| 150–75 μm | 0.8 | 1 max | | 56 max | 28 | | |
| 75–50 μm | 6.0 | 99 | | 40 max | 70 | | |
| < 50 μm | 93.1 | | | | | | |

Source: Ref. 1.

TABLE 9.2 Typical Composition (wt%) in Various Commercially Produced Glass Fibers

| Letter designation | A | Cemfil | C | D[a] | E | R | S |
|---|---|---|---|---|---|---|---|
| $SiO_2$ | 71.8 | 71 | 65 | 74.5 | 55.2 | 60 | 65 |
| CaO | 8.8 | | 14 | 0.5 | 18.7 | 25 | |
| $Al_2O_3$ | 1 | 1 | 4 | 0.3 | 14.8 | | 25 |
| $B_2O_3$ | | | | 22 | 7.3 | | |
| MgO | 3.8 | | | | 3.3 | 6 | 10 |
| $Na_2O$ | 13.6 | | 11 | 1 | 0.3 | 9 | |
| $K_2O$ | 0.6 | | | 0–1.3 | 0.2 | | |
| $Fe_2O_3$ | 0.5 | | | | 0.3 | | Trace |
| $F_2$ | | | | | 0.3 | | |
| $ZrO_2$ | | 16 | | | | | |
| $Li_2O_3$ | | | | | | | |
| $SO_3$ | | | | | 0.1 | | |

[a]Reprinted from Ref. 8 with permission from Elsevier Science—NL, Sara Burgerhartstraat 25, Amsterdam, The Netherlands.
*Source:* Ref. 1.

corrosiveness of E-glass in the molten state, dense chrome oxide is the best refractory material for side walls. The refractory is backed by zircon blocks and another layer of clay blocks. The floor of the furnace is tilted to allow free flow of the glass to the exit channel.

Natural gas is the least expensive source of fuel to melt glass, and energy constitutes a large portion of the cost of making glass fibers. The melting temperature of the E-glass in the furnace is typically in the range of 2500°F when it leaves the melter and enters into a narrow channel forehearth. The forehearth is a supply channel for conveying liquid glass into bushing to produce fibrous form. The T- and H-shaped forehearths are the most widely used. The bushing is the equivalent of the spinneret of the synthetic fiber industry. But, unlike the latter, the bushing is normally rectangular. Again, because of the requirement of high temperature and corrosion resistance, the alloy of platinum and rhodium works the best. With tens and hundreds of bushings used in a large-fiber glass plant, the cost of precious metal becomes a substantial investment. A bushing contains a multiplicity of small holes up to several thousands. The bushing is further heated electrically for better temperature control and uniformity. The molten glass flows through the individual tips by gravity. The design of bushing and geometry of nozzle is each company's proprietary technology. It is still very much more an art than a science.

## Glass Fiber–Reinforced Polypropylene

Once the liquid glass leaves the tip of the nozzle, it is quenched rapidly by circulating air and spraying water. Unbeknownst to people outside the glass producer is the importance of nozzle shields, which space between rows of nozzles. They absorb radiant heat from the nozzle and glass and serve as a heat sink. The glass should be fully solidified when it reaches the size applicator about a few feet lower from the bushing. At this step the organic size is applied to the glass surface by either roller or belt applicator. Because the glass composition is commonly an E-glass, how compatible each glass product is to the resin matrix entirely depends on the size of the glass surface. The chemistry of the size and its importance to the interfacial bonding will be discussed in Sections 9.2.3 and 9.2.4. Common chemicals and several specific size formulations are given in chapter 6 of Ref. 1.

*Fabrication of Product Forms.* Once the individual glass filament, which almost always contains lubricants, is wetted by water and size, it can be gathered through split shoes. The size of the glass strand is determined by how the bushing is split. Micarta or graphite is the common material for gathering shoes. For multiply-split cake, the shoe can be a flat laminate stationed horizontally under the size applicator or it can be a rotatable grooved disk.

The attenuated glass fibers are traditionally wound on a collet. The collet is a cylinder powered by the winder. Today the cylinder is usually long enough to accommodate two cakes. The fibers are actually wound on the paper or plastic tube that is slid onto the collet. The function of the tube is to facilitate doffing and further handling of the forming cakes. The rotation speed of the collet is the key parameter to the determination of fiber diameter in the final form. The other very important part of the winder is the traverse. The traverse places individual strands on the collet in such a fashion that the cake has even buildup and the strands can be easily unwound after drying. The design and operating parameters of traverse—notably strand split efficiency and size migration—are critical to the product.

Because all size systems are water based, the wet cakes contain about 10–20 wt% water. They then go through a drying process to remove water and, in many instances, to cure the size. The hot-air oven, operating up to 300°F, is commonly used. The profile and duration of drying temperature depend on the size used and the product requirements. The drying time is typically very long, commonly up to 15 hr. Dielectric drying is an alternative to hot-air drying. The advantage of the former is a shorter drying time and more uniform drying.

Many different forms of fiberglass products occur after the drying of the cakes. Direct roving gets its name from selling the packages directly. The major markets for direct rovings are in filament winding, pultrusion, and woven fabric. When the direct rovings are sized for thermoplastics compatibility, they are sold to long fiber–reinforced thermoplastic compounders. More details about this process are discussed in Section 9.4.2. Assembled rovings require more assembly work from cakes to roving package. In the fabrication area, the glass strands from rows of forming cakes are pulled together by another winder. In this way, each assembled roving contains multiple strands, up to 100 or more. The end count depends on the yield of the roving and the number of splits in the forming cake. Assembled rovings serve the compression molding market and the open-mold market. The latter is dominated by the spray-up process. An example of closed-mold process is sheet molding compounds (SMCs). Ironically, the reinforced thermoplastics industry began with feeding assembled rovings to the extruder [3]. In the late 1960s and early 1970s, the thermoplastics compounders in the developed world moved from rovings to chopped strands.

Chopping continuous glass fibers from forming cakes is done in the secondary operation. Again, glass strands from rows of cakes are pulled by a chopper. The chopped length varies from $\frac{1}{8}$ to 1 in. For thermoplastic resins, $\frac{1}{8}$- to $\frac{1}{4}$-in. strands are produced. Very recently, one glass producer also introduced the 4-mm chop length. For thermoset resins, $\frac{1}{4}$ and $\frac{1}{2}$ in. are more common in the bulk molding compounds (BMCs) application. The thermoplastic BMC does use these types of chop length in the process. The important criteria for a good chopped glass strand are low fuzz and consistent flowability. Chopped glass strands can also be produced by a direct chop process. Basically, the chopper is installed under the bushing. After the glass fibers are coated with the size, they are pulled directly into the chopper. The glass fibers are collected while they are still wet. This type of product is sold to make roofing mat or gypsum. One exception is the Taffen process for making thermoplastic stampable sheet. For these market applications, large fiber diameters from 13 to 17 μm are acceptable, compared with the normal 10–13 μm made for short-fiber-reinforced thermoplastics. The direct chop process can be operated efficiently and profitably under these conditions.

Forming cakes or assembled rovings can also be chopped to make the chopped strand mat. In this case, the chopped length is much longer at 1–4 in. Once the mat is formed, binder is applied to provide strength for handling. The binder content can be as high as 6% by weight of the glass mat. The general application of chopped strand mat is in the open-mold process. One unique use is the manufacture of inexpensive printed circuit boards by a continuous laminating process.

The continuous strand mat provides much stronger reinforcement to the composites than the chopped strand mat. The mat can be formed directly under the bushings or from forming cakes. In either case, glass strands are continuously placed onto a moving conveyer in a random pattern. Additional binder is applied as with the chopped strand mat. The product is widely used in the reaction injection molding (RIM) application. The Azdel stampable sheet is made from this process.

# Glass Fiber–Reinforced Polypropylene

Scrap or virgin fibers can be milled to less than 20:1 aspect ratio. The milled fibers are widely used as reinforcing fillers [9, pp. 212–214]. Figure 9.1 shows the flow chart of the glass fiber production for the various product forms mentioned in this section.

## 9.1.3 The Future of the Glass Fiber Industry

The glass fiber industry output has grown from a few tons to more than 3 billion pounds volume globally in the last 60 years. The projected growth is still above the gross domestic product (GDP). Because of the nature of the business, major

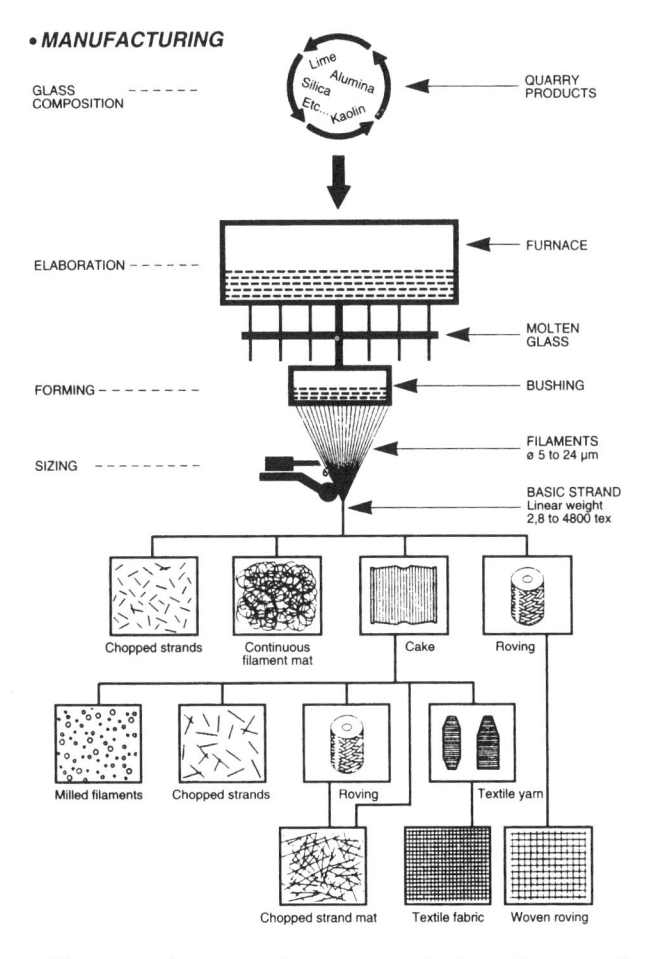

**FIGURE 9.1** The manufacturing flow chart of glass fiber reinforcements. (Courtesy of Vetrotex International.)

capital investment is needed to expand any production facility. In the meantime, all glass suppliers are achieving incremental improvement to the production volume. Like any industry, to survive global competition, productivity increases are the top priority. The low inflation rate of the last 15 years has also put tremendous pressure on cost reduction. Bushing design plays a big part of the technology innovation [10]. Larger and larger bushings are built to increase throughput. High-throughput bushing for small fiber diameter presents particular challenges to the industry. Better quality control was in every glass producer's mind in the 1980s. After a decade's effort, glass fiber suppliers in the developed countries have achieved remarkable progress, All major producers are ISO certified. The quality standard also applies to every manufacturing plant within the company. To be a global supplier is the game of the 1990s. Owens Corning 144A for polypropylene reinforcement is publicized as the first glass fiber to achieve global standardization [11].

In response to consumer awareness and stringent government regulations, glass fiber manufacturers have spent millions of dollars to minimize waste. Reducing furnace emission is no doubt the number one challenge to all glass makers [12]. Major pollutants from a glass furnace are dust, sulfur oxides, and nitrogen oxides. The use of oxygen combustion is gaining popularity because the process decreases the emission of nitrogen oxide to the environment [13]. The oxyfuel firing also provides better glass temperature control and homogeneity of the glass fibers. Fluorine-free glass fibers are produced for the sole purpose of eliminating fluorine pollution [14]. Boron-free glass fibers can minimize the air pollutants in manufacturing [15]. The company (Owens Corning) already converted the entire plant at Guelph, Ontario to this boron-free formulation [16]. Dust control is also a concern for the furnace rebuild [17]. The organic size is normally applied on the glass surface at less than 2% by weight. Its waste has to be treated to meet environmental guidelines [18]. Unlike the flat glass industry, which can recycle 100% of its waste glass, the recycling of forming and fabrication wastes remains a difficult task.

Although plastics compounders are mostly concerned with the processibility of the glass fibers and short-term and long-term composite properties, the above-mentioned programs are necessary steps that will enable glass manufacturers to grow into the 21st century.

## 9.2 MECHANISM OF REINFORCEMENT TO POLYPROPYLENE

### 9.2.1 Physical Properties of Glass Fibers

The mechanical, physical, thermal, and electrical properties of two common glass fiber reinforcements for plastics, E and R, are listed in Table 9.3 [19]. The tensile

## Glass Fiber–Reinforced Polypropylene

**TABLE 9.3** Properties of Glass Fiber Reinforcements for Plastics

| | Properties | Units | E Glass | R Glass |
|---|---|---|---|---|
| Main properties of fiber from various types of glass (measured on new untreated filament) | Ultimate tensile strength | psi × $10^3$ | 493 | 638 |
| | Young's modulus | psi × $10^6$ | 10.5 | 12.7 |
| | Elongation at break | % | 4.4–4.5 | 5.2 |
| | Poisson's ratio | — | 0.22 | — |
| General properties of various types of glass | Physical properties | | | |
| | Specific gravity (in bulk) | | 2.60–2.82 | 2.55 |
| | Specific gravity (on filaments) | | 2.50–2.59 | 2.53 |
| | Hardness (Mohs) | | 6.5 | |
| | Thermal properties (glass in bulk) | | | |
| | Coefficient of thermal expansion (1°F × $10^{-6}$) | | 2.80 | 2.30 |
| | Optical properties | | | |
| | Refractive index: @25°C | | 1.550–1.566 | 1.541 |
| | UV transmission | | Opaque | |
| | Electrical properties (glass in bulk) | | | |
| | D-C volume resistivity ($\log_{10}$ 150–400°C) | ohm/cm | 17.7–10.4 | |
| | Dielectric constant at $10^6$ Hz (disk 30 mm diameter, thickness 3 mm) | | 6.5–7.0 | 6.0–8.1 |
| | Chemical properties | | | |
| | Alkalinity ($Na_2O$ equivalent) | % | 0.3 | 0.4 |
| | Solvent resistance | | Good | Good |
| | Alkali resistance | | Good except hydrofluoric acid | Good except hydrofluoric acid |
| | Acid resistance | | | |

*Source*: Ref. 19.

strength given in this table is the virgin fiber strength. In a separate publication [20], the nonimpregnated glass strand has 35–46% of virgin strength. The loss of strength of the fiber is due to deterioration from various steps in the production. This value is slightly lower than that measured by Cheng et al. [21], who reported a mean tensile strength of 2.22 GPa for the water-sized E-glass fiber. When the glass fiber is treated with γ-aminopropyltriethoxysilane (γ-APS), the mean tensile strength did not change. But the Weibull modulus became slightly larger. Glass fiber is a perfectly elastic material with 100% elastic recovery at room temperature. The softening point of E-glass fiber is above 1500°F. This temperature is beyond any reaction and processing temperature of polymer-based composites. However, in the friction market this has become an important consideration [22].

The excellent electrical insulation property of E-glass has found many applications in this field. Fabrics made of textile glass yarn are the backbone of the printed circuit board of the modern electronic industry. Polybutene terephthalate, polyphenylene sulfide, and liquid crystal polymer reinforced with chopped glass fiber are also widely molded into electronic components [23].

The diameter of glass filaments is commonly produced at 10–25 μm. In the glass fiber industry, a letter is assigned to a range of fiber diameter (Table 9.4). For glass rovings, the strand yield is represented by the yard per pound. The smaller the number of yield is, the heavier the glass strand is. There is an inverse relationship to a more common metric unit of TEX (g/1000 m) used in the textile

TABLE 9.4 Letter Designation of Glass Fiber Diameter

| \multicolumn{3}{c}{Inches × $10^{-6}$} | | \multicolumn{3}{c}{Micrometers[a]} |
|---|---|---|---|---|---|---|
| Min. | Nom. | Max. | Letter | Min. | Nom. | Max. |
| 35 | 37.5 | 40 | G | 8.89 | 9.50 | 10.12 |
| 40 | 42.5 | 45 | H | 10.12 | 10.78 | 11.43 |
| 45 | 47.5 | 50 | J | 11.43 | 12.07 | 12.70 |
| 50 | 52.5 | 55 | K | 12.70 | 13.34 | 13.97 |
| 55 | 57.5 | 60 | L | 13.97 | 14.61 | 15.24 |
| 60 | 62.5 | 65 | M | 15.24 | 15.88 | 16.51 |
| 65 | 67.5 | 70 | N | 16.51 | 17.15 | 17.78 |
| 70 | 72.5 | 75 | P | 17.76 | 18.42 | 19.05 |
| 75 | 77.5 | 80 | Q | 19.05 | 19.69 | 20.32 |
| 80 | 82.5 | 85 | R | 20.32 | 20.96 | 21.59 |
| 85 | 87.5 | 90 | S | 21.59 | 22.23 | 22.86 |
| 90 | 92.5 | 95 | T | 22.86 | 23.50 | 24.13 |
| 95 | 97.5 | 100 | U | 24.13 | 24.77 | 25.40 |

[a]One micrometer = 39.37 × $10^{-6}$ in.
*Source:* Ref. 19.

## Glass Fiber–Reinforced Polypropylene

industry. The strand designations and its conversion to TEX are listed in Table 9.5. In contrast to the composite properties, the tensile strength of glass fiber itself is independent of the fiber diameter [24]. The tensile strength of glass fiber is also isotropic because the breaking strengths of a glass fiber in tension and in torsion are equivalent. In the finished product, silane, film-former, lubricant, and the gage length all have profound influence to the recorded tensile strength of glass fibers [25]. A coating from the first three components improves the strength, whereas longer gage length reduces the test values. At 0.25, 1.0, and 3.0 gage lengths, the

TABLE 9.5 Fiberglass Strand Designations and Conversion Table

**Strand Designations**
- Filament diameter—denoted by letter (G, K, M, etc.)
- Filament count—number of filaments per strand or bundle
- Strand yield—yards per pound/100
    Example: ECK30.5
    E = "E" glass
    C = Continuous filament
    K = "K" fiber (13.3 μm)
    30.5 = 3050 yards per pound
- Filament diameter (in.) $= \dfrac{(0.622)}{(\sqrt{YN})}$
    Y = strand yardage in yards/lb
    N = number of filaments per strand

**Metric Designation (TEX)**
   TEX = g/1000 m
   $\text{TEX} = \dfrac{496.238}{\text{yards/lb}}$

**Basic Glass Designation**
Standard Products

| Yield | Tex[a] | Tex[b] |
|---|---|---|
| 62 | 8004 | 8065 |
| 113 | 4392 | 4425 |
| 160 | 3102 | 3125 |
| 165 | 3008 | 3030 |
| 207 | 2397 | 2359 |
| 225 | 2206 | 2225 |
| 250 | 1985 | 2000 |
| 450 | 1103 | 1152 |
| 675 | 735 | 741 |
| 900 | 551 | 556 |

[a]Conversion factor = 496,238.
[b]Common conversion factor = 500,000.
*Source:* Ref. 19.

tensile strength of glass fiber decreases from 332 to 238 to 182 ksi for one type of glass and from 233 to 143 to 98 ksi in the second sample. The strength is the test value of the weakest spot on the glass surface. The dependence of fiber strength on the gage length is also confirmed in recent articles [26,27].

### 9.2.2 Surface Chemistry of Glass Fibers and Its Adhesion to Polymers

Although the composition of E-glass fiber has been well publicized in the literature, the actual composition at the surface of glass fiber has gained very little attention. In an earlier paper, Wong [28] found more silica concentration on the surface than in the bulk. A more complete investigation was published by Rastogi et al. [29] in 1976. The E-glass surface was analyzed by Auger electron spectroscopy (AES) and electron spectroscopy for chemical analysis (ESCA) electron spectroscopy. From AES the E-glass surface showed lower concentrations in magnesium, boron, and calcium but higher in fluorine, silicon, and aluminum. The results from ESCA were not as consistent as those from AES. There was still an agreement by both techniques: silicon is high and calcium and boron are low. Wang and Jones [30] examined the surface composition of heat-cleaned E-glass slide by X-ray photoelectron spectroscopy more recently. The calcium element was very rich in the bulk but less concentrated on the surface (Table 9.6). The measurement depended on the take-off angle.

The virgin or heat-cleaned E-glass fiber has a very smooth surface under scanning electron microscopy (SEM) (Fig. 9.2). This is further confirmed by a recent topographic study [31] by atomic force microscopy (AFM). Some bumps in the form of crates are observed and can be as high as 700 nm. The glass fiber with coupling agent shows a grooved surface. This is due to its tearing off from

TABLE 9.6 E-Glass Fiber Surface Composition

|    | Wong [28] Bulk | Wong [28] Surface AES | Rastogi, Rynd, Stassen [29] Bulk | Rastogi, Rynd, Stassen [29] Surface AED | Rastogi, Rynd, Stassen [29] ESCA | Wang et al. [30] Bulk | Wang et al. [30] Surface XPS |
|----|------|------|------|------|------|------|------|
| Si | 18.8 | 24.3 | 18.6 | 24.1 | 29.6 | 22.3 | 26.1 |
| Al | 5.8  | 7.0  | 6.1  | 8.4  | 5.0  | 7.4  | 7.9  |
| Mg | 2.3  | 0.6  | 2.2  | 0.7  | 1.8  | 0.4  | —    |
| Ca | 6.4  | 1.9  | 6.3  | 1.8  | 3.4  | 16.4 | 8.5  |
| B  | 4.5  | 3.4  | 4.1  | 3.0  | 0    | —    | —    |
| F  | 0.32 | 1.4  | 0.4  | 1.2  | 0.4  | —    | —    |
| O  | 61.8 | 61.2 | 61.8 | 61.1 | 59.9 | 49.6 | 57.5 |

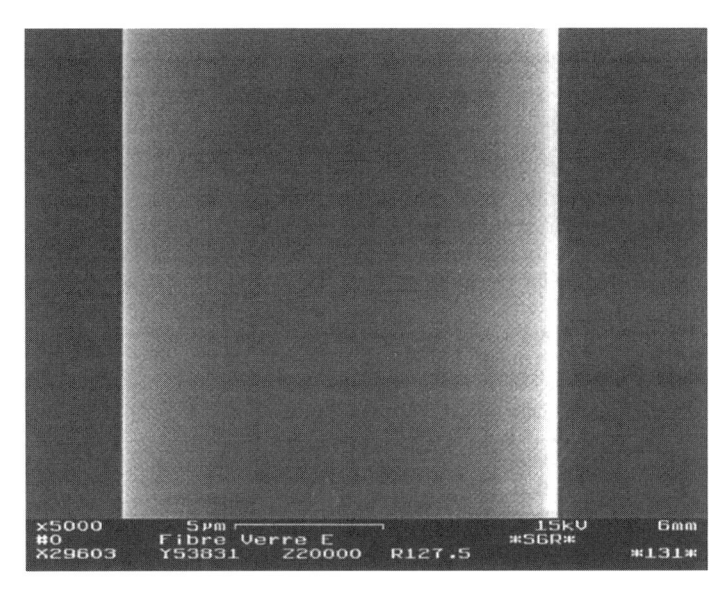

FIGURE 9.2 SEM micrograph of the Virgin E-glass fiber. (Courtesy of Vetrotex International.)

adjacent fibers. The groove depth and width are 0.33 and 3.4 nm for A1120. The atomic force microscopy (AFM) image of the glass fiber coated with A1100 reveals small smooth droplets with occasional big blisters. A line along the fiber is also present from the cyclization of the aminosilane. Figure 9.3 also shows organized zigzag lines by AFM. With complete size on the glass fiber, there are no torn-off areas on the surface The hypothesis is that the silane layer is covered by the film former and lubricants. At high resolution, a vein structure of the size can be seen in Fig. 9.4. This kind of thin vein and tree limb surface appearance of a complete sizing is also reported [31].

Water is the worst enemy for any glass-fiber-reinforced plastic (GFRP). The damage is done at the interface rather than to the glass reinforcement. When E-glass fiber is aged for 1 week at 50°C and 90% relative humidity, only a slight increase of surface area is observed [32]. It can correspond to an increase of roughness and the appearance of hydrophilic sites. The thermoanalysis curve of E-glass fiber is given in Fig. 9.5. The weight loss between 20°C and 400°C amounts to only 0.01%. The change in slope around 450°C can be attributed to a loss of water provided by other oxides. The aged E-glass has the same curve, meaning that the fiber is not chemically altered by moisture. From the steep slope of the water adsorption–desorption isotherms of an E-glass fiber, the affinity of the water is high (Fig. 9.6). A hysteresis occurs during the desorption that shows

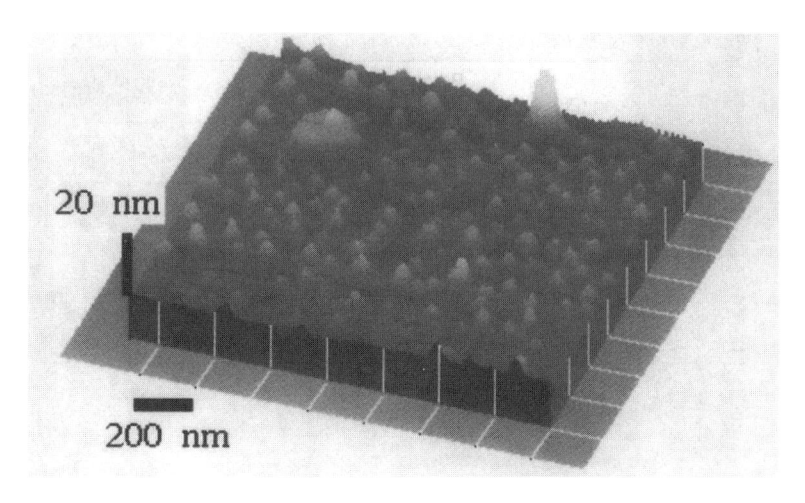

**FIGURE 9.3** The E-glass fiber surface sized with A-1100 silane by atomic force microscopy. (Courtesy of Vetrotex International.)

an important part of irreversibility. The second adsorption isotherm is very close to the adsorption isotherm on the aged glass fiber. This leads to the conclusion that a few hours in high humidity is as damaging as a week.

The surface energetics and wettability theory of adhesion is the most widely used hypothesis to explain the intermolecular and interatomic forces on the surface energies of the polymer and the glass. For those forces to have any measurable value, the polymer must come into intimate contact with the glass (i.e., the surface of the glass must be completely wetted by the polymer). Interfacial energy, which is usually smaller in magnitude than surface energy, is found when dissimilar materials are in contact with each other. This energy arises because the intermolecular forces in one medium are not necessarily matched by the intermolecular forces in another.

For liquid, the surface energy and surface tension are numerically the same. It is usually expressed by the symbol $\gamma$ and in units of millijoules per square meter. The room temperature surface energy of water is about 72 mJ/m$^2$. For a solid surface, the surface energy is a change in a free energy function with a change in the surface area of a material. There is no trustworthy method for measuring the surface energy of most solids. However, the value for most inorganic solids is thought to be in the hundreds of mJ/m$^2$. The wettability theory of adhesion is related to the study of contact angles of liquids on solid surfaces. The energy balance at the point of contact between the liquid and solids can be written as [33]:

$$\gamma_{SA} = \gamma_{SL} + \gamma_{LA} \cos \theta \tag{9.1}$$

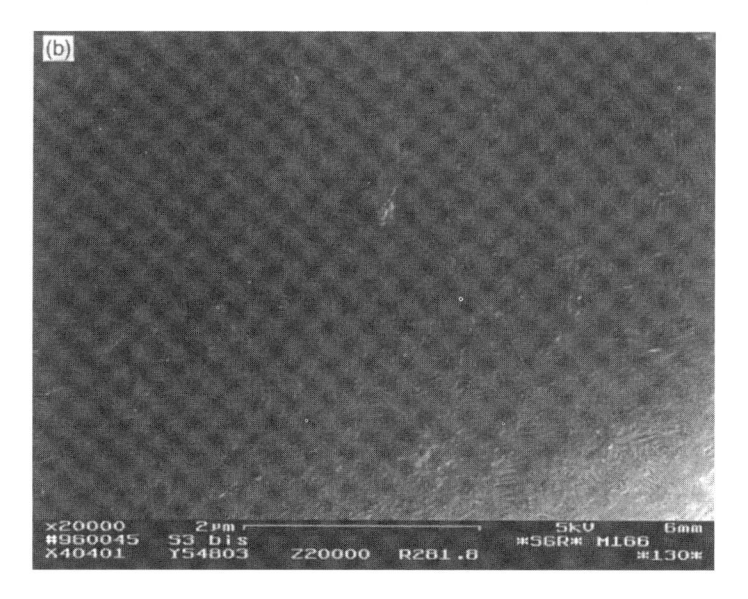

FIGURE 9.4 SEM micrographs of the glass fiber with polypropylene size. (Courtesy of Vetrotex International.)

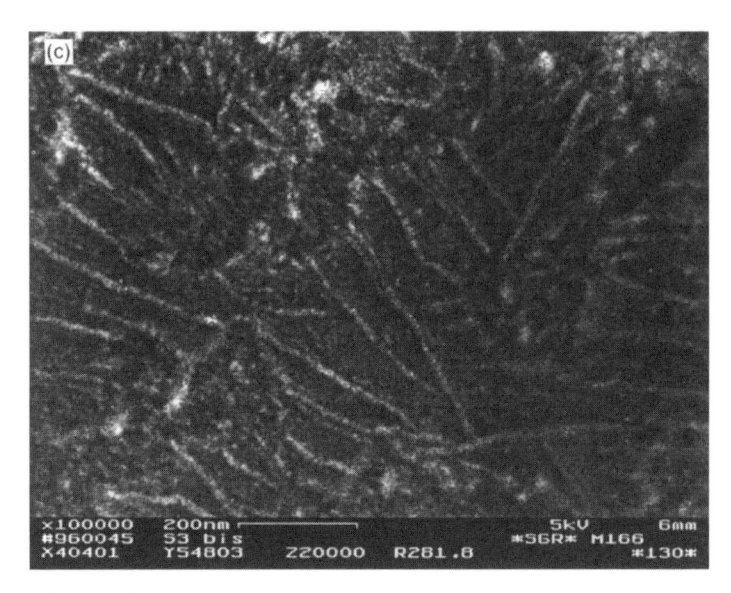

**FIGURE 9.4** (*continued*)

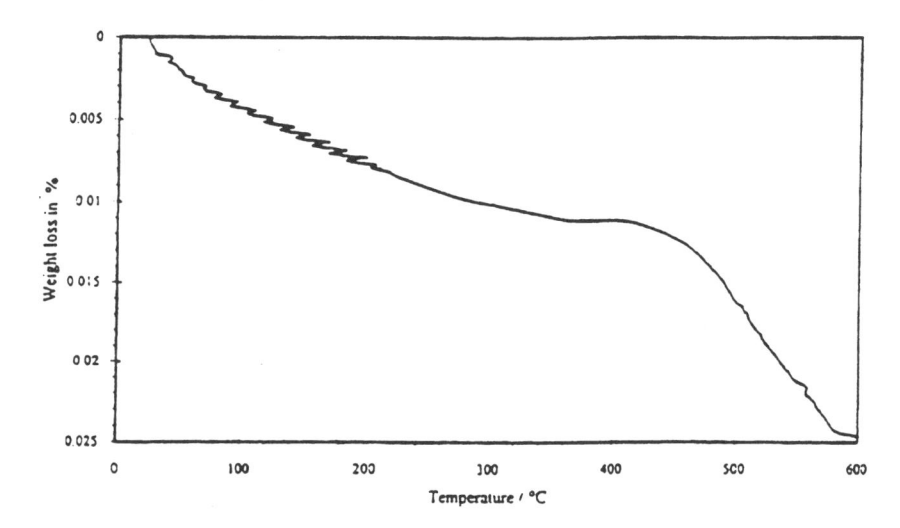

**FIGURE 9.5** Weight loss (in %) versus temperature of Virgin E-glass fiber. (Reproduced from Ref. 32 with permission.)

# Glass Fiber–Reinforced Polypropylene

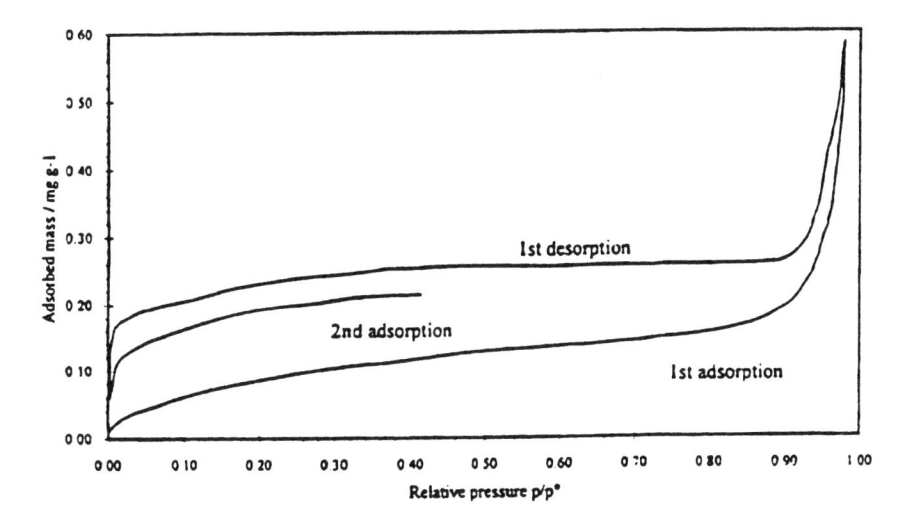

**FIGURE 9.6** Adsorption–desorption isotherm of water on E-glass fibers. (Reproduced from Ref. 32 with permission.)

$\gamma_{SA}$ is the surface tension of solid–air, $\gamma_{SL}$ is the surface tension of solid–liquid, $\gamma_{LA}$ is the surface of liquid–air, and $\theta$ is the contact angle.

The work of adhesion is defined by the Dupre equation:

$$W = \gamma_{SA} + \gamma_{LA} - \gamma_{SL} \tag{9.2}$$

Combining these two equations, the unknown solid surface tension is eliminated:

$$W_{SL} = \gamma_{LA} * (1 + \cos\theta) \tag{9.3}$$

This Young-Dupre equation allows one to determine the work of adhesion from a simple measurement of the contact angle and the liquid surface tension. Much of the experimental work in adhesion science has centered around the relationship of these two parameters.

As discussed above, to make glass-reinforced polypropylene (PP) composites, the PP resin has to wet and to encapsulate glass fibers. At the wetting stage, the adsorption or adhesion of liquid on solid consists of both dispersive and nondispersive interactions. The dispersive forces are London forces, which exist universally. The nondispersive forces are Lewis acid–base or electron acceptor–donor interaction, except saturated hydrocarbons. By the measurement of specific retention volume of several kinds of probe molecules, the dispersive component of the surface free energy of the bare E-glass fiber was determined to be 40–49 mJ/m$^2$ [34,35]. Most silane treatments reduce the dispersive component of the surface energy of glass fiber. In a similar work, the dispersive component of surface free energy is a constant for glass fibers treatment with various sizes [36]. The hydroxylated glass has the highest polar component of surface free energy and the highest total surface free energy. Thus, this water-sized glass fiber is expected to be wetted much more easily than sized fibers. Polypropylene is a nonpolar polymer. The homopolymer does not readily wet the glass fibers. The contact angle of molten PP on glass cloth decreases from 38° to 22° with the increase in temperature from 200°C to 250°C [37]. It was postulated that appearance of carbonyl and hydroxyl groups in the PP molecules results from thermal oxidation. The data match the contact angle measured by Lellig and Ondracek [38]. In their work, the dispersive surface energy parameter of PP at 533 K is 32.7 mJ/m$^2$. The polar surface energy parameter is merely 0.2 mJ/m$^2$. The time dependence of wetting angles of the PP melt shows a linear line in a double logarithmic expression. For a highly polar polymer, such as polyamide, macroscopic cohesion failure occurs inside the glass, whereas the joint strength of the nonpolar PP polymer to the glass is so weak that it cannot be measured.

Four techniques were used to characterize the surface of bare (sample 1) and sized glass fibers [39]. The sizes include cationic silane (sample 2), polyethylene dispersion (sample 3), silane/polyurethane dispersion (sample 4), and silane/epoxy dispersion (sample 5). Sample 2 shows highest surface free energy by capillary rise method. There is almost 30 mJ/m$^2$ difference between samples 2 and 4. However, the difference in surface free energy among five samples is nonexistent by inverse gas chromatography. Samples 1 and 2 have an acidic surface character by potential measurement, whereas samples 3 and 5 are basic. Sample 4 exhibits basic group from the aminosilane and acidic character from the polyurethane film former and emulsifying agent in the dispersion.

Chemically modified PP is commonly used to improve the interfacial adhesion between PP and glass fibers. It does not alter the wetting kinetics of untreated glass fibers [40]. In either case, both types of PP can wet bare glass fibers relatively quickly. With aminosilane on the glass surface, the wetting by homopolypropylene remained unchanged. It is with chemically modified PP and aminosilane-treated glass fiber that the wetting tension is drastically lowered. The amino group of the silane and the carboxylic groups of the grafted PP has a strong interaction. The goal of better adhesion is achieved.

### 9.2.3 Adhesion Promoter

Glass is an inorganic material, whereas the polymer is organic. These two materials are naturally incompatible and do not form hydrolytically stable bonds. A coupling agent is necessary for the chemical reaction to occur between the two materials. Coupling agents contain chemical functional groups that can react with silanol groups on glass. At the other end of coupling agent, the organofunctional group is to react with the polymer. Covalent bonds, which are assumed to form, lead to the strongest interfacial bond. In the dawn of the fiberglass-reinforced plastics industry, the chromium complex Volan was proven to be the most effective coupling agent. In the late 1950s and 1960s, organofunctional silanes began to displace Volan in many applications. Various silanes were developed for specific resin systems with much more superior composite properties than Volan [41,42]. A complete review of adhesion promoters, including nonsilanes, was written by Sathyanarayana and Yaseen [43].

Because all fiberglass sizes are water based, alkoxysilanes have to be hydrolyzed before being applied to glass surfaces to function as coupling agents. The hydrolysis is for trialkoxysilanes to react with excess water to form silane triols:

$$R'Si(OR)_3 + \text{excess } H_2O \rightarrow R'Si(OH)_3 + 3ROH$$

The mechanism and rate of hydrolysis of various silanes have been the subject of intense study in the last four decades. The kinetic model for hydrolysis of mono-, di-, and trifunctional silane esters in aqueous and aqueous-organic solutions can be simply described by the rate of silane ester disappearance:

$$-d[S]/dt = k_{\text{spon}}[H_2O]^n[S] + k_H[H^+][H_2O]^m[S] + k_{HO}[HO^-]^o[H_2O]^p[S] + k_B[B][H_2O]^q[S]$$

while S is the silane ester and B any basic species other than hydroxide anion [44]. The model shows that the rate of hydrolysis highly depends on the pH of the solution. In addition, the alkyl substituents on silicon were also found to have a significant effect on the rates of aqueous hydrolysis of alkyltrialkoxysilanes under basic and acidic conditions. The pH–rate profile for the hydrolysis of γ-glycidoxypropyltrimethoxysilane is shown in Ref. 31. For a more widely used silane for thermoplastics, the γ-APS shows the highest deposition rate at the natural pH around 10–11 by Fourier transform infrared spectroscopy (FTIR) [45]. In the pH range of 2–7, the bands in the spectra at 1610 and 1505 $cm^{-1}$ were assigned to the deformation mode as a protonated amine. Then change occurred at pH about 8. A new peak began to appear at 1570 $cm^{-1}$ from pH 8–10. At pH 10, the peaks were clearly defined at 1575 and 1505 $cm^{-1}$ as the aminobicarbonate salt. At pH 12, the amine group is essentially free protonation. The bands at 1575 and 1485 $cm^{-1}$ disappeared, leaving a prominent peak at 1592 $cm^{-1}$. The structure of γ-APS hydrolyzate changed from $NH_3^+Cl^-$ to $NH_3^+(HCO^-)$ and to $NH_2$. The pH-dependent deposition rate of γ-APS is also confirmed by Wang and Jones [30]. At 1.5% silane concentration, E-glass slide picked up three times the γ-APS at the natural pH than at either pH 2 or 12.

Freshly prepared silane solution is predominantly monomer silanols. After the first and second alkoxy groups are hydrolyzed, condensation reaction to form siloxanols follows:

$$R'Si(OH)_3 + R'Si(OH)_3 \leftrightarrow R'Si(OH)_2OSi(OH)_2R' + H_2O$$

Dimers, trimers, and low molecular weight oligomers are always present unless in a very dilute solution. The fact that the hydrolysis and condensation occur concurrently makes it very difficult to identify the numerous structures of condensed species and to monitor their formation with time. A kinetic model of condensation of silanol to disiloxanol can be expressed as the rate of silanol disappearance:

$$-d[S']/dt = k_H[H^+][S']^2 + k_{HO}[HO^-][S']^2 + k_B[B][S']^2$$

where $S'$ represents the silanol. The neutral organofunctional trialkoxysilanes hydrolyze rapidly in mildly acidic water and then condense slowly to siloxanols. In contrast, the aminosilane at their normal pH in water hydrolyzes and condenses very quickly.

In the case of hydrolyzed γ-methacryloxypropyltrimethoxysilane, the monomer portion was only 7% from $^1$HNMR integration [46]. The dimer, trimer, tetramer, and pentamer had more than 10% concentration. Cyclic trimer and tetramer were actually the majority of the species at 22%.

Besides the concentration and pH of the silane solution, drying condition is another significant factor that can affect the performance of the coupling agent. The reaction of silanols with siliceous surface to form the Si-O-Si covalent bond occurs at the drying stage when the byproduct, water, is removed. When E-glass fibers were treated with 20% by weight γ-APS aqueous solution, various drying conditions yielded somewhat different infrared spectra. The increased amount of free amine was observed by the drying in nitrogen. Whereas in a carbon dioxide or air atmosphere, the amine group formed an amine bicarbonate structure [47]. The entire reaction flow chart of alkoxysilanes [48] is presented in Fig. 9.7.

In the late 1970s and early 1980s, Ishida and Koenig at the Department of Macromolecular Science of Case Western Reserve University used FTIR and Raman spectroscopy to conduct a series of studies on the hydrolysis and structure of silanes and its interactions with silica–E-glass surface [45,47,49–59]. They were able to obtain direct evidence on the molecular level of hydrolysis and reformation during drying of the Si-O-Si bonds of the polyvinylsiloxane on E-glass fibers. The infrared band at 893 cm$^{-1}$ was assigned to the Si-O stretching mode of the silanol group. During the drying, this peak decreased in intensity while new peaks appeared at 1170 and 1080 cm$^{-1}$. These two peaks arose from the chemical reaction of the silane oligomer with the glass surface (i.e., Si-O-Si covalent bond). They also found that the coupling agent on the glass fiber reacted much more quickly than without glass at room temperature. The hydrolyzed

# Glass Fiber–Reinforced Polypropylene

**Hydrolysis**

$$RSi(OCH_3) + 3H_2O \rightarrow RSi(OH)_3 + 3CH_3OH$$

**Condensation**

$$2RSi(OH)_3 \rightarrow \begin{array}{c} R \quad R \quad R \\ | \quad | \quad | \\ HO\text{-}Si\text{-}O\text{-}Si\text{-}O\text{-}Si\text{-}OH \\ | \quad | \quad | \\ OH \quad OH \quad OH \end{array} + 2H_2O$$

OH OH OH
| | |
Substrate

**Hydrogen Bonding**

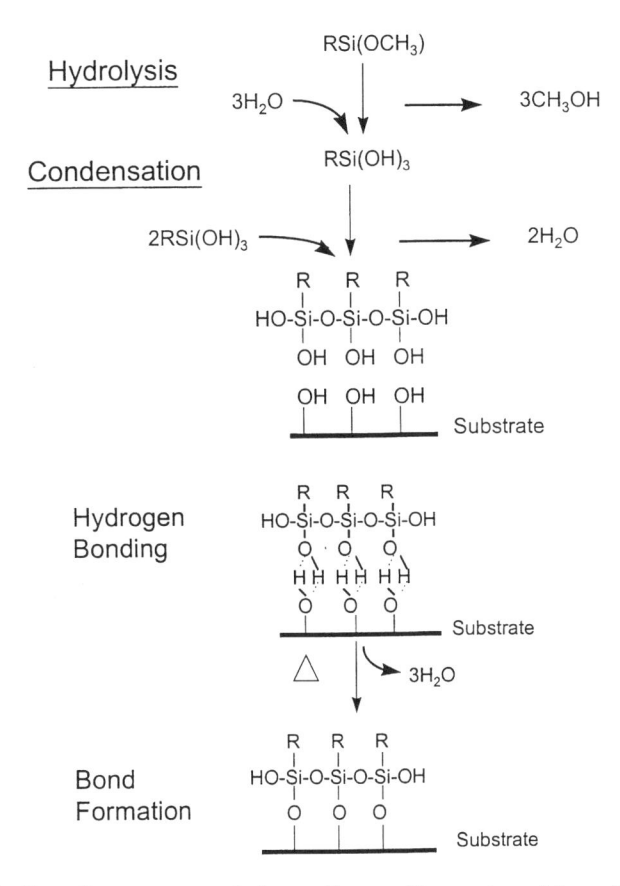

R  R  R
|  |  |
HO-Si-O-Si-O-Si-OH
|  |  |
O  O  O
H H  H H  H H
O  O  O
|  |  |
Substrate

Δ ↓ → 3H₂O

**Bond Formation**

R  R  R
|  |  |
HO-Si-O-Si-O-Si-OH
|  |  |
O  O  O
|  |  |
Substrate

FIGURE 9.7  Reaction process of alkoxysilanes. (Reproduced from Ref. 48 with permission.)

γ-APS existed in two structural forms: a chelate ring and a nonring extended structure. On the glass surface, a multiply hydrogen bonded was proposed where the amino group formed an intramolecular ring structure. Unfortunately, most of their study of silane-polymer reaction were on thermoset resins. A complete review was given by Ishida in 1984 [60].

In the last two decades, Dr. Lawrence Drzal of Michigan State University also published a series of articles on the interface of composites. More recently, Wang and Jones of the University of Sheffield UK used X-ray photoelectron spectroscopy (XPS) and secondary ion mass spectrometry to study the reaction of silane and glass [21,30,61–63]. The γ-APS coating on the glass surface is at least 3 nm thick. The very outer physisorbed layer can be easily removed by warm water and consists mainly of small molecular oligomers. The interfacial layer is more hydrolytically stable. The stability comes from a higher cross-link density within the polymeric deposit and/or the presence of a strongly chemisorbed layer. The thickness of this layer is likely less than 1 nm.

The thickness of silane layer on the glass surface also depends on the concentration of the silane solution [64]. At concentrations from 1 to 30 wt%, the thickness of the aminosilane increased from 1.0 to 4.8 μm on a 200-μm glass fiber. However, the interfacial shear strength peaked at 20% concentration. Using pyrolysis–gas chromatography/FTIR to analyze glass fiber treated with 2% γ-APS, Nishio et al. [65] determined that most silane vaporized from glass surface between 400°C and 500°C. The decomposition occurred at the Si-C or N-C bonds. Multilayer silane structure on the glass surface was reconfirmed. However, the fixing ratio never reached 1.0. This implied that even at very low silane concentrations, not all silane molecules were chemically bound to the glass surface.

A new alkoxysilane to silica surface adhesion mechanism was proposed by Dubois and Zegarski [66]. The silica surface was dehydroxylated to less than 0.2 OH group/nm. It was found that model silanes and organosilanes could still form the stable chemical bond to the silica surface.

The selection of silane for glass fiber size has tremendous impact on the properties of polypropylene composites. Plueddemann [41, p. 211] showed that the flexural strength of GFRP is twice as strong with cationic styrylsilane as with diamine silane. Four different silanes were used to treat glass microspheres to test their coupling effectiveness with polypropylene [67,68] (Table 9.7). However, when Thomason and Schoolenberg [69] applied six commonly used coupling

TABLE 9.7 Effect of Silane Type on Mechanical Properties of PP/E-Glass Bead Composites

| Silane type | Tensile strength | Flexural strength | Flexural modulus | Izod impact | Charpy impact |
|---|---|---|---|---|---|
| Control (none) | 100 | 100 | 100 | 100 | 100 |
| Vinylbenezylamine[a] | 130 | 107 | 123 | 113 | 115 |
| Ionomer (Zn$^{++}$)[b] | 141 | 110 | 136 | 94 | 121 |
| Carboxy[c] | 121 | 102 | 111 | 114 | 129 |
| Amino[d] | 145 | 117 | 141 | 99 | 129 |

[a]Dow Corning Z-6032 in G.P. PP.
[b]Dow Corning 7169-45B in carboxylated PP.
[c]Dow Corning 7169-45B in G.P. PP.
[d]Dow Corning Z-6020 in G.P. PP.
*Source:* Refs. 67 and 68.

# Glass Fiber–Reinforced Polypropylene

agents to the glass fiber and tested the interfacial strengths by single-fiber pull-out, the interfacial shear strength (IFSS) values were very close (Table 9.8). The use of silane did not significantly improve the adhesion over bare glass.

When the level of the silane increased from 1 to 2.5 wt%, the interfacial bond strength increased from 32 to 46 MPa [70]. This was confirmed by the SEM study of 30 wt% glass flake–reinforced polypropylene [71]. The glass flakes were treated with 0–0.8 wt% $N$-β-($N$-vinylbenzylamino)ethyl-γ-aminopropyltrimethoxysilane (Dow Corning Z6032). The amount of PP remaining on the flake surface after fracture increased with the higher level of silane. It was also shown that silane altered the matrix morphology by inducing nucleation at the filler–matrix interface. The effect of coupling agent on the fracture performance occurred at very low concentration. Tensile strength of the 30% glass fiber–reinforced polypropylene increased almost 30% from 0 to 0.3% silane content. Additional silane up to 0.9 wt% had a negligible effect on the tensile strength. The tensile modulus remained constant at all levels of silane.

A recent trend is to supply water-based prehydrolyzed silane [72]. The advantage is simple to use for the formulators. At least one supplier has claimed to achieve equivalent performance to regular silane [73]. Reducing or totally eliminating volatile organic content (VOC) is in every silane user's mind [74]. Ethanol is considered a nonpolluting solvent, whereas methanol or isopropanol are pollutants. Silylated polymers are available on the market. The trend is to extend this technology to lubricants and nucleating agents. Polymeric silane was found to have a greater effect on the deposition of the silane [75]. When hydrolyzed silanes bond to the glass surface, it is desirable to exclude water completely from the glass surface after drying. Hydrolysis and thermal resistance of silane-glass bond needs to be further improved.

TABLE 9.8 Effect of Silane Type on the Interfacial Strength by Single Fiber Pullout (IFSS)

| Silane type | Fiber diameter (μm) | IFSS (MPa) |
| --- | --- | --- |
| Control (none) | 10.9 | 3.6 |
| Amino | 18.0 | 4.5 |
| Methacryl | 18.5 | 3.5 |
| Epoxy | 15.7 | 3.1 |
| Chloro | 16.8 | 3.7 |
| Propyl | 16.9 | 3.1 |
| Vinyl | 9.6 | 3.9 |

Source: Ref. 69, © 1994, with permission from Elsevier Science.

## 9.2.4 Role of Glass Fiber Size for Polypropylene

Although it is proven that silane forms the covalent bond to the glass surface and provides the interfacial strength to the composites, glass fiber producers must use many other ingredients in their size formulations. These can generally be categorized as film formers for binding the filaments together for downstream processing; lubricants for reducing fuzz and improving flow; antistatic agents to eliminate static generation and fuzz accumulation; plasticizers to improve flexibility and control solvent solubility; and nucleating agents to improve composite properties and other miscellaneous chemicals such as acid for pH adjustment, antioxidant for nonyellowing, and surfactant for size stability. Common ingredients are listed on pages 261–266 of ref. 1. On page 278 of ref. 1, two specific size formulations for the reinforcement of polypropylene are given. In both cases, a polypropylene emulsion is used as the film former along with a second film former, i.e., urethane or epoxy.

The effect of polypropylene film former and the addition of polyurethane or epoxy film former has been studied by Mader et al. [76] (Table 9.9). Fiber glass sized with aminosilane (A1100) with or without film former(s) was tested in homopolypropylene and a maleic anhydride grafted polypropylene. The results show that when changing resin matrix from nonpolar homopolypropylene to maleated polypropylene, there is a 45% increase in IFSS on bare fibers alone. With so much emphasis on the mechanism of coupling agents, the A1100 silane does not react to the homopolymer. The epoxy film former does nothing on the interface of glass–homopolypropylene but actually deteriorates the IFSS in the chemically modified polypropylene. Polypropylene dispersion is the best film former for polypropylene resin. The diffusion theory of adhesion is overlooked by many researchers. Although glass and polymer are dissimilar materials, the coating on the glass surface plays an important role in interfacial adhesion. The adhesion between the size on the glass surface and the resin matrix is maximized when the solubility parameters of the two are matched. It is some-

TABLE 9.9  Effect of Fiber Glass Sizing Agents on the IFSS

| Silane | Film former | HomoPP IFSS (MPa) | Coupled PP IFSS (MPa) |
|---|---|---|---|
| None  | None              | 5.9 | 8.6  |
| A1100 | None              | 5.2 | —    |
| A1100 | Epoxy             | 6.0 | 6.6  |
| A1100 | PP/PUR dispersions | 7.2 | 11.9 |
| A1100 | PP dispersion     | 7.2 | 12.4 |

*Source:* Ref. 76, © 1996, with permission from Elsevier Science.

## Glass Fiber–Reinforced Polypropylene

times described as "like dissolves like." Most of the polypropylene dispersions are made from chemically modified polypropylene. It is no wonder that the combination of silane and polypropylene dispersion works the best in the coupled polypropylene system.

This is further confirmed by Hoecker and Karger-Kocsis [77] from single-fiber pull-out and microdroplet method. The IFSS of unsized glass and homopolymer is 5.7 MPa, same as in Table 9.8. The effect of size and coupled polypropylene is higher. The maximal IFSS is reported at 20.2 MPa.

A more recent work [78] by Nippon Electric Glass (NEG) extended the film former to include acrylstyrene (Table 9.10). The glass is sized with aminosilane and three different types of film former: acrylstyrene copolymer emulsion, polypropylene homopolymer, and maleic anhydride–modified polypropylene dispersion. The interesting point is that even with a not-so-effective film former such as acrylstyrene, whose IFSS with homopolypropylene is 0.9 MPa, the coupled polypropylene resin matrix can still provide reasonably strong interfacial strength. This is different from the epoxy film former discussed earlier. Higher kneading can slightly improve the IFSS of the best system but has negligible effect on nonoptimized system.

What has been missing is the straight comparison of the effect of silane and film former on the final composite properties. The recent work at Vetrotex CertainTeed involves the testing of a commercial polypropylene reinforcement 968 (sample A) in polypropylene, Montell Profax 6523. Sample B is the E-glass sized with polypropylene dispersion only and sample C is with silane only. Sample D is water-sized E-glass fiber at same fiber diameter of 13.3 μm. The results (Table 9.11) clearly demonstrate the advantage of a suitable film former not only to the processibility but also to the final composite properties. From the SEM micrographs, the clean pull-out of glass fibers from the fractured surface in samples C and D indicate that bare glass and silane-treated glass have almost no bonding to polypropylene resin (Fig. 9.8), whereas the glass fiber sized with polypropylene film former (sample B) shows little resin attachment. Sample A is the optimized size system. The unsized and silane-only glass fibers cannot be handled by the industrial material handling equipment. Without a film former, the

TABLE 9.10 Effect of Film Formers on the IFSS

| Silane | Film former | HomoPP IFSS (MPa) | Coupled PP IFSS (MPa) |
|---|---|---|---|
| Amino | Acrylstyrene | 0.9 | 6.2 |
| Amino | Homo-PP | 1.4 | 13.9 |
| Amino | Modified PP | 7.6 | 16.1 |

Source: Ref. 78.

TABLE 9.11 Influence of Glass Fiber Size on the Mechanical Properties of GFRP

| Glass size system | A<br>Silane + PP | B<br>PP dispersion | C<br>Silane | D<br>Unsized |
| --- | --- | --- | --- | --- |
| % Glass in Profax 6523 | 22.9 | 22.2 | 23.7 | 26.5 |
| Tensile properties | | | | |
| Strength (MPa) | 67.8 | 58.2 | 35.9 | 32.9 |
| Modulus (GPa) | 3.87 | 3.58 | 2.70 | 3.40 |
| Elongation (%) | 2.5 | 2.4 | 3.6 | 4.4 |
| Flexural properties | | | | |
| Strength (MPa) | 103 | 95 | 61 | 60 |
| Modulus (GPa) | 3.90 | 4.08 | 2.54 | 4.58 |
| Izod impact | | | | |
| Notched (kJ/m$^2$) | 12.3 | 9.82 | 7.87 | 4.11 |
| Unnotched (kJ/m$^2$) | 28.2 | 24 | 22.7 | 32.3 |

fuzz generation and fiber breakage are simply too big a problem for compounders to process the glass fibers.

Crystalline carboxylated polypropylene, such as Hercoprime G, and the carboxylated amorphous polypropylene, such as Epolene E-43, can be used as the base polymer to make polypropylene dispersion [79]. To make polypropylene emulsion, the polymer is heated to 170–175°C under 100–120 psig superatmospheric pressure. Triton X-100, an ethoxylated alkylphenol, or Igepal CO 630, an ethoxylated nonylphenol, can be used as the surfactant. The acid-modified polyolefin must be neutralized by a base such as diethylethanolamine to maintain the stability of the dispersion. The surfactant is mixed with the molten polyolefin and then the base is metered in. After the mixture reacts and mixes well, quenching water is added to rapidly cool the mixture. One of the commercially available emulsions of the carboxylated amorphous polypropylene is RL-5440 and that of carboxylated crystalline polypropylene is RL-5140. Both are from the Procter Division of National Starch and Chemicals. Chemical Corporation of America also has a line of polyemulsion with a wide selection of nonionic or cationic acrylic acid and maleic anhydride grafted polypropylene emulsions.

The other film formers listed in these patents include polyurethane dispersions, a polyurethane-modified epoxy dispersion, and a polyvinyl acetate emulsion. The choices of silanes are γ-aminopropyltriethoxysilane (A1100 or Z6011), $N$-β-aminoethyl-γ-aminopropyltrimethoxysilane (A1120 or Z6020), and γ-glycidoxypropyltrimethoxysilane (A187 or Z6040).

Isotactic or syndiotactic polypropylene has a strong tendency to crystallize during cooling. It is therefore more difficult to emulsify these types of poly-

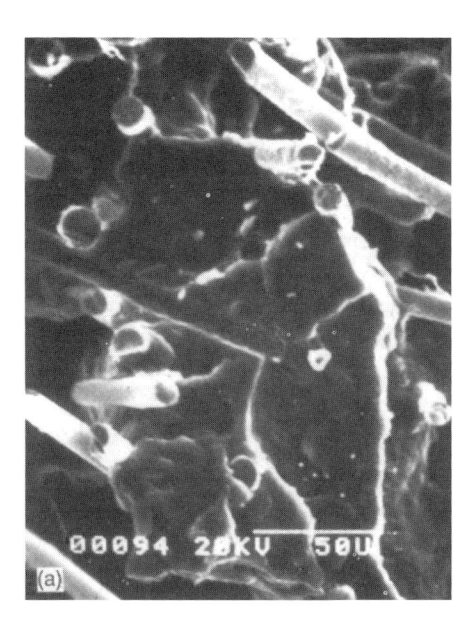

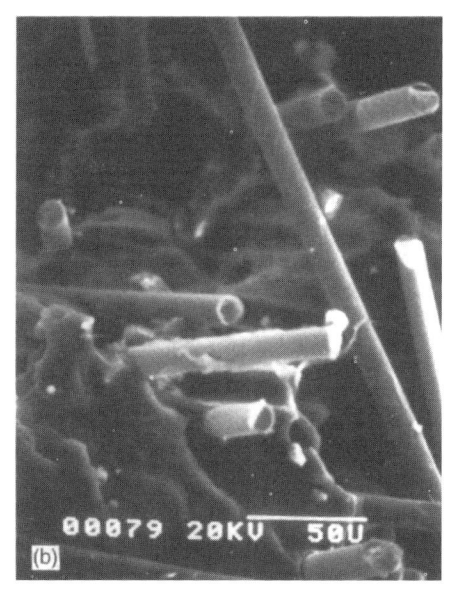

FIGURE 9.8 SEM micrographs of the fractured surface of 20 wt% GFRP. (a) Sample A, (b) sample B, (c) sample C, and (d) sample D. (Courtesy of OSi/Witco.)

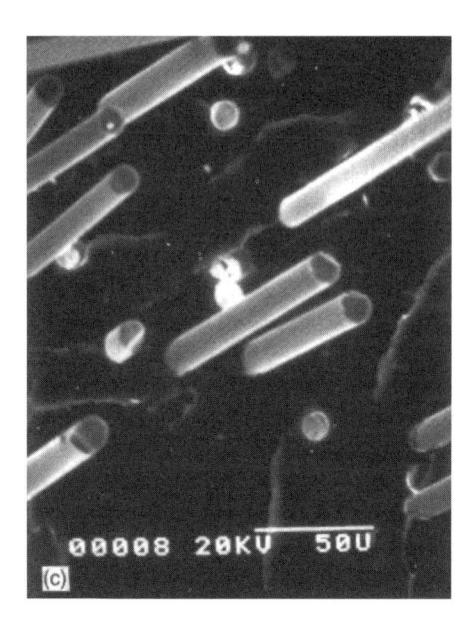

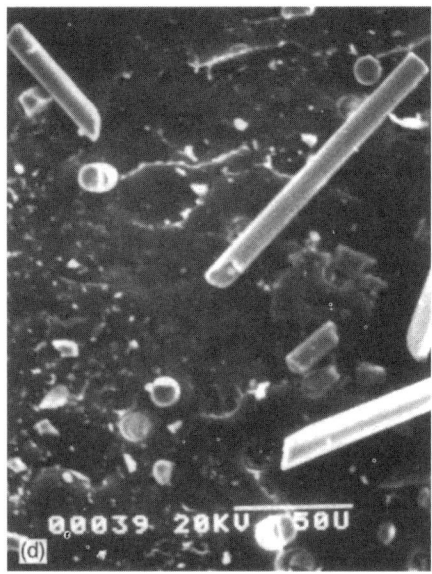

FIGURE 9.8 (continued)

propylene. It is even more difficult to emulsify such polymers of high molecular weight. Two recent U.S. patents disclose the glass fiber sizing containing such emulsion [80,81]. Fatty acids can fluidize the polyolefins and form a first mixture at a temperature higher than the melting point of the polymer. Adding to the first mixture are an inorganic or organic base, water, and an emulsifying agent. The second mixture is stirred under pressure and then cooled down to solid state. The aqueous polypropylene emulsion made by this method is very stable by itself. However, a monoamine stabilizing agent is needed for such an emulsion to be formulated with aminosilane.

The effects of nucleating agents have been studied by postcoating glass fibers with sodium salt of methylene *bis*(2,4-*di*-*t*-butylphenol) acid phosphate (M1.NA.11 by Adela Argus), which yields the α-crystalline form or quinacridone pigment (E3B by Hoechst Celanese), which yields the β-crystalline form at the glass–polypropylene interface [82]. The α (monoclinic) phase is the preferred form in melt-crystallized isotactic polypropylene. Any significant quantity of β (hexagonal) form has to be introduced through the proper nucleating additive. The growth of α and β transcrystallinity from a glass fiber has the same form as in the polypropylene bulk resin [83]. The α-form lamella is found to lay edge-on (i.e., with the lamella surface parallel normal to the composite). The β-form lamella lays flat-on (i.e., with the lamellar surface parallel to the composite) (Fig. 9.9). For glass fibers coated with M1.NA.11, the individual damaged region of the single-fiber composite specimen consists of an interlamellar crack within the α-transcrystalline zone, which subsequently propagates through spherulites in the polypropylene resin. In contrast, for glass fibers coated with E3B, the damage is limited to the β-transcrystalline regions exclusively and no longer gives the tree appearance. The β form is significantly tougher than the α form due to its propensity for uniform intralamillar deformation.

A more recent work through thermodynamic theory leads to a megacoupled polypropylene composite [84,85]. The data show that this system is approaching the theoretical maximum strength of combined reinforcement and polypropylene resin matrix. More is discussed in Chapter 12.

### 9.2.5 Testing of Interfacial Strength of the Composites

The ultimate strength of a composite material ($\sigma_{cu}$) can be simply expressed as the sum of the strengths of each component:

$$\sigma_{cu} = v\sigma_{fu} + (1-v)\sigma_m \tag{9.4}$$

where $\sigma_{fu}$ is the tensile strength of the reinforcing fiber, $v$ is volume fraction of the fiber, and $\sigma_m$ is the resin matrix stress at failure. This model provides an adequate prediction of a continuous fiber–reinforced composite. However, when the fibers are of finite length, stress is transferred from the matrix to the fiber by a

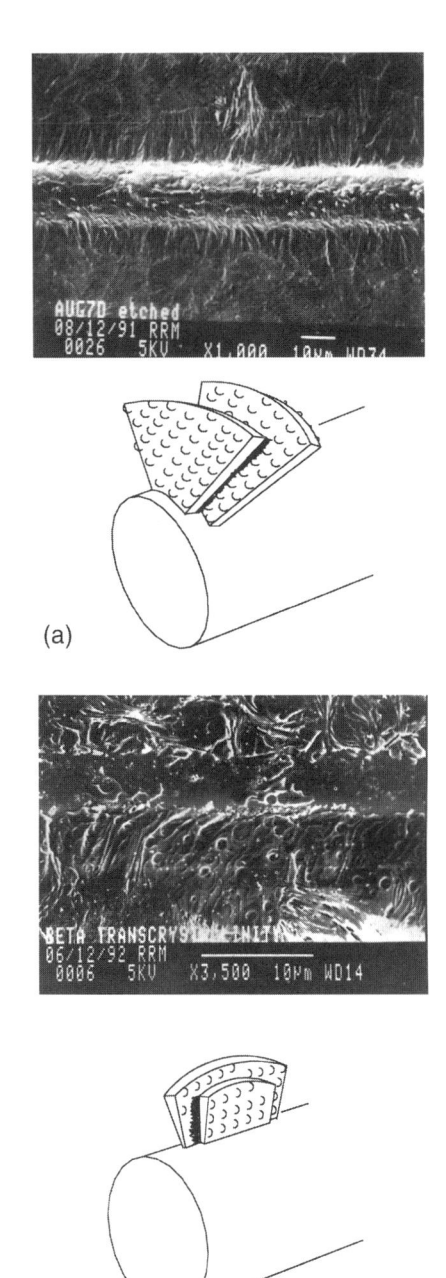

FIGURE 9.9 SEM view of transcrystallinity from glass fiber and illustration of (a) edge-on lamellae and (b) flat-on lamellae. (Reproduced from Ref. 83 with permission.)

shear transfer mechanism. This gives rise to the concept of critical fiber length ($L_c$). For fibers whose length are shorter than $L_c$, stress transfer to the fracture–stress level will not occur. The critical fiber length $L_c$ can be calculated from three parameters:

$$L_c = \frac{d\sigma_{fu}}{2\tau_u} \tag{9.5}$$

where $d$ is the fiber diameter and $\tau_u$ is the shear strength of the interface, which is often assumed to be the shear strength of the resin matrix.

In an injection-molded short-fiber-reinforced thermoplastic, there is a distribution of fiber lengths. Not all fibers are oriented in one direction. Taking these factors into consideration, Bowyer and Bader [86] proposed the following equation to predict the theoretical strength of the short-fiber-reinforced composite:

$$\sigma_{cu} = C\left[\sum_i \tau_i l_i v_i/d + \sum_j \sigma_{fu} v_j(1 - L_c/2l_j)\right] + (1-v)\sigma_m \tag{9.6}$$

where $C$ is the fiber orientation factor, $l_i$ and $l_j$ are the subcritical and supercritical fiber lengths, respectively, and $v_i$ and $v_j$ are the fiber volume fractions of the subcritical and supercritical fiber lengths, respectively.

It is generally recognized that in the fiber-reinforced polymeric matrix composites there is a separate interphase region between the bulk resin and solid fibers. The interphase is one of the most important parameters governing the composite performance. In the short term, the interphase influences the mode of failure and the toughness of the composites. In the long term, the interface determines by and large the wet strength retention against water, solvent, and thermal aging. Among the many methods to characterize the interfacial strength, fiber pull-out, microbond, single-fiber fragmentation, and microindentation techniques are most widely used (Fig. 9.10). In fact, these four methods were chosen in a multinational round-robin program for evaluating the micromechanical properties of fiber–matrix interface involving 12 international renowned research laboratories [87].

The single-fiber pull-out test measures the force required to pull the fiber out of resin matrix. A single fiber is embedded vertically in a block of resin to a controlled depth. The free fiber end is gripped in a tensile machine and then pulled from the resin block. In a typical plot of pull-out force $F_p$ versus displacement $x$ (Fig. 9.11), three sequential mechanisms are observed. The first peak, where $F_p$ increases linearly up to a maximal value of $F_d$, is attributed to debonding and frictional resistance to slipping. The subsequent lower peak and sometimes a few oscillations are due to friction coupled with stick and slip of the fiber. The last curve represents the force sustained by friction that gradually

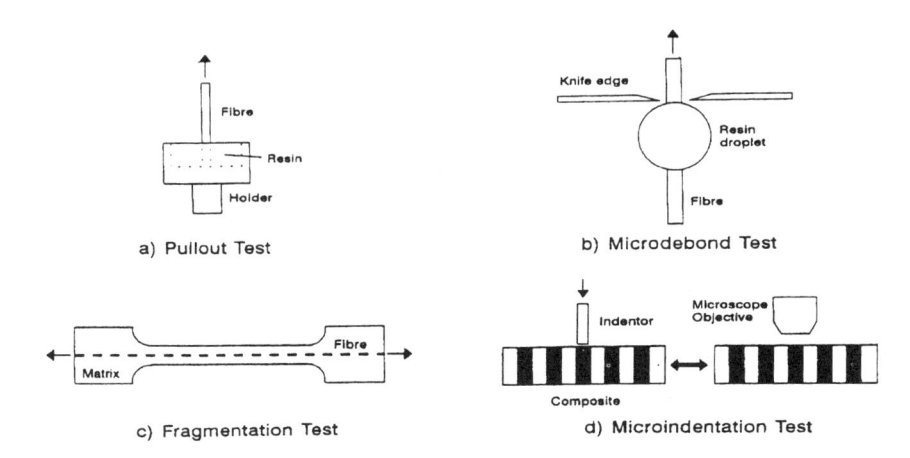

**FIGURE 9.10** Four common micromechanical test methods. (Reprinted from Ref. 87, Copyright 1993, with permission from Elsevier Science.)

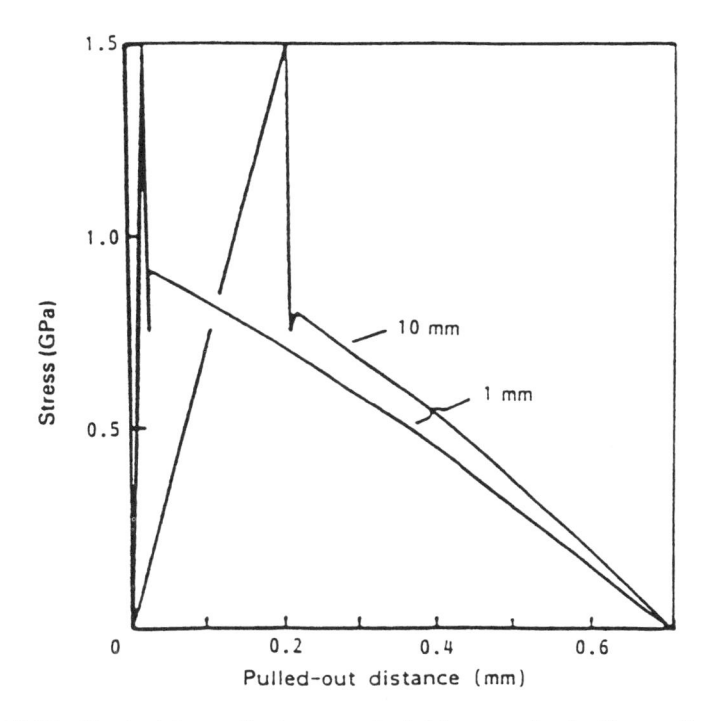

**FIGURE 9.11** Typical force-displacement plot from a single fiber pullout test. (Reproduced from Ref. 88 with permission.)

decreases as the fiber is continuously pulled out. This test method is best suited to low interfacial strength system. Short embedment fiber length causes sample preparation to be very difficult and large data scatter. Alignment of the fiber with the loading axis is a major source of error.

In the microbond method, a droplet of resin is placed on the fiber and solidified in position. The specimen is placed in a tensile tester so that one fiber end is gripped and the resin droplet is placed between two knife edges. The fiber is pulled against the knife edge and load introduced through the resin droplet. The debonding force is a function of the embedded length. The sample preparation is the simplest and quickest. For fibers of very small diameter, the maximum embedded length is less than 1 mm. The small microdrop size makes the failure process difficult to record. The meniscus formed on the fiber makes the embedded length determination difficult. All of these contribute to large data error.

The essential aspect of the single-fiber fragmentation test is that a single fiber is carefully aligned down the center of a "dog bone" mold. The mold is then filled with the resin. The coupon is stretched under tensile load. As the coupon extends, the fiber will fail repeatedly into shorter and shorter lengths until a saturation level is achieved and the fiber no longer fails. At the end of the test, the fragment lengths are measured. As discussed in Section 9.2.1, the strength of the fiber depends on the gauge length. The Weibull-Poisson model is often used to obtain the fiber fracture strength. The failure strain of the matrix must be larger (more than three times) than the failure strain of the fiber to promote multi-fragmentation of the fiber. The fully elastic model of the resin matrix is over-simplified. Recent progress on acoustic emission to monitor the failure process has expanded the composite system beyond transparent matrices. The most appealing advantage of this test method is that it replicates the events in situ in the composite.

The microindentation test is run on an individual selected fiber oriented normal to a polished cross-section of a high-fiber volume fraction composite. Instead of stretching fibers in a tension field, a stepwise compressive load is applied to the fiber through an indentor. The load/displacement response is recorded until debonding occurs. The diameter of the fiber and the distance to the nearest neighboring fiber are also recorded. A simplified axisymmetrical finite element model is used and is independent of the tensile strength of the fiber. The debonding force is dominated by the interfacial shear stress component that reaches maximal value at a distance of approximately one-half the fiber diameter below the contact surface. Glass and carbon fibers are the suitable fibers for this test method, although splitting of the fiber is still observed frequently.

Carbon fibers in the cured epoxy was the composite system for the international round-robin program on interfacial test method. The comparison of four test methods is given in Table 9.12. The results indicate that within each

TABLE 9.12 Interfacial Shear Strength of Carbon Fiber/Epoxy by Various Test Methods

| Methods | Size | No. of labs | IFSS (MPa) | SD (MPa) | CV (%) |
|---|---|---|---|---|---|
| Pullout | None | 3 | 64.6 | 8.2 | 13 |
|  | Yes | 3 | 84.1 | 19.4 | 23 |
| Microbond | None | 4 | 48.3 | 14.1 | 29 |
|  | Yes | 4 | 69.7 | 19.7 | 28 |
| Fragmentation | None | 6 | 23.8 | 6.6 | 28 |
|  | Yes | 7 | 47.3 | 15.4 | 33 |
| Indentation | None | 2 | 47.8 | 0.5 | 1 |
|  | Yes | 3 | 49.5 | 9.1 | 19 |

*Source:* Ref. 87, © 1993, with permission from Elsevier Science.

laboratory the reproducibility is good, with coefficient of variation (CV) less than 10%. However, the CV among the laboratories can be as high as 25%. The indentation method has the smallest percentage of CV because two of the laboratories used the same commercial equipment. The sized fibers show 20 MPa higher IFSS than untreated fibers in pull-out, microbond, and fragmentation methods. The microindentation test method is insensitive to the surface treatment of the fiber.

Yue and colleagues [89–91] examined the many factors that affect the test results of the single-fiber pull-out. The interfacial shear stress in the glass-polypropylene composite system concentrates at the emergent end of the fiber because of the smaller ratio of modulus of the fiber to that of the matrix, 60 to 1.4 GPa, respectively. However, this is true only when the radius of the PP matrix block is 20 times that of the glass fiber. At very small amounts of PP, the interfacial shear stress at the embedded end of the fiber is actually larger than that at the emergent end. Those two values are equal when the radius of the PP matrix block is eight times that of the glass fiber. The pull-out force increases linearly with the embedded fiber length and interfacial frictional stress. The effect of Poisson shrinkage of fiber becomes significant only when the fiber diameter is large and fiber embedded length exceeds 5 mm. Specimen failure by fiber fracture and matrix yielding occurs more often in fixed-bottom than in restrained-top pull-out testing.

The crystallization of the isotactic polypropylene, particularly transcrystallinity at the glass fiber interphase region, has been investigated by many researchers [92–102]. The presence of glass fibers in polypropylene composite decreases the crystalline time versus neat resin itself because of the nucleating agent role of the fibers. However, the growth rate of the crystallites is the same in the bulk as in the polymer adjacent to the fibers. The transcrystalline occurs only

## Glass Fiber–Reinforced Polypropylene

when a shear stress is applied to the glass fiber below the melting temperature of the polypropylene and at very rapid cooling rate (above 100°C/min). Lower molecular weight or higher melt flow of the polypropylene increases the level of stress necessary to give stress-induced transcrystallization. This is also confirmed by the actual measurement of interfacial shear strength of glass-polypropylene from a modified single-fiber pull-out test [77]. The explanation is that the amorphous phase of the polypropylene is attributing to the better wetting and adhesion. In another study of the effect of test temperature on interfacial shear strength by the single-fiber pull-out method, Schulz et al. [103] found that IFSS decreases with the increase in test temperature from −30°C to +70°C. Larger diameter of the glass fiber, from 10 to 17 μm, increases the IFSS slightly at −30°C and 25°C. At 70°C, the IFSS is weaker with the 17-μm glass fiber than with the 10-μm glass fiber.

Most work on single-fiber fragmentation tests has been done on thermoset resin matrix with a carbon fiber [104–108]. In the round-robin program, the number of fragments varied from 87 to 528. The mean fragment length is more than 800 μm for the untreated fiber and about 400 μm for the sized fiber. Good bonding, as expected, decreases the critical fiber length. Considering a perfect bond at the fiber–matrix interface, the critical fiber aspect ratio, critical fiber length over the fiber diameter ($L_c/d$), is related to the square root of the ratio of fiber to matrix modulus. The interfacial shear strength increases linearly with the shear modulus of the matrix. The fiber length distribution can be measured by the acoustic emission (AE) to overcome the limitation of optical procedures [109,110]. Netravali and Sachse [110] obtained excellent agreement between the AE and optical method for E- and S-2 glass and AS-4 graphite fibers in epoxy resin.

Fiber lengths between 5 and 10 mm are conveniently selected for the microindentation test [111]. For a carbon fiber/epoxy system, as the fiber volume fraction increases from 10 to 50 vol%, the indentation displacement distance decreases from 44 to 36 μm but the interfacial shear strength increases from 33 to 46 MPa. When the interphase-to-matrix modulus ratio increase from 1.0 to 7.5, the interfacial shear stress increases by only 10%. Likewise, interphase thickness and fiber diameter have marginal effects on the interfacial shear stress. Three types of thermoplastic polymers (polyester, polyamide, and polypropylene) were tested for their interfacial shear strength to the glass fiber by Desaeger and Verpoest [112]. The IFSS of thermoplastic polyester-glass fiber at 59 MPa is stronger than that of polyamide. For polyamide polymer, a higher IFSS is obtained from well-polished cross-section, pyramidal tip, and good adhesion from the fiber coating. The interphase adhesion is so weak between polypropylene and glass that debonding occurs during the specimen preparation. An excellent SEM picture was included to demonstrate the separation of PP from the glass fiber.

Other test methods for interface adhesion include short-beam shear and dynamic mechanical analysis [88,113,114].

## 9.3 INFLUENCE OF GLASS FIBER IN COMPOSITE PROPERTIES

### 9.3.1 Effect of Glass Fiber Parameters

*Initial Chopped Length of the Glass Fibers.* In the reinforced thermoplastic extrusion compounding and injection molding, the lengths of glass fibers are deteriorated to a few hundred micrometers. The initial chopped length of the fiber has no effect on the final composite properties [115,116]. In Schweizer's study, a homopolypropylene Profax 6523 was used as the resin matrix. The reinforcement was a K-fiber in lengths of $\frac{1}{8}$, $\frac{3}{16}$, and $\frac{1}{4}$ in. Table 9.13 gives the average fiber length and corresponding mechanical properties after extrusion compounding and injection molding. The data support the above-mentioned conclusion. The average length of the glass fiber decreases by 0.005–0.0026 in. in the injection molding process. Three different input lengths of glass fibers at 3.2, 4.5, and 6.4 mm were also compounded in PP [117]. The final average lengths were indifferent. The glass fiber attrition during extrusion and injection molding is discussed in the next two sections.

The input strand length influences the glass fiber-reinforced polypropylene compression-molded laminates (GMT) [118–121]. Chopped lengths at 0.1, 0.8,

TABLE 9.13 Effect of Input Strand Length on the Average Glass Fiber Length and Polypropylene Composite Properties

| Factor | Input strand length (in.) | | |
|---|---|---|---|
| | $\frac{1}{8}$ | $\frac{3}{16}$ | $\frac{1}{4}$ |
| WAFL after extrusion compounding (in.) | 0.030 | 0.030 | 0.030 |
| WAFL in injection molded bars (in.) | 0.025 | 0.026 | 0.027 |
| In Profax 6523 composites | | | |
| Glass content (%) | 19.9 | 20.3 | 20.2 |
| Tensile strength (kpsi) | 6.1 | 6.1 | 5.7 |
| Tensile modulus (kpsi) | 0.70 | 0.68 | 0.70 |
| Flexural strength (kpsi) | 9.8 | 9.7 | 9.5 |
| Notched Izod impact strength (ft-lb/in.) | 1.6 | 1.7 | 1.8 |
| Unnotched Izod impact strength (ft-lb/in.) | 4.7 | 4.6 | 4.4 |

WAFL, weight average fiber length.

## Glass Fiber–Reinforced Polypropylene

3, 4.5, 6, and 12 mm were investigated. The tensile and flexural strengths increase with very short fiber length and then level off at lengths greater than 3 mm. The Charpy impact strength also increases with chopped length up to 6 mm. At very short fiber lengths, the impact strength is approaching that of neat polypropylene. This clearly indicates that when the fiber length is shorter than the critical fiber length, there is not much reinforcement effect from the glass fibers.

*Fiber Diameter.* It is generally known in the composites industry that smaller glass fiber diameter improves composite strength. The change in properties varies with each resin matrix [122]. In a recent work, the commercially available 968 PP glass was produced at 10, 11, 12.5, 13.3, 14, and 16 µm in diameter. They were compounded in Montell Profax 6523 at 10%, 20%, and 30% by weight. The tensile and impact strengths are listed in Table 9.14. It can be clearly seen that the tensile and unnotched Izod impact strengths and tensile strain of the GFRP decreases with the increase of fiber diameter at all three glass contents (Figs. 9.12 and 9.13). The notched Izod impact strength also decreases

TABLE 9.14 Effect of Glass Fiber Diameter on the Mechanical Properties of the GFRPP

| | Tensile strength in kpsi (ASTM D690) | | | | | |
|---|---|---|---|---|---|---|
| % Ash | 10 | 11 | 12.5 | 13.3 | 14 | 16 µm |
| 30 | 11.1 | 11.2 | 10.7 | 10.4 | 10.2 | 10.1 |
| 20 | 10.1 | 9.74 | 9.38 | 9.24 | — | 8.92 |
| 10 | 7.81 | 7.52 | 7.58 | 7.40 | 7.34 | 7.24 |

| | Notched Izod impact strength in ft-lb/in. (ASTM D690) | | | | | |
|---|---|---|---|---|---|---|
| % Ash | 10 | 11 | 12.5 | 13.3 | 14 | 16 µm |
| 30 | 1.32 | 1.32 | 1.29 | 1.23 | 1.22 | 1.19 |
| 20 | 1.22 | 1.31 | 1.31 | 1.34 | — | 1.26 |
| 10 | 0.92 | 0.95 | 0.98 | 0.91 | 1.02 | 0.95 |

| | Unnotched Izod impact strength in ft-lb/in. (ASTM D690) | | | | | |
|---|---|---|---|---|---|---|
| % Ash | 10 | 11 | 12.5 | 13.3 | 14 | 16 µm |
| 30 | 7.47 | 7.28 | 6.72 | 6.57 | 6.45 | 5.68 |
| 20 | 7.55 | 7.31 | 6.75 | 6.94 | — | 6.52 |
| 10 | 7.64 | 7.17 | 7.23 | 7.16 | 7.12 | 6.94 |

Glass fiber: 968 Chopped Strand from Vetrotex CertainTeed; Resin Matrix: Profax 6523 from Montell.

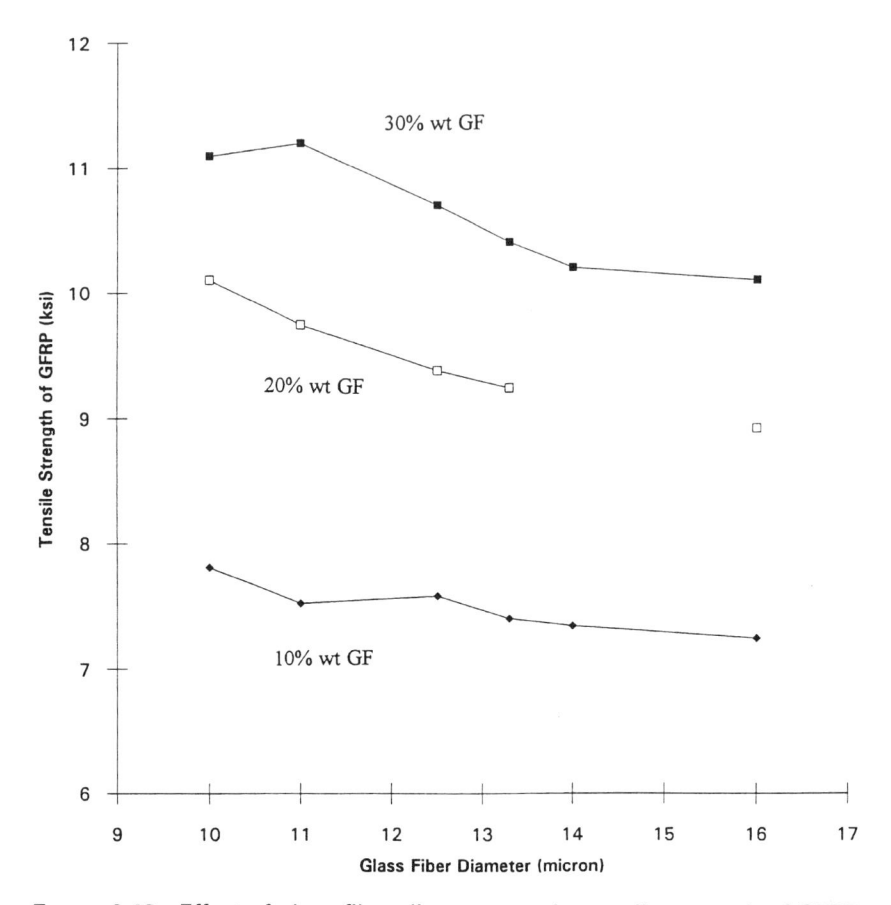

**FIGURE 9.12** Effect of glass fiber diameter on the tensile strength of GFRP.

with larger fiber diameter at 30 wt% glass loading. At 10 and 20 wt% glass contents, the fiber diameter has an insignificant effect on the notched Izod impact strength.

The small gain in the selective mechanical properties sometimes cannot make up the higher cost of finer diameter of fiberglass. More important to the molders is the increased melt viscosity associated with smaller fiber diameter (Fig. 9.14). At 20 wt% of glass content, the melt flow decreases from 1.9 to 1.35 g/10 min.

*Total Glass Content.* From Table 9.14 and other testings, it can be concluded that tensile strength and flexural modulus of the GFRP increase with higher glass content (Figs. 9.15 and 9.16). Notched Izod impact strength

# Glass Fiber–Reinforced Polypropylene

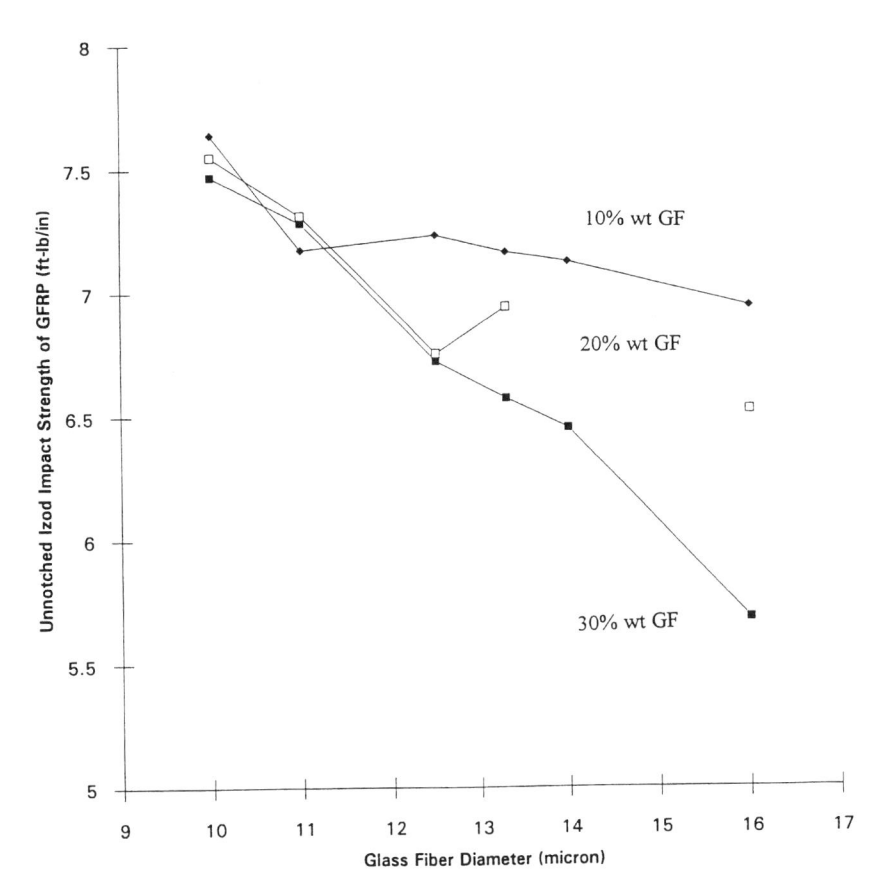

**FIGURE 9.13** Effect of glass fiber diameter on the unnotched Izod impact strength of GFRP.

increases with the glass content and peaks at either 20 or 30 wt% depending on the fiber diameter. Unnotched Izod impact strength decreases with more glass fibers in the composites. A more comprehensive study was published by researchers at Ferro Corporation [123]. Melt flow decreases with more glass fibers from 0 to 40% in both homo- and copolymer polypropylene. Tensile/flexural strengths and tensile/tangent modulus values increase linearly with the glass content. Manual and instrumented notched Izod impact strengths peak at 30 wt% ash content. Unnotched Izod is insensitive to the glass content. Reduction of melt viscosity and fiber length distribution with the increase of glass fiber concentration is also reported by Barbosa and Kenny [124].

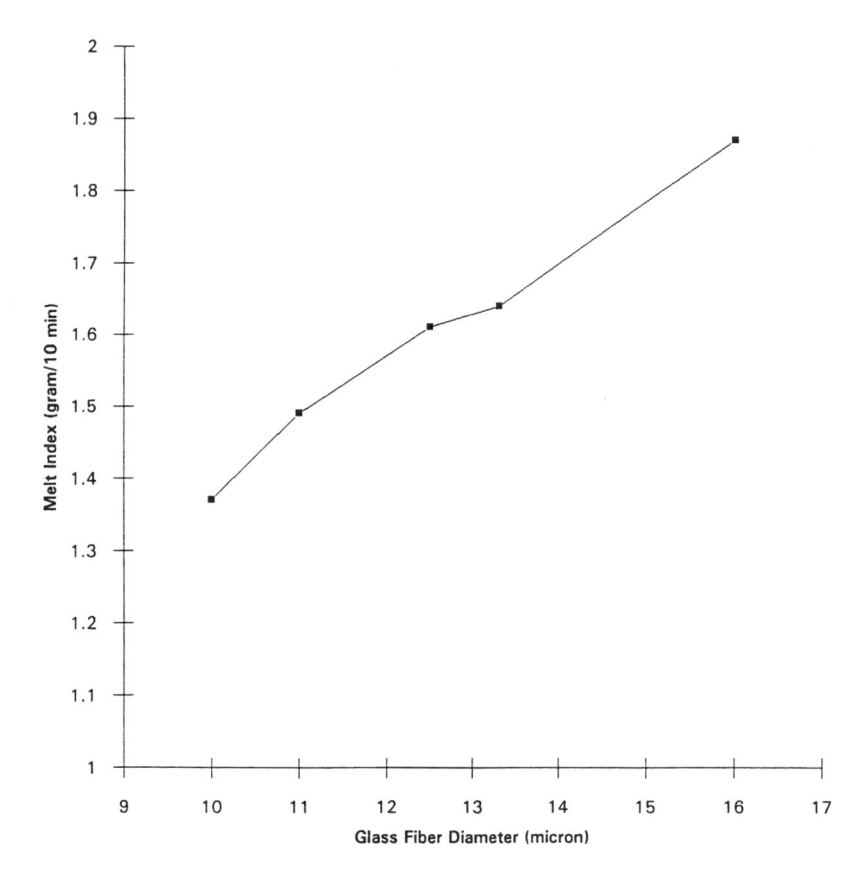

**FIGURE 9.14** Effect of glass fiber diameter on the melt index of 20 wt% GFRP.

The higher impact strength versus glass content is also reported by Karger-Kocsis [125] in both T- and L-notched specimens. This contributes to the fact that the fiber avoidance cracking path, mostly due to fiber debonding and pull-out process, increases with increasing glass content.

More torque is required to compound glass fiber in polypropylene than the neat resin itself because of higher viscosity with the filled resin [126]. With as little as 10% w/w of glass fibers, the initial torque increases from 10 to 20 Nm. The torque decreases rapidly in the first few minutes of mixing, an indication of fiber breakage. For 10 and 25 wt% glass content, the torque levels off after 5 min of mixing, but for 40 and 60 wt% glass contents, the torque continues to decrease with time. This is because higher glass content leads to higher melt temperature. Thermal degradation of the polypropylene and/or size is progressively removed from the fibers, providing internal lubrication. The viscosity of the compound

# Glass Fiber–Reinforced Polypropylene

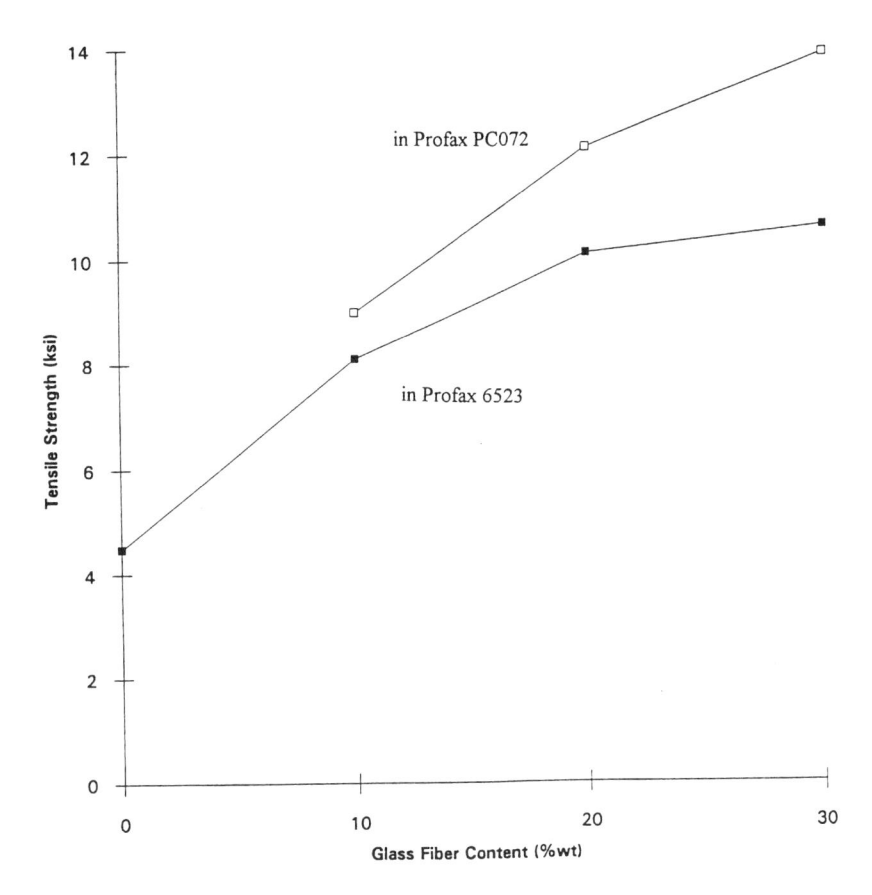

**FIGURE 9.15** Tensile strength vs. glass contents in homo- and coupled-PP composite systems.

increases with any quantity of glass fibers [127]. The differential is larger at low shear rate and diminishes at higher shear rate. The pressure drop in the mold also increases with the increase of glass content [128].

In the glass fiber–reinforced polypropylene mat (GMT), the tensile and notched Charpy impact strengths increase linearly with the glass content up to 60% by weight [121]. The addition of glass fibers severely reduces the strain to failure. Higher fiber content results in a higher concentration of failure initiation sites and thus a lower tensile strain. Similarly, the flexural strength also increases with the glass content linearly up to 25 wt%. There is a steep rise in the flexural strength between 25% and 30% fiber content. The flexural strength at 60% fiber content is actually lower than that at 50 wt% because of high void content.

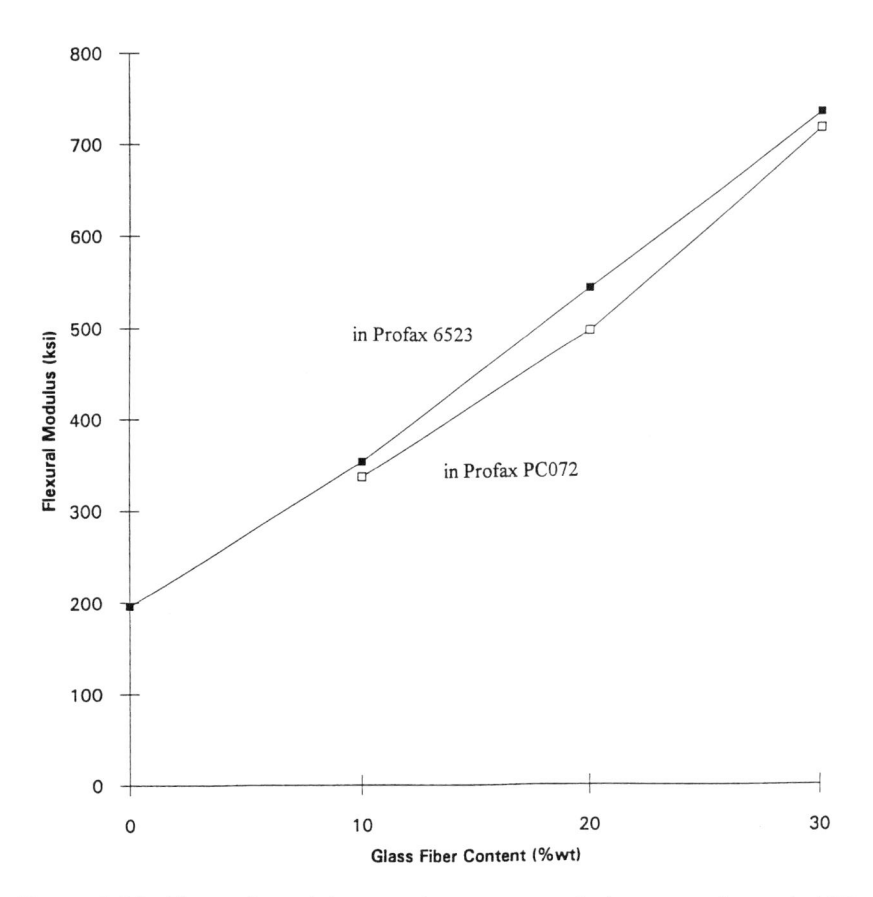

**FIGURE 9.16** Flexural modulus vs. glass contents in homo- and coupled-PP composite systems.

## 9.3.2 Extrusion Compounding of Glass-Reinforced Polypropylene

To get the maximal reinforcement properties out of the glass fibers in the polypropylene composites, the resin needs to be modified with coupling agents. The PP chemical coupling agent technology has been discussed in detail in Chapter 3. Compounders can purchase commercially available coupling agents and mix them in small quantity (1–10 wt%) with polypropylene during the extrusion with glass fibers. Many PP compounders also develop their own proprietary coupling technology. In this case the additive (peroxide, acrylic acid, or maleic anhydride) is added in the feeding and mixing sections of the extruder before the introduction of glass fibers into the melt [129]. The reactive extrusion of coupled polypropylene with glass fibers has been reviewed in Chapters 3 and 10. Even so, the glass fibers receive great damage in the extruder. In Table 9.13, the average fiber length in the compound is down to 0.03 in. from the original 0.125–0.25 in. Unlike the initial input length, there is a distribution of fiber length in the compound or the composite. The fiber length distribution (FLD) depends on many factors in the process and at the end on the measurement techniques. Assuming 0.03 in. in length and 13.3 µm in diameter, the total number of glass fibers in one single $\frac{1}{8}$-in. pellet with 10% by weight of glass fiber is about 7000 individual fibers. The number of fibers manually measured to obtain the FLD is around 200–500 counts. Using image analysis, up to 3000 fibers can be measured within a reasonable time [115,116,130]. High-temperature ashing is the simplest method to retrieve glass fiber fragments, but it makes glass fiber very brittle and subject to further damage during handling. The use of solvent to dissolve the organic polymer is very tedious. Low-temperature plasma oxidation process is the preferred method. A mixture of methyltrimethoxysilane with Cat-X at pH 4 provides good dispersion for fiber length measurement.

The temperature curve of the extruder depends on the viscosity of the polymer. Being more viscous than nylon or polybutylene terephthalate (PBT), polypropylene resin heats up much faster than the latter two in the compression zone [131]. The introduction of glass fibers downstream reduces the melt temperature by 5°C. The viscosity of the compound increases once glass fibers are incorporated, and this leads to a sharp rise of melt temperature in the metering zone. The glass FLD at various sections in a single-screw extruder showed rapid reduction at the end of the melting zone [117]. When the 3-mm chopped fiber was preblended and fed into the resin hopper, the number average fiber length only decreased to 2.73 and 2.30 mm in the first two sections in the compression zone. This indicates that fiber bending between solid polymer pellets does not play a large role in fiber fracture. Real breakage to the fibers takes place in the melting zone. The number average fiber length is 1.58 and 0.75 mm in sections 3 and 4, respectively. At the end of section 4, there is no 3-mm fiber left. At sections 5 and 6 in the mixing zone, the number average fiber length is 0.55 and 0.425 mm, respectively. The mean values of the two are indifferent, indicating that not much additional damage occurs in the last section. The only difference is the absence of any percentage of fibers longer than 0.15 mm.

The fiber length degradation as a function of mixing time from 0 to 40 min in the polypropylene melt was carried out on the plasticorder torque rheometer by Fisa [126]. The transformation of the 3-mm input strand length into short broken fragments in the final compound occurred via two processes overlapping in time. First, the strand bundles are filamentized into individual fibers. The study showed a significant fiber degradation at this stage. The higher the glass concentration, the shorter time it happens. The individual fibers are further broken down into small fragments as a result of shear stress in the melt. This fiber–melt interaction was confirmed by the continuing decrease of fiber length with mixing time at very low glass concentration. Increasing PP resin viscosity shows a strong decline of fiber length at 40% glass content. With low fiber concentration (2 wt%) and high melt flow rate PP (12 g/10 min), only half of the fibers break in 2 min. The effect of mixer rotor speed from 15 to 90 rpm on the final fiber length is negligible for 40 wt% glass content. At 2 wt% glass content, the fiber length decreases with the increase of rotor speed. It is also confirmed that there is no effect of the screw speed on the final fiber length because fibers are all degraded to its final length, given enough mixing time [117].

In the twin-screw extruder, the kneading disk region is the most important for compounding efficiency. The effects of screw speed, barrel temperature, and fill factor on torque, pressure, and polypropylene resin temperature were measured in this region in a corotating twin-screw extruder by Shimizu et al. [132,133]. Resin temperature increases with the rotational speed and barrel temperature but is independent of fill factor. Torque and pressure increase with higher fill factor but decrease with higher barrel temperature. When compounding 10 wt% of glass fibers in polypropylene, the average fiber length decreases with the residence and rotation speed. In only 60 sec all fibers have been fractured. The fiber length distribution changes very little with the barrel temperature, fill factor, and amount of glass fibers from 5 to 15 wt%. The combination of rotational speed, residence time, and shear stress is the determining condition to control the final fiber length after compounding.

### 9.3.3 Injection Molding of GFRP Composites

The advances of injection molding of fiber-reinforced engineering thermoplastics have been critically reviewed in recent articles [134–136]. Gas injection molding, coinjection molding, lost/fusible core injection molding, and push–pull injection molding are cited as the most significant developments of the last two decades. The lost/fusible core technology has gained much publicity due to a dedicated process for dedicated parts, such as air intake manifold for automotive application [137]. The push–pull injection molding overcomes the biggest drawback of fiber-reinforced plastics with poor weldline strength. When the glass fiber–reinforced polypropylene compound is melted and injection molded into the final part, the glass fibers are subject to another degradation in the process. Higher screw speed and screw-back time caused greater damage to the glass fiber and lower tensile and flexural strengths of the composite [138]. Higher barrel temperature decreases flexural strength and modulus, Izod impact strength, mold shrinkage, $\beta$-crystal content, and thickness of skin layer of the test specimen [139]. Higher injection pressure lowers the tensile strength, but higher injection speed increases the orientation of fibers and improves the mechanical properties [140]. Preheating

## Glass Fiber–Reinforced Polypropylene

the compounds can preserve fiber length during the injection molding process [141].

The melt flow during cavity filling in the injection molding process induces the orientation effect of glass fibers. In planes parallel to the cavity wall, the skin layers contain the fibers aligned to the flow direction. The core layer in the center of the part contains randomly oriented fibers. Many articles have been published that measured planar and three-dimensional orientation of glass fibers in injection-molded parts [142–145]. For a rectangular plaque, the dependence of fiber orientation on glass content and injection speed was studied [146]. The fibers become more aligned as the flow progresses from the gate to the midstream and until the flow front is reached. At the flow front, the fibers become more random because of the fountain effect. The fiber orientations are similar for plaques 1 and 2.54 mm thick [147]. However, at 5.08 mm the flow is no longer quasi-unidirectional. At high injection speed into this thick plaque, all fibers are aligned in the $z$ direction, perpendicular to the flow plane. The ultimate tensile strength decreases with higher concentration of core region in the specimen. The fracture toughness of 30% glass fiber–reinforced polypropylene is investigated with six different gate geometries and by the specimen locations [148]. The toughness increases with the direction of melt flow. The highest value is at the end of cavity wall due to the presence of a higher percentage of fibers aligning normal to the fill direction. The corner gate and off-center angle gate give higher toughness values than center edge–gated molding, which is a standard practice for any test bar. The fracture toughness in the transverse direction is higher from the specimen closer to the cavity wall than those from the center of the plaque. This is due to the fibers that align parallel to the walls. In the transverse direction, the toughness is lower from corner gate and off-center angle gate, and all values are higher than in the flow direction. The tensile modulus of injection-molded plaque has the same fiber orientation effect with 20 and 40 wt% glass fiber in polypropylene [149]. In the normal direction, the tensile modulus is strongest at the end of fill. For specimens parallel to the injection direction, the center of the plaques has the lowest tensile modulus, showing again a strong orientation at the cavity wall. The two-dimensional mapping of tensile strength was also presented in the same conference and is discussed in detail in Chapter 12 [150].

It is a well-known fact that weldlines are unavoidable in most injection-molded parts of even moderate complexity. The presence of a second material and the orientation effect of glass fibers in the melt flow exaggerates the weakness of the molded parts. Because of the formation of fountain flow, the glass fibers tend to align perpendicular to the flow front. The weldline strength is independent of cavity shape and part thickness [151]. The fiber content and fiber length distribution have predominate effects on the weldline strength. Table 9.15 compares the weldline strengths of short and long glass fiber–reinforced polypropylene. The use of more glass fiber supposedly improves the composite

TABLE 9.15 Weldline Strength of Short and Long Glass Fiber–Reinforced Polypropylene Composites

| Glass content (%) | Pellet form | No weldline (MPa) | With weldline (MPa) | Weldline factor (%) |
|---|---|---|---|---|
| 0  | —           | 35 | 30 | 86 |
| 20 | Short fiber | 65 | 30 | 47 |
| 30 | Short fiber | 67 | 23 | 34 |
| 30 | Long fiber  | 70 | 17 | 24 |
| 40 | Long fiber  | 66 | 12 | 18 |
| 50 | Long fiber  | 64 | 10 | 16 |

Source: Ref. 152, © 1995, with permission of John Wiley & Sons, Inc.

strength. However, if there is a weldline present in the finished part, the ultimate failure strength is actually weaker with higher glass content.

Adequate venting, an increase of holding time, appropriate gating, and sequential filling can improve weldline strength [152,153]. The push–pull type of injection molding process offers the best solution to this problem. The principle of this process is to oscillate the melt to provide better entanglement of the two flow fronts. The injection molding machine has two injection units that are controlled independently of each other. The cavity is filled simultaneously by both units via two separate gates. After the flow fronts meet, the control electronic permits one unit at a time to be retracted, whereas the second continues to inject the molten material and reverses the flow of the molten core of the other front. Unfortunately, the skin layers have solidified before and during the pressure cycling. Continuous improvement in process technology is necessary to further improve the weldline strength.

## 9.4 MARKET AND APPLICATION OF GFRP

### 9.4.1 Short Glass Fiber–Reinforced Polypropylene

In 1996 [154], the composite industries registered more than 3 billion pounds production volume and the sales of glass fiber reinforcement exceeded more than 1 billion pounds for the third consecutive year in the United States (Fig. 9.17). Of the 1 billion pounds, 280 million pounds was chopped strands for thermoplastic reinforcement. In the United States, 10% of the chopped strands were used in polypropylene resin matrix. In 1995, 470,000 tons of glass fibers was processed by the plastics industry in Europe [155]. A total of 130,000 tons was consumed by thermoplastics compounders. The economic recovery since 1991 in the United States has led to a tremendous growth of thermoplastic composites [156] (Fig. 9.18). In the last 20 years, the transportation segment leads in the growth

# Glass Fiber–Reinforced Polypropylene 357

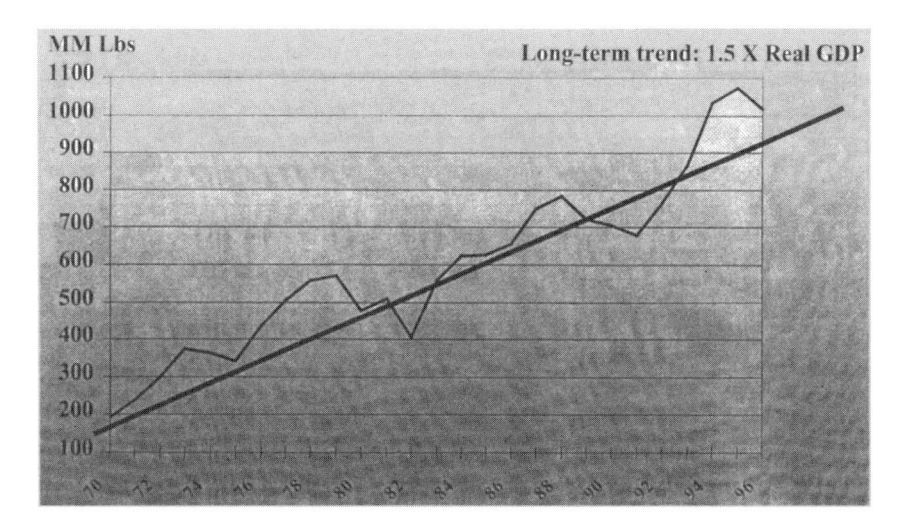

**FIGURE 9.17** The FRP shipment in the United States from 1970 to 1996. (Reproduced from Ref. 154 with permission.)

of GFRP [154] (Fig. 9.19). The sale of glass fibers to the marine market has gradually decreased. The GFRP market of Japan in 1990 is less than half of that of the United States [157]. A large part (51%) of the GFRP consumption in Japan is used for building, construction, and related areas. The usage in marine and transportation is very small at 8.8% and 5.5%, respectively. In both the United

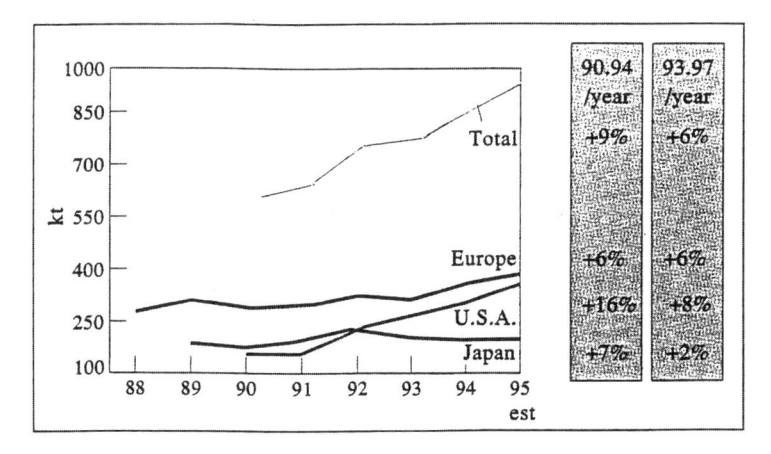

**FIGURE 9.18** Worldwide composites market for thermoplastics. (Reproduced from Ref. 156 with permission.)

(a)

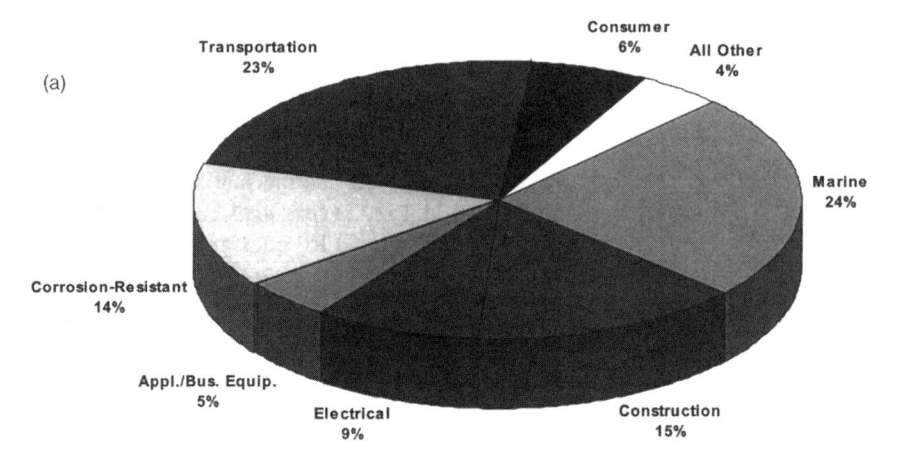

- Transportation 23%
- Consumer 6%
- All Other 4%
- Marine 24%
- Construction 15%
- Electrical 9%
- Appl./Bus. Equip. 5%
- Corrosion-Resistant 14%

**TOTAL MARKET SIZE: 556.3 Million Pounds**

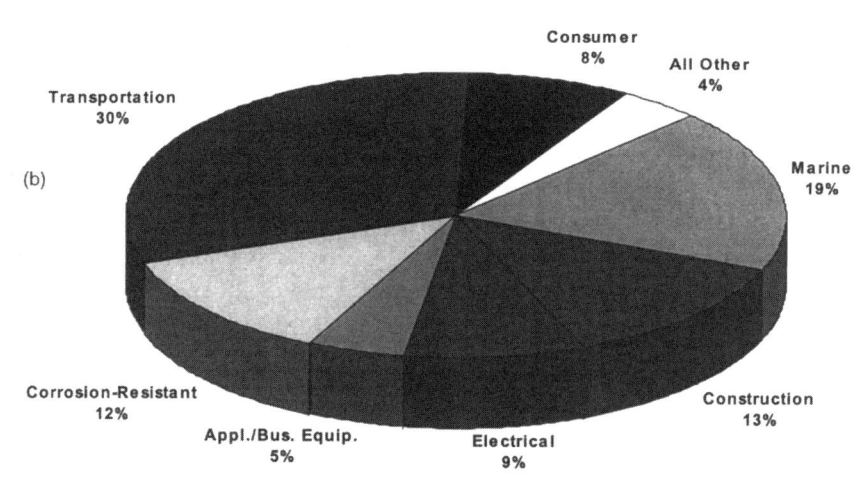

(b)

- Transportation 30%
- Consumer 8%
- All Other 4%
- Marine 19%
- Construction 13%
- Electrical 9%
- Appl./Bus. Equip. 5%
- Corrosion-Resistant 12%

**TOTAL MARKET SIZE: 784.8 Million Pounds**

FIGURE 9.19 The changing market FRP industry in the United States from (a) 1978, (b) 1988, to (c) 1996. (Reproduced from Ref. 154 with permission.)

States and Europe, the leading reinforced thermoplastics are polyamides (PA6,6 and PA6), thermoplastic polyesters (PBT and PET), and polypropylene. The sales of GFRP in Europe are stagnating because of the reduced price difference between GFRP and glass fiber–reinforced polyamide (GFRPA) in recent years [158]. Glass fiber–reinforced polypropylene has certain advantages over other

## Glass Fiber–Reinforced Polypropylene

(c)

Consumer 7%
All Other 5%
Transp. 37%
Marine 12%
Construction 12%
Electrical 12%
Appl./Bus. Equip. 4%
Corrosion-Resistant 11%

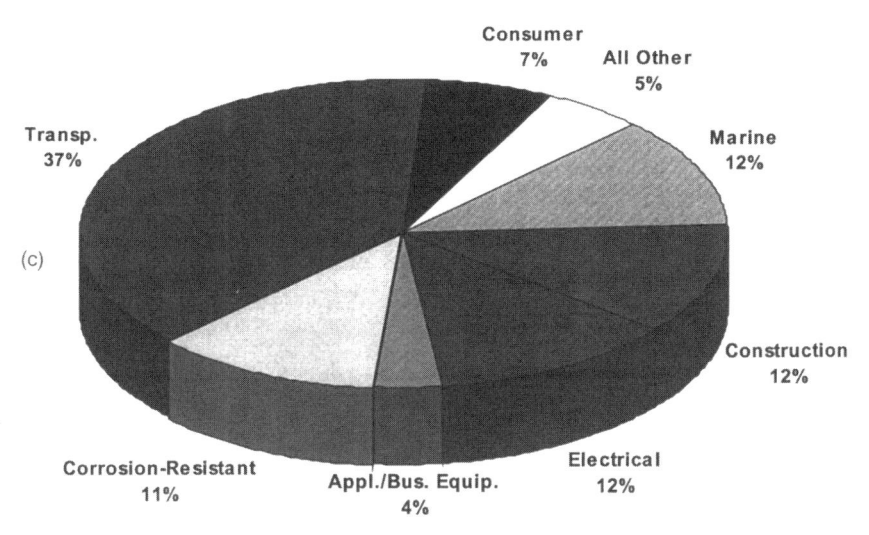

**TOTAL MARKET SIZE: 1,017 Million Pounds**

FIGURE 9.19 (*continued*)

engineering plastics, the main ones being low cost and low specific gravity. The limitation is its thermal stability.

The overview of the polypropylene composite industry was discussed in Chapter 1. Several new trends can enhance the GFRP properties and open doors to new applications. The breakthrough in metallocene catalysts can make possible new combinations of properties. Metallocene technology can also be applied to impact modifiers to further improve the impact strength of GFRP by 1 ft-lb/in. [159]. Polyolefin alloys is another field under intensive study. The mechanical properties have been optimized for glass fiber–reinforced PP/low-density polyethylene (LDPE) [160,161]. The rheological and mechanical properties of glass fiber–reinforced PP/ethylene-propylene-diene monomer (EPDM) blends are published by Gupta and colleagues [162,163]. The blends of polypropylene and polyamide are particularly interesting because of the low cost of PP and mechanical and thermal properties of PA. The compatibility of PP and PA has been investigated [164]. Lower molecular weight nylon favors the compatibility in the GFRP/PA composites [165]. At 20 wt% glass fiber and a 50 : 50 ratio of PP to PA, the composite properties depend strongly on the maleic anhydride content of the compatibilizer [166]. However, it is not known if a PP or a PA glass fiber was used in the study. The most comprehensive work on the GFRP/PA system using either a PP-compatible or a PA-compatible glass fiber was done by Perwuelz et al. [167]. With a PP-compatible glass fiber, the heat deflection temperature (HDT) does not increase until PP is less than 30% of the resin matrix. Also, at 30% PP content, the composite has the lowest ultimate tensile stress. The Izod and instrumented impact strengths decrease with lower PP content. With a PA-compatible fiberglass, the HDT rises sharply between 0 and 50% PA content. The ultimate tensile strength increases linearly with the decrease of PP content. The notched Izod impact strength is almost independent of the PP/PA ratio. At 30% PP, the composite has the highest unnotched Izod and instrumented impact strength.

Glass fiber–reinforced isotactic polypropylene is considered the first-generation GFRP [168]. The composite containing metallocene-based resin is considered the second-generation GFRP because of higher impact and flexibility. Glass fiber–reinforced PP/PA alloy is the third-generation GFRP for its superior strength and toughness. A growing application of thermoplastic resins is in the blow molding of engineering parts. The best known example is the high-density polyethylene (HDPE) fuel tank. This product impressively demonstrates the complicated shape that can be obtained by blow molding. Pegugorm uses blow-molded glass-reinforced PP for spoilers application [169]. Shredder housing can be made from Hostalen PPG 1022 through a double-walled blow molding process [170]. Another process gaining popularity is the in-line extrusion. The injection of a liquid lubricant at the surface of a diverging die can change the fiber orientation in the glass fiber–reinforced polypropylene tube and increases the resistance of internal pressure of the tube [171].

Continuous improvement on the performance of glass fiber–reinforced polypropylene has been gaining ground to replace metal and other more expensive composite materials. Examples can be found in the weed trimmer from Black and Decker, the housing and adapter plate for pumps of Hayward pool products, air-cleaner housings, and trays for light trucks [172]. The impact grade is suitable for hockey skate components, automotive fender liners, and ammunition boxes.

### 9.4.2 Long Glass Fiber–Reinforced Polypropylene

The concept of long-fiber compounds is not new. Because the mechanical properties of the discontinuous fiber–reinforced composites depends on the effective length of reinforcing fibers, every attempt is made to preserve the fiber length during composite fabrication. The Verton products were introduced by ICI in the United Kingdom in the mid-1980s. In the United States they are now sold by LNP Engineering Plastics.

Most long glass fiber–reinforced polypropylene (LGFRP) on the market is produced by melt impregnation process. A simplified diagram is given in Fig. 9.20a [173]. Long continuous glass rovings are pulled through a crosshead die that is filled with the molten polymer from an extruder. After the glass

# Glass Fiber–Reinforced Polypropylene

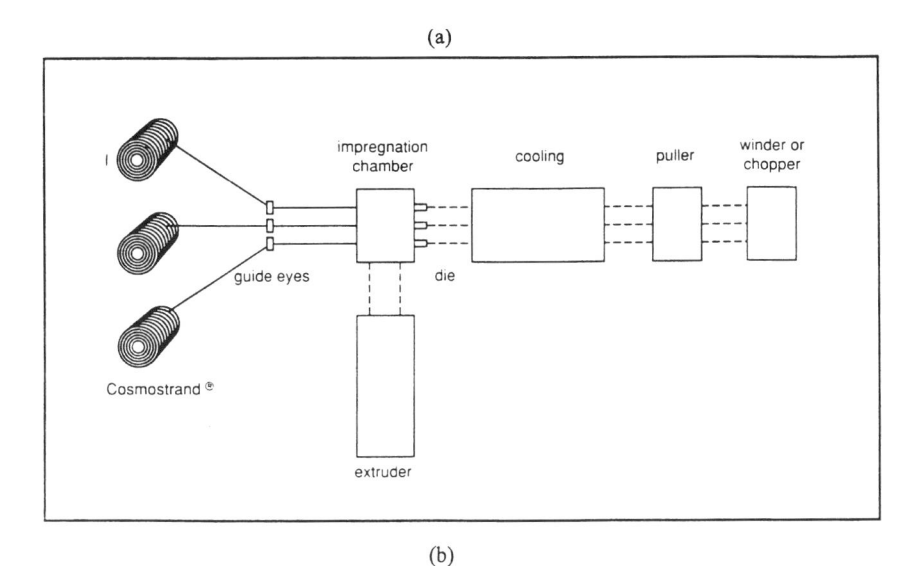

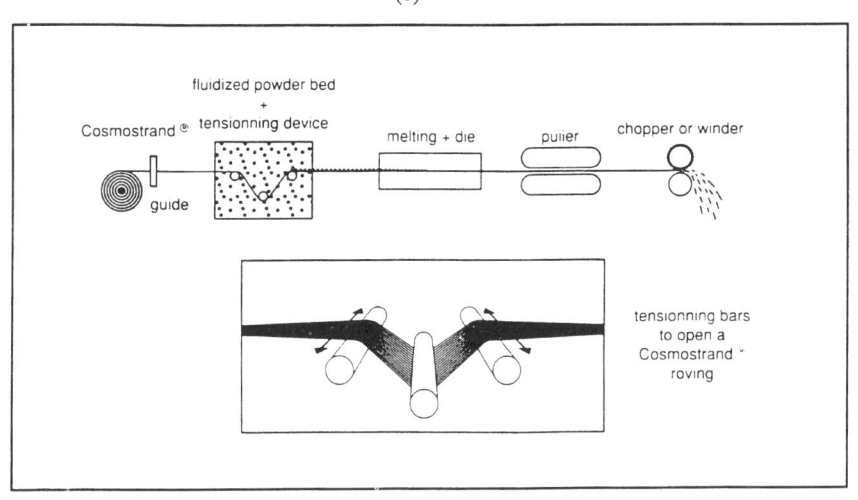

**FIGURE 9.20** Illustration of long glass fiber–reinforced polypropylene compounds by (a) melt and (b) powder impregnation process. (Reproduced from Ref. 173 with permission.)

filaments are impregnated with the resin, they are cooled and chopped into $\frac{1}{4}$- or $\frac{1}{2}$-in. pellets. Because the fiber length is the chopped length of the pellets, the LGFRP compounds avoid the fiber degradation associated with extrusion compounding of SGFRP. Typical glass rovings, such as Vetrotex P319, are 17 µm in diameter and 1200 Tex in linear weight [174]. They are produced by the direct roving process as discussed in Section 9.1.2. Therefore, each roving package contains several thousand individual filaments depending on the size of bushing each company uses.

Although this process resembles the wire coating process, the key for good LGFRP pellets is to fully impregnate all several thousand individual filaments, and every company regards their process as a proprietary technology. Many patents have been issued [175–181], but the detail mechanism has rarely been discussed in publications. Opening up the glass roving for resin penetration can be achieved by the concave and convex pins inside the crosshead die [182,183] (Fig. 9.21). Longer impregnation time and contact surface, higher melt temperature and melt flow of PP, more pins in the die, and higher roving tension and vacuum all improve the degree of impregnation [184]. Preheating the rods has very little effect, and higher isotropic pressure of molten PP actually has a negative effect on the degree of impregnation. Longer residence time is counterproductive for the already slow process. Productivity improvement has been claimed in a recent article with higher line speed [185].

The degradation of glass fiber in the screw is less for the long fiber pellets. Interesting work was done to freeze the screw, and the fiber length distribution was measured on each channel [186,187]. With short-fiber pellets, most fiber degradation occurs between the 7th and the 12th screw channels. There is very little change in the FLD after that. The fiber attrition is also pronounced between the 7th and 12th screw channels for the long-fiber pellets. However, a small but

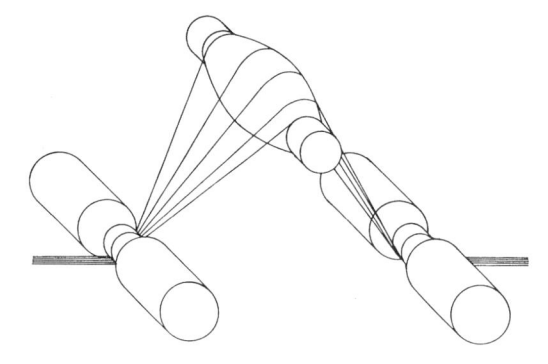

FIGURE 9.21 Concave and convex pins in the crosshead die for maximal spreading of filaments. (Reproduced from Ref. 182 with permission.)

significant portion of fibers is still intact at the original length. The reduction of FLD continues in the compression zone between the 13th and 19th zone.

Another factor to the maximum usage of LGFRP is in the injection molding process. Higher processing temperature and generous gate configuration are recommended. The barrel temperature should be 10–20°C higher than the comparable short-fiber compounds [188]. Round or fan gates with large cross-section are best with the gate diameter no less than 1.5–2 mm. The diameter of the runner should be 1.5 times the wall thickness of the molded parts. Longer fibers from the larger gate are found mainly in the core section of the molded plate [189]. This effect arises from the shear associated with the molding-filling process. The higher fiber content allows longer fibers to be retained. This would suggest that the flow properties deviate more from Newtonian toward pseudoplastic behavior, so the retained fiber lengths increase. The level of fiber attrition in LGFRP is consistently less than for nylon resin matrix. The fatigue behavior of long glass fiber–reinforced alloys of PP and PA is lower than that of LGFRP or LGFRPA [190]. The decrease is large when nylon is blended with PP and smaller when PP is blended with nylon. Comparing the composite properties, the long-fiber polypropylene compounds can easily double the flexural modulus and Izod impact strengths over the corresponding short-fiber polypropylene compounds [191]. The improvement of falling-dart impact energy by long glass fibers is, however, only 20–40% [192]. The long fiber length plays less of a role in this multiaxial impact loading condition because other factors such as resin ductility, composite stiffness, and compressive properties contribute to the total impact energy to a greater extent.

Long-fiber pellets or prepregs can also be produced by the powder impregnation process [193–195], as shown in Fig. 9.20b. In North America, the materials are available from Baycomp. Examples of LGFRP application include Proton battery tray, Quicksilver prop wrench, Mercury marine drain plug, Pompanette boat ladder, Fill-rite dry disconnect, Dosmatic metering pump, Ryobi drive wheel [196], and the reverse cap of the handle of Snap-on Tools [197]. Gas injection molding has also been applied to the molding of LGFRPs to produce gear levels [198]. In Vetrotex CertainTeed, when ergonomic requirements call for a light-weight load-bearing tray, a 30 wt% long glass fiber–reinforced heat-stabilized polypropylene is selected (Fig. 9.22). Previously, a thermoset BMC tray was used. The LGFRP molded tray is 50% lighter without sacrificing any other service parameter.

### 9.4.3 Stampable Polypropylene Composites

The joint venture between General Electric and PPG in 1986 to form the Azdel Corporation has created a unique composite market of its own. The most publicized application is for bumper beams [198]. Azdel composite sheets were

**FIGURE 9.22** Composite tray molded with 30% long glass fiber–reinforced polypropylene. (Courtesy of Vetrotex CertainTeed.)

first introduced in the Corvette bumper system in 1984. Honda began to use it on the 1990 models [199]. In 1992, the worldwide sales of glass mat thermoplastic was projected at more than 60,000 tons [200]. Half of the sales were in the United States (32,700 tons). European sales, at 15,400 tons, were slightly more than that in Japan at 12,000 tons. In addition, Europe and especially Japan were projected to experience higher growth than the United States.

In the United States, Exxon was the only other GMT producer until their merge with Azdel in 1997 [201]. In Europe, Symalit, a subsidiary of Shell, and Elastogran Kunststoff-Technik, a part of BASF, have 40% each of the European market share [202]. Although Azdel expanded the capacity into Europe [203], BASF is conducting a feasibility study regarding entry into the U.S. market [204]. By their estimate, the worldwide GMT market was 198 million pounds in 1996, with U.S. consumption at 72 million pounds.

The Azdel process is shown in Fig. 9.23a. The most detailed study on GMT was done by Stokes [205–208]. A single 230 × 405 mm P100 plaque was divided into 288 specimens for testing. There is a 30% variation in density, ranging from 1.11 to 1.32 g/mL. The Young's modulus can vary by a factor of 2, ranging from 476 to 1411 ksi. The left and right modulus also exhibit large-scale fluctuations along a specimen. An appropriate mean modulus would be a face modulus that measures the average tensile response across the 12.7-mm face width. The tensile

# Glass Fiber–Reinforced Polypropylene

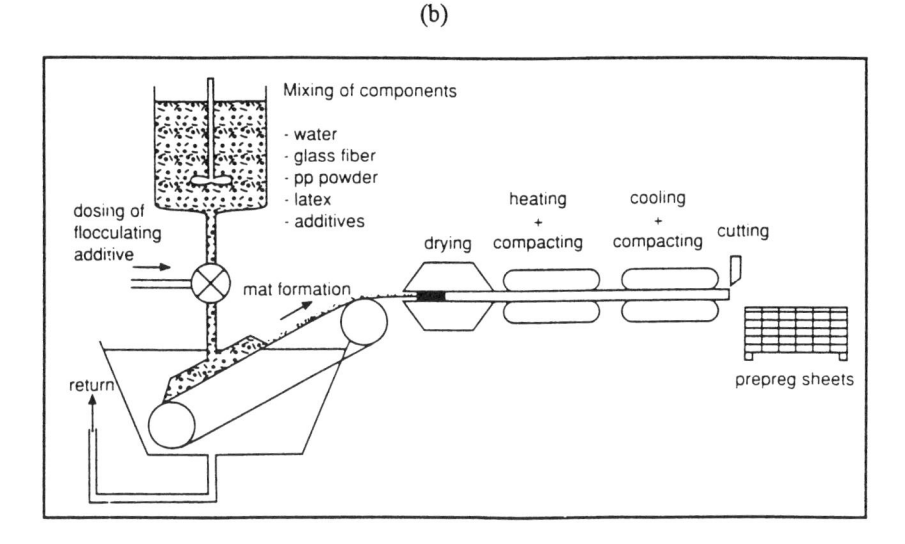

**FIGURE 9.23** Stampable glass fiber mat–reinforced polypropylene sheet by (a) melt impregnation and (b) wet/paper machine process. (Reproduced from Ref. 173 with permission.)

strength can vary between 12 and 18 ksi. The sheet is stiffer and stronger in the cross-machine direction than in the machine direction. The general guidelines of processing conditions are to preheat materials at 199–227°C [209]. The press speed should be 8.5–25.4 mm/sec and the pressure should be 12.8 MPa to obtain optimal performance of a bumper beam. The new generation of GFRP bumpers will feature thinner walls, multiple fixation point, and zero gap [210]. Higher productivity can come from longer service life and lower costs for molds and from reducing processing temperature and shorter mold cycle. Improved paintability with water-based paints is required. The incorporation of polyether amines into polypropylene can provide significantly better paintability [211]. The Azdel Plus combines the GMT with unidirectional roving for additional strength in the longitudinal direction [212]. It is used in the 1997 Firebird/Camaro rear bumper beam to replace the RIM. Adding a low molecular weight polypropylene coupling agent can further improve the strength of the GMT [213]. The tensile and flexural strengths peak at 7 wt% of the coupling content. However, the Charpy impact strength decreases with the higher concentration of the coupling agent. Figure 9.24 compares the stiffness and toughness of glass fiber–reinforced thermoplastics by various methods. GMT made of continuous fiberglass strand mat offers superior impact strength.

The schematic diagram of the Taffen process is shown in Fig. 9.23b. Exxon purchased the patent from Arjomari-Prioux and built the plant in Lynchburg, Virginia in 1991 [214]. At the time it was sold to Azdel, it had an annual capacity of 16.5 million pounds. The process is simulated by Battelle Lab with three different grades of polypropylene [215]. The impact strength of PP-based GMT compares favorably with PET, PPS, and nylon-6,6. A structural battery tray for GM's EV1 electric vehicles weighing 33 pounds is molded from Taffen [216,217]. The part won the Grand Award for innovative plastics use by the

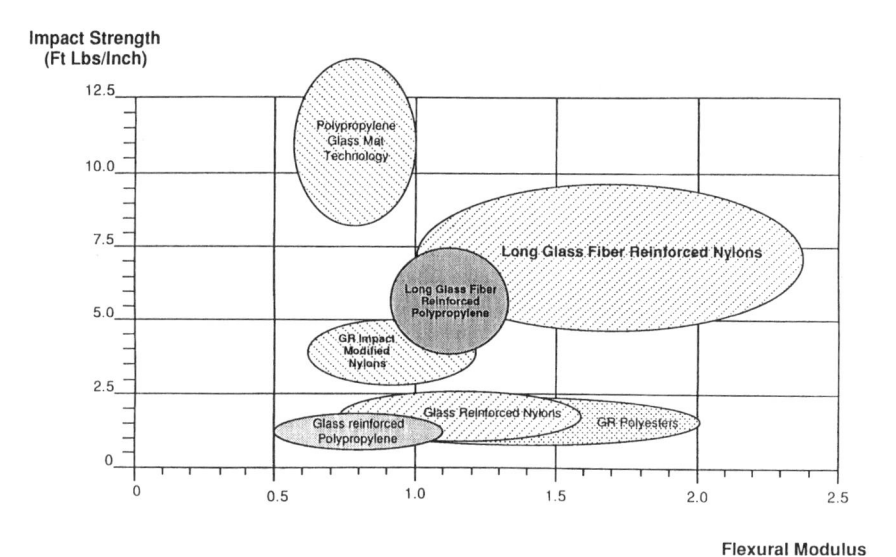

FIGURE 9.24 Crystalline-reinforced engineering thermoplastics. Stiffness–toughness positioning chart. (Courtesy of DSM.)

# Glass Fiber–Reinforced Polypropylene

SPE in 1996. It is also possible to run glass rovings through a slurry bath and consolidate them into a unidirectional prepreg [218].

Another interesting long glass fiber–reinforced thermoplastic BMC operation was developed by Composite Products [219,220]. Basically, it is an on-site extrusion and molding operation. The long glass fibers, up to 50 mm long, are fed downstream to the extruder. The compound exits the open die as a log and is stationed in the accumulator until compression molded. Thermoplastic SMC process has been recently patented by ThermoComp [221].

The most exciting and revolutionary development of glass fiber–reinforced polypropylene composite was announced in 1995 [222]. A unique manufacturing process combines glass and polypropylene fibers as they are being fiberized into commingle rovings [223–226]. The patented process was developed by Vetrotex in Europe and commercialized in the United States by Vetrotex CertainTeed under the trade name Twintex. Uniform commingling of the fiberglass and polypropylene fibers in Twintex minimizes the distance the molten PP is required to flow and allows the material to be molded under very low pressure. The same property allows the PP to wet the glass fibers even at very high percentages of glass. The weight percent glass in PP is typically 60–70%. The commingled rovings can be woven into fabrics of many variations including plain, twill, or satin weave; knitted; unidirectional; oriented; and three-dimensional. The comolding of Twintex and a GMT stampable sheet or a thermoplastic BMC combines the flow properties of the random fiber molding materials with the mechanical properties of the continuous fiberglass. The synergistic effect is shown in Table 9.16 [227]. The applications of Twintex include firefighters' helmets and door panels [228] (Fig. 9.25). The direct composite process is also patented by other glass fiber manufacturers [229,230].

## 9.4.4 Recycling of GFRP

In the United States, plastics constitute roughly 7% by weight or about 16% by volume of all municipal solid waste (MSW) [231]. The major method for MSW disposal is landfilling, which accounts for 73%. The rest is incinerated (14%), recycled (11%), or composted (2%). Plastics recycling is a higher priority in Europe and Japan due to the shortage of landfill capacity. In 1990, Japan incinerated 65% of all MSW; 23% was landfilled and 5% was recycled. The reinforced plastics account for 4% of all plastics production in the United States. In the automotive segment, 56% of the composites are thermoset matrix whereas 44% are thermoplastic. The advantage of thermoplastic over thermoset plastics is in the ability to be recycled. However, the glass fiber–reinforced thermoplastics suffer the same fade as thermoset material due to continuous degradation of the reinforcing fibers from repeated processing. The use of virgin glass fibers in waste PP to make prime compounds is more feasible [232,233]. Compression

TABLE 9.16 Mechanical Properties of Twintex, GMT, and Comolded Panels

| | \multicolumn{6}{c}{Laminate construction} |
|---|---|---|---|---|---|---|
| | GMT | TwintexTM | G-T | G-T-G | T-G-T | T-G-T |
| Ply thickness, mm | 4 | 4 | 2/2 | 1/2/1 | 1/2/1 | 0.5/2/0.5 |
| Glass content, wt% | 40 | 60 | 52 | 52 | 50 | 42 |
| Glass content, vol% | 19 | 34 | 27 | 27 | 26 | 20 |
| Flexural strength, MPa | 147 | 300 | 200 | 200 | 280 | 204 |
| Flexural modulus, GPa | 6 | 12 | 8.7 | 7.5 | 11.3 | 9.3 |
| Elongation, % | 3.3 | 2.5 | 2.7 | 3.5 | 3 | 4.5 |
| Unnotched Charpy, $kJ/m^2$ | 82 | 220 | 185 | 130 | 143 | 141 |
| Unnotched Charpy, $J/cm^3$ | 2 | 8.9 | 4.8 | 3.4 | 6.8 | 4.4 |
| Notched Izod, $kJ/m^2$ | 76 | 224 | 125 | 118 | 118 | 139 |
| Falling-dart impact, J | 35 | 100 | — | — | 45 | — |
| Tensile strength, MPa | 70 | 240 | — | — | 185 | 141 |
| Tensile modulus, GPa | 6.5 | 13 | — | — | 9 | 7.5 |

*Source:* Ref. 227.

molding is less damaging to the fibers; 95% of Huffy Sports' product line of home backboard systems are made of recycled PP and reground glass fibers [234]. They are mixed in the extruder, but the board is compression molded. Several recycling processes are found to be useful. The pyrolysis of composites involves decomposition of organic matrix at high temperature in the absence of oxygen. The reaction products are olefins and other hydrocarbons that can be

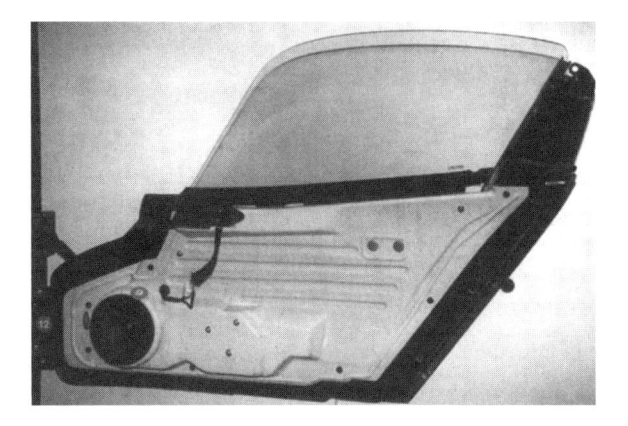

FIGURE 9.25 Lightweight door module with compression-molded Twintex. (Courtesy of Vetrotex CertainTeed.)

used as fuels or as feedstocks for petrochemicals. Recycling through hydrolysis is better suited for recovering the monomers of resins manufactured by polyaddition and polycondensation reactions, such as polyethylene, polyester, and polyamides. Regrind as filler is the method of choice for thermoset composites. Although it is technically feasible, the cost is not competitive to the virgin material. Ford's recycling projects make the automaker the world's largest user of postconsumer nylon [235]. However, it is still a distant goal to reclaim 100% of the glass fiber–reinforced composite materials.

## ACKNOWLEDGMENT

I wish to express my gratitude to Messrs. Michel Arpin and Jean-Philippe Gasca of Vetrotex International and Paul Lucas of Vetrotex France for their technical assistance. Appreciation is also extended to Messrs. Fred Krautz and Bo Dismukes and Ms. Catherine Gillis of Vetrotex CertainTeed for their valuable input.

## REFERENCES

1. KL Loewenstein. The Manufacturing Technology of Continuous Glass Fibres, 3rd ed. Amsterdam: Elsevier, 1993.
2. JG Mohr, WR Rowe. Fiber Glass. New York: Van Nostrand Reinhold, 1978.
3. JR Gonterman, WW Wolf. The Technology of Glass Fibers. Alfred, NY: Proceedings of 1st International Conference on Advances in the Fusion of Glass by American Ceramic Society, 1988, pp. 7.1–7.15.
4. WW Boeschenstein. They didn't know what they couldn't do. Glass Industry 76:26–47, 1995.
5. TF Starr. Glass-Fibre Databook. London: Chapman & Hall, 1993.
6. P Morgan. Glass Reinforced Plastics. 3rd Impression. London: Iliffe & Sons Ltd., 1954, p. 3.
7. JJ Svec, LB Streight-Schulz. Computers automate fiber glass production. Ceramic Industry 130:38–41, 1980.
8. P Gupta. Fibre Reinforcement for Composite Materials. Composite Materials Series Vol. 2. New York: Elsevier, 1988.
9. VI Kostikov, ed. Fibre Science and Technology. London: Chapman & Hall, 1995.
10. J Batten. Innovative bushing technology. Glass 72:423–424, 1995.
11. Owens-Corning News. New polypropylene chopped strand for thermoplastics has improved properties and may increase throughput. February 8, 1993.
12. G Tackels. Furnace emission challenge to French glassmakers. Glass 70:137–141, 1993.
13. K. Aydin, A Akinci. Application of oxy-fuel firing to an E-glass furnace. Glass Technol 34:256–258, 1993.
14. First Fluorine-Free Fibres. Glass 68:34, 1991.

15. AdvantexTM Glass Fibers. Owens-Corning Publication Number 5-PL-21433, February, 1997.
16. Composites '97. Plastics Technol 42:32–37, 1997.
17. BQ Kinsman. Environmental and health aspects of glass furnace repairs. Glass Technol 31:197–207, 1990.
18. LV Toropina, EP Pun'ko, LA Budnik, VM Dyaglev, YM Rassadin, NI Parafenko. Purification of wastewaters from glass fibre production. Fibre Chem 28:115–117, 1996.
19. A Handbook of FRP Products, Processes and Design Procedures. Valley Forge, PA: Vetrotex CertainTeed Corporation, 1991.
20. The Glass Fibre for Performance. Vetrotex International, 1997.
21. T-H Cheng, FR Jones, D Wang. Effect of fibre conditioning on the interfacial shear strength of glass-fibre composites. Compos Sci Tech 48:89–96, 1993.
22. J Bijwe. Composites as friction materials: recent developments in nonasbestos fiber reinforced friction materials—a review. Polym Compos 18:378–396, 1997.
23. Facts & Figures. Ticona, September, 1997.
24. FW Preston. The shoe on the other foot. Ceramic Bull 33:356–358, 1954.
25. WA Fraser, FH Ancker, AT DiBenedetto, B Elbibli. Evaluation of surface treatments for fibers in composite materials. Polym Compos 4:238–248, 1983.
26. B Yavin, HE Gallis, J Scherf, A Eitan, RD Wagner. Continuous monitoring of the fragmentation phenomenon in single fiber composite materials. Polym Compos 12:443–446, 1991.
27. WA Curtin. Determining fiber strength vs. gage length. Polym Compos 15:474–478, 1994.
28. R Wong. Recent aspects of glass fiber-resin interfaces. J Adhesion 4:171–179, 1972.
29. AK Rastogi, JP Rynd, WN Stassen. Investigations of Glass Fiber Surface Chemistry. 31st Annual Conference of the Reinforced Plastics Division, SPI, Washington DC, 1976, Section 6-B, pp. 1–8.
30. D Wang, FR Jones. Surface analytical study of the interaction between γ-amino propyl triethoxysilane and E-glass surface. J Mater Sci 28:2481–2488, 1993.
31. A El Acharl, A Ghenaim, V Wolfe, C Caze, E Carlier. Topographic study of glass fibers by atomic force microscopy. Textile Res J 66:483–490, 1996.
32. P Trens, R Denoyel, E Guilloteau. Evolution of surface composition, porosity, and surface area of glass fibers in a moist atmosphere. Langmuir 12:1245–1250, 1996.
33. NK Adam. The Physics and Chemistry of Surfaces. New York: Dover, 1968.
34. K Tsutsumi, Y Abe. Determination of dispersive and nondispersive components of the surface free energy of glass fibers. Colloid Polym Sci 267:637–642, 1989.
35. K Tsutsumi, T Ohsuga. Surface characterization of modified glass fibers by inverse gas chromatography. Colloid Polym Sci 268:38–44, 1990.
36. BK Larson, LT Drzal. Glass fibre sizing/matrix interphase formation in liquid composite moulding: effects on fibre/matrix adhesion and mechanical properties. Composites 25:711–721, 1994.
37. LV Zaborskaya, VA Dovgyalo, OR Yurkevich. Study of the wetting of reinforcing fibers by melts of thermoplastics. Fibre Chem 27:125–129, 1991.
38. K Lellig, G Ondracek. Glass and polymer: wetting and adhesion. Glastech Ber Glass Sci Tech 89:357–367, 1996.

39. AK Bledzki, J Lieser, G Wacker, H Frenzel. Characterization of the surface of treated glass fibres with different methods of investigation. Compos Interfaces 5:41–53, 1997.
40. K Grundke, P Uhlmann, T Gietzelt, B Redlich, HJ Jacobasch. Studies on the wetting behaviour of polymer melts on solid surfaces using the Wilhelmy balance method. Colloid Surface A 116:93–104, 1996.
41. EP Plueddemann. Silane Coupling Agents, 2nd ed. New York: Plenum Press, 1991, pp. 17–18.
42. PG Pape, EP Plueddemann. History of Silane Coupling Agents in Polymer Composites. 192nd Annual Meeting, American Chemical Society, Div. of the History of Chemistry, Anaheim, 1986, Paper No. 30, pp. 1–35.
43. MN Sathyanarayana, M Yaseen. Role of promoters in improving adhesion of organic coatings to a substrate. Progr Organic Coat 26:275–313, 1995.
44. FD Osterholtz, ER Pohl. Kinetics of the hydrolysis and condensation of organofunctional alkoxysilanes: a review. J Adhesion Sci Tech 6:127–149, 1992.
45. S Naviroj, SR Culler, JL Koenig, H Ishida. Structure and absorption characteristics of silane coupling agents on silica and E-glass fibre dependence on pH. J Colloid Interface Sci 97:308–317, 1984.
46. S Savard, LP Blanchard, J Leonard, RE Prud'Homme. Hydrolysis and condensation of silanes in aqueous solutions. Polym Compos 5:242–249, 1984.
47. S Naviroj, JL Koenig, H Ishida. Molecular structure of an aminosilane coupling agent as influenced by carbon dioxide in air, pH, and drying conditions. J Macromol Sci Phys B 22:291–304, 1983.
48. GL Witucki. A silane primer: chemistry and applications of alkoxy silanes. J Coatings Tech 65:57–60, 1993.
49. H Ishida, JL Koenig. Fourier transform infrared spectroscopic study of the silane coupling agent/porous silica interface. J Colloid Interface Sci 64:555–564, 1978.
50. H Ishida, JL Koenig. Fourier transform infrared spectroscopic study of the structure of silane coupling agent on E-glass fiber. J Colloid Interface Sci 64:565–576, 1978.
51. H Ishida, JL Koenig. An investigation of the coupling agent/matrix interface of fiberglass reinforced plastics by Fourier transform infrared spectroscopy. J. Polym Sci Polym Phys Ed 17:615–628, 1979.
52. H Ishida, JL Koenig. Molecular organization of the coupling interphase of fiberglass reinforced plastics. J Polym Sci Polym Phys Ed 17:1807–1813, 1979.
53. H Ishida, JL Koenig. Effect of hydrolysis and drying on the siloxane bonds of a silane coupling agent deposited on E-glass fibers. J Polym Sci Polym Phys Ed 18:233–237, 1980.
54. H Ishida, JL Koenig. A Fourier transform infrared spectroscopic study of the hydrolytic stability of silane coupling agents on E-glass fiber. J Polym Sci Polym Phys Ed 18:1931–1943, 1980.
55. C-H Chiang, H Ishida, JL Koenig. The structure of γ-aminopropyltriethoxysilane on glass surface. J Colloid Interface Sci 74:396–404, 1980.
56. H Ishida, JL Koenig, B Asumoto, ME Kenney. Application of UV resonance Raman spectroscopy to the detection of monolayers of silane coupling agent on glass surfaces. Polym Compos 2:75–80, 1981.

57. C-H Chiang, JL Koenig. Comparison of primary and secondary aminosilane coupling agents in anhydride-cured epoxy fiberglass composites. Polym Compos 2:192–198, 1981.
58. H Ishida, S Naviroj, SK Tripathy, JJ Fitzgerald, JL Koenig. The structure of an aminosilane coupling agent in aqueous solutions and partially cured solids. J Polym Sci Polym Phys Ed 20:701–718, 1982.
59. C-H Chiang, JL Koenig. Spectroscopic characterization of the matrix-silane coupling agent interface in fiber-reinforced composites. J Polym Sci Polym Phys Ed 20:2135–2143, 1982.
60. H Ishida. A review of recent progress in the studies of molecular and microstructure of coupling agents and their functions in composites, coatings and adhesive joints. Polym Compos 5:101–123, 1984.
61. FR Jones. Interfacial Aspects of Glass Fiber Reinforced Plastics. Proceedings of the International Conference on Interface Phenomena in Composite Materials, September 1989, pp. 25–32.
62. D Wang, FR Jones, P Denison. Surface analytical study of the interaction between γ-amino propyl triethoxysilane and E-glass surface. Part 1. Time-of-flight secondary ion mass spectrometry. J Mater Sci 27:36–48, 1992.
63. D Wang, FR Jones. TOF SIMS and XPS study of the interaction of silanized E-glass with epoxy resin. J Mater Sci 28:1396–1408, 1993.
64. CY Yue, MY Quek. The interfacial properties of fibrous composites. Part III. Effect of the thickness of the silane coupling agent. J Mater Sci 29:2487–2490, 1994.
65. E Nishio, N Ikuta, T Hirashima, J Koga. Pyrolysis-GC/FT-IR analysis for silane coupling treatment of glass fibers. Appl Spectrosc 43:1159–1164, 1989.
66. LH Dubois, BR Zegarski. Bonding of alkoxysilanes to dehydroxylated silica surface: A new adhesion mechanism. J Phys Chem 97:1665–1670, 1993.
67. G. Smith. An Investigation of Experimental Silane Coupling Agents and Their Effects on Glass Microspheres in Polypropylene Homopolymer. ANTEC, SPE, Dallas, 1990, pp. 1946–1948.
68. E Galli. Update: Coupling agents. Plastics Compound 14:50–56, 1991.
69. JL Thomason, GE Schoolenberg. An investigation of glass fibre/polypropylene interface strength and its effect on composite properties. Composites 25:197–203, 1994.
70. MJ Folkes, WK Wong. Determination of interfacial shear strength in fibre-reinforced thermoplastic composites. Polymer 28:1309–1314, 1987.
71. J Denault, T Vu-Khanh. Role of morphology and coupling agent in fracture performance of glass-filled polypropylene. Polym Compos 9:360–367, 1988.
72. B Arkles. Trends in Silane Coupling Agent Technology. Adhesion Coupling Agent Technol 97, Boston, 1997.
73. B Arkles, JR Steinmetz, I Zazyczny, M Zolotnitsky. Stable, Water-Borne Silane Coupling Agents. 46th Annual Conference, Composite Institute, SPI, Washington, DC, 1991, 2-D:1–7.
74. A Diwanji. Next Generation Coupling Agents for Glass Fibers. Adhesion Coupling Agent Technol 97, Boston, 1997.

75. B Arkles, J Steinmetz, J Hogan. Polymeric Silanes: An Evolution in Coupling Agents. 42nd Annual Conference, Reinforced Plastics/Composite Institute, SPI, Cincinnati, 1987, 21-C:1–4.
76. E Mader, H-J Jacobasch, K Grundke, T Gietzelt. Influence of an optimized interphase on the properties of polypropylene/glass fibre composites. Composites, Part A, 27:907–912, 1996.
77. F Hoecker, J Karger-Kocsis. On the effects of processing conditions and interphase of modification on the fiber/matrix load transfer in single fiber polypropylene composites. J Adhesion 52:81–100, 1995.
78. H Hamada, M Hiragushi, A Hamamoto, T Sekiya, N Nakamura, M Sugiyama. Effect of Interfacial Properties on Mechanical Properties in Glass Fiber Reinforced Polypropylene Injection Molding, ANTEC, SPE, Toronto, 1997, pp. 2519–2523.
79. CS Temple. Glass Fibers for Reinforcing Polymers. U.S. Patent 5130197, July 14, 1992.
80. M Arpin, P Petit, A Vagnon. Finish Composition for Coating and Protecting a Reinforcing Substrate. U.S. Patent 5389440, February 14, 1995.
81. J-P Gasca, G Tardy. Glass Fibers for Reinforcing Organic Matrices. U.S. Patent 5470658, November 28, 1995.
82. HD Wagner, A Lustiger, CN Marzinsky, RR Mueller. Interlamellar failure at transcrystalline interfaces in glass/propylene composites. Compos Sci Tech 48: 181–184, 1993.
83. A Lustiger, CN Marzinsky, RR Mueller, HD Wagner. Morphology and damage mechanisms of the transcrystalline interphase in polypropylene. J Adhesion 57:1– 14, 1995.
84. HG Karian. Thermodynamic Probe on Inter-molecular Coupling in the Interphase Region of Glass Fiber Reinforced Polymer Composites. ANTEC, SPE, Boston, 1995, pp. 1665–1669.
85. HG Karian. Thermodynamic probe of intermolecular coupling in the interphase region of glass fiber reinforced polymer composites. J Vinyl Additive Tech 1:264– 268, 1997.
86. J Bowyer, MG Bader. On the reinforcement of thermoplastics by imperfectly aligned discontinuous fibers. J Mater Sci 7:1315–1321, 1972.
87. MI Pitkethly, IP Favre, U Gaur, I Jakubowski, SF Mudrich, DL Caldwell, LT Drzal, M Nardin, HD Wagner, L Di Landro, A Hampe, JP Armistead, M Desaeger, I Verpoest. A round-robin programme on interfacial test methods. Compos Sci Tech 48:205–214, 1993.
88. PJ Herrera-Franco, LT Drzal. Comparison of methods for the measurement of fibre/matrix adhesion in composites. Composites 24:2–27, 1992.
89. CY Yue, WL Cheung. Interfacial properties of fibre-reinforced composites. J Mater Sci 27:3843–3855, 1992.
90. CY Yue, WL Cheung. Interfacial properties of fibrous composites. Part 1. Model for the debonding and pull-out process. J Matr Sci 27:3173–3180, 1992.
91. CY Yue, HC Looi, MY Quek. Assessment of fibre-matrix adhesion and interfacial properties using the pull-out test. Int J Adhesion Adhesives 15:73–80, 1995.

92. MJ Folkes, ST Hardwick. The molecular weight dependence of transcrystallinity in fibre reinforced thermoplastics. J Mater Sci Lett 3:1071–1073, 1984.
93. MJ Folkes, ST Hardwick. Direct study of the structure and properties of transcrystalline layer. J Mater Sci Lett 6:656–658, 1987.
94. E Devaux, B Chabert. Non-isothermal crystallization of glass fibre reinforced polypropylene. Polym Commun 31:391–394, 1990.
95. E Devaux, B Chabert. Nature and origin of the transcrystalline interphase of polypropylene/glass fibre composites after a shear stress. Polym Commun 32:464–468, 1991.
96. E Devaux, JF Gerard, P Bourgin, B Chabert. Two-dimensional simulation of crystalline growth fronts in a polypropylene/glass-fibre composite depending on processing conditions. Compos Sci Tech 48:199–203, 1993.
97. JL Thomason, AA van Rooyen. Transcrystallized interphase in thermoplastics composites. Part I. Influence of fibre type and crystallization temperature. J Mater Sci 27:889–896, 1992.
98. JL Thomason, AA van Rooyen. Transcrystallized interphase in thermoplastics composites. Part II. Influence of interfacial stress, cooling rate, fibre properties and polymer molecular weight. J Mater Sci 27:897–907, 1992.
99. J Varga, J Karger-Kocsis. Direct evidence of row-nucleated cylindritic crystallization in glass fiber–reinforced polypropylene composites. Polym Bull 30:105–110, 1993.
100. J Varga, J Karger-Kocsis. The occurrence of transcrystallization or row-nucleated cylindritic crystallization as a result of shearing in a glass-fiber-reinforced polypropylene. Compos Sci Tech 48:191–198, 1993.
101. F Hoecker, J Karger-Kocsis. Effects of matrix microstructure on the interfacial strength in GF/PP examined by a single fiber pull-out technique. Mekhanika Kompozitmykh Materialov 29:723–733, 1993.
102. B Monasse. Polypropylene nucleation on a glass fibre after melt shearing. J Mater Sci 27:6047–6052, 1992.
103. E Schulz, G Kalinka, S Meretz, A Hampe. Pull-out of single fibres: a method of determining interfacial strength. Kunststoffe Plastics Europe 86: 25–26, 1996.
104. L Dilandro, AT DiBenedetto, J Groeger. The effect of fiber-matrix stress transfer on the strength of fiber-reinforced composite materials. Polym Compos 9: 209–221, 1988.
105. EIM Asloun, M Nardin, J Schultz. Stress transfer in single-fibre composites: effect of adhesion, elastic modulus of fibre and matrix, and polymer chain mobility. J Mater Sci 24:1835–1844, 1989.
106. V Rao, LT Drzal. The dependence of interfacial shear strength on matrix and interphase properties. Polym Compos 12:48–56, 1991.
107. SF Zhandarov, EV Pisanova, VA Dovgyalo. Fragmentation of a single filament during tension in a matrix as a method of determining adhesion. McKhanika Kompozitnykh Materialov 28:384–403, 1992.
108. J-P Favre, M-H Auvray, A Mavel. Fibre/resin interaction in fragmentation tests: coaxial vs. simple specimens comparison. J Microsc 185:102–108, 1997.
109. M Takeshima, J-I Yamaki. Estimation of adhesion between fiber and matrix by the acoustic emission method. Jpn Polym Sci Tech 6:490–503, 1977.

## Glass Fiber–Reinforced Polypropylene

110. AN Netravali, W Sachse. Some remarks on acoustic emission measurements and the single-fiber-composite test. Polym Compos 12:370–373, 1991.
111. H Ho, LT Drzal. Evaluation of interfacial mechanical properties of fiber reinforced composites using the microindentation method. Composites 27:961–971, 1996.
112. M Desaeger, I Verpoest. On the use of the micro-indentation test technique to measure the interfacial shear strength of fibre-reinforced polymer composites. Compos Sci Tech 48:215–226, 1993.
113. PS Chua. Dynamic mechanical analysis studies of the interphase. Polym Compos 8:308–313, 1987.
114. JL Thomson. A note on the investigation of the composite interphase by means of thermal analysis. Compos Sci Tech 47:87–90, 1992.
115. RA Schweizer. Glass Fiber Length Degradation in Thermoplastics Processing. 36th Annual Conference, Reinforced Plastics/Composite Institute, SPI, Washington, DC, 1981, 9-A:1–4.
116. RA Schweizer. Glass fiber length degradation in thermoplastics processing. Polym Plast Tech Eng 18:81–91, 1982.
117. R von Turkovich, L Erwin. Fiber fracture in reinforced thermoplastic processing. Polym Eng Sci 23:743–749, 1983.
118. JL Thomason, MA Vlug. The influence of fibre length and concentration on the properties of glass fibre-reinforced polypropylene. Part 1. Tensile and flexural modulus. Composites A 27:477–484, 1996.
119. JL Thomason, WM Groenewoud. The influence of fibre length and concentration on the properties of glass fibre-reinforced polypropylene. Part 2. Thermal properties. Composites A 27:555–565, 1996.
120. JL Thomason, MA Vlug, G Schippar, HGLT Krikor. The influence of fibre length and concentration on the properties of glass fibre-reinforced polypropylene. Part 3. Strength and strain at failure. Composites A 27:1067–1074, 1996.
121. JL Thomason, MA Vlug. The influence of fibre length and concentration on the properties of glass fibre-reinforced polypropylene. Part 4. Impact properties. Composites A 28:277–288, 1997.
122. EC Hsu, CS Temple. Fiber Glass Reinforced Thermoplastics: Effect of Reinforcement Parameters on Composite Properties. 36th Annual Conference, Reinforced Plastics/Composite Institute, SPI, Washington, DC, 1981, 9-F:1–6.
123. X Leguet, M Ericson, D Chundury, G Baumer. Filled and Reinforced Polypropylene Compounds as Alternatives to Engineering Resins. ANTEC, SPE, Toronto, 1997, pp. 2117–2130.
124. SE Barbosa, JM Kenny. Rheology of thermoplastic matrix short glass fiber composites. J Vinyl Additive Tech 1:269–272, 1995.
125. J Karger-Kocsis. Microstructural aspects of the fatigue crack growth in polypropylene and its chopped glass fiber reinforced composites. J Polym Eng 10:97–121, 1991.
126. B Fisa. Mechanical degradation of glass fibers during compounding with polypropylene. Polym Compos 6:232–240, 1985.
127. B Chun, C Cohen. Glass fiber-filled thermoplastics. I. Wall and processing effects of rheological properties. Polym Eng Sci 25:1001–1007, 1985.

128. JP Greene, JO Wilkes. Numerical analysis of injection molding of glass fiber reinforced thermoplastics. Part 1. Injection pressures and flow. Polym Eng Sci 37:590–602, 1997.
129. RJ Nichols, F Kheradi. Reactive Modification of Polypropylene with a Non-Intermeshing Twin Screw Extruder. 3rd Chemical Congress of North America, Toronto, June 1988.
130. LC Sawyer. Determination of fiberglass lengths: Sample preparation and automatic image analysis. Polym Eng Sci 19:377–382, 1979.
131. K Stade. Techniques for compounding glass fiber-reinforced thermoplastics. Polym Eng Sci 17:50–57, 1977.
132. Y Shimizu, S Aral, T Itoyama, H Kawamoto. Experimental analysis of the kneading disk region in a co-rotating twin screw extruder. Part 1. Flow characteristics of the kneading disk region. Adv Polym Tech 15:307–314, 1996.
133. Y Shimizu, S Aral, T Itoyama, H Kawamoto. Experimental analysis of the kneading disk region in a co-rotating twin screw extruder. Part 2. Glass-fiber degradation during compounding. Adv Polym Tech 16:25–32, 1997.
134. PG Kelleher. Report on the state of the art: Injection molding of fiber reinforced thermoplastics. Part I. Materials and processes. Adv Polym Tech 10:219–230, 1990.
135. PG Kelleher. Report on the state of the art: Injection molding of fiber reinforced thermoplastics. Part II. Design and manufacture. Adv Polym Tech 10:277–284, 1990.
136. PG Kelleher. Advances in injection molding of fiber-reinforced thermoplastics during 1991. Part I. Materials and processing. Adv Polym Tech 11:305–313, 1992.
137. E Schmachtenberg, M Polifke. From an Intake Manifold to a Pump Casing: Fusible Core Technology Makes its Way into New Applications. Kunststoffe Plast Europe March 1996, pp. 16–17.
138. RE Richards, D Sims. Glass-filled thermoplastics: Effect of glass variables and processing on properties. Composites 3:214–220, 1971.
139. M Fujiyama. Structures and properties of injection moldings of glass fiber-filled polypropylene. Int Polym Proc VIII:245–254, 1993.
140. SF Xavier, D Tyagi, A Misra. Influence of injection-molding parameters on morphology and mechanical properties of glass fiber-reinforced polypropylene composites. Polym Compos 3:88–96, 1982.
141. HJ Wolf. Screw plasticating of discontinuous fiber filled thermoplastics: mechanisms and prevention of fiber attrition. Polym Compos 15:375–383, 1994.
142. G Fisher, P Schwarz, U Mueller, U Fritz. Measuring spatial fiber orientation—a method for quality control of fiber reinforced plastics. Adv Polym Tech 10:135–141, 1990.
143. S Toll, PO Anderson. Microstructural characterization of injection-moulded composites using image analysis. Composites 22:298–306, 1991.
144. H Yaguchi, H Hojo, DG Lee, EG Kim. Measurement of planar orientation of fibers for reinforced thermoplastics using image processing. Int Polym Proc X:262–269, 1995.
145. L Averous, J Quantin, A Crespy. Evolution of the three-dimensional orientation distribution of glass fibers in injected isotactic polypropylene. Polym Eng Sci 37:329–337, 1997.

## Glass Fiber–Reinforced Polypropylene

146. JP Greene, JO Wilkes. Numerical analysis of injection molding of glass fiber reinforced thermoplastics. Part 2. Fiber orientation. Polym Eng Sci 37:1019–1035, 1997.
147. M Sanoui, B Chung, C Cohen. Glass fiber-filled thermoplastics. II. Cavity filling and fiber orientation in injection molding. Polym Eng Sci 25:1008–1016, 1985.
148. S Hashemi, M Koohgilani. Fracture toughness of injection molded glass fiber reinforced polypropylene. Polym Eng Sci 35:1124–1132, 1995.
149. SE Barbosa, JM Kenny. Analysis of the Relationship Between Processing Conditions—Fiber Orientation—Final Properties in Short Fiber Reinforced Polypropylene. ANTEC, SPE, Toronto, 1997, pp. 1855–1859.
150. HG Karian. Two-Dimensional Mapping of Tensile Strength for Injection Molded Composites of Glass Fiber Reinforced Polypropylene. ANTEC, SPE, Toronto, 1997, pp. 1832–1836.
151. S Fellahi, A Meddad, B Fisa, BD Favis. Weldlines in injection-molded parts: a review. Adv Polym Tech 14:169–196, 1995.
152. PJ Cloud, F McDowell, S Gerakaris. Reinforced thermoplastics: understanding weld-line integrity. Plastics Tech 22:48–51, 1976.
153. W Michaeli, S Caluschia. Procedures for Increasing the Weldline Strength of Injection Molded Parts. ANTEC, SPE, Detroit, 1993, pp. 534–542.
154. CM Gillis. U.S. Economic Outlook and Composites Industry Forecast. Adhesion Coupling Agent Technology 97, Boston, 1997.
155. C den Beston. Reinforcements. Kunststoffe Plast Europe, 86:21–22, July 1996.
156. Glass Makers Outline European Priorities. Reinforced Plastics, 39:22–26, 1995.
157. S Umekawa, S Momoshima. Composites in Japan. Composites Eng 2:677–690, 1992.
158. E Seiler. Polypropylene Still has Scope for Innovation. Kunststoffe Plast Europe 85:30–33, 1995.
159. D Chundry, S Edge, B MacIver, J Vaughn. Polyolefin alloys, blends and compounds. Plastics Formulating & Compounding 2:18–24, 1996.
160. F Avalos, M Arroyo, JP Vigo. Optimization of a short glass fiber filled composite based on polyolefin blends, I. Tensile and flexural behavior. J Polym Eng 9:157–170, 1990.
161. F Avalos, M Arroyo, JP Vigo. Optimization of a short glass fiber filled composite based on polyolefin blends. II. Impact behavior. J Polym Eng 9:157–170, 1990.
162. AK Gupta, PK Kumar, BK Ratnam. Glass-fiber-reinforced polypropylene/EPDM blend. I. Melt rheological properties. J Appl Polym Sci 42:2595–2611, 1991.
163. AR Gupta, KR Srinivasan, PK Kumar. Glass fiber reinforced polypropylene/EPDM blends. II. Mechanical properties and morphology. J Appl Polym Sci 43:451–462, 1991.
164. FP La Mantia. Blends of polypropylene and nylon 6: Influence of the compatibilizer, molecular weight, and processing conditions. Adv Polym Tech 12:47–59, 1993.
165. J-Y Wu, W-C Lee, W-F Kuo, H-C Kao, M-S Lee, J-L Lin. Effects of molecular weights and compatibilizing agents on the morphology and properties of blends containing polypropylene and nylon-6. Adv Polym Tech 14:47–58, 1995.

166. S Kenig, A Silberman. Structure—Properties Relationships of Short Fiber Reinforced Polypropylene-Polyamide Blends. ANTEC, SPE, Boston, 1995, pp. 2160–2164.
167. A Perwuelz, C 'Caze, W Piret. Morphological and mechanical properties of glass-fiber-reinforced blends of polypropylene/polyamide 6,6. J Thermoplastics Composite Materials 6:176–189, 1993.
168. P Dave, P Chundury, G Baumer, L Overley. Performance-Property of Novel Glass Fiber Reinforced Polypropylene Compounds and their Applications. ANTEC, SPE, Indianapolis, 1996, pp. 1826–1830.
169. W Ast. Blow moulding of engineering parts. Kunststoffe German Plastics 80:3–10, 1990.
170. W Ast. Blow moulding of engineering parts. Kunststoffe German Plastics 81:8–14, 1991.
171. C Ausias, M Vincent, I Jarrin. Optimization of the extrusion process for glass-fiber-reinforced tubes. J Thermoplastics Composite Materials 8:435–448, 1995.
172. T Stevens. Polypropylene: Not just a cheap plastic. Mechanical Eng 112:30–43, 1990.
173. Reinforcements for polypropylene: Technology developments. Reinforced Plastics 36:17–21, 1992.
174. Customer Acceptance Standards, RO99 1200 P319, Vetrotex International S.A., Reference 194/02, January 1, 1997.
175. M Glemet, G Cognet. Process for the Manufacture of Sections of Thermoplastic Resin Reinforced with Continuous Fibers. U.S. Patent No. 4883625, Nov. 28, 1989.
176. DE Woodmansee, BM Kim. Method and Apparatus for Coating Fibers with Thermoplastics. U.S. Patent No. 5006373, Apr. 9, 1991.
177. C Koppernses, S Nolet, JP Fanucci. Method and Apparatus for Wetting Fiber Reinforcements with Matrix Materials in the Pultrusion Process Using Continuous Inline Degassing. U.S. Patent No. 5073413, Dec. 17, 1991.
178. PJ Bates, J-M Charrier. Apparatus for Continuously Coating Fibers. U.S. Patent 5133282, July 28, 1992.
179. LV Montsinger. Apparatus and Method for Forming Fiber Filled Thermoplastic Composite Materials. U.S. Patent No. 5176775, January 5, 1993.
180. LV Montsinger. Apparatus and Method for Forming Fiber Filled Thermoplastic Composite Materials. U.S. Patent 5447793, September 5, 1995.
181. T Asai, T Ohara, T Tanaka, S Hashizume. Method of Manufacturing Continuous Fiber-Reinforced Thermoplastic Prepregs. U.S. Patent 5529652, June 25, 1996.
182. J-M Charrier, PJ Bates, D Guillon, G Zanella. The Effect of Pin Shape on Spreading Roving Filaments for a Thermoplastic Pultrusion Process. 45th Annual Conference, Composites Institute, SPI, Washington, DC, 1990, 18-A, pp. 1–4.
183. PJ Bates, H Ripert, J-M Charrier. Transverse Permeability of Fiber Rovings to Fluids. ANTEC, SPE, Detroit, 1992, pp. 1709–1711.
184. P Peltonen, K Lahteenkorva, EJ Paakkonen, PK Jarvela, P Tormala. The influence of melt impregnation parameters on the degree of impregnation of a polypropylene/glass fibre prepreg. J Thermoplastic Composite Materials 5:318–343, 1992.
185. LV Montsinger. MTI Long Fiber Compounding Productivity Improvement. ANTEC, SPE, Toronto, 1997, pp 2335–2338.

186. VB Gupta, RK Mittal, PK Sharma, G Mennig, J Wolters,. Some studies on glass fiber-reinforced polypropylene. Part I. Reduction in fiber length during processing. Polym Compos 10:8–15, 1989.
187. VB Gupta, RK Mittal, PK Sharma, G Mennig, J Wolters. Some studies on glass fiber-reinforced polypropylene. Part II. Mechanical properties and their dependence on fiber length, interfacial adhesion, and fiber dispersion. Polym Compos 10:16–27, February 1989.
188. M Zettler, E Doring. Latest developments in the processing and application technology of long fibre reinforced thermoplastics. Kunststoffe German Plastics 79:19–22, 1989.
189. R Bailey, H Kraft. A study of fibre attrition in the processing of long fibre reinforced thermoplastics. Int Polym Proc 2:94–101, 1987.
190. T Harmia. Fatigue behavior of neat and long glass fiber (LGF) reinforced blends of nylon 66 and isotactic PP. Polym Compos 17:926–936, 1996.
191. LNP Technical Data. Verton MFX Series, May 9, 1997.
192. LW Glenn, HC Kim, DE Miller, CS Ellis. Toughness of Long Glass Fiber Re-in-forced Thermoplastics. ANTEC, SPE, Toronto, 1997, pp. 2325–2329.
193. L Ye, V Klinkmuller, K Friedrich. Impregnation and consolidation in composites made of GF/PP powder impregnated bundles. J Thermoplastic Composite Materials 5:32–48, 1992.
194. M Ostgathe, Ch Mayer, M Neitzel. Thermoplastic Composite Sheets from Powder. Kunststoffe Plast Europe 86:13–15, 1996.
195. D Cutolo, E Zoppi. Process of Making a Composite Article. U.S. Patent 5614139, Mar. 25, 1997.
196. Verton Structural Composites. LNP Engineering Plastics. Bulletin 270–196, 1996.
197. Glass-reinforced PP gives new twist to screwdriver. Modern Plastics p. 82, Vol. 73, December 1996.
198. A Lucke. Thermoplastics with backbone. Kunststoffe Plast Europe 87:7–9, 1997.
199. Honda literature. Bumper systems. Automotive composites. A design and manufacturing guide 1:44–47, 1997.
200. Special report—GMT. Reinforced Plastics 36:17, 1992.
201. Azdel to Buy Exxon's Composites Business. Plastics News p. 7, February 17, 1997.
202. Reinforced PP making headway in cars. Reinforced Plastics 36:22–24, 1992.
203. Capacity for Making Fiber Glass-Reinforced Thermoplastic Sheet is Doubled by Azdel Inc., Shelby, NC, USA. Reinforced Plastics Newsletter 18:4, February 21, 1994.
204. BASF, Owens Corning May Form Alliance. Plastics News p. 3, March 24, 1997.
205. VK Stokes. Random glass mat reinforced thermoplastic composites. Part I. Phenomenology of tensile modulus variations. Polym Compos 11:32–44, 1990.
206. VK Stokes. Random glass mat reinforced thermoplastic composites. Part II. Analysis of model materials. Polym Compos 11:45–55, 1990.
207. VK Stokes. Random glass mat reinforced thermoplastic composites. Part III. Characterization of the tensile modulus. Polym Compos 11:342–353, 1990.
208. VK Stokes. Random glass mat reinforced thermoplastic composites. Part IV Characterization of the tensile strength. Polym Compos 11:354–367, 1990.

209. HF Giles. Effect of Processing Conditions on Thermoplastic Glass Reinforced Bumper Beam Performance. ANTEC, SPE, Toronto, 1997, pp. 1976–1978.
210. A new generation of bumpers. Kunststoffe Plast Europe 84:11–12, 1994.
211. RJ Clark. Polyether Amine Modification of Polypropylene Paintability Enhancement. ANTEC, SPE, Boston, 1995, pp. 3306–3310.
212. More Than US $10 per Unit is Saved. Reinforced Plastics Newsletter 21(9), April 28, 1997.
213. WJ Cantwell, W Tato, HH Kausch, R Jacquemet. The influence of a fiber-matrix coupling agent on the properties of a glass/polypropylene GMT. J Thermoplastic Composite Materials 5:304–317, 1992.
214. P Gerault, M Gognelin, P Fredesucci. Papermaking Process and Composition for the Production of Tridimensional Products Containing Thermoplastics Resin and Reinforcing Fibers. U.S. Patent 4929308, May 29, 1990.
215. DM Bigg, DF Hiscock, JR Preston, EJ Bradbury. The properties of wet-formed thermoplastic sheet composites. Polym Compos 14:25–34, 1993.
216. Composite Battery Tray Wins SPE Auto Award. Modern Plastics p. 13, December 1996.
217. GM EV1 battery tray. Automotive Engineering 105:55, 1997.
218. R Dyksterhouse, JA Dyksterhouse. Production of Improved Preimpregnated Material Comprising a Particulate Thermoplastic Polymer Suitable for Use in the Formation of a Substantially Void-Free Fiber Reinforced Composite Article. U.S. Patent 5128198, July 7, 1992.
219. RC Hawley. Extruder Apparatus and Process for Compounding Thermoplastic Resin and Fibres. U.S. Patent 5165941, November 24, 1992.
220. FS Deans, JU Raisoni. Long-Fiber Thermoplastic Direct Melt Phase Molding. ANTEC, SPE, New Orleans, 1993, pp. 1314–1315.
221. DR Fitchmun. Thermoplastic Thermoformable Composite Material and Method of Forming Such Material. U.S. Patent 5604020, February 18, 1997.
222. A Revolutionary New Class of Reinforced Thermoplastics is Created. Reinforced Plastics Newsletter 19(6)1, March 20, 1995.
223. G Roncato, R Federowsky. Process and Device for Producing a Yarn or Ribbon Formed from Reinforcement Fibers and a Thermoplastic Organic Material. U.S. Patent 5011523, April 30, 1991.
224. G Roncato, R Fedorowsky, P Boissonnat, D Loubinoux. Apparatus for Manufacturing a Composite Strand Formed of Reinforcing Fibers and of Organic Thermoplastic Material. U.S. Patent 5316561, May 31, 1994.
225. G Roncato, R Fedorowsky, P Boissonnat, D Loubinoux. Apparatus for Manufacturing a Glass and Organic Composite Strand, Including a Blowing Device. U.S. Patent 5328493, July 12, 1994.
226. D Loubinoux, G Roncato. Method of and an Apparatus for Forming a Composite Thread Including Stretching of Thermoplastic Filaments. U.S. Patent 5425796, June 20, 1995.
227. S Osten, C St. John, D Guillon, G Zanella, T Renault. Compression Molding of Twintex and Random Fiber Thermoplastic Molding Materials. ANTEC, SPE, Toronto, 1997, pp. 2432–2436.

228. Moldable PP/glass fabric makes headway in composites. Plastic Technology 43: 15, 1997.
229. AB Woodside, JV Gauchel, LJ Huey, DL Shipp, R Macdonald, PW Woodside, DB Mann. Contact Drying of Fibers to Form Composite Strands. U.S. Patent 5626643, May 6, 1997.
230. WH Kielmeyer, MD Peterson, DR Larratt. Method of and Apparatus for Forming Composite and Other Fibers. U.S. Patent 5639291, June 17, 1997.
231. JM Henshaw, W Han, AD Owens. An overview of recycling issues for composite materials. J Thermoplast Compos Mater 9:4–20, 1996.
232. A Adewole, K Dackson, M Kolkowicz. The effect of GFR (glass-fiber-reinforcement) on the composition–property balance of compatibilized recycled polyolefinic blends. J Thermoplast Compos Mater 8:272–292, 1995.
233. M Vinci, FP La Mantia. Properties of filled recycled polypropylene. J Polym Eng 16:203–215, 1996/7.
234. Recycled Plastic Scores Big. Plastics Technol p. 84, March 1997.
235. D Loepp. Automakers ignore recycling. Plastics News, March 3, 1997, p. 10.

# 10

# Functionalization and Compounding of Polypropylene Using Twin-Screw Extruders

**Thomas F. Bash**
Ametek Westchester Plastics, Nesquehoning, Pennsylvania, U.S.A.

**Harutun G. Karian**
RheTech, Inc., Whitmore Lake, Michigan, U.S.A.

## 10.1 INTRODUCTION

Polypropylene (PP) has been modified commercially via reactive extrusion since the 1960s [1]. The earliest work, with the goal of reducing the melt viscosity, involved controlled degradation of the high molecular weight fraction of the molecular weight distribution. This was done because polypropylene produced with Ziegler-Natta catalysts came out of the reactor with very broad molecular weight distributions [1], where the high molecular weight species caused processing problems in downstream equipment.

Shortly after controlled degradation technology was developed, pioneer workers explored polypropylene functionalization reactions in extruders [1]. The early work involved grafting of maleic anhydride onto polypropylene to improve adhesion properties. Later work involved grafting other monomers, and even polymers, onto polypropylene in twin-screw extruders.

Chapter 3 provides a thorough discussion of the chemistry and free-radical reaction mechanisms involved in producing functionalized polypropylene. However, the actual sequences of physical and chemical changes during the extrusion process are still much less understood by scientist and manufacturing compounder alike. In a sense, the shear field within the kneading elements of the twin screw remains very much a "black box" to the processor who seeks higher throughput rates with maximum yield of good-quality products.

Bartilla et al. [2] point out that the properties of a modified polymer blend are a combined function of the individual ingredients in regard to type, percentage content, and actual blending technique used. In the case of corotating twin-screw extrusion, they identify the melting zone as having a primary significance in attaining desirable blend properties. The axial location of conversion of solid polymer resin in the feedstream to molten mass is associated with an abrupt rise in melt viscosity that provides sufficient shearing of the material. Hence, the melting process entails both homogenization and desired or undesired high shearing, i.e., much like any functionalization process based on peroxide initiation and grafted monomer addition. The desired end of such a process is incorporation of monomer ingredient at high percentage yield. The beta scission of the polypropylene backbone leading to vis breaking and melt flow increase is an example of undesirable side reaction. Likewise, the homopolymerization of monomer into oligomers is a source of waste because monosubstitution of functional groups is preferred in enhancing adhesion of the polymer backbone to sized fiber glass and various surface-treated filler reinforcements. Because of the superposition of residence time and shear deformation aspects of this type of reactive extrusion process, the functionalization of molten polypropylene resin is representative of compounding methodology to modify polypropylene resin in general.

This chapter first reviews the history of polypropylene modification using twin-screw extruders. The discussion moves to the topic of the underlying principles of current twin-screw extrusion technology. In particular, corotating, intermeshing, twin-screw extrusion is described in terms of fundamental relationships between modular extruder screw design, operating extruder conditions, and the physicochemical processes involved in the grafting of functional groups onto the polypropylene backbone.

## 10.2 POLYPROPYLENE MODIFICATIONS USING TWIN-SCREW EXTRUDERS

### 10.2.1 Controlled Degradation of Polypropylene in an Extruder

As mentioned above, controlled degradation of polypropylene was one of the earliest reactive extrusion technologies developed. A fairly substantial literature on the subject has appeared in the past 15 years. The mechanism of free-radical attack of the polymer backbone is now thought to be well known [3–6]. The following sequence of free-radical reactions in the extruder provide a summary of key process steps involved in grafting functional groups onto chains of the polypropylene molecules.

The first step of the mechanism involves generation of *peroxy radicals* through thermal decomposition of a peroxide:

$$ROOR \rightarrow 2RO^{\cdot}$$

The radical then attacks the backbone of the polymer, abstracting a tertiary *hydrogen*:

$$RO^{\cdot} + {\sim\sim}-CH(CH_3)-CH_2-CH(CH_3)-CH_2-{\sim\sim} \xrightarrow{-ROH} {\sim\sim}-C^{\cdot}(CH_3)-CH_2-CH(CH_3)-CH_2-{\sim\sim}$$

The backbone then degrades by beta scission:

$${\sim\sim}C^{\cdot}(CH_3)-CH_2-CH(CH_3)-CH_2-{\sim\sim} \rightarrow {\sim\sim}C(CH_3)=CH_2 + {\cdot}CH(CH_3)-CH_2-{\sim\sim}$$

The polymeric radicals then terminate by disproportionation:

$${\sim\sim}-CH_2-CH^{\cdot}(CH_3) + {\cdot}CH(CH_3)-CH_2-{\sim\sim} \rightarrow {\sim\sim}-CH_2-CH_2(CH_3) + CH(CH_3)=CH-{\sim\sim}$$

This basic mechanism has been used to model the degradation reactions of polypropylene in an extruder [5–7].

A closed-loop extruder control scheme loosely based on this mechanism has been described by Curry et al. [8]. They used an on-line rheometer at the discharge of a ZSK 30 twin-screw extruder to generate a signal related to melt flow index to a programmable logic controller (PLC). The PLC compared this value with a set point and manipulated the ratio of feeders, where one feeder contained a master batch of polypropylene and peroxide. Their control scheme featured a process lag time on the order of 4–7 min.

When controlled rheology technology was in its infancy, peroxides were not used to generate a source of radicals. Instead, very high extruder barrel temperatures (up to 420°C) in the feed area were used, which promoted oxidative degradation from the oxygen in the air [1]. A reverse temperature profile was used so that the discharge temperature was lower than the feed temperature.

The influence of screw design on the thermal and peroxide-induced degradation of polypropylene was studied by Kim and White [7]. They used a Japan Steel Works TEX-30 intermeshing twin-screw extruder operated in both corotating and counterrotating modes. The counterrotating screw elements had thin flights and were not completely self-wiping. The melting profiles were determined from screw pull after dead stopping the extruder screw and rapidly cooling the barrel to freeze the steady-state holdup material.

Four different screw configurations (A–D) were used for both modes of operation, with the screw designs differing in the number and location of kneading elements:

1. Screw design A consisted of only forward conveying bushing screw elements.
2. Screw design B used one kneading block just before the die.
3. Screw design C had the kneading block closer to the feed hopper.
4. Screw design D used three kneading blocks along the screw length.

Onset of melting occurred particularly where there was a restriction to flow or increased holdup in a kneading block. For example, for screw design A, melting was delayed to just before the die.

The thermal degradation reactions were carried out at a barrel temperature of 230°C, whereas peroxide-induced degradation reactions were catalyzed by injecting 2,5-dimethyl-2,5-*bis*(*t*-butylperoxy)hexane at rates of 0.05, 0.1, and 0.2 phr (per hundred parts resin) into the feed hopper of the extruder. The polypropylene resin was supplied by Himont and had a melt flow index of 0.14 g/10 min.

The degree of degradation was found to vary with the screw design and screw speed and was different between corotating and counterrotating modes. The counterrotating mode resulted in greater levels of degradation for the thermal degradation reaction, and the extent of degradation was found to increase with increasing screw speed for all screw designs. In the corotating mode, the extent of degradation decreased with increasing screw speed, except for screw design D with the three kneading disk blocks. This difference between corotating and counterrotating performance was explained by the greater shear stresses and increased viscous heating experienced when operating in counterrotating mode.

The performance of the two rotation modes also differed for peroxide-induced degradation. When a small quantity of peroxide was used (0.05 phr), the results were similar to the thermal degradation study. The extruder, when run in counterrotating mode, achieved greater levels of degradation at all screw speeds studied. When the peroxide level was increased to 0.2 phr, the extent of the degradation reaction became high and insensitive to the mode of operation. This result was explained as an increased significance of mixing when peroxide is used, which narrows the difference in performance between the corotating and counterrotating modes of operation.

### 10.2.2 Functionalization of Polypropylene in an Extruder

The development of extruder-based polypropylene functionalization technology followed that of controlled degradation technology. The early work related to grafting of monomers, such as acrylic acid (AA) and maleic anhydride (MA), onto polyolefin backbones [1]. Initial studies were done with single-screw extruders, whereas more recent work has concentrated on using twin-screw extruders because of their superior mixing performance. A survey of the early work is presented in Brown [9].

Polypropylene is functionalized to improve or enhance one or more properties. Carboxylation of polypropylene adds an acid group to the polymer that improves adhesion to many substrates and imparts compatibility with many other polymers. Polypropylene is commonly carboxylated by grafting onto the backbone either AA or MA. The grafting reaction mechanisms are quite different for AA and MA. Glicidyl methacrylate (GMA) is another monomer that can be grafted onto the polypropylene backbone.

*Acrylic Acid–Grafted PP via Melt Extrusion.* The melt phase grafting reaction of acrylic acid onto polypropylene proceeds by a free radical mechanism [10]. Radicals generated by thermal decomposition of an initiator abstract hydrogen from the polypropylene backbone and initiate homopolymerization of acrylic acid. Acrylic acid also adds to the sites on the backbone, with the result that the product contains acrylic acid grafted polypropylene (AA-g-PP) and poly(acrylic acid) homopolymer.

An example of grafting acrylic acid onto polypropylene in an extruder was presented by Chiang and Yang [11]. They dissolved the acrylic acid and benzoyl peroxide initiator in acetone, mixed the solution in with polypropylene powder, and then allowed the acetone to evaporate. The mixture was then reacted in an unspecified extruder at a temperature of 200°C to prepare the AA-g-PP. Grafting efficiencies were found to decrease as the amount of acrylic acid was increased, resulting in relatively greater production of poly(acrylic acid) homopolymer. The AA-g-PP was then blended in a Brabender Plasti-Corder with a mixture of a silane coupling agent and mica filler. The AA-g-PP blends showed enhanced mechanical and thermal properties in comparison with ungrafted polypropylene. An example of preparing acrylic acid–grafted polypropylene by injecting the reactants in a twin-screw extruder is discussed in the next section.

*Maleic Anhydride Grafted PP via Melt Extrusion.* The grafting of maleic anhydride onto polyolefins proceeds by a different mechanism, which has been extensively studied by Gaylord and coworkers [10,12,13]. Unlike acrylic acid and other vinyl monomers, maleic anhydride does not undergo radical-induced homopolymerization under standard conditions. It does undergo homopolymerization under conditions of high radical concentration. The radical-induced reaction of maleic anhydride with polypropylene results in polymer containing individual grafted maleic anhydride units at the chain ends, along with degraded polypropylene.

The mechanism of maleic anhydride grafting is related to the mechanism of maleic anhydride homopolymerization in that the reactive species is an excited dimer of maleic anhydride [11]. Because the excited dimer is capable of abstracting hydrogen from the backbone of polypropylene, maleic anhydride contributes to the degradation of polypropylene. Gaylord [10] and Gaylord and Mishra [12] showed that the addition of a small amount of an inhibitor for maleic anhydride homopolymerization such as an *N,N*-dialkylamide significantly reduces the amount of degradation of the polypropylene.

An example of grafting of maleic anhydride onto polypropylene in a twin-screw extruder was described by Wong [14]. Polypropylene was coated with 1.1 wt% maleic anhydride and fed to a 53-mm twin-screw extruder at 30 kg/hr. Four temperature control zones were used, and styrene monomer at 1.4 wt% was injected into the second zone. They obtained 0.8 wt% bound maleic anhydride on the product, and the melt flow index was 17°/min. When 500 ppm Lupersol 130 catalyst was added and no styrene was injected, the melt flow index increased to 258°/min, indicating significant degradation. No grafting was obtained when an antioxidant was added to the feed.

The effect of screw configuration on the grafting reaction of maleic anhydride onto polypropylene was studied by Kim and White [7]. This investigation used the *same* screw designs A–D that were used to characterize controlled degradation of PP discussed above in Section 10.2.1. They fed maleic anhydride and 2,5-dimethyl-2,5-*bis*(*t*-butylperoxy)hexane peroxide together as a solution in a nonreacting solvent, such as acetone, to a 30-mm twin-screw extruder. The extruder was run in both the corotating and counterrotating modes using the screw designs.

The relative levels of grafting achieved were determined for each screw design in corotating and counterrotating modes and as a function of screw speed. In both the corotating and counterrotating modes, increasing the number of kneading blocks increased the level of grafting achieved. This is expected because the screw designs with more kneading blocks melted the polymer earlier and had greater residence time in the melt phase for reaction. The incremental change in grafting level along the screw root profile was determined by postanalysis of axial segments of the frozen mass of reacted polymer taken from the extruder after dead stop of the drive motor, cool-down, and screw-pull experiments.

In comparing the modes of operation, Kim and White concluded that corotation resulted in higher levels of maleation for each screw configuration tested, even though melting occurred earlier in the counterrotation mode. This difference in performance between the two modes of operation was explained as a difference in mixing efficiency. The maleation reaction required intimate mixing of the maleic anhydride, peroxide, and polymer in the melt phase. The results of this study indicated that the corotating mode of operation resulted in superior mixing of the reactants in the melt phase.

*Glicidyl Methacrylate Grafted PP via Melt Extrusion.* More recently, Cartier and Hu [15] used direct in-line characterization of the reactive extrusion process of grafting of GMA monomer while minimizing homopolymerization of the monomer and undesired molecular changes of the original polypropylene resin. They studied the combination of physical (plastification) and chemical (fast-grafting kinetics) on the grafted GMA yield. These authors point out that virtually all investigations cited in the literature analyzed efficacy of grafting yield and changes in molecular architecture (degradation or cross-linking) at the *die exit* only.

Like Bartilla et al. [2], Cartier and Hu [15] identified the plastification zone as the focal point of the reactive extrusion as a function of material, machine design, and process variables:

1. Material variables: melting point, heat capacity, and enthalpy of polymer resin
2. Machine variable: screw design
3. Process variables: barrel temperature profiles, screw speed, and feed rate

By using a ZSK 30 corotating twin-screw extruder, they used top ports at Kombi 4 and 6 positions to sample molten mass before and after compounding in an arrangement of kneading elements. The neutral 90° staggered kneading disks followed by reverse-screw bushing provides sufficient working volume to attain the required grafting reaction promoted by plastification of either high-density polyethylene (HDPE) or polypropylene resin and simultaneous free-radical mechanism for GMA addition. By using Fourier transform infrared analysis, they determined grafting yield. Their results are as follows:

1. *Material-barrel temperature*: HDPE powder melted before the kneading block because of the heat transfer from the 200°C barrel, whereas the porous PP pellets were only partially molten before the kneading section. Hence, the HDPE conversion to grafted GMA occurred in the upstream screw-conveying section.

2. *Screw speed–feed rate*: Plastification was more sensitive to screw speed than feed rate. Lower grafted GMA yield was attributed to decreased barrel heating of the polymer as the main source of heat for plastification.

The direct in-line evidence for stepwise grafting along the extruder screw during the actual reactive extrusion process provides fundamental insights into the controlling mechanism for enhanced grafting of monomer onto the PP, i.e., what happens before and after given sections of kneading elements? The process conditions promoting the plasticification of the PP resin are identified as the key to effective grafting efficiency. Compared with the in-line method used by Cartier and Hu, the postanalytical technique used by Kim and White provides a similar but much more *blurred* indication of grafting efficiency because of the long delay of quenching the molten mass into a frozen form.

### 10.2.3 Single-Stream Polypropylene Functionalization–Glass Fiber Incorporation

The compounding of glass fibers into modified polypropylene yields improved mechanical properties for many applications, as discussed elsewhere in this book (see Chapters 3, 9, 12, and 13). Usually, this is accomplished by two separate extrusion steps. First, polypropylene is functionalized to give a chemical coupling concentrate in an upstream extrusion process. Then, in a second extrusion process, a level of chemical coupling agent (AA-g-PP or MA-g-PP) is dry blended with PP resin followed by glass fiber addition to give PP composite. Twin-screw extruders, and some specially designed single-screw extruders, provide an option of downstream addition of fiberglass to minimize attrition of the glass fiber length.

A consecutive sequence of three process steps was attained by Nichols and Kheradi [16] in a single-extrusion process by using various mixing zones in a twin-screw extruder:

1. Polypropylene resin was melted.
2. Acrylic acid was grafted onto the PP backbone.
3. Downstream addition of fiberglass was carried out.

A general process schematic illustrating the feeds and product streams of the process described by Nichols and Kheradi is shown in Fig. 10.1. A counter-rotating, nonintermeshing, twin-screw extruder was used, with an overall length/diameter $(L/D)$ ratio of $55:1$, of which the last $6:1$ $L/D$ comprised a single-screw discharge pump.

The initiator used was 2,5-dimethyl-2,5-*bis*(*t*-butylperoxy)hexane, which was injected into the extruder through a specially designed injection valve. The injection valve was mounted in a water-cooled adapter that featured thermal pins

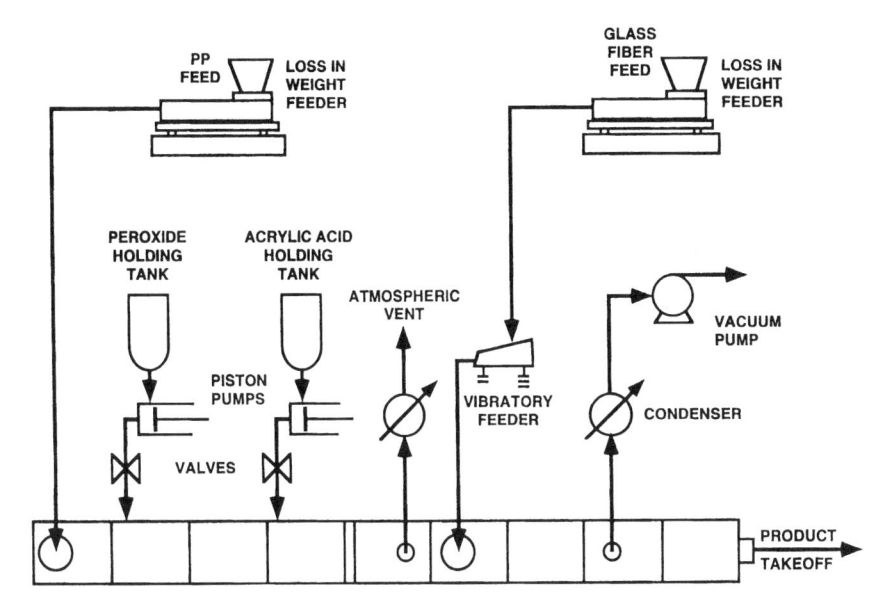

FIGURE 10.1 Flow diagram of combined acrylic acid grafting-glass fiber addition process. (From Ref. 16, courtesy of Welding Engineers.)

to ensure cooling down to the injection entry location at the barrel inner surface. The same kind of injection valve and adapter was used for injection of the acrylic acid. Cooling the injection valves prevented the peroxide from decomposing before entering the extruder and prevented the acrylic acid from homopolymerizing. The piping to the injection valves was also traced with cooling lines.

The initiator was injected into the second barrel where it mixed with the unmelted polypropylene in the solids conveying zone. In the melting section, a process temperature of about 200°C resulted in an initiator half-life on the order of 5 sec. This is about the residence time in the melting section, so that when acrylic acid was injected into barrel 4, polypropylene radicals would already have been formed. The grafting reaction of acrylic acid was essentially completed within one barrel section. An atmospheric vent was used in the fifth barrel to remove unreacted acrylic acid volatiles and entrapped air.

The glass fibers that were used were $\frac{3}{16}$-in.-long standard grade from Owens Corning fiberglass and were fed into the sixth barrel. The glass fibers were compounded into the melt, and a devolatilization zone with a vacuum vent was used in the eighth barrel. Samples were collected for analysis of the extruded polypropylene resin, resin plus 30% glass fibers, and grafted resin plus 30% glass fibers. The samples were analyzed for tensile, flexural, impact strength, and heat distortion temperature in accordance with standard ASTM test methods.

The tensile, flexural, and impact properties of glass fiber–reinforced polypropylene are a function of fiber loading and length distribution, as discussed by Ramos and Belmontes [17]. They compared glass-filled polypropylene blends, at various glass loading levels, prepared on a two-roll mill and a twin-screw extruder. They found that the longer glass fibers in the blends made on the twin-screw extruder resulted in superior tensile and flexural properties. Nichols and Kheradi did not analyze their samples for glass fiber length distribution, but it was expected that the length distribution would be the same for the modified and unmodified samples.

In Chapter 3, mechanical properties of glass fiber–reinforced polypropylene are improved by addition of a coupling agent. Likewise, the in situ generation of modified polypropylene by grafting acrylic acid onto the polymer backbone in the Nichols and Kheradi investigation yields enhanced tensile strength compared with unfilled polypropylene and uncoupled composite (Fig. 10.2).

## 10.3 TWIN-SCREW EXTRUDER TECHNOLOGIES

Thiele [18] stressed the utility of continuous-single and twin-screw extruders in comparison with Banbury batch-type mixers on the basis of the working volume, i.e., small working mass in the extruder channels in continuous process equip-

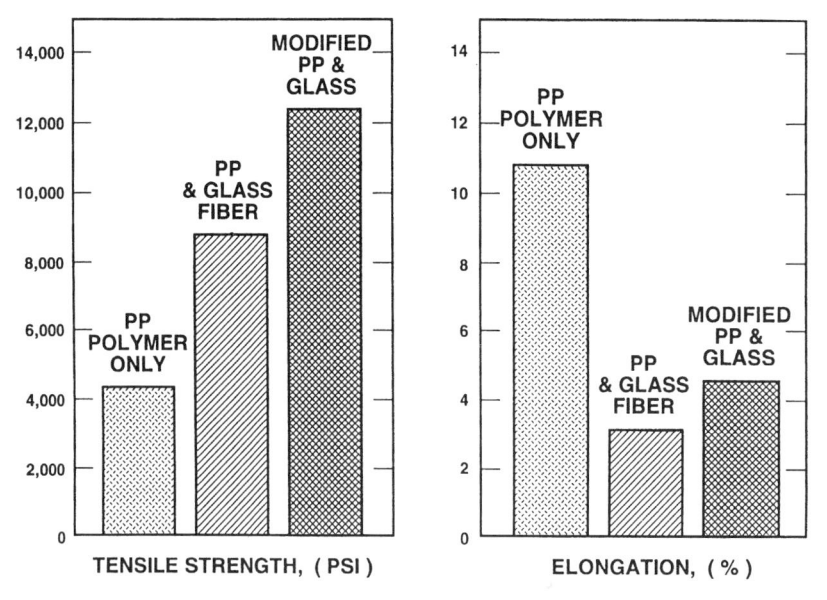

FIGURE 10.2 Tensile properties of polypropylene. (From Ref. 16, courtesy of Welding Engineers.)

ment versus a large amount of blended material per cycle in batch-type mixers. With a relatively smaller working volume in an extruder, there is more enhanced intimate mixing on the microscopic scale in a shorter period of residence time.

Given that "small working mass" is a more economical means of reactive extrusion, the distinction between different types of extrusion equipment branches off into a host of processing attributes. Because of increasing trends in the late 1990s to replace single extruders by twin-screw extruders, our attention is now directed at twin-screw methodology to modify and reinforce polypropylene resins into a wide spectrum of composite materials.

White et al. [19] made a critical assessment of the technological development of twin-screw extruder design and concepts of flow mechanisms. The 194 references cited in their review article provide a very comprehensive geneology of twin-screw equipment designs from the 19th century to the present state of the art. This includes details concerning tangential counterrotating, intermeshing counterrotating, and intermeshing corotating twin-screw machines.

### 10.3.1 Intermeshing Corotating and Counterrotating Twin-Screw Extruders

Nonintermeshing counterrotating machines have the oldest technology [19]. The Farrel continuous mixer (FCM) remains today as a good compromise of the Banbury batch mixer but with continuous mixing capability with the "small mass" concept. The incorporation of various fillers into polypropylene (talc, calcium carbonate, etc.) at 30–80 wt% levels is still accomplished via the FCM.

The behavior of a counterrotating twin-screw extruder with nonintermeshing screw design (as described in Ref. 16) is in marked contrast to that of intermeshing-type design for rotation in the same direction (corotation) or in opposite directions (counterrotation). Sakai et al. [20] provide experimental evidence comparing the extrusion performance of the two types of intermeshing twin-extruder equipment. Their results combined with comments by Thiele are given in Table 10.1. It is apparent from a comparison between the two types of compounding equipment that corotating twin-screw extrusion offers more versatile options for both reactive extrusion and incorporating a wide spectrum of ingredients to modify polypropylene to meet increasing demands in the marketplace.

### 10.3.2 Composite of Modular Elements

With the availability of slip-on sections of screw bushings and kneading blocks, the processor is equipped with a wide variety of options to tailor a screw design that suits the compounding process needed to attain desired product properties. This building block approach is described by White [21] as being a composite system that is influenced by the performance of individual modules in the

TABLE 10.1  Comparison Between Corotating and Counterrotating Intermeshing Twin-Screw Extruders

| | Corotating | Counterrotating |
|---|---|---|
| | $L/D$ ratio up to 60:1, screw speed up to 1400 rpm | $L/D$ ratio usually < 24:1, screw speed usually < 150 rpm |
| Distributive mixing | Good to excellent<br>This is due to high degree of radial mixing between pairs of intermeshing elements with material transport from one shaft to the other at a given axial location | Poor |
| Dispersive mixing | Good | Excellent<br>The movement of material between the calender gap of the two screws is similar to Banbury-type mixing |
| Flow in longitudinal direction | Open system | Closed system |
| Material conveying mechanism | Viscous drag at barrel surface | Positive displacement |
| Flow stability (surging) | Fair | Excellent |
| Pressure generation | Moderate | High |
| Venting capability | Fair to good | Fair |
| Feeding capability | Good | Good |
| Molten flow pattern in 8-figure cross-section | Flow around two screws just like one oval single-screw axis | Because of restraints on closed C chambers, material is forced through calender gap between screws |
| Attrition of glass fibers | Lower degree of glass fiber break-up | Higher break-up of glass fibers with lower mechanical properties |
| Melt purge properties of $TiO_2$ cleaning time characteristics | Slower with self-wiping effect—much broader residence time distribution | Rapid due to sharper residence time distribution |
| Length of melting region | Longer in continuous screw bushing | Shorter length |
| Mean residence time | Same for equal free volume | Same for equal free volume |
| Relative operational screw speed | High | Low |
| Relative throughput rate | Higher at same torque | Lower at same torque |

Source: Refs. 18 and 20.

composite modular twin-screw extruder. A consistent procedure of how to effectively obtain modular design with minimum trial and error has been the subject of many investigations [21–29].

The central theme of modular design is the same for any effective screw configuration. There must be a sufficient degree of conveying mechanism to keep control of the mixing process for a good admixture of dispersive and distributive mixing that is appropriate for a given compounded material. This is accomplished by an appropriate arrangement of forward (right-hand) and reverse (left-hand) screw bushings with given lead length and staggered arrays of kneading disks in blocks with given design features (offset angle, orientation, and thickness). In addition, neutral kneading blocks (disks arranged at 90° offset angle) are generally placed between forward–reverse melt pumping sections to provide a means of controlling the holdup volume of a given mixing zone.

*Compounding Powdered Polyvinyl Chloride Blends.* To exemplify the need for controlled extrusion processing for thermally sensitive polymer resins, let us consider the situation of compounding polyvinyl chloride (PVC) powdered blends into pellet form. The compounding of PVC resin is best accomplished via a slowly rotating screw at about 40–60 rpm for an intermeshing counterrotating extruder. By acting as a positive displacement pump, the discharge screws (closed system of C-shaped chambers) can generate high die head pressures, particularly for rigid PVC resins. This attribute combined with operating at low screw speed prevents excessive heating of PVC to minimize thermal degradation.

In comparison, the corotating twin-screw extruder requires a more delicate balance of radial–axial mixing and melt pumping through the die head due to thermal sensitivity of the PVC resin. Consequently, powdered PVC densification into pellets provided a good model for studying the effects of screw design for a compounding process with severe constraints on both residence time and degree of shear mixing of ingredients.

A variety of blends ranging from flexible to rigid PVC base resin were compounded via a 50-mm intermeshing corotating twin-screw extruder [22]. An array of screw design configurations consisted of a block of six kneading disks arranged in forward staggered offset angles of 30, 45, 60, and 90° followed by a pair of neutral elements of wide or narrow thickness. The extruder was equipped with a barrel value [23] that provided external control of the holdup volume. The split-barrel feature of the twin-screw extruder permitted *immediate* inspection of the steady-state material profile in the extruder barrel after immediate dead stop of the drive motor.

The powdered PVC feed went through a distinct sequence of process steps along the barrel length that are described as a CDFE mechanism: *C*ompaction, *D*ensification, *F*usion, *E*longation. By visual assessment of steady-state mass and axial melt temperature–pressure measurements, this investigation provided a means for characterizing the dynamic effects of screw design. In particular, the location of the compressed PVC plug is just before onset of plastification (continuous viscous mass). This is the *fulcrum point* for the appropriate dynamic balance between solids pressure buildup during compression of the PVC powder and the drag–pressure flow characteristics of the downstream plasticated mass.

The so-called melting length $L$ (PVC does not melt in this process but becomes plasticated) was defined as the axial distance between the generated compressed plug of PVC powder and the discharge end of the mixing section. The magnitude of $L$ increased with the offset angle between adjacent kneading disks (i.e., 90° or neutral elements). This increased length corresponded to excess heat rise of the PVC, leading to onset of thermal degradation.

These results stress that at the heart of any intermeshing corotating twin-screw extrusion process is the melting or plastification step [19]. By viscous drag of the highly viscous polymer resin at its melting point, the generated shear stress at the extruder barrel is maximal for intense dispersive mixing of ingredients. It is primarily at this point in the compounding process that there is effective reduction in domain size of any aggregates of filler and any other ingredient present in the feed stream.

### 10.3.3 Recent Trends in High-Volume Compounding Equipment

In lieu of the existing trends in the industry to high-volume equipment at minimal space requirement, there is an inherent need to consider any shear rate constraints on modular design to optimize compounding extrusion processes at high screw speeds. The key to effective modular screw design is how to best use the maximum viscous stresses generated at the onset of the melting or plastification process for the most effective size reduction of aggregates into finely dispersed particles.

With a desire for greater throughput rates for minimally sized equipment, a recent review article of existing trends in the compounding equipment industry [24] indicates a move primarily to corotating intermeshing equipment. A growing list of manufacturers featuring this type of equipment includes Farrel Corp., Welding Engineers-Toshiba, Davis Standard, Berstorff Corp., and Werner & Pfleiderer. As noted in Table 10.1, corotating twin-screw equipment favors high screw speeds at given motor loading. With higher quality gear boxes and better machine design, the energy density in a relatively small-volume extruder is intensified; hence, the notion of small mass shearing takes a whole new meaning at 600–1400 rpm for 8000–15,000 lb/hr for units having 100–133 mm bore diameter.

Obviously, the functionalization of molten polypropylene requires much longer residence times than attained by such high-volume equipment. Nevertheless, the *fundamental principles* are the same for a wide range of extrusion processes, i.e., chemical coupling via separate reactive extrusion steps and typical composite compounding processes involving the incorporation of chemical coupling agents, fillers, fibers, flame retardants, and other additives to modify the physicochemical properties of virgin polypropylene resins.

In the next section, general concepts of kneading block flow are described that relate machine design to viscous drag of molten polymer. This fundamental approach provides insights into effective arrangement of kneading elements for general use of twin-screw extrusion equipment featuring the corotating type of methodology.

## 10.4 FUNDAMENTAL CONCEPTS OF MODULAR COROTATING EXTRUSION

### 10.4.1 Newtonian Flow Model

Huneault et al. [25] describes the flow patterns in the channels of a twin-screw extruder as being very complex in nature. This is due to a nonuniform shear deformation of molten mass in the stress field of varying types of flow streamlines in axial and radial directions. For a bilobe-shaped kneading disk, there are three flow channels in all. Fundamentally, molten flow characteristics are a function of localized buildup of back pressure in an array of kneading blocks against a given downstream flow restriction, the molten material viscosity $\eta$, and the holdup volume with axial length $L$ of the molten mass in a given section of screw bushing or kneading disk elements. These authors conclude that there are dispersive or distributive types of mixing depending on the flow dynamics for a given kneading element.

Even though polypropylene composites exhibit non-Newtonian flow behavior with shear-dependent melt viscosity, fundamental concepts of flow mechanisms can be visualized more easily in the limit of Newtonian flow characteristics. Consequently, both White [21] and Todd [26] have made an analogy between drag flow in a given kneading block of elements with the flow characteristics of a conventional single-screw extruder with the following simple expression:

$$Q = AN - B\Delta P/\eta L \tag{10.1}$$

where $A$ and $B$ are geometrical parameters corresponding to drag and pressure flow, respectively. The throughput rate $Q$ is directly dependent on the screw speed $N$ and back pressure $\Delta P$. The pressure flow term is inversely proportional to the melt viscosity $\eta$ and backup length $L$.

In Fig. 10.3, the forward arrangement of bilobe disks at an offset angle of 30° generates a tip-to-tip arrangement that is much like a continuous single-screw helix. This is the basis for representing molten flow in the intermeshing kneading disks by a single-screw model depicted by Eq. (10.1).

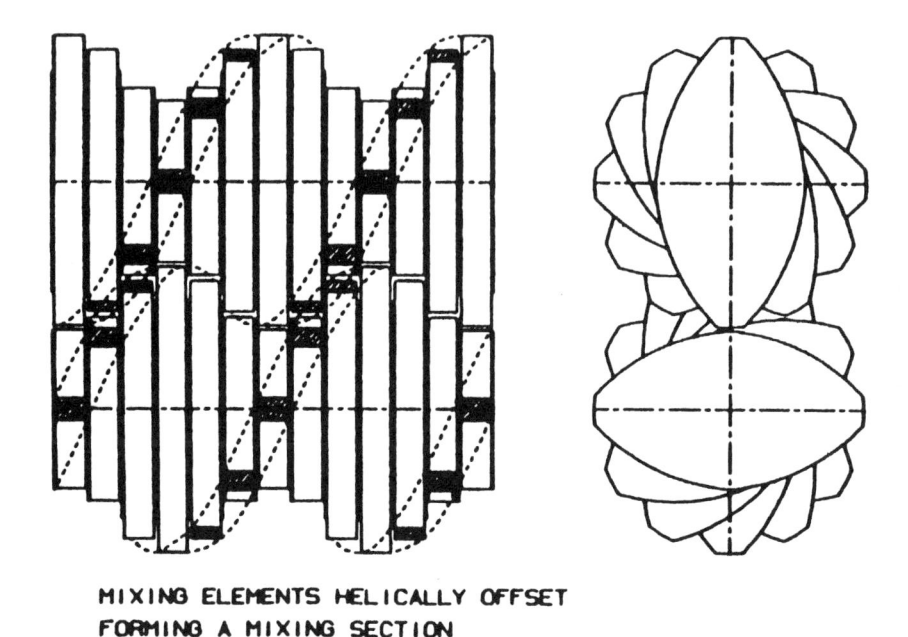

MIXING ELEMENTS HELICALLY OFFSET
FORMING A MIXING SECTION

**FIGURE 10.3** Helical arrangement of kneading disks in staggered array with off-set angle of 30°. (From Ref. 29, courtesy of APV.)

Please note that in Fig. 10.3, the repeating sequence of kneading disks for 180° rotation is 7 for the 30° staggered offset angle. The corresponding sequence is 5 and 4 for 45° and 60° offset angles, respectively. To facilitate matching of neighboring bilobe orientation, groups of kneading elements in manufactured block lengths consist of 7, 5, and 4 for the corresponding offset angles of 30°, 45°, and 60°, respectively.

White et al. [21,27,28] used Eq. (10.1) to determine the backup length $L$ of given kneading block design with given throughput rate $Q$, screw speed $N$, fluid viscosity η, and axial back pressure due to a downstream restriction. The various options for flow restriction are several left-hand screw bushing sections, a left-hand kneading block, an in-line barrel valve with variable setting [23], or simply the back pressure due to an array of capillary holes in the die head at the exit of the extruder.

### 10.4.2 Determination of Drag–Pressure Flow Parameters

Todd [26] characterized the flow behavior of a polybutene fluid in both rotating and static kneading disk arrangements of a 50-mm corotating twin-screw extruder. Using a viscous drag flow model, the values of $A$ and $B$ parameters for an array of kneading disk designs were determined by Karian (private communication) and verified later by Loomans et al. [29].

To visualize backup lengths as a function of kneading disk arrangements, screw speed, throughput rate, and flow restriction, the extruder barrel consisted of a transparent plastic. The blocks of kneading disk elements consisted of various sets of staggered elements with an array of offset angles (30, 45, or 60°) and disk thickness $W$ ($D/8$, $D/4$, or $D/2$ for $D$ equal to 50 mm or about 2-in. bore diameter). By using polybutene as the model fluid having about 1000 poise viscosity at room temperature, the pumping characteristics were measured by regulating the die head pressure via a regulating value at the end of the extruder length. A liquid-type pressure gauge gave accurate measurements of fluid pressure. After analyzing measured throughput rates versus measured die head pressure at given screw speed and backup length, the parameter values of $A$ and $B$ were empirically determined.

By using a flat plate model of drag and pressure flow in blocks of kneading disks, Loomans et al. [29] used Cad-Cam analysis to define flow patterns along the downstream flow channels of the unwrapped moving barrel surface. Figures 10.4 and 10.5 represent flow patterns as a function of staggered offset angle and kneading disk thickness, respectively.

In the flat plate model, the barrel moves over the unwrapped channels of the stationary shafts. By using simple vector analysis, the magnitude and direction is determined for forward $U_f$ and reverse $U_r$ flow vectors that are axial components of the peripheral velocity $U(\pi DN)$ due to rotation speed $N$ of the intermeshing twin screws.

The angular directions of forward and reverse rotation vectors were dependent on the given kneading disk design (i.e., staggered angle and disk thickness). This effort led to the following expression for $A$ in terms of rotation speed components and the cross-sectional areas of flow channels in the forward $A_f$ and reverse directions $A_r$. Figure 10.6 provides cross-sectional areas of neighboring kneading disks that are offset at various staggered angles.

The following mathematical expression used the above geometrical results to theoretically compute the values for $A$ for different kneading disk geometries:

$$A = \text{(volumetric flow rate} \div \text{screw speed)} \\ = (\pi D/2U)(3A_f U_f - 3A_r U_r) \tag{10.2}$$

The factor 3 in Eq. (10.2) comes from the three flow channels due to twin lobe-shaped kneading disks.

Tables 10.2 and 10.3 provide a summary of the above two approaches to determine the magnitude of drag flow parameter $A$. The corresponding measured pressure flow parameter $B$ is given in Table 10.4.

| OFFSET ANGLE | 30° | 45° | 60° |
|---|---|---|---|
| Unwrapped Block of Kneading Disks with Disk Thickness of D/4<br><br>Forward and Reverse Flow Channels | | | |
| Down Channel Flow Patterns | | | |
| Projections of Forward-Reverse Velocity Components Along Axial Length of Screw With Rotation Speed U | | | |

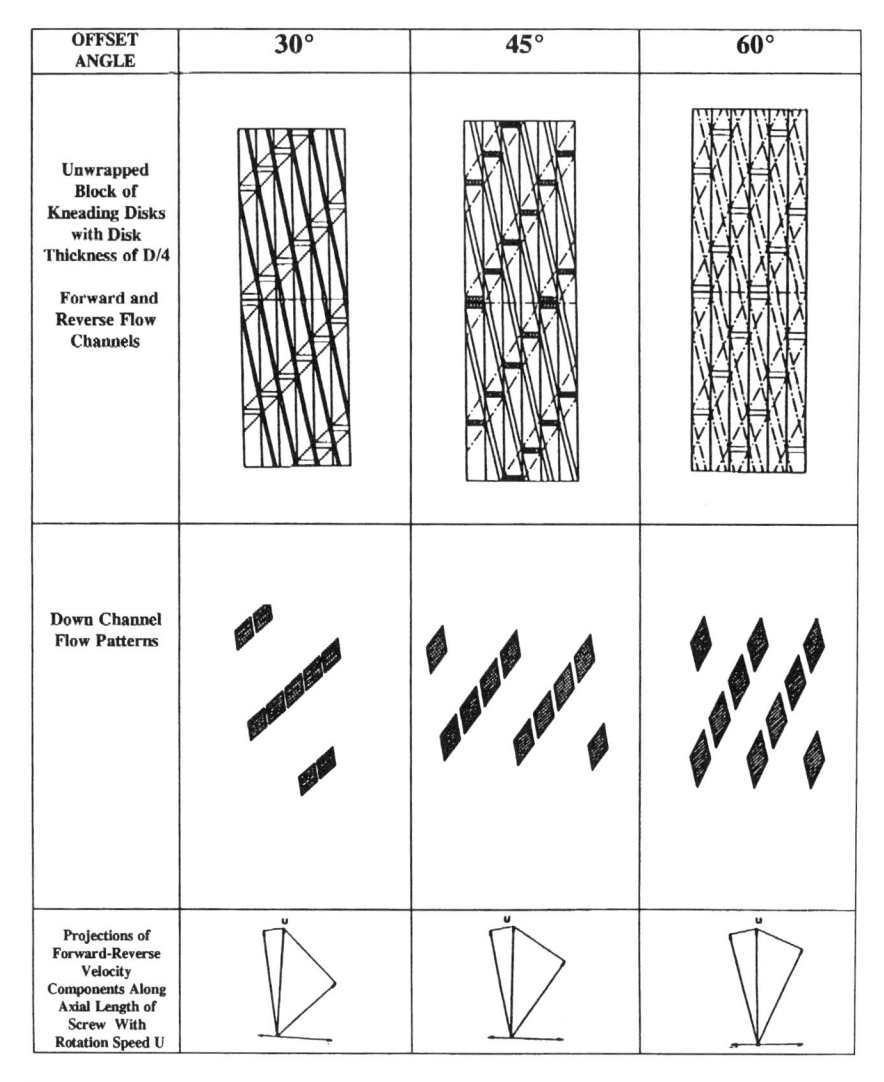

**FIGURE 10.4** Flow patterns in unwrapped array kneading disks with different off-set angle. (From Ref. 29, courtesy of Mr. Ben Loomans, APV.)

### 10.4.3 Applications of Flow Characteristics

*Analysis of Drag-Pressure Flow Parameters for Kneading Block Design.* For a given kneading element design, the drag flow parameter $A$ has a positive value for forward conveying of fluid that corresponds to right-handed offset angles for most European-designed gear boxes. For reverse or left-handed

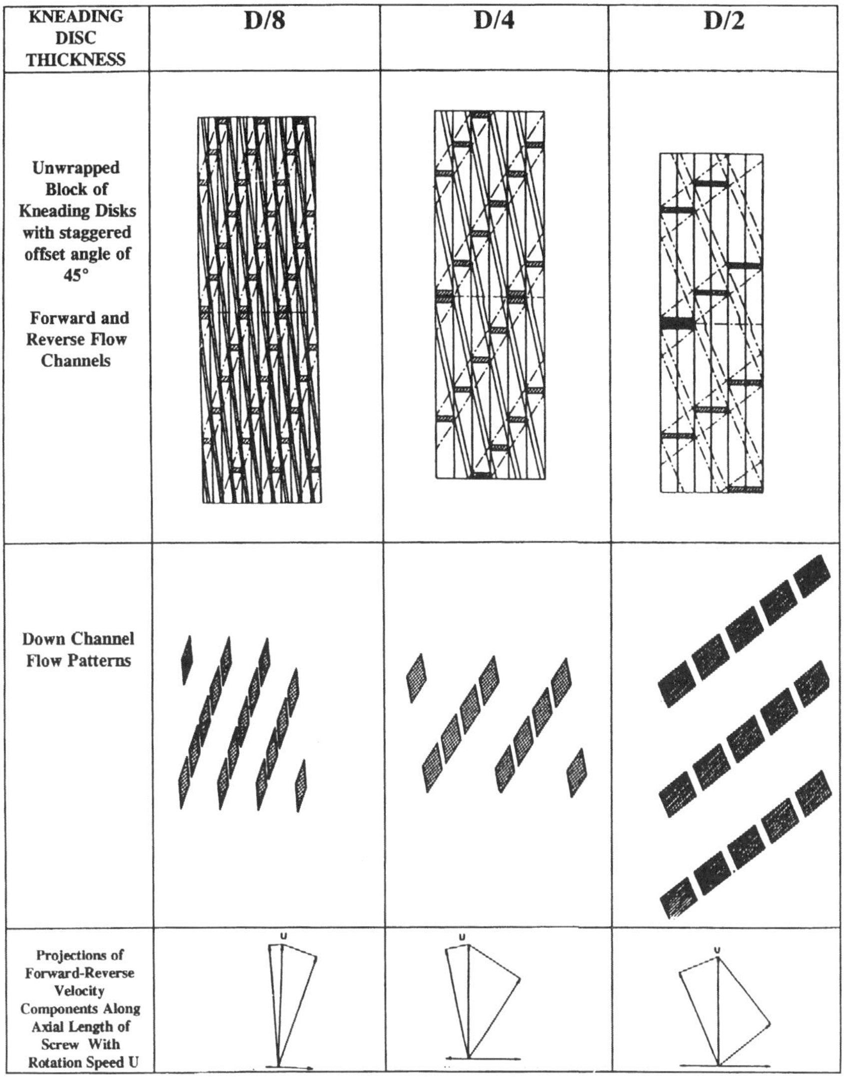

# Use of Twin-Screw Extruders

| KNEADING DISC THICKNESS | D/8 | D/4 | D/2 |
|---|---|---|---|
| Unwrapped Block of Kneading Disks with staggered offset angle of 45° Forward and Reverse Flow Channels | | | |
| Down Channel Flow Patterns | | | |
| Projections of Forward-Reverse Velocity Components Along Axial Length of Screw With Rotation Speed U | | | |

FIGURE 10.5  Flow patterns of unwrapped array kneading disks with different thickness. (From Ref. 29, courtesy of Mr. Ben Loomans, APV.)

| CROSS-SECTIONAL VIEW OF FLOW CHANNELS | Staggered Offset Angle | Full Channel Area in$^2$ | Forward F Channel Area $A_f$ (in$^2$) | Reverse R Channel Area $A_r$ (in$^2$) |
|---|---|---|---|---|
| 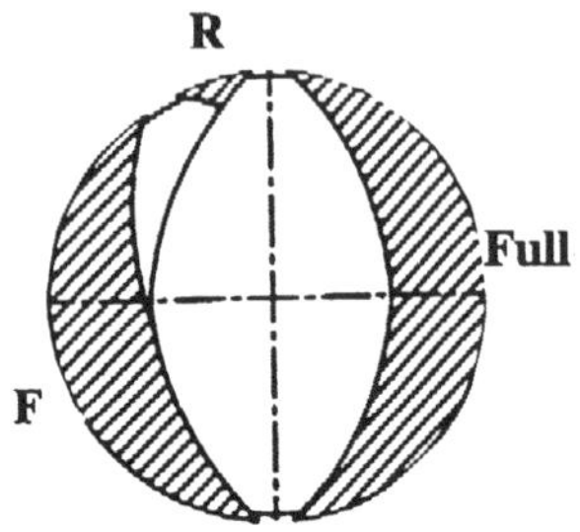 | 30 | 0.7919 | 0.6094 | 0.0247 |
| 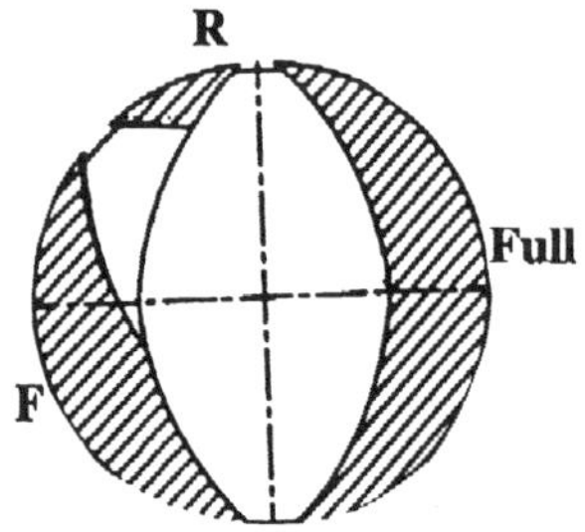 | 45 | 0.7919 | 0.5192 | 0.0663 |
| 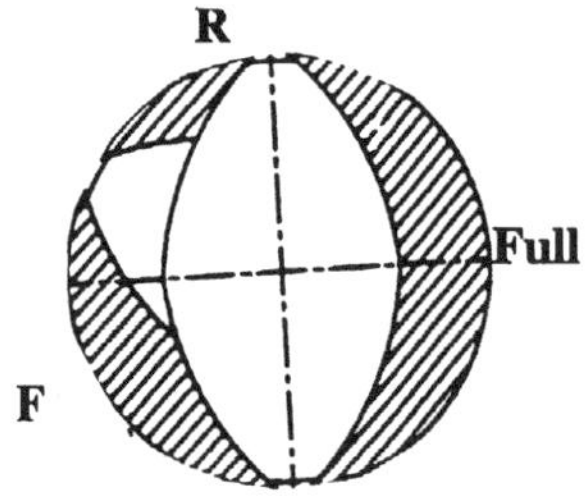 | 60 | 0.7919 | 0.4310 | 0.1224 |

FIGURE 10.6 Cross-sectional view of flow channels. (From Ref. 29, courtesy of Mr. Ben Loomans, APV.)

elements, the value of $A$ is negative. In the special case of neutral elements with 90° staggered arrangement of adjacent kneading disks, the value of the drag parameter is zero.

Trend data in Figs. 10.7–10.9 yield notions concerning the relationship between the staggered arrangement of bilobe-shaped kneading elements and the degree of mixing required at a given axial location.

## Use of Twin-Screw Extruders

**TABLE 10.2** Drag and Pressure Flow Parameters

| Kneading disk design features | | Drag flow parameter $A$, in.$^3$ (cm$^3$) | | Pressure flow parameter $B$, in.$^4$-sec (cm$^4$-sec) | |
|---|---|---|---|---|---|
| Offset angle | $W$ | Measured [26] | Calculated [29] | Dynamic measurement [26] | Static measurement [26] |
| 30 | $D/4$ | 3.118 (51.1) | 2.823 (46.3) | 0.01220 (0.508) | 0.01281 (0.533) |
| 45 | $D/8$ | 1.141 (18.7) | 1.349 (22.1) | 0.00476 (0.198) | 0.00466 (0.194) |
| 45 | $D/4$ | 1.898 (31.1) | 2.090 (34.2) | 0.00836 (0.348) | 0.00812 (0.338) |
| 45 | $D/2$ | 2.222 (36.4) | 2.153 (35.3) | 0.01449 (0.603) | |
| 60 | $D/8$ | 0.348 (5.7) | 0.782 (12.8) | 0.00548 (0.228) | |
| 60 | $D/4$ | 1.092 (17.9) | 1.319 (21.6) | 0.00879 (0.366) | 0.00853 (0.353) |
| 60 | $D/2$ | 1.397 (22.9) | 1.578 (22.9) | 0.01170 (0.487) | |
| 90 | $D/4$ | 0.000 | 0.000 | | 0.01307 (0.429) |

Figure 10.7 shows the angular dependence of drag flow on the staggered arrangement of kneading disk elements having a thickness of $D/4$. A comparison between computed and measured values of $A$ is quite good. The given plots indicate that lower staggered angles favor greater drag flow of molten polymer.

Figure 10.8 provides comparison between measured and calculated values for parameter $A$ versus kneading disk thickness for 45° and 60° offset angles.

The corresponding plots of measured values of $B$ pressure flow parameter versus kneading disk thickness are shown in Fig. 10.9. The increase in $B$ with disk thickness implies greater shear work using an array of wide kneading disks. Conversely, the narrower kneading disk elements are more prone to enhancing distribution of ingredients at minimal shear deformation.

*Computation of the Axial Fluid Pressure Profile for Modular Design.* A series of pressure transducers can be installed in the extruder barrel to determine the axial fluid pressure profiles located at various positions in the kneading blocks. Without the option of a transparent plastic barrel, one can measure the buildup of fluid pressure before downstream flow restriction and thereby determine the backup length $L$.

TABLE 10.3  Computation of Drag Flow Parameter

| Offset angle | W | $\theta_f$ (degrees) | $\theta_r$ (degrees) | $U_f/U$ | $U_r/U$ | $A_f$ (in.$^2$) | $A_r$ (in.$^2$) | Calculated A (in.$^3$) | Calculated A (cm$^3$) |
|---|---|---|---|---|---|---|---|---|---|
| 30 | D/4 | 43.68 | 10.81 | 0.499 | 0.184 | 0.6094 | 0.0247 | 2.823 | 46.3 |
| 45 | D/8 | 17.67 | 6.06  | 0.289 | 0.105 | 0.5192 | 0.0663 | 1.349 | 22.1 |
| 45 | D/4 | 32.50 | 11.98 | 0.453 | 0.203 | 0.5192 | 0.0663 | 2.090 | 34.2 |
| 45 | D/2 | 51.85 | 23.00 | 0.486 | 0.360 | 0.5192 | 0.0663 | 2.153 | 35.3 |
| 60 | D/8 | 13.42 | 6.81  | 0.226 | 0.118 | 0.4310 | 0.1224 | 0.782 | 12.8 |
| 60 | D/4 | 25.52 | 13.42 | 0.389 | 0.226 | 0.4310 | 0.1224 | 1.319 | 21.6 |
| 60 | D/2 | 43.68 | 25.52 | 0.499 | 0.389 | 0.4310 | 0.1224 | 1.397 | 22.9 |

Source: Ref. 29.

# Use of Twin-Screw Extruders

TABLE 10.4  Axial Fluid Pressure Profile Calculation for Modular Design

|  | Kneading section 1 | Kneading section 2 | Kneading section 3 |
|---|---|---|---|
| Orientation direction of staggered arrangement of kneading disks | Right handed (forward) | Neutral | Left handed (reverse) |
| Number $n_i$ of disk elements per section | 5 | 2 | 5 |
| Disk thickness $W_i$, fraction of $D$ | $D/4$ | $D/4$ | $D/8$ |
| cm | 1.27 | 1.27 | 0.635 |
| in. | 0.50 | 0.50 | 0.25 |
| Drag flow parameter $A_i$ (cm$^3$) | +31.1 | 0.0 | −18.7 |
| Pressure flow parameter $B_i$ (cm$^4$) | 0.348 | 0.429 | 0.198 |
| Fluid pressure $P(L)$, kPa |  |  |  |
| Beginning of section | 0 | 44.6 | 40.7 |
| End of section | 44.6 | 40.7 | 0.0 |
| Net change | +44.6 | −3.9 | −40.7 |
| Kneading block length $L$ (in.) | 2.5 | 1.0 | 1.25 |
| Drag flow term kPa $\eta N n_i W_i A_i B_i$ | 56.8 | 0.0 | −30.0 |
| Pressure flow term $-\eta Q n_i W_i / B_i$ | −12.2 | −3.9 | −10.7 |
| $\Delta P_i$ | +44.6 | −3.9 | −40.7 |

Combined axial fluid pressure = [drag term] − [pressure term] = $[N n_i W_i(A_i/B_i)]$ − $[Q n_i(W_i/B_i)]$
Extruder: bore diameter $(D) = 50$ mm
Process conditions: Throughput rate $(Q) = 6.66$ cm$^3$/sec
Screw speed $(N) = 60$ rpm $= 1$ rps
Fluid viscosity $= 1,000$ poise $= 100$ Pa/sec.

As the fluid flows through the restriction region, the fluid pressure drops to zero. The slope of this dropoff is proportional to the characteristics of the given restriction, for example, die pressure is predictable for given fluid viscosity, capillary die hole geometry, and throughput rate. In the more practical situation, the flow restriction is located at the end of a given mixing zone. Different arrangements of kneading block designs may be stacked in succession as a composite of modular characteristics necessary for dynamic balance of viscous forces in conveying and back-pressure flow directions.

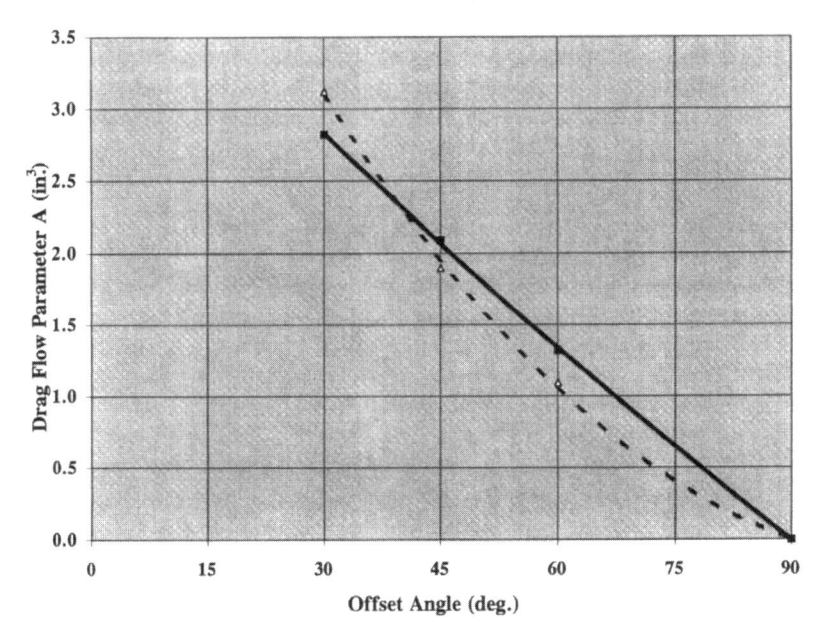

**FIGURE 10.7** Drag flow parameter A versus offset angle. —■—, Measured data (26); – –△– –, calculated data [29]. (From Refs. 26 and 29.)

Under isothermal conditions (constant fluid viscosity), this buildup is expressed in terms of summation of individual kneading characteristics using the Newtonian model:

$$L = \sum_i L_i \qquad (10.3)$$

$$P_{\text{restriction}} = \sum_i \Delta P_i \qquad (10.4)$$

$$P_{\text{restriction}} = \eta \left[ N \sum_i (W_i A_i / B_i) - Q \sum_i (W_i / B_i) \right] \qquad (10.5)$$

To demonstrate the utility of knowing the magnitude of $A$ and $B$ parameter values for a given set of kneading disk designs, let us consider the flow dynamics of a modular screw design consisting of the following arrangement of three sections of kneading blocks for a bore diameter of 50 mm with a screw speed ($N$) of 60 rpm (1 rps) and a throughput rate ($Q$) of 6.66 cm$^3$/sec:

1. Five forward kneading disks
2. Two neutral kneading disks
3. Five reverse kneading disks

## Use of Twin-Screw Extruders

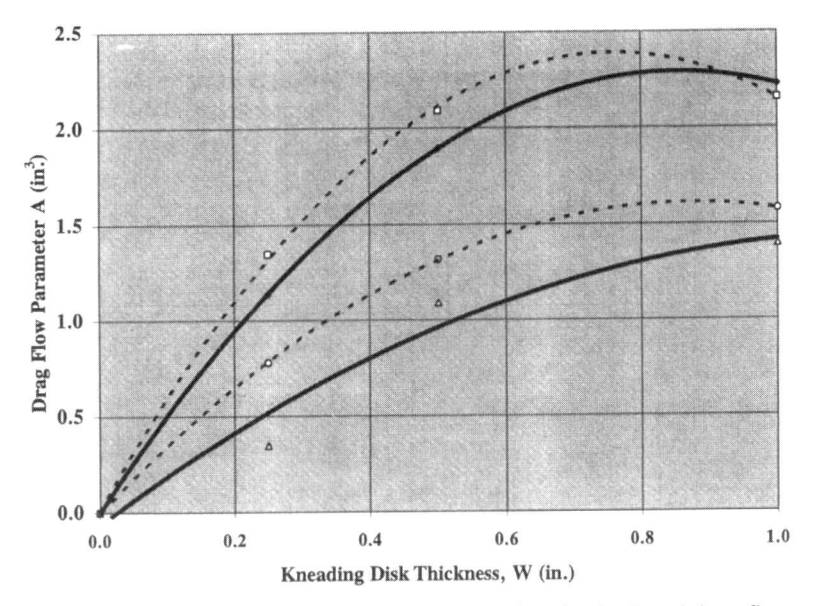

**FIGURE 10.8** Comparison between measured and calculated drag flow versus kneading disk thickness. △, measured values—60° offset angle [26]; O, calculated values—60° offset angle [29]; ♦, measured values—45° offset angle [26]; □, calculated values—45° offset angle [29]. (From Refs. 26 and 29.)

Table 10.4 provides a summary of design-process input data with the computation of axial pressure via Eq. (10.5). Figure 10.10 depicts the buildup of melt pressure in section I of the screw design followed by incremental decrease in pressure value as one encounters sections II and III. Ultimately, by decompression, the fluid pressure drops to zero at 4.75-in. axial length. The accompanying Fig. 10.11 provides a side by side comparison of the axial pressure change per individual kneading disk element.

*Notions of Distributive Mixing for Kneading Block Design.* The flow characteristics of kneading block design correlate well with recent numerical analyses of corotating intermeshing twin-screw extrusion models. For example, Huneault et al. [25] developed a relationship between normalized recirculation and disk thickness. Recirculation number is the average number of times that fluid goes through a given disk. The increase in disk thickness was found to reduce recirculation but to increase shear rate in the kneading block. Consequently, wider kneading disks are best positioned in the solid–melt transition position where shear stress is maximal and opportunity for dispersive mixing exists. Conversely, narrow kneading disks feature lower shear rate combined with a very high degree of recirculation patterns, giving enhanced distributive mixing.

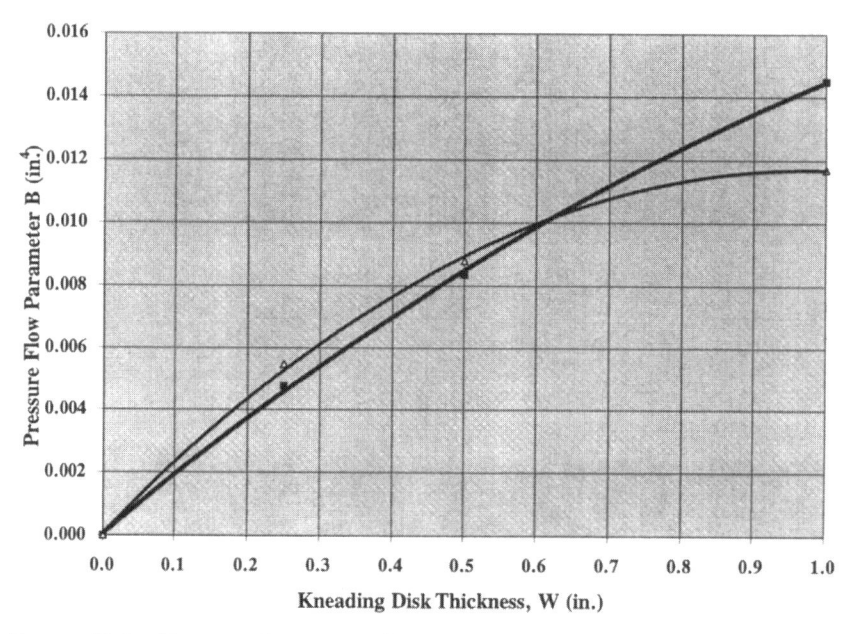

**FIGURE 10.9** Measured pressure flow parameter versus kneading disk thickness. △, 60° offset angle; ■, 45° offset angle. (From Ref. 26.)

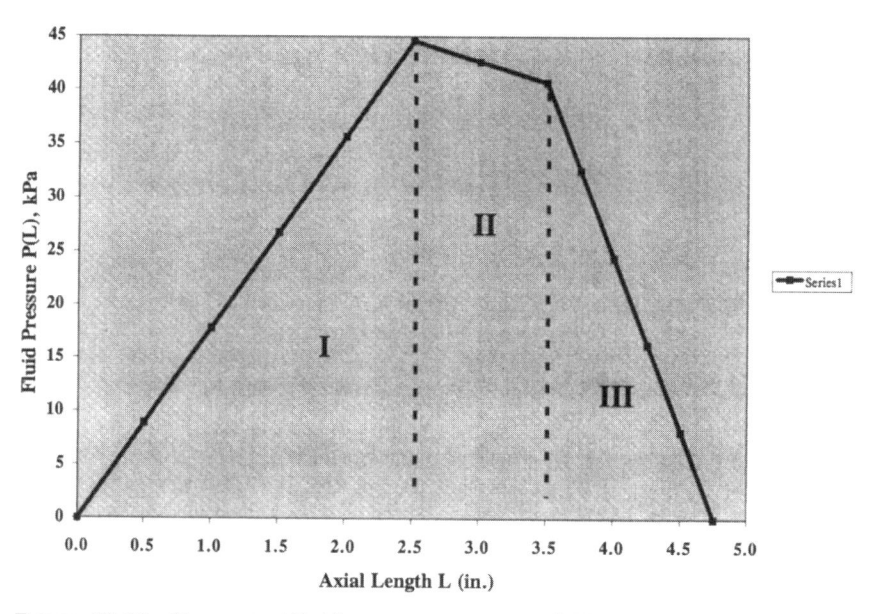

**FIGURE 10.10** Computed fluid pressure versus axial length profile: I, pressure buildup along right-handed kneading block I; II, pressure decrease along neutral kneading block II; III, pressure decrease along left-handed kneading block III.

# Use of Twin-Screw Extruders 409

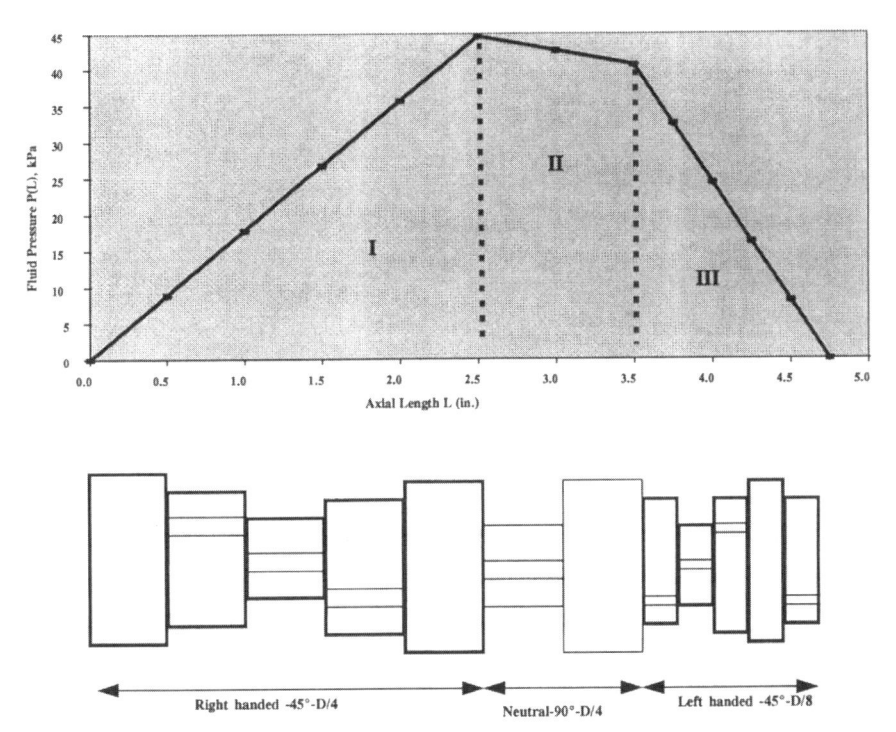

**FIGURE 10.11** Side by side comparison pressure plot—modular screw design.

The same authors assessed the influence of stagger angle between the kneading elements on the flow. Their conclusions mirror the empirical data obtained via plastic barrel extrusion studies.

The relationship between kneading disk design and the degree of distributive mixing was characterized by Karian (unpublished results reported in Ref. 23). Aggregates of titanium dioxide particles are relatively easy to disperse in polyolefin resin. However, the dispersed particles can reagglomerate if there is inadequate distributive mixing to keep the particles apart. An array of staggered kneading blocks featuring an array of offset angles and disk thicknesses were used to compound a 50 wt% concentrate of titanium dioxide ($TiO_2$) in linear low-density polyethylene (LLDPE) using a 50-mm-diameter corotating twin-screw extruder. About a 10:1 let-down of the compounded pellets with virgin LLDPE was blow molded into thin film as a means of allowing visual assessment of the degree of dispersion due to given kneading block design. The number of $TiO_2$ specks observed per gram of film was plotted as a function of offset angle for three sets of disk thicknesses. Figure 10.12 provides a comparison of the results.

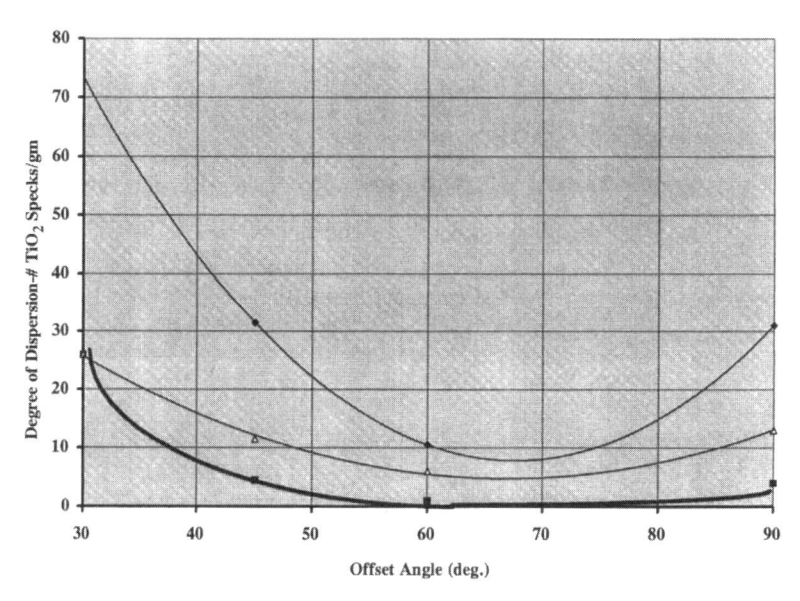

**FIGURE 10.12** Degree of dispersion of $TiO_2$ particles versus offset angle. ■, $D/8$ disk thickness; △, $D/4$ disk thickness; ◆, $D/2$ disk thickness. (From Ref. 23.)

The plot corresponding to the thinnest disk ($D/8$) depicts the overwhelming effect of distributive mixing. A minimum number of $TiO_2$ specks/g corresponds to a kneading block design having the most effective distributive mixing: 60° offset angle with $D/8$ thickness.

## 10.5 CONCLUDING REMARKS

We used simple Newtonian models to help the reader grasp some key fundamental concepts of molten flow in the channels of intermeshing corotating twin-screw extruders. Hopefully, notions of dispersive and distributive mixing can be better understood in terms of the dynamic balance of conveying action of rotating screws acting against the back pressure of various flow restrictions. As an application of these conceptual models, the modification of polypropylene resin by reactive extrusion and general compounding of fillers or fibers may reach a common ground for general considerations of composite modular design.

## REFERENCES

1. RC Kowalski. Fit the reactor to the chemistry. In: Reactive Extrusion: Principles and Practice. New York: Hanser Publishers, 1992, pp. 7–32.

2. T Bartilla, D Kirch, J Nordmeier, E Promper, T Strauch. Physical and chemical changes during the extrusion process. Adv Polym Technol 6:339–387, 1986.
3. C Tzoganakis, J Vlachopoulos, AE Hamielec. Controlled degradation of polypropylene. Chem Eng Progr 84(11):47–49, 1988.
4. M Xanthos. Process analysis from reaction fundamentals. In: Reactive Extrusion: Principles and Practice. New York: Hanser Publishers, 1992, pp. 33–53.
5. R Lew, P Cheung, ST Balke. Reactive extrusion of polypropylene elucidating degradation kinetics. In: T Provder, ed. Computer Applications in Applied Polymer Science II. ACS Symposium Series 404, 1989, pp. 507–520.
6. AE Hamielec, PE Gloor, S Zhu. Kinetics of free radical modification of polyolefins in extruders: chain scission, crosslinking and grafting. Can J Chem Eng 69:611–618, 1991.
7. BJ Kim, JL White. Thermal/peroxide induced degradation and maleation of polypropylene by reactive extrusion. Int Polym Proc 10(3):213–220, 1995.
8. J Curry, S Jackson, B Stoehrer, A van der Veen. Free radical degradation of polypropylene. Chem Eng Progr 84(11):43–46, 1988.
9. SB Brown. Reactive extrusion: a survey of chemical reactions of monomers and polymers during extrusion processing. In: Reactive Extrusion: Principles and Practice. New York: Hanser Publishers, 1992, pp. 75–199.
10. NG Gaylord. Reactive extrusion in the preparation of carboxyl-containing polymers and their utilization as compatibilizing agents. In: Reactive Extrusion: Principles and Practice. New York: Hanser Publishers, 1992, pp. 55–71.
11. WY Chiang, WD Yang. Polypropylene composites. I. Studies of the effect of grafting of acrylic acid and silane coupling agent on the performance of polypropylene-mica composites. J Appl Polym Sci 35:807–823, 1988.
12. NG Gaylord, MK Mishra. Nondegradative reaction of maleic anhydride and molten polypropylene in the presence of peroxides. J Polym Sci Polym Lett 21:23–30, 1983.
13. NG Gaylord, R Mehta. Peroxide-catalyzed grafting of maleic anhydride onto molten polyethylene in the presence of polar organic compounds. J Polym Sci A Polym Chem 26:1189–1198, 1988.
14. CS Wong. U.S. Patent 4,857,254. DuPont Canada, 1989.
15. H Cartier, G-H Hu. Plastification or melting: a critical process for free radical grafting in screw extruders. Polym Eng Sci 38:177–185, 1998.
16. RJ Nichols, F Kheradi. Reactive modification of polypropylene with a non-intermeshing twin screw extruder. Proceedings of 3rd Chemical Congress of North America, Toronto, Canada, 1988.
17. MA Ramos, FA Belmontes. Polypropylene/low density polyethylene blends with short glass fibers. II. Effect of compounding method on mechanical properties. Polym Compos 12(1):1–6, 1991.
18. W Thiele. Operating characteristics of intermeshing co-rotating and counterrotating extruders in reactive processing. Proceedings of the Second International Congress on Compatibilizers and Reactive Polymer Alloying. Compally '90. New Orleans: Shotland, 1990, pp. 149–163.
19. JL White, W Szylowski, K Min, M-H Kim. Twin screw extruders; development of technology and analysis of flow. Adv Polym Technol 7:295–332, 1987.

20. T Sakai, N Hashimoto, N Korayashi. Experimental Comparison Between Counter-Rotation and Co-Rotation on the Twin Screw Extrusion Performance. Los Angeles: SPE Preprints Antec '87, 1987, pp. 146–151.
21. JL White. Twin Screw Extrusion: Technology and Principles. New York: Hanser Publishers, 1990, p. 247.
22. HG Karian. Co-rotating twin screw compounding studies of PVC formulations. J Vinyl Technol 7:154–159, 1985.
23. D Gelok. Practical compounding of fire retardants: controlled shear in twin-screw compounders. In: Fire Retardant Engineering Polymers. San Antonio: Fire Retardant Chemicals Association, 1989, pp. 161–181.
24. What's New in Compounding Extrusion. Plastics Formul Compound 2:16–30, 1996.
25. MA Huneault, MF Champagne, A Luciani. Polymer blend mixing and dispersion in the kneading section of a twin-screw extruder. Polym Eng Sci 36:1694–1706, 1996.
26. DB Todd. Society of Plastics Engineers (SPE) Preprints, Antec '89, New York, 1989, p. 168.
27. JL White, Z Chen. Simulation of non-isothermal flow in modular co-rotating twin screw extrusion. Polym Eng Sci 34:229–237, 1994.
28. JL White, S Montes, JK Kim. Experimental study and practical engineering analysis of flow mechanisms and starvation in a modular intermeshing corotating twin screw extruder. Kunstoffe 43:20–25, 1990.
29. BA Loomans, JE Kowalczyk, JW Jones. Flow distribution in co-rotating twin shaft continuous mixers. American Defence Preparedness Association (ADPA) Compatibility and Processing Symposium, New Orleans, April 1988.

# 11

## Engineered Interphases in Polypropylene Composites

**Josef Jancar**
Technical University Brno, Brno, Czech Republic

## 11.1 INTRODUCTION

There is now more than 30 years of research and plenty of published data on the role of interphase layer in the mechanical response of filled semicrystalline thermoplastics, such as polypropylene (PP). Intuitively, it has been believed that this thin layer is responsible to a great extent for the variations of the mechanical properties of filled PP described in the literature. However, a clear interpretation of the experimental data has not occurred until the late 1980s and beginning of the 1990s. In the last two decades, a significant growth in the science and technology of composite materials has been achieved. It began with an interest in the properties of primary components (i.e., fillers and polymer matrices). As more knowledge has been accumulated, it became apparent that a third entity (the region in an imminent vicinity of the surface of a reinforcement) plays a profound role in the behavior of composites. This region was termed interface/interphase. As the importance and widespread use of particulate composites grew, interphase/interface phenomena have also been declared important for this class of materials. To estimate the extent of the internal surface, it seems worthwhile to note that in 1 kg of a common PP filled with 50 wt% of microground $CaCO_3$ with an average particle size of 5 μm interface area equals about three football fields.

### 11.1.1 Definition of Concepts

The *interface* is commonly defined as a perfect two-dimensional (2-D) mathematical surface dividing two distinguished phases or components in a composite. Interface is characterized by an abrupt change in properties and, frequently, in chemical composition. This 2-D surface does not have any physical properties itself. For the purpose of investigation of stress transfer from the matrix to the reinforcement, one can assume that all stress transfer phenomena take place at the interface, which is then characterized by a single property (i.e., interfacial shear strength). This approach is frequently used when there is an emphasis on a chemical bond between the constituents as a primary parameter controlling the mechanical response of a composite.

The *mesophase* concept was developed by Theocaris [1] and Lipatov [2]. In general, constituents of a composite material are insoluble in one another. Most frequently, inorganic filler or fibrous reinforcement form a discontinuous phase dispersed in a continuous polymer matrix. Ideally, there is a sharp 2-D interface between the two constituents. In reality, however, around an inorganic inclusion a complex situation exists due to imperfect bonding, surface topology of the filler, stress gradients, voids, and microcracks. Moreover, a presence of a solid in the polymer melt or monomer in the course of solidification or cure facilitates physical changes in the morphology of the polymer phase in a region near solid surface. This results, most probably, in a restriction of molecular mobility of molecules in this region especially in the case of strong interfacial interactions. Hence, it appears reasonable to expect that this layer, termed *mesophase*, possesses physical properties different from those of the polymer bulk.

The *interphase* is a three-dimensional (3-D) layer in the immediate vicinity of filler surface, possessing physical properties different from those of the two main phases or components in a composite (i.e., matrix and filler). For the purpose of this chapter, the term interphase is limited to the layers introduced on the filler surface intentionally in a controlled manner—engineered interphase layers (EILs). In these layers, a gradient of chemical composition can also exist as well as a gradient of physical properties. The pivotal problem is therefore obtaining a definition and an evaluation of an interphase thickness and its properties, namely, stiffness and fracture toughness. Interphase behavior plays a paramount role in the ability to transfer loads from the matrix to the reinforcements, hydrolytic stability of the material, and fracture behavior of a particulate composite.

## 11.1.2 Importance of Interphase Design

Treatment of the stress transfer phenomena as an interphase-governed problem is much closer to reality. However, the problem is substantially more complicated than the interface approach. Here one can assume that the stress transfer phenomena are concentrated in the interphase. A full set of physical properties, including elastic moduli, Poisson's constant, and fracture toughness, is then needed to characterize the interphase. The complexity of mathematical treatment of the interphase phenomena was the primary cause of the first closed form solutions of this problem being published only in the last decade [3,4]. Most attention has been paid to the use of numerical methods.

New concepts combining micromechanical models with the macromechanics of composite bodies could account for experimental data and predict limits of mechanical properties. Proposed models were used as the link between micro- and macromechanics of the composite body. In the calculations, in addition to properties of the matrix and the filler, properties and spatial arrangement of the interphase have been included [5]. This model allows for a prediction of the structure–property relationships in PP filled with randomly distributed core-shell inclusions with EIL shell. This is of a pivotal importance in an attempt to develop and manufacture materials "tailored" to a particular end-use application.

An interphase formation, its morphogenesis, and final structure are all extremely important for the resulting physical properties of the interphase and thus for the behavior of any composite material. The problem of interphase formation is, however, very complex and in many aspects still not completely understood. The rapid advances in the technology and application of a wide range of matrices of different polarities and chemical reactivity have led to a resurgence of interest in optimizing the interfacial bonding and the performance of the interphase region in recent years. Some of the latest results have shown that the physicomechanical properties of the interphase are the primary factors controlling mechanical response of particulate composites under common conditions, especially their fracture and impact behavior [6–9].

In the field of rigid inclusion-filled polymers, EILs formed by silane, titanate, and low molecular weight fatty acid agents are essential to good performance, easy processing, and excellent appearance. Apparently, most understanding of the phenomena, related to the effects these substances have on the interface/interphase in filled polymers, has been acquired from technological rather than fundamental grounds [10]. Since the 1960s, there has been a substantial effort to produce thin interphases of elastomers on the surface of common reinforcements. The major thrust in these investigations was a hypothesis that the presence of a thin, highly flexible layer on the surface of reinforcing fibers will contribute to an enhancement of composite toughness without jeopardizing its strength and stiffness.

In these studies, an advantage of "in situ" formation of chemically distinguished interphase was explored. Elastomers such as ethylene-propylene rubber (EPR) and ethylene-propylene-diene monomer (EPDM) were chemically modified to introduce polar groups onto their backbone chains. Most frequently, maleic anhydride and acrylic acid were used as the grafting comonomers [11–13]. In the process of melt mixing, carboxyl groups from the grafted comonomer react with hydroxyls, amines, or other suitable reactive groups present on the filler surface. Often the resulting bond is a mixture of contributions from several types of interactions ranging from covalent bonds through hydrogen bonds to electrostatic interactions. Hence, the term *acid–base interaction* is commonly used to describe interfacial bonding in these systems. Interphases formed in this fashion were about 500 nm thick, which was about one order of magnitude thicker than the silane or oligomer interphases, commonly 20–100 nm thick [14].

Further expansion of the use of high-volume polymeric materials, such as PP, in engineering applications (automotive, home appliances, mass transportation vehicles, and construction industries) is dependent on the ability to enhance both their stiffness and toughness. Fire safety concerns bring about an additional requirement of reduced flammability, especially for those used in the mass transport and construction industries. This extends the requirements placed on the properties of an interphase.

### 11.1.3 Attainment of Optimum Stiffness–Toughness Balance

The enhancement of PP stiffness is commonly achieved by incorporating rigid fillers or reinforcements, whereas the enhancement of fracture resistance is achieved through a blending with other polymers, mainly elastomers. Particulate pulverized fillers are used in polymers to reduce cost and modify physical properties, processability, and appearance. Carbonates ($CaCO_3$), metal oxides ($TiO_2$), metal hydroxides [e.g., $Al(OH)_3$ and $Mg(OH)_2$], carbon black, and silicates (mica) are among the most common fillers used. Glass, carbon, graphite, and organic fibers, such as Kevlar, ultrahigh molecular weight polyethylene (UHMW PE), and polyimide fibers are among the reinforcements most frequently used to enhance stiffness and strength of a material. Various EILs of prescribed properties, chemical structure, and thickness have to be used in composites containing these fillers and reinforcements. This methodology provides the means to tailor the stiffness–toughness balance according to the requirements of the end user [14].

Halogenated hydrocarbons, phosphorus-based flame retardants, and inorganic fillers such as alumina trihydrate, $Al(OH)_3$, and magnesium hydroxide, $Mg(OH)_2$, are used in high-volume polymers to minimize the evolution of dense smoke and acidic poisonous gases during combustion [15]. Incorporation of

Mg(OH)$_2$ into polypropylene resin also increases the ignition temperature, even at long exposition times [16]. However, filler loading in the range of 60–80 wt% (0.3 < $v_f$ < 0.5), necessary to achieve the required nonflammability, reduces the yield strength and fracture toughness of the composite material.

One of the shortcomings most adversely affecting the utilization of PP filled with flame retardant magnesium hydroxide [Mg(OH)$_2$] is the steep reduction of subambient fracture toughness compared with that of neat polypropylene. Generally, an increase in the resistance to crack initiation and growth can be achieved by modifying the stress state within the specimen, altering the size of the crack tip plastic zone, and inducing favorable mechanisms of matrix failure [17]. The most common method of obtaining these changes is to "rubber toughen" the matrix phase by the addition of elastomer inclusions. Under proper processing conditions, the presence of elastomer particles will reduce the matrix yield strength, thereby increasing the size of crack tip plastic zone and producing changes in the state of stress at the crack tip.

As was shown earlier [18], thermoplastic composites consisting of PP filled with rigid inorganic particles covered with EILs can be designed to be both stiffer and tougher than the neat PP resin. In addition to the effects of filler volume fraction and particle shape, interphase thickness, elastic moduli relative to those of the matrix, and adhesion to the matrix bulk play a crucial role in controlling the composite response to mechanical loading. Thickness of one-step, melt-mixing, deposited EIL is a result of a frozen dynamic equilibrium between thermodynamic and shear forces in the melt during the mixing procedure. The thermodynamic forces are determined by the surface free energy of components, and the system tends to acquire a morphology with minimal total free energy. In a majority of the PP composites, the free energy of the filler is significantly greater than that of soft and rigid interphases used, which is again greater than that of the PP bulk [19]. Thus, the morphology possessing the lowest free energy occurs when filler particles are encapsulated by the EIL and these complex core-shell inclusions are embedded in the PP. The shear forces in the melt are controlled by the relative viscosity, temperature, particle size and shape, and other rheological parameters [20]. Viscous drag flow patterns tend to remove the EIL from the filler surface in the course of melt mixing, leading to only partial coverage of the filler particles. Hence, a random distribution of partially covered core-shell inclusions in the PP matrix is the morphology resulting from an uncontrolled mixing process. The reproducibility of such a random morphology is somewhat low, resulting in only a limited applicability of such a process on an industrial scale.

Increasing polarity of the EIL via grafting of maleic anhydride (MA), acrylic acid (AA), and other reactive polar groups containing species onto EIL backbone molecules increases significantly the surface free energy of the respective component and thus affects the resulting deposition of EIL in the course of mixing via modification of the thermodynamic forces of attraction between solid surface and these molecules. At some concentration of the polar groups and under a well-defined processing window, the thermodynamic forces become dominant in controlling the deposition of EILs and formation of strongly adhering EIL dominates its dewetting in the course of melt mixing. This makes the process sufficiently reproducible and capable of utilization on an industrial scale in engineering compounds according to the requirements of the end user efficiently. Figure 11.1 compares the effects of EIL polarity on the type and uniformity of morphology of Mg(OH)$_2$-filled PP.

Kolarik et al. [21–23], Stamhuis [24], and Jancar and DiBenedetto [25] showed that in the two limiting cases (i.e., rigid and soft EILs), compounded PP composites exhibited substantial differences in both stiffness and fracture toughness at the same filler and elastomer concentrations. A model predicting upper and lower limits of elastic moduli was proposed recently by Jancar and DiBenedetto [25]. Although a large number of experimental data on yielding and fracture behavior of ternary composites and blends have been published, no

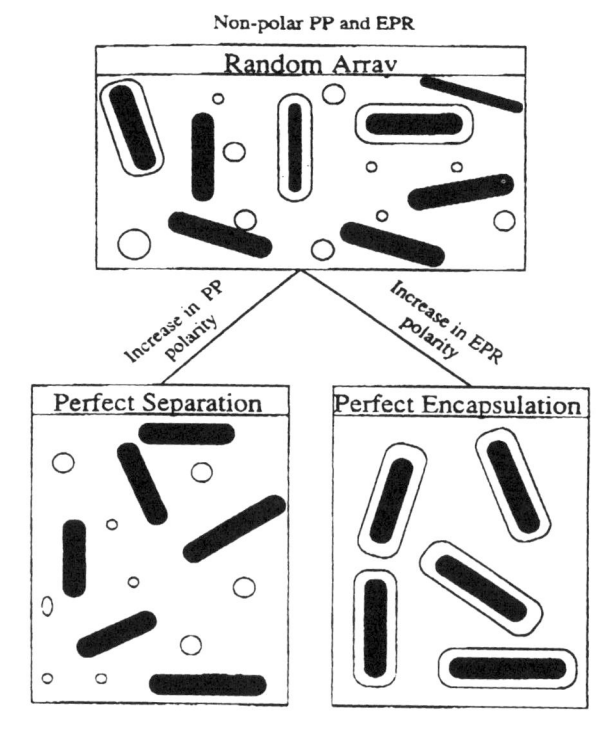

FIGURE 11.1 Effects of EIL polarity on morphology of covered core shell inclusions in PP matrix. (From Ref. 30, courtesy of Chapman & Hall.)

clear elucidation of the failure mechanics and mechanisms has been reported in the literature. An attempt to model and predict the yield strength and fracture toughness under impact loading has been proposed by Jancar et al. [5,29,30] for PP containing rigid inclusions of various partible shapes and rigid or soft EILs.

Unlike the case of toughening binary blends, where the primary variables controlling the fracture resistance are elastomer volume fraction and the size and distribution of elastomer inclusions [26–28], the toughness of composites with EILs is extremely sensitive to the thickness and elastic moduli of the interphase [6,7,31]. A lower limit of strain energy release rate, $G_c$ (or fracture toughness, $K_c$), for materials with rigid EIL can be calculated using linear elastic fracture mechanics (LEFM) and Irwin's concept of the plastic zone [32] along with knowledge of the matrix yield strength [33,34]. When the adhesion between rigid filler and PP is poor or in the case of inclusions with soft EIL, the character of the stress field in the vicinity of heterogeneities is altered favorably. Thereby, this changes the main failure mechanism from unstable shear banding, nucleated at the filler surface, to extended shear yielding. As was shown in Chapter 6, LEFM and Irwin's concept of plastic zone, along with the Nicolais-Narkis model for yield strength of these materials, can be used to predict an upper limit of $G_c$ (or $K_c$) for composites filled with soft EIL–covered inclusions.

The aim of this chapter is to review current state of the art in the application of composite models in prediction of tensile modulus, yield strength, and critical strain energy release rate of PP composites with engineered interphases.

## 11.2 INTERPHASE FORMATION AND STRUCTURE

### 11.2.1 Chemically Formed EILs

*Silane and Titanate EILs.* Silane coupling agents have been used in composites since the 1940s. Most of these compounds contain hydrolyzable substituents such as chlorine or alkoxy groups. Most commercial silanes contain one to three alkoxy substituents and one alkyl chain attached to a single silicon atom. A typical structure of a common silane is $RSiX_3$. The functionality $X_3$ interacts with reactive groups on the substrate surface, most often with hydroxyls. Figure 11.2 provides a list for a number of silanes structures. Most commercially available silanes contain three ethoxy or methoxy groups and a linear chain end capped with a reactive group such as amino, vinyl, or methacryloxy. In some instances, chlorine is used instead of ethoxy groups, especially in vinyl-terminated silane agents for use with unsaturated polyester matrices.

Titanates, though very similar in structure, have the silicon atom replaced with titanium, which somehow alters both reactivity of these agents and properties of the resulting interphase. However, there are no principal differences in the chemistry of silanes and titanates. Therefore, discussion in this section is

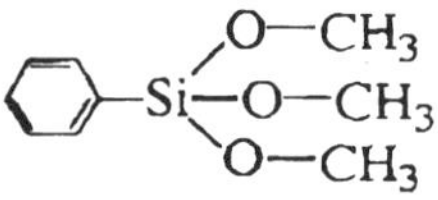

phenyltrimethoxysilane: C₆H₅–Si(O–CH₃)₃

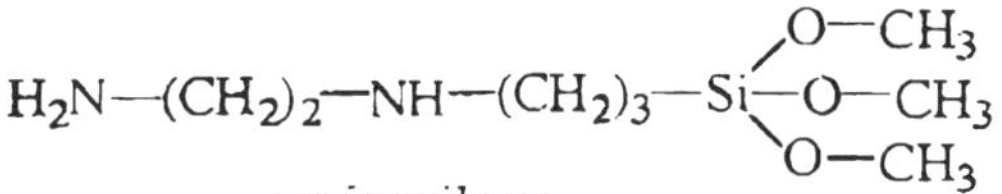

aminosilane: H₂N—(CH₂)₂—NH—(CH₂)₃—Si(O—CH₃)₃

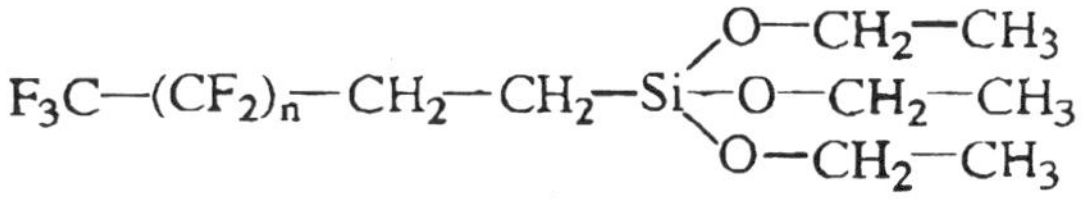

fluoroalkylsilanes: F₃C—(CF₂)ₙ—CH₂—CH₂—Si(O—CH₂—CH₃)₃

| | |
|---|---|
| Vinyldimethylchlorosilane (VDMCS) | CH₂=CH(CH₃)₂SiCl |
| Vinylmethyldichlorosilane (VMDCS) | CH₂=CH(CH₃)SiCl₂ |
| Vinyltrichlorosilane (VTCS) | CH₂=CHSiCl₃ |
| Octenyldimethylchlorosilane (OEDMCS) | CH₂=CH(CH₂)₆(CH₃)₂SiCl |
| Octenyltrichlorosilane (OETCS) | CH₂=CH(CH₂)₆SiCl₃ |
| Ethyltrichlorosilane (ETCS) | CH₃CH₂SiCl₃ |
| γ-Methacryloxypropyltrichlorosilane | CH₂=C(CH₃)COO(CH₂)₃SiCl₃ |

FIGURE 11.2 Assortment of silane additives.

exemplified by silanes as an example. Specific reference to titanates is made only when a substantial difference exists. In general, titanates exhibit lower thermal and hydrolytic stability compared to silanes.

The traditional method of applying silanes has been to use aqueous solvent systems directly in the course of fiber spinning. The water in the aqueous solution hydrolyzes the silane into silanols before immersion of the inorganic reinforcement. Hydroxyls from the inorganic surface can also serve to hydrolyze silane agents [35–37]. The silanol moieties readily condense to form siloxane bonds that can result in a linear, branched, or cross-linked polysiloxane structure. Reproducible coverage using the above-described method is difficult to obtain because of two competing condensation reactions (i.e., condensation on the surface and in the solution). A development of silylation techniques from organic solvents has come about as a result of attempts to obtain more reproducible siloxane layers or as a route to silane monolayers [38,39].

One, two, or three hydrolyzable siloxane groups are present on a common silane coupling agent. It is clear that in the case of monofunctional silanes, only a monomolecular layer can be obtained on the solid surface containing bonded or physisorbed hydroxyls [40]. Similarly, it is more probable that the difunctional silanes will form a monomolecular coverage rather than forming linear chains. Hence, only the trifunctional silanes are capable of forming three- dimensional siloxane multilayer on the solid surface when deposited from an aqueous solution [41]. (See Section 9.2.3 in Chapter 9 for more details.)

Thorough investigations of the 3-D siloxane coating by Schrader et al. [42] revealed that even after extensive washing in boiling water, a thin siloxane multilayer remained tenaciously bonded to the solid surface. Using a radioisotope labeling technique with γ-aminopropyltriethoxysilane and various washing procedures, three separate siloxane regions were identified. The first layer, most distant from the solid surface, could be easily removed by simple water wash and consisted of several layers of weakly physisorbed, hydrolyzed silane coupling agent. Rinsing with benzene followed by water wash resulted in removal of the second layer consisting most probably of linear or branched polymeric siloxane without a significant number of cross-links. After all the washing procedures used, a third and strongly bonded layer of 3-D polysiloxane network remained on the glass surface. It is now generally accepted that the original concept of a siloxane monolayer is not correct and that this strongly bonded cross-linked coating contains several tens of layers. Even though most of the studies were conducted on an E-glass surface, their validity remains principally intact for all sorts of inorganic glasses and silicates. As long as there is strongly adsorbed water on the surface allowing for silane hydrolysis, one will obtain qualitatively the same results.

DiBenedetto and Scola [43] have shown there are three distinctive regions of the interphase formed by γ-aminopropyltriethoxysilane on the surface of S2-glass. The layer most distant from the S2-glass surface was found to consist of a high molecular weight polysiloxane, a middle region containing mostly siloxane oligomers, and the region closest to the glass surface had structure of high molecular weight polysiloxane with slightly differing chemical composition compared with that of the most distant region. These results were confirmed by Pawson and Jones [44], who also concluded that a removal of the outer layer is essential for exposing the reactive groups of silanes capable of a reaction with unsaturated polyester matrix used.

A great deal of understanding of the relations between the way of silane deposition and the topology of siloxane layers on glass and metal surfaces has been acquired by Bascom [45], Lee [46], Boerio and Cheng [47], DiBenedetto [48], DiBenedetto and Lex [49], and Wu [50]. It has been found that the siloxane layers deposited from nonpolar solvents were thicker and more resistant to removal by both organic solvents and water, whereas interphases deposited from polar solvents were fairly thin and relatively easily desorbed in polar liquids. Scanning electron microscopy (SEM) observation revealed that the surface coverage is not uniform in thickness and siloxane forms an island-like irregular deposits. It is highly probable that despite the existence of these deposits, a very thin layer of siloxane covers uniformly the glass surface. The observed deposits are related to an adsorption of siloxane microgels formed in the solution in the course of silane deposition. However, there are no direct experimental proofs for such an explanation in the literature.

In addition, Lee [46], Boerio and Cheng [47], and Urban and Koenig [51] studied the orientation of alkyl chains containing reactive groups attached to the siloxane bonds with respect to the inorganic surface. It has been determined that the orientation depends on the carbon chain length, type of the reactive group, and extent of surface coverage. The most complete study of the effect of alkyl chain length on the orientation of silane monolayers with respect to the aluminum surface was published by Cave and Kinloch [52]. Their findings are discussed more thoroughly in Section 11.4.

Titanium coupling agents also contain hydrolyzable groups capable of bonding to inorganic surfaces and reactive groups capable of forming chemical linkages to the matrix. Interestingly, there is evidence that unlike silanes, titanate coupling agents form thinner interphases. Arroyo et al. [53] found that titanium coupling agents form preferentially monomolecular layers on the glass surface. Most frequently, titanates containing long hydrocarbon chains were used as processing aids [54]. The major limitation was their relatively poor thermal stability, allowing for processing only up to 200°C [55]. This shortcoming of titanates was partially overcome by substituting long hydrocarbon chains in their molecule by pyrophosphato groups. The hydrolytic stability of titanate interphases remains unclear because of contradictory results published. Yang et al. [56] showed that titanate coupling agents are relatively susceptible to hydrolysis, whereas Gomes [57] found that titanate interphases were more stable in a moist environment than silane-based interphases.

*EILs Formed via Grafting Oligomers.* Recently, a new approach to deposit low surface free energy EILs onto glass fiber reinforcement has been developed by Ranade and DiBenedetto [58] dealing with techniques of a deposition of oligomers of polysulfone and polycarbonate on the E-glass fiber surface. In this approach, a suitable intermediate, such as silicon tetrachloride ($SiCl_4$) or titanium tetrachloride ($TiCl_4$), reacts with surface hydroxyls of glass in a nonaqueous environment. During this reaction, hydrochloric acid gas is evolved, which can be used to detect the reaction. Spacing of surface hydroxyls and the small size of the intermediates results in a forming of primarily di- and trichloro species. This procedure is well known and used in producing thin layers onto glass sheets [59,60]. The chlorine atoms, remaining on the pretreated glass surface, are then used to react with hydroxyls from polysulfone or polycarbonate oligomers. It is clear that any trace of water in the reacting system will cause a severe reduction in an extent of grafted oligomers due to reactive chlorine elimination. After the reaction is completed, remaining weakly physisorbed oligomers are rinsed off and the treated glass is dried.

It was found that increasing the molecular weight of the *bis*phenol A polycarbonate oligomers has led to an increase in the interphase thickness. Apparently, the thickness, as determined by thermogravimetric analysis (TGA), under a very simplified assumption of a uniform coverage corresponded well with the average end-to-end distance calculated using a freely joint segments model over the entire $M_w$ interval ranging from 1300 to 20,000 described by DiBenedetto et al. [61]. The interphase thickness has been between 3 nm ($M_w = 1300$) and 14 nm ($M_w = 20,000$). In this technique, $SiCl_4$ intermediate appeared to yield better results than $TiCl_4$ [58,61,62].

*Elastomeric EILs.* Since the 1960s, there has been a substantial effort to produce thin interphases of elastomers on the surface of common reinforcements. The major thrust in these investigations was a hypothesis that the presence of a thin highly flexible layer on the surface of reinforcing fillers and fibers will contribute to an enhancement of composite toughness without jeopardizing its strength and stiffness [63–65]. This concept was not very successful in continuous fiber–reinforced composites and laminates mostly due to very tedious and time-consuming procedures to deposit uniform elastomeric layers onto the small-diameter fibers. It seems that a more plausible application of the above hypothesis was found in particulate and short fiber–reinforced thermoplastic composites [25–28,66].

Here an advantage of in situ forming of chemically distinguished soft EILs has been explored. Elastomers such as EPR and EPDM are chemically modified to introduce polar groups onto their backbone chains. Most frequently, maleic anhydride and acrylic acid are used as the grafting comonomers [21]. In the process of melt mixing, carboxyl groups from the grafted comonomer reacts with hydroxyls, amines, or other suitable reactive groups presented on the filler surface. Often the resulting bond is a mixture of contributions from several types of interactions ranging from covalent bonds to hydrogen bonds to electrostatic interactions. Hence, a term acid–base interaction is commonly used to describe bonding in these systems [67–69]. Pfeiffer [70] described process in which thick flexible interphases were deposited onto glass surfaces to modify toughness of a short glass fiber (SGF) reinforced composite.

## 11.2.2 Nonchemically Originated EILs in PP

During solidification from melt, gradients of mass transport occur near the filler surface caused by potential field at the surface of the solid inclusion [2]. As a result, both supermolecular structure and chain mobility of PP in a thin layer near the filler surface is altered compared with the bulk, with the extent of such a change proportional to the surface activity of the filler and with relation to the topology of the surface. Surface affinity is usually defined as the ratio between adhesive and cohesive energies of the filler and matrix, and, generally, more surface active fillers cause greater changes in PP morphology. The thickness of such a layer ranges from 10 to several hundred nanometers [1,2] and has a more or less amorphous character.

In addition to the formation of such an interphase, fillers affect morphology of PP in the bulk. Many fillers act as primary nucleation agents, increasing the number of spherulites and reducing their mean diameter without any significant modification of the crystallization kinetics. A slight decrease in the degree of crystallinity by about 10% has been observed using X-ray diffraction measurements. Interpretation of these measurements is partly obscured by peak overlaps from the PP and the filler and by reduced perfection of the crystallites formed in filled PP. Up to particulate filler content of about 20 vol%, $\alpha$-form lamellae are formed similar to unfilled PP. Thickness of the lamellae is also reported to be reduced in filled PP [71]. Above 20 vol% of the filler, spherulite morphology of PP disintegrates into system of lamellae oriented more or less randomly with respect to filler surface.

*Transcrystalline EILs.* Neat isotactic polypropylene (iPP) crystallized from melt exhibits spherulitic morphology of the crystalline phase [72,73]. In some cases and under very specific conditions, cylindrites, axialites, quadrites, hedrites, and dendrites of iPP may be formed [74]. In general, crystallization from quiescent melts results in spherulitic morphology, whereas crystallization from melts subjected to mechanical loads results in cylindrites [75]. Crystalline supermolecular structure caused by oriented crystal growth from heterogeneous surfaces is commonly termed transcrystallinity [76].

Transcrystalline morphology is formed when crystallization takes place on the solid surface of fillers or reinforcements. Transcrystallization takes place when the density of the crystal nuclei is substantially greater on the surface of solid inclusions than in the melt bulk [77]. Because polyhedral spherulites cannot develop due to restricted lateral growth on the solid surface, crystallites are allowed to grow only in stacks perpendicularly to the surface plane [78]. In the case when only one crystal form occurs in a polymer, Keller [79] confirmed that the microstructure of transcrystalline layer and bulk crystalline phase is identical. For PP, however, the situation is more complicated by the polymorphism so that one crystal form can exist in the transcrystalline layer and another in the polymer bulk. The nature of nucleation of the transcrystalline layer is still somewhat controversial with experimental results supporting hypotheses based on both the chemical and topological nucleation [80].

The strong effect of the surface topology on the epitaxial type of transcrystallization of iPP has been described by Hobbs [81,82]. The primary factors affecting transcrystallization on the filler surface include temperature gradient near the filler surface, chemical composition of the filler surface, crystalline morphology of the filler surface, surface energy of the filler, adsorption of nucleants presented in the polymer, epitaxy, and topography of the filler [83–86].

The molecular weight of the PP used has been found to play a major role in developing transcrystallinity in filled and fiber-reinforced PP [87,88]. There is also a large body of literature showing effects of shear flow on the transcrystallinity nucleation, suggesting a strong role of composite processing on the development of transcrystallinity in PP composites. It has been shown that, using various combinations of crystallization conditions, one can obtain different crystal forms in the transcrystalline layer and in the bulk in a controlled manner [89–95].

A transcrystalline layer possesses different mechanical properties. Moreover, unlike PP in bulk, this layer is strongly anisotropic. This is expected to be extremely important for determining the mechanical properties and hydrolytic stability, especially in the case of glass fiber–reinforced PP.

Mechanical properties of transcrystalline PP and bulk spherulitic morphology PP differ significantly as shown by Folkes and Hardwick [96]. One has to keep in mind that these data were measured in an unconstrained plane stress state of stress, thus inevitably giving lower values of yield strength and larger values of elongation to break. The Young modulus of transcrystalline layer was found to be 40% greater than that for the fine spherulitic morphology (bulk); both the tensile and shear yield strength was about 30% greater than those for bulk PP, whereas the elongation to break of 4% was measured for the transcrystalline layer and about 300% for the bulk PP. This results in the energy to failure of about $28\,\text{kJ/m}^2$ for transcrystalline PP compared with about $49\,\text{kJ/m}^2$ for the bulk PP.

The increase in elastic modulus for the transcrystalline layer is in agreement with the findings of other researchers [85,86]. These values were measured in the direction perpendicular to the transcrystallinity growth direction, or, in other words, the loading direction has been parallel to the surface of the solid inclusion. One can expect that due to the orthotropic symmetry of the properties of crystalline PP lamellae, different set of properties will be measured when loaded in the direction perpendicular to the solid inclusion surface. In addition, the elongation to break and the yield strength were measured in an unconstrained transcrystalline layer under the plane stress state of stress, which is far from the plane strain state of stress in the highly constrained transcrystalline interphase at the filler surface. This objection has been partly eliminated by Folkes and Hardwick [96], who measured properties of a "laminate" formed by transcrystalline layers and thin glass sheets.

It has been found that the presence of transcrystalline region increases ductility of glass fiber–reinforced PP [76]. However, these findings contradict results of Folkes and Hardwick [96]. It seems that the role of transcrystalline interphase on the mechanical properties of filled and reinforced PP is very matrix-fiber specific. High shear strength of the aramid fiber–PP transcrystalline region [97] is in contrast to very low shear strength of the transcrystalline interphase in glass fiber–PP composite [98].

Direct measurements of the interfacial shear strength provide somewhat contradictory results. On one hand, Elemendorp and Schoolenberg [99] reported substantial increase of the average shear strength of the transcrystalline interphase compared with the same composite without the transcrystalline interphase, whereas Hoecker and Karger-Kocsis [98] have not observed any significant effect of the transcrystalline interphase on the interfacial shear strength in PP/glass fiber composite.

## 11.3 ELASTIC MODULI OF POLYPROPYLENE COMPOSITES WITH INTERPHASES

### 11.3.1 General Concepts

*Phenomenology of Elasticity.* In general, Hooke's law is the basic constitutive equation giving the relationship between stress and strain. Generalized Hooke's law is often expressed in the following form [100]:

$$\sigma_{ij} = C_{ijkl}\varepsilon_{kl} \tag{11.1}$$

where $\sigma_{ij}$, $C_{ijkl}$, and $\varepsilon_{kl}$ are tensor of stresses, tensor of elastic constants, and tensor of deformations, respectively. Equation (11.1) has been derived for materials with energy drive behavior, such as metals. Elasticity of polymers contains substantial entropic contribution even for temperatures below $T_g$. Hence, elasticity of polymeric materials deviates somewhat from behavior of ideally elastic solids in several ways. First, deformation of a polymer is strongly dependent on its thermal and strain history and test conditions such as strain rate, temperature and pressure. As a result, time or frequency should be added as a variable into the constitutive Eq. (11.1). In other words, deformation response of polymeric solids is of a viscoelastic nature. Second, unlike in metals, not all deformation within the elastic region of polymer behavior can be recovered (duration of the test becomes a test variable). Third, unlike in ideal elasticity, the constitutive elasticity equations for polymers are generally nonlinear. Fourth, stress and strain used in Hooke's law [Eq. (11.1)] are defined for polymers in their original sense only at very small deformations, generally well below 0.5%. For larger deformations, these parameters must be defined in a more general fashion. Last but not least, polymers can become anisotropic or even orthotropic upon drawing, which greatly complicates the mathematical form of the generalized Hooke's law that has to be used. Particulate-filled and short fiber–reinforced PP, though anisotropic on a microscale, can be considered quasi-isotropic and homogeneous on the macroscopic scale. Thus, for the purpose of phenomenological description of the mechanical response of PP filled with inclusions covered in engineered interphases, one can consider these materials as isotropic and homogeneous with the elastic constants calculated using existing composite models [123,124].

*Polymer Elasticity on the Molecular Level.* Attempts to establish transformation algorithms between phenomenological and microscopic theories have met with complications. Unlike in metals, substantial entropic contribution to small deformation response of solid polymers results in very strong dependence of the elastic moduli on temperature with abrupt changes in moduli over very narrow (10–20°C) temperature intervals. Thus, one has to keep in mind that molecular interpretations of elasticity differ in temperature regions above and below $T_g$ [122]. In addition, the underlying difficulty is attributed to the need to distinguish between response of glassy and semicrystalline polymers principally. Semicrystalline polymers must be considered two-phase solids with substantial difference between the response of the amorphous and crystalline phases.

The relationships between constitution, configuration, and attainable conformations of individual macromolecules and the elastic behavior of a solid glassy polymer varies with changing the temperature, substantially. In general, three distinct regions are recognized—glassy, glass transition, and rubbery (leathery). As a result of the long-range order existing in the crystalline domains of semicrystalline polymers, elastic behavior of semicrystalline thermoplastics is far less dependent on temperature, compared with glassy thermoplastics. Unlike glassy thermoplastics and thermosets, no direct correlation has been found so far between molecular parameters and elastic properties of semicrystalline polymers. Voight-Martin et al. [125] clearly demonstrated that elastic response of semicrystalline thermoplastics is controlled directly by their morphology rather than by their molecular structure. The influence of chain constitution, configuration, molecular weight, and molecular weight distribution on polymeric elasticity is always indirect in effect.

The molecular theory of polymer viscoelasticity rests on the work of Bueche [101–104], Rouse [105], and Zimm [106,107] investigating the behavior of diluted solutions of linear polymers. The molecular theory of viscoelasticity has not found much success in describing viscoelasticity of solid polymers over the whole temperature interval, and thus a modified theory of rubber elasticity has to be used to describe the elastic response of amorphous polymers above their $T_g$. On the other hand, it seems reasonable to expect that well below the $T_g$, the entropic character of polymer elasticity yields to more energy-driven elasticity of polymer glasses. Frozen-in conformations below $T_g$ allow us to assume that the energy-drive elasticity corresponds to a Young modulus for a glassy polymer that is lower in value than for an ideal polycrystalline polymer discussed below. Various linear glassy polymers such as polycarbonate (PC), polymethyl methacrylate (PMMA), polyvinyl chloride (PVC), and polystyrene (PS) exhibit Young's moduli within the interval 2–4 GPa, suggesting that the disordered supermolecular structure with no translational symmetry of potential fields resists external forces only to the limit given by intermolecular interactions. The expression for the Young modulus of elasticity for a system on $N$ macromolecules each consisting of $z$ equivalent segments can be expressed as follows [108]:

$$E_r(t) = 3NKT \sum_{p=1}^{z} \exp\left(-\frac{t}{\tau_{max}} p^2\right) \tag{11.2}$$

where $NKT$ is the elastic constant of the molecular "spring" at temperature $T$, Poisson's ratio is assumed to be 0.5, and $\tau_{max}$ is the maximal relaxation time of the molecular dash pots. The pure elastic modulus is equal to the $E_r$ at $t = 0$:

$$E_1 = 3zNKT \tag{11.3}$$

where $zN$ is equivalent segments in cubic centimeters. Qualitatively, Eq. (11.2) predicts an increase of the elastic moduli with increasing polymer density, segment stiffness, and size. In addition, strong intermolecular interactions result in higher elastic moduli, in general.

Molecular interpretation of elasticity of semicrystalline polymers is complicated even more by their two-phase morphology and the complex manner in which amorphous and crystalline domains are interconnected. Generally, behavior of the amorphous fraction resembles behavior of glassy polymers, and elasticity of crystalline regions is of the same nature as elasticity of "atomic" crystals. Internal segment rotations are restricted in the polymer crystal, and the internal segment mobility consists of simple vibrations about the equilibrium position, resulting in energy-driven ideal elasticity. Moreover, elastic properties of crystals are strongly anisotropic. For the PE monocrystal, the Young modulus in the direction of polymer chains is about 300 GPa, whereas the Young modulus measured in the perpendicular direction reaches only about 3 GPa [100]. Similar values for the longitudinal and transverse elastic moduli have been found for other semicrystalline thermoplastics such as PP and PA. The random spatial orientation of crystalline domains within a semicrystalline polymer solid is the reason why the elastic moduli of common semicrystalline plastics are even lower than the transverse moduli of the monocrystals. Preferential orientation of the crystalline regions and orientation of tie molecules within the amorphous domains in highly drawn polymer fibers is the reason for these fibers to reach elastic moduli of the order of 200 GPa, which is about 70% of the longitudinal moduli values for a monocrystal.

To interpret the large difference between the longitudinal and transverse moduli, one has to keep in mind that in the direction of chains, covalent bond angles and their lengths would have to be changed to extend the long chain length requiring large forces. On the other hand, substantially lower force is necessary to move the chains relatively to each other (i.e., against the weak van der Waals intermolecular interactions). Because the long-chain molecules in polyamides are interconnected with stronger hydrogen bonds, transverse modulus of polyamide 6 (PA6) is about three times greater than that of PE, when measured in dry conditions. On the other hand, the planar zig-zag conformation of the extended PE chain contributes to its higher longitudinal elastic modulus compared to helix-shaped PP chains.

In the PP monocrystal, helical configuration of the long-chain molecules, built in the lamellae topology, results in lowering the longitudinal elastic moduli compared with planar zig-zag configuration of PE molecules. The need for high forces to change the covalent bond angles and lengths is reduced for the much lower force demanding rotational twisting/rocking movement of the segments. As a result, PP crystals exhibit longitudinal elastic modulus of only about 40 GPa. The transverse elastic modulus is not affected by the helix configuration of the chain because the intermolecular forces in PP are of the same strength as in the PE and thus PP has transverse modulus of about 2–3 GPa. In semicrystalline polymers exhibiting stronger intermolecular interactions, such as polyvinyl alcohol and polyamides, transverse elastic moduli of up to 9 GPa have been observed experimentally [100].

In reality, the morphology of a polycrystalline thermoplastic consists of spherulites that holds for common polymer types (e.g., PP, PE, PA6, PA6,6, and polyether-ether ketone (PEEK) crystallized under common conditions). Some semicrystalline polymers and moderately filled composites of the above resins may exhibit lamellar crystalline morphology without any spherulitic order. As a result of random orientation of individual crystallites in spherulites and the manner of their connectivity, the elastic modulus of about 10 GPa has been extrapolated for a hypothetical ideal polycrystalline PE containing no amorphous phase from the dependence of the elastic modulus of PE on the degree of crystallinity. Presence of amorphous phase reducing the content of crystalline phase results in a further reduction of the overall elastic modulus of the semicrystalline polymers compared with ideal monocrystals [109].

## 11.3.2 Composites with Spherical Inclusions

Quantitative prediction of the dependence of elastic moduli on the properties of constituents and the material composition for multicomponent polymeric materials has been the center of attention of both scientific and engineering communities since the advent of composites. Several models have been developed over the years capable of predicting elastic moduli, e.g., shell models [110–112,115–118], self-consistent models [117], variational methods [113,114], and semiempirical approaches [119]. The accuracy of the prediction varies for the respective models depending on the set of assumptions put into the foundations of the theories. For the purpose of this chapter, continuum-based models are only discussed assuming spherical inclusions and uniformly thick interphase layer of known bulk and shear moduli and total volume fraction of inclusions.

Generally, the presence of rigid inclusions ($G_f \gg G_m$) of volume fraction $v_f$ in a matrix polymer causes an increase of the elastic modulus of the composite, $G_c$, resulting from a matrix volume exclusion of an effective volume fraction ($Av_f$) [71,123]:

$$G_c = 1.4 G_m \left\{ \frac{1}{1 - Av_f} \right\} \tag{11.4}$$

with

$$A = \left(1 - \frac{1.4 G_m}{G_f}\right) + \frac{v_i}{v_f}\left(1 - \frac{G_m}{G_i}\right) \tag{11.5}$$

Coefficient $A$ depends on the relative elastic moduli of the matrix and filler, $G_m/G_f$, and matrix and interphase layer, $G_m/G_i$, and on the relative volume fraction of the interphase and filler, $v_i/v_f$. Equation (11.5) was simplified from a more general expression assuming values of Poisson's ratio 0.28 for the filler, 0.35 for the matrix, and 0.30 for the composite. For rigid interphase ($G_m < G_i$), $A > 0$ and elastic modulus of the composite is increasing with the interphase volume fraction, $v_i$, at constant $v_f$ and with $v_f$ at constant interphase thickness. On the other hand, for soft and compressible interphase ($G_m > G_i$), $A < 0$ resulting in a reduction of the elastic modulus with increasing $v_i$ at constant filler loading or with increasing $v_f$ at constant interphase thickness. In Eq. (11.5), however, $G_i$ remains a semiempirical fitting parameter because very few direct data on the elastic moduli of thin interphases constrained on rigid surfaces are available in literature.

The Kerner-Nielsen equation [119] is a useful alternative to Eqs. (11.4) and (11.5) when the elastic modulus of an interphase is substantially greater than that of the matrix. Most often, interphase elastic modulus is assumed to be equal to that of the filler resulting in a slightly overestimated elastic modulus of the compound. The effect of an interphase is then introduced in the Kerner-Nielsen equation substituting an effective filler volume fraction, $v_{\text{eff}} = (v_f + v_i)$, for the filler volume fraction, $v_f$ [71]:

$$G_c = G_m \frac{1 + AB v_f^{\text{eff}}}{1 - B\psi v_f^{\text{eff}}} \tag{11.6}$$

# Engineered Interphases in PP Composites

In Eq. (11.6), $A$ and $B$ are parameters dependent of the elastic constants of the components:

$$A = \frac{8 - 10\nu_m}{7 - 5\nu_m} \tag{11.7}$$

$$B = \frac{\frac{G_f}{G_m} - 1}{\frac{G_f}{G_m} + A} \tag{11.8}$$

The following expression for the $\psi$ term in Eq. (11.6) indicates that there exists a maximal volume fraction, $v_f^{max}$, which rigid inclusions of a given shape and size distribution can occupy. Similarly, a maximum effective filler volume fraction has to be used in calculating $\psi$ [119]:

$$\psi = 1 + \frac{1 - (v_f^{eff})^{max}}{[(v_f^{eff})^{max}]^2} v_f^{eff} \tag{11.9}$$

To account for the presence of an interphase, two simplifying assumptions had to be made: the elastic moduli of the interphase and the filler are assumed to be equal and $(v_f^{eff})^{max}$ can reach a value of 1 when all the polymeric phase consists of an interphase.

## 11.3.3 Composites with Nonspherical Reinforcement

By analogy, the Halpin-Tsai equation can be used to predict elastic moduli of the composite containing randomly oriented anisometric (nonspherical) inclusions with engineered interphases [30]. This is a plausible model of the 3-D randomly oriented $Mg(OH)_2$ platelets or 2-D randomly oriented SGF in the PP matrix. In the first step, the modified Halpin-Tsai equation is used to calculate the longitudinal and transverse moduli of the composites with unidirectionally oriented nonspherical inclusions:

$$E_c^L = E_m \frac{1 + \eta_L \xi_L v_f^{eff}}{1 - \eta_L v_f^{eff}} \tag{11.10}$$

and

$$E_c^T = E_m \frac{1 + \eta_T \xi_T v_f^{eff}}{1 - \eta_T v_f^{eff}} \tag{11.11}$$

where

$$\eta_L = \frac{\dfrac{E_f^{eff}}{E_m} - 1}{\dfrac{E_f^{eff}}{E_m} + \xi_L}$$

$$\eta_T = \frac{\dfrac{E_f^{eff}}{E_m} - 1}{\dfrac{E_f^{eff}}{E_m} + \xi_T}$$

(11.12)

and

$$\xi_L = 2\frac{L}{t} \text{ for platelet-shaped inclusions}$$

$$\xi_L = 2\frac{L}{D} \text{ for SGF}$$

$$\xi_T = 2 \text{ for square platelets}$$

$$\xi_T = 2 \text{ for SGF}$$

Unlike the case of spherical inclusions where the rigid inclusions cause stiffening of the composite by excluding volume of a deformable matrix, the presence of an interphase layer affects the true reinforcing efficiency of the inclusions. Hence, the effective filler modulus of the inclusions have to be calculated as a function of interphase thickness and elastic modulus. This can be done effectively using a simple rule of mixture:

$$E_f^{eff} = \xi E_f v_f' + E_i v_i \qquad (11.14)$$

where $\xi$ is the reinforcing efficiency, $E_f$ and $E_i$ are the filler and interphase elastic moduli, respectively. In Eq. (11.14), $v_f'$ is the filler volume fraction in the complex core-shell inclusion. The reinforcing efficiency can be calculated using numerical methods such as the finite element analysis (FEA). $\zeta$ is the ratio between the average stress, $[\sigma_f^{cs}(L)]$, in an inclusion of the required aspect ratio and the maximal attainable stress, $[\sigma_f(L \gg L_c)]$, in the center of a long inclusion ($L \gg L_c$) without an interphase layer. Both stresses are calculated at an arbitrary strain assuming linear elastic behavior [120]:

$$\zeta = \frac{[\sigma_f^{cs}(L)]}{[\sigma_f(L \gg L_c)]} \qquad (11.15)$$

This approach has been proven valid for the PP filled with platelet-shaped and fibrous inclusions encapsulated in an elastomer interphase [30]. Needless to say, however, the boundary conditions and the assumptions used to derive this model restrict its prediction power to the quantitative prediction of the upper and lower limits of the elastic modulus. The upper limit is achieved for ideal cases of perfectly adhering rigid interphase, whereas the lower limit is achieved for either perfectly adhering soft interphase or for nonadhering interphase. In reality, measured values lay between these two limiting values for filler volume fractions up to 0.5. The observed deviation from the ideal case is due to the fact that in the course of composite preparation, defects of the morphology are inevitably introduced. Most commonly, these defects result in values closer to the lower limit [30,124] due to lower interphase stiffness.

## 11.4 YIELDING OF POLYPROPYLENE COMPOSITES WITH ENGINEERED INTERPHASES

### 11.4.1 General Concepts of Yielding in Particulate-Filled Thermoplastics

*Deformation Response of Semicrystalline Thermoplastics.* The importance of spherulites and spherulitic morphology for the mechanical response of semicrystalline thermoplastics rests on experimental evidence that spherulite sizes fix the scale for coordinated local mobility of an assembly of segments (i.e., for the extent of plastic deformation) [121]. The deformation response of semicrystalline polymers depends also on the morphology of the crystalline phase, especially on the size and shape of spherulites and location and density of the molecules. Polymers with small uniform spherulites prepared by quenching polymer melts exhibit more ductile behavior, whereas morphologies with coarse spherulites resulting from slow crystallization fail in rather brittle manner [127]. In the latter case, cracks propagate generally along the boundaries of large spherulites. When cracks meet spherulites head-on at boundary intersections, they tend to propagate through these spherulites along surfaces of locally radial lamellae. However, this behavior is dependent on the ease of interlamellar cleavage. Polymers with fine spherulitic morphology exhibit generally plastic yielding, although not always in the same manner. In the case of lower crystallinity, these polymers undergo affine deformation, whereas specimens with high crystallinity generally yield in a heterogeneous manner, forming a neck [128]. Observed elongations within the neck up to 500% or more are facilitated by ease of chain extension from neatly folded conformations, and the localized character of the deformation is largely attributable to the manner in which chain-folded crystals respond to stress.

According to current studies, local heating and consequent melting of crystallites seems to play a part only during deformation at high strain rates, especially during impact loading. The interpretation of the differences between deformation response of fine and coarse spherulitic structures is based on the fact that it is the spatial arrangement of crystalline and amorphous domains as well as the density of extra- and intraspherulitic tie molecules bridging the amorphous regions and allowing for the load transfer between crystallites that controls the mechanical response of these complex heterogeneous solids. At very small strains below 1% and, at $T > T_g$ of the amorphous fraction, deformation inside the solid semicrystalline polymer is confined to disordered amorphous regions due to their low stiffness above their $T_g$. As a result, elastic modulus of semicrystalline polymers increases with increasing degree of crystallinity and with increasing spherulite size. Inside the spherulites, rotation of lamellae occurs within this deformation region. The resulting orientation of lamellae depends on their position within the spherulite in respect to the equator-pole orientation of the external deformation [70,71,79,84,100,129–137].

At larger strains, after disintegration of spherulites, crystalline lamellae undergo plastic deformation processes manifested by macroscopic yielding. Lamellae, under tension in their own planes, fracture almost immediately with the microcracks being bridged by fine microfibrils 10–30 μm in diameter formed by extended chains pulled out of their folded conformations [100]. As deformation continues, other molecules become extended and feed into these microfibrils, with large elongations being achieved before original crystals are consumed [79]. As a result, morphology comprising of stacks of crystalline lamellae oriented obliquely to the tensile stress axis is formed. The total fibrillation can eventually consume the whole crystal when strained at extremely low strain rates and molecules are sufficiently long or there are stronger intermolecular attractions than the weak van der Waals forces [71,84,87]. Small spherulitic structures have more evenly distributed stresses, resulting in more even loading of individual chains. At very large deformations with draw ratios greater than 5, highly oriented fibrillar structure is formed consisting of microfibrils clumped together into coarser fibrils bridged by unoriented microfibrils crossing longitudinal voids between them [100].

*Phenomenology of Yielding of Thermoplastics.* Similar to metals, large number of thermoplastics can develop macroscopic plastic deformation as a response to mechanical loading. The onset of macroscopic plastic deformation is commonly termed the yield point. The viscoelastic-plastic nature of mechanical response of polymers to mechanical loads makes identification of the onset of true plastic deformation very difficult. Very often, rather arbitrary definitions are used. One has to keep in mind that polymers in general show necking, cold drawing, brittle fracture, and large homogeneous deformations depending on the exact loading conditions (strain rate, sample size and shape, mode of loading, temperature, pressure). Due to the phenomenological nature of the yield point, this fact is a general one, with no respect to chemical and physical structure [136].

The phenomenological analysis of the stress state at yielding results in the conclusion that the shear yielding is caused by deviatoric components of the stress tensor. This distinguishes shear yielding from crazing caused by triaxial stress state [137]. The shear yielding deformation occurs without significant change in volume by changing the shape of the deformed volume element. The *intrinsic* yield point is independent of test geometry used and is always followed by strain softening. All isotropic glassy polymers exhibit the strain softening. Some of them, however, orientate very rapidly, and the resulting strain hardening can obscure this process even under compressive or pure shear loading conditions. Similarly, some semicrystalline polymers do not exhibit strain softening due to the onset of orientational strain hardening, masking any possible signs of strain softening under usual test conditions.

In standardized tests, $\sigma$–$\varepsilon$ plots are most commonly obtained from the testing machines with the first maximum corresponding to an instability in deformation and neck formation. This point is often called *extrinsic* yield point, and it depends on the loading geometry. Quite often, extrinsic and intrinsic yield points are very close to each other, and the extrinsic yield point is referred to as the yield point in most literature on deformation behavior of polymers.

Because the test geometry plays a substantial role in the determination of the yield point, it appears reasonable to comment on some of the most common tests. The geometrical features of tensile tests promote extrinsic instabilities causing necking; thus, they are less favorable for investigations of true yielding behavior of polymers compared with more favorable compressive loading. Compressive loading eliminates the presence of triaxial tension sufficient to cause craze development, and compression also promotes the occurrence of an intrinsic yield point. If tested under uniaxial tensile loading, necking and cold drawing may result, which are accompanied by a nonuniform distribution of stress and strain along the length of test specimen. In this case, most deformation is concentrated within the neck.

It is always very useful to be able to predict at what level of external stress and in which directions the macroscopic yielding will occur under different loading geometries. Mathematically, the aim is to find functions of all stress components that reach their critical values equal to some material properties for all different test geometries. This is mathematically equivalent to derivation of some plastic instability conditions commonly termed the yield criteria. Historically, the yield criteria derived for metals were applied for polymers. Later these criteria were modified as knowledge about the differences in deformation behavior of polymers compared with metals has been acquired.

The simplest yield criterion is that of Tresca [136], who states that yielding will occur when the maximal shear stress on any plane in the tested solid reaches its critical value:

$$|\sigma_1 - \sigma_2| = 2\tau_y = \sigma_y \qquad (11.16)$$

where $\sigma_1$, $\sigma_2$ are principal stresses, $\tau_y$ and $\sigma_y$ are the yield stress in pure shear and the yield stress in pure uniaxial tension, respectively (materials properties). More accurate and most widely used is the von Mises yield criterion because the third principal stress component, $\sigma_3$, can also be taken into consideration [136]:

$$(\sigma_1 - \sigma_2)^2 + (\sigma_2 - \sigma_3)^2 + (\sigma_3 - \sigma_1)^2 = 6\tau_y^2 \tag{11.17}$$

Lower values of the yield stress measured in tension compared with those measured in compression suggest that the effect of pressure, which is important for polymers, is not accounted for in this criterion. Hence, appropriate correction has to be made to account for effect of external pressure. The most frequent version of pressure-dependent yield criterion is the modified von Mises criterion [136]:

$$(\sigma_1 - \sigma_2)^2 + (\sigma_2 - \sigma_3)^2 + (\sigma_3 - \sigma_1)^2 = 6(\tau_y - \mu p)^2 \tag{11.18}$$

where $\mu$ is the friction coefficient of the polymer and $p$ is the hydrostatic pressure.

In the case of compressive loading, the Coulomb yield criterion is often used in the form:

$$\tau = \tau_c + \sigma_N \tan \phi \tag{11.19}$$

where $\tau$ is the shear stress, $\sigma_N$ is the normal stress, and the angle $\phi$ is determined as

$$\theta = \frac{\pi}{4} + \frac{\phi}{2} \tag{11.20}$$

where $\theta$ is the angle between a normal to the yield plane and the direction of applied external loading.

The measured yield strength of polymers depends substantially on test temperature and strain rate, expressing the underlying viscoelastic nature of solids made of long-chain molecules. Similarly to the elastic modulus, the yield strength, $\sigma_y$, increases with decreasing temperature and increasing strain rate, $\dot{\varepsilon}$. Arrhenius type of analysis, expressed in the Eyring model of flow, is usually used to describe this dependence. Based on Eyring theory, the thermal and strain rate dependence of the yield strength for polymers sufficiently below their softening point ($T_g$, $T_m$) is expressed as:

$$\sigma_y = \frac{E^*}{\gamma} + \frac{RT}{\gamma} \ln \frac{\dot{\varepsilon}}{A} \tag{11.21}$$

where $A$ is a constant for a given polymer, R is the universal gas constant, and $\gamma$ is the activation volume. Despite very good linearity in the $\sigma_y/T$ versus $\log \dot{\varepsilon}$ plot, the proposed interpretations of the true physical meaning of the activation energy, $E^*$, and the activation volume, $\gamma$, has still not been unambiguously accepted. Experimental and theoretical evidences suggest that plastic flow is a cooperative motion of a large number of equivalent chain segments. This might be why $E^*$ does not correspond to any molecular relaxation energy and $\gamma$ is substantially greater than the size of the statistical random link by 2–20 times.

*Semicrystalline Polymers.* A two-phase model is often considered for analysis of mechanical response of semicrystalline polymers. This model simply assumes presence of crystalline domains of different shape and orientation in an amorphous matrix. One can imagine that for relatively low degree of crystallinity, such a model is close to reality; however, this concept is far from reality for most commercially important semicrystalline polymers exhibiting degree of crystallinity greater than 40%. Above its $T_g$, the amorphous "matrix" is assumed to have a rubber-like character; below $T_g$ it behaves as a glass. The crystalline regions are supposed to undergo shear flow by slip, twinning, or martensitic transformation. The deformation of crystalline domains is directionally dependent, and rotations of crystallites before their yielding are assumed to take place. Unlike in the Bowden's model of homogeneous yielding of glassy polymers, true dislocations exist in the PE crystals. It has been found that the plastic deformation of thin PE crystals can be initiated by thermal activation of screw dislocations with the Burgers vectors parallel to the chain direction in chain-folded crystallite.

## 11.4.2 Heterogeneous Yielding

In practice, homogeneous plastic deformation of polymers occurs very rarely. Most frequently, surface defects, local changes in the specimen cross-section, and so on cause localization of strain, leading to a more rapid increase of local plastic deformation compared with the rest of the solid. There are principally two reasons for such a plastic instability to occur. The first reason is a geometrical inhomogeneity in the form of variation in the cross-section or shape of the solid resulting in necking when subject to uniaxial tension. The second reason consists of a strain softening taking place nonuniformly in a localized volume of the solid. If strain softening is followed by strain hardening as commonly observed for semicrystalline thermoplastics, stabilization of the localized plastic deformation can take place, also resulting in necking.

As mentioned above, necking is a typical macroscopic manifestation of a geometrical inhomogeneity. This instability occurs because of the stress being concentrated at some point of the solid. The yield stress is therefore attained at this point earlier than in the rest of the solid. Resulting localized plastic flow causes orientation of molecules in the neck region, resulting in a strain hardening. This allows for stabilization of the neck and its extension through the whole solid. Increase in test temperature helps the neck development. A more fundamental reason for inhomogeneous plastic deformation to occur is strain softening, which may occur after the yield point. Strain softening is an intrinsic property of polymers. It can be described as a substantial reduction of a resistance to plastic deformation after reaching the yield point. If this occurs locally, the material in this region will undergo even larger deformation than the rest of the specimen. In some cases, orientation of molecules followed by strain hardening can stabilize this process and lead to a neck development and extension. Types of inhomogeneous deformation are discussed in more detail by Kinloch and Young [128].

Crazing and shear banding are the most common forms of inhomogeneous plastic deformation in glassy polymers. Shear banding develops in glassy polymers in the case when crazing is suppressed. This occurs mostly in the case of compressive loading. As mentioned previously, strain softening is an intrinsic property of glassy polymers, and its localization due to molecular heterogeneities is almost inevitable, resulting in heterogeneous plastic deformation.

A gross localization of plastic deformation in craze fibrils, bridging the craze surfaces, makes crazes precursors of brittle fracture. Shear yielding in the form of a quasihomogeneous bulk process can contribute substantially to the crack resistance of a polymeric solid. On the other hand, however, localized shear yielding in the form of shear microbands is believed to be a precursor of brittle fracture in many semicrystalline and glassy thermoplastics. The nature of developed shear bands depends on test conditions and the type of polymer. In PS, sharp fine texture of shear bands exists, hence the term "microshear bands." Microshear bands are usually 0.5–1 μm thick and are oriented in an angle of 38 ° to the direction of compressive load. On the other hand, shear bands observed in PMMA are wide with poorly defined boundaries resembling broad diffuse zones, termed "diffuse shear bands." More importantly, however, the strain in microshear bands reaches several hundred percent, whereas that in the diffuse shear bands is only a few percent above the strain level in the rest of the material. Some indications suggest that the diffuse shear bands are viscoelastic rather than plastic. Other glassy thermoplastics form shear bands intermediate between microshear and diffuse shear bands.

Strain rate, test temperature, and thermal history of the specimen all affect the appearance of shear bands in a particular polymer. The differences in morphology of shear bands were proposed to be due to different rates of strain softening and the rate sensitivity of the yield stress. Microshear bands tend to develop in polymers with small deformation rate sensitivity of $\sigma_y$ and when relatively large inhomogeneities existed in the specimen before loading.

Interestingly, the contribution of diffuse shear bands to the total deformation of the specimen is large, despite relatively low deformation existing in them. On the other hand, large plastic deformation in microshear bands does not contribute so substantially to the total deformation of the specimen. It appears that a small deformation over a large volume has much greater effect than a large deformation in a small volume. It was also found that the diffuse shear bands are formed by very fine deformation bands of the width of 50 nm. Unlike the 500- to

1000-nm microshear bands, these fine deformation elements of diffuse shear bands are not continuous. On the other hand, the shear deformation in these submicroshear elements was of the order of the deformation found in microshear bands.

Localization of shear deformation into narrow bands is now believed to be a precursor of brittle failure in thermoplastics and in filled polymers. The experimental evidence for this hypothesis lies in observations of microvoid formation at the intersections of shear bands. It is proposed that stress raisers, such as air bubbles, flaws, defects, molecular inhomogeneities, and filler particles, enable microshear banding to occur at loads well below the bulk yield stress. When two microshear bands intersect, microfibrils from each of these bands are subject to further straining. In this extremely strained area, chain disentanglement or, eventually, even chain scission take place preferentially in comparison with the rest of the shear bands. This process results in a generation of microvoids that act as nuclei for crack growth initiation. An additional mode of failure resulting from the presence of sharp shear microbands is the rupture of microfibrils in these bands. Kinloch and Young [128] summarized the observed sequence of failure processes leading to a brittle fracture of glassy polymers in the following steps: strain softening in the plane strain conditions causes microshear bands to develop, growing shear bands intersect, microvoids form at the intersection of shear bands nucleating a craze structure, and craze breakdown and crack initiation occurs. The presence of sharp notches, thick specimens, low temperature, thermal annealing, and high strain rates promotes this type of fracture.

The two-phase morphology of semicrystalline polymers causes a substantially different response to mechanical excitations, resulting in an inhomogeneous plastic flow on microscopic and macroscopic levels different from those observed in glassy thermoplastics. Moreover, the deformation of crystalline regions is anisotropic, resulting in slips and twinning in only certain crystallographic directions. It is common, however, that no shear bands develop when the morphology of a semicrystalline polymer is satisfactorily isotropic. Interactions of localized shear bands, resulting in formation of microcracks, are also believed to be a precursor of brittle fracture in semicrystalline polymers. On the other hand, large kink bands are observed in deformed oriented semicrystalline polymer fibers. It is thought that these kinks are formed by combination of a slip in crystalline regions and plastic flow in the amorphous regions. The angle of the kinks in respect to chain orientation depends on the chemical structure of the long-chain molecules and the conditions of the test.

Semicrystalline polymers deform elastically by affine deformation up to the relative deformation of about 1%. The two-phase morphology and relatively low shear strength of lamellae results in an onset of plastic deformation at very low strains. In a macroscopically quasi-isotropic polymer, orthotropic lamellae are randomly oriented in respect to the external loading direction between the two extreme orientations (i.e., parallel and perpendicular) due to their radial growth pattern in a spherulite and due to their helix topology. In the course of plastic deformation, rotation and shear deformation of lamellae occurs within the crystallites and bunches of lamellae followed by increasing inclination of chain orientation with respect to the lamella plane. Eventually, lamellae disintegrate into small blocks of folded chain morphology joined by tie molecules. As a result of lamellae breakage, the concentration of tie molecules increases with increasing deformation. Blocks of folded chains held together by the tie molecules form microfibrils with orthotropic symmetry of properties caused by orientation of chains in crystalline blocks and orientation of extended tie molecules in the amorphous phase. The diameter of these fibrils range from 10 to 30 nm, and their length may be on the order of micrometers. The nature of this deformation process dictates the tie molecules to be concentrated on the surface of individual microfibrils with a portion of them bridging the gap between neighboring microfibrils and ensuring integrity of larger bunches of microfibrils-fibrils. Orientation of molecules in the amorphous phase is one of the most profound changes of morphology occurring in microfibrils compared with an undeformed polymer.

### 11.4.3 Yielding in a Vicinity of an Inclusion

Investigations of the localized yielding caused by stress concentrations at the vicinity of solid or soft inclusions are of extreme importance for the understanding of the processes underlying the global yielding of a particulate filled or fiber reinforced PP. In general, when a heterogeneous material is subject to external loading, the mismatch between the elastic properties of the continuous phase and the inclusions results in stress concentration in the region near the inclusion surface [138]. Localized plastic deformation of the matrix material originates first in the place of the largest stress concentration. The mechanisms of plastic deformation in thermoplastics can be either shear yielding or crazing or their combination. The onset of the macroscopic plastic deformation in a particulate-filled composite can be interpreted as a percolation threshold or, in other words, as formation of the first path of completely yielded material throughout the whole specimen cross-section formed by connecting the zones of locally plastically deformed material. The understanding of the way in which the localized plastic deformation occurs is of pivotal importance [139–143].

Most experimental studies of localized plastic deformation in the vicinity of an inclusion have been performed using either a glass bead or an elastomeric inclusion. The actual shape of the local stress fields as the magnitude of external stress amplification are strongly affected by the shape of the inclusion and the degree of adhesion between the matrix and the inclusion. Despite complications caused by these factors, the principal understanding of the localized yielding has been gained in the course of investigating systems containing only one spherical inclusion. In general, two conically shaped surfaces of highly plastically deformed matrix are formed in system containing the well-adhering inclusion. Plastic deformation starts in this case at the poles of the spherical inclusion. This case also equals the behavior of a matrix containing one rigid inclusion covered by a well-adhering rigid interphase layer. On the other hand, a poorly adhering rigid inclusion of diameter greater than the critical value initiates matrix dewetting at the poles first and then development of plastic deformation near inclusion equator. Similar results with minor modifications are valid for inclusions with well-adhering soft interphases, for soft inclusions, and for holes [144–146].

The effect of particle size on the localized plastic deformation has been studied extensively by Vollenberg and colleagues [147,148], who based their interpretation of the effect of inclusion size and interfacial adhesion on the yielding behavior of thermoplastics on the competition between dewetting and shear yielding. If the dewetting stress indirectly proportional to the square root of the particle diameter was greater than the stress needed for localized yielding or crazing, the latter processes were dominant even in the case of poorly adhering inclusions. They found that particles smaller than 4 µm with no adhesion to the matrix fulfilled this condition, resulting in local behavior resembling behavior of a matrix containing well-adhering particles. If the dewetting stress was lower than the yield strength, dewetting occurred first. This was also the case of inclusions with soft well-adhering interphases, even though in this case interphase cavitation is another mechanism capable of relieving local stress concentrations.

Guild and Young [149] and Jancar et al. [150] used numerical methods (FEA) to assess the stress fields in the vicinity of spherical inclusions and predict the onset of local yielding. The latter paper investigates the effect of adhesion on the local stress fields in the two ideal limiting cases of perfect adhesion and zero adhesion modeling and the effect of rigid and soft interphase on the onset of yielding in filled PP, respectively. Matonis and Small [151] and Matonis [152] derived analytical expressions for the components of the local stress field in the vicinity of a solid spherical inclusion encapsulated in perfectly adhering soft interphase layer. Application of their results for an elastomer interphase layer on a rigid filler in PP matrix yielded the conclusion that for elastomer interphase thickness greater than 1% of the spherical inclusion diameter, the core-shell inclusion behaves principally as an elastomeric inclusion of the volume equal to the sum of the volumes of the core and the shell components. In other words, covering the rigid filler such as $CaCO_3$ with a uniformly thin elastomer interphase of the total volume equal to about 2–4 vol% of the filler, one can replace up to 98% of the elastomer with substantially cheaper rigid filler and still obtain the same toughening effect as with the elastomer alone. Needless to say, however, deposition of uniformly thin elastomeric interphases on the particulate fillers or fibrous reinforcements is a tedious task that has not been reduced to an industrial practice yet. Interesting experimental data on the effect of rigid interphase on the stress field around an inclusion have been published by Kendall and Sherliker [153] for PE filled with carbon black.

Jancar et al. [154] attempted to calculate the effect of a soft interphase on the stress field around and in the platelet-shaped and fibrous inclusions of small aspect ratio. Because of the presence of a shear component of the stress in the interphase, a transfer of a portion of the load from the matrix to the core-shell inclusion is possible, even when the interphase layer has modulus of elasticity substantially lower than the matrix. At least five to six times thicker soft interphase compared with spherical inclusion is necessary to reduce the reinforcing efficiency of platelets with aspect ratio of 5 to a negligible value. Above the elastomer interphase volume fraction equal to about 12 vol% of the inclusion, the elastic modulus of the complex core-shell inclusion equals that of the PP matrix.

### 11.4.4 Global Yielding as a Percolation Threshold

The yield strength of thermoplastics filled with particulate filler depends strongly on the structural variables such as concentration and shape of the filler particles, morphology of the matrix, interphase thickness and properties, and the matrix–interphase and interphase–filler adhesion. The spatial arrangement of the filler particles and agglomeration also have great influence. In addition, mode of loading and the state of stress within the polymeric solid have great influence on the yielding behavior of filled PP. There have been a number of attempts to interpret yield strength of filled PP in terms of the structural parameters. Despite indisputable practical success of these models, most of them lack the desired predicting power because semiempirical parameters are used to fit the experimental data [155–159]. To overcome this deficiency, Jancar et al. [154] proposed a model to predict the yield strength of particulate-filled thermoplastics using only the structural parameters and the properties of the constituents measured in independent experiments. This was achieved by considering the onset of macroscopic global yielding as a percolation threshold [160]. This model has later been extended to describe yielding behavior of PP filled with core-shell inclusions.

Consider a particulate-filled thermoplastic consisting of a random distribution of uniform size spherical inclusions of diameter $d$. Although they are randomly distributed on the macroscopic scale, one may imagine that on a microscopic scale there is some degree of order in the arrangement of particles. Thus, we visualize the composite structure as a random array of ordered cells consisting of a small number of particles in simple geometrical arrays. Assume each cell consists of particles in a cubic arrangement in a matrix with a center-to-center distance $a$. The material then consists of a large number of these cells with a random distribution of center-to-center distances.

Upon applied stress load, there will be a stress concentration around each particle in the cell. If the matrix has a yield point, there will be a zone of yielded matrix around each of adjacent particles in the given cell. The rest of the matrix remains in an elastic state. If particles are far enough apart (low filler volume fraction), the localized behavior will occur independent of the neighbors. If adjacent particles are close enough to allow overlapping of their respective yielded microzones, the matrix within the cell will fully yield. The yielding behavior of the composite will then be a combination of the localized microcracking and shear yielding with the ultimate deformation response dependent on the volume fraction of the cells undergoing the various failure mechanisms. The model for interacting stress concentration fields used by Sjoerdsma [139] in utilizing the percolation concept to predict the ductile–brittle transition was used to predict the yield strength of the material.

The shape and size of the localized yielded microzones (Fig. 11.3) have been calculated [150] using a finite element analysis software, and von Mises stresses were compared with the independently measured yield strength of the matrix. Perfect adhesion at all the interfaces and rigid interphase were assumed. Following the lead of Sjoerdsma, the yielded area fraction, $f_y$, has been expressed as

$$f_y = F(c) v_f^2 \tag{11.22}$$

where $F(c)$ is proportional to the average yielded area per particle:

$$F(c) = \frac{9c^4}{\pi} \left( \frac{\sqrt{(1-c^{-2})}}{2c(c^{-2} - 0.25) \arccos\left(\frac{1}{c}\right)} + \frac{1}{16} \sin\left[4 \arccos\left(\frac{1}{c}\right)\right] \right)$$

The physical meaning of $F(c)$ is the rate of increase in connectivity of yielded microzones around individual particles as a function of their size. The larger $F(c)$, the lower $v_f^{crit}$ or, in other words, the larger the yielded microzones around individual particles, the sooner the maximum yield strength is achieved. In agreement with qualitative expectations, it appears that $F(c)$ is, for a given combination of matrix and filler, a function of matrix ductility. It was observed that $F(c)$ increases with increasing matrix ductility. Equation (11.22) provides means to calculate $F(c)$ and numerical methods such as FEA can be used to estimate size of the plastic microzones $c$.

In an uniaxial tensile test, we assume that the composite yield strength $\sigma_{yc}$ is proportional to the fraction of the yielded material:

$$F(c) v_f^2 = \frac{\sigma_{yc} - \sigma_{ym}^*}{\sigma_{yc}^{max} - \sigma_{ym}^*} \tag{11.23}$$

LOW VOLUME FRACTION
GOOD ADHESION

LOW VOLUME FRACTION
NO ADHESION

HIGH VOLUME FRACTION
GOOD ADHESION

HIGH VOLUME FRACTION
NO ADHESION

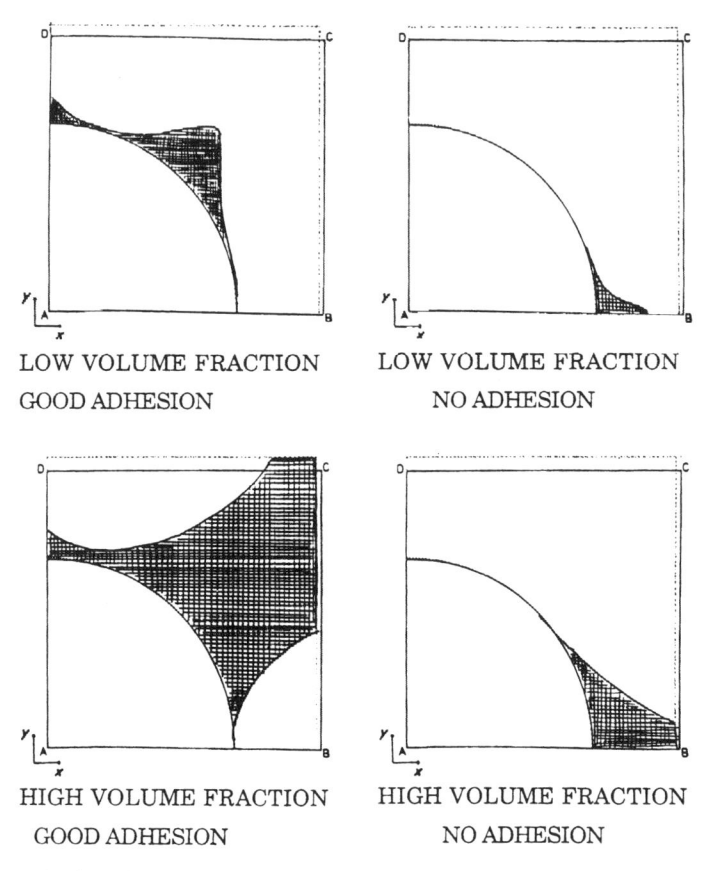

**FIGURE 11.3** Localized yielding from FEM calculations: volume fraction–level of adhesion. (From Ref. 150, courtesy of SPE.)

where $\sigma_{ym}^*$ is the matrix yield strength in the presence of a very small percentage of the filler (generally slightly less than the yield strength of the unfilled matrix). Equation (11.23) can be rearranged into a more useful form:

$$\sigma_{yc} = \sigma_{ym}^*[1 + 0.33F(c)v_f^2] \tag{11.24}$$

for $F(c)$ greater or equal to $v_f^2 > 0$ and

$$\sigma_{yc} = 1.33\sigma_{ym}^* \tag{11.25}$$

for $1/F(c) < v_f^2$. Equations (11.24) and (11.25) represent the upper bound of the composite yield strength. There is a turning point on the concentration dependence of the yield strength at a critical filler volume fraction given by:

$$v_f^{crit} = F(c)^{-1/2} \tag{11.26}$$

The model predicts an increase of $\sigma_{yc}$ with increasing $v_f$ below $v_f^{crit}$ and a constant maximum value equal to $1.33\sigma_{ym}$ above $v_f^{crit}$. This upper limiting value is based on the elastic stress concentration needed to yield the matrix in the direction of 45–50° with respect to the applied tensile stress. Above $v_f^{crit}$, the yielded microzones interconnect throughout the whole specimen cross-section, i.e., $F(c)v_f^2 = 1$, providing a yield strength of the material independent of $v_f$. The upper limiting value of $1.33\sigma_{ym}$ is independent of the matrix yield strength and depends only on the ratio of filler and matrix moduli. A reduction in matrix yield strength increases the size of the yielded microzone around individual particles, leading to a steeper increase in $\sigma_{yc}$ with $v_f$ and a lower critical volume fraction of the filler. The ductility of a polymer continuum can be increased either by raising the temperature or by introducing subcritical size stress concentrators.

In the case of core-shell inclusions with soft interphase thicker than 1% of the particle diameter, the yield strength can be calculated using the simple idea of reduction of the matrix cross-section as the cause of stress concentration as proposed for example by Nicolais and Narkis [161] in the case of nonadhering particles embedded in a polymer matrix. The lower bound for the composite yield strength can be written as:

$$\sigma_{yc} = \sigma_{ym}^*(1 - 1.21v_f^{2/3})S \tag{11.27}$$

where the strength reduction factor, $S$, can be calculated using the finite element analysis and in general varies between 1.0 and 0.2. Figure 11.4 brackets the concentration dependence of the composite yield strength between an upper limit of perfect adhesion and lower zero adhesion limit. Untreated calcium carbonate (as received with no stearic acid addition) exhibited brittle failure at 40–50 wt% loading limiting the use of this model. At 80°C, the more ductile matrix appears to show a greater degree of localized shear yielding. This may explain the slightly higher values of the relative tensile yield strength compared with that measured at 23°C.

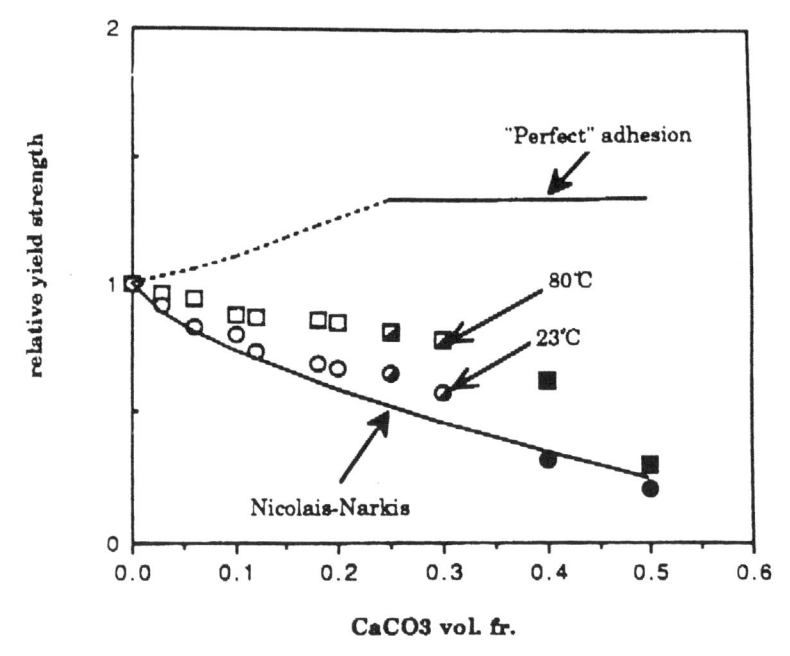

**FIGURE 11.4** Concentration dependence of composite yield strength. Upper boundary: 0.0–0.2 vol% Eq. (11.24) at $F(c) = 15$; 0.25–0.5 "perfect" adhesion. Lower boundary: Nicolais-Narkis data [161] for "no" adhesion. (From Ref. 150, courtesy of SPE.)

## 11.5 LOCALIZED PLASTIC DEFORMATION AND BRITTLE FRACTURE IN FILLED POLYPROPYLENE WITH ENGINEERED INTERPHASES

### 11.5.1 Mechanisms of Localized Plastic Deformation and Brittle Fracture in Thermoplastics

*Localized Shear Banding.* Shear yield in a form of homogeneous bulk process can contribute substantially to the crack resistance of a polymeric solid. On the other hand, localized shear yielding in the form of shear microbands is believed to be a precursor of brittle fracture in many semicrystalline thermoplastics. Localized shear yielding is thus a major contributor to the initiation of cracks along with crazing, which, however, is believed not to operate in semicrystalline thermoplastics.

CRACK INITIATION. Shear yielding is frequently considered synonymous with ductile failure. This is more or less true in the case of large-volume shear yielding. However, localization of shear deformation into narrow bands, so-called shear banding, and the interactions of shear bands are now believed to be precursors of brittle failure in thermoplastics and in filled polymers [128]. The experimental evidence for this hypothesis lies in observations of microvoid formations at the intersections of shear bands. It is proposed that stress raisers such as air bubbles, flaws, filler particles, and molecular inhomogeneities enable microshear banding to occur at the external loads well below the bulk yield stress. When two microshear bands intersect, microfibrils from each of these bands are subject to further straining. In this extremely strained area, chain disentanglement or, eventually, even chain scission take place preferentially compared with the rest of the shear bands. These processes will result in a generation of microvoids that can act as nuclei for crack growth initiation. An additional mode of fracture resulting from the presence of sharp shear bands is the rupture of microfibrils in these bands.

This latter mode of fracture occurred in notched PS loaded in tension. Notches or other severe flaws give rise to coarse shear bands, and Wu and Li [162] reported the sequence of events occurring during failure of interacting shear microbands in compression test in PS. Chau and Li [163] reported on the effect of rupture of narrow shear bands under tensile loading in PS. They generated shear bands using a sharp notch and compressive load. After shear bands had developed, the portion of the specimen containing notch was removed and the surface was polished. The specimen was then subjected to tensile loading. A typical feature of the fracture surface formed by crack initiated from the intersection of shear microbands is a presence of sheets of highly strained plastically deformed materials that extend throughout the whole specimen cross-section.

More frequently, however, fracture does not initiate from fracturing the fibrils in microshear bands but rather from the microvoids formed by interacting shear microbands. It is now accepted that the microvoids, formed by the interaction of shear bands, expand plastically and groups of them stabilize to form a craze structure with the craze plane perpendicular to the maximum principal tensile stress. Crack initiation and growth result from propagation and breakdown of the craze structure [164]. These microvoids are 10–30 nm in diameter. At temperatures below $T_g$, these microvoids grow more elliptical, concentrating stresses in the surrounding matrix that caused additional nuclei to form under triaxial state of stress. These studies were originally done in PS; however, it has been shown later than similar mechanisms also act in PMMA, PVC, PC, and other glassy polymers [165–167]. These investigations were carried out at relatively low strain rates and under plane strain conditions.

CRACK PROPAGATION. Shear yielding, involving large shear deformations, either can be a mechanism of brittle fracture when highly localized or can cause large energy dissipation and consequent ductile failure when extended over a large volume of the material ahead of the crack tip. The fracture becomes more ductile as the extent of shear yielding at the crack tip increases, causing an effective blunting of the crack and consequent increase in apparent toughness. The limiting case occurs when the whole ligament thickness is yielded before crack initiation [17].

Just why the mechanism of brittle fracture occurs in some polymers by crazing and in others by localized shear yielding is not well understood. Both mechanisms are favored when severe localization of strain occurs. This is exemplified in the case when no severe defects exist (e.g., for glassy polymers exhibiting strain softening). Under tension loading, crazing is the most favorable failure mechanism in glassy polymers. Cross-linking the macromolecules inhibits crazing in glassy polymers and leads to brittle failure due to localized shear yielding. Typical examples are epoxy, phenolic resins, and unsaturated polyesters. Using the two-phase model of semicrystalline polymers considering crystalline domains as effective cross-links connected by bunches of tie molecules, the localized shear yielding mechanism of brittle fracture can be expected in these polymers.

However, crazing and shear yielding are not mutually exclusive deformation processes. Both mechanisms were observed simultaneously in many polymers [168–173]. It is natural to expect that an interaction between crazes and shear bands will occur as they grow through the specimen cross-section. Some of the initial models of interactions between crazes and shear bands proposed termination and stabilization of crazes by existing shear bands [171]. This idea was based on the fact that craze growth is stabilized when a molecular orientation exists at the tip of a growing craze. This occurs when the maximum principal tensile stress in oriented specimens is parallel to the orientation direction. However, subsequent investigations did not confirm this hypothesis, suggesting that the only possibility for a shear band to terminate craze growth is in the case when the shear band is formed at the craze tip [172,173]. The interaction between a craze and an existing shear band would not lead to a stress relief at the craze tip because of already large strain in the shear band. A more probable result of this interaction would be a premature craze breakdown and brittle crack nucleation.

Ductile failure is mostly associated with extensive shear yielding. However, as will be shown later, multiple stable crazing as observed in rubber-modified polymers can lead to the same ductile failure mode. There is very little understanding of which molecular parameters favor ductile fracture. Quenching of the melt seems to favor ductile fracture. Typical test variables leading to a ductile fracture are plane stress, elevated temperature, low strain rates, absence of aggressive environments, and lack of sharp defects. Ductile fracture was studied in many polymers such as PC, polyphenylene sulfide (PPS), and polyphenylene oxide (PPO) using thin precracked sheets [174–178]. Dugdale-type plastic zone shape is generally observed in this type of fracture, and the experimentally determined size and shape are in a very good agreement with the calculated dimensions. Additionally, substantial lateral contractions were observed in the area ahead of the crack tip. The maximum displacements in the Dugdale zones were of the order of the thickness of the polymer sheets. The shear bands became more localized when the polymer was annealed below its $T_g$ due to an increase in the yield stress and strain softening. An average extension ratio in these shear zones was found to be of the order of 1.5–2.5 and was, in general, proportional to the maximum predicted extension ratio. Donald and Kramer [176–178] proposed a mathematical form for this observation:

$$\lambda_{sz} = 0.6\lambda_{max} = 0.6\frac{l_e}{\langle R_{rms}^2 \rangle_e^{1/2}} \qquad (11.28)$$

where entanglements are assumed to act as permanent cross-links with no slippage or chain scission occurring. The $l_e$ is the chain contour length between entanglements, and the denominator expresses the root mean square end-to-end distance of a chain of molar mass $M_e$. Equation (11.28) can be interpreted such that the extension ratio within the shear yield zone is governed by the maximal extension ratio achievable by the entanglement network of the polymer [128].

The postyield ductile fracture of polymers was extensively investigated [170–183], resulting in a conclusion that the crack is initiated from cavities growing from defects in the drawn material. These cavities have a rhombic shape with the long and short diagonals perpendicular and parallel to the draw direction. These cavities were observed in PVC, PE, polyethylene terephthalate (PET) at room temperature, and in PC, PMMA, polyether sulfone (PES), and PS at elevated temperatures. At slow strain rates, the growth of these cavities in a plastically deformed material loaded in tension is stable until the critical size is reached, resulting in an unstable catastrophic failure.

*Crazing.* Crazes develop mostly in glassy polymers by a coalescence of microvoids formed at points of high dilatational stress concentrations such as defects, flaws, surface scratches, sudden changes in cross-section, and molecular heterogeneities. The major difference between craze and crack is the ability of craze to support load. This is allowed by the presence of highly plastically deformed fibrils bridging the craze. Gross localization of plastic deformation in craze fibrils makes crazes precursors of brittle fracture. Low temperature, high strain rates, aggressive environments, and plane strain conditions promote crazing. Despite reports suggesting the presence of crazes in PP, no conclusive evidence has been presented for such a hypothesis. In any case, crazes could develop only in amorphous regions in PP.

*Interactions Between Crazes and Shear Bands.* Localized shear yielding and crazing are competing mechanisms of brittle fracture under tensile loading. Two principal interactions can occur between craze and shear band in an early mechanism proposed by Bucknall et al. [171], Newman and Wolock [169], and

Jacoby and Camer [170]. Later, however, Donald et al. [172,173] showed that interaction A cannot effectively lead to a stabilization of craze and only interaction B will stabilize craze growth and contribute to toughening of the polymeric material.

Mechanism A consists of interaction between approaching craze and oriented molecules in the preexisting shear band. It was suggested that the molecular orientation within the shear band alters completely the stress field at the craze tip and larger external stress is necessary to propagate craze through the shear yielded band. Additionally, changes in craze structure will lead to its stabilization. Mechanism B assumes stabilization of a growing craze only in the case when a new shear band forms at the tip of this craze, effectively relieving triaxial dilatational stresses in this region. This mechanism was supported by experimental observations. The same authors concluded that when a growing craze meets existing shear band, the stress at the craze tip can be relieved only by the craze breakdown leading to a crack initiation. At this time, however, it is not clear if the outcome of such an interaction is independent of craze size, number and proximity of crazes, and density of shear bands.

## 11.5.2 Delocalization of Shear Banding at Particles with Interphases

There are several mechanisms capable of stabilizing crack growth before it becomes unstable, causing a catastrophic failure or arresting already supercritical cracks (i.e., cracks proceeding catastrophically). Among the primary processes of crack stabilization are extensive large-volume yielding, reduction of the crack driving force by localized redistribution of stresses around inclusions, and reduction of the overall yield strength of the material. The most important parameters controlling which of the mechanisms will take major role in this process at given external conditions ($T$, $\dot{\varepsilon}$, etc.) are stiffness of the interphase relative to that of the matrix and the filler, fracture toughness of the interphase relative to that of matrix, thickness of the interphase relative to the size of solid inclusions, and shape and spatial arrangement of the inclusions.

The deflecting of the crack path, which can effectively increase fracture resistance of the fibrous composite, plays a much less important role in particulate-filled composites because of relatively small size of inclusions relative to the length of the crack front. The most important mechanism of crack stabilization is reduction of the crack driving force by localized redistribution of stresses around inclusions and reduction of the overall yield strength of the material leading to extensive plastic deformation ahead of the running crack. Large crack tip plastic zones are characteristic of composites with soft interphases when interphase thickness exceeds 19% of the particle diameter. Matonis and

## Engineered Interphases in PP Composites

Small [151] and Matonis [152] showed analytically that the presence of a soft layer on the spherical particle alters the stress fields in the surrounding matrix substantially. For thickness greater than 10% of the particle diameter, the stress field in the matrix is similar to that of a soft inclusion or a hole with the maximal stress concentration at the inclusion equator. Similar results were confirmed numerically by Broutman and Paniza [7], Guild and Young [149], and Jancar et al. [150].

As a result, localized shear yielding occurs at low levels of external loadings, causing plastic deformation of the matrix near the particles in the body well ahead of the crack tip. These yielded microzones interconnect with increasing load, forming eventually an uninterrupted plastically deformed zone at the crack tip. This modifies the fracture mode from a brittle to a quasi-ductile or ductile tearing, leading to a substantial increase in the apparent fracture toughness or impact strength. One has to bear in mind, however, that the change in fracture mode is the primary cause of the observed steep toughness rise and that there is no direct correlation to the material morphology. As was shown in Sections 11.2 and 11.3, elastic modulus and yield strength of the composite are the material parameters controlling this transition and directly connected with composite structure.

The state of adhesion between the interphase and matrix affects somewhat the stress field in the matrix or, in other words, the external load level at which the above-described processes develop. In the commercially most important case of soft interphases, a good interphase–matrix adhesion only changes the cavitation site from the interfacial dewetting at interface between matrix and the interphase to the interphase cavitation. Among the systems exhibiting this kind of behavior are the nylon/maleine anhydride-grafted EPR (MEPR) blends [4] and vinyl-siloxane layers and vinyl ester matrices in glass-containing composites [121].

Interfacial adhesion between the interphase and matrix becomes substantially more important in the case of rigid interphases, such as those formed by maleated PP (see Chapter 3). In these cases, the interphase itself is involved in the deformation process and can be drawn plastically absorbing substantial amount of deformation energy. On the other hand, these processes are important only for composites containing reinforcements such as fibers or flakes. In the case of irregular, approximately spherical inclusions, rigid interphase will not deform before the matrix deformation even if an ideal bond exists between the two. Such an hard interphase contributes only to an effective increase in the particle diameter.

*Delocalization of Shear Banding.* As discussed in Chapter 6, delocalization of shear banding is a desirable process capable of enhancing crack resistance of these thermoplastics. It can be achieved in many otherwise brittle polymers by incorporating a uniformly dispersed secondary discontinuous component. This approach has been shown to be especially effective when the elastic modulus of the dispersed inclusions is substantially lower than that of the matrix.

The effect of secondary inclusions on the delocalization of shear banding is based on the concept of modifying the local stress fields and achieving favorable distribution of stress concentrations in the matrix due to the presence of inclusions. This leads to reduction of the external load needed to initiate plastic deformation of the polymer. As a result, plastically deformed matter is formed at the crack tip, effectively reducing the crack driving force. Above approximately 20 vol% of the elastomer inclusions, cracks become effectively blunted and the originally brittle failure is transformed into the plastic hinging.

The location of the maximum stress concentration moves away from the particle surface with increasing inclusion volume fraction and its actual position depends on the spatial packing of the inclusions. In Chapter 6, it was shown that not much of a difference is seen in the shape of the stress field around an elastomer inclusion, a void, and a rigid inclusion with no adhesion to the matrix.

## 11.6 EFFECT OF INTERPHASES ON COMPOSITE FRACTURE TOUGHNESS

### 11.6.1 Brittle Fracture in PP

*Phenomenology.* Brittle fracture is considered the most dangerous type of failure in any engineering or load-bearing application of polymers. Flaws of various kinds in any polymeric solid cause localized stress concentration, resulting in local overloading of the material and consequent fracture. The stress concentration factor associated with a microcrack of length $2a$ having a very small radius $r$ at the tip, positioned perpendicularly to the applied stress direction, was expressed as $2(a/r)^{1/2}$. The inverse relationship of $r$ is a serious source of stress risers in a mold design. Consequently, it is of pivotal importance to determine the critical value of the external stress, $\sigma_c$, causing a brittle crack to propagate catastrophically through the specimen cross-section. In Chapter 6 (Section 6.5.1), an expression is derived for the modified Griffith criterion that relates $\sigma_c$ to materials characteristics and flaw size for brittle glass. The following expression is a function of the critical surface crack depth $a_c$ within an infinitely thick part with opening $2ka$. Other parameters include critical strain energy release rate $G_c$, the surface energy $\gamma$, and tensile modulus of elasticity $E$:

$$\sigma_c = [E(2\gamma + G_c)/(ka_c)]^{1/2} \qquad (11.29)$$

In common thermoplastics, surface free energy $\gamma$ is two to three orders of magnitude lower than $G_c$ and thus can be neglected [184]. Therefore, Eq. (11.29) is simplified to

$$\sigma_c \approx [EG_c/ka_c]^{1/2} \qquad (11.30)$$

At small and moderate strain rates, a fracture in semicrystalline polymers occurs in the process of cold drawing before or during deformation strengthening. In the course of cold drawing, the primary macroscopically isotropic properties of a semicrystalline polymer are altered into the orthotropic ones due to the transformation of the spherulitic morphology into the microfibrilar one. Microfibril ends form point vacancies or dislocations representing the weakest points of this morphology and are most probably the loci for crack initiation. These defects grow with increasing strain either perpendicularly to the draw direction by a fracture of neighboring microfibrils or parallel to the draw direction up to the locus of a similar defect in the structure of the neighboring microfibril, allowing expansion of the defect in the direction perpendicular to the draw direction. The former mechanism of crack growth is more common in polymers with strong intermolecular interactions (nylons), whereas the latter is common in polymers with weak intermolecular interactions (polyolefins). Both processes proceed initially in a stable manner resulting in an increased stress concentration in the vicinity of the growing defect. As soon as the growing crack reaches its critical size, catastrophic failure occurs, resulting in the failure of the solid. Fracture surfaces resulting from the two mechanisms of crack initiation differ substantially. Smooth glass-like fracture surfaces are observed for the first mechanism, whereas rough surfaces covered by ends of fractured microfibrils are characteristic for the latter one.

It has been shown that in many cases, especially at low temperature fractures, crack initiation described above contributes to the polymer toughness far less than deformation processes accompanying its propagation. All three stages of the failure process (i.e., initiation, propagation, and termination (important for very thin specimens)) dissipate mechanical energy stored in the material during its loading to the point of failure. The distribution of the strain energy dissipated during the individual stages of the fracture depends on both structural variables (glassy, semicrystalline, filled, elastomer modified, molecular weight, etc.) and test conditions (strain rate, temperature, specimen geometry, etc.). The structural variables control the nature and character of the actual dissipative mechanisms on the microscale (shear yielding, crazing, shear banding, etc.), whereas the test conditions control the relative importance of the respective microdeformation processes (operates the process for which the test conditions are more favorable). Unlike in the case of the elastic modulus, fracture is strongly dependent on the state of stress at the crack tip and ahead of it throughout the fracturing solid. Both structural variables, expressed in terms of material properties ($E$, $\sigma_y$, $G_c$, $\nu$) and test conditions (temperature and strain rate dependence of the material properties, specimen geometry, crack tip radius), are bound together in controlling the state of stress.

The major contributing dissipative process at the crack tip is shear yielding in semicrystalline thermoplastics. Frequently, a transition in major deformation mechanism from shear yielding to shear banding or vice versa is accompanied by a sudden change in the measured crack resistance. This phenomenon is often termed ductile-brittle transition (DBT).

The large practical importance of this phenomenon has led to many hypotheses and theories attempting to interpret the experimental data and predict effects of both types of variables on the position of DBT and the magnitude of toughness change. Some explanations of DBT have been proposed using the Ludwik-Davidenkov-Orowan hypothesis that the brittle fracture occurs when the yield stress as a material property exceeds a critical value. It assumes that brittle fracture (brittle strength) and yielding (yield strength) are independent processes with different dependencies of the respective strengths on the strain rate and temperature. Generally, yield strength is expected to depend on both of these variables in a more profound fashion than brittle strength. It is argued that whichever process can occur at lower external stress will be the operative one. Thus, the intersection of yield and brittle stress defines position of the DBT where conditions for both types of failure possess the same probability. Because this intersection must depend on $T$ and $\varepsilon$ mostly due to the strong dependence of $\sigma_y$ on these variables, DBT shifts to higher temperatures with increasing strain rate. At very high strain rates, heat formed during yielding is not conducted away due to the poor heat conductivity of polymers, causing local isothermic-adiabatic transition and preventing any strain hardening; thus, the ductile fracture prevails. In agreement with this, increase in fracture resistance has been observed at moderately high strain rates (lower than ballistic strain rates).

It is generally accepted that the most effective dissipative processes are those involving large-volume plastic deformations of the fracturing solid before crack initiation. Large plastic deformations also take place in shear bands and large crazes; however, extreme localization of plastic deformations into small volumes of the material leads to macroscopically brittle failure initiated from these areas of large but localized plastic flow. This is even more pronounced during straining of preexisting cracks or notches at high strain rates or at low temperatures. An obvious method of increasing the amount of dissipated energy is an expansion of the volume of a polymer involved in shear yielding or crazing, or, in other words, delocalization of plastic deformation. This is most effectively achieved by incorporating a secondary component of prescribed elastic properties, inclusion size, and interfacial adhesion to the matrix. Elastomers, thermoplastic inclusions, and, in some instances, rigid inorganic particles and reinforcements are the most frequently used secondary "toughening agents." The two primary deformation processes encountered in thermoplastics (shear yielding, crazing) are greatly altered by incorporation of secondary particulate matter into thermoplastic matrices. In addition, completely new deformation/dissipative processes may be introduced due to the presence of particulate inclusions both soft and rigid. The new dissipative processes and redistribution of the contributions from the existing ones often arises from the presence of the secondary component (filler particles, elastomer inclusions) and from interactions between the host polymer and the secondary particles (interfacial cavitation, particle deformation, etc.).

*Molecular Fracture of Polypropylene.* Because an understanding of the importance of any one process contributing to the failure in thermoplastics and the control over these processes is only partly attainable, the knowledge and understanding of the nature of endurance limits is extremely important for successful use of plastics and especially engineered thermoplastics [185]. In terms of the failure type, polymer fracture may occur as a rapid extension of an initial defect, plastic flow of the matter, and the thermally activated flow of the macromolecules. In all of these cases, fracture is a localized phenomenon characterized by a large inhomogeneity of deformations.

The fracture processes on the molecular level are defined in terms of localized physical rearrangements of chains realized by segment rotation, cavitation, and slip. In some instances, especially in highly oriented fibrillar polymers, at low temperature and high strain rates, primary bond breakage may be involved in rupture. This process is commonly termed *chain scission*. The relative importance of the particular process in the overall balance of mechanical energy dissipation depends on various parameters such as molecular weight, presence and type of side groups, and intra- and intermolecular interactions, and it is often affected by morphological factors such as the crystalline superstructure. In addition, the state of stress, loading geometry, strain rate, and temperature are the most important external variables affecting the nature of the fracture.

The molecular interpretation of macroscopic fracture is obscured due to a large difference in scales between individual long-chain molecules and macroscopic polymer solid subjected to mechanical loads. In addition, it is worthwhile to note that most of the results providing background for understanding the fracture on the molecular level were obtained using highly oriented polymeric fibers; thus, the concepts can be translated into the investigations of brittle fracture of quasi-isotropic polymeric solids only with a great caution.

It is now generally accepted that the fracture of thermosets and rubbery materials consists of breaking the primary chemical bonds in their backbone chains. The role of bond breaking appears much less important for thermoplastics. Electron paramagnetic resonance spectroscopy experiments have shown that in quasi-isotropic PE and PP specimens broken at low temperatures, less than 1% of the primary bonds existing on the crack plane were actually broken.

Disentanglement of the long-chain molecules was the primary molecular process allowing crack to propagate. Voids and other crack nuclei can be formed in thermoplastics by thermal motion of chain segments without breaking any primary load-bearing bonds. Thus, localized chain scission and bond strength determine the fracture resistance of polymers only in the case when no flow occurs. One can expect that loading of individual chains up to the bond strength can take place in thermosets with little or no flow allowed and, in some instances, in highly crystalline oriented thermoplastics, strained at extremely high strain rates or at very low temperatures.

To load the chain up to the strength of primary bonds, intermolecular interactions of the order of those encountered in a crystal and sufficiently large elastic modulus of the chain segments must exist. An alternative way of loading chains in a glassy polymer is via the force of inertia (i.e., by stress wave propagation). At room temperature, thermoplastic fibers (PA-6, PET, aromatic polyimide, etc.) behave "classically" in the range of strain rates from 0.1 to 140 $sec^{-1}$. After reducing the test temperature below $-67°C$, a decrease in the strength with increasing strain rate was observed for strain rates above 30 $sec^{-1}$. It has to be borne in mind that the strain rates are considered in the axial direction of the macromolecules.

Upon loading to a high stress level, chains have to relieve the built-in stresses. The mechanisms available to relieve the built-up stress include chain slip, conformation change, and chain scission. Generally, in thermoplastics, the relaxation times needed to relax strained chains at room temperature vary from $10^{-3}$ to $10^2$ sec. Thus, the relaxations are faster than the mechanical excitations at slow tests, allowing for a relief of a substantial portion of the load in the duration of the test. Rapid loading during an impact test or at the tip of a running crack can bring more nonextended chain segments to high stress levels and, eventually, to chain scission, compared with slow loading, within the same strain interval.

The effect of change in conformation of the chain can be illustrated by the behavior of PE. The transformation of the four gauche conformations within a PE segment of length 5 nm corresponds to an increase in length by 0.25 nm (i.e., by 5%). This reduces the axial elastic forces in the chain by $0.05 \times E$, which equals about 10 GPa. If the maximal static loading in the PE chain produces axial stress of 7.5 GPa, this conformation change fully relieves this load and completely unloads the chain. The rates of conformational changes are relatively high and tend to increase with the applied external load.

The final mechanism of stress relief is the thermomechanically activated chain scission. Primary bond breakage can be homolytic, ionic, or via a degrading chemical reaction. It is worthwhile to note that the relative slippage of chains, microfibrils, and fibrils reduces or prevents the mechanical scission of chains in quasi-isotropic polymeric solids. In other words, chain scission is an important mode of fracture only in highly oriented semicrystalline thermoplastics.

In PE, a draw ratio of 5 has to be achieved before any measurable bond breakage occurs. The fact that semicrystalline polymers do not exhibit significant chain scission for draw ratios below 3 is in an agreement with Peterlin's model of plastic deformation. The interspherulitic tie molecules undergo chain scission first, whereas the intraspherulitic and intercrystallite tie molecules are not overloaded in the first step. They become crucial in the later stages of deformation in formation of microfibrilar structure. Chain scission was observed in semicrystalline polymers for draw ratios ranging from 3 to 10. At these draw ratios, the presence of laterally rigid crystalline regions permits static transition of large axial stresses into chains, resulting in attainment of critical stress before macroscopic fracture.

From numerous experimental observations, it is generally accepted that the morphology of a fracture surface is not simply formed by rupture of molecular chains across a fracture plane. There is always a plastic deformation at the crack tip, resulting in large plastic flow of the chains within a layer of finite thickness from the fracture surface. Intuitively, one can expect dependence of the extent of plastic deformation on the same external parameters that affect yielding (i.e., temperature and strain rate). Generally, the higher the possibility of intersegmental motion, the larger the extent of plastic flow. Localization of fracture processes in an immediate vicinity of the crack plane emphasizes local stochastic behavior of chains during the fracture.

A hydrocarbon chain is in a constant thermal motion, and without external force field the chains fluctuate around the most stable position given by the distribution of possible conformations at the temperature. The action of external forces at the ends of a molecule causes displacements of chains from their equilibrium conformations and evokes retractive forces. For a hydrocarbon chain of $M_w = 14{,}000$, extended length 125.5 nm, and the end-to-end distance $r = 7$ nm, the maximal exerted force is 10 MPa. The level of forces exerted by the random coil macromolecules are much lower than the theoretical strength of the primary bonds. The presence of strong intermolecular interactions, such as hydrogen bonds in polyamides, affects the retractive force substantially, causing a restriction of the number of possible chain conformations. In addition, the transitions between the conformations can be dependent on each other, resulting in a cooperative character of these rearrangements. Theoretical calculations performed for extended PE and PA chains at room temperature provided small retractive forces in comparison with the ultimate forces that can be supported by the chain.

The characteristic feature of the loading of a chain in the semicrystalline thermoplastic is the periodicity of the potential field. For the ideal PE lamella, the limiting axial stress was estimated at about 7.5 GPa. In other words, by being built in the crystalline lamella, the long-chain molecules in semicrystalline thermoplastics can be loaded to a very high level of stress, which are generally two orders of magnitude greater than those achievable in glassy thermoplastics. It has been found that the mechanical excitation decays in the direction of lamella thickness. For an ideal PE crystal, the penetration length of the mechanical excitation into the lamella was found to be about 5 nm. As a result, if the tie molecule ends in the lamella or continues and ends in an amorphous region, the axial forces to which it can be exposed are very small. Such a tie molecule contributes very little to the strength of a semicrystalline solid.

Addition of strong intermolecular interactions such as hydrogen bonding between long-chain molecules, e.g., in polyamides, creates a relatively inhomogeneous potential field because these attractions are stronger at C=O and N-H sites than at $CH_2$ sites. It was suggested that the hydrogen bonds are responsible for the difference between the cohesion energies for PA-6 (18.4 kcal/mol) and PE (6.3 kcal/mol). These calculations provide the limiting stress to pull a tie molecule from an ideal PA-6 crystal of the order of 22.4 GPa (i.e., about three times that of PE). Despite relatively crude assumptions and simplifications used in the original model, the results hold even when more sophisticated approaches were applied.

### 11.6.2 Fracture Toughness of Particulate-Filled PP

Considerable experimental data can be found in the recent literature dealing with fracture toughness of particulate-filled polypropylene. However, there is little analysis of the effect of filler on the measured values of strain energy release rate ($G'_c$). Effects of the specimen geometry and mode of loading are often neglected, and observed changes are attributed only to the change in composite structure, mainly to the filler volume fraction, particle shape, and filler surface treatment. Needless to say, the phenomenological character of the term *fracture toughness* results in a lack of structural information. Hence, one has to be very careful when trying to interpret directly the changes in fracture toughness in terms of structural variations. It seems more correct to interpret fracture toughness in terms of structural variations. It seems more correct to interpret fracture toughness in terms of other mechanical properties, namely, elastic modulus, yield strength, and Poisson ratio. Application of existing models connects the changes in these parameters to the variations in composite structure.

In Chapter 6, Charpy notched impact strength measurements are analyzed for calcium carbonate–filled polypropylene. It has been shown that the effect of filler volume fraction on the strain energy release rate $G'_c$ of particulate- filled polypropylene can be described by calculating the contributions from mixed modes of fracture using basic principles of LEFM. Quantitative results for strain energy release rate, obtained using a model combining contributions from the plane stress and plane strain regions, are in good agreement with the experimental data at low filler volume fractions for both soft and rigid interphases.

Above $v_f = 0.2$, fracture toughness in the case of soft interphase is determined by the ratio between the square of the relative yield strength and the relative modulus of the composite. A lower bound for $G'_c$ in good agreement with experimental data has been calculated. The reduction in material yield strength leads to a higher extent of plastic deformation in the front of the crack and thus to an increase in the strain energy release rate for crack propagation and an increase in Charpy notched impact strength.

## 11.7 BALANCING COMPOSITE PROPERTIES AND DESIGN RECOMMENDATIONS

It has been shown that, based on a combination of numerical method and percolation concept, a quantitative model can be developed capable of reasonably accurate predictions of compositional dependencies of elastic moduli, yield strength, and fracture toughness for PP filled with inclusions of various shapes. The quantitative predictions were derived for the two limiting cases of soft and rigid engineered interphases and perfect adhesion at all interfaces. This, along with other simplifications, allowed calculations of dependencies of the upper and lower limits of the mechanical properties on filler volume fraction, interphase thickness and modulus, particle shape, and properties of the constituents. Good quantitative agreement has been found between theory and experimental data in composition regions satisfying the model assumptions. The theoretical predictions were found less accurate when larger deviations from the assumed ideal morphology occurred. Despite all that, conceptual models can be used to design filled polypropylenes according to the requirements of the end user in a practical choice of filler volume fraction interval.

### REFERENCES

1. PS Theocaris. Mesophase concept in composite materials. Berlin: Springer-Verlag, 1987, p. 3.
2. Yu S Lipatov. Physical chemistry of polymer composites (Engl. Trans.). Int Polym Sci Technol Monograph: #2, Moscow, 1977.
3. A DiAnselmo, J Jancar, AT DiBenedetto, JM Kenny. Composite Materials. London: Elsevier, 1992, p. 49.
4. SM Connelly. Ph.D. Thesis, University of Connecticut, 1993.
5. J Jancar, AT DiBenedetto, A DiAnselmo. Polym Eng Sci 32:1394–1399, 1994.
6. MY He, JW Hutchinson. Int J Solid Struct 25:1053, 1989.
7. JL Broutman, J Paniza. Polym Eng Sci 14:1254, 1974.
8. SM Connelly, AT DiBenedetto, M Accorsi. Proc. 5th Eur. Conf on Compos Mater, Lyon, 1993, p. 699.
9. S Ranade, AT DiBenedetto. Proc. Adhesion '96. London: Cambridge VSP Publ, 1996, Vol. II, p. 714.

10. JE Castle, JF Watts. Surface analytical techniques for studying interfacial phenomena in composite materials. In: FR Jones, ed. Interfacial Phenomena in Composite Materials. Oxford, UK: Butterworths, 1989, pp. 3–6.
11. B Pukanszky, F Tudos, J Jancar, J Kolarik. J Mater Sci Lett 6:345, 1989.
12. J Jancar. J Mater Sci 24:3947–3955, 1989.
13. J Jancar. J Mater Sci 24:4268–4274, 1989.
14. J Jancar, AT DiBenedetto, A DiAnselmo. Chem Papers 50:187–203, 1996.
15. PN Hornsby, CL Watson. Plast Rubb Proc Appl 6:169, 1986.
16. J Rychly, K Vesely, E Gal, M Kummer, J Jancar, L Rychla. Polym Degr Stabil 30:57–72, 1990.
17. JG Williams. Polymer Fracture. Chichester: Horwood, 1984, p. 100.
18. J Jancar, AT DiBenedetto. Proc. IUPAC 1992, Prague. Macromolecules 1992. Berlin: VSP Publ, 1993, pp. 399–409.
19. B Pukanszky, E Fekete. In: J Jancar, ed. Inorganic Fillers in Thermoplastics. Munich: Springer-Verlag, 1998.
20. L Delamare, B Vergnes. Computation of the morphological changes of a polymer blend along a twin-screw extruder. Polym Eng Sci 36:1685–1693, 1996.
21. J Kolarik, J Jancar. Polymer 33:4961–4967, 1992.
22. J Kolarik, F Lednicky. In: B Sedlacek, ed. Polymer Composites. Berlin: de Gruyter, 1986, p. 537.
23. J Kolarik, J Jancar, F Lednicky, B Pukanszky. Polym Commun 31:201, 1990.
24. JE Stamhuis. Polym Compos 9:280, 1988.
25. J Jancar, AT DiBenedetto. J Mater Sci 29:4651–4658, 1994.
26. J Jancar, A DiAnselmo, AT DiBenedetto. Polymer 34:1684, 1993.
27. J Jancar, J Kucera. Polym Eng Sci 30:707, 1990.
28. J Jancar, J Kucera. Polym Eng Sci 30:714, 1990.
29. J Jancar, A DiAnselmo, AT DiBenedetto. Polym Eng Sci 33:559–563, 1993.
30. J Jancar, AT DiBenedetto. J Mater Sci 30:2438–2445, 1995.
31. J Jancar. J Polym Sci Polym Symp 178:236, 1995.
32. FA McClintock, GR Irwin. Fracture Toughness Testing. Philadelphia: ASM, 1965, p. 85.
33. PL Fernando, JG Williams. Polym Eng Sci 20:215, 1980.
34. M Bramuzzo. Polym Eng Sci 29:1077, 1989.
35. M Hair, W Hertl. J Phys Chem 73:2372, 1969.
36. EP Pluedemann. Silane Coupling Agents. New York: Plenum Press, 1982, p. 49.
37. LV Phillips, DM Hercules. In: DE Leyden, ed. Silanes, Surfaces and Interphases. New York: Gordon and Breach, 1986, p. 235.
38. RT Morrison, RN Boyd. Organic Chemistry. Boston: Allyn & Bacon, 1983.
39. HA Clar, EP Pluedemann. Mod Plast 409(6):133, 1963.
40. P Lex, Ph.D. Thesis, University of Connecticut, Storrs, 1988.
41. SW Morral, DE Leyden. In: DE Leyden, ed. Silanes, Surfaces and Interfaces. New York: Gordon and Breach, 1982, p. 501.
42. ME Schrader, I Lerner, FJ D'Oria. Mod Plast 45:195, 1967.
43. AT DiBenedetto, DA Scola. J Coll Interf Sci 74:150, 1980.
44. D Pawson, FR Jones. In: FR Jones, ed. Interfacial Phenomena in Composite Materials '89. London: Butterworths, 1989, p. 188.

45. W Bascom. Macromolecules 5:792, 1972.
46. LH Lee. J Coll Interf Sci 27:751, 1968.
47. FJ Boerio, SY Cheng. J Colloid Interf Sci 68:252, 1979.
48. AT DiBenedetto. Pure Appl Chem 57:1659, 1985.
49. AT DiBenedetto, P Lex. Polym Eng Sci 29:543, 1989.
50. S Wu. Surface and Interfacial tensions of polymers, oligomers, plasticizers and organic pigments. In: J Brandrup, EH Immergut, eds. Polymer Handbook. New York: Wiley, 1989.
51. MW Urban, JL Koenig. Appl Spectrosc 40:513, 1986.
52. NG Cave, AJ Kinloch. Polymer 33:1162, 1992.
53. M Arroyo, A Iglesias, F Perez. J Appl Polym Sci 30:2475, 1985.
54. CD Han, C Sanford, HJ Yoo. Polym Eng Sci 18:849, 1978.
55. SJ Monte, G Sugerman. In: A Whelan, JL Craft, eds. Developments in Plastics Technology. New York: Elsevier, 1985, p. 86.
56. CO Yang, JF Moulder, WJ Fataley. Adhes Sci Technol 2:11, 1988.
57. J Gomes. Ph.D. Thesis, University of Connecticut, 1989.
58. S Ranade, AT DiBenedetto. Proc. EURADH '96, Cambridge, 1996, p. 717.
59. JB Peri. J Phys Chem 70:2937, 1966.
60. JB Peri, AL Hensley Jr. J Phys Chem 72:2926, 1968.
61. AT DiBenedetto, SJ Huang, D Birch, J Gomez, WC Lee. Compos Struct 27:73, 1994.
62. T Perez. M.S. Thesis, University of Connecticut, 1995.
63. Q Fu, G Wang. Polyethylene toughened by rigid inorganic particles. Polym Eng Sci 32:94–97, 1992.
64. KU Schaefer, A Theisen, M Hess, R Kosfeld. Properties of the interphase in ternary polymer composites. Polym Eng Sci 33:1009–1021, 1993.
65. H Kitamura. In: T Hayashi, K Kawata, S Umekawa, eds. Progress in Science and Engineering of Composites. Tokyo: Japan Society for Compos. Mater., 1982, p. 1787.
66. JF Gerard, B Chabert. Macromol Symp 108:137, 1996.
67. E Mader, HJ Jacobasch, K Grundke, T Gietzelt. Influence of an optimized interphase on the properties of polypropylene/glass fibre composites. Composites 27A:907–912, 1996.
68. E Mader, KH Freitag. Interface properties and their influence on short fibre composites. Composites 21:397–402, 1990.
69. HP Schreiber. Proc. EURADH '96, Cambridge, 1996, p. 432.
70. DC Pfeiffer. J Appl Polym Sci 24:1451–1455, 1979.
71. J Jancar. Deformation behavior of polypropylene composites. Ph.D. Thesis, Inst. Macromol. Chem. Czech Academy of Sciences, Prague, 1987.
72. AJ Keller. Polym Sci 15:31, 1955.
73. HD Keith. Morphology of polymers. In: MB Bever, ed. Encyclopedia of Materials Science and Engineering. New York: Pergamon Press, 1986, p. 3110.
74. MJ Folkes. Interfacial crystallization of polypropylene in composites. In: J Karger-Kocsis, ed. Polypropylene: Structure, Blends and Composites. London: Chapman and Hall, 1995, 3, p. 340.
75. E Devaux, JF Gerard, P Bourgin, B Chabert. Compos Sci Technol 48:199, 1993.

76. D Campbell, MM Qayyum. J Polym Phys Ed 18:83, 1980.
77. E Devaux, B Chabert. Polym Commun 31:391, 1990.
78. M Fujiyama, T Wakino. J Appl Polym Sci 42:9, 1991.
79. A Keller. In: A Cifferi, IM Ward, eds. Ultrahigh Modulus Polymers. London: Elsevier, 1977.
80. A Galeski. In: J Karger-Kocsis, ed. Polypropylene: Structure, Blends and Composites. London: Chapman & Hall, 1996, p. 116.
81. YS Hobbs. Nature 234:12, 1970.
82. YS Hobbs. Nature 239:28, 1972.
83. JD Hoffmann, GT Davies, JI Lauritzen. In: NB Hannay, ed. Treaty on Solid State Chemistry. New York: Plenum Press, 1976, p. 497.
84. DG Bassett. Principles of Polymer Morphology. Cambridge: Cambridge University Press, 1981, p. 16.
85. F Rybnikar. Macromol Sci Phys B 19:1, 1981.
86. F Rybnikar. J Appl Polym Sci 27:1479, 1982.
87. L Mandelkern. Crystallization and melting of polymers. In: G Allen, ed. Comprehensive Polymer Science. Oxford: Pergamon Press, 1989.
88. M Fujiyama, TJ Wakino. Appl Polym Sci 43:57, 1991.
89. J Varga. Crystallization, melting and supermolecular structure of isotactic PP. In: J Karger-Kocsis, ed. Polypropylene: Structure, Blends and Composites. London: Chapman & Hall, 1995, p. 56.
90. B Lotz, AJ Lovinger, RS Cais. Macromolecules 21:2375, 1988.
91. Y Chatani, H Maruyama. J Polym Sci Polym Phys 29:1649, 1991.
92. Y Chatani, H Maruyama, K Noguchi. J Polym Sci Polym Lett 8:393, 1990.
93. AJ Lovinger, D Davis, B Lotz. Macromolecules 24:552, 1991.
94. AJ Lovinger, B Lotz, D Davis, FJ Padden Jr. Macromolecules 26:3494, 1993.
95. J Rodriguez-Arnold, A Zhang, ZD Che, AJ Lovinger. Polymer 35:1884, 1994.
96. MJ Folkes, ST Hardwick. J Mater Sci Lett 3:1071, 1984.
97. GE Schoolenberg, AA Von Rooyen. Transcrystallinity in fiber reinforced thermoplastic composites. In: I Verpoest, F Jones, eds. Proc. Conf. Int. Phenomena in Composite Materials (IPCM '91). Oxford: Butterworth-Heinemann, 1991, p. 111.
98. F Hoecker, J Karger-Kocsis. Polym Bull 707, 1993.
99. JJ Elemendorp, GE Schoolenberg. Some wetting and adhesion phenomena in polypropylene composites. In: J Karger-Kocsis, ed. Polypropylene-Structure Blends and Composites. London: Chapman & Hall, 1995, p. 257.
100. J Kolarik. High Modulus Polymer Fibers and Fibrous Composites. Prague: Academia Press, 1987, p. 10.
101. F Bueche. J Chem Phys 20:1959, 1952.
102. FJ Bueche. J Chem Phys 22:603, 1954.
103. F Bueche. J Appl Phys 26:738, 1955.
104. F Bueche. J Chem Phys 25:599, 1956.
105. PE Rouse. J Chem Phys 21:1272, 1953.
106. BH Zimm. J Chem Phys 18:830, 1950.
107. BH Zimm. J Chem Phys 24:269, 1956.
108. AV Tobolski. Properties and Structure of Polymers. New York: Wiley, 1960, p. 166.
109. JC Halpin, JL Kardos. Polym Eng Sci 16:344, 1976.

## Engineered Interphases in PP Composites

110. JD Eshelby. Proc Soc Lond A241:376, 1957.
111. RM Christensen, H Lok. J Mech Phys Solids 27:2234, 1975.
112. B Budiansky. J Mech Phys Solids 13:223, 1965.
113. Z Hashin, S Shtrikman. J Mech Phys Solids 10:335, 1962.
114. Z Hashin, S Shtrikman. J Mech Phys Solids 11:127, 1963.
115. CTD Wu, RL McCollough. In: GS Hollister, ed. Developments in Composite Materials. London: Applied Science Publ, 1977.
116. T Pakula, M Kryszewski, J Grebowicz, A Galeski. Polym J 6:94, 1974.
117. R Hill. J Mech Phys Solids 13:223, 1965.
118. YA Dzenis. Mekh Kompoz Mater (Russ) 1:14, 1986.
119. TB Lewis, LE Nielsen. J Appl Polym Sci 14:1449, 1970.
120. L DiLandro, AT DiBenedetto, J Groeger. Polym Compos 9:209, 1988.
121. K Dijkstra. Deformation behavior of toughened nylon 6. Ph.D. Dissertation, Univ. Twente, The Netherlands, 1993.
122. JD Ferry. Polymer Viscoelasticity. New York: Wiley, 1965.
123. B Pukanszky, F Tudos. In: M Lewin, ed. Proc. IUPAC Int. Symp. on Polymers for Advanced Technologies, Jerusalem 1987, p. 792.
124. JL Kardos. In: H Ishida, G Kumar, eds. Molecular Characterization of Composite Interfaces. New York: Plenum Press, 1985, p. 1.
125. IG Voight-Martin, EW Fischer, L Mandelkern. J Polym Sci Polym Phys 18:2347, 1980.
126. J Jancar, AT DiBenedetto. Proc. ANTEC '93, SPE, New Orleans 1993, p. 1698.
127. K Friedrich. Ph.D. Dissertation, Univ. Bochum, 1980.
128. AJ Kinloch, RJ Young. Polymer Fracture. London: Elsevier, 1983, p. 128.
129. AS Argon. Phil Mag 28:839, 1973.
130. NJ Brown. J Mater Sci 18:2241, 1983.
131. EJ Kramer. J Polym Sci Polym Phys 13:509, 1975.
132. PB Bowden, RJ Young. J Mater Sci 9:2034, 1974.
133. TM Lin, IM Harrison. Polym Mater Sci 59:430, 1988.
134. AN Gent, SJ Madon. Polym Sci Polym Phys 27:1529, 1989.
135. JA Roetling. Polymer 6:311, 1965.
136. DL Hatl. J Appl Polym Sci 12:1653, 1968.
137. C Bauwens-Crowet, JC Ots, JC Bauwens. J Mater Sci 9:1197, 1974.
138. MEJ Dekkers, D Heikens. In: H Ishida, JL Koenig, eds. Proc 1st Int. Conf. on Compos. Interfaces. New York: North Holland, 1986, p. 161.
139. SD Sjoerdsma. Polym Commun 30:106, 1989.
140. J Jancar, A DiAnselmo, AT DiBenedetto. Polym Eng Sci 32:1394, 1993.
141. J Jancar, AT DiBenedetto. Sci Eng Compos Mater 3:217, 1994.
142. J Jancar, AT DiBenedetto. J Mater Sci 30:1601, 1995.
143. J Jancar. Macromol Symp 108:163, 1996.
144. MEJ Dekkers, D Heikens. J Mater Sci 20:3865, 1985.
145. MEJ Dekkers, D Heikens. J Mater Sci 20:3873, 1985.
146. MEJ Dekkers, D Heikens. J Mater Sci 18:3281, 1983.
147. PHT Vollenberg, D Heikens. In: H Ishida, JL Koenig, eds. Proc 1st Int. Conf. on Compos. Interfaces. New York: North Holland, 1986, p. 171.
148. PHT Vollenberg, D Heikens, HCB Ladan. Polym Compos 9:382, 1988.

149. FJ Guild, RJ Young. J Mater Sci 24:2454, 1989.
150. J Jancar, A DiAnselmo, AT DiBenedetto. Polym Eng Sci 32:1394, 1992.
151. VA Matonis, NC Small. Polym Eng Sci 9:91, 1969.
152. VA Matonis. Polym Eng Sci 9:100, 1969.
153. K Kendall, FR Sherliker. Br Polym J 12:111, 1980.
154. J Jancar, A DiAnselmo, AT DiBenedetto. Polym Eng Sci 32:1394, 1992.
155. B Pukanszky, F Tudos, F Kelen. In: B Sedlacek, ed. Polymer Composites. Berlin: de Gruyter, 1986, p. 167.
156. B Pukanszky, B Turczanyi, F Tudos. In: H Ishida, ed. Proc. 2nd Int. Conf. on Interfaces in Polym. Compos. New York: Elsevier, 1988.
157. J Jancar, J Kucera. Polym Eng Sci 30:707, 1990.
158. J Jancar, J Kucera. Polym Eng Sci 30:714, 1990.
159. B Pukanszky, F Tudos, J Jancar, J Kolarik. J Mater Sci Lett 8:1040, 1989.
160. J Jancar, AT DiBenedetto. J Mater Sci 30:160, 1995.
161. L Nicolais, M Narkis. Polym Eng Sci 11:194, 1971.
162. JCB Wu, JCM Li. J Mater Sci 11:434, 1976.
163. CC Chau, JCM Li. J Mater Sci 16:1858, 1981.
164. S Wellinghoff, E Baer. J Macromol Sci B4:1195, 1977.
165. L Camwell, D Hull. Phil Mag 27:1135, 1973.
166. NJ Mills. J Mater Sci 11:363, 1976.
167. I Narisawa, M Ishikawa, H Ogawa. J Mater Sci 15:2059, 1980.
168. RP Kambour. J Polym Sci Macromol Rev 7:1, 1973.
169. SB Newman, I Wolock. J Res Natl Bur Stds 58:339, 1957.
170. G Jacoby, C Cramer. Rheol Acta 7:23, 1968.
171. CB Bucknall, D Clayton, WE Keast. J Mater Sci 7:1443, 1972.
172. AM Donald, EJ Kramer. J Mater Sci 17:1871, 1982.
173. AM Donald, EJ Kramer, RP Kambour. J Mater Sci 17:1739, 1982.
174. NJ Mills. Eng Fract Mech 6:537, 1974.
175. HF Brinson. Exp Mech 27:72, 1970.
176. AM Donald, EJ Kramer. Polymer 23:1183, 1982.
177. AM Donald, EJ Kramer. Polym Sci Polym Phys 20:899, 1982.
178. AM Donald, EJ Kramer. J Mater Sci 17:1765, 1982.
179. PL Cornes, RN Haward. Polymer 15:149, 1974.
180. PL Cornes, K Smith, RN Haward. Polym Sci Polym Phys 15:955, 1977.
181. N Walker, RN Haward, JN Hay. J Mater Sci 14:1085, 1979.
182. N Walker, RN Haward, JN Hay. J Mater Sci 16:817, 1981.
183. N Walker, JN Hay, RN Haward. Polymer 20:1056, 1979.
184. N Brown. Yield behavior of polymers. In: W Brostow, RD Corneliussen, eds. Failure of Plastics. New York: Hanser Publ, 1986, pp. 121–130.
185. HH Kausch. Polymer Fracture. Berlin: Springer, 1978, p. 9.

# 12

## Mega-Coupled Polypropylene Composites of Glass Fibers

**Harutun G. Karian**
RheTech, Inc., Whitmore Lake, Michigan, U.S.A.

Material specifications for under-the-hood automotive applications require retention of mechanical properties for materials subjected to changes in ambient temperatures ($-40°C$–$175°C$) for prolonged periods. Besides temperature, the prolonged service lifetime of a molded part reflects material durability to extremes in environmental conditions that include moisture, thermo-oxidative degradation, exposure to sunlight, and creep resistance under high static stress loads.

Traditionally, material design requirements that suit such demanding end-use applications have been limited within the domain of engineering plastics based on polyamide 6 or 66, polyester alloys, and polyacetal-type resins. However, as described in Chapter 9, glass fiber-reinforced polypropylene (GFRP) composites continue to gain a market share in automotive molded parts.

The underlying cause for increased opportunities for chemically coupled GFRP composites in the realm of engineering plastics is efficient mechanical stress transfer from the polymer matrix to the load-bearing glass fiber reinforcement via strong adhesive bonds at the polymer–fiberglass interface. There is a general consensus that macroscopic mechanical behavior is dependent on the degree of chemical coupling at the molecular scale in the region of the matrix–glass fiber interface.

Many factors are responsible for closing the performance gap between composites of engineering plastics and polypropylene (PP) resins. Enhanced mechanical properties of GFRP composites are primarily due to practical utilization of fundamental concepts of the design of engineered interphases of GFRP composites as described in Chapter 11. Furthermore, because polyamide resins are quite hydroscopic, there is significant loss in mechanical properties of corresponding glass fiber–reinforced composites at elevated temperatures. Economically, the lower raw material cost of polypropylene resins combined with lower density yields favorable cost-to-volume for molded parts.

In this chapter, recent improvements in the degree of chemical coupling are described in the context of improved interphase design. The resulting development of mega-coupled GFRP composites [1] are described that exhibit significantly higher mechanical properties approaching the "perfect coupling" limit. As a result, mega-coupled GFRP composites can be truly described as engineering plastics.

## 12.1 CONTRIBUTING FACTORS TO STRENGTH OF GFRP

The underlying factors that determine the mechanical strength of GFRP composites are many and interrelated:

1. Interfacial adhesion models
2. Concept of interphase microstructure
3. Glass fiber length and orientation

### 12.1.1 Interfacial Adhesion

Because the polymer matrix is inherently nonpolar and therefore incompatible with untreated glass fibers, it is necessary to add a type of coupling agent. Chapter 3 reviews chemical coupling technology. The spectrum of chemical coupling agents consist of a variety of functional groups grafted onto polypropylene resins by a separate reactive extrusion step, described in Chapter 10 (i.e., maleic anhydride, acrylic acid, himic anhydride, fumeric acid, and other derivatives of dicarboxylic acids). This type of functionality promotes interfacial adhesion by bonding with basic groups adhering to the glass fiber surface from surface treatment via glass fiber manufacture.

The polymer chain extending from the reactive end of the coupling agent provides the means for physical entanglement with the polypropylene matrix to provide sufficient anchor in the stress transfer mechanism. Hence, the chemical coupling agent acts as a compatibilizer bridge (depicted in Fig. 12.1) between the nonpolar polypropylene matrix and the ionic substrate of glass fiber surfaces.

# Mega-Coupled PP Composites of Glass Fibers

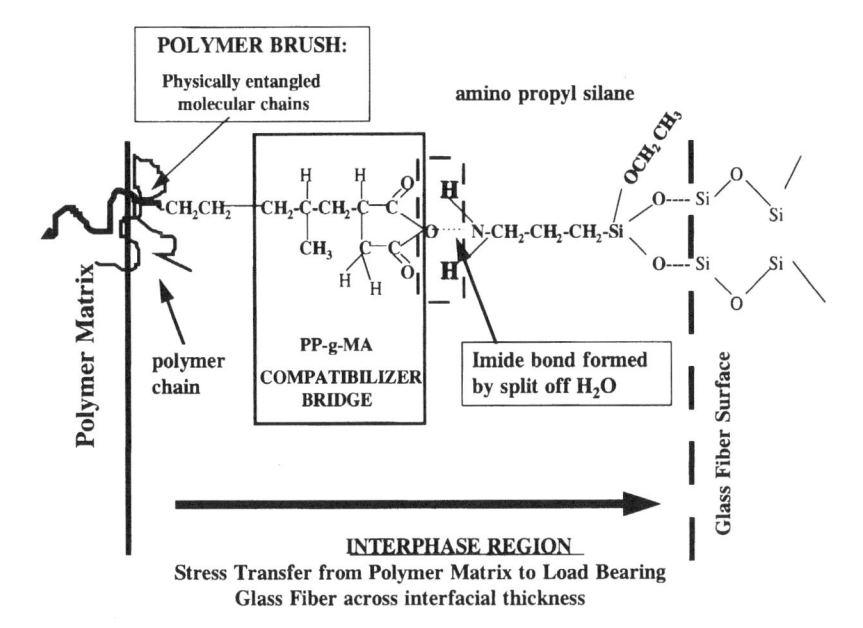

**FIGURE 12.1** Chemically coupled reinforcement. (Reprinted courtesy of Chapman & Hall from Pritchard [8].)

## 12.1.2 Interphase Microstructure

Efficient stress transfer from the polymer matrix to the load-bearing glass fibers implies greater involvement of the polymer matrix than just adhesion at the boundary of the fiber–polymer matrix interface. Drzal [2] attributed "good" interfacial adhesion to the existence of interphase microstructure in a distinct three-dimensional region of space surrounding each glass fiber with interphase thickness $\Delta r_i$ as shown in Fig. 12.2. Presence of a sufficiently thick interphase is a "necessary" criterion for attaining desired strength, stiffness, and impact resistance of polymer composites. The magnitude of $\Delta r_i$ is dependent on the degree of chemical coupling between the glass fibers and polymer matrix.

Scanning electron microscopy (SEM) photomicrographs of fracture surfaces (provided by Dr. K. McLoughlin, Aristech, Monroeville, PA) make a comparison between uncoupled (Fig. 12.3) and chemically coupled (Fig. 12.4) composites of 30 wt% GPRP. Figure 12.3 exhibits distinctly bare glass fibers having no visible adhesion to the polymer matrix. However, the surface of glass fibers in Fig. 12.4 is covered by a thin layer of bonded chemical coupling agent (2 wt% Unite MP-1000). Similar direct observations of the interphase region are given by Okawara et al. [3].

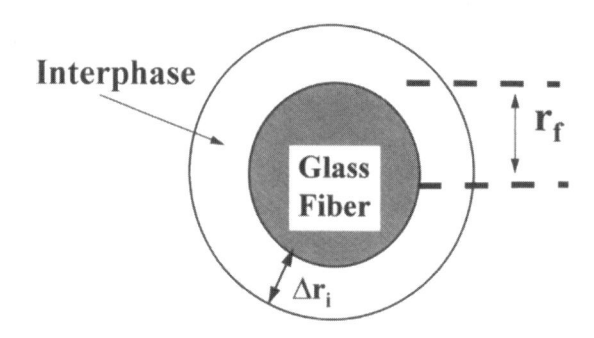

FIGURE 12.2 Geometry of interphase.

FIGURE 12.3 SEM micrograph of GFRP: uncoupled–no coupling agent. (Provided by Dr. K. McLouglin, Aristech, Monroeville, PA.)

## Mega-Coupled PP Composites of Glass Fibers

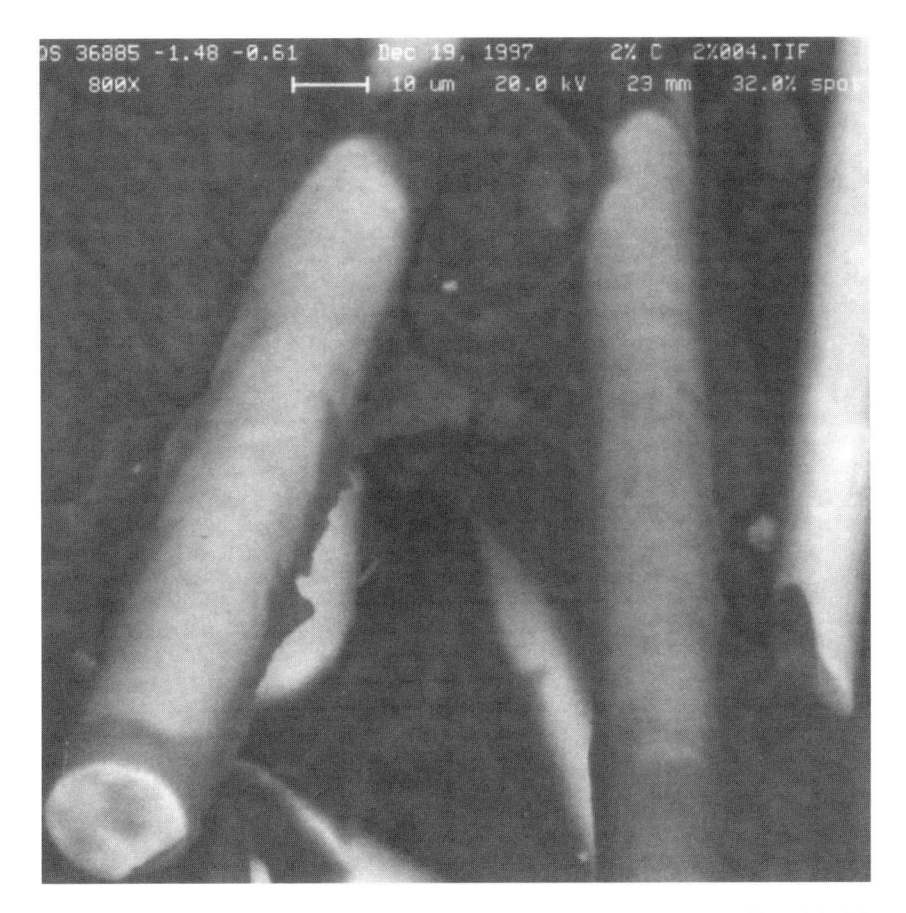

**FIGURE 12.4** SEM micrograph of GFRP: 2 wt% Unite MP-1000. (Provided by Dr. K. McLouglin, Aristech, Monroeville, PA.)

### 12.1.3 Glass Fiber Length and Orientation

An ideal composite consists of perfect bonding at the fiber–matrix interface and uniaxial alignment of continuous or very long glass fibers with respect to the applied tensile stress load. However, Folkes [4] described the microstructure of injection-molded short fiber–reinforced thermoplastics as being much more complex than this simple uniaxial model. As the result of inhomogeneous flow patterns of molten polymer due to different types of mold designs, glass fiber orientation is dependent on localized direction angles with respect to stress field in three-dimensional space. Furthermore, the composite strength and stiffness depend on the cumulative effects of extrusion and injection molding processes [5] on the degree of fiber length attrition, the resultant distribution of glass fiber lengths, an interfacial adhesion with the polymer matrix.

Karian et al. [6] described the effects of short glass fibers on the mechanical properties of thermoplastic composites. They reviewed various conceptual models used to predict the tensile strength of composites in terms of measurable parameters related to glass fiber reinforcement, such as fiber glass loading level, glass fiber length distribution, load bearing strength of glass fibers, and an assessment of glass fiber orientation due to the given injection molding process. In particular, a modified rule-of-mixture model [7] was highlighted due to its relative simplicity and amenability to correlating measured tensile strength $\sigma_c$ values with fiber length determinations:

$$\sigma_c = v_f \sigma_f [1 - L_c/2L_{ave}] C_o + v_m \sigma_m \tag{12.1}$$

where $L_c$ is the critical fiber length, $L_{ave}$ is the average fiber length, $v_f$ and $v_m$ are volume fractions for the fiber glass and polymer matrix, $\sigma_f$ and $\sigma_m$ are the corresponding strengths of glass fiber and polymer resin, and $C_o$ is the fiber orientation factor with values ranging from 0 (completely random) to 1 (completely oriented to load direction).

More recently, an empirical model was proposed by Fu and Lauke [5] that provides guidelines for attaining desired short GFRP composite strength: optimal fibre length distribution, definition of required fiber orientation distribution, and necessary interfacial adhesion. They propose the following general trends as useful rules of thumb in analyzing glass fiber length distribution data:

1. Tensile strength increases rapidly with increase of the average fiber length as the percentage of minimal fiber lengths approach the critical limiting value $L_c$ (fiber length that is dependent on the interfacial adhesion).
2. As the average fiber length increases to $>5L_c$, the tensile strength values reach a plateau.
3. Composite strength increases with decrease in $L_c$, i.e., increase in interfacial adhesion strength $\tau$ by virtue of the fundamental relationship associated with the well known Kelley-Tyson expression [8] based on a shear-lag theory for metal filaments:

$$\tau = \sigma_f r_f / L_c \tag{12.2}$$

where $r_f$ is the radius of individual filaments that are formed into glass fiber strands.

## 12.2 INTERPHASE DESIGN FEATURES

Figure 12.1 portrays the combination of key building blocks required for efficient stress transfer from the polymer matrix to load-bearing glass fibers: sizing chemistry (aminopropylsilane and film formers), compatibilizer bridge by incorporation of maleic anhydride–grafted PP (MA-g-PP), and physical entanglement of polymer chains in a polymer brush. Recent research cited below describes the effects of these building blocks on interphase design.

### 12.2.1 Sizing Chemistry

Wu et al. [9] stated that the properties of the fiber–matrix interphase in glass fiber–reinforced composites can play a dominant role in dictating the end-use performance of molded parts. They identify sizing chemistry on the fiberglass surface as being the first building block of the "interphase." The glass surface composition varies as the result of changes in the fiber-forming process. The first step of the process is extrusion of molten glass at > 1000°C through the fiberglass bushing to produce filaments ranging from about 9 to 17 μm diameter for commercial grades of fiberglass for thermoplastic composites [6]. After being quenched by a water spray to reduce temperature of the glass filaments to < 100°C, 800–1000 of these filaments are surface treated by a sizing application in aqueous media. The coated filaments are formed into fiberglass strands followed by a series of secondary process steps (e.g., rolled into rovings, dried, chopped in-line or off-line). Chapter 9 discusses glass fiber manufacture in detail.

The "sizing" solutions provide a variety of attributes [6]: lubrication, bundle integrity to abrasion and fiber bending, compatibility with the modified polymer matrix, and durability to environmental conditions of the molded part. For fiberglass appropriate for chemically coupled polypropylene, the amount of sizing chemistry coated on the glass fiber surface is about 0.7–1.2 wt%. As a qualitative determination of the actual amount of sizing, the dried fibers are put into a muffle furnace or microwave oven to burn off the sizing to obtain the loss on ignition.

### 12.2.2 Compatibilizer Bridge

Felix and Gatenholm [10] characterized the effects of compatibilizer molecular weight on the mechanical response of interphase design of cellulose fiber–reinforced polypropylene composites. The cellulose fiber surface was treated with three different compatibilizers having a range of molecular weights: alkyl succinic anhydride, Epolene 43, and Hercuprime G. The latter two compatibilizers are MA-g-PP.

Titrimetric measurements were used to estimate the average number of carboxylic groups grafted onto each compatibilizer molecular chain. This information combined with molecular weight yielded an estimate of theoretical length of stretched chains attached at one end to the cellulose fiber and fully extending into the surrounding space occupied by the polymer matrix. As described in Chapter 11, dynamic mechanical measurements can be used to determine interphase thickness. These authors [10] used this technique for the given array of compatibilizers.

Table 12.1 gives a comparison between data relating to compatibilizer molecular weight and measured tensile yield strength for the 30 vol% loading of cellulose fiber. For a given molecular weight, the interfacial thickness was found to be much thicker than the fully extended compatibilizer chain. This observation led to the inference that the compatibilizer chains are entangled with the surrounding polymer matrix having a "brush-like" structure with restricted mobility of a portion of matrix polymer chains.

### 12.2.3 Polymer Brush Model

Figure 12.5 depicts a cartoon of chemical coupling chains bonded at one end to the glass fiber sizing groups with the free end intermingled with molecules of the surrounding polymer matrix. Felix and Gatenholm used a similar interaction model to explain the long-range nature of chemical coupling in cellulose-reinforced polypropylene composites. In general, the existence of a interphase with a three-dimensional microstructure is supported by the various analytical probes, particularly the thermodynamic probe via dynamic scanning calorimetry (DSC) [1].

At sufficiently high concentrations of compatibilizer bonded to the glass fiber surface, there is severe crowding of molecular chains to cause them to assume a perpendicular arrangement and stretch away from the glass fiber surface to satisfy steric hindrance. The free molecular ends become miscible with the polypropylene matrix in an entangled polymer brush structure. An interphase design consisting of Hercuprime G yielded the best mechanical properties for the

TABLE 12.1 Summary of Felix and Gatenholm Results

| Compatibilizer | Molecular weight | Theoretical stretched chain length (nm) | Interphase thickness at 20°C (nm) | Tensile yield strength (MPa) |
|---|---|---|---|---|
| Alkyl succinic anhydride | 350 | 2–3 | 220 | 24 |
| Epolene 43 | 4,500 | 7–30 | 400 | 34 |
| Hercuprime G | 39,000 | 20–300 | 600 | 40 |

Source: Refs. 10 and 11.

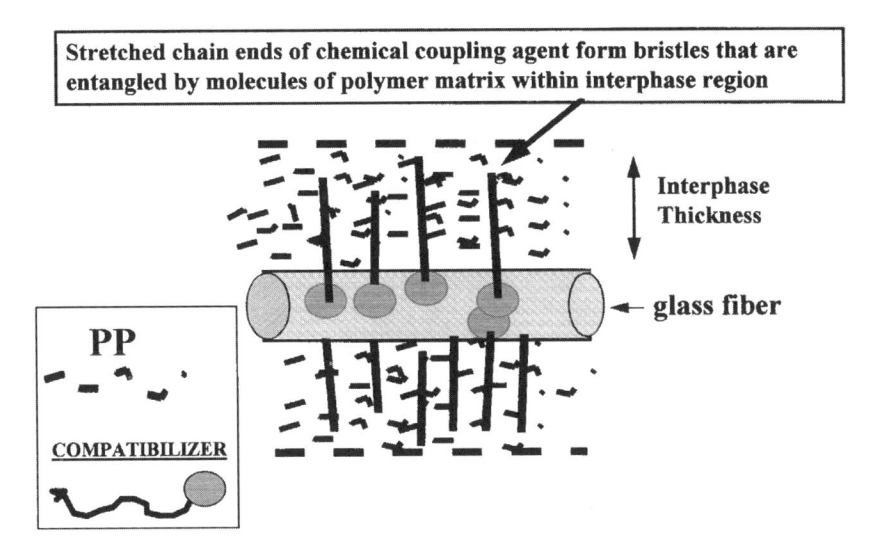

FIGURE 12.5  Polymer brush model.

cellulose fiber–reinforced polypropylene composites [11]. This enhanced mechanical response was correlated with the greatest interphase thickness due to extended molecular chains of compatibilizer having the highest molecular weight.

### 12.2.4 Characterization of Chemical Coupling Agent

Based on the findings of Felix and Gatenholm [10], the particular choice of chemical coupling agent should include consideration of its molecular weight combined with effective incorporation of grafted maleic anhydride functionality. In the polymerization of polypropylene resins and the compounding of polypropylene composites, the measurement of melt flow rate (MFR) is used as a key process control variable. In particular, the MFR of polypropylene resin is inversely related to its molecular weight (i.e., a high MFR infers low molecular weight and correspondingly low melt viscosity).

The reactive extrusion process for grafting maleic anhydride monomer (MAH) onto polypropylene backbone involves a balance of simultaneous reaction steps as described in Chapter 3:

1. Peroxide initiator is used to abstract hydrogen atoms attached to tertiary carbon atoms along the polypropylene molecular chain. This initial reaction step generates a number of free-radical reactive sites.
2. Addition of MAH monomer to the reactive sites.

3. For unreacted sites, there can be chain scission or visbreaking of the polypropylene molecular chain with accompanying increase in MFR.

In Chapter 3, the direct relationship between MFR and amount of grafted MFR is described as a consequence of the peroxide-initiated reaction kinetics. The inference here is that one should seek the highest concentration of grafted MAH for maximal molecular weight of MA-g-PP. Hence, one can envision the inherent difficulty of attaining the desired balance of MFR and grafted MAH content.

The spectrum of commercial chemical coupling agents based on MA-g-PP is depicted in Table 12.2.

### 12.2.5 Effect of Chemical Coupling Level in Tensile Behavior

Karian and Wagner [12] assessed interfacial adhesion in chemically coupled 30 wt% GFRP. The chemical coupling agent used was acrylic acid grafted polypropylene (AA-g-PP) at 0–15 wt% addition levels. The commercially sized glass fibers had a filament diameter of 13 µm. Figures 12.6 and 12.7 are SEM photomicrographs of fracture surfaces due to 2 and 15 wt% levels of AA-g-PP, respectively. Figure 12.8 is a plot of ultimate tensile strength versus chemical coupling concentration. The observed tensile behavior represents a sequence of failure mechanisms depending on the levels of compatibilizer.

At low chemical coupling concentrations (0–5 wt% AA-g-PP), tensile failure is due to fiber pullout of fibers or debonding from the polymer matrix due to relatively weak adhesion at the fiber–polymer matrix interface. From 5 to 8 wt% concentrations of chemical coupling agent, there is a progressive increase in interfacial adhesion in the fiber–glass interphase region up to the saturation limit of available sites on the crowded glass fiber surface. At about the 8 wt% level, tensile failure occurs mainly in the weaker polymer matrix. Consequently, the corresponding interfacial shear strength value at the saturation limit is equivalent to the yield strength of the polypropylene matrix (e.g., 32.8 MPa). The accompanying ultimate tensile strength values reach a plateau limit of 91–92 MPa for $\geq 8$ wt% AA-g-PP levels.

A similar trend for a plateau limit in tensile behavior was also described by Jancar and Kucera [13] for calcium carbonate–reinforced composites of polypropylene using maleic anhydride to promote interfacial adhesion. Likewise, they observed convergence of interfacial shear strength value to that of the polypropylene yield strength.

Mader and Freitag [14] used the single-fiber pullout method to characterize the interfacial shear strength of chemically coupled GFRP composites having the polymer matrix also modified by acrylic acid–grafted polypropylene. By using glass fibers having diameters between 30 and 50 µm, the greatly reduced surface-

TABLE 12.2  Commercial Grades of Maleic Anhydride–Grafted Polypropylene

| Manufacturer | Product name | Form | Total grafted monomer content (wt%) | Melt flow rate g/10 min @230°C, 2.16 kg [ASTM D-1238] | Molecular weight |
|---|---|---|---|---|---|
| Uniroyal | PB 3001 | Pellet | 0.075 | 5 | 320,000 |
|  | PB 3002 | Pellet | 0.15 | 7 | 261,000 |
|  | PB 3150 | Pellet | 0.40 | 50 | 139,000 |
|  | PB 3200 | Pellet | 0.80 | 90–120 | 110,000 |
| Nippon Hydrazine | FTH 100 | Powder | 3.0 | High |  |
| Hercules | Hercuprime G | Powder | 3.9 | High | 39,000 |
| Hoechst Celanese | Hostaprime HC5 | Powder | 4.0 | High | 30,000 |
| DuPont | Fusabond P: |  |  |  |  |
|  | MD-9508 | Pellet | 0.1 | 30 | 160,000 |
|  | MZ-135D | Pellet | 0.1 | 12 | 200,000 |
| Eastman Kodak | P-1824-003 | Pellet | 0.2 | 13 | 200,000 |
| Exxon | POX1-1015 | Pellet | 0.3 |  |  |
| Elf Atochem NA | Orevac CA 100 | Pellet | 1.1 | High |  |
| Honan Petrochemical | PH-200 | Powder | 4.2 | High | 41,000 |
| Sanyo Chemical | Youmex 1001 | Pellet | 4.0 | High | 15,000 |
|  | Youmex 1010 | Pellet | 9.6 | High | 4,000 |
| Aristech | Unite MP 320 | Pellet | 0.2 | 22 | 180,000 |
|  | Unite MP 620 | Pellet | 0.4 | 34 | 150,000 |
|  | Unite MP 880 | Pellet | 0.8 | 200 |  |
|  | Unite MP 1000 | Pellet | 1.2 | 1000 | 100,000 |

*Source:* Ref. 6, Courtesy Chapman & Hall.

**FIGURE 12.6** SEM photograph of fiber matrix interface at fracture surface: 2 wt% AA-g-PP. (Courtesy Mr. J. Helms, Thermofil, Inc.)

to-volume ratio for available sites for chemical coupling leads to the corresponding decrease in load bearing strength to applied tensile stress.

### 12.2.6 Optimal Interphase Design Attributes

*Direct Fiber-Melt Wetting Experiments.* Mader et al. [15] developed a new method to quantitatively optimize interphase design to enhance properties of GFRP composites. Direct fiber-melt wetting experiments were conducted under conditions comparable with processing conditions to rectify shortcomings in describing the crucial role played by film formers in interphase formation [16]. They used $\zeta$ potential measurements to characterize the combined effect of acid–base interactions and physical interactions in the interphase region between sized glass fibers and modified polypropylene matrix.

The wetting kinetics of sized glass fibers by melt changed due to acid–base interactions and compatibility between the film former and surrounding bulk polymer matrix. The highest interfacial shear strength values corresponded to the

# Mega-Coupled PP Composites of Glass Fibers

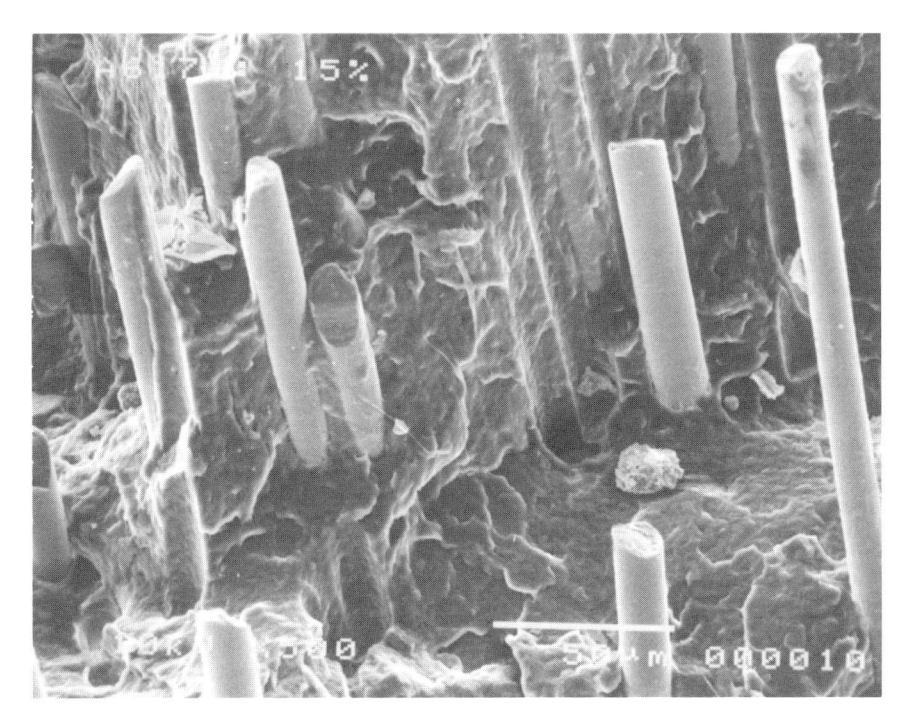

**FIGURE 12.7** SEM photograph of fiber matrix interface at fracture surface: 15 wt% AA-g-PP. (Courtesy Mr. J. Helms, Thermofil, Inc.)

combination of polypropylene film formers in the glass fiber sizing chemistry and modified polypropylene matrix containing 2 wt% Uniroyal Polybond 3150 (maleic anhydride–grafted polypropylene).

*Effects of Compounding Extrusion.* Hamada et al. [17] conducted a systematic study of interphase design for an array of 10 materials as a function of kneading condition, binder, and matrix resin for 20 wt% GFRP with chopped short glass fibers having 3 mm length and filament diameter of 13 µm. The polymer matrix options consisted of unmodified or modified polypropylene (added level of MA-g-PP). The sizing chemistry for surface treatment of the glass fibers consisted of aminosilane coating combined with an array of binder options to glue together separate glass filaments into bundles of glass fiber strands for compounding extrusion. The variety of film formers consisted of maleic anhydride–modified polypropylene emulsion, acrylstyrene copolymer emulsion, unmodified polypropylene emulsion, and chlorinated polypropylene emulsion.

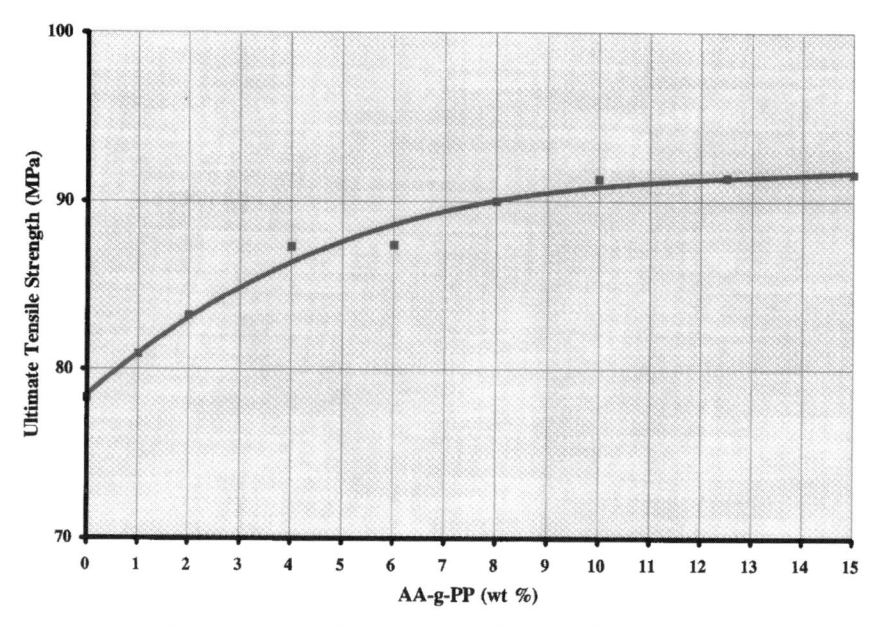

**FIGURE 12.8** Ultimate tensile behavior. (Data taken from Ref. 12.) ■, Measured tensile strength values.

The array of materials were compounded via a twin-screw extruder having options for low and high kneading conditions.

As a matter of characterizing the mechanical response to a given interphase design, molded dumbbell-shaped tensile test specimens were used to measure tensile strength and determine tensile fatigue behavior. The corresponding analyses of glass fiber distribution in molded test specimens yielded calculated interfacial shear strength values. A fiber–matrix model was defined as the interphase region as bounded by two regimes of interactions: interface 1 and interface 2. Interface 1 relates to the chemical bond formed by the acid–base interaction of the maleic anhydride end of the modified polypropylene additive and the binder chemistry on the glass fibers. Interface 2 relates to the miscibility or compatibility between binder and matrix resin. These two interfaces bound the interphase region given in Fig. 12.2.

In all binder–matrix combinations, the high kneading condition for compounding extrusion gave greater interfacial strength compared with low kneading condition. The optimal interphase design consisted of an extrusion process with the high kneading condition combined with maleic anhydride–modified emulsion (binder) and chemically coupled polypropylene matrix compositions. This resulted in maximal tensile strength and greatest resistance to tensile fatigue.

These compounding extrusion results agree with the findings via the direct fiber-melt wetting method [15] to probe the kinetics for the formation of optimal interphase structure. Thereby, optimal interphase design is now fundamentally linked to the relationship between the effects of physical and chemical interactions that influence interfacial shear strength and the macromechanical properties of composite materials.

## 12.3 METHODS TO PROBE THE INTERPHASE MICROSTRUCTURE

Despite SEM evidence for significant microstructure development by highly chemically coupled composites, actual quantitative determination of interphase thickness remains a rather controversial topic of research. Pukanszky [18] cited values from the literature ranging from 1 nm to several millimeters depending on the type of interfacial interaction and method used to probe the interphase microstructure.

Schaefer et al. [19] studied the interphase microstructure of ternary polymer composites consisting of polypropylene, ethylene-propylene-diene monomer (EPDM), and different types of inorganic fillers (e.g., kaolin clay and barium sulfate). They used extraction and dynamic mechanical methods to relate the thickness of absorbed polymer coatings on filler particles to mechanical properties. The extraction of composite samples with xylene solvent for prolonged periods indicated that the bound polymer thickness around filler particles increased from 3 to 12 nm between kaolin to barium sulfate filler types. Solid-state nuclear magnetic resonance (NMR) analyses of the bound polymer layers indicated that EPDM was the main constituent adsorbed to the filler particles. Without doubt, the existence of an interphase microstructure was shown to exist and have a rather sizable thickness. They proceeded to use this interphase model to fit a modified van der Poel equation to compute the storage modulus $G'(T)$ and loss modulus $G''(T)$ properties.

Perwuelz et al. [20] determined that heat distortion temperatures of glass fiber–reinforced composites of polypropylene and polyamide-66 (PA-66) blends were dependent on the given sizing chemistry on glass fiber reinforcement. By using boiling xylene extraction to dissolve the polypropylene ingredient, these workers inferred that the glass fibers sized for polypropylene were indeed coated by the same polymer. Conversely, PA-66-sized glass chemistry led to a preferred coating by polyamide polymer. Although the extraction method aids in the identification of the polymer coating surrounding glass fiber reinforcement, the actual extent of the interphase thickness is somewhat blurred due to the inherent complexity of three-dimensional polymeric network structures as described here.

Perhaps that attribute explains why methods based on mechanical response yield much greater interphase thickness.

Theocaris [21] developed a method for estimating interphase thickness in composites via dynamic mechanical measurements. This type of method provides a more fundamental relationship between interphase thickness and mechanical properties attributed to the interphase structure. Chapter 11 discusses this technique to probe the interphase microstructure.

More recently, Karian [1] used DSC methodology to determine the magnitude of interfacial thickness and correlate calculated values with measured tensile strength for a spectrum of composite materials having varying degrees of chemical coupling. This thermodynamic probe of the interphase microstructure is based on the Lipatov model [22] of the interphase microstructure.

## 12.4 THERMODYNAMIC PROBE OF INTERPHASE MICROSTRUCTURE

Pukanszky [18] defines the interphase as the immobilized polymer layer. Increased interphase thickness is deemed equivalent to increased filler content. Consequently, both strength and stiffness of the composite increases with the relative thickness of the interphase. Therefore, thermal and mechanical properties of GFRP composites are clearly related to the characteristics of an interphase having a three-dimensional microstructure.

Lipatov [22] investigated the effects of interphase thickness on the calorimetric response of particulate-filled polymer composites. Based on experimental evidence, his analysis led to the conclusion that the interphase region surrounding filler particles had sufficient thickness to give rise to measurable calorimetric response. The proposed existence of a thick interphase region correlates with limitations of molecular mobility for supermolecular structures extending beyond the two-dimensional filler boundary surface.

The basis for a thermodynamic probe to describe interphase microstructure is founded on fundamental concepts of the abrupt change in molecular mobility at the glass transition temperature. If filler–polymer matrix interactions were limited to a thin interfacial boundary layer, it would not be possible to observe any difference in glass transition temperature $T_g$ behavior for the composite materials having different levels of filler loading or glass fiber reinforcement.

Instead, the thermal behavior of polymer composites undergoing a glass transition is a *cooperative process* that is characteristic of macroscopic behavior due to a large number of molecules spanning a region of space beyond that of a boundary layer. Similarly, the observed differences in heat distortion temperature due to glass fiber sizing [20] could not be of any consequence if limited to a few molecular layers surrounding the fibers.

Besides observed shifts in glass transition temperature, the accompanying change in the specific heat $\Delta C_p(T)$ versus temperature behavior is indicative of the degree of interphase interactions between the glass fiber reinforcement and the surrounding polymer matrix.

### 12.4.1 Lipatov Model

Lipatov [22] analyzed specific heat data for an array of filled polymer composites. He characterized the interactions due to the existence of the interphase region surrounding filler particles as a function of filler content. The fact that the magnitude of the specific heat jump at the glass transition temperature decreases with increase in filler content is indicative of exclusion of a certain portion of macromolecules in the polymer matrix from participating in the cooperative process of glass transition.

The ratio of molecules in the immobilized layer of molecules in the region of the glass fiber surface to the remaining polymer matrix has an empirical relationship with the corresponding specific heat jump attributed to the unfilled polymer resin:

$$\lambda = 1 - \Delta C_p^f / \Delta C_p^0 \qquad (12.3)$$

where $\Delta C_p^f$ is the composite heat capacity jump at $T_g$ and $\Delta C_p^0$ is the corresponding heat capacity jump of the unfilled polymer at the glass transition temperature.

By multiplying the volume fraction of filler $v_f$ by $\lambda$, one obtains the volume fraction of the immobilized polymer. In comparison, $1 - v_f$ is the volume fraction of the entire polymer matrix. The radius of the filler particle is $r_f$. As depicted in Fig. 12.2, the radial thickness of the interphase layer is expressed as $r_f + \Delta r_i$.

The resulting empirical expression for particulate filler reinforcement for the Lipatov model can be used to calculate the effective interphase thickness $\Delta r_i$ using given values of filler volume fraction, filler particle radius, and measured calorimetric evaluation of $\lambda$:

$$(r_f + \Delta r_i)^3 / r_f^3 - 1 = \lambda v_f / [1 - v_f] \qquad (12.4)$$

Theocaris [21] developed an analogous expression for the interphase of unidirectional glass fiber-reinforced epoxies using similar terminology as given for Eq. (12.4):

$$(r_f + \Delta r_i)^2 / r_f^2 - 1 = \lambda v_f / [1 - v_f] \qquad (12.5)$$

For the given geometry, $r_f$ is radius of the glass filament with $v_f$ as volume fraction loading. In the case of GFRP composites, the relationship between weight percentage fibreglass loading and volume fraction terms is given in Table 12.3. The value of the interphase thickness $\Delta r_i$ can be computed from DSC

TABLE 12.3 Volume Fraction Terms of Glass Fiber–Reinforced Polypropylene Composites

| Fiberglass loading (wt%) | Volume fraction ($v_f$) | $v_f/[1 - v_f]$ |
|---|---|---|
| 20 | 0.084 | 0.0917 |
| 30 | 0.135 | 0.1561 |
| 40 | 0.194 | 0.2401 |
| 50 | 0.230 | 0.2987 |

measurement of heat capacity jump values for the composite and unfilled polypropylene resin materials.

Lipatov's equations have recently been used by Silberman et al. [23] to characterize the effects of different pigments on the crystallization of polypropylene resin. They defined the immobilized boundary layer as the amorphous phase absorbed on the solid filler surface.

A practical application of Lipatov's model was quantitative evaluation of the geometry of the interfacial layer in polyimide coatings [24].

## 12.4.2 Effect of Glass Fiber Content on Composite Heat Capacity

Figures 12.9 and 12.10 are plots of heat capacity as a function of glass fiber content and temperature. These plots were based on data obtained by Karian [1] using DSC methodology to investigate glass fiber–reinforced composites of polypropylene.

Figure 12.9 features three heat capacity versus temperature plots for homopolymer polypropylene resin, 50 wt% glass fiber–reinforced polypropylene, and glass fiber alone. The abrupt increase in the heat capacity at the glass transition temperature or melting temperature is attributed to molecular mobility or free volume. Even though the polypropylene resin was highly isotactic, Fig. 12.9 shows that the polypropylene melting point is not at a precise temperature. Instead the melt transition is really over a broad temperature range (e.g., 150–175°C).

Hsieh and Wang [25] characterized a variety of factors that influence the heat capacity for glass fiber–reinforced composites of molten polypropylene resins (e.g., temperature, pressure, molecular weight of the polypropylene resin, and glass fiber loading level). They proposed an additivity rule to predict the molten heat capacity of GFRP composites. This additivity rule is expressed as a

# Mega-Coupled PP Composites of Glass Fibers

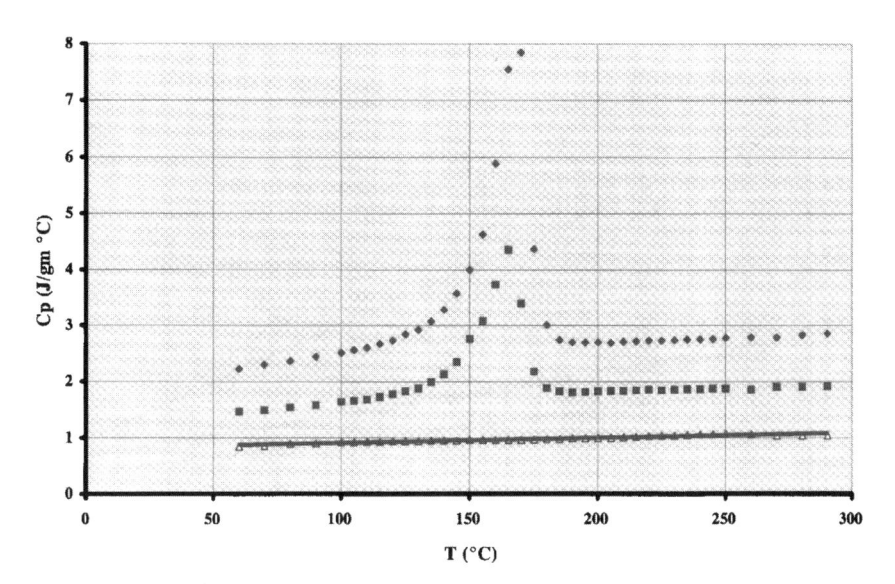

**FIGURE 12.9** Heat capacity versus temperature. ♦, 0 wt% fiberglass; ■, 50 wt% fiberglass; △, 100 wt% fiberglass.

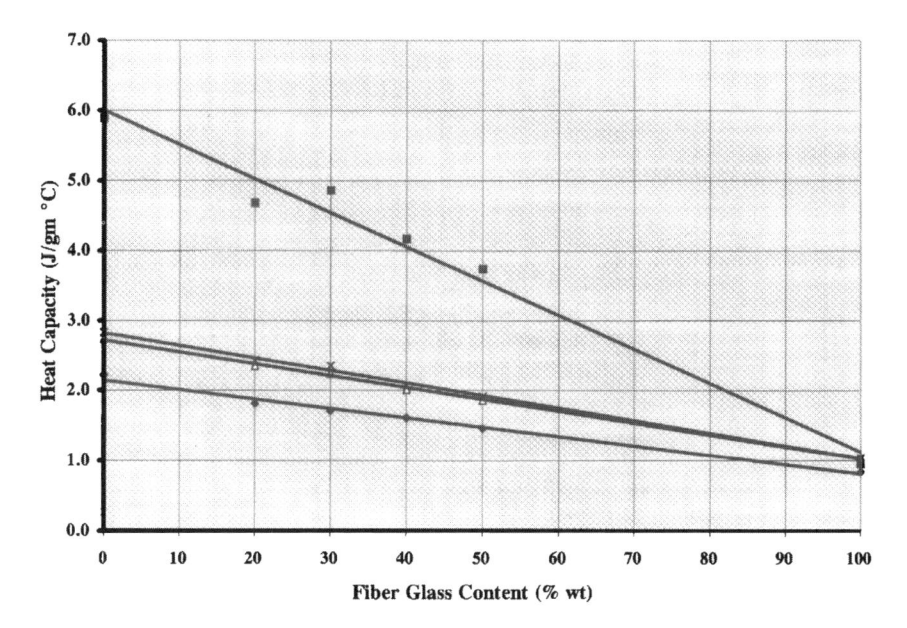

**FIGURE 12.10** Heat Capacity Plots for polypropylene composites (similar plots given in Ref. 1). ♦, 60°C; ■, 160°C; △, 230°C; ×, 280°C.

linear combination of weighted contributions of heat capacities for polypropylene and fibreglass ingredients:

$$C_p(\text{composite}) = (1 - W)C_p(\text{PP}) + WC_p(\text{glass fiber}) \tag{12.6}$$

where $W$ is the weight fraction of glass fiber. Furthermore, Fig. 12.10 provides experimental evidence for application of the additivity rule for both solid and molten composite material over a wide temperature range of 60–280°C [1].

The linear decrease of heat capacity with increase in fiberglass content during the melt transition process is analogous to decreased heat capacity jump at the glass transition temperature. In both melting and glass transition processes, increased glass fiber loading tends to immobilize a greater portion of the polymer matrix. In that sense, the observed additivity rule for solid–melt transition is mirrored by Eq. (12.5), depicting the effects of glass fiber content.

### 12.4.3 Measurement of Heat Capacity Jump at Glass Transition

As reported by Karian [1], a series of GFRP composites were compounded via a twin-screw extruder. Various levels of chopped short glass fibers were fed downstream via a side feeder. The melt extrudate was quenched in a water bath and strand pelletized. The compounded pellets were molded into ASTM tensile bars. The tensile bar specimens were used to measure tensile strength at 5 mm/min crosshead speed. Small 5- to 20-mg samples were taken from the center cut of the molded tensile bars for DSC measurement of heat capacity versus temperature.

*DSC Method.* The sample was loaded into the DSC cell and first annealed by heating to about 290°C followed by slow cooling of the molten mass. Upon cooling to about −40°C, the temperature was heated at 10°C/min to 60°C. This temperature range brackets the onset of glass transition for the polypropylene ingredient. By using sapphire standard of known heat capacity, measurement of the change in heat flow for the DSC cell leads to determination of heat capacity as a function of cell temperature.

*Heat Capacity Jump for Unfilled Polypropylene Resin.* The corresponding heat capacity jump $\Delta C_p^0$ for unfilled polypropylene resin is shown in Fig. 12.11. The observed glass transition temperature range is about −15° to −5°C. Before and after onset of glass transition, heat capacity versus temperature behavior is linear but with different slopes. The measurement of heat capacity jump $\Delta C_p$ is simply the difference of heat capacity values defined by end points A and B for the two linear regions shown in Fig. 12.11. The resulting value of $\Delta C_p^0$ is 0.105 J/g °C.

# Mega-Coupled PP Composites of Glass Fibers

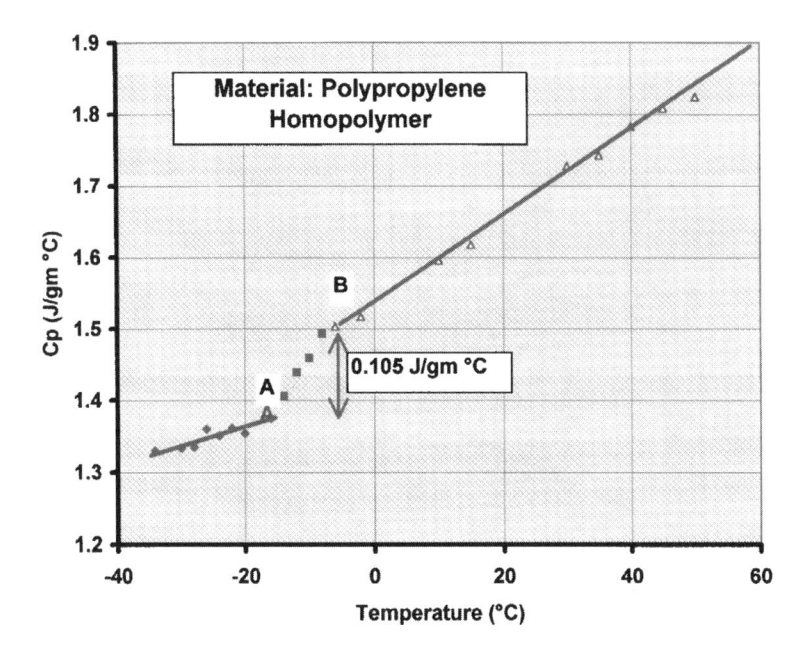

**FIGURE 12.11** Heat capacity jump for polypropylene resin. (Data from Ref. 1.) ♦, $< T_g$; ■, glass transition; △, $> T_g$.

*Heat Capacity Jump for Given Interphase Design.* By using the DSC method to evaluate $\Delta C_p^f$ for a series of short GFR polypropylene composites, the calorimetric response has been used to gauge the relative effectiveness for an array of interphase designs given in Table 12.4.

At 10 wt% fiber glass content, the heat capacity versus temperature plots for three types of composite material are given in Fig. 12.12. The significant reduction in heat capacity jump value at the glass transition temperature for P7 interphase design is indicative of a greater portion of the polymer matrix that is immobilized by a more extensive polymer brush structure, i.e., a significant

**TABLE 12.4** Interphase Design Features

| Composite material | Fiberglass content (wt%) | Chemical coupling level | Degree of entanglement |
|---|---|---|---|
| P  | 0, 10, 20, 30    | None | Very low |
| P1 | 10               | Low  | Low |
| P6 | 10, 20, 30, 40   | High | Medium |
| P7 | 10, 20, 30, 40   | High | High |

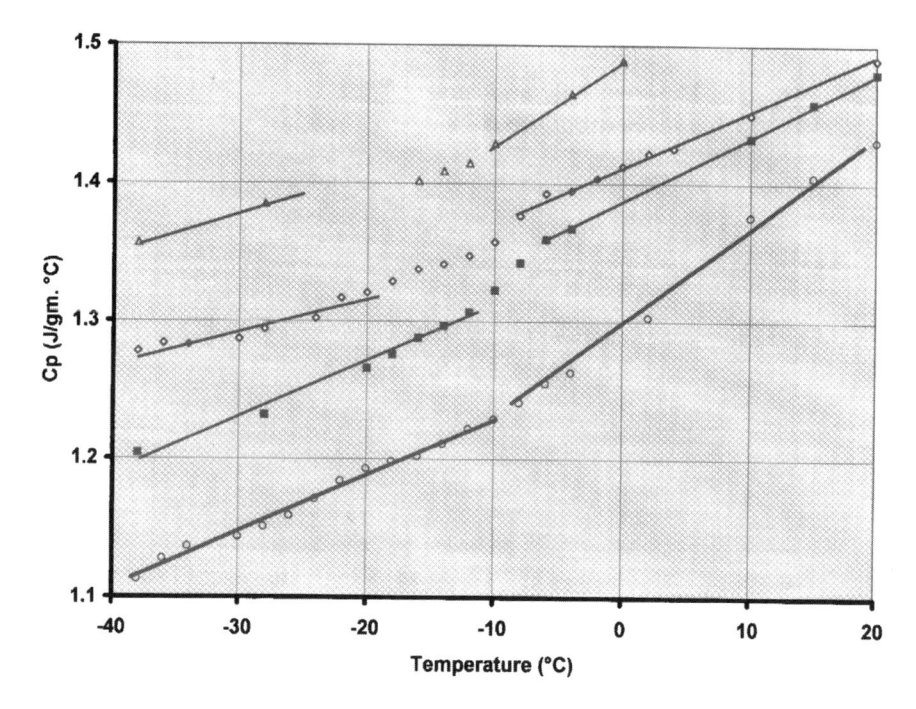

**FIGURE 12.12** Heat capacity jump for 10 wt% fiberglass loading (DSC method described in Ref. 1). ◇, $\Delta C_p^f = 0.064$ J/g °C; ■, P1: $\Delta C_p^f = 0.063$ J/g °C; ○, P7: $\Delta C_p^f = 0.010$ J/g °C; △ P6: $\Delta C_p^f = 0.038$ J/g °C.

portion of the polymer matrix is not available to participate in the glass transition process.

The addition of a greater amount of fiberglass combined with enhanced polymer brush structure development leads to rather profound results for the 40 wt% glass fiber–reinforced polypropylene composites Figure 12.13 depicts a significant reduction of the heat capacity jump value from 0.041 to 0.000 J/g °C in going from P6 to P7 interphase designs. In other words, there is no observed glass transition temperature for the P7 material. The corresponding tensile strength values of 108 MPa (P6) and 121 MPa (P7) are macroscopic consequences of the relative degree of intermolecular entanglement present in the given interphase designs.

### 12.4.4 Evaluation of Interphase Thickness

The evaluation of interphase thickness of the immobilized polymer layer surrounding glass fibers provides quantitative means for gauging the relative

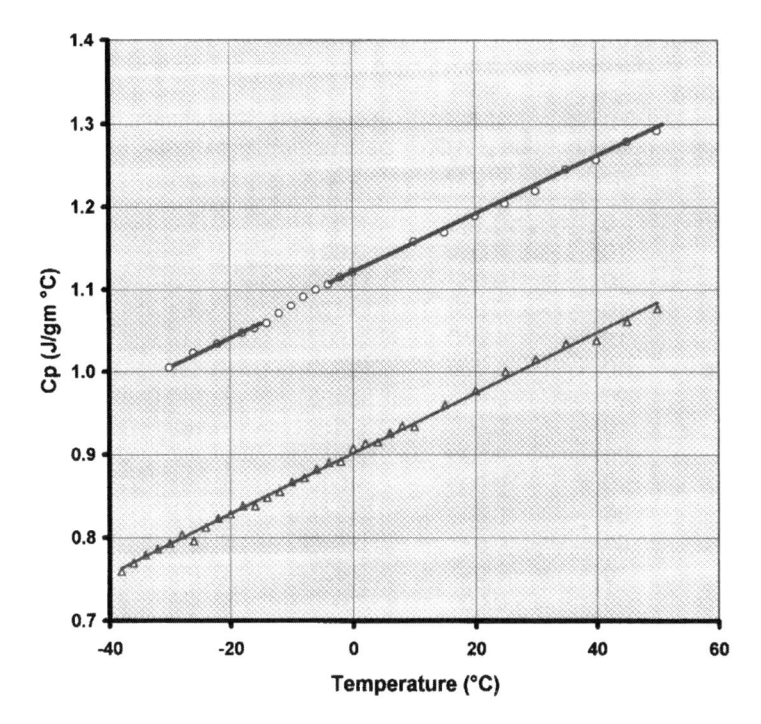

**FIGURE 12.13** Thermal behavior of 40 wt% GFRP composites (data obtained by DSC method described in Ref. 1). ○, P6: $\Delta C_p^f = 0.064$ J/g °C; △, P7: $\Delta C_p^f = 0.000$ J/g °C.

effectiveness of given interphase design. After obtaining necessary input data from DSC measurements, the following sequence of calculations is followed:

1. Equation (12.3) was used to compute the interaction parameter λ from known heat capacity jump values for filled and unfilled polypropylene resin.
2. Then for a given volume fraction of glass fiber loading, Eq. (12.5) was used to compute the interphase thickness $\Delta r_i$ for each interphase design.

Table 12.5 provides a summary of the combined results of DSC and tensile strength measurements that characterize a set of interphase designs. Folkes [27] stressed the need to do further research of this type to develop quantitative relationships between factors defining microstructure and measured macroproperty values. The resulting structure–property correlations can be used as effective

TABLE 12.5 Characterization of Interphase Designs

| Material–interphase design | Fiberglass content (wt%) | Heat capacity jump $\Delta C_p$ (J/g °C) | Interaction parameter $\lambda$ | Interphase thickness $\Delta r_i$ (μm) | Ultimate tensile strength (MPa) |
|---|---|---|---|---|---|
| Unfilled PP resin | 0 | 0.105 | 0.000 | 0.000 | 36.0 |
| P | 10 | 0.064 | 0.391 | 0.051 | 51.1 |
| P1 | 10 | 0.063 | 0.400 | 0.053 | 54.3 |
| P6 | 10 | 0.038 | 0.638 | 0.084 | 58.6 |
| P7 | 10 | 0.010 | 0.907 | 0.118 | 65.3 |
| Perfect coupling | 10 | 0.000 | 1.000 | 0.131 | 70.3 |
| P | 20 | 0.053 | 0.496 | 0.146 | 69.3 |
| P6 | 20 | 0.040 | 0.619 | 0.182 | 82.0 |
| P7 | 20 | 0.020 | 0.810 | 0.237 | 88.9 |
| Perfect coupling | 20 | 0.000 | 1.000 | 0.291 | 93.1 |
| P | 30 | 0.069 | 0.343 | 0.172 | 75.8 |
| P6 | 30 | 0.037 | 0.648 | 0.321 | 90.5 |
| P7 | 30 | 0.017 | 0.838 | 0.412 | 107.3 |
| Perfect coupling | 30 | 0.000 | 1.000 | 0.489 | 110.3 |
| P6 | 40 | 0.057 | 0.457 | 0.348 | 102.8 |
| P7 | 40 | 0.016 | 0.848 | 0.633 | 116.1 |
| Perfect coupling | 40 | 0.000 | 1.000 | 0.740 | 120.7 |

*Source:* Ref. 1.

research tools to develop cost-effective composites of reinforced thermoplastics in "critical" load bearing situations.

## 12.5 MEGA-COUPLED POLYPROPYLENE COMPOSITES

P6-type interphase design features conventional chemical coupling (as described in Chapter 3) to promote good mechanical properties for GFRP composites A unique combination of proprietary ingredients in P7-type interphase design (base polypropylene resin, type and amount of added chemical coupling agent, and appropriate sizing chemistry on the glass fiber surface) yield GFRP composites having *mega*-coupled interphase design. The nomenclature term *mega* is used to distinguish a more effective compatibilizer bridge (Fig. 12.1) than that attained by chemical Coupling alone for P6 interphase design.

Besides the formation of strong chemical bonds at the glass fiber surface, mega-coupled interphase design features *physical entanglement* of molecular chains into a polymer brush structure (Fig. 12.5). Notions of enhanced crystallinity in a polymer matrix with incorporation of glass fibers [28,29] provide an explanation for the generation of a high degree of intermolecular entanglement attributed to mega-coupling.

Pukanszky et al. [28] pointed out that mechanical properties of both unfilled and reinforced polypropylene composites are dependent on the degree of crystallinity in the base polymer resin. They developed empirical relations between Young's modulus and crystallization characteristics of molten polypropylene resins upon cooling. DSC measurement of the peak temperature of crystallization and the heat of crystallization correspond to lamella thickness and degree of crystallinity, respectively. The preponderance of successive methyl groups on the same side of the polymer backbone provides a maximal number of sites for promoting intermolecular entanglement and intermingling of crystallite layers throughout the bulk polymer matrix.

As described in Section 12.2.2, the stretched molecular chains of the chemical coupling adduct are bonded at one end to the glass fiber surface with the free end oriented in the normal direction. Milner [30] defined this unique arrangement of long-chain molecules as *bristles* of a polymer brush.

### 12.5.1 Tensile Strength–Interphase Thickness Relationship

Data given in Table 12.5 were used to generate Figs. 12.14 and 12.15. Figure 12.14 exemplifies the relationship between the interphase thickness of various interphase designs and tensile strength response as a function of fiberglass content. At constant levels of tensile strength, P7 composites require about 10 wt% less fiberglass loading than P6 type.

The corresponding relationship between interphase thickness and fiberglass content for the series of interphase designs is given in Fig. 12.15. Table 12.6 exemplifies the effects due to increased degree of entanglement in interphase design for 10 wt% fiberglass loading.

The entire set of tensile strength and interphase thickness values from Table 12.5 coalesce into a single master plot shown in Fig. 12.16. This is powerful evidence that the thermodynamic probe of the microstructure for various interphase designs appears to provide a fundamentally meaningful correlation between structure and properties.

### 12.5.2 Effect of Interphase Design on Composite Strength and Stiffness

Table 12.7 provides typical mechanical properties for a set of 10–40 wt% GFRP composites having mega-coupled (P7) type interphase design. In particular, these materials exhibit excellent strength and stiffness. Even though these types of short-term properties are often used to gauge the relative effectiveness of different interphase designs, the design engineer is more concerned with composite performance during the service lifetime of a molded part. In Chapter 13, methods

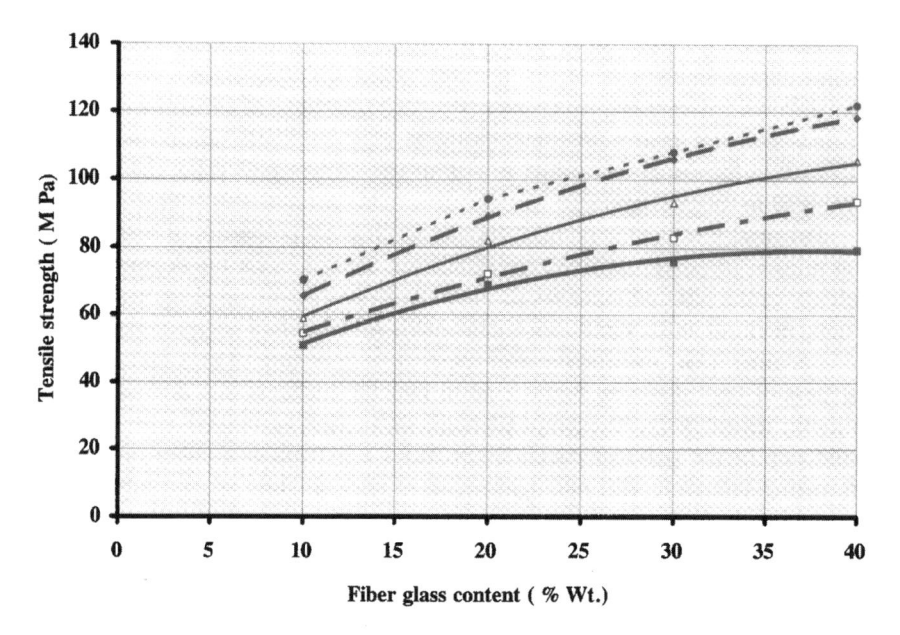

**FIGURE 12.14** Tensile strength versus fiberglass level for chemical coupling spectrum. (Reprinted courtesy of Chapman & Hall from Pritchard [6].) ■, P; □, P1; △, P6; ◆, P7; ●, perfect coupling.

are discussed for the characterization of long-term creep and fatigue behavior as better measures of performance criteria for interphase design. In particular, accelerated creep evaluations at elevated temperatures allow for quicker assessment of long-term durability.

*Tensile Creep Behavior.* The characterization of the tensile behavior of P6-20FG-0600 and P7-20FG-0600 demonstrates differences between short-term and long-term determination of composite strength under constant load. Tensile creep determination was at a chamber temperature setting of 80°C with an applied load of 80 kg (about 20.6 MPa tensile stress). In Fig. 12.17, P6-type interphase design exhibits an abrupt yield-to-failure after about 30 hr time. The corresponding P7-type composite exhibits only gradual increase in axial elongation after 45 hr duration. In comparison, the difference in ultimate tensile strength values (from Table 12.5) is relatively small, e.g., 82.0 MPa (P6) and 88.9 MPa (P7).

*Flexural Fatigue Behavior.* Particularly in automotive applications, the design engineer selects materials that will endure periodic fluctuations of both tensile- and flexural-type loads for the service lifetime of the molded part. Under-

# Mega-Coupled PP Composites of Glass Fibers

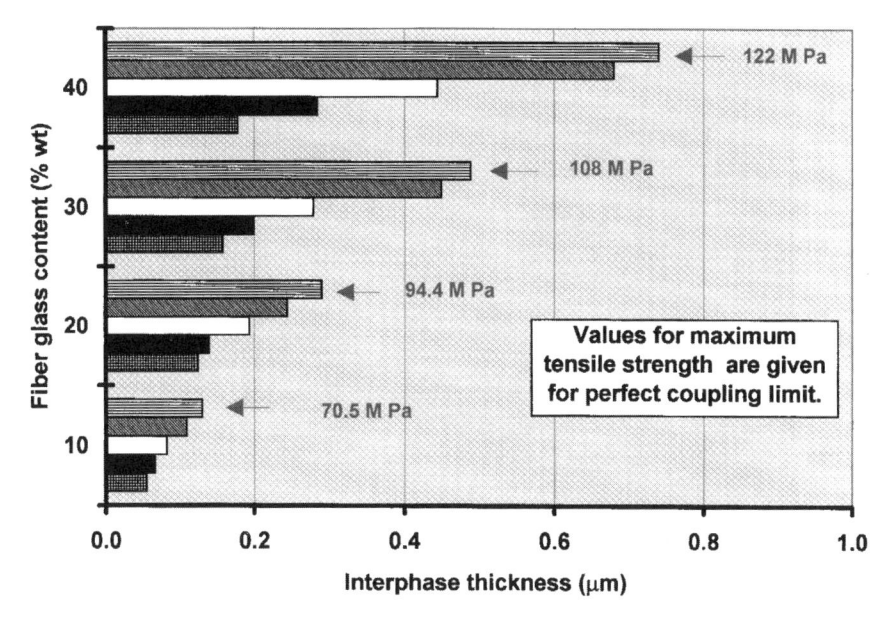

**FIGURE 12.15** Interphase thickness versus fiberglass content. (Reprinted courtesy of Chapman & Hall from Pritchard [6]. ☐, P; ■, P1; ☐, P6, ▨, P7; ▤, perfect coupling.

**TABLE 12.6** Characterization of Interphase Designs for 10 wt% Fiberglass Loading

| Material | Heat capacity jump @ $T_g$ $\Delta C_p$ (J/g °C) | Interaction parameter $\lambda$ | Interfacial thickness $\Delta r_i$ (μm) | Measured tensile strength (MPa) |
|---|---|---|---|---|
| Unfilled PP resin | 0.105 | 0.000 | 0.000 | 36.0 |
| P | 0.064 | 0.391 | 0.051 | 51.1 |
| P1 | 0.063 | 0.400 | 0.053 | 54.1 |
| P6 | 0.038 | 0.638 | 0.084 | 58.6 |
| P7 | 0.01 | 0.907 | 0.118 | 65.3 |
| Perfect coupling | 0.000 | 1.000 | 0.131 | 70.3 |

*Source:* Ref. 1.

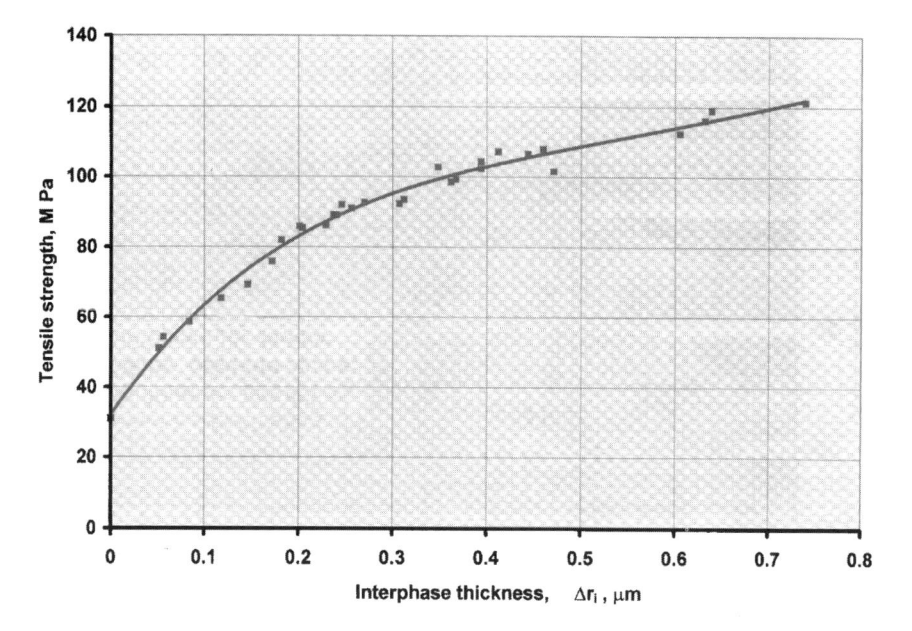

FIGURE 12.16 Correlation between tensile strength and interphase thickness. (Reprinted courtesy of Chapman & Hall from Pritchard [6].) ■, Measured tensile strength–calculated interphase thickness value.

TABLE 12.7 Summary of P7-XXFG-0600 Properties

|  | P7-10FG-0600 | P7-20FG-0600 | P7-30FG-0600 | P7-40FG-0600 |
|---|---|---|---|---|
| Fiberglass content XX (wt%) | 10 | 20 | 30 | 40 |
| Tensile strength (MPa) | 65.3 | 88.9 | 107.3 | 116.1 |
| Flexural strength (MPa) | 95 | 126 | 154 | 159 |
| Flexural modulus (GPa) | 2.898 | 4.347 | 6.348 | 8.211 |
| HDT-264 (°C) | 146 | 154 | 157 | 159 |
| Notches impact (J/m) | 48.0 | 72.0 | 77.3 | 64.0 |
| Unnotched impact (J/m) | 506.4 | 607.6 | 611.9 | 421.1 |
| Specific gravity | 0.98 | 1.04 | 1.13 | 1.23 |

Products manufactured by Thermofil, Inc., Brighton, Michigan.

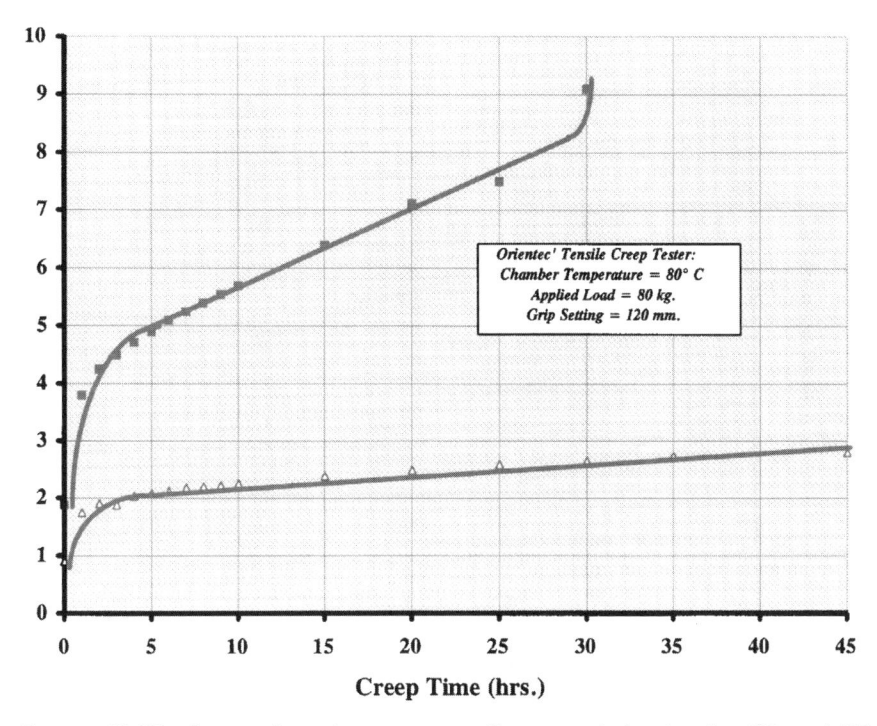

**FIGURE 12.17** Comparison between tensile creep behavior for P6 and P7 composites. (Courtesy of L. Ochs, Thermofil, Inc., Brighton, MI.) △, P7-20FG-0600; ■, P6-20FG-0600.

the-hood automotive conditions include extremes in ambient temperature combined with possible localized vibrations from the engine and rotation of molded parts (cooling fans) under continuous load. Although deformation will occur due to creep, there is concern regarding fatigue type failure after repeated fluctuations of the applied load.

The flexural fatigue behavior of a series of composite materials in Fig. 12.18 provides vivid evidence for mega-coupled composites exhibiting the type of performance usually attributed to engineering resins. The corresponding S-N plots for 20 wt% GFR composites having P6 and P7 interphase design provide a practical determination of stiffness for molded parts subjected to a range of stress levels for periodic cantilever bending.

As was determined for molded part strength in Fig. 12.17, the P7-type composite exhibits enhanced performance compared with P6-type composite. Likewise, 40 wt% GFR PA-66 composites show proportionally similar trends but with much greater stress for fluctuating cantilever bending mode compared with

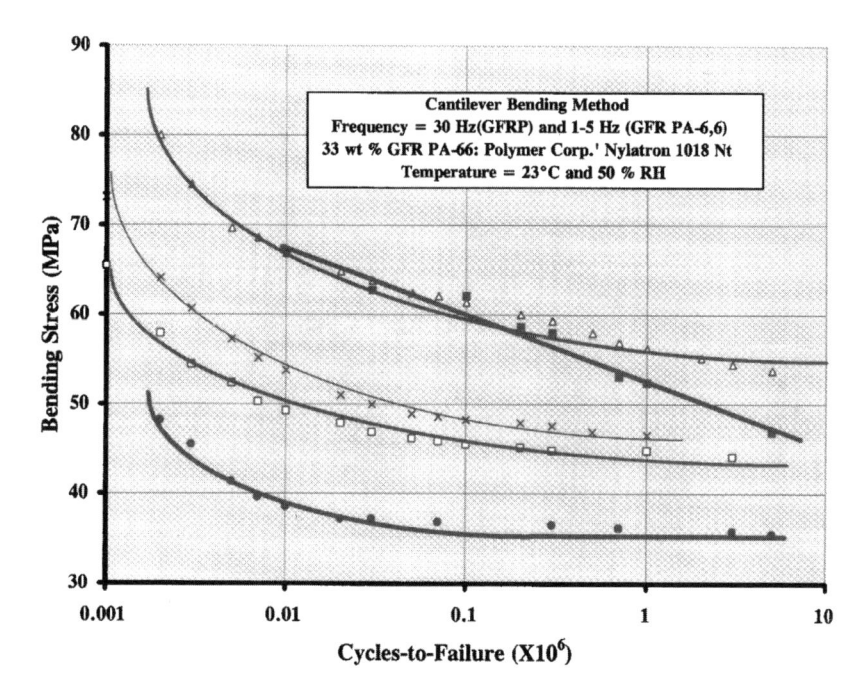

FIGURE 12.18 Flexural fatigue plots for GFRP and GFR PA-66. (GFRP measurements courtesy of Amoco Polymers, Inc., 1995; GFR PA-66 data from Modern Plastics Encyclopedia 1986–1987, p. 615, McGraw-Hill.) ●, P6-20FG-0600; ×, P6-40FG-0600; □, P7-20FG-0600; △, P7-40FG-0600; ■, GFR PA-66;.

20 wt% glass fiber loading. Furthermore, P7-40FG-0600 product (manufactured by Thermofil, Inc., Brighton, MI) exhibit better durability to flexural fatigue compared with GFR PA-66 composite that contains about 1.64 wt% absorbed moisture due to the 50% relative humidity (RH) conditioning. This type of behavior is indicative of the mega-coupled GFRP functioning as an engineering plastic.

*Comparison Between Mega-Coupled GFRP and Engineering Plastics.* Long-term tensile creep determinations over the period of 1000 hr are given in Fig. 12.19. The observed trends in axial elongation versus time indicate yield-to-failure for Minlon 22C (glass-mineral composite of nylon-66 manufactured by DuPont) and 30 wt% GFRP with P7 interphase design. Even though 33 wt% GFR PA-66 initially shows less axial elongation than 40 wt% GFRP (P7), the onset of yielding appears to be closing the performance gap after 1000 hr of tensile creep.

The effect of moisture level on flexural modulus for PA-66-based composites (Fig. 12.20) is a dictating factor influencing long-term stiffness of corre-

## Mega-Coupled PP Composites of Glass Fibers

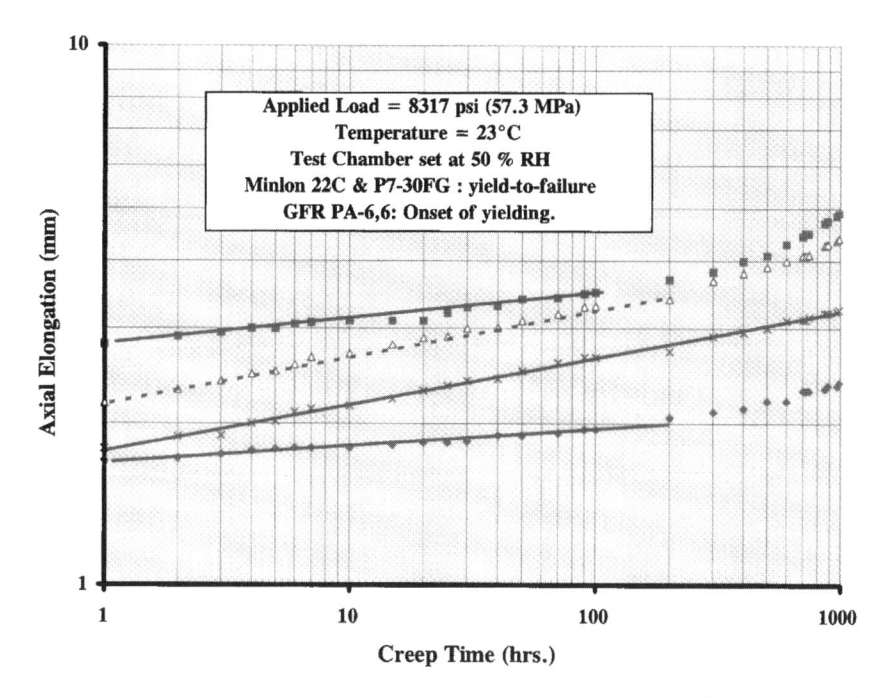

**FIGURE 12.19** Comparison between tensile creep behavior for P7 vs. polyamide composites. ◆, 33 wt% GFR PA-66; ■, Minlon 22C; △, P7-30FG; ×, P7-40FG.

sponding molded parts. Polypropylene resins, in comparison, are hydrophobic with little absorption of water molecules into the composite. That is, the flexural modulus would not be greatly influenced by direct water contact in end-use applications.

The plots shown in Fig. 12.21 exhibit trends in regards to base polymer type, interphase design, and ambient temperature:

1. *Base polymer type*—Because of the influence of moisture, polypropylene-based composites exhibit much greater stiffness than hydroscopic polyamide resins preconditioned at 50% RH before testing of flexural modulus.
2. *Interphase design*—The sequence of best to worst stiffness up to about 120°C is P7 > P6 > PA-66 > Minlon 22C.
3. *Temperature*—For continuous-use temperatures ≤140°C, 40 wt% P7 GFRP provides the broadest anticipated performance regarding resistance to static flexural loads.

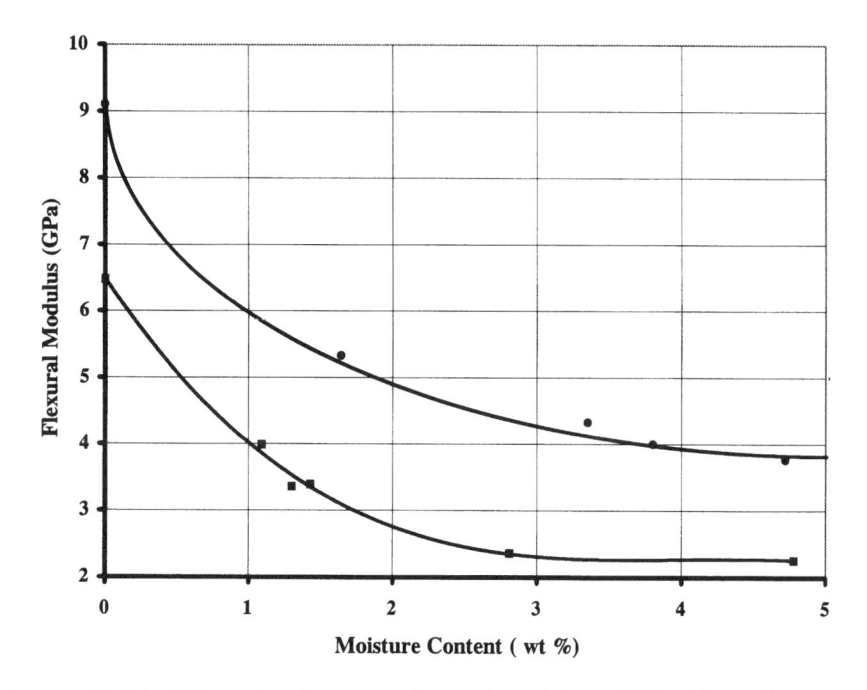

**FIGURE 12.20** Effect of moisture on flexural modulus of PA-66-based composites. ●, 33 wt% GFR PA-66; ■, Minlon 22C.

### 12.5.3 Perfect Coupling Limit

Folkes [4,27] suggested the possibility of transcrystallinity playing a key role in determining the mechanical properties of GFR thermoplastics. If the transcrystalline layer is sufficiently thick, a certain portion of the polymer matrix would form a cylindrical sheath that surrounds each glass fiber. At high enough glass fiber loading levels, impingement of neighboring cylindrical sheaths would propagate throughout the entire bulk polymer matrix with accompanying enhancement of mechanical properties. When the entire polymer matrix is entirely immobilized by total entanglement with molecular chains of the chemical coupling agent anchored to the glass fiber surface, one attains the perfect coupling limit.

At the perfect coupling limit, there is no observed glass transition temperature with $\Delta C_p^f = 0$ as shown in Fig. 12.13. This thermodynamic condition is represented by setting the interaction parameter $\lambda = 1$ in Eq. (12.3).

Upon substitution of $\lambda = 1$ into Eq. (12.5) and rearrangement of terms, the expression for maximal interphase thickness is obtained:

$$\Delta r_i(\max) = r_f\{[1/(1 - v_f)]^{1/2} - 1\} \tag{12.7}$$

## Mega-Coupled PP Composites of Glass Fibers

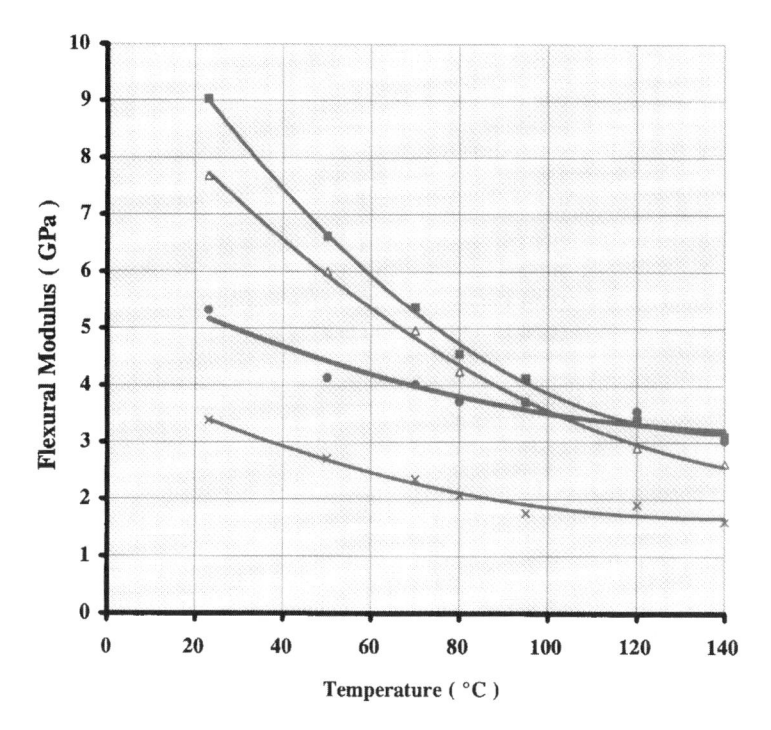

FIGURE 12.21 Thermal behavior of flexural modulus. (Courtesy Mr. R. Burton, Thermofil, Inc., Brighton, Ml.) △, P7-40FG-0600; ■, P7-40FG-0600; ●, 33 wt% GFR PA-66; ×, Minlon 22C.

Using the corresponding values for $r_f$ (6.5 μm) and $v_f$ (from Table 12.3), the calculated value of interphase thickness corresponds to the perfect coupling limit for a given fiberglass loading.

Once the maximum interphase thickness value is computed, Fig. 12.16 can be used to make an estimate of ultimate tensile strength in the perfect coupling limit. Table 12.8 summarizes the values obtained for $\Delta r_i$ (max) at various fiberglass levels. These numbers are remarkable in two respects. First, the corresponding measured values of tensile strength for P7 composites are quite close in magnitude to that predicted at the perfect coupling limit. Second, the computed value of 16,000 psi (110.3 MPa) for chemically coupled 30 wt% GFRP is very close to the maximal theoretical value of 15,820 psi (109.1 MPa) obtained by Schweizer (R. A. Sehweizer, private communication, Owens Corning Fiberglass, 1994) using a modified version of the Kelley-Tyson equation to empirically fit fiber length distribution data.

TABLE 12.8  Estimate of Tensile Strength: Perfect Coupling Limit

| Fiberglass level | | Maximum interphase thickness (μm) | Ultimate tensile strength | |
|---|---|---|---|---|
| wt% | $v_f$ | $\Delta r_i$ (max) | psi | MPa |
| 10 | 0.039 | 0.131 | 10,200 | 70.3 |
| 20 | 0.084 | 0.291 | 13,500 | 93.1 |
| 30 | 0.135 | 0.489 | 16,000 | 110.3 |
| 40 | 0.194 | 0.740 | 17,500 | 120.7 |

*Source:* Ref. 1.

## 12.6  QUANTITATIVE ASSESSMENT OF ANISOTROPIC TENSILE BEHAVIOR

For measurement of tensile strength, molded dumbbell-shaped tensile specimens are either end or side gated. In the case of short GFRP, the alignment of glass fibers are essentially in the longitudinal flow direction during the injection-molding process. The molded test specimen is subjected to an applied load along the same axial direction (0°). Hence, the measured tensile strength represents the maximal attainable value for a given composite material. For a series of composite materials measured in the same manner, one can obtain preliminary estimates of relative strength. Likewise, tensile creep determination using the same type of test specimens will exhibit the best strength for axially oriented short fibers in the molded tensile specimen.

Unfortunately, tensile strength evaluation depends on the direction of applied load with respect to a fixed axis in the molded test specimen. Therefore, the tensile behavior of short GFRP composites is anisotropic. Consequently, uniaxial values of tensile strength are of little practical use in product design.

### 12.6.1  Effect of Load Angle–Glass Fiber Orientation

The cartoon illustrated in Fig. 12.22 emphasizes that tensile strength progressively decreases from an *upper bound* longitudinal limit value at $\Theta = 0°$ ($\sigma_{0°}$) to a *lower bound* transverse limit value at $\Theta = 90°$ ($\sigma_{90°}$).

Folkes [4] stressed the shortcomings of using tensile strength values for product design at either extreme of measurement:

1.  At upper bound longitudinal limit at $\Theta = 0°$, values for tensile strength overestimate the safe working stress load necessary to suit end-use applications.

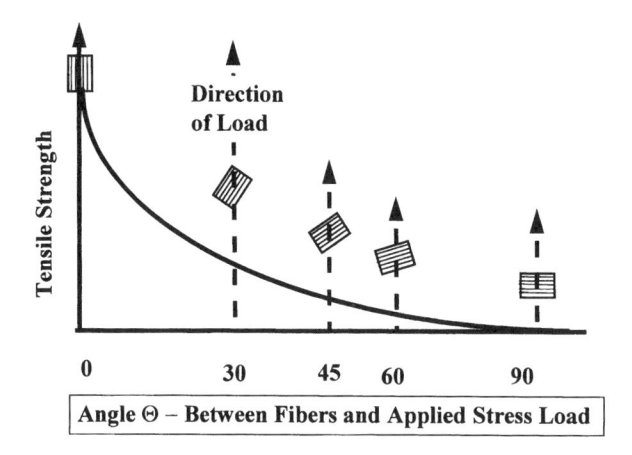

FIGURE 12.22 Angular dependence of tensile strength.

2. At lower bound transverse limit at $\Theta = 90°$, this is uneconomic over design of the molded part.

## 12.6.2 Experimental Methodology

*Overview of Two-Dimensional Mapping of Tensile Behavior.* To quantitatively determine tensile strength constraints on product design, there is a need to obtain a two-dimensional mapping of tensile strength [26] for general assessment of anisotropic behavior. For a coordinate reference frame defined as $(X, Y)$, the goal would be to develop an analytical expression or graphical representation having the general form:

$$\sigma = \text{function } (X, Y) \quad (12.8)$$

For a given molded part, one can use the resulting $X, Y$ mapping to determine the magnitude and location of the minimum or weakest tensile strength value. If one transforms the $(X, Y)$ mapping into an expression that explicitly defines the angular dependence of tensile strength, a failure mode criterion can be defined. Then dimensional deformations under applied stress can be anticipated for a given glass fiber orientation.

*Experimental Method.* Table 12.9 provides a summary of mechanical properties for four different types of composite materials that were used in an investigation [26] to obtain a two-dimensional mapping of tensile strength for molded rectangular plaques having the dimensions $200 \times 100 \times 3.2$ mm. The 30 wt% GFR composites bracket a wide spectrum of chemical coupling and

TABLE 12.9 Mechanical Properties of 30 wt% GFRP Composites

| Material | Long GFRP | P7-30FG-A729H | P6-30FG-A729G | P-30FG-0100 |
|---|---|---|---|---|
| Type of glass fiber reinforcement | Long glass fiber | Short glass fiber | Short glass fiber | Short glass fiber |
| Base PP resin | Homopolymer | Homopolymer | Homopolymer | Homopolymer |
| Degree chemical coupling | High | Mega | High | None |
| Impact modifier | No | Yes | Yes | No |
| Melt flow rate 230°C–2.16 kg | 0.73 | 4.4 | 4.2 | 1.9 |
| Tensile strength (MPa) | 116.2 | 99.4 | 90.9 | 72.9 |
| Flexural strength (MPa) | 164.8 | 151.9 | 142.4 | 98.2 |
| Flexural modulus tangent (GPa) | 6.01 | 6.43 | 5.76 | 5.04 |
| HDT-1.82 MPa (°C) | 159 | 156 | 153 | 151 |
| Notched impact (J/m) | 160 | 124 | 139 | 63 |
| Unnotched impact (J/m) | 794 | 706 | 747 | 246 |

Tensile and flexural values measured at cross-head speed of 5.0 mm/min.
Source: Ref. 26.

feature short/long glass fiber reinforcement. The long GFRP composite was Celstrand PPG-30-02-4 manufactured by Polymer Composites, Inc. (Winona, MN), whereas the remaining three materials, P7-30FG-A729H, P6-30FG-A729G, and P-30FG-0100, were manufactured by Thermofil, Inc. (Brighton, MI).

Side-gated rectangular plaques were molded via a 150-ton HPM injection-molding machine. By evaluating tensile strength for test specimens taken from a plaque at various horizontal ($X$) and vertical ($Y$) positions, the effects of skewed glass fiber orientation angles versus loading level were characterized for each material.

To obtain dumbbell-shaped test specimens having edges that are minimally perturbed by the cutting technique, a Dewes-Gumbs die (DGD) expulsion press was used to punch out sections of dumbbell-shaped type I ASTM tensile specimens along the longitudinal mold flow direction ($Y$) and transverse ($X$) direction. Photographs in Fig. 12.23a and b show the apparatus used and how dumbbell-shaped specimens were obtained.

Figures 12.24 and 12.25 indicate specimen locations in the $Y$ and $X$ directions, respectively. A total of seven tensile test specimens were obtained (i.e., three along the $Y$ axis and four along the $X$ axis). Tensile strength was measured at a crosshead speed of 5 mm/min at 23°C and 50% relative humidity. To maintain consistency between measurement of two sets of specimens having different lengths due to the $X$ and $Y$ dimensions of the plaque, the span was maintained at 6.35 cm.

The array of measured tensile strength values were plotted in Figs. 12.26 and 12.27 to give longitudinal and transverse trendlines, respectively.

### 12.6.3 Tripartite Model of Stowell-Liu

The tripartite model proposed by Stowell and Liu [31] has been used to analyze tensile behavior in a two-dimensional mapping (26) of injection-molded rectangular plaques. This model divides the angular dependence of tensile strength [4,32] into three possible failure mechanisms: shear failure at the matrix–fiber interface, fiber failure in tension, and matrix failure in tension.

The first type of failure mode is *shear failure at the matrix–fiber interface*. The corresponding expression

$$\sigma = \tau/[\sin \Theta \cos \Theta] \tag{12.9}$$

for a range of angles defined by the given range of $\Theta_1 \leq \Theta \leq \Theta_2$. By inclusion of a term $\tau$ for interfacial shear strength, the tensile strength is directly dependent on the nature of the adhesive bond between the chemical coupling agent and glass

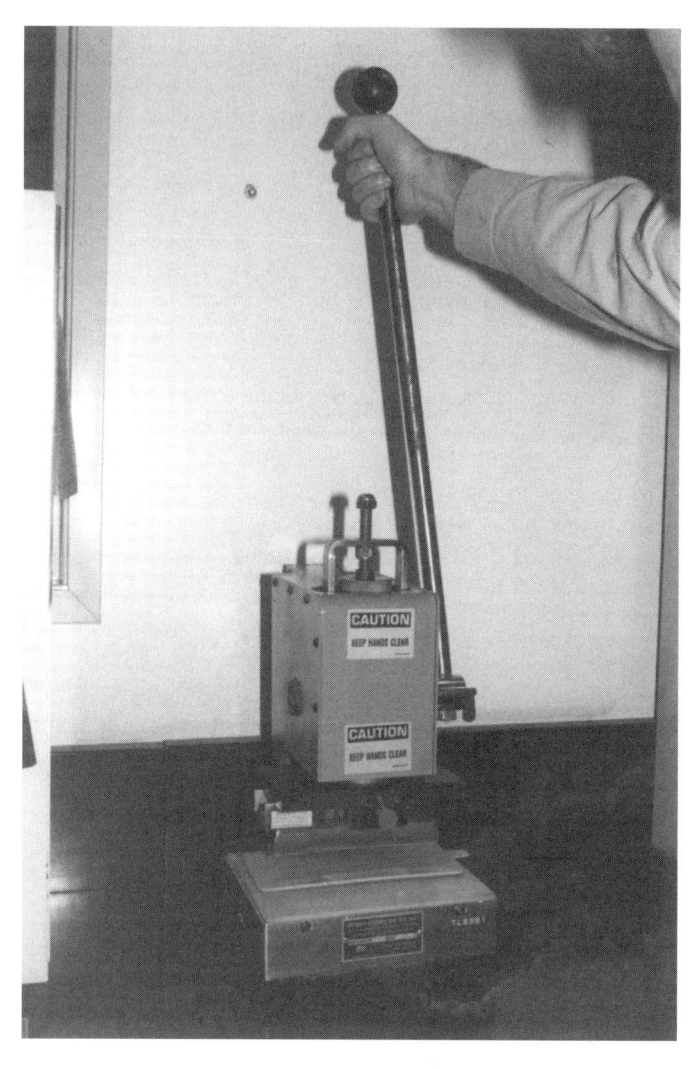

FIGURE 12.23 (a) DGD expulsion press. (Photographs courtesy Ms. D. Chrysler, Thermofil, Inc., Brighton, MI.)

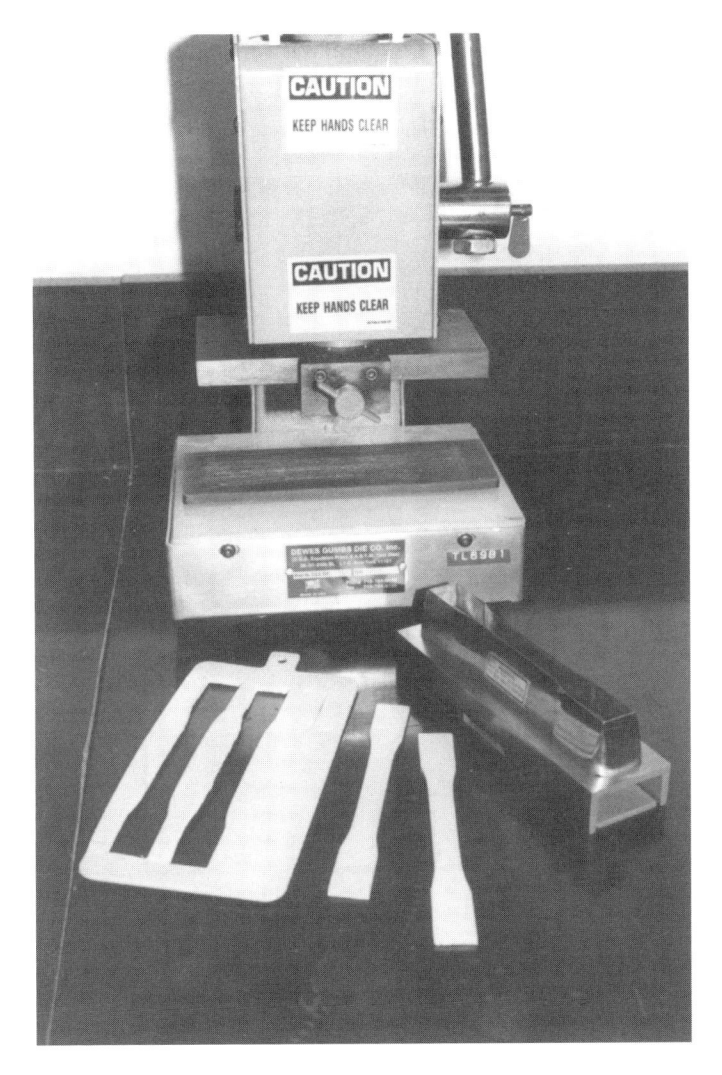

FIGURE 12.23 (b) Test specimens via DGD expulsion press. (Photographs courtesy Ms. D. Chrysler, Thermofil, Inc., Brighton, MI.)

**Location of Longitudinal Tensile Strength**

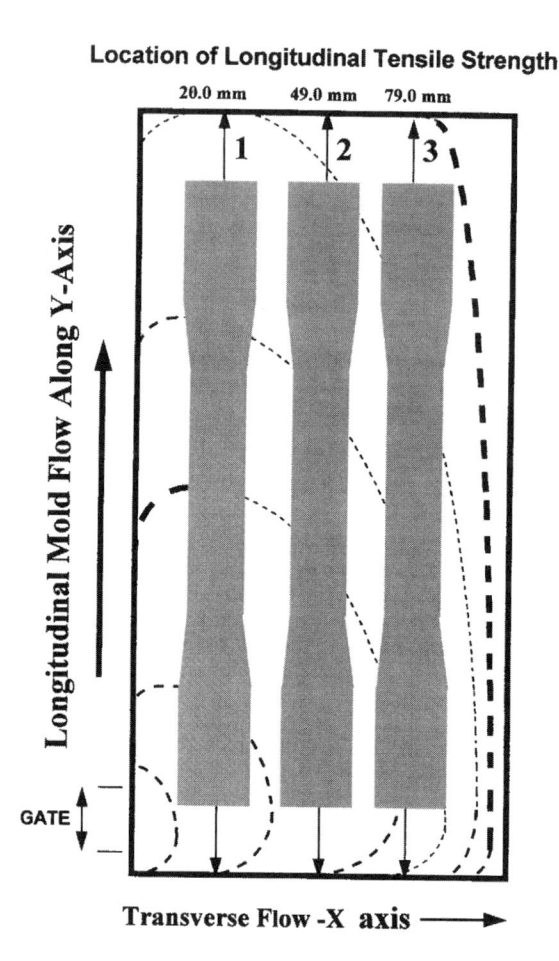

**FIGURE 12.24** Location of test specimens along X axis.

fiber sizing chemistry. The limiting angle values $\Theta_1$ and $\Theta_2$ were defined by Lees [32] in terms of tensile stress and interfacial shear values:

$$\Theta_1 = \tan^{-1}[\tau/\sigma_{0°}] \qquad (12.10)$$

$$\Theta_2 = \tan^{-1}[\sigma_{90°}/\tau] \qquad (12.11)$$

The second type of failure mode is a *fiber failure in tension* mechanism of the tripartite model for orientation angles in the $\Theta$ range of $0 < \Theta < \Theta_1$:

$$\sigma = \sigma_{0°}/\cos^2 \Theta \qquad (12.12)$$

# Mega-Coupled PP Composites of Glass Fibers

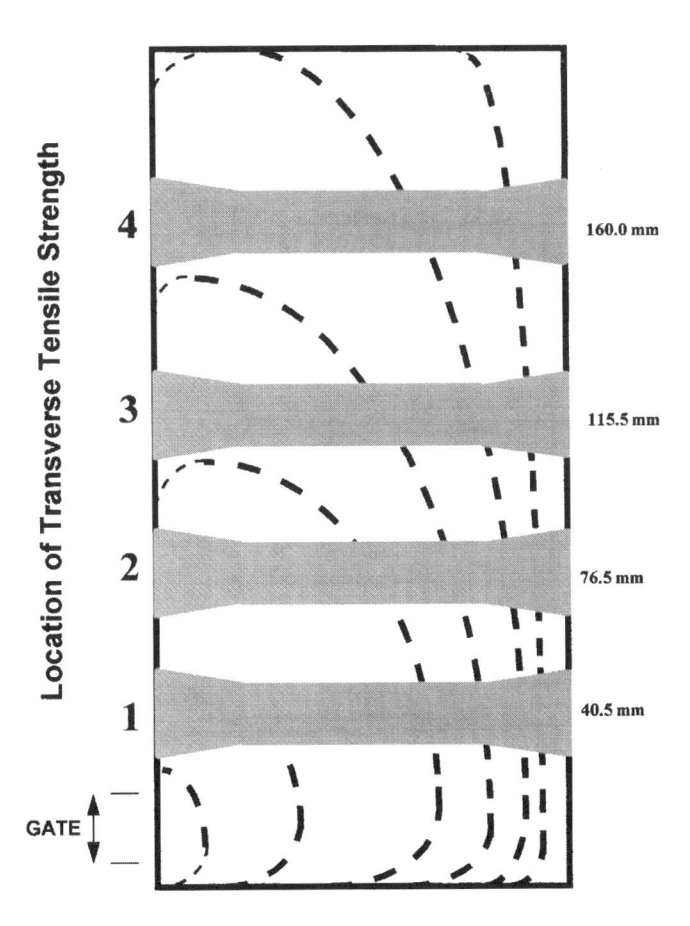

**FIGURE 12.25** Location of test specimens along Y axis.

where $\sigma_{0°}$ corresponds to the measured tensile strength values given in Table 12.9. If the composite strength is enhanced by mega-coupling with intermolecular entanglement giving strength to the polymer matrix, the weakest link is the near-uniaxial alignment of individual glass filaments that carry the brunt of the applied stress load.

A third possibility for tensile behavior is the *failure in the base polymer matrix* in tension. The localized tensile strength approaches the lower bound limit value:

$$\sigma = \sigma_{90°}/\sin^2 \Theta \tag{12.13}$$

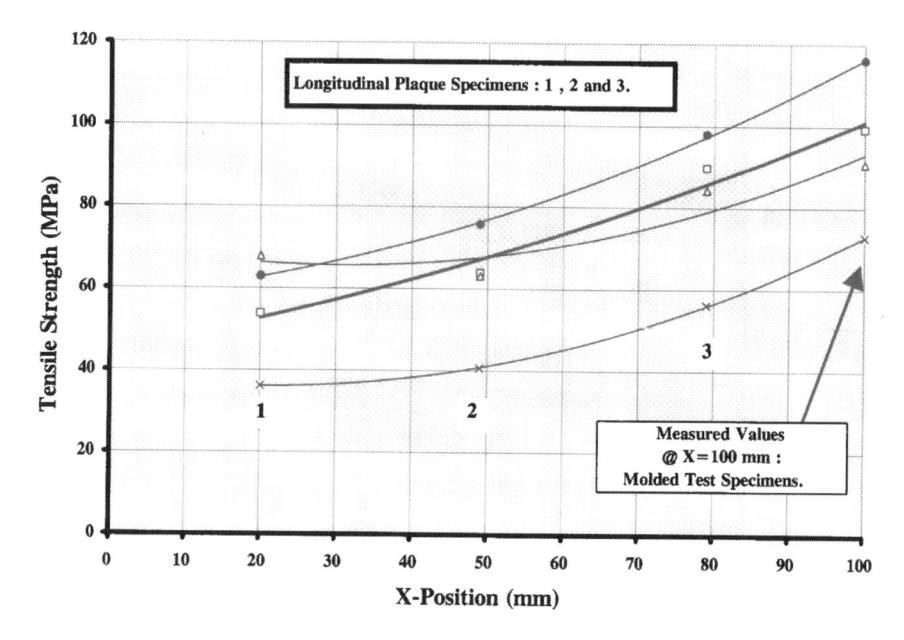

**FIGURE 12.26** Longitudinal tensile strength profile for 30 wt% GFRP composites. (Data given in Table 12.10.) ●, Long GFRP; □, P7-30FG-A729H; △, P7-30FG-A729G; ×, P-30FG-0100.

with the range of angles being $\Theta_2 \leq \Theta \leq 90°$, where $\sigma_{90°}$ is the lower bound strength measured perpendicular to the longitudinal flow direction.

Figure 12.28 shows the result of the transformation of $\sigma(X, Y)$ data into a single tensile strength versus $\Theta$ plot for each material. The angular-dependent tensile strength plots were generated by a sequence of five steps:

1. The measured tensile strength $\sigma_{0°}$ using a molded ASTM tensile specimen was used to define the boundary value at $\Theta = 0$, i.e., $\sigma(0) = \sigma_{0°}$.
2. In Fig. 12.27, each material exhibits a parabolic curve suggesting transverse tensile behavior typical of the shear failure in the fiber–matrix interface. Based on Lees [32], the minimal tensile strength value for each curve approximates $\sigma(45°)$ for the given composite material.
3. By substituting the value of $\sigma(45°)$ into Eq. (12.9), the effective interfacial shear strength value was obtained: $\tau = \sin(45°)\cos(45°)\sigma(45°) = (0.707)(0.707)\sigma(45°) = \sigma(45°)/2$.
4. Because the calculated value of $\Theta_1$ tends to be about 9–10°, even tensile strength values obtained along the longitudinal direction in

# Mega-Coupled PP Composites of Glass Fibers

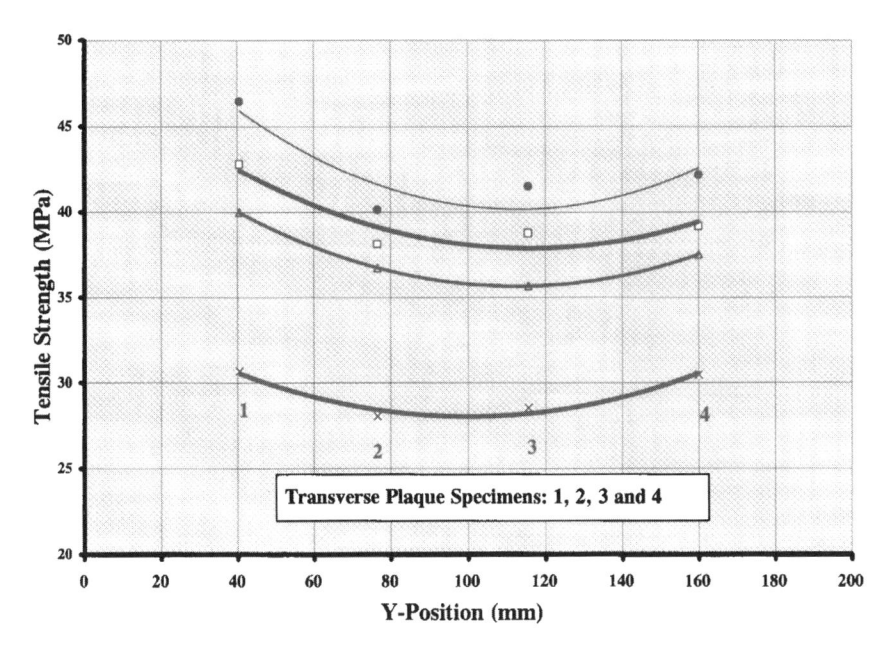

**FIGURE 12.27** Cross-sectional tensile strength profile for 30 wt% GFRP composites. (Data given in Table 12.11.) ●, Long GFRP; □, P7-30FG-A729H; △, P7-30FG-A729G; ×, P-30FG-0100.

Fig. 12.26 at positions 1, 2, and 3 are assumed to be attributed to the failure mechanism at the matrix–fiber interface. Therefore, for any given value of tensile strength σ for specimens cut from the molded plaque in both $X$ and $Y$ directions, an estimate of the corresponding orientation angle $\Theta$ can be computed using a derivation of Eq. (12.9) based on the trigonometric identity $\sin 2\Theta = 2\sin\Theta\cos\Theta$:

$$\sigma = \tau/[\sin\Theta\cos\Theta] = \tau/[\sin 2\Theta]/2 \qquad (12.14)$$

$$\Theta = \{\sin^{-1}[2\tau/\sigma]\}/2 \qquad (12.15)$$

5. Finally, Fig. 12.28 features a smooth curve of tensile strength versus $\Theta$ values obtained by Eq. (12.14) combined with the intercept value at 0° and the plateau region estimated between 45° and 90° having a constant value of $\sigma(\Theta) = \sigma(45°)$.

Tensile values at given specimen locations and estimates of localized glass fiber orientation angles $\Theta$ are given for each composite material in Tables 12.10 and

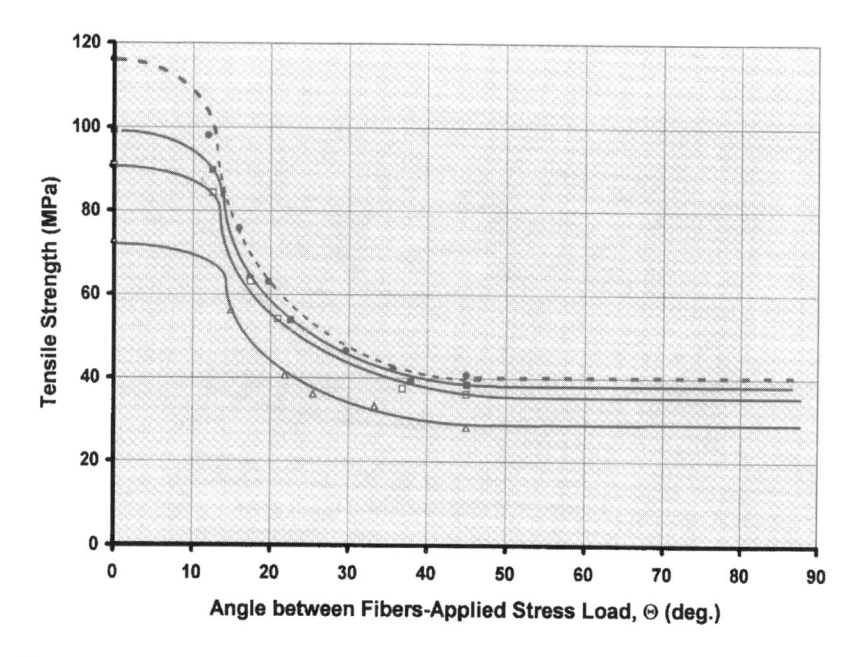

FIGURE 12.28 Tensile strength–angular behavior. (Data given in Tables 12.10 and 12.11.) ●, Long GFRP; ■, P7-30FG-A729H; □, P7-30FG-A729G; △, P-30FG-0100.

12.11. Table 12.10 features characteristics of longitudinal flow in the $Y$ direction, whereas Table 12.11 provides analysis of transverse flow in the $X$ direction.

The nearly overlapping plots in Fig. 12.28 for the long glass fiber–reinforced and short glass fiber–reinforced mega-coupled polypropylene composite (P7-30FG-A729H) indicate the same degree of mechanical anisotropy for the two materials. These analytical results suggest similar mechanical response for any molded product design subjected to stress loads of varying magnitude and direction.

## 12.7 CONCLUDING NOTIONS OF FIBER REINFORCEMENT

In this concluding section, the combined effects of mega-coupling and impact modification on fiber–fiber interactions are described in terms of observed creep resistance at extremes of elevated temperature, SEM evidence, and impact resistance.

TABLE 12.10 Longitudinal Tensile Strength Measurements

Strength in Y direction—angle between fibers and stress load

| Material | Specimen 1 20 mm along X axis MPa | Θ | Specimen 2 49 mm along X axis MPa | Θ | Specimen 3 79 mm along X axis MPa | Θ | Standard injection-molded specimen 100 mm along X axis MPa | Θ |
|---|---|---|---|---|---|---|---|---|
| Long GFRP Celstrand PPG30-02-4 | 63.0 | 19.7 | 75.8 | 15.9 | 98.1 | 12.0 | 116.2 | 0.0 |
| P7-30FG-A729H | 53.8 | 22.5 | 63.8 | 17.3 | 89.8 | 12.5 | 99.4 | 0.0 |
| P6-30FG-A729G | 54.0 | 20.9 | 63.1 | 17.4 | 84.3 | 12.6 | 90.9 | 0.0 |
| P-30FG-0100 | 36.2 | 25.4 | 40.6 | 21.8 | 56.2 | 14.9 | 72.9 | 0.0 |

Source: Ref. 26.

TABLE 12.11 Transverse Tensile Strength Measurements

Strength in X direction—angle between fibers and stress load

| Material | Interfacial shear strength $\tau$ (MPa) | Specimen 1 40.5 mm along Y axis MPa | $\Theta$ | Specimen 2 76.5 mm along Y axis MPa | $\Theta$ | Specimen 3 115.5 mm along Y axis MPa | $\Theta$ | Specimen 4 160.0 mm along Y axis MPa | $\Theta$ |
|---|---|---|---|---|---|---|---|---|---|
| Long GFRP Celstrand PPG30-02-4 | 20 | 46.5 | 29.6 | 40.1 | 45–90° plateau | 41.5 | 45–90° plateau | 42.2 | 35.7 |
| P7-30FG-A729H | 19 | 42.8 | 31.3 | 38.1 | 45–90° plateau | 38.8 | 45–90° plateau | 39.2 | 37.9 |
| P6-30FG-A729G | 18 | 40.0 | 32.1 | 36.7 | 45–90° plateau | 35.7 | 45–90° plateau | 37.5 | 36.9 |
| P-30FG-0100 | 14 | 30.7 | 32.9 | 28.0 | 45–90° plateau | 28.5 | 45–90° plateau | 30.5 | 33.3 |

Tensile strength orientation angle $\Theta = 45°$.
Source: Ref. 26.

## 12.7.1 Attributes of Stampable Polypropylene Composites

Silverman characterized flexural creep resistance at 120°C [33] and room temperature impact resistance [34] for test specimens from injection-molded composites of chopped fibers versus machined specimens from compression-molded continuous long GFR polypropylene sheets. His concepts of glass fiber reinforcement in stampable composites tend to mirror many of the attributes related to mega-coupling of short GFRP.

Stampable composites feature "fiber"-dominated strength and modulus in both longitudinal and transverse directions that give general isotropic behavior for applied stress loads. In other words, the thermally softened polymer matrix does not hinder fiber–fiber interactions necessary for maintaining strength and modulus. Compared with short GFRP, the role of the polymer matrix in any mechanical failure mechanism is deemed insignificant [34]. Consequently, this type of reinforcement exhibits enhanced creep resistance [33] at elevated temperatures encountered in under-the-hood automotive applications. The combination of isotropic behavior and creep resistance is attributed to the randomness of glass fiber orientation and exceptionally long lengths of continuous fibers that are embedded in the polypropylene resin with little breakage due to the compression molding process.

In addition to the characterization of flexural creep, Silverman [34] determined the corresponding impact resistance for injection-molded and stampable GFRP composites. Retardation of crack propagation in the stampable polypropylene sheet composite was attributed to the onset of fiber debonding and interply delamination. Hence, the long fibers exhibited significant impact strength due to this combination of impact energy dissipation.

## 12.7.2 Flexural Creep Behavior of Mega-Coupled GFRP

Even with a high degree of chemical coupling to promote efficient stress load transfer from the polypropylene matrix to the load-bearing glass fibers, it is intuitive that the softening of the polymer matrix at elevated temperatures leads to certain constraints on long-term retention of mechanical properties. However, Figs. 12.17 and 12.21 exhibit enhanced tensile and flexural behavior for mega-coupled polypropylene composites at elevated temperatures.

Figure 12.29 describes flexural creep behavior at the chamber temperature of 120°C for injection-molded specimens of 40 wt% mega-coupled GFRP composite subjected to three levels of stress load: 12.5, 18.7, and 21.9 MPa. Other than a higher fiberglass level, this material has essentially the same composition as P7-30FG-A729H used in the previous investigation of two-dimensional mapping of tensile behavior [26]. For comparison, a creep plot for machined specimens of compression-molded 40 wt% stampable polypropylene sheet subjected to 24.1 MPa stress load [33] is given in Fig. 12.29.

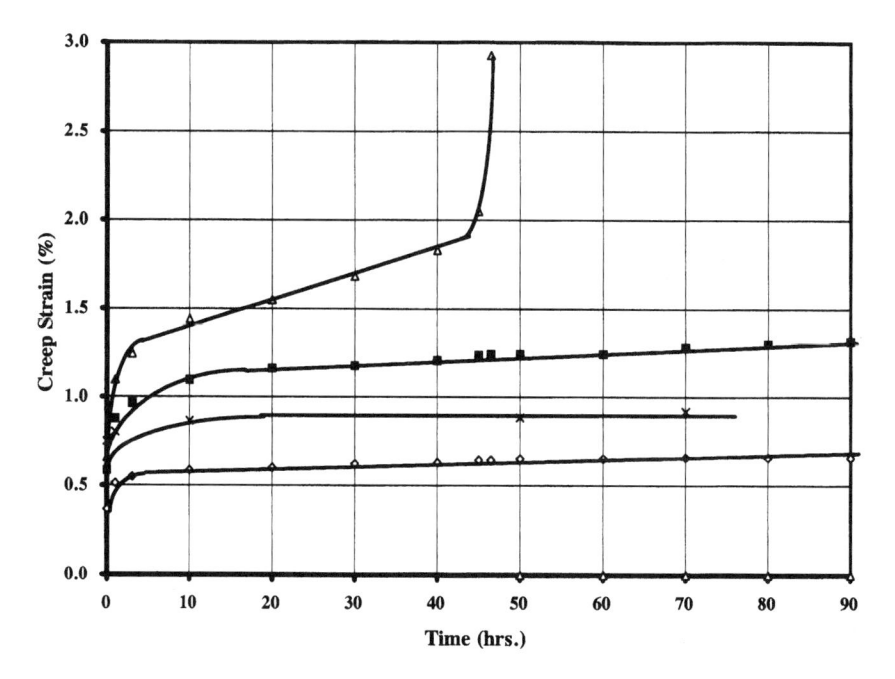

**FIGURE 12.29** Flexural creep behavior of 40 wt% GFRP composites at 120°C. (Measurements by Mr. B. Gardner, Thermofil, Inc., Brighton, MI.) ◇, P7-12.5 MPa; ■, P7-18.7 MPa; △, P7-21.9 MPa; ×, Stampable PP-24.1 MPa.

At 21.9 MPa stress load, the mega-coupled composite exhibits a much steeper strain versus time behavior than stampable polypropylene composite and fails after about 46.5 hr. However, this creep behavior is exceptional in comparison with the corresponding uncoupled GFRP, which fails almost immediately as reported by Silverman [33]. This trend indicates that the mega-coupled stress transfer mechanism promotes fiber–fiber interactions that tend to offset the inherent softening of the base polypropylene resin at 120°C.

### 12.7.3 Effects of Impact Modification

In short glass fiber reinforcement, the enhancement of impact resistance is dependent on the incorporation of rubbery type polyolefin ingredient. An impact modifier concentrate was developed to permit stepwise addition to mega-coupled composites to control the degree of material toughness. By limiting the level of impact modification, P7-30FG-A729H exhibited a good balance of mechanical properties, given in Table 12.9.

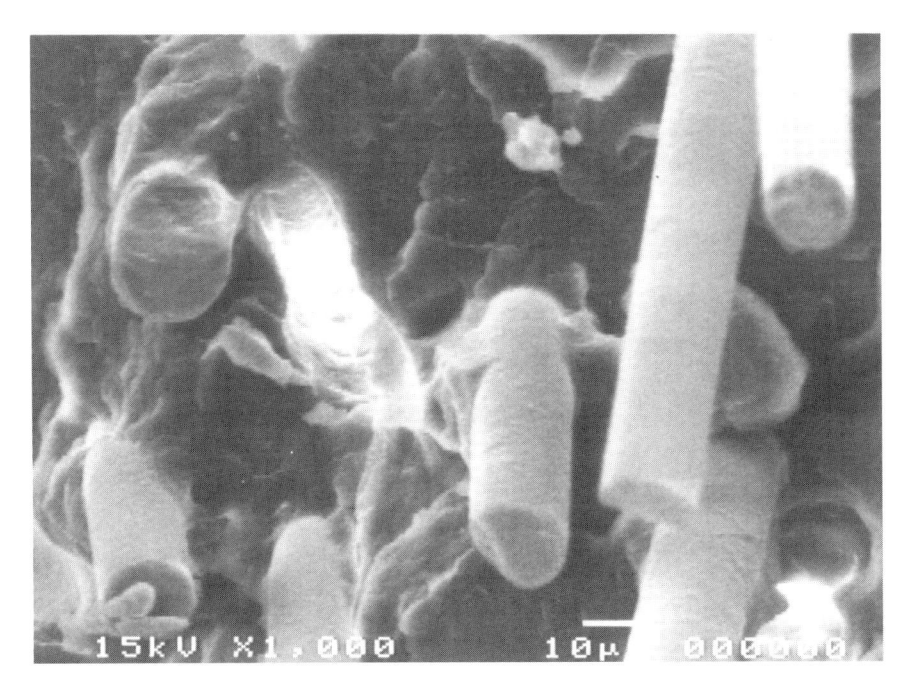

FIGURE 12.30 SEM microphotograph of P7-30FG-A729H. (Courtesy of Mr. L. Ochs, Thermofil, Inc., Brighton, MI.)

From SEM analysis of the fractured cross-section of a tensile specimen, the microstructure of P7-30FG-A729H composite is shown in Fig. 12.30. Mega-coupling is manifested by a thick polymeric coating on each glass fiber, whereas the web-like connection between neighboring fibers is due to impact modification of the polypropylene matrix. Consequently, this type of interphase design features an effective conduit from the modified polymer matrix for both stress load transfer and impact energy dissipation between glass fibers.

To determine the upper limit of impact modification without mega-coupling, the polypropylene homopolymer ingredient was replace entirely by the impact modifier concentrate to give P6-30FG-0877A product (manufactured by Thermofil, Inc.). This supertough composite contained the same chemical coupling package. The resulting notched Izod value increased about twofold from 124 to 267 J/m. The SEM photograph of the impact modifier dominated microstructure is shown in Fig. 12.31. Compared with Fig. 12.30, the high level of impact modifier leads to thick layers of a stringy polyolefin wrapped around each glass fiber and a significant network of stringy polymeric bonds between neighboring fibers to give the observed supertough impact strength.

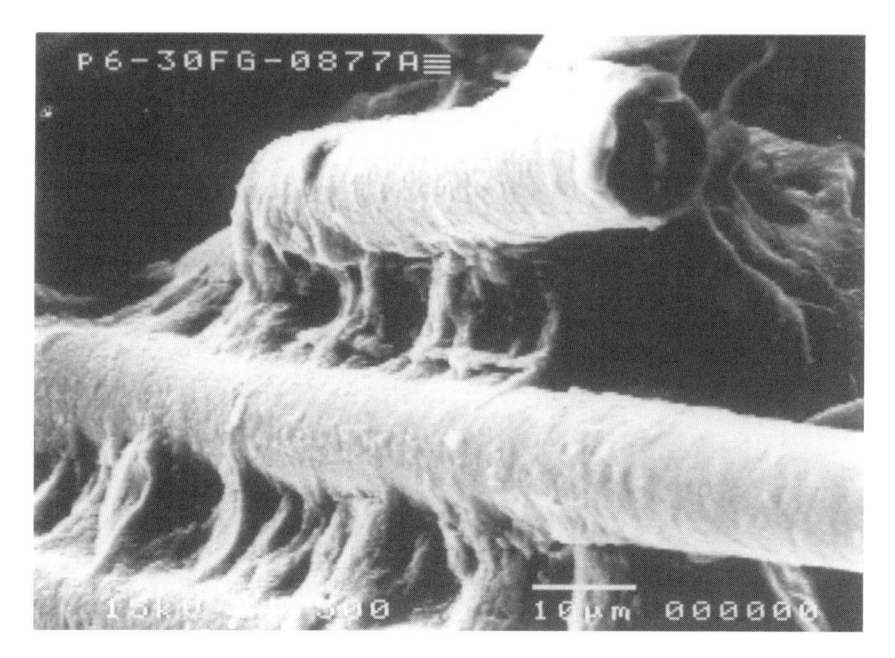

**FIGURE 12.31** SEM microphotograph of P6-30FG-0877A. (Courtesy of Mr. L. Ochs, Thermofil, Inc., Brighton, MI.)

Even at low impact modifier content, these connecting links between glass fibers impart both impact resistance and transverse strength to P7-30FG-A729H composite.

## REFERENCES

1. HG Karian. Designing interphases for "mega-coupled" polypropylene composites. Plastics Eng 52(1):33–35, 1996.
2. LT Drzal. Fiber-matrix interphase structure and its effect on adhesion and composite mechanical properties. In: H Ishida, ed. Controlled Interphases in Composite Materials. New York: Elsevier, 1990, pp. 309–311.
3. A Okawara, S Yonemori, A Kitsunezuka, H Nishimura. Direct observation of interfaces between glass fibers and polymers in glass fiber reinforced plastics. In: H Ishida, ed. Controlled Interphases in Composite Materials. New York: Elsevier, 1990, pp. 61–64.
4. MJ Folkes. Short-Fibre Reinforced Thermoplastics. 1st ed. New York: Research Studies Press, 1982, pp. 143–149.

5. SY Fu, B Lauke. Effects of fiber length and fiber orientation distributions on the tensile strength of short-fiber-reinforced polymers. Compos Sci Technol 56:1179–1190, 1996.
6. HG Karian, RW Smearing, H Imajo. Short glass fibers: their effects on the mechanical properties of thermoplastics. In: G Pritchard, ed. Plastics Additives: An A-Z Reference. 1st ed. London: Chapman Hall, 1998, pp. 226–240.
7. Z Yu, J Brisson, A Ait-Kadi. Prediction of mechanical properties of short kevlar fiber-nylon 6,6 composites. Polym Compo 4(4):238–248, 1983.
8. A Kelley, WR Tyson. Tensile properties of fibre-reinforced metals: copper tungsten copper/molybdenum. J Mech Phys Solids 13:329, 1965.
9. HF Wu, DW Dwight, NT Huff. Effects of silane coupling agents on the interphase and performance of glass-fiber-reinforced polymer composites. Compos Sci Technol 57:975–983, 1997.
10. IM Felix, P Gatenholm. Interphase design in cellulose fiber/polypropylene composites. ACS Polym Mat Sci Eng Div Preprints, Washington DC, Fall 1992, pp. 315–316.
11. JM Felix, P Gatenholm. Formation of entanglements at brushlike interfaces in cellulose-polymer composites. J Appl Polym Sci 50:699–708, 1994.
12. HG Karian, HR Wagner. Assessment of interfacial adhesion in chemically coupled glass fiber reinforced polypropylene. SPE ANTFC Preprints, New Orleans, 1993, pp. 3449–3455.
13. J Jancar, J Kucera. Yield behavior of PP/CaCO$_3$ and PP/Mg(OH)$_2$ composites. II. Enhanced interfacial adhesion. Polym Eng Sci 30(12):714–720, 1990.
14. F Mader, KH Freitag. Interface properties and their influence on short fibre composites. Composites 21(5):397–402, 1990.
15. E Mader, HJ Jacobasch, K Grundke, T Gietzelt. Influence of an optimized interphase on the properties of polypropylene glass fibre composites. Composites 27A:907–912, 1996.
16. BJR Scholtens, JC Brackman. Presented at ICCI-V Conf., Goteborg, Sweden, June 20–23, 1994.
17. H Hamada, M Hiragushi, A Hamamoto, T Sekiya, N Nakamura, M Sugiyama. Effects of interfacial properties on mechanical properties in glass fiber reinforced polypropylene injection moldings. Antec 1997 Preprints, Toronto, 1997, pp. 2519–2523.
18. B Pukanszky. Particulate-filled polypropylene: structure and properties. In: J Karger-Kocsis, ed. Polypropylene Structure, Blends and Composites, Vol. 3, Composites. London: Chapman & Hall, 1995, pp. 21–35.
19. KU Schaefer, A Theisen, M Hess, R Kosfeld. Properties of the interphase in ternary polymer composites. Polym Eng Sci 33(16):1009–1021, 1993.
20. A Perwuelz, C Caze, W Piret. Morphological and mechanical properties of glass-fiber-reinforced blends of polypropylene polyamide 6.6. J Thermoplast Compos Mater 6:176–189, 1993.
21. PS Theocaris. On the evaluation of adhesion between phases in fiber composites. Colloid Polym Sci 262(12):929–938 1984.

22. YS Lipatov. Physical Chemistry of Filled Polymers. English translation by RJ Mosely Int Polym Sci Tech Monograph No. 2. Moscow (1977).
23. A Silberman, E Raninson, I Dolgopolsky and S Kenig. The effect of pigments on the crystallization and properties of polypropylene. Polym Adv Technol 6:643–652, 1995.
24. VM Startsev, MR Kiselev, NF Chugunova. Geometry and nature of the interfacial layer in polyimide coatings. Polym Eng Sci 32(2):81–84, 1992.
25. KH Hsieh, YZ Wang. Heat capacity of polypropylene composite at high pressure and temperature. Polym Eng Sci 30(8):476–479, 1990.
26. HG Karian, K Stoops. Two-dimensional mapping of tensile strength for injection molded composites of glass fiber reinforced polypropylene. Polym Compos 19(1):1–6, 1998.
27. MJ Folkes. Interfacial crystallization of polypropylene in composites. In: J Karger-Kocsis, ed. Polypropylene Structure, Blends and Composites, Vol. 3, Composites. 1st ed. London: Chapman & Hall, 1995 pp. 340–370.
28. B Pukanszky, I Mudra, P Stanier. Relation of crystalline structure and mechanical properties of nucleated polypropylene. J Vinyl Additive Tech 3(1):53–57, 1997.
29. RA Phillips, MD Wolkowicz. Structure and morphology. In: EP Moore, ed. Polypropylene Handbook, 1st. ed. New York: Hanser Publishers, 1996, pp. 113–118.
30. ST Milner. Polymer brushes. Science 251:905–913, 1991.
31. EZ Stowell, TS Liu. On the mechanical behavior of fiber-reinforced crystalline materials. J Mech Phys Solids 9:242–260, 1961.
32. JK Lees. A study of the tensile modulus of short fiber reinforced plastics. Polym Eng Sci 8(3):186–201, 1968.
33. EM Silverman. Fiber reinforcement effects on elevated temperature performance of reinforced thermoplastics. Tech. Proc., 39th Annual Conference, RP/C Institute, SPI, Section 13-G, 1984, pp. 1–6.
34. EM Silverman. Effect of glass fiber length on the creep and impact resistance of reinforced thermoplastics. Polym Compos 8(1):8–15, 1987.

# 13

## Characterization of Long-Term Creep–Fatigue Behavior for Glass Fiber–Reinforced Polypropylene

**Les E. Campbell**
Owens Corning Fiberglas, Anderson, South Carolina, U.S.A.

### 13.1 INTRODUCTION

Although flexural, tensile, and impact ties are commonly used to characterize glass-reinforced thermoplastics (GRTPs), the time scale for measurement is always short. The results give little information about the changes that may occur in GRTPs under continual stress or fluctuating stress for long periods of time. In practical applications, these periods may vary from hours to years. The changes in the thermoplastic may be large, and this deformation is one of the most important limiting factors in their application to load-bearing units.

Thus, the characterization of creep and fatigue is a good predictive estimate of actual service lifetime of thermoplastic molded parts. As a result, there is growing interest in the performance of glass fiber–reinforced polypropylene (GFRP) for creep and fatigue. More applications are being designed for parts "under load." Because much of the interest in creep and fatigue performance of GFRP is for under-the-hood applications, all testing in this investigation was measured at elevated temperature, i.e., 80°C.

An investigation has been undertaken to better understand the influence of glass fiber characteristics on tensile creep and fatigue in polypropylene. Specifically, the characteristics include the physical attributes and chemical composition of sizing on the glass filaments. The physical aspects of the glass fiber discussed in this chapter are filament diameter and retained filament length in the molded parts. It is known that the selected glass fiber manufacturing process has an effect on composite properties of thermoplastic parts. For this reason, generating creep and fatigue data for different glass fiber manufacturing processes was of interest. The influence of sizing composition was of interest in this investigation. It is the major aspect of glass fibers that influences composite properties.

All creep and fatigue data were generated from high chemically coupled polypropylene injection molded parts. For lab compound, the composition of the molded parts was limited to resin, coupler, and glass fiber. No other additives were added to the compound, and no attempt was made to maximize the composite properties and creep or fatigue. Commercial compounders have optimized their compounds to achieve significantly higher creep and fatigue values as compared with somewhat simple lab compound. Some data that are discussed were generated from commercial compound.

## 13.2 EXPERIMENTAL

Compounded materials consisted of 5 wt% Polybond 3002 (Uniroyal) added to polypropylene homopolymer (Montel Profax 6523) to promote interfacial adhesion. Except for glass content studies, composites had a 30 wt% fiber glass loading. Blends were prepared and compounded in HPM single-screw extruder, 2.5-in, diameter screw with 30 : 1 $L/D$ ratio. The compound was dried for 4 hr at 80°C before molding in a Cincinnati 150-ton molding machine with 8-oz shot size.

Creep can be defined as deformation that occurs over time when the material is subjected to constant stress at constant temperature. In this investigation, the creep was measured in tension. The molded tensile bars with 0.5-in. taper were placed in an Instron 1331 servohydraulic testing machine, in load control, using a fixed mean level of 120 kg and an amplitude of zero. The elevated test temperature of 80°C is achieved using a Thermotron environmental chamber. Testing is controlled by an IBM-compatible PC running Instron MAX software. Failure times (hours to creep rupture) were averaged for three specimens.

Fatigue strength is a measure of the magnitude of fluctuating stress required to cause failure in a test specimen after a specific number of loading cycles. The tension fatigue testing was performed on an Instron 1331 servohydraulic testing machine in load control, using a sinusoidal waveform. The ratio ($R$) of the minimal to maximal stress on each cycle is 0.05. The test frequency is 6 Hz.

## Creep–Fatigue Behavior for GFRP

Testing is controlled by an IBM-compatible PC running Instron MAX software. Three stress levels were often tested: 8400, 8900, and 10,000 psi. The cycles to failure were averaged for three specimens.

Figures 13.1 and 13.2 show the tensile creep and fatigue equipment with the clamping devices enclosed in the environment chamber.

### 13.3 PHYSICAL ASPECTS OF THE GLASS FIBER

It is known in the industry that filament diameter of GRTPs is a factor in the performance of thermoplastic composites. The micronage of the filaments in polypropylene is less significant than for other polymers such as polyamide. However, initial review of data suggested that some less significant factors for composite properties (flexural, tensile, and impact) may be quite significant for creep and fatigue. For this reason, it was decided to include the significance of filament micronage upon creep and fatigue.

A route to maximizing creep values is to retain longer filament length in the molded parts. Experiments were undertaken in an attempt to affect the length of the filaments in the molded parts. Data presented in this section for retained filament length are an average of 400 measurements per each of two tensile bars.

FIGURE 13.1 Tensile creep equipment. (Reprinted courtesy of Owens Corning).

**FIGURE 13.2** Tensile fatigue equipment. (Reprinted courtesy of Owens Corning).

### 13.3.1 Filament Diameter

The change in strength properties (flexural, tensile, and impact) for filament diameter is well known. An investigation comparing filament diameter change with strength properties and tensile fatigue found that the change in fatigue is much greater than for commonly tested strength properties. In a highly coupled lab compound, the change in tensile strength is linear at approximately 1.5%/μm over the range 11–20 μm. The change in unnotched Izod was found to be approximately 1.3% μm.

Over this same micronage range, Fig. 13.3 shows the dramatic change for tensile fatigue and indicates the change is nonlinear. In this case, the stress was 8.4 kpsi, which is 65% of the ultimate breaking strength. As the filament diameter approaches 20 μm, the loss in tensile fatigue is so large as to indicate that such a large filament diameter may have little commercial value for underload applications.

### 13.3.2 Retained Filament Length

In short-fiber GRTP compounding and injection molding, the filament length degradation is significant and well documented. One approach to improving

# Creep–Fatigue Behavior for GFRP

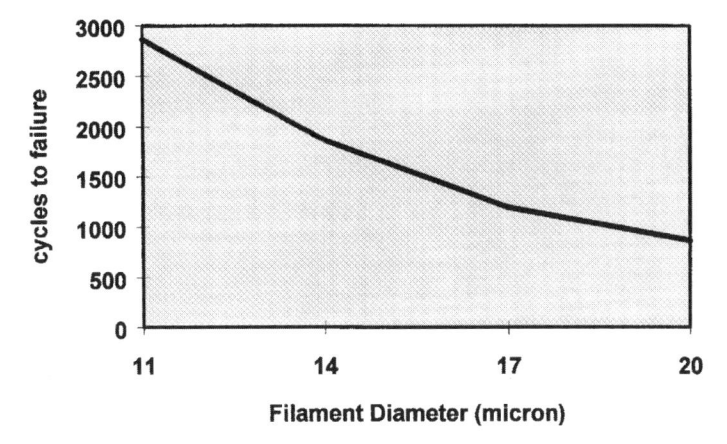

FIGURE 13.3 Micronage versus tensile fatigue.

composite performance is increasing the retained filament length in molded parts. This is difficult due to the high shear forces that somewhat fragile glass filaments are exposed to during compounding and injection molding processes. A typical filament degradation for a 4.5-mm-length input fiber will be reduced to approximately 0.5 mm in polypropylene-molded parts. There are numerous factors that will affect the final retained length, but typically it will vary by no more than approximately 10% around 0.5 mm.

There are several approaches that can be pursued in an attempt to increase the retained filament length. In practice, experience indicates that attempts to significantly increase the retained length in injection-molded parts is very difficult and generally not achievable without major changes in the processing equipment.

These observations are normally measured with strength properties such as impact, tensile, and flexural. Generally, the effect of very small changes in retained length on creep and fatigue is unknown. Historically, to lengthen retained filament length 10–20% would be considered a significant accomplishment. Such a change in length would increase the strength properties of molded parts at most maybe 10%, for example. Review of historical observations and data suggested that rather minor reductions in retained filament length could cause significant reduction in tensile creep.

The retained glass filaments can be influenced through changes in fiber-producing processing or by modifications of the sizing chemistry. Whatever approach is chosen to change the filament length, the length change has been confirmed to be quite small. The corresponding influence on tensile creep has a major effect. Table 13.1 shows the effect of 10% change in average retained filament length on strength properties and tensile creep. Interestingly, when

TABLE 13.1 Retained Filament Length in Molded Parts for 30 wt% GFRP

| Fiber | Mean length (μm) | Tensile strength (kpsi) | Unnotched Izod (ft-lbF/in.) | Tensile creep rupture time (hr) |
|---|---|---|---|---|
| A | 527 | 11.87 | 11.85 | 1.8 |
| B | 574 | 12.42 | 11.68 | 5.0 |
| C | 550 | 12.55 | 12.75 | 6.8 |
| D | 587 | 12.69 | 13.11 | 8.2 |
| E | 584 | 12.66 | 12.94 | 6.3 |

average retained filament is tested for significance, it is not statistically due to large standard deviations.

### 13.3.3 Glass Content

The influence of glass content on the performance of thermoplastic molded parts is well documented and is significant. Glass content is another aspect of molded parts in which the tensile fatigue was dramatically affected, much more so than for strength properties (tensile, impact, and flexural). In Table 13.2, the tensile strength values decrease by 43% when the glass content is reduced from 30 to 10 wt% for a highly coupled compound. Extrapolation of this observed trend of glass content versus fatigue performance to unreinforced polypropylene suggests very limited utility for underload applications as low fiber glass loadings. For example, 10 wt% GFRP composites exhibited instantaneous failure for tensile fatigue determination at applied stress corresponding to 65% ultimate tensile strength. However, there are available commercial compounds that suit many underload applications with less than 30 wt% fiber glass loading. This is due to an effort by compounders to optimize interphase design to achieve acceptable creep and fatigue performance (see Section 13.7 and Chapter 12).

TABLE 13.2 Glass Content versus Tensile Fatigue

| Glass content (wt%) | Cycles to failure at stress = 8.4 kpsi | Tensile strength (kpsi) |
|---|---|---|
| 30 | 1870 | 12.8 |
| 20 | 470 | 10.4 |
| 10 | 0 | 7.3 |

## 13.4 EFFECT OF FIBER MANUFACTURING PROCESSING ON PERFORMANCE

### 13.4.1 In-Line Chopped-Strand Production

One manufacturing method to produce glass chopped strands is a process known as "in-line" processing (CRATEC in Owens Corning). In this process, the fiber forming, chopping, size application, drying, and packaging are accomplished in a continuous process. This section compares the creep and fatigue performance of chopped strands made via this process versus off-line, also described by some people as "two-step," process. Generally, this in-line processing produces lower composite properties than the off-line process. Preliminary investigation in cooperation with a major compounder indicated that the in-line (CRATEC) process produced significantly lower tensile creep performance than off-line chopped strands. One purpose of the investigation was to confirm this observation and gain knowledge about possible routes to achieving equivalent performance with in-line chopped to off-line chopped strands.

### 13.4.2 Off-Line, Two-Step, Chopped-Strand Production

Another manufacturing route common to the industry in producing chopped strands for thermoplastics is a two-step process, known also as off-line in Owens Corning. In this process, the fiber-forming and sizing application is performed as a separate operation from the chopping and drying. The sized glass strands are wound onto doffs in fiber forming and then transferred to a separate operation where the doffs are either dried before chopping or chopped wet and then dried.

Historically, this method usually produces higher composite properties than the in-line process. Differences in composite properties performance between these two processes depend to a great extent on the matrix being reinforced. Polypropylene is one of the least affected by the two processes. In cooperation with a major compounder, it was observed that the processes affect creep and fatigue performance greatly, far more than measured for flexural, tensile, and impact strengths. In Fig. 13.4, the increase in tensile strength for l44A made off-line is typically 5% in a chemically coupled compound. The differential for Izods was 3–5% for the two processes as shown in Fig. 13.5.

When the chopped strands produced by these two described processes were tested for creep and fatigue, the difference in performance was very significant. Figure 13.6 shows the difference in tensile creep for lab made compound to be 270% and the same chopped strand in an optimized commercial compound to be even greater, at 700% in Fig. 13.7.

Indeed, the investigation confirmed numerous times that the in-line process for producing chopped strand resulted in greatly lower tensile creep and fatigue performance. There is a far greater differential of long-term properties than that

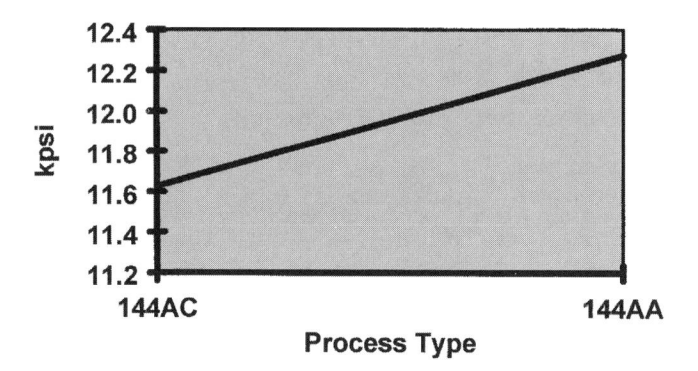

FIGURE 13.4  Tensile strength versus process type.

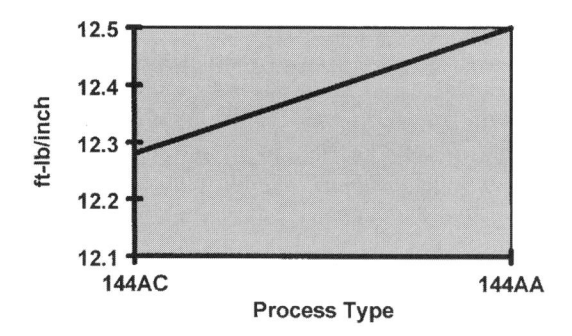

FIGURE 13.5  Unnotched Izod impact versus process type.

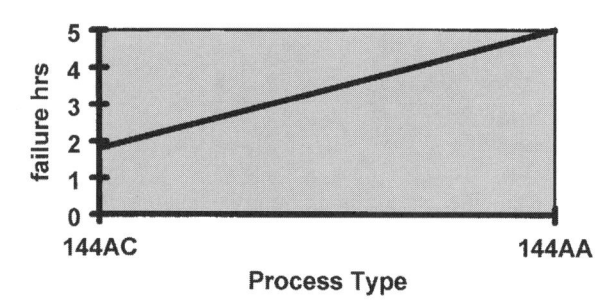

FIGURE 13.6  Tensile creep rupture for lab compound.

## Creep–Fatigue Behavior for GFRP

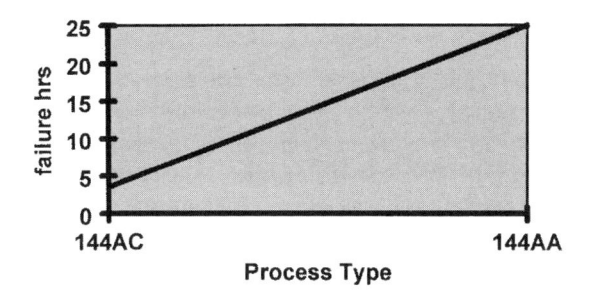

**FIGURE 13.7** Tensile creep rupture for commercial compound.

associated with short-term determination of typical mechanical strength properties (e.g., tensile and flexural measurements). After having confirmed this performance difference, the task was to find a solution to this problem. One route to the solution lies in reducing damage done to the filaments during the chopping process.

The challenge was to discover methods through process modifications and sizing chemistry to achieve equivalent performance of in-line produced chopped strand as off-line chopped strand. This was accomplished by process improvements and by changing the sizing chemistry significantly. The focus of this effort involved reducing the damage to the filaments during the chopping process and the development of new sizing, which is further discussed in Section 13.5. This effort has led to a new generation of chopped strand in which the tensile creep and fatigue are at least equivalent or superior to 144A made in-line.

Figure 13.8 shows the benefit of this recent development in processing improvement and chemistry change. In Fig. 13.8, 144AC was the in-line chopped strand, 144AA is the same sizing chemistry applied off-line, and product C is new-generation sizing applied in-line.

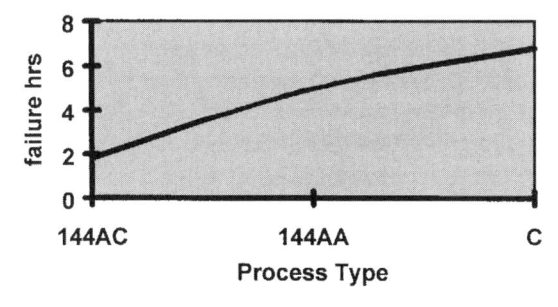

**FIGURE 13.8** Tensile creep behavior versus process type.

## 13.5 INFLUENCE OF SIZING ON PERFORMANCE

The significance of sizing in the composite properties of thermoplastics is well known and discussed at much length in the previous chapters. Glass fiber producers are constantly striving to improve performance by influencing interfacial adhesion and interphase microstructure. A large portion of this continual effort involves discovery of new components composing the sizing chemistry.

It is common practice in the glass fiber reinforcement industry to tailor the sizing composition for a specific type of matrix. The reinforcement of polypropylene dictates the unique chemistry of the sizing composition. Over the years, glass fiber producers have participated in improving composite properties of GFRPs by continually developing better sizings. Each evolution in next-generation sizing and reinforcement may be considered to be significant when composite properties are improved by about 10% on average.

Earlier investigations into understanding the influence of glass fiber on creep and fatigue suggested that changes in processing or sizing composition could cause very significant change in creep values. A new sizing composition that affects composite properties, such as tensile strengths and impacts, by 5–10% could affect creep significantly more.

The greater the freedom of the molecules to move, the more extensive the creep. The focus of the effort is to increase the wetting of the filament surface and increase the toughness of the polypropylene film formers of the interface. One approach taken in this regard is to increase the molecular weight of the film formers. Another approach is to affect the spherulites' structure after nucleation of the polypropylene interphase.

Three sizing compositions were studied to compare the improvement in composite properties such as flexural, tensile, and impact versus creep and fatigue properties. The sizing designated as (A) in this discussion is Owens Corning 144A. The major film former in this sizing is a low molecular weight maleated polypropylene wax. In Fig. 13.9, the change from a rather low molecular

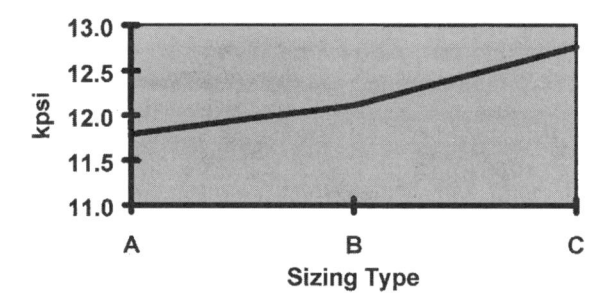

FIGURE 13.9 Tensile strength versus sizing type.

## Creep–Fatigue Behavior for GFRP

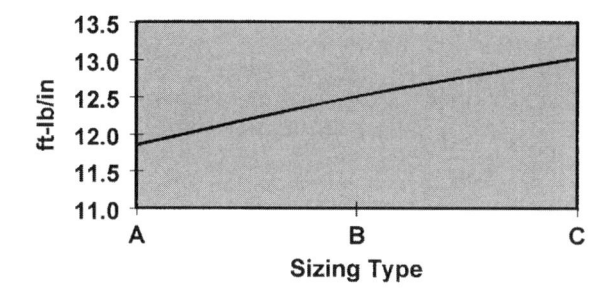

**FIGURE 13.10** Unnotched Izod impact versus sizing type.

weight (MW) film former to progressively higher MW film former results in 8–10% improved tensile strength. Figure 13.10 describes a 10% improvement in unnotched Izod from sizing (A) to sizing (C). This amount of change in product evolution is often considered significant. There are other component changes involved in these sizings as well that contribute to improved performance.

However, when evaluating the change in sizing composition on creep and fatigue, the improvement in tensile creep and fatigue was much larger in magnitude. The increase for tensile creep was nearly 300% for a lab compound and 600% for an optimized commercial compound for the same three chopped strands. These two creep observations are shown in Fig. 13.11. Figure 13.12 shows the corresponding change in tensile fatigue of 133% for the lab compound.

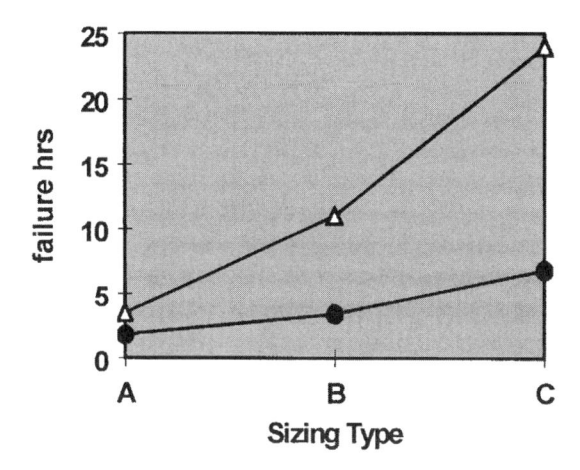

**FIGURE 13.11** Tensile creep versus sizing type. △, Commercial compound; ●, Lab compound.

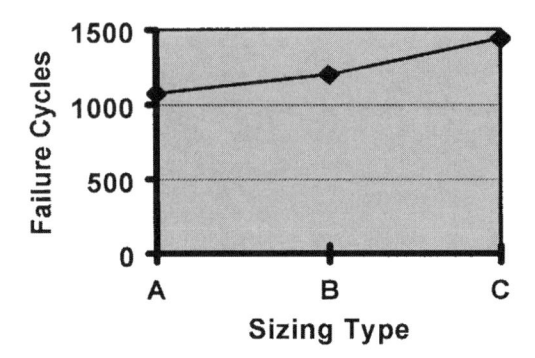

**FIGURE 13.12** Tensile fatigue versus sizing type at 8.4$^{kpsi}$.

In summary, this investigation found that the chemistry of the sizing can have very significant benefit on underload properties. This finding indicates the possibility that even greater creep and fatigue performance is possible with future glass fiber development. The ability to effectively increase creep resistance would be a valuable option to parts application designers for glass fiber reinforced polypropylene (GFRP).

## 13.6 CORRELATION BETWEEN TENSILE CREEP AND TENSILE FATIGUE

There was interest in determining whether any correlation existed between tensile creep and fatigue. If there is any correlation, it may suggest that the failure mechanisms are similar. Future investigations will focus on developing a conceptual model to describe failure mechanisms for tensile creep and fatigue processes. A clear understanding of a common failure mode could be quite beneficial in future development for increasing creep or fatigue performance, allowing the designers more application opportunities in GFRPs.

Figure 13.13 is a correlation for 8.4 kpsi stress for tensile fatigue versus tensile creep. For several years, testing for fatigue was done at three stress levels, representing 60%, 70%, and 80% of ultimate tensile strength (UTS). At 80% of UTS, 10 kpsi, the average cycles to failure for tensile fatigue was approximately 21% of 8.4 kpsi applied stress of lab compound. In commercial compound, the 80% UTS tensile fatigue values were just 5–10% of 60% UTS (8.4 kpsi) values. The fact that there is some correlation between creep and fatigue allowed us to select an appropriate stress level for screening purposes because testing for creep and fatigue can be very time consuming. This approach helps to reduce the time required for characterizing creep and fatigue properties of optimized commercial compound having enhanced long-term durability. A choice of a sufficiently high UTS stress is appropriate for development work as an effective screening tool.

# Creep–Fatigue Behavior for GFRP

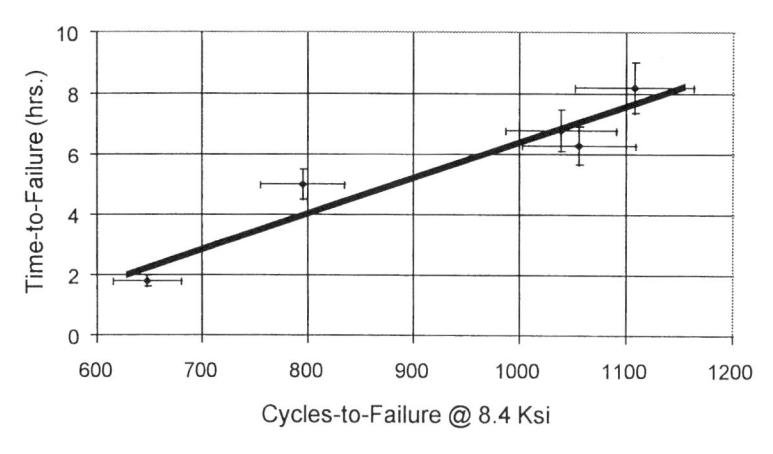

**FIGURE 13.13** Correlation tensile creep–fatigue for GR-polypropylene. (Courtesy of Owens Corning).

## 13.7 DETERMINATION OF CREEP RUPTURE ENVELOPES FOR GFRP

Creep rupture strength determination is the long-term analogue of short-term measurement of tensile strength. The effective uniaxial strength of a GFRP composite depends on the duration of a stress load applied to a given cross-sectional area of a molded part subjected to an ambient temperature encountered in the end-use application. Depending on strength limits imposed by stress–temperature–time relationships, a creep rupture envelope can be defined as a criterion for strength. For example, an envelope of limiting values can be defined by plotting creep rupture time versus applied stress load. The design engineer can then use the resulting master plot to predict maximal stresses below which the molded part suits material requirements for suitable service lifetime.

By deriving fundamental relationships for service lifetime as a function of stress and temperature, accelerated tensile creep experiments can be effectively used to optimize interphase design for GFRP composites. In previous sections of this chapter, long-term determinations of tensile creep and fatigue characteristics were shown to be quite useful for optimizing both sizing chemistry and glass fiber manufacturing processes.

In this section, tensile creep methodology is used to define the limits to creep resistance for a series of mega-coupled GFRPs described in Chapter 12: P7-20FG-0600, P7-30FG-0600, and P7-40FG-0600. Analyses of the creep rupture times for given sets of load and temperature give a definition of the creep rupture envelopes near the perfect coupling limit.

### 13.7.1 Method

A four-station Orientec creep tester is depicted in Fig. 13.14a–c. It was used to measure axial elongation versus time for various sets of applied loads ranging from 120 to 220 kg. For a dumbbell-shaped molded ASTM molded tensile specimen, this corresponds to a range of tensile stresses of about 4435–8132 psi or 30.6–56.1 MPa.

The 10:1 cantilever arm is loaded with a variety of stacked weights as shown in Fig. 13.14c. By requiring only 10% of the actual stretching force applied at the free end of the tensile specimens at the other end of the lever arm, this mechanical advantage permits easy application of loads by hand. The test temperature was precisely controlled at 50–80°C in an efficient air-heated chamber shown in Fig. 13.14b.

### 13.7.2 Creep Deformation Behavior

Tensile creep deformation is characterized by measurement the relative change in uniaxial length $\ell(t)$ from the initial 118.5-mm grip distance $\ell_0$ for prolonged periods. The continuous pen recorder generates a plot of axial displacement $\Delta\ell$ versus creep time.

Total creep strain $\varepsilon_T$ is equivalent to change in axial length $\Delta\ell$ divided by this grip distance

$$\varepsilon_T(t) = [\ell(t) - \ell_0]/\ell_0 = \Delta\ell(t)/\ell_0 \tag{13.1}$$

Generally, the corresponding creep strain versus time plots feature a sequence of three stages [1] of axial deformation for a test specimen under constant stress load:

1. *First-stage deformation: elastic response.* The instantaneous axial elongation of the tensile specimen is attributed to dissipation of elastic strain that is equivalent to the ratio of the applied tensile stress and tensile modulus, that is the amount of strain predicted by the short-term characteristics of stress–strain encountered in the measurement of tensile strength.
2. *Second-stage deformation: creep strain increase.* Depending on the magnitude of the applied stress and chamber temperature, the increase in strain with time will proceed at a relatively slow rate. Generally, the logarithmic plot of strain due to creep versus time will be linear for a duration defined by third-stage characteristics of the given GFRP composite.
3. *Third-stage deformation: creep routine.* The critical third stage of tensile creep depends on the relative ductility of the GFRP composite. If the homopolymer polypropylene in the polymer matrix is highly

# Creep–Fatigue Behavior for GFRP

**FIGURE 13.14** Tensile creep equipment. (Courtesy of Thermofil, Inc.) (a) Four-station oriented creep tester; (b) temperature-controlled test chamber; (c) weight stack to cantilever arm.

crystalline or isotactic, ultimate failure will occur suddenly at a critical threshold limit of strain, (e.g., 2–3% strain as determined by Cessna [2]). However, incorporation of even a small amount of impact modifier will lead to a gradual change in slope of the linear logarithmic plot with onset yielding just before creep rupture.

The design engineer of plastic component parts needs more than just short-term stress–strain data for anticipating long-term deformation behavior. For example, the useful service lifetime of the molded part is curtailed by onset of *excessive creep deformation*, leading to ultimate creep rupture. Consequently, there have been significant efforts by researchers to correlate creep strain with stress–temperature–time parameters that suit end-use constraints imposed on material design.

Trantina [3] analyzed the creep behavior of talc-filled polypropylene composites under constant uniaxial stress loads. He proposed a general expression for the strain $\epsilon_c$ attributed to creep alone as a superposition of time ($t$), stress ($\sigma$), and temperature ($T$) dependence. The model equation is written as a product of three separate functions for each parameter:

$$\epsilon_c(t) = \epsilon_T(t) = \epsilon_0 = f_1(t) f_2(\sigma) f_3(T) \tag{13.2}$$

Time-dependent creep strain is obtained by correcting the total strain $\epsilon_T$ for the first-stage deformation $\epsilon_0$ due to the instantaneous elastic response to the applied stress load:

$$\epsilon_0 = \sigma/[\text{tensile modulus}] \tag{13.3}$$

Based on empirical evidence, Trantina [3] defined functions of time and stress as power law expressions. An Arrhenius-type relationship is used to describe temperature dependence. Hence, Eq. (13.2) is written as

$$\epsilon_c = C_1 t^n \sigma^m \exp(-C_2/T) \tag{13.4}$$

where $C_1$ is a material constant and $C_2$ is an activation energy term for creep deformation. The values of the power law indices $n$ and $m$ are the corresponding slopes of log-log plots of strain versus time and stress, respectively.

### 13.7.3 Analytical Expression of Creep Rupture

By assuming that creep rupture occurs at a constant creep strain limit [2], Eq. (13.4) can be transformed into an expression of creep rupture time $t_{CR}$ as a function of stress and temperature:

$$t_{CR} = \text{constant} \exp(C_2/T) \sigma^{-m} \tag{13.5}$$

## Creep–Fatigue Behavior for GFRP

Because the applied tensile stress σ is directly proportional to the given weight $L$ acting on cantilever arm ($\sigma = L/\text{cross-functional area}$), Eq. (l3.5) can be expressed in terms of test chamber temperature $T$ and load $L$:

$$\log t_{CR} = A1 + A2/(T + 273.2) + A3 \log L \tag{13.6}$$

where A1, A2, and A3 are constants. A1 is a material parameter, whereas A2 and A3 correspond to temperature and stress load dependence. The temperature $T(°C)$ is the test chamber setting. $L$ is the kilogram weight that is hand loaded to the lever arm (Fig. 13.14c).

The form of Eq. (13.6) is essentially identical to a model for predicting the service lifetime of polyethylene pipes [4]. The particular σ term is the hoop stress that is related to internal pressure of the pipe. The Arrhenius expression is based on an activated rate-process theory of failure under creep loading conditions.

Within certain limits of accuracy, accelerated tensile creep determinations can be used to determine the unknown constants in Eq. (13.6). These constants are the coefficients of regression analysis for a set of test data featuring variation of temperature and load settings.

### 13.7.4 Analysis of GFRP Composites

Figures 13.15 and 13.16 depict the relationship between applied stress load and measured creep rupture times for a series of composite materials at 50°C and 80°C, respectively. The inherent relationship between creep resistance and fiberglass content is displayed in Fig. 13.15, i.e., higher glass fiber reinforcement leads to improved load bearing capability for the given composite materials.

As described in Chapter 12, mega-coupled composites exhibit tensile behavior that acts like 10 wt% more glass fiber reinforcement attained by conventional chemical coupling. This enhanced performance is demonstrated in Fig. 13.16 in regard to load bearing capability for a constant creep rupture time of 30 hr. P6-20FG-0600 having conventional chemical coupling is compared with mega-coupled products of P7-10FG-0600 and P7-20FG-0600. With just 10 wt% fiber glass loading, P7-10FG-0600 has load bearing capability nearly equal to that of P6-20FG-0600. Furthermore, P7-20FG-0600 can endure a load of about 130 kg, or more than 60% higher than that of the P620FG-0600.

Tables 13.3–13.5 provide a summary of regression analyses of P7-20FG-0600, P7-30FG-0600, and P7-40FG-0600. Table 13.6 summarizes the values of the three coefficients Al, A2, and A3 for the three composite materials. The regression analysis statistics indicate a very good fit of observed creep rupture data to Eq. (13.6).

The thermal behavior of tensile strength values in Fig. 13.17 tend to mirror the effects of temperature on long-term creep behavior. Obviously, load bearing capability to resist creep is greatly reduced by elevation of ambient temperature.

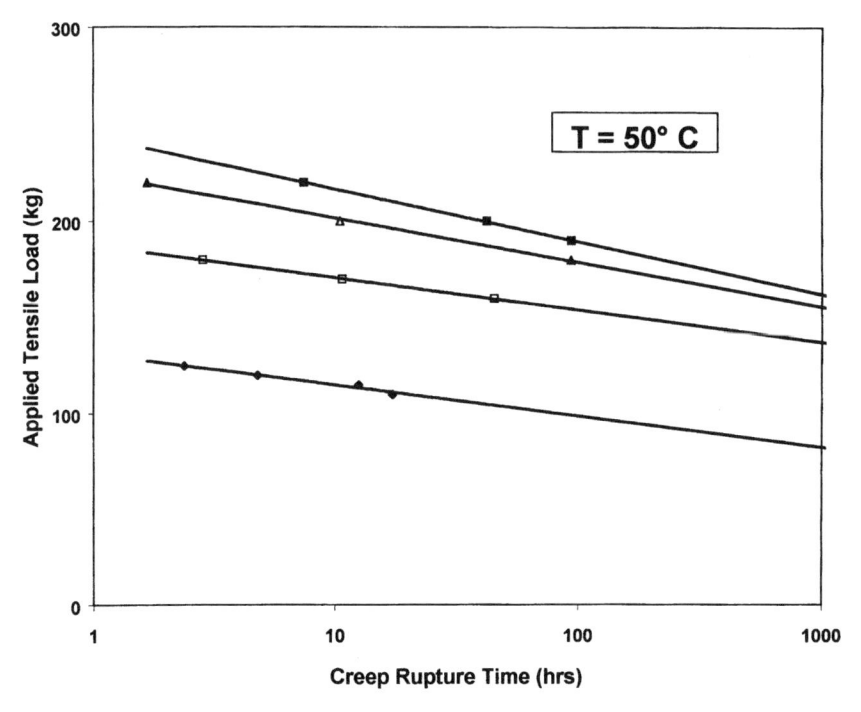

**FIGURE 13.15** Tensile creep rupture time–load plots at 50°C. ◆, P7-10FG-0600; □, P7-20FG-0600; , P7-30FG-0600; ■, P7-40FG-0600.

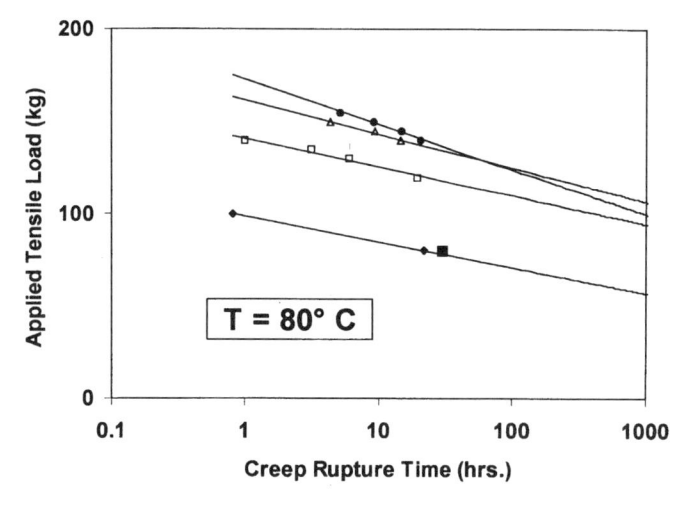

**FIGURE 13.16** Tensile creep rupture time–load plots at 80°C. ◆, P7-10FG-0600; □, P7-20FG-0600; △, P7-30FG-0600; ●, P7-40FG-0600; ■, P6-20FG-0600.

# Creep–Fatigue Behavior for GFRP

**TABLE 13.3** Summary of Regression Analysis of Tensile Creep Data: P7-20FG-0600

| Chamber temp. $T$ (°C) | Axial load $L$ (kg) | Tensile stress $S$ (psi) | Creep rupture time $t$ (hr) | Independent variables $x1$ $1/T$ (°K) | $x2$ $\log(L)$ | Response function $Y = (x1, x2)$ Time $t$ (hr) Log($t$)-meas | Calc. | Calculated load (kg) for creep rupture $t = 1000$ hr |
|---|---|---|---|---|---|---|---|---|
| 50 | 160 | 5914 | 45.16 | 0.003095 | 2.2041 | 1.6548 | 38.29 | 136 |
| 50 | 170 | 6284 | 10.70 | 0.003095 | 2.2304 | 1.0294 | 11.18 |  |
| 50 | 180 | 6653 | 2.88  | 0.003095 | 2.2553 | 0.4594 | 3.50  |  |
| 65 | 135 | 4990 | 40.00 | 0.002958 | 2.1303 | 1.6021 | 47.17 | 116 |
| 65 | 140 | 5175 | 27.36 | 0.002958 | 2.1461 | 1.4371 | 22.53 |  |
| 65 | 145 | 5360 | 13.68 | 0.002958 | 2.1614 | 1.1361 | 11.05 |  |
| 65 | 150 | 5544 | 5.12  | 0.002958 | 2.1761 | 0.7093 | 5.55  |  |
| 65 | 160 | 5914 | 1.48  | 0.002958 | 2.2041 | 0.1703 | 1.50  |  |
| 80 | 130 | 4805 | 6.00  | 0.002832 | 2.1139 | 0.7782 | 5.22  | 100 |
| 80 | 135 | 4990 | 3.08  | 0.002832 | 2.1303 | 0.4886 | 2.43  |  |
| 80 | 140 | 5175 | 1.00  | 0.002832 | 2.1461 | 0.0000 | 1.16  |  |
| 80 | 120 | 4435 | 19.48 | 0.002832 | 2.0792 | 1.2896 | 26.55 |  |

TABLE 13.4 Summary of Regression Analysis of Tensile Creep Data: P7-30FG-0600

| Chamber temp. $T$ (°C) | Axial load $L$ (kg) | Tensile stress $S$ (psi) | Creep rupture time $t$ (hr) | Independent variables $x1$ $1/T$ (°K) | Independent variables $x2$ $\log(L)$ | Response function $Y = (x1, x2)$ Time $t$ (hr) Log($t$)-meas | Response function $Y = (x1, x2)$ Time $t$ (hr) Calc. | Calculated load (kg) for creep rupture $t = 1000$ hr |
|---|---|---|---|---|---|---|---|---|
| 50 | 200 | 7392 | 10.48 | 0.003095 | 2.3010 | 1.0204 | 11.06 | 159 |
| 50 | 220 | 8132 | 1.68  | 0.003095 | 2.3424 | 0.2253 | 1.72  |     |
| 50 | 180 | 6653 | 93.40 | 0.003095 | 2.2553 | 1.9703 | 86.51 |     |
| 65 | 180 | 6653 | 3.04  | 0.002958 | 2.2553 | 0.4829 | 2.81  | 133 |
| 65 | 170 | 6284 | 10.76 | 0.002958 | 2.2304 | 1.0318 | 8.57  |     |
| 65 | 165 | 6099 | 19.32 | 0.002958 | 2.2175 | 1.2860 | 15.36 |     |
| 65 | 160 | 5914 | 26.00 | 0.002958 | 2.2041 | 1.4150 | 28.01 |     |
| 80 | 150 | 5544 | 4.28  | 0.002832 | 2.1761 | 0.6314 | 4.29  | 113 |
| 80 | 145 | 5360 | 9.28  | 0.002832 | 2.1614 | 0.9675 | 8.32  |     |
| 80 | 140 | 5175 | 14.46 | 0.002832 | 2.1461 | 1.1602 | 16.50 |     |

# Creep–Fatigue Behavior for GFRP

**TABLE 13.5** Summary of Regression Analysis of Tensile Creep Data: P7-40FG-0600

| Chamber temp. T (°C) | Axial load L (kg) | Tensile stress S (psi) | Creep rupture time t (hr) | Independent variables x1 1/T (°K) | x2 log(L) | Response function Y = (x1, x2) Time t (hr) Log(t)-meas | Calc. | Calculated load (kg) for creep rupture t = 1000 hr |
|---|---|---|---|---|---|---|---|---|
| 50 | 220 | 8132 | 7.44  | 0.003095 | 2.3424 | 0.8716 | 8.54  | 165 |
| 50 | 200 | 7392 | 42.00 | 0.003095 | 2.3010 | 1.6232 | 40.91 |     |
| 50 | 190 | 7023 | 93.70 | 0.003095 | 2.2788 | 1.9717 | 95.04 |     |
| 60 | 210 | 7762 | 1.76  | 0.003002 | 2.3222 | 0.2455 | 1.98  | 144 |
| 65 | 190 | 7023 | 3.72  | 0.002958 | 2.2788 | 0.5705 | 3.54  | 135 |
| 65 | 180 | 6653 | 10.00 | 0.002958 | 2.2553 | 1.0000 | 8.62  |     |
| 65 | 175 | 6468 | 16.04 | 0.002958 | 2.2430 | 1.2052 | 13.69 |     |
| 65 | 170 | 6284 | 23.68 | 0.002958 | 2.2304 | 1.3744 | 22.05 |     |
| 80 | 140 | 5175 | 20.64 | 0.002832 | 2.1461 | 1.3147 | 26.42 | 112 |
| 80 | 145 | 5360 | 14.80 | 0.002832 | 2.1614 | 1.1703 | 14.84 |     |
| 80 | 150 | 5544 | 9.12  | 0.002832 | 2.1761 | 0.9600 | 8.50  |     |
| 80 | 155 | 5729 | 5.12  | 0.002832 | 2.1903 | 0.7093 | 4.96  |     |

TABLE 13.6  Statistical Analysis of Tensile Creep Data

| | A1 | A2 | A3 | | Regression statistics | | |
| | Intercept material constant | x1 Temperature parameter | x2 Tensile load parameter | | | | |
|---|---|---|---|---|---|---|---|
| Material | | | | Multiple R | R square | SE | Observ. |
| P7-20FG-0600 | 14.61256394 | 10,256.04522 | −20.3129384 | 0.98924442 | 0.97860453 | 0.0890043 | 12 |
| P7-30FG-0600 | 12.42081132 | 10,847.99066 | −19.52694264 | 0.99422777 | 0.98848887 | 0.06046796 | 10 |
| P7-40FG-0600 | 7.229676437 | 10,402.28137 | −16.43307774 | 0.99352367 | 0.98708929 | 0.06038549 | 11 |

$\text{Log}(t) = A1 + A2/(T + 273.2) + A3*(\log L)$, where $t$ = creep rupture time (hr), $L$ = applied tensile load (kg), and $T$ = test chamber temperature (°C).

### Creep–Fatigue Behavior for GFRP

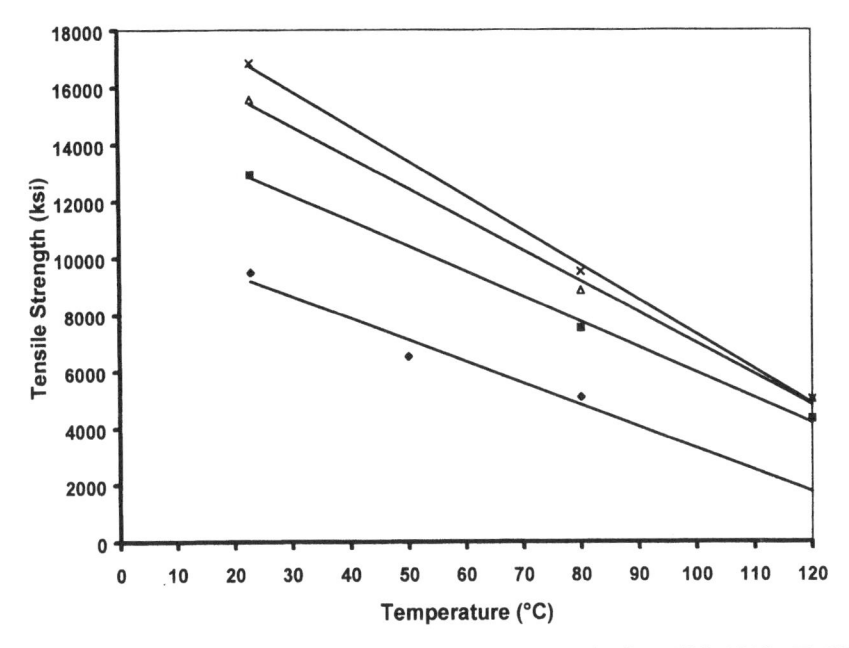

**FIGURE 13.17** Tensile strength–temperature plots. ◆, P7-10FG-0600; ■, P7-20FG-0600; △, P7-30FG-0600; ×, P7-40FG-0600.

To prevent unrealistically short creep rupture times, one needs to maintain stress loads below about 60–70% of the measured tensile strength at the same test chamber temperature. This is an important criterion for selecting appropriate weights for accelerated tensile strength determination.

Equation (13.6) was used to develop a set of creep rupture envelopes to predict attainable creep rupture time at given load $L$ for three mega-coupled composites at 80°C as shown in Fig. 13.18. With measured data only at creep rupture times <100 hr for accelerated tensile creep determinations, we can make estimates for anticipated service lifetime as a function of applied stress load that is indicated by the weight $L$.

#### 13.7.5 Comparison Between Performance of Dry Blends and Compounded Material

Cessna [2] used a stress–time superposition model to analyze creep data for chemically coupled GFRP. Instead of compounding PC-072 powder (containing a portion of Hercuprime G) with glass fibers in an extruder, he dry blended ingredients and directly injection-molded test specimens with glass rovings. This

[FIGURE: Creep Rupture Time t (hrs) vs Applied Tensile Load L (kg), log-linear plot]

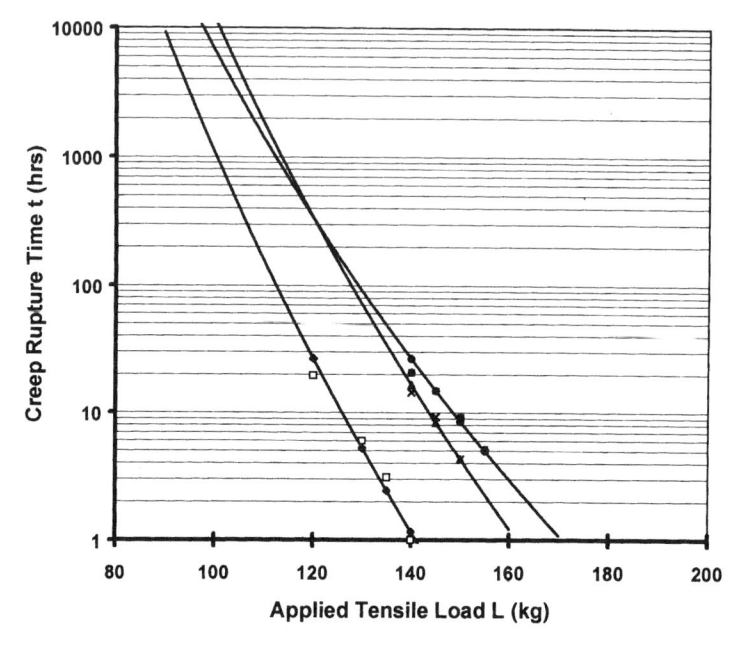

**FIGURE 13.18** Analysis of tensile creep at 80°C. ♦, 20 wt% calc; □, 20 wt% meas; △, 30 wt% calc; ×, 30 wt% meas; ●, 40 wt% calc; ■, 40 wt% meas.

type of sample preparation produces essentially a long glass–reinforced composite having nearly perfect tensile behavior in the uniaxial direction. The inference is that the tensile creep determination would also reflect the effects of perfect coupling. The corresponding creep rupture data for 1000 hr duration was published in *Modern Plastics Encyclopedia* [5] for this series of composites subjected to 23°C and 80°C temperatures. These data were used to gage the relative effectiveness of the corresponding mega-coupled composites in Fig. 13.19.

The closeness in creep rupture envelopes of stress–temperature for 1000 hr duration for the two sets of data shown in Fig. 13.19 provides additional verification of the perfect coupling behavior of mega-coupled composites described in Chapter 12.

## ACKNOWLEDGMENTS

We express special thanks to Dr. Harry Karian, Thermofil, Inc. for providing Section 13.7; to Chris Gill and Don Wise, Jr., Owens Corning, Technical Center, Granville, Ohio; and to Eric Vickery, Owens Corning, Anderson, South Carolina.

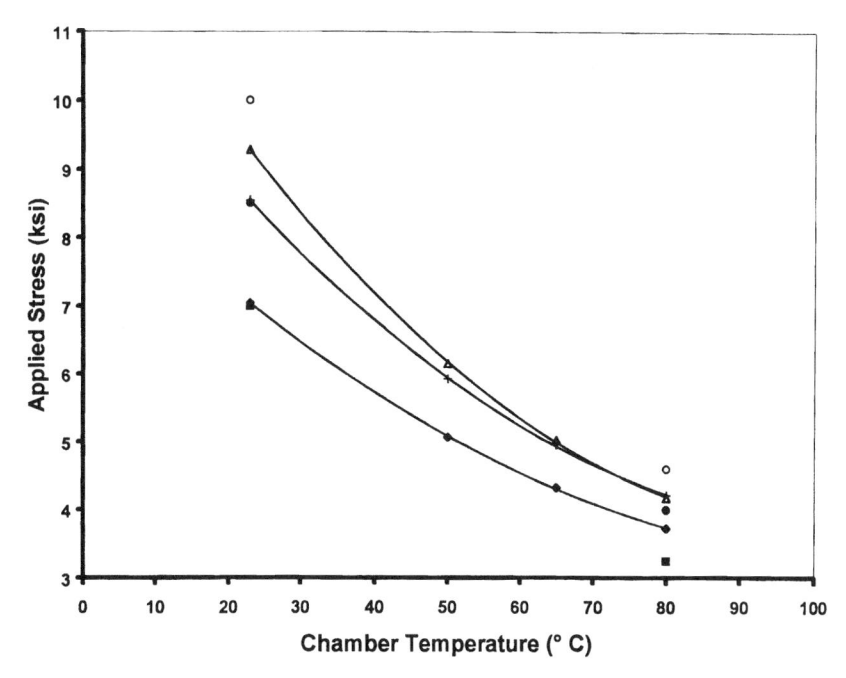

FIGURE 13.19  Creep rupture strength versus temperature at 1000 hr. ◆, P7-20FG-0600; +, P7-30FG-0600; △, P7-40FG-0600; ■, PC-072/20; ●, PC-072/30; ○, PC-072/40.

## REFERENCES

1. Modern Plastics Encyclopedia. New York: McGraw-Hill, Inc., 1986–1987, pp. 405–407.
2. LC Cessna Jr. Stress-time superposition of creep data for polypropylene and coupled glass-reinforced polypropylene. Polym. Eng Sci 11:211–218, 1971.
3. GS Trantina. Creep analysis of polymer structures. Polym Eng Sci 26:776–780, 1986.
4. AW Birley, B Haworth, J Batchelor. Physics of Plastics. New York: Hanser Publishers, 1992, pp. 363–366.
5. Modern Plastics Encyclopedia. New York: McGraw-Hill, 1988, p. Cr-20.

# 14

## Mica Reinforcement of Polypropylene

**Levy A. Canova**
Franklin Industrial Minerals, Kings Mountain, North Carolina, U.S.A.

### 14.1 INTRODUCTION
#### 14.1.1 Markets

Mica is combined with polyolefin resins to make composites that compare favorably to widely used engineering resins These reinforced compounds are now replacing acrylonitrile-butadiene-styrene (ABS), glass fiber–reinforced polypropylene, various grades of nylon and polyester resins, and steel in many applications.

Polypropylene composites have many advantages over metal such as corrosion resistance, low weight, easily molded complicated shapes, and low cost. A 1979 article [1] compared the cost of polypropylene composites to that of sheet at equal stiffness (Table 14.1). Mica filled (40 wt%) polypropylene was the only composite with a relative cost less than that of steel (0.98) for a part with stiffness equal to steel. The relative thickness of the composite was 2.88 and the relative weight was 0.45. Cost calculations were based on mica at $0.18/lb and polypropylene at $0.31/lb. The relative cost of a 40% glass-filled polypropylene composite was 2.48 and its relative weight was 0.49. The relative cost of a 40 wt% filled talc/polypropylene composite was 1.32 and the relative weight was 0.61.

TABLE 14.1 Composite Costs Relative to Steel at Equal Stiffness Composite Molded at Equal Stiffness to Steel Sheet

| Material | Density relative to steel | Relative weight | Relative thickness | Relative cost |
|---|---|---|---|---|
| Steel | 1.000 | 1.00 | 1.00 | 1.00 |
| Polyester | 0.199 | 0.57 | 2.84 | 3.68 |
| Polypropylene | 0.155 | 0.55 | 4.77 | 1.09 |
| PP/40 wt% mica | 0.157 | 0.45 | 2.88 | 0.98 |
| PP/40 wt% glass | 0.157 | 0.49 | 2.10 | 2.48 |
| PP/40 wt% talc | 0.157 | 0.61 | 3.88 | 1.32 |
| PP/40 wt% CaCO$_3$ | 0.157 | 0.66 | 4.21 | 1.43 |
| Nylon | 0.145 | 0.80 | 5.55 | 6.04 |
| Nylon/33 wt% glass | 0.176 | 0.56 | 3.31 | 4.52 |
| Nylon/40 wt% mineral | 0.191 | 0.70 | 3.68 | 3.93 |

*Source:* Ref. 1.

Mica-reinforced polypropylene composites are meeting the growing demand for larger, thinner, and stronger parts; reduced weight; better surface appearance; higher heat stability; and good recyclability. These requirements for higher performance plastics together with enhancement with surface treatments and additives are playing a major role in the increasing use of mica-reinforced composites. Virtually every market is being affected by these compounds. Automotive, construction, and appliance markets are some of the larger growth areas.

### 14.1.2 Mica Types

The two principal minerals in the mica group used as reinforcements for polypropylene are muscovite mica and phlogopite mica. Chemical formulas for these mica varieties are shown below:

$$\text{Muscovite:} \quad KAl_2(Si_3O_{10})(OH_3F)_2$$
$$\text{Phlogopite:} \quad KMg_3(AlSi_3O_{10})(OH_3F)_2$$

The above formulas are idealized representations of the elemental composition of these mica types. Naturally occurring muscovite and phlogopite mica contain small amounts of other elements, including iron, magnesium, and titanium, that are bound up in the internal structure of the minerals. Table 14.2 gives the chemical assay of muscovite and phlogopite mica.

TABLE 14.2 Composition of Muscovite and Phlogopite Mica

| Elemental assay | Muscovite mica, North Carolina | Phlogopite mica, Canada | Phlogopite mica, Finland |
|---|---|---|---|
| $SiO_2$ | 47.9 | 40.7 | 42 |
| $Al_2O_3$ | 33.1 | 15.8 | 11 |
| $Fe_2O_3$ | 0.7 | 1.2 | 4.7 |
| FeO | 2.0 | 7.8 | 5.5 |
| MgO | — | 20.6 | 24 |
| $K_2O$ | 9.8 | 10 | 10.4 |
| $Na_2O$ | 0.8 | 0.5 | 0.1 |
| $TiO_2$ | 0.7 | 0.4 | 0.4 |

*Source:* Ref. 39.

## 14.1.3 Mica Morphology and Effects

Mica is a generic term for any of several complex hydrous aluminosilicate minerals characterized by their platy shape and pronounced basal cleavage. This permits splitting or delamination into extremely thin high-aspect-ratio particles that are tough and flexible. An example of a highly delaminated mica product can be seen in Fig. 14.1, which is a photograph of muscovite mica obtained with a scanning electron mictoscope. Flake glass is the only other platy reinforcement with fairly high aspect ratio. Because of its brittle nature, flake glass is much more difficult to compound and mold than mica and is much more expensive.

The unique platy shape of mica particles is very beneficial in polypropylene composites. Because the length and width dimensions are very similar, uniform shrinkage in molded parts essentially eliminates warpage. Glass fibers and acicular minerals like wollastonite have a very large difference in length-to-width dimensions. Because of this, the fibers align in the direction of flow during injection molding. This orientation causes significant differences in shrinkage in the transverse and flow directions with part warpage as the end result. Addition of mica to glass fiber–reinforced polypropylene composites results in a significant reduction of part warpage. In addition to equal shrinkage, total shrinkage is also decreased as measured by the coefficient of linear thermal expansion (CLTE). This is quite important when one needs to produce a product that maintains its dimensions over a wide range of use temperatures.

Another benefit resulting from the thin platy shape of mica is its ability to reduce flow of gases and liquids through polypropylene parts. This reduction in permeability is important in various end applications that require good barrier properties.

**FIGURE 14.1** Scanning electron microscopy (SEM) of mica. Muscovite, H-360.

### 14.1.4 Aspect Ratio and Effects

For mica minerals, aspect ratio is defined as the average ratio of the average diameter of all particles to the average thickness of all particles. Until recently, it was impossible to accurately determine aspect ratios of different products. Attempts at predicting aspect ratio were made by measurement of diameters and thickness of individual particles using scanning electron microscopy. It is now possible to make this measurement with modern particle size measurement equipment. Research is currently underway to determine and develop a correlation of the aspect ratio of mica in processed polypropylene composites to observed mechanical properties.

Aspect ratio is a very important characteristic of mica products because it has a significant effect on the increase in flexural modulus obtained with a given loading level of mica. High-aspect-ratio mica produces the largest increase in flexural modulus.

Higher amounts of iron in phlogopite mica bond the individual layers of mica together with relatively high forces. Because of this, hooks of phlogopite mica are more difficult to delaminate than books of muscovite mica, and resulting aspect ratios are somewhat lower.

## 14.1.5 Color Variations and Effects

Muscovite mica is silvery white to off-white. When incorporated into polypropylene, the composite typically is a very light beige because of the refractive index difference between mica and the polymer. Phlogopite mica is a bronze to dark brown or black mineral. The color of mica is important when one wants to produce a colored composite. It is possible to add colorants to composites containing muscovite mica to produce a variety of colored products. The dark color of phlogopite mica inhibits color change when colorants are used. For this reason, phlogopite mica is primarily limited to color-insensitive applications.

## 14.1.6 Other Important Mica Properties

Muscovite mica is inert to strong acids and bases and all solvents. It can only be dissolved by hot hydrofluoric acid. It can be used in polypropylene composites that are designed for corrosive applications. However, phlogopite mica will dissolve in strong acids.

The relatively large amounts of iron in phlogopite mica (Table 14.2) are likely the cause of the poorer heat aging characteristics of phlogopite-filled polypropylene relative to that of muscovite mica.

Mica provides better scratch and mar resistance than other minerals. The ability to improve scratch resistance is especially useful for automotive interior applications where scratch and mar resistance has been a problem with talc-filled polyolefins. Considerable research is presently underway to take advantage of this characteristic of mica.

Muscovite mica is also an excellent electrical insulator (high dielectric properties) and provide both thermal and sound dampening properties. When used in low molecular weight polymers, sound dampening is enhanced greatly. Muscovite mica also shields out ultraviolet (UV) radiation below 300 nm. Blocking of UV radiation below 300 nm is likely one of the factors causing enhanced weatherability of exterior architectural coatings. This phenomenon may also increase the life of polypropylene composites exposed to sunlight.

Mica minerals are quite soft and nonabrasive, so that equipment wear is minimal. The hardness of mica minerals on the mohs scala of hardness varies from 2.0 to 2.5 Mohs for muscovite mica arid 2.5 to 3.0 Mohs for phlogopite mica.

Mica products are also very stable at high temperatures. Thermogravimetric analysis shows that both muscovite and phlogopite mica are thermally stable to over 450°C, which is far above the processing temperatures for polyolefin composites. Table 14.3 lists some of the more important physical properties of mica.

TABLE 14.3  Properties of Mica

| Property | Muscovite | Phlogopite |
|---|---|---|
| Density, g/cm$^3$ | 2.8–3.2 | 2.6–3.2 |
| Hardness, Mohs | 2.0–3.2 | 2.5–3.0 |
| Thermal conductivity Perpendicular to cleavage | 0.67 | 0.67 |
| Specific heat (25°C) | 0.206–0.209 | 0.206–0.209 |
| Modulus of elasticity, MPa | $1.723 \times 10^5$ | $1.723 \times 10^5$ |
| Tensile strength, MPa | 255–296 | 255 296 |
| Refractive index, $\eta_D$ | | |
| $\alpha$ | 1.552–1.570 | 1.43–1.63 |
| $\beta$ | 1.582–1.607 | 1.57–1.69 |
| $\gamma$ | 1.588–1.611 | 1.57–1.69 |
| Melting point | Decomposes | Decomposes |
| Water constitution, wt% | 4.5 | 3.2 |
| Power factor at 25°C, 1 MHz | 0.01–0.02 | 0.3 |
| Dielectric constant | 8.5–9 | 5–6 |
| Dielectric strength, 2°C, V/μm | 235–118 | 165–83 |
| Resistivity, Ω-cm | $10^{12}-10^{15}$ | $10^{10}-10^{13}$ |
| Coefficient of expansion per °C, perpendicular to cleavage | | |
| 20–100°C | $(15-25) \times 10^{-6}$ | $1 \times 10^{-6} - 1 \times 10^{-3}$ |
| 100–300°C | $(15-25) \times 10^{-6}$ | $2 \times 10^{-4} - 2 \times 10^{-2}$ |
| 300–600°C | $(16-36) \times 10^{-6}$ | $1 \times 10^{-5} - 3 \times 10^{-3}$ |
| Parallel to cleavage | | |
| 0°C–200°C | $(8-9) \times 10^{-6}$ | $(13-14.5) \times 10^{-6}$ |

Source: Ref. 40.

### 14.1.7  Production

In the United States, muscovite mica occurs in commercial quantities in North Carolina, South Carolina, Georgia, South Dakota, and New Mexico. Phlogopite mica does not occur in the United States in commercial amounts. The main sources of phlogopite mica are Canada and Finland.

Mica ores contain from 15% to 80% mica and other minerals, such as silica, feldspar, and kaolin. Mica producers use a variety of purification processes to purify mica before grinding into finished products.

Mica is reduced in size by two basic comminution processes: dry grinding and wet grinding. A third process uses a screening procedure to produce "flake" mica products from unground mica. In dry grinding, particle attrition is caused by contacting mica particles at high velocity with other mica particles or with metal or ceramic surfaces. The former procedure, called jet milling, is used to produce

# Mica Reinforcement of Polypropylene

the smallest particle size products with $D(v, 0.5)$ of 8–20 μm. The $D(v, 0.5)$ number is a 50% percentile value obtained with modern laser particle size measurement equipment. That is, 50% of the particles are smaller and 50% are larger than the median micron size. Contact with metal produces intermediate particle-sized products with $D(v, 0.5)$ of 50 μm and higher. Dry grinding results in much less delamination and therefore lower aspect ratios than wet grinding.

Wet grinding is used when highly delaminated mica products with $D(v, 0.5)$ of 20–50 μm are desired. Particles produced during wet grinding are not only very thin (0.1–1 μm); they are also very clean as a result of the washing action of water. Wet grinding is a more costly process than dry grinding because of lower production rates and higher capital costs.

A sieving process is used to produce "flake" products, although all micas are, of course, flake products. This process is used with both phlogopite and muscovite mica. The $D(v, 0.5)$ for flake mica products varies from 30 to more than 1000 μm.

After grinding, various forms of screening and air classification are used to produce commercial mica products. Shipments are made in 50-lb bags or bulk containers such as pneumatic trucks, rail cars, Gaylord boxes, or fabric tote bags.

## 14.1.8 Analysis and Specifications

Mica products are usually analyzed by screening to determine particle distributions. Specifications typically include minimal and maximal values for the amount of mica passing or retained on screens of varying mesh sizes. Another value often used for a material specification is loose bulk density. Laser particle size analysis is also being used to some extent to define mica products. One should remember that most other minerals are measured by an instrument called a sedigraph, which measures the size of particles based on their settling rate. This procedure can be used with very small low-aspect-ratio minerals such as calcium carbonate, talc, silica, very-small-particle mica, and aluminum trihydrate but cannot be used for analysis of most mica products. Sedigraph particle size values for very-small-mica products are about one-half to one-third of the values obtained by laser diffraction.

## 14.2 PROCESSING

### 14.2.1 Handling

Mica products for reinforcing polypropylene vary in mean size from about 8 to 250 μm or higher. Because of the platy nature of the particles, mica tends to "rathole" or "bridge" during discharge from a hopper into a transfer line. This is less of a problem with products that have fairly large mean particle sizes (i.e., 200 μm and higher). Proper hopper design is essential to avoid the erratic flow problems created by this phenomenon. Using a hopper with moving sides is one approach that has been successful in providing an even flow of mica from the hopper. A loss-in-weight system with a special hopper can control the bridging problem [2]. The hopper has flexible vinyl sides that are massaged by paddles to keep the mica flowing at a steady rate.

### 14.2.2 Compounding Equipment

*Equipment.* Single-screw extruders may be used for preparation of compounds containing up to 20 wt% mica. Twin-screw extruders are generally needed for efficient compounding of higher amounts of mica.

*Processing.* Even though mica is somewhat flexible, it is possible to significantly change the size of relatively large mica particles during melt compounding. With twin-screw compounding and throat addition of mica, one can expect a 266-µm mica in 30 wt% filled polypropylene to be reduced to about 100 µm. Table 14.4 shows the effects of twin-screw compounding on mica products with median particle sizes that vary from 20 to 266 µm. As expected, mica with small median size is not degraded nearly as much as mica with large flakes. Size reduction is not desired when one is trying to maintain the aspect ratio to obtain the highest possible flexural modulus. The effects of flake aspect ratio on flexural properties of mica-reinforced plastics has been described thoroughly [3].

One should compound mica in the manner that glass fiber is processed to minimize particle breakage. This can be accomplished by adding mica downstream through a side feed port to the resin after it has been softened by heating. This procedure avoids many of the high shear forces that break mica flakes when

TABLE 14.4  Compounding Effect on Mica Particle Size

| Mica product | Initial particle size D(v, 0.5) (µm) | Mica size after processing, PP-CP | Mica size after processing, TPO[a] |
|---|---|---|---|
| C-3000 | 24.0 | 21.9 | 23.1 |
| WG-325 | 37.0 | 27.7 | 30.1 |
| 4K | 50.0 | 32.8 | 39.5 |
| L-140 | 86.0 | 40.2 | 45.4 |
| L-135 | 266.0 | 96.3 | 115.0 |
| Barrel temp., °C | — | 195 | 185 |

[a] 30 wt% mica, two passes through a 17:1 L/D, corotating, 35-mm conical twin-screw extruder; mica and resin were premixed and added at the upstream port. 15 wt% Polysar 306 rubber in Profax grade 6301 polypropylene homopolymer.
*Source:* Ref. 41.

mica and solid resin pellets are fed together through the extruder throat. It is helpful to keep the rear zone temperature higher than the front zone to decrease the resin viscosity at the mica addition point if mica is added at this location.

Another helpful procedure is to preheat the mica before adding it to the resin. This results in lower energy consumption, reduced equipment wear, and improved product quality according to a 1994 article in *Plastics Technology* [4]. This preheating procedure is covered in U.S. Patent 4,980,390 and was available for license when the article was published.

A paper presented at the 1990 SPE ANTEC [5] discussed techniques for compounding highly filled polymers with corotating twin-screw extruders. The fusion time as measured in a torque rheometer of 40 wt% filled polypropylene with melt flow index (MFI) of 15 was 2.4 min for glass fibers, 8 min for mica, and 13.5 min for talc. For a 30 wt% loading of mica, fusion time was reduced to about 5 min.

The bulk density of mica also effects compounding rates. Higher bulk density mica products are wet out by resin much faster than low bulk density mica. This occurs because it is more difficult to remove adsorbed air from smaller particles in low bulk density fillers.

Extrusion rates are also increased by use of surface-treated mica [6]. The extrusion rate increased 20% for a 50 wt% filled polypropylene homopolymer compound using surface-modified mica.

Because of the stiffening effects of mica on extruded compound, less cooling is needed before pelletizing.

### 14.2.3 Molding

*Equipment.* Polypropylene composites are usually molded by an injection molding machine equipped with a reciprocating screw. Compression molding is used only rarely. If compression molding is used, however, much higher flexural modulus values can be obtained due to retention of the aspect ratio of the mica.

A technical bulletin produced by Marietta Resources International [7] was the source for data in Table 14.5, which table shows that precompounding mica into polypropylene reduces properties versus injection molding of a dry blend of large particle size mica in polypropylene. Flexural modulus was increased from 1.44 to 1.65 GPa for a 50 wt% filled polypropylene. Because of the difficulty in feeding small particle size mica and the high air content, dry-blend compounding/molding should be used only for coarser grades of mica. Use of high-intensity compounding as in a Banbury mixer should be avoided to maintain aspect ratio.

Molding screws that contain sections for intensive mixing may cause significant particle size reduction of the larger sizes of mica. Additional shear should always be avoided when possible. Use of hot runner systems will usually

TABLE 14.5  Injection Molding with Dry Blends vs. Compounds

|  | Unfilled PP[a] | 40 wt% 325-H mica (fine) || 50 wt% 60-H mica (coarse) ||
| Property | | Powder blend | Com-pounded | Powder blend | Com-pounded |
| --- | --- | --- | --- | --- | --- |
| Tensile strength, MPa | 28.5 | 33.8 | 35.0 | 35.7 | 28.9 |
| Flexural modulus, GPa | 1.37 | 7.03 | 6.64 | 11.4 | 9.92 |
| Flexural strength, MPa | 28.8 | 57.3 | 58.4 | 63.2 | 51.6 |

[a]Hercules Inc., Profax PCO-72 (maleic anhydride-modified polypropylene).
*Source:* Ref. 7.

help minimize flake breakage. In general, the same processing procedures used to minimize breakage of glass fiber during the injection molding process should be used when molding mica-reinforced polypropylene.

Gate size should be selected according to the size of the mica flakes to enable good flow and to prevent additional breaking of the flakes.

*Processing.* When molding, both the material and mold temperatures should be kept as high as possible without adversely affecting part finish. Keeping the melt temperature 50–100°F hotter than the temperature normally used for the unfilled resin is recommended. Typical cylinder temperatures for 30–50 wt% filled polypropylene are 400–440°F. Melt temperatures should be obtained primarily by use of external heating rather than increased back pressure on the screw. This will help minimize degradation of the mica flakes. Also, use the minimal injection speed that will produce a good part. Screw speed and back pressure should be kept at minimal levels as long as a homogeneous melt can be obtained.

Mold temperature should be 85–160°F, and hot runner temperature should be 40–80°F higher than the cylinder temperature.

*Mold Design.* Molds should be designed to minimize flow and weld lines. The gate should be a wedge-shaped fan or tab to minimize flow problems. Also, gates should be located away from flow obstructions. There should be only one gate if possible and it should be in the thicker section of the part. Space between the end of the gate and the runner system should be minimized. This space should also be wedge shaped. Sometimes, cold plugs in the runner system are helpful in producing parts with good finish.

## 14.2.4 Loading Level Effects on Melt Index

Contrary to what one might believe, the melt flow of mica-reinforced polypropylene is not decreased substantially when one increases the mica loading level. In a recent paper [8], the melt flow of 20, 30, and 40 wt% filled polypropylene homopolymer decreased only slightly with higher levels of mica. The melt flow decreased from 2.9 g/10 min for 20 wt% mica to 2.6 g/10 min for 30 wt% and 1.8 g/10 min for 40 wt% mica.

Melt flow for polypropylene composites containing 3.2-mm chopped glass fiber decreased from 1.7 g/10 min for a 20 wt% loading to 1.3 g/10 min with a 40 wt% loading.

It is sometimes advisable to use additives in the formulation to increase melt flow when high mica loadings are used or when parts are being molded. Very high molecular weight silicones are now available, and products such as Struktol TR-016 from Struktol Corporation provide excellent increases in melt flow. In one experiment in Franklin Industrial Minerals Applications Lab (L. Canova, unpublished data), the melt flow of a polypropylene copolymer containing 20 wt% mica increased from 17.1 to 27.5 g/10 min when 2 wt% Struktol TR-016 was used to replace 3 wt% Unite MP-880.

## 14.3 MICA VERSUS OTHER MINERALS AND GLASS REINFORCEMENTS

Table 14.6 shows that in a 40 wt% filled polypropylene compound, mica increases flexural modulus much more than other minerals and glass products. The mica

TABLE 14.6  Mica versus Other Reinforcements

| 40 wt% filler in polypropylene homopolymer[a] | Tensile strength (psi) | Flexural modulus (psi) | Izod impact (ft-lb/in.) | HDT (°F@264 psi) |
|---|---|---|---|---|
| Control, 100 wt% PP | 4580 | 240 | 2.90 | 150 |
| HiMod-360 mica | 4435 | 1110 | 0.39 | 238 |
| L-135 mica | 3710 | 1010 | 0.46 | 244 |
| Wollastonite | 3690 | 740 | 0.41 | 190 |
| Glass flake, 1/64" | 3370 | 710 | 0.43 | 183 |
| Milled glass, 1/8" | 3580 | 670 | 0.42 | 200 |
| Talc | 4220 | 660 | 0.44 | 181 |
| Calcium carbonate | 3250 | 430 | 0.41 | 157 |

[a]Profax 6501, MI = 4.
*Source:* Ref. 42.

compound had a flexural modulus of 7.2 versus 5.1 GPa for wollastonite, 4.9 GPa for glass flake, 4.6 GPa for milled glass, 4.5 GPa for talc, 3.0 GPa for calcium carbonate, and 2.6 GPa for silica. Mica has been reported to provide the highest stiffness and heat distortion temperature (HDT) relative to other mineral fillers in the *Handbook of Reinforcements for Plastics* [9]. The mica-reinforced compound had a heat deflection temperature of 91°C versus 93°C for milled glass, 84°C for flake glass, 83°C for talc, 69°C for calcium carbonate, and 68°C for silica. Gardner impact strength of the mica-filled polypropylene was lower than observed with most other reinforcements. Calcium carbonate and talc had Gardner impact values of 0.68 and 0.46 J versus 0.33 J for the mica-filled compound.

A comparison of the properties of surface-treated phlogopite mica with talc, calcium carbonate, glass, and unfilled resin is given in a technical bulletin produced by Marietta Resources International [10]. Table 14.7 shows that 40 wt% mica-filled polypropylene (Profax 6523) had a higher flexural modulus than 30 wt% glass fiber–filled polypropylene (1.04 × $10^6$ psi versus 0.93 × $10^6$ psi). The compound filled with 40 wt% surface-modified mica also had higher Izod impact than the 40 wt% talc-filled compound (0.65 versus 0.5 ft-lb/in.).

TABLE 14.7 Surface-Modified Phlogopite Mica vs. Other Minerals for Polypropylene Reinforcement

| Mechanical properties of filled polypropylene[a] | None 0.0 wt% | Talc 40 wt% | CaCO$_3$ 40 wt% | Glass fiber 30 wt% | Mica Suzorex 200-NP, 40 wt% |
|---|---|---|---|---|---|
| Tensile strength, psi | 4930 | 4270 | 2770 | 6340 | 6190 |
| Flexural strength, psi | 4450 | 6420 | 4720 | 10,060 | 9320 |
| Flexural modulus, $10^3$ psi | 193 | 676 | 421 | 933 | 1040 |
| Izod impact, ft-lb/in., 22°C | 0.45 | 0.45 | 0.75 | 0.79 | 0.65 |
| HDT at 264 psi, °F | 136 | 162 | 183 | 257 | 226 |
| CLTE, 73–163°F, per °F × $10^{-5}$ | 7.4 | 4.0 | 5.3 | 2.0 | 3.3 |
| Mold shrinkage, % (parallel to flow) | 2.0 | 1.2 | 1.4 | 0.3 | 0.8 |

[a]Profax 6523 polypropylene homopolymer.
*Source:* Ref. 10.

## 14.4 EFFECTS ON COMPOSITE PROPERTIES

### 14.4.1 Effects on Flexural Modulus

As one might expect, increasing the amount of different mica products in polypropylene composites increases flexural modulus significantly. See Tables 14.8, 14.9, and 14.10 for the changes in flexural modulus obtained with different amounts of mica in polypropylene homopolymer, copolymer, and thermoplastic polyolefin (TPO) formulations. Flexural modulus increases more in polypropy-

TABLE 14.8  Effect of Mica Concentration on Flexural Modulus of Polypropylene Homopolymer Composites

| Muscovite mica, grade | Profax 6501 homopolymer | | | |
|---|---|---|---|---|
| | 12 wt% | 20 wt% | 30 wt% | 40 wt% |
| C-3000 | 360 | 460 | 620 | 835 |
| HiMod-270 | 410 | 570 | 810 | 1080 |
| WG-325 | 370 | 480 | 700 | 930 |
| 4K | 360 | 470 | 690 | 900 |
| L-140 | 410 | 560 | 780 | 1010 |
| L-135 | 410 | 530 | 750 | 1010 |
| PP, 100 wt% | 240 | 240 | 240 | 240 |

Flexural modulus (kpsi), mica in polypropylene homopolymer.
*Source:* Ref. 43.

TABLE 14.9  Effect of Mica Concentration on Flexural Modulus of Polypropylene Copolymer Composites

| Muscovite mica, grade | Profax 8623 copolymer | | |
|---|---|---|---|
| | 12 wt% | 20 wt% | 30 wt% |
| PM-325 | 290 | 310 | 420 |
| C-3000 | 270 | 280 | 380 |
| HiMod-270 | 270 | 355 | 495 |
| WG-325 | 300 | 360 | 480 |
| 4K | 280 | 340 | 460 |
| L-140 | 250 | 325 | 470 |
| L-135 | 250 | 310 | 490 |
| PP, 100 wt% | 160 | 160 | 160 |

Flexural modulus (kpsi), mica in polypropylene copolymer.
*Source:* Ref. 41.

TABLE 14.10 Effect of Mica Concentration on Flexural Modulus of TPO Composites

| | TPO[a] | | |
|---|---|---|---|
| Muscovite mica, grade | 10 wt% | 20 wt% | 30 wt% |
| PM-325 | 200 | 260 | 290 |
| C-3000 | 230 | 270 | 360 |
| HiMod-270 | 260 | 350 | 430 |
| WG-325 | 240 | 320 | 390 |
| 4K | 230 | 300 | 380 |
| L-140 | 250 | 370 | 430 |
| L-135 | 240 | 295 | 380 |
| PP, 100 wt% | 155 | 150 | 140 |

[a]TPO made with Profax 6501 and constant 15 wt% Polysar 306. Flexural modulus (kpsi), mica in TPO.
Source: Ref. 41.

lene homopolymer than in the other systems. For 30 wt% loadings, the flexural modulus increase in homopolymer with L-135 mica was 212% versus 94% in copolymer and 97% in the TPO system.

## 14.4.2 Effects on HDT

Effects of different loading levels of several mica products on HDT are shown in Tables 14.11, 14.12, and 14.13. Increasing the amount of mica always increases HDT of the composite. The same mica products that gave the largest increases in

TABLE 14.11 Effect of Mica Concentration on Heat Deflection Temperature of Polypropylene Homopolymer Composites

| | Profax 6501 homopolymer | | | |
|---|---|---|---|---|
| Muscovite mica, grade | 12 wt% | 20 wt% | 30 wt% | 40 wt% |
| C-3000 | 163 | 183 | 200 | 225 |
| HiMod-270 | 171 | 184 | 222 | 225 |
| WG-325 | 169 | 186 | 218 | 242 |
| 4K | 162 | 180 | 212 | 235 |
| L-140 | 184 | 205 | 218 | 247 |
| L-135 | 173 | 182 | 212 | 244 |
| PP, 100 wt% | 150 | 150 | 150 | 150 |

HDT (°F) @ 264 psi, mica in polypropylene homopolymer.
Source: Ref. 43.

## Mica Reinforcement of Polypropylene

TABLE 14.12 Effect of Mica Concentration on Heat Deflection Temperature of Polypropylene Copolymer Composites

| Muscovite mica, grade | Profax 8623 copolymer | | |
|---|---|---|---|
| | 12 wt% | 20 wt% | 30 wt% |
| PM-325 | 138 | 146 | 148 |
| C-3000 | 131 | 140 | 154 |
| HiMod-270 | 146 | 152 | 171 |
| WG-325 | 136 | 142 | 156 |
| 4K | 144 | 153 | 163 |
| L-140 | 137 | 143 | 164 |
| L-135 | 135 | 145 | 165 |
| PP, 100 wt% | 139 | 139 | 139 |

HDT (°F) at 264 psi, mica in polypropylene comopolymer.
Source: Ref. 41.

flexural modulus also produced the highest increases in HDT. With 30 wt% L-135 mica, HDT increased 41% with the homopolymer, 19% with the copolymer, and 20% in the TPO composites. Lowest increases were obtained with C-3000, which has the lowest aspect ratio of the products tested.

### 14.4.3 Effects on Flexural Strength

Flexural strength improvements, with several mica products are listed in Tables 14.14, 14.15, and 14.16. Mica was much less effective in increasing flexural

TABLE 14.13 Effect of Mica Concentration on Heat Deflection Temperature of TPO Composites

| Muscovite mica, grade | TPO[a] | | |
|---|---|---|---|
| | 10 wt% | 20 wt% | 30 wt% |
| PM-325 | 131 | 134 | 130 |
| C-3000 | 136 | 147 | 156 |
| HiMod-270 | 144 | 157 | 171 |
| WG-325 | 137 | 145 | 157 |
| 4K | 132 | 143 | 159 |
| L-140 | 136 | 143 | 165 |
| L-135 | 134 | 146 | 160 |
| TPO, 100 wt% | 138 | 137 | 133 |

[a]TPO made with Profax 6501 and constant 15 wt% Polysar 306. (HDT) (°F) at 264 psi, mica in TPO.
Source: Ref. 41.

TABLE 14.14 Effect of Mica Concentration on Flexural Strength of Polypropylene Homopolymer Composites

| Muscovite mica, grade | Profax 6501 homopolymer | | | |
|---|---|---|---|---|
| | 12 wt% | 20 wt% | 30 wt% | 40 wt% |
| C-3000 | 7910 | 8170 | 8430 | 8420 |
| HiMod-270 | 8030 | 8270 | 8590 | 8080 |
| WG-325 | 7910 | 8210 | 8560 | 8240 |
| 4K | 7620 | 7770 | 8220 | 8000 |
| L-140 | 8250 | 8625 | 8410 | 7980 |
| L-135 | 7770 | 7740 | 7670 | 7190 |
| PP, 100 wt% | 7360 | 7360 | 7360 | 7360 |

Flexural strength, psi, mica in polypropylene homopolymer.
Source: Ref. 43.

strength than in increasing flexural modulus. For homopolymer, copolymer, and TPO composites, loadings of 30 wt% C-3000 mica increased flexural strength 14.5%, 16.6%, and 21.6%, respectively. The higher aspect ratio micas, such as HiMod-270. L-140, and L-135, gave less improvement than C-3000 except in the TPO formulation. In TPO. 30 wt% HiMod-270 improved flexural strength 30% versus 21.6% for C-3000.

TABLE 14.15 Effect of Mica Concentration on Flexural Strength of Polypropylene Copolymer Composites

| Muscovite mica, grade | Profax 8623 copolymer | | |
|---|---|---|---|
| | 12 wt% | 20 wt% | 30 wt% |
| PM-325 | 5290 | 5260 | 5360 |
| C-3000 | 5190 | 5350 | 5570 |
| HiMod-270 | 4960 | 5030 | 5220 |
| WG-325 | 5060 | 5280 | 5460 |
| 4K | 5000 | 5095 | 5190 |
| L-140 | 4660 | 4870 | 5030 |
| L-135 | 4510 | 4630 | 4760 |
| PP, 100 wt% | 4775 | 4775 | 4775 |

Flexural strength, psi, mica in polypropylene copolymer.
Source: Ref. 41.

# Mica Reinforcement of Polypropylene

TABLE 14.16  Effect of Mica Concentration on Flexural Strength of TPO Composites

| Muscovite mica, grade | TPO[a] 10 wt% | 20 wt% | 30 wt% |
|---|---|---|---|
| PM-325 | 5210 | 5200 | 5090 |
| C-3000 | 5280 | 5260 | 5280 |
| HiMod-270 | 5680 | 5910 | 5640 |
| WG-325 | 5170 | 5020 | 5080 |
| 4K | 5240 | 5210 | 5130 |
| L-140 | 5140 | 5280 | 5135 |
| L-135 | 5245 | 4990 | 4730 |
| TPO, 100 wt% | 4790 | 4550 | 4340 |

[a]TPO made with Profax 6501 and constant 15% Polysar 306.
Flexural strength, psi, mica in TPO.
Source: Ref. 41.

## 14.4.4  Effects on Izod Impact

Tables 14.17, 14.18, and 14.19 show that addition of mica to polypropylene compounds has a negative effect on Izod impact properties. In each of the three resin systems, Izod impact decreases steadily with increasing amounts of mica. In homopolymer composites (Table 14.17), the largest particle size mica product, L-135, gives the highest Izod impact values. With 30 wt% L-135 mica in

TABLE 14.17  Effect of Mica Concentration on Izod Impact of Polypropylene Homopolymer Composites

| Muscovite mica, grade | Profax 6501 homopolymer 12 wt% | 20 wt% | 30 wt% | 40 wt% |
|---|---|---|---|---|
| C-3000 | 1.05 | 0.77 | 0.55 | 0.42 |
| HiMod-270 | 1.15 | 0.78 | 0.53 | 0.40 |
| WG-325 | 0.90 | 0.71 | 0.49 | 0.41 |
| 4K | 1.00 | 0.81 | 0.61 | 0.49 |
| L-140 | 1.18 | 0.93 | 0.60 | 0.45 |
| L-135 | 1.05 | 1.09 | 0.82 | 0.62 |
| PP, 100 wt% | 2.94 | 2.94 | 2.94 | 2.94 |

Izod impact, ft-lb/in., mica in polypropylene homopolymer.
Source: Ref. 43.

TABLE 14.18 Effect of Mica Concentration on Izod Impact of Polypropylene Copolymer Composites

|  | Profax 8623 copolymer |  |  |
| --- | --- | --- | --- |
| Muscovite mica, grade | 12 wt% | 20 wt% | 30 wt% |
| PM-325 | 9.7 | 7.0 | 4.3 |
| C-3000 | 6.8 | 5.5 | 3.7 |
| HiMod-270 | 12.1 | 7.5 | 2.8 |
| WG-325 | 9.6 | 6.6 | 4.0 |
| 4K | 9.4 | 6.8 | 2.8 |
| L-140 | 5.7 | 3.7 | 2.0 |
| L-135 | 5.9 | 3.5 | 1.9 |
| PP, 100 wt% | 15.4 | 15.4 | 15.4 |

Izod Impact, ft-lb/in., mica in polypropylene copolymer.
Source: Ref. 41.

homopolymer. Izod impact is 0.82 ft-lb/in., whereas the smaller C-3000 mica product gives an Izod impact of 0.55 ft-lb/in.

The smallest mica products give the best Izod impacts in copolymer and TPO composites. For 30 wt% filled copolymer and TPO, the Izod impact with PM-325 was 4.3 and 2.7 ft-lb/in., respectively. Izod impact values of 1.9 and 1.8 ft-lb/in. were obtained when 30 wt% of the larger L-135 mica was used.

TABLE 14.19 Effect of Mica Concentration on Izod Impact of TPO Composites

|  | TPO[a] |  |  |
| --- | --- | --- | --- |
| Muscovite mica, grade | 10 wt% | 20 wt% | 30 wt% |
| PM-325 | 6.4 | 4.0 | 2.7 |
| C-3000 | 6.4 | 3.7 | 2.5 |
| HiMod-270 | 4.6 | 2.4 | 1.4 |
| WG-325 | 7.7 | 4.1 | 2.2 |
| 4K | 4.9 | 3.0 | 2.1 |
| L-140 | 4.3 | 2.5 | 1.5 |
| L-135 | 4.3 | 2.9 | 1.8 |
| TPO, 100 wt% | 13.5 | 15.0 | 15.3 |

[a]TPO made with Profax 6501 and constant 15 wt% Polysar 306.
Izod impact, ft-lb/in., mica in TPO.
Source: Ref. 41.

TABLE 14.20  Effect of Mica Concentration on Gardner Impact of Polypropylene Homopolymer Composites

| Muscovite mica, grade | Profax 6501 homopolymer ||||
|---|---|---|---|---|
| | 12 wt% | 20 wt% | 30 wt% | 40 wt% |
| C-3000 | 6.7 | 5.5 | 4.6 | 2.8 |
| HiMod-270 | 7.9 | 5.7 | 3.4 | 2.0 |
| WG-325 | 6.9 | 5.5 | 3.4 | 3.0 |
| 4K | 7.3 | 5.1 | 3.7 | 2.8 |
| L-140 | 6.8 | 5.2 | 3.7 | 2.6 |
| L-135 | 4.8 | 4.6 | 4.2 | 3.6 |
| PP, 100 wt% | 16.4 | 16.4 | 16.4 | 16.4 |

Gardner impact, in.-lb, mica in polypropylene homopolymer.
*Source:* Ref. 43.

## 14.4.5  Effects on Gardner Impact

As noted for Izod impact above, Gardner impact properties also decrease steadily with increasing amounts of mica in the composite (Tables 14.20–14.22). The micas that provide highest increases in flexural modulus and HDT give the lowest Gardner impact properties. For 30 wt% filled systems. the relatively large L-135 gave Gardner impact values of 4.2, 14.8, and 3.7 in.-lb in homopolymer, copolymer, and TIPO, respectively. In similar systems with the small particle size C-3000, Gardner impact values were 4.6, 39.9, and 7.8 in.-lb. The higher

TABLE 14.21  Effect of Mica Concentration on Gardner Impact of Polypropylene Copolymer Composites

| Muscovite mica, grade | Profax 8623 copolymer |||
|---|---|---|---|
| | 12 wt% | 20 wt% | 30 wt% |
| PM-325 | 149.0 | 139.0 | 13.5 |
| C-3000 | 57.7 | 45.5 | 39.9 |
| HiMod-270 | 238.0 | 88.0 | 5.3 |
| WG-325 | 55.5 | 36.1 | 35.9 |
| 4K | 224.0 | 108.0 | 4.6 |
| L-140 | 74.8 | 54.4 | 15.2 |
| L-135 | 61.1 | 47.8 | 14.8 |
| PP, 100 wt% | NB | NB | NB |

Gardner impact, in.-lb, mica in polypropylene copolymer.
*Source:* Ref. 41.

TABLE 14.22  Effect of Mica Concentration on Gardner Impact of TPO Composites

| Muscovite mica, grade | TPO[a] | | |
| --- | --- | --- | --- |
| | 10 wt% | 20 wt% | 30 wt% |
| PM-325 | 50.2 | 28.4 | 16.9 |
| C-3000 | 40.6 | 16.9 | 7.8 |
| HiMod-270 | 14.0 | 10.2 | 5.5 |
| WG-325 | 152.5 | 20.7 | 6.3 |
| 4K | 53.2 | 22.7 | 6.9 |
| L-140 | 127.8 | 21.0 | 6.9 |
| L-135 | 28.2 | 7.9 | 3.7 |
| TPO, 100 wt% | NB | NB | NB |

[a]TPO made with Profax 6501 and constant 15 wt% Polysar 306.
Gardner impact, in.-lb, mica in TPO.
*Source:* Ref. 41.

Gardner impact values obtained in polypropylene copolymer are quite significant and would direct one to this system when high Gardner impact is required.

## 14.5 EFFECTS OF PARTICLE SIZE ON PERFORMANCE PROPERTIES

Mica particle size has significant effects on most composite properties. Table 14.23 shows these effects for a 30 wt% filled polypropylene copolymer composite. It is not possible to get good trends from this table based on particle size only. These products were prepared by different methods including wet grinding, dry grinding, and screening of spiral mica. These production processes provide different degrees of and different aspect ratios for particles of similar diameter.

Processing into polypropylene compounds tends to make the median particle sizes more similar. Before processing, the median size varied from 22.1 to 266 μm. After processing, the median size varied from 18.5 to 96.3 μm. Relatively small mica products give lower flexural modulus and HDT except for the very high aspect ratio HiMod-270, which gives the highest flexural modulus, 495 kpsi, and highest HDT, 171°C.

Mica products with smallest particle size, PM-325 and C-3000, provide highest flexural strength and Izod impact values. For example, C-3000 and HiMod-270 at 30 wt% loading in the copolymer give flexural strength values of 5570 and 4760 psi, respectively. Izod impact for the C-3000 composite was 3.7 versus 1.9 ft-lb/in. for the HiMod-270 composite.

TABLE 14.23  Mica Particle Size Effects on Polypropylene Composite Properties

| Muscovite mica, grade | Initial D(v, 0.5) (μm) | Processed D(v, 0.5) (μm) | Flexural modulus (kpsi) | Flexural strength (psi) | HDT at 264 psi (°F) | Izod impact (ft-lb/in.) | Gardner impact (in.-lb) |
|---|---|---|---|---|---|---|---|
| PM-325 | 22.1 | 18.5 | 420 | 5360 | 148 | 4.3 | 13.5 |
| C-3000 | 24.0 | 21.9 | 380 | 5570 | 154 | 3.7 | 39.9 |
| HiMod-270 | 34.6 | 26.4 | 495 | 5220 | 171 | 2.8 | 5.3 |
| WG-325 | 37.0 | 27.7 | 480 | 5460 | 156 | 4.0 | 35.9 |
| 4K | 50.0 | 32.8 | 460 | 5190 | 163 | 2.8 | 4.6 |
| L-140 | 86.0 | 40.2 | 470 | 5030 | 164 | 2.0 | 15.2 |
| L-135 | 266.0 | 96.3 | 490 | 4760 | 165 | 1.9 | 14.8 |

Effects of mica particle size on 30 wt% filled polypropylene copolymer, Profax 8623.
Source: Ref. 41.

Gardner impact differences are mixed with no clear trends. The highest Gardner impact (39.9 in.-lb) was obtained with C-3000 and the lowest with 4K (4.6 in.-lb).

## 14.6 WARPAGE CONTROL

### 14.6.1 Warpage Causes

Part warpage is caused by differential shrinkage within the molded part when the part is removed from the mold. Many factors are involved. Nonuniform part cooling, high internal stresses, molecular orientation of polymeric chains, packing of polymers during the injection or packing phase of the molding cycle, and orientation of fillers are some of the major conditions that cause parts to warp.

Differential shrinkage can be divided into three categories: area shrinkage, average of directional shrinkage values for a given area of the part; cross-sectional shrinkage, shrinkage across the part thickness; and directional shrinkage, differences in shrinkage between parallel and perpendicular directions [11]. These issues have been the subject of numerous theoretical and empirical investigations [12–14].

### 14.6.2 Methods of Reducing Warpage

Warpage can be minimized using a number of different approaches that include mold design (e.g., uniform wall section, radiused corners, smooth melt flow), process conditions (e.g., injection pressure, mold temperature, filling rate, cooling rate), and material formulation. In fiber-filled composites, addition of a flat platelet mineral such as mica can result in more uniform shrinkage in the flow and cross-flow directions, resulting in less warpage.

### 14.6.3 Mica Combinations with Glass Fiber

Blends of mica and glass fiber have been used to provide enhanced mechanical properties and reduced warpage in polypropylene [15–19]. A paper presented at the 1997 SPE ANTEC discussed the effect of combinations of glass fiber and mica on the physical properties and dimensional stability of injection-molded polypropylene composites [8]. Warpage was least with high levels of mica and highest with high levels of glass fiber with no mica present. Table 14.24 shows the warpage range measured in millimeters at six different locations on the test plaque for different combinations of glass fiber and mica. One of the conclusions of this paper was that "dimensional stability or warpage is directly related to the mica content. Higher levels of mica result in flatter more dimensionally stable parts." A technical brochure produced by Microfine Minerals [20] contained a study of warpage reduction by mica in polypropylene. It was stated that "a large

# Mica Reinforcement of Polypropylene

TABLE 14.24 Warpage of Fiberglass–Mica Polypropylene Composites: Warpage Range[a] for Combinations of Fiberglass and Muscovite Mica

| Polypropylene Profax 6523 (wt%) | Mica L-135 (wt%) | Fiberglass (wt%) | Warpage range (mm) |
|---|---|---|---|
| 80 | 0 | 20 | 13.1 |
| 80 | 10 | 10 | 11.0 |
| 60 | 0 | 40 | 8.6 |
| 65 | 7.5 | 27.5 | 7.4 |
| 70 | 15 | 15 | 7.3 |
| 60 | 20 | 20 | 3.1 |
| 80 | 20 | 0 | 2.1 |
| 70 | 30 | 0 | 1.7 |
| 60 | 40 | 0 | 1.1 |

[a]Difference between minimal and maximal value, 152 × 228 × 4 mm plaque.
*Source:* Ref. 8.

reduction in warpage can be achieved by only a small replacement of the glass. In this case, a replacement of one sixth of the glass was required to give a 60% reduction in warpage."

## 14.7 EFFECTS OF ANNEALING ON COMPOSITE PROPERTIES

It is widely known that conditions used in injection molding can have significant effects on molded-in stresses. Locked-in stresses are created in composites by use of excessive molding pressures, low melt temperature, uneven cooling, and cooling too quickly. The effect of varying molding stresses may be eliminated by annealing in accordance with supplier recommendations [21]. The required temperature is related to the manufacturing process and the phase transition in the base resin. In film applications, it is easy to see the effect of reducing internal stresses because the film shrinks considerably after annealing.

It has been theorized that the internal stresses in molded plastics and composites are the result of freeze-up of the molded part before the molecules have a chance to arrange themselves in the most relaxed configuration. This frozen condition would prohibit movement of polypropylene molecules into positions for better bonding. Less intermolecular bonding should result in lower flexural and tensile properties clue to the relative ease of separating the molecules. The specimen size will also be somewhat larger because of suboptimal association, which translates to more space between molecules. With built-in stresses, one would expect an exaggerated change in specimen dimensions during the heat deflection test as the specimen shrinks at higher temperatures and the resin molecules become more highly bonded. The observed heat deflection of the specimen would be a combination of shrinkage and actual bending under the applied load.

Heat treatment, often referred to as annealing, will decrease the viscosity of the composite and increase movement of the individual molecules because of the higher kinetic energy. This increased mobility of the polymer molecules allows the molecules to move into a more highly bonded lower energy arrangement with corresponding increases in tensile and flexural properties. Preshrinkage of test specimens is expected to produce significant increases in HDT because the shrinkage caused by stress relief has already occurred.

### 14.7.1 Effects of Annealing Temperature

A paper given to the 48th Annual SPE Conference [22] reported that annealing composites composed of HiMont's Profax 6523 and 40 wt% mica at 150°C for 15 min gave excellent increases in HDT. The HDTs for various annealing temperatures are shown in Table 14.25. Obviously, the annealing temperature must be changed depending on the softening point of the specific resin in the composite.

### 14.7.2 Annealing Effects on Flexural Strength

HAR-160 mica was modified with a proprietary silane coupling agent and tested in polypropylene. Table 6 of the previously referenced paper [22] gives data for several types of mica that had different particle size distributions. In general, the flexural strength of composites containing untreated mica decreases with increas-

TABLE 14.25 Effect of Annealing Temperature on Heat Deflection Temperature

| Annealing temp. (°C) | HDT after 5 min (°C) | HDT after 15 min (°C) | HDT after 30 min (°C) | HDT after 60 min (°C) |
|---|---|---|---|---|
| 100 | 114 | 117 | 117 | 117 |
| 125 | 123 | 127 | 128 | 129 |
| 150 | 129 | 131 | 129 | 130 |

Annealing effects on HDT, 1820-kPa load, 40% HAR-160 mica in Profax 6523 polypropylene.
Source: Ref. 22.

# Mica Reinforcement of Polypropylene

ing mica loading. With treated mica, flexural strength increases with increasing amounts of mica.

For composites containing untreated mica, annealed specimens with HAR-160 mica had flexural strength increases of 14.6%, 12.4%, and 9.1% for 20, 30, and 40 wt% loadings, respectively, relative to the flexural strength of unannealed specimens. Surface-treated HAR-160 mica gave flexural strength increases of 14.2%, 10.5%, and 8.3% for 20, 30, and 40 wt% loadings compared with unannealed specimens. In both cases, there is a trend toward smaller flexural strength increases in annealed specimens with higher loading levels of mica. Flexural strengths for 40 wt% filled composites containing untreated and treated mica varied from 56.5 to 87.4 MPa, respectively, in annealed specimens.

## 14.7.3 Annealing Effects on Flexural Modulus

Data for composites containing HAR-160 mica is shown in Table 14.26. In composites containing untreated mica, annealed specimens with HAR-160 mica had flexural modulus increases of 27.0%, 20.8%, and 15.7% for 20, 30, and 40 wt% loadings, respectively, when compared with unannealed specimens. The largest increases were obtained with the lowest amount of mica because there are more polypropylene molecules available to reorient to form more highly bonded configurations. Composites containing surface-treated HAR-160 mica had flexural strength increases of 31.6%, 18.0%, and 20.0% for 20, 30, and 40 wt%

TABLE 14.26  Annealing Effects on Flexural Strength and Modulus

| HAR-160 mica (wt%) | Flexural strength (MPa) Not annealed | Annealed | Property change (%) | Flexural modulus (GPa) Not annealed | Annealed | Property change (%) |
|---|---|---|---|---|---|---|
| | | | *Plain Mica* | | | |
| 20 | 57.0 | 65.3 | 14.6 | 3.7 | 4.7 | 27.0 |
| 30 | 56.3 | 63.3 | 12.4 | 5.3 | 6.4 | 20.8 |
| 40 | 51.8 | 56.5 | 9.1 | 7.0 | 8.1 | 15.7 |
| 50 | 48.3 | 51.3 | 6.2 | 8.9 | 10.6 | 19.1 |
| | | *Surface-Modified Mica–KMG Silane* | | | | |
| 20 | 65.7 | 75.0 | 14.2 | 3.8 | 5.0 | 31.6 |
| 30 | 75.3 | 83.2 | 10.5 | 6.1 | 7.2 | 18.0 |
| 40 | 80.7 | 87.4 | 8.3 | 8.0 | 9.6 | 20.0 |
| 50 | 79.8 | 83.8 | 5.0 | 11.2 | 12.0 | 7.1 |

HAR-160 mica in Profax 6523 polypropylene.
*Source:* Ref. 22.

loadings compared with unannealed specimens. In each case, there is a trend toward lower flexural modulus increases in annealed specimens with higher loading levels of mica. Flexural modulus values for 40 wt% filled composites containing untreated and treated mica varied from 8.1 to 9.6 GPa, respectively, in annealed specimens.

### 14.7.4 Annealing Effects on Heat Deflection Temperature

Heat deflection temperature data for composites containing HAR-160 mica are shown in Table 14.27. In composites containing untreated mica, annealed specimens with HAR-160 mica had HDT increases of 20.9%, 20.3%, and 21.2% for 20, 30, and 40 wt% loadings, respectively, compared with unannealed specimens.

Composites containing surface-treated HAR-160 mica exhibited HDT increases of 27.6%, 23.7%, and 17.1% for 20, 30, and 40 wt% loadings, respectively, when compared with unannealed specimens. It is interesting that the HDT increases obtained with untreated mica did not change with increased loading levels of mica, whereas the level of increase in the composites made with surface-treated mica decreased with increasing loading. The HDT values for 40 wt% filled composites containing untreated and treated mica varied from to 103°C to 130°C, respectively, in annealed specimens.

TABLE 14.27 Annealing Effects on HDT and Izod Impact

| HAR-160 mica (wt%) | HDT (°C) Not annealed | HDT (°C) Annealed | Property change (%) | Izod impact (J/m) Not annealed | Izod impact (J/m) Annealed | Property change (%) |
|---|---|---|---|---|---|---|
| | | | *Plain Mica* | | | |
| 20 | 67 | 81 | 20.9 | 19.8 | 23.5 | 18.7 |
| 30 | 74 | 89 | 20.3 | 21.4 | 23.5 | 9.8 |
| 40 | 85 | 103 | 21.2 | 20.3 | 20.3 | 0.0 |
| 50 | 99 | 113 | 14.1 | 20.8 | 20.3 | −2.4 |
| | | | *Surface-Modified Mica–KMG Silane* | | | |
| 20 | 76 | 97 | 27.3 | 18.2 | 17.6 | −3.3 |
| 30 | 97 | 120 | 23.7 | 21.9 | 20.3 | 7.3 |
| 40 | 111 | 130 | 17.1 | 24.0 | 20.8 | −13.3 |
| 50 | 118 | 130 | 10.2 | 20.3 | 21.4 | 5.4 |

HAR-160 mica in polypropylene homopolymer.
*Source:* Ref. 22.

## 14.7.5 Effects of Annealing on Izod Impact Strength

Izod impact and HDT data for composites containing HAR-160 mica are shown in Table 14.27. In composites containing untreated mica, annealed specimens with HAR-160 mica had Izod impact increases of 18.7%, 9.8%, and 0.0% for 20, 30, and 40 wt% loadings, respectively, when compared with values for unannealed specimens. In contrast, composites containing surface-modified HAR-160 mica had decreasing Izod impact values of 3.3%, 7.3%, and 13.3% for 20, 30, and 40 wt% loadings, respectively. Decreases observed with surface-modified mica may be the result of additional bonding between mica and polypropylene that would produce a more brittle material with less ability to absorb impact energy. Another possibility is that the surface treatment process left latent free radical generators on the mica that could cause additional cross-linking during compounding and molding. Izod impact values for 40 wt% filled composites containing untreated and treated HAR-160 mica varied from 20.3 to 20.8 J/m, respectively, in annealed specimens. The average Izod impact strength of unfilled polypropylene was 15.5 J/m for unannealed specimens and 17.6 J/m for annealed specimens.

In summary, heat treatment of mica-reinforced polypropylene composites can produce substantial performance improvements. Annealing may be imperative when highest mechanical and thermal properties are required.

## 14.8 EFFECTS OF SURFACE TREATMENTS ON COMPOSITE PROPERTIES

### 14.8.1 Interfacial Considerations

Mica is a hydrophilic (high surface energy) mineral because of the surface silanol (Si-OH) groups. Actual surface energy is 2400–5400 mJ/m$^2$. Polypropylene is of course a very hydrophobic resin with low surface energy that is not easily wet by water or any high-energy substance.

Good dispersion of mica into hydrophobic resins like polypropylene is difficult because of the hydrophillic nature of the mica surface. Poor interaction at the surface of mica and polypropylene results in lowered mechanical and thermal properties. These property reductions are caused by the absence of good interfacial bonding and also by the presence of air and water in the interfacial microvoids. Treatment of mica with organofunctional silanes will place a hydrophobic surface on mica that can form weak bonds with polypropylene. This phenomena is described in many papers and bulletins [23–27]. In one bulletin [27], coupling agents PC-1A and PC-1B (produced by OSi Specialties) were used in a 4:1 ratio. Table 14.28 lists the changes in mechanical properties obtained with mica pretreated with 1 and 2 wt% of a PC-1A/PC-1B blend.

TABLE 28 Effect of PC-1A/PC-1B Treatment Level on Mechanical Properties of Mica/Polypropylene Composites

| Property | Untreated mica | PC-1A/PC-1B 1 wt% | PC-1A/PC-1B 2 wt% |
|---|---|---|---|
| Tensile strength at yield, psi (MPa) | 4330 (29.85) | 5840 (40.26) | 5870 (40.47) |
| Elongation at yield, % | 5.5 | 7.5 | 9.5 |
| Flexural modulus (tan) psi (MPa × $10^3$) | 870 (6.0) | 850 (5.86) | 950 (6.55) |
| Flexural strength at yield, psi (MPa) | 7520 (51.85) | 10,280 (70.88) | 10,510 (72.46) |
| Izod impact strength, ft-lb/in. | 0.5 | 0.4 | 0.6 |
| Unnotched izod impact, ft-lb/in. | 2.3 | 4.0 | 6.3 |

40 wt% WG-2 muscovite mica in Profax 6523 pm homopolymer.
Source: Ref. 27.

TABLE 14.29 Surface Treatments and Additives Studied

| Material | Manufacturer | Level (wt%) |
|---|---|---|
| Aminosilane, A-1100 | Union Carbide | 1.0 |
| Azidosilane, AZ-Cup N | Hercules | 1.0 |
| Byk-W-980 | Byk Chemie | 1.0 |
| Cavco Mod NP | Cavedon Chemical Co. | 1.0 |
| Epolene E-43 wax | Eastman Chemical Products | 2.0 |
| Proprietary treatment | KMG Minerals | 1.0 |
| Neoalkoxy titanate, LICA-01 | Kenrich Petrochemicals | 1.0 |
| Neoalkoxy zirconate, LZ-12 | Kenrich Petrochemicals | 1.0 |
| Ucarsil PC-1A/PC-1B 3:1 ratio | Union Carbide | 1.0 |
| Ucarsil PC-1A/PC-1B 4:1 ratio | Union Carbide | 1.0 |
| Silane, code 4A-2167 | PCR Inc. | 2.0 |
| Silane, code 4B-2167 | PCR Inc. | 2.0 |
| Silane code 4-2167 | PCR Inc. | 2.0 |
| Polybond 1001-40MF | BP Performance Polymers | 16.7 |
| Polypro PC-072 PM | Himont, USA | 60.0 |

Source: Ref. 28.

# Mica Reinforcement of Polypropylene

## 14.8.2 Materials and Experimental Procedures

A study of the effects of 11 surface-treating agents and three modified polypropylene resins on the strength properties of mica-reinforced polypropylene was presented in a paper at the 1990 annual SPI conference [28]. HAR-160 mica, the mica used in the previous section on annealing, was also used in this study. The prolyproylene used in the study was Profax 6523 pm, which has a nominal melt index of 4. Surface treatments and treatment levels are listed in Table 14.29.

HAR-160 mica was modified with the various treatment chemicals in a Patterson Kelly blender to coat 1 wt% of the active treating material on the surface. Treated mica products were dried at 100°C for 2 hr before use. The two maleic anhydride modified polypropylene additives, Polybond 1001 and Polypro PC-072, were added to the composite formulation before compounding. Test specimens containing 40 wt% mica were produced by compounding in a twin-screw extruder followed by injection molding.

## 14.8.3 Results of Surface Treatment Study

Composite properties are listed in Table 14.30. Graphs for flexural strength, flexural modulus, heat deflection temperature, and Izod impact strength are shown in Figs. 14.2–14.5.

## 14.8.4 Effects of Treatments and Additives on Flexural Strength

Highest increases in flexural strength when comparing untreated mica with treated mica were obtained when HAR-160 was modified with PCR-1 and KMG silanes. PCR silane, code 4A-2167, increased the flexural strength from 7260 to 12,000 psi (a 65% increase), whereas the KMG silane increased flexural strength to 11.860 psi (63% increase). See Fig. 14.2 for a graph of all flexural strength data.

## 14.8.5 Effects of Treatments and Additives on Flexural Modulus

Flexural modulus changes are shown in Fig. 14.3. Performance improvements were not as dramatic as those obtained for flexural strength but were still substantial. Highest increases in flexural modulus were obtained with PC-072 and KMG, OSC 3:1, and A-1100 silanes. Flexural modulus increased from 980,000 to 1,140,000 psi with PC-072 (16% increase) and to 1,070,000 (9% increase) with the three varieties of silane-modified mica.

Effects of different loading levels of plain HAR-160 on the flexural modulus of polypropylene composites were reported. The derived equation for flexural modulus was flexural modulus $= 965 + 30.15 \times$ wt% HAR-160 mica in

TABLE 14.30 Effects of Surface Treatments and Additives on Composite Properties: Mechanical Properties of Polypropylene[a] Composites Containing 40 wt% HAR-160 Muscovite Mica

| Treatment or additive (wt%) | Flexural strength (psi) | Flexural modulus (kpsi) | Tensile strength (psi) | Izod (ft-lb/in.) | HDT@264 psi (°F) |
|---|---|---|---|---|---|
| Untreated HAR-160 | 7260 | 980 | 4130 | 0.39 | 199 |
| PC-072, 60 | 9700 | 1140 | 5920 | 0.30 | 225 |
| KMG, 1.0 | 11,860 | 1070 | 6940 | 0.43 | 241 |
| PC-1A/PC-1B, 1.0, 3:1 | 10,630 | 1070 | 6350 | 0.39 | 230 |
| A-1100, 1.0 | 9720 | 1070 | 5900 | 0.39 | 224 |
| PC-1A/PC-1B, 1.0, 4:1 | 10,670 | 1050 | 6370 | 0.38 | 229 |
| PCR, 4A-2167, 2.0 | 12,000 | 1050 | 6720 | 0.42 | 236 |
| AZ-Cup N, 1.0 | 11,110 | 1030 | 6390 | 0.41 | 228 |
| P-Bond 1001, 16.7 | 10,650 | 1010 | 6470 | 0.42 | 240 |
| PCR, 4-2167, 2.0 | 11,770 | 1000 | 6670 | 0.42 | 235 |
| LZ-12, 1.0 | 7630 | 980 | 4510 | 0.40 | 209 |
| PCR, 4B02167, 2.0 | 11,630 | 970 | 6520 | 0.42 | 229 |
| Epolene E-43, 2.0 | 8390 | 930 | 5260 | 0.42 | 224 |
| Lica-01, 1.0 | 7150 | 920 | 4100 | 0.42 | 196 |
| BYK-W-980, 1.0 | 7190 | 910 | 4000 | 0.45 | 182 |
| Cavco Mod NP, 1.0 | 7110 | 820 | 4090 | 0.43 | 207 |

[a]Himont's Profax 6523 pm, polypropylene homopolymer, MI = 4.0.
Source: Ref. 28.

the composite. This equation was an excellent fit for the data as shown by a correlation coefficient of nearly 1.0.

### 14.8.6 Effects of Treatments and Additives on Heat Deflection Temperature

Heat deflection temperature changes are shown in Fig. 14.4. Highest heat deflection temperatures were obtained with KMG silane treatment and by addition of an acrylic acid–modified polypropylene. The 241°F and 240°F heat deflection temperatures are approximately 20% higher than observed with untreated mica.

### 14.8.7 Effects of Treatments and Additives on Izod Impact Strength

Changes in Izod impact strength are shown in Fig. 14.5. Treatments and additives had little effect on Izod impact values, which varied from 0.30 ft-lb/in. with PC-072 to 0.45 ft-lb/in. with BYK-W-980. Ineffectiveness of additives and surface

# Mica Reinforcement of Polypropylene

**TREATMENT AND ADDITIVE EFFECTS ON FLEXURAL STRENGTH**
**40% HAR-160 Mica in Profax 6523 Polypropylene**

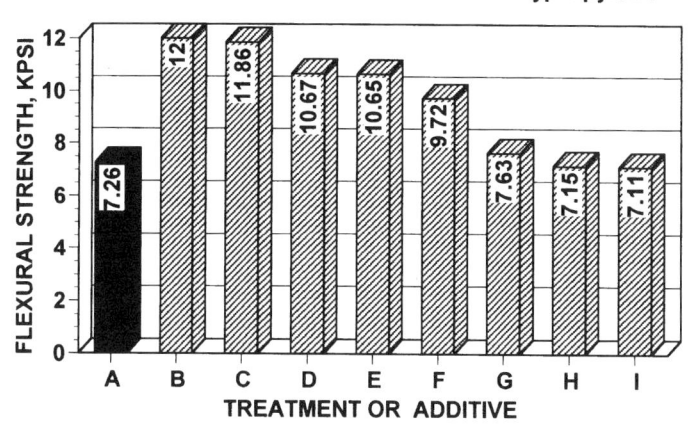

FIGURE 14.2  Treatment and additive effects on flexural strength. 40% HAR-160 mica in Profax 6523 polypropylene. A, untreated mica; B, PCR silane, Code 4A-2167, 2%; C, KMG silane, 1%; D, 3:1; OSC silane, 1%; E, Polybond 1001, 16.7%; F, aminosilane, A-1100, 1%; G, neoalkoxy zirconate, LZ-12, 1%, H, neoalkoxy titanate, LICA-01, 1%; I, Cavco Mod NP, 1%.

**TREATMENT AND ADDITIVE EFFECTS ON FLEXURAL MODULUS**
**40% HAR-160 Mica in Profax 6523 Polypropylene**

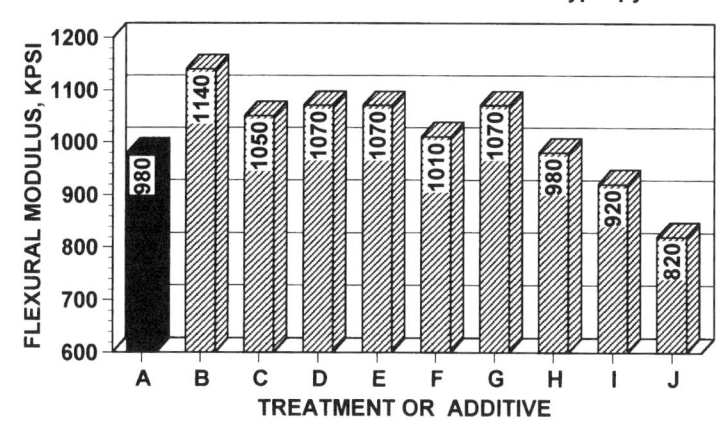

FIGURE 14.3  Treatment and additive effects on flexural modulus. 40% HAR-160 mica in profax 6523 polypropylene. A, untreated mica; B, Polypro PC 072 pm, 60%; C, PCR silane, Code 4A-2167, 2%; D, KMG silane, 1%; E, 3:1 OSC silane, 1%; F, Polybond 1001, 16.7%; G, aminosilane, A-1100, 1%; H, neoalkoxy zirconate, LZ-12, 1%; I, neoalkoxy titanate, LICA-01, 1%, J, Cavco Mod NP, 1%.

**TREATMENT AND ADDITIVE EFFECTS ON HDT**
**40% HAR-160 Mica in Profax 6523 Polypropylene**

Bar chart values (HDT, 264 psi load, F): A, 199; B, 240; C, 236; D, 241; E, 230; F, 225; G, 224; H, 209; I, 196; J, 207.

FIGURE 14.4 Treatment and additive effects on HDT. 40% HAR-160 mica in Profax 6523 polypropylene. A, Untreated mica; B, Polybond 1001, 16.7%; C, PCR silane, Code 4A-2167, 2%; D, KMG silane, 1%; E, 3:1 OSC silane, 1%; F, Polypro PC 072 pm, 60%; G, amino silane, A-1100, 1%; H, neoalkoxy zirconate, LZ-12, 1%; I, neoalkoxy titanate, LICA-01, 1%; J, Cavco Mod NP, 1%.

**TREATMENT AND ADDITIVE EFFECTS ON IZOD IMPACT STRENGTH**
**40% HAR-160 Mica in Profax 6523 Polypropylene**

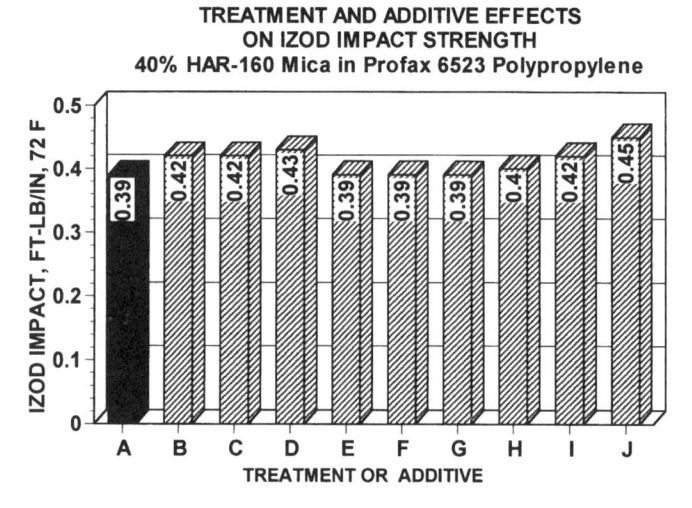

Bar chart values (IZOD IMPACT, FT-LB/IN, 72 F): A, 0.39; B, 0.42; C, 0.42; D, 0.43; E, 0.39; F, 0.39; G, 0.39; H, 0.4; I, 0.42; J, 0.45.

FIGURE 14.5 Treatment and additive effects on Izod impact. 40% HAR-160 mica in profax 6523 polypropylene. A, untreated mica; B, Polybond 1001, 16.7%; C, PCR silane, Code 4A-2167, 2%; D, KMG silane, 1%; E, 3:1 OSC silane, 1%; F, Polypro PC 072 pm, 60%; G, aminosilane, A-1100, 1%; H, neoalkoxy zirconate, LZ-12, 1%; I, neoalkoxy titanate, LICA-01, 1%; J, BYK W-980, 1%.

treatments on Izod impact suggests that the main cause of low Izod impact values is the high notch sensitivity of polypropylene.

### 14.8.8 Effects of Surface Treatment on Long-Term Heat Aging

Organosilicone chemicals were shown to increase the long-term heat aging characteristics of mica/polypropylene composites [29]. An article [29] compared the effects of plain mica with azidosilane and Ucarsil PC-1A/lB silane (1.5 wt%/0.5 wt%) treatments. Table 14.31 shows that with phlogopite mica, Ucarsil PC-1A/1B treatment provided greater crazing resistance than the azidosilane when specimens were aged at 150°C. The time to 10% crazing was 41–54 days for the Ucarsil-treated mica versus 26–30 days for the azidosilane treatment and 26–34 days for mica without any treatment. It should be noted that the azidosilane product is no longer available.

### 14.8.9 Effect of Surface Treatment on Charpy Impact Strength

Talc is sometimes preferred to mica because of higher impact properties. Information published by Briggs et al. [30] shows that use of 30 wt% mica pretreated with PC-1A/PC-1B coupling agents gives composites with higher flexural modulus and unnotched Charpy impact than composites containing 40 wt% talc. Data from this study are shown in Table 14.32.

TABLE 14.31 Azidosilane and Ucarsil Silane Effects on Heat Aging of Mica/Polypropylene Composites: Long-Term Heat Aging Efffects on Crazing of Mica[a]–Polypropylene[b] Composites Held at 150°C

|  | 20 wt% | | | 40 wt% | | |
| --- | --- | --- | --- | --- | --- | --- |
| Factor | None | Azido-silane | PC-1A/PC-1B 1.5 wt%/0.5 wt% | None | Azido silane | PC-1A/PC-1B 1.5 wt%/0.5 wt% |
|  | *Time to Crazing (days)* | | | | | |
| Tensile bars | 21 | 21 | 34 | 26 | 26 | 47 |
| Flexural modules bars | 26 | 26 | 34 | 34 | 30 | 54 |
| Gardner discs | 19 | 19 | 30 | 26 | 26 | 41 |

[a]Suzorite 60 HK.
[b]Profax 6523 polypropylene homopolymer, MI = 4.0.
*Source:* Ref. 44.

TABLE 14.32 Impact Comparisons for Talc- and Surface-Modified Mica in Polypropylene

|  | Tensile strength (MN/m$^2$) | Flexural strength (MN/m$^2$) | Notched Charpy (kJ/m$^2$) | Unnotched Charpy (kJ/m$^2$) |
|---|---|---|---|---|
| Homopolymer |  |  |  |  |
| 40 wt% talc | 34.0 | 53.7 | 2.35 | 11.1 |
| 30 wt% mica | 40.7 | 64.2 | 1.60 | 9.95 |
| 30 wt% mica, OSC[a] | 45.6 | 71.2 | 1.97 | 13.6 |
| Copolymer |  |  |  |  |
| 40 wt% talc | 27.3 | 44.2 | 3.67 | 16.1 |
| 30 wt% mica | 30.6 | 48.3 | 3.48 | 12.4 |
| 30 wt% misa, OSC* | 35.7 | 57.5 | 3.11 | 18.4 |

[a]Mica treated with Ucarsil PC-1A/PC-1B silanes.
*Source:* Ref. 30.

## 14.8.10 Summary

The best surface treatments for mica are the organofunctional silanes made by PCR Inc., by Witco's Osi silicones division, and the proprietary material made by KMG Minerals. (Note: KMG Minerals was acquired by Franklin Industrial Minerals in 1994.) More will be said about the use of anhydride–modified polypropylene additives in Section 14.9. One must decide whether to use a surface-modified mica or an additive such as maleic anhydride–modified polypropylene when increased performance properties are needed. Because of the manufacturing and chemical costs associated with producing surface-modified mica, modified mica products are considerably more expensive than plain mica. And maleic anhydride–modified polypropylenes are likewise more expensive than polyprol pylene.

A third choice used by some compounders is addition of the silane to the extruder or to the mica just before it enters the extruder. This requires good control of addition rates and provision for venting evolved volatile material. The best choice can only be made after the relative costs of these approaches are known. The amounts of surface treatment can be varied but usually range from 0.4 to 2.0 wt%. The amount of maleic anhydride–modified polypropylene used in the formulation can also be adjusted over a fairly wide range. The decision as to which approach is better depends entirely on the particular performance properties required because this will determine the amount of surface-treated mica or maleic anhydride–modified polypropylene in the compound. The best way to make an informed decision is to work closely with the suppliers of

# Mica Reinforcement of Polypropylene

surface-treated mica and maleic anhydride–modified polypropylene to find the best cost/performance balance.

## 14.9 INCREASING PERFORMANCE PROPERTIES WITH ADDITIVES

Reinforcement of polypropylene with mica provides outstanding flexural modulus and heat deflection temperature, minimizes warpage, and gives good dimensional stability. Often, however, impact properties are decreased. A paper given to the SPE Polyolefins X Conference in 1997 [31] focused on improvement of the impact–stiffness balance of mica-reinforced polypropylene by use of additives.

### 14.9.1 Use of Maleic Anhydride-Modified Polypropylene

Anhydride-modified polypropylene is used widely to improve mechanical properties of mica-polypropylene composites. Many excellent papers discuss the use of anhydride-modified polypropylene for property enhancement [32–36]. These additives contain polar carboxyl groups on the resin surface that wet out the hydrophillic mica surfaces, whereas the nonpolar portion of the molecule bonds well via van der Waals forces with polypropylene. The paper [31] gave results of tests conducted with seven different maleic anhydride–modified polypropylene homopolymers and six anhydride-modified polypropylene copolymers and rubbers.

### 14.9.2 Other Additives Studied

Three other additives, commonly referred to as processing aids, were tested to see if they would improve impact properties while maintaining most of the original stiffness [31]. Processing aids function by improving movement of the melted polymer against itself and the machine. They can also serve as mineral wetting agents, melt promoters, and impact modifiers.

An excellent paper [37] compares the benefits of a fatty acid amide derivative used with talc-filled polypropylenes. The most effective additives and specific resins used are identified in Table 14.33.

### 14.9.3 Experimental

Polypropylene homopolymer, polypropylene copolymer, and a polypropylene elastomer blend (TPO) were melt compounded with 2.0, 4.0, and 6.0 wt% additive. Additional tests with additive levels of 0.5, 1.0, and 1.5 wt% were made with Struktol TR-016. HiMod-360 mica was used at 30 wt% levels in all formulations, and there was a constant 0.2 wt% level of a stabilizer package in each compound. The maleic anhydride–modified polypropylene most effective in

TABLE 14.33 Property Changes (% Change) with Most Effective Additives for PP-HP, PP-CP, and TPO Resins

| Additive | Flexural modulus | Izod impact | Unnotched Izod | Gardner impact | HDT | Flexural strength | Tensile strength |
|---|---|---|---|---|---|---|---|
| Polypropylene homopolymer | | | | | | | |
| Struktol TR-016, 1 wt% | −6 | +90 | +115 | +88 | −12 | −14 | −14 |
| Struktol TR-016, 2 wt% | −15 | +150 | +200 | +221 | −24 | −17 | −18 |
| Polybond 3150, 6 wt% | +2 | −4 | +33 | +43 | +23 | +26 | +38 |
| Unite MP-1000, 6 wt% | 0 | +6 | +115 | +55 | +23 | +26 | +38 |
| Polypropylene copolymer | | | | | | | |
| Struktol TR-016, 2 wt% | +73 | +37 | +44 | +272 | −13 | −10 | −10 |
| Struktol TR-016, 4 wt% | +11 | +43 | +133 | +201 | −14 | −9 | −13 |
| Unite MP-880, 6 wt% | +11 | −61 | −23 | +37 | +36 | +53 | +56 |
| Fusabond P MZ-109D, 4 wt% | +26 | −57 | +47 | +23 | +15 | +46 | +84 |
| TPO | | | | | | | |
| Struktol TR-016, 0.5 wt% | +5 | +55 | +127 | +376 | −7 | −7 | −10 |
| Struktol TR-016, 2 wt% | +7 | +70 | +196 | +67 | −15 | −8 | −8 |
| Unite MP-1000, 6 wt% | +38 | −25 | +145 | +190 | +37 | +55 | +50 |

Struktol TR-016 is produced by Struktol Co. of America; Unite MP-1000 and MP-880 are produced by Aristech Chemical Corp; Fusabond P, Grade MZ-109D is produced by DuPont Canada, Inc.
Source: Ref. 31.

increasing performance properties was Unite MP-1000, but many other similar products made by a variety of companies performed nearly as well. To select the very best product, one must consider the relative costs of such additives and required use levels.

### 14.9.4 Impact Improvements in Polypropylene Homopolymer

*Izod Impact.* Best increases in notched Izod impact properties for polypropylene homopolymer composites were obtained with 1 wt% Struktol TR-016 while maintaining good flexural modulus. The notched Izod values increased from 25.6 to 48.6 J/m, and there was only a 5% decline in flexural modulus (5.72–5.44 GPa).

*Unnotched Izod Impact.* All anhydride additives worked well in increasing unnotched Izod impact properties. Unnotched Izod impact was improved 115% (144.2 to 309.7 J/m) with 6 wt% Unite MP-l000, flexural modulus was not effected, and heat deflection temperature increased from 100.6°C to 123.9°C. Addition of 2 wt% Struktol TR-016 gave the highest unnotched Izod value (432.5 J/m), but flexural modulus was decreased about 14%.

*Gardner Impact.* Addition of 6 wt% Unite MP-1000 gave a 55% increase in Gardner impact (0.42–0.65 J) and HDT increased 23% (100.6–123.9°C). As observed for unnotched Izod changes in the previous paragraph, 2 wt% Struktol TR-016 gave the best Gardner impact (1.35 J), but flexural modulus decreased by 13% and HDT decreased by 24%. Figure 14.6 shows the Izod, notched Izod, and Gardner impact changes in polypropylene homopolymer obtained with the best additives.

### 14.9.5 Impact Improvements in Polypropylene Copolymer

*Izod and Unnotched Izod.* For polypropylene copolymer. higher levels of Struktol TR-016 were needed to reach the best impact properties. Substantial increases of Izod and unnotched Izod impact properties were obtained with 4 wt% Struktol TR-016. (See Fig. 14.7 for a graph of Izod impact changes.) Izod impact was improved almost 50% (160.2–229.6 J/m) and unnotched Izod was improved from 304 to 710 J/m (130% increase). This was accomplished while also increasing flexural modulus from 3.34 to 3.72 GPa.

*Gardner Impact.* A 270% increase (1.45–5.4 J) in Gardner impact strength of a polypropylene copolymer compound was obtained by addition of 2 wt% Struktol TR-016. Flexural modulus was also increased about 10% (3.34–3.79 GPa). Unite MP-880 produced a 40% increase in Gardner impact while also increasing flexural and tensile strength about 50%. However, Unite MP-880

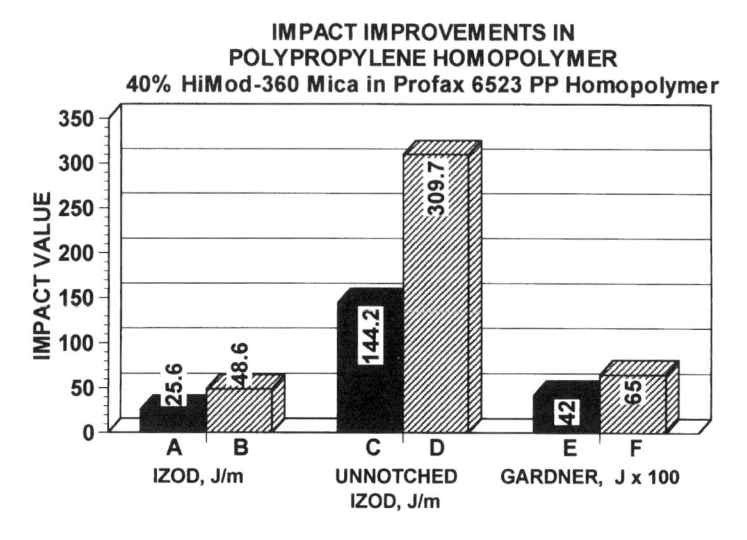

**FIGURE 14.6** Impact improvements in polypropylene homopolymer. 30% HiMod-360 mica in Profax 6501 homopolymer. A, Untreated mica; B, 1% Struktol TR-016; C, 6% Unite MP-1000; D, 6% Unite MP-1000.

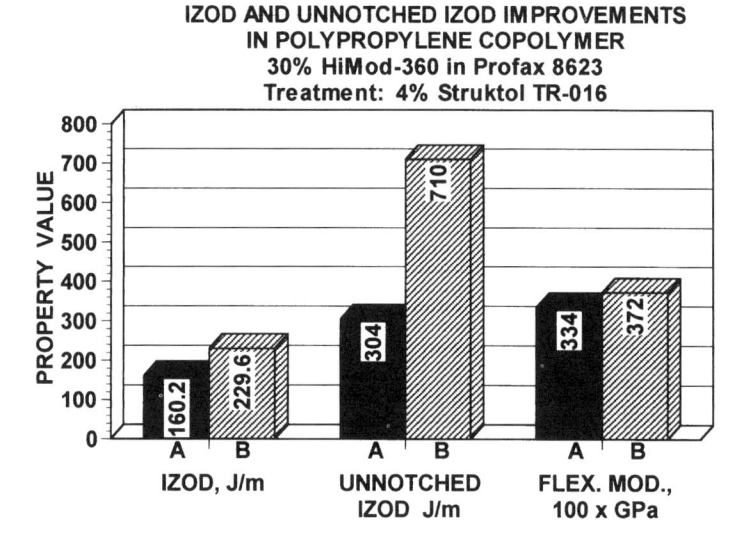

**FIGURE 14.7** Izod and unnotched Izod improvements in polypropylene copolymer. 30% HiMod-360 mica in profax 8623 copolymer. (From Ref. 30.) A, No additive; B, 4% Struktol TR-016.

# Mica Reinforcement of Polypropylene

**GARDNER IMPACT IMPROVEMENTS IN POLYPROPYLENE COPOLYMER**
**30% HiMod-360 Mica in Profax 8623 Polypropylene**

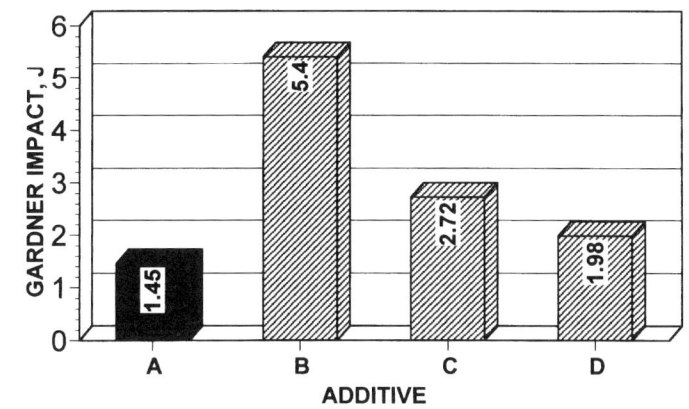

A: 1.45; B: 5.4; C: 2.72; D: 1.98

**FIGURE 14.8** Gardner impact improvements in polypropylene copolymer. 30% HiMod-360 mica in Profax 8623 copolymer. (From Ref. 30.) A, No additive; B, 2% Struktol TR-016; C, 1% Struktol TR-016; D, 6% Unite MP-880.

caused a 60% decrease in Izod impact. Figure 14.8 is a graph of Gardner impact changes observed in polypropylene copolymer. See Table 14.34 for flexural strength, flexural modulus, tensile strength, and HDT improvements obtained with additives.

### 14.9.6 Impact Improvements in a TPO Formulation

TPO compounds containing 30 wt% HiMod-360 mica, 15 wt% Polysar EPM 306 (p), and polypropylene homopolymer were compounded to contain 2, 4, and

**TABLE 14.34** Flexural Strength, Flexural Modulus, Tensile Strength, and HDT Improvements in Polypropylene Copolymer

| Additive | Flexural strength (MPa) | Flexural modulus (GPa) | Tensile strength (MPa) | HDT at 1.82 MPa (°C) |
|---|---|---|---|---|
| None | 36.52 | 3.34 | 18.7 | 78.9 |
| Unite MP-880, 6 wt% | 55.81 | 4.48 | 29.1 | 107.2 |
| Unite MP-1000, 6 wt% | 55.19 | 3.03 | 29.1 | 108.3 |
| Fusabond P MZ-109D, 6 wt% | 50.02 | 3.72 | 26.9 | 86.1 |

30 wt% HiMod-360 mica in Profax 8623.
*Source:* Ref. 31.

6 wt% additives, with the balance being the polypropylene homopolymer Profax Grade 6501.

*Izod Impact.* Struktol TR-016 again provided the highest increase of Izod impact strength. Addition of 2 wt% of this additive gave an Izod impact change from 106.8 to 181.6 J/m. Flexural modulus increased from 2.89 to 3.1 GPa.

*Unnotched Izod and Gardner Impact.* The best improvements in unnotched Izod and Gardner impact properties were obtained by addition of 6 wt% Unite MP-1000 to the TPO formulation. This additive increased the unnotched Izod impact 146% (256.3–630.1 J/m) while simultaneously increasing HDT from 75.6°C to 103.3°C. Gardner impact was increased by 190% (145–5.4 J) and flexural modulus was increased by about 13%. Use of Struktol TR-016 provided a 375% increase in Gardner impact (0.7–3.7 J), but HDT decreased from 75.6 to 71.1°C. See Fig. 14.9 for a graph of Izod, unnotched Izod, and Gardner impact properties.

### 14.9.7 Summary

Changes in impact, flexural modulus, and heat deflection temperature are shown in Table 14.35. With Struktol TR-016, all three impact properties studied were

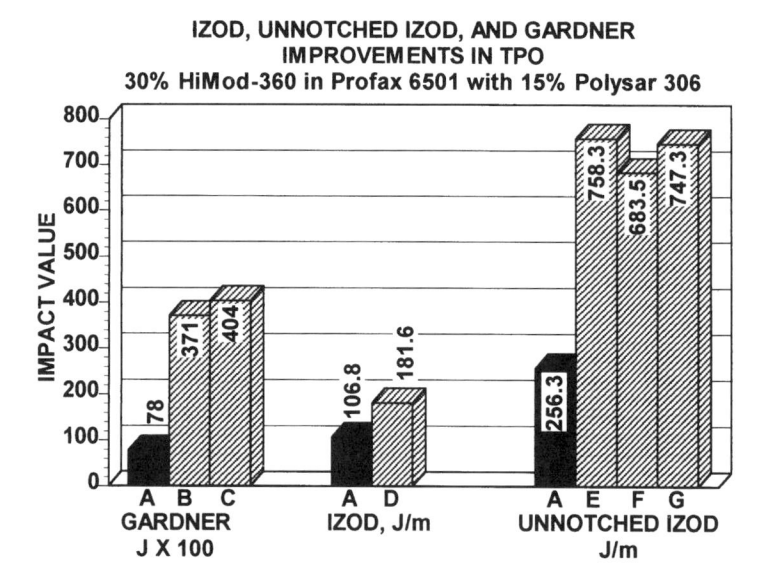

**FIGURE 14.9** Izod, unnotched Izod, and Gardner impact improvements in TPO. 30% HiMod-360 mica in TPO. (From Ref. 30.) A, No additive; B, 0.5% Struktol TR-016; C, 4% Unite MP-880; D, 2% Strucktol TR-016; E, 2% Struktol TR-016; F, 1% Struktol TR-016; G, 4% Unite MP-880.

## Mica Reinforcement of Polypropylene

TABLE 14.35 Effects of Best Additives on Flexural Modulus, HDT, and Impact Properties in Three Polypropylene Systems

| Additive | Izod impact (J/m) | Unnotched Izod (J/m) | Gardner impact (J) | Flexural modulus (GPa) | HDT at 1.82 MPa (°C) |
|---|---|---|---|---|---|
| Homopolymer[a] | | | | | |
| No additive | 25.6 | 144.2 | 0.42 | 5.72 | 100.6 |
| Struktol TR-016, 1 wt% | 48.6 | 309.7 | 0.79 | 5.44 | 88.9 |
| Struktol TR-016, 2 wt% | 64.1 | 432.5 | 1.35 | 4.96 | 76.7 |
| Unite MP-880, 6 wt% | 25.1 | 277.7 | 0.41 | 5.58 | 120.6 |
| Unite MP-1000, 6 wt% | 27.2 | 309.7 | 0.65 | 5.72 | 123.9 |
| Fusabond PMZ-109D, 6 wt% | 25.6 | 240.3 | 0.35 | 5.65 | 115.6 |
| Copolymer[b] | | | | | |
| No additive | 160.2 | 304.4 | 1.45 | 3.34 | 78.9 |
| Struktol TR-016, 2 wt% | 218.9 | 437.9 | 5.40 | 3.79 | 68.3 |
| Struktol TR-016, 4 wt% | 229.6 | 710.2 | 4.36 | 3.72 | 67.8 |
| Unite MP-880, 6 wt% | 62.5 | 235.0 | 1.98 | 4.48 | 107.2 |
| Fusabond P MZ-109 D, 4 wt% | 61.9 | 224.3 | 0.94 | 3.72 | 92.8 |
| TPO[c] | | | | | |
| No additive | 106.8 | 256.3 | 0.78 | 2.89 | 75.6 |
| Struktol TR-016, 0.5 wt% | 165.5 | 582.1 | 3.71 | 3.03 | 70.6 |
| Struktol TR-016, 2 wt% | 181.6 | 758.3 | 1.30 | 3.10 | 64.4 |
| Unite MP-1000, 6 wt% | 80.1 | 630.1 | 2.26 | 4.00 | 103.3 |

[a]Pro-fax 6501.
[b]Pro-fax 8623.
[c]Profax 6501 plus Polysar EPM 306 (15 wt% of total formulation).
30 wt% HiMod-360 mica in PP-HP, PP-CP, and TPO.
*Source:* Ref. 31.

improved, but there were small decreases in heat deflection temperature, tensile strength, and flexural strength.

Maleic anhydride–modified polypropylene consistently decreased Izod impact but provided large increases in Gardner and unnotched Izod impact. Heat deflection temperature as well as tensile strength and flexural properties were also increased substantially.

The information presented in this section shows that polypropylene composites respond differently to additives. An additive that works best for a polypropylene homopolymer may not be the best additive for a TPO or polypropylene copolymer formulation. There is no short cut to laboratory trials to determine the best combination of materials for the best cost/performance result.

## 14.10 MAXIMIZATION OF TPO PERFORMANCE PROPERTIES—EXPERIMENTAL DESIGN

### 14.10.1 Introduction

Earlier sections in this chapter discussed efforts to increase performance properties of mica-reinforced polypropylene with specific procedures (i.e., annealing, surface treatment of mica and use of additives). This final section is based primarily on a paper presented at the TPOs in Automotive '97 Conference in Novi, Michigan [38]. The objective was to learn how to optimize performance properties and to develop equations that would predict performance properties of mica-reinforced TPO composites.

### 14.10.2 Experimental Design and Materials

An experimental design was produced using Design Expert software from Stat-Ease, Inc. Statistically designed experiments were produced D-optimally for the mixture variables. Nineteen different formulations and six replicates were needed to develop reliable predictive equations. Table 14.36 identifies the five formulation components and gives their concentration ranges. Selection of materials was based on earlier studies by Franklin Industrial Minerals indicating that they would be expected to provide the best balance of impact and flexural modulus. After completing the original set of experiments, predicted property values were tested by preparation of compounds and checking of their properties.

TABLE 14.36 Experimental Design: Formulation Components and Concentrations Materials Used in Experimental Design

| Product | Source | Amount (wt%) |
|---|---|---|
| HiMod-360 mica | Franklin Industrial Minerals | 10.0–40.0 |
| Pro-fax 6501 PP | Montell N.A. | 40.0–76.5 |
| Exact 4033 | Exxon | 0.0–20.0 |
| Unite MP-1000 | Aristech Chemical Corp. | 0.0–20.0 |
| Struktol TR-016 | Struktol Co. | 0.0–1.5 |

*Source:* Ref. 38.

### 14.10.3 Best Impact–Stiffness Balance

The reader is referred to Table 14.37 for a summary of the formulations that produced the best impact properties while maintaining a flexural modulus of 3.5 GPa. Use of nearly 40 wt% mica, 17.1–18.2 wt% Exact 4033, and 1.5 wt% Struktol TR-016 produced the best flexural modulus–impact balance for Izod and maximal load (Dynatup) properties. A similar composition that contained 3 wt% Unite MP-1000 gave the best Gardner impact–flexural modulus balance. The best balance of unnotched Izod and failure energy (Dynatup) with flexural modulus was obtained with 76.5 wt% polypropylene, 20.8 wt% mica, 1.2 wt% Exact 4033, and 1.5 wt% Struktol TR-016.

### 14.10.4 Best Tensile and Flexural Properties

The reader is referred to Table 14.38 for formulations that give maximal predicted tensile strength and flexural properties. Maximum predicted tensile strength,

TABLE 14.37 Formulations for Best Impact Properties with Flexural Modulus at 3.5 GPa

| | Izod impact and maximal load (wt%) | Gardner impact (wt%) | Unnotched Izod and failure energy (wt%) |
|---|---|---|---|
| Formulation: | | | |
| Mica, HiMod-360 | 40.0 | 38.1 | 20.8 |
| Pro-fax 6501 | 40.3 | 40.4 | 76.5 |
| Exact 4033 | 18.2 | 17.1 | 1.2 |
| Unite MP-1000 | 0.0 | 3.0 | 0.0 |
| Struktol TR-016 | 1.5 | 1.4 | 1.5 |
| Properties: | | | |
| Flexural strength, MPa | 26.9 | 34.5 | 47.6 |
| Flexural modulus, GPa | 3.5 | 3.5 | 3.5 |
| Tensile strength, MPa | 15.4 | 18.2 | 23.3 |
| Tensile modulus, GPa | 3.5 | 3.3 | 2.8 |
| Izod, J/m | 217.5 | 138.6 | 157.5 |
| Unnotched Izod, J/m | 339.0 | 268.8 | 452.0 |
| Gardner, J | 6.4 | 6.5 | 1.7 |
| Failure energy, J | 7.1 | 7.0 | 9.7 |
| Maximal load, N | 1140.0 | 996.0 | 677.0 |
| HDT @ 1.82 MPa, °C | 69.9 | 76.1 | 76.7 |
| Melt flow, g/10 min | 2.7 | 3.4 | 3.2 |
| Mold shrinkage, % | 0.9 | 0.8 | 1.7 |

*Source:* Ref. 38.

TABLE 14.38 Formulations for Maximal Predicted Tensile and Flexural Properties: HiMod-360 Mica in Polypropylene

|  | Maximum tensile strength | Maximum tensile modulus | Maximum flexural strength | Maximum flexural modulus |
|---|---|---|---|---|
| Property value | 44.3 MPa | 5.1 GPa | 70.0 MPa | 6.4 GPa |
| Formulation (in wt%): |  |  |  |  |
| HiMod-360 | 40.0 | 40.0 | 32.6 | 40.0 |
| Pro fax 6501 | 47.3 | 52.4 | 60.4 | 53.0 |
| Exact 4033 | 5.7 | 0.0 | 0.0 | 0.0 |
| Unite MP-1000 | 7.0 | 7.0 | 7.0 | 7.0 |
| Struktol TR-016 | 0.0 | 0.6 | 0.0 | 0.0 |

*Source:* Ref. 38.

44.3 MPa, was obtained with 40 wt% mica, 47.3 wt% polypropylene, 5.7 wt% Exact 4033, and 7 wt% Unite MP-1000.

Maximal predicted tensile modulus, 5.1 GPa, was obtained with 40 wt% mica, 52.4 wt% polypropylene, 7 wt% Unite MP-l000, and 0.6 wt% Struktol TR-016.

Maximal predicted flexural strength, 70.0 MPa, was obtained with 32.6 wt% mica, 60.4 wt% polypropylene, and 7 wt% Unite MP-1000.

Maximal predicted flexural modulus, 6.4 GPa, was obtained with 40 wt% mica, 53 wt% polypropylene, and 7 wt% Unite MP-1000.

### 14.10.5 Best Izod and Unnotched Izod Impact

See Table 14.39 for formulations that give the best Izod and unnotched Izod properties. Maximal predicted Izod impact strength, 261.6 J/m, was obtained with 10 wt% mica, 68.5 wt% polypropylene, 20 wt% Exact 4033, and 1.5 wt% Struktol TR-016.

Maximal predicted unnotched Izod impact strength, 190 J/m, was obtained with 10 wt% mica, 68.7 wt% polypropylene, 19.8 wt% Exact 4033, and 1.5 wt% Struktol TR-016.

### 14.10.6 Best Gardner and Instrumented Dart Impact Properties

The reader is referred to Table 14.40 for formulations that give the highest Gardner and instrumented Dart impact properties. Maximal predicted Gardner impact, 14.3 J, was obtained with 10 wt% mica, 67 wt% polypropylene, 20.0 wt% Exact 4033, 1.5 wt% Unite MP-1000, and 1.5 wt% Struktol TR-016.

# Mica Reinforcement of Polypropylene

TABLE 14.39 Formulations for Maximum Predicted Izod and Unnotched Izod Properties: HiMod-360 Mica in Polypropylene

|  | Maximal Izod | Maximal unnotched Izod |
|---|---|---|
| Property value | 261.6 J/m | 1,290 J/m |
| Formulation (in wt%) |  |  |
| HiMod-360 | 10.0 | 10.0 |
| Pro-fax 6501 | 68.5 | 68.7 |
| Exact 4033 | 20.0 | 19.8 |
| Unite MP-1000 | 0.0 | 0.0 |
| Struktol TR-016 | 1.5 | 1.5 |

Source: Ref. 38.

Highest predicted maximal load, 1755 N, was obtained with 10 wt% mica, 68.5 wt% polypropylene, 20 wt% Exact 4033, and 1.5 wt% Struktol TR-016.

Maximal predicted failure energy, 15.9 J, was obtained with 10 wt% mica, 68.5 wt% polypropylene, 20 wt% Exact 4033, and 1.5 wt% Struktol TR-016.

## 14.10.7 Best HDT, Melt Index, and Mold Shrinkage

The reader is referred to Table 14.41 for formulations that give the best HDT, melt index, and mold shrinkage values. Maximal predicted heat deflection temperature, 143°C, was obtained with 40 wt% mica, 53 wt% polypropylene and 7 wt% Unite MP-1000.

TABLE 14.40 Formulations for Maximum Predicted Gardner and Instrumented Impact Properties: HiMod-360 Mica in Polypropylene

|  | Maximum Gardner impact | Instrumented impact maximal load | Instrumented impact maximal failure energy |
|---|---|---|---|
| Property value | 14.3 J | 1755 N | 15.9 J |
| Formulation (in wt%) |  |  |  |
| HiMod-360 | 10.0 | 10.0 | 10.0 |
| Prof-fax 6501 | 67.0 | 68.5 | 68.5 |
| Exact 4033 | 20.0 | 20.0 | 20.0 |
| Unite MP-1000 | 1.5 | 0.0 | 0.0 |
| Struktol TR-016 | 1.5 | 1.5 | 1.5 |

Source: Ref. 38.

TABLE 14.41 Formulations for Best Predicted HDT, Melt Index, and Minimum Mold Shrinkage[a] Properties: HiMod-360 Mica in Polypropylene

|  | Maximal HDT | Maximal melt flow | Maximal mold shrinkage |
|---|---|---|---|
| Property value | 143°C | 9.5 g/10 min | 0.5% |
| Formulation (in wt%) |  |  |  |
| HiMod-360 | 40.0 | 10.0 | 40.0 |
| Pro-fax 6501 | 53.0 | 76.5 | 40.6 |
| Exact 4033 | 0.0 | 9.1 | 12.4 |
| Unite MP-1000 | 7.0 | 4.4 | 7.0 |
| Struktol TR-016 | 0.0 | 0.0 | 0.0 |

[a]Parallel to flow direction.
*Source:* Ref. 38.

Maximal melt flow, 9.5 g/10 min, was obtained with 10 wt% mica, 76.5 wt% polyproyylene, 9.1 wt% Exact 4033, and 4.4 wt% Unite MP-l000.

Minimal mold shrinkage, 0.5%, was obtained with 40 wt% mica, 40.6 wt% polypropylene, 12.4 wt% Exact 4033, and 7 wt% Unite MP-1000.

### 14.10.8 Summary

Predictive equations derived from all generated data are listed in Table 14.42. These 12 equations can be used to predict properties of composites with any combination of ingredients as long as the concentrations remain within the experimental design space.

It should be noted that there are literally hundreds of different polypropylene resins available for use in producing mica-reinforced composites. There are also many different types and suppliers of anhydride-modified polypropylene that can be used successfully, and there are many types of elastomer suitable for use in TPO formulations. The above information is provided to show the importance of the use of additives when designing mica-reinforced TPO composites. Well-planned experimental designs are needed to determine optimal formulations.

## 14.11 SUMMARY AND CONCLUDING COMMENTS

New developments in resin technology (i.e., metallocene catalysts) continue to give us better resins for use as toughening agents and new basic polyolefin resins. Better additives, better surface treatments, and mica products with higher reinforcement potential are also being developed.

This chapter has presented information to help plastic engineers understand the basics of formulating and processing mica-reinforced polypropylene. Hope-

# Mica Reinforcement of Polypropylene

TABLE 14.42 Predictive Equations

| Property | Predictive equation |
|---|---|
| Tensile strength, MPa | 0.33[M] + 0.26[P] − 1.52[E] + 21.77[U] + 4.83[T] + 0.02[M][E] − 0.18[M][U] − 0.25[M][T] = 0.017[P][E] − 0.23[P][U] − 0.21[E][U] − 0.85[U][T] |
| Tensile modulus, GPa | 0.12[M] + 0.014[P] − 0.009434[E] + 1.22[U] − 36.46[T] − 0.0005279[M][P] − 0.001567[M][E] − 0.013[M][U] + 037[M][T] − 0.012[P][U] + 0.36[M][T] − 0.012[P][U] + 0.37[P][T] − 0.14[E][U] + 0.37[E][T] + 0.34[U][T] |
| Flexural strength, MPa | − 0.19[M] + 0.42[P] − 0.48[E] + 1.03[U] + 3.38[T] + 0.014[M][P] + 0.018[M][E] + 0.071[M][U] − 0.24[M][T] − 0.77[U][T] |
| Flexural modulus, GPa | 0.15[M] + 0.007248[P] + 0.016[E] + 0.027[U] − 0.037[T] − 0.003984[M][E] |
| Logit (Izod Impact), Y | 0.07[M] + 0.012[P] + 0.094[E] + 0.41[U] − 27.68[T] − 0.003591[M][P] + 0.27[M][T] − 0.008971[P][U] + 0.31[P][T] − 0.007727[E][U] + 0.32[E][T] |
| Izod impact, J/m | $[(262.7 \times e^Y) + 21.428]/(1+e^Y)$ |
| Unnotched Izod, J/m | 53.3[M] + 16.86[P] + 35.58[E] − 7.54[U] + 55.7[T] − 1.28[M][P] − 1.57[M][E] − 27.61[U][T] |
| Gardner impact, J | 0.8[M] + 0.17[P] + 2.35[E] − 17.16[U] + 5.88[T] − 0.017[M][P] − 0.047[M][E] + 0.18[M][U] − 0.024[P][E] + 0.18[P][U] − 0.075[P][T] + 0.17[E][U] − 0.37[U][T] |
| Maximum load, N | 51.46[M] + 15.75[P] + 117[E] + 23.03[U] + 810.74[T] − 0.91[M][P] − 2.77[M][E] − 3.35[M][T] − 0.75[P][E] − 11.04[P][T] − 1.67[E][U] − 47.44[U][T] |
| Logit (failure energy), Y | − 0.06[M] − 0.002092[P] + 0.062[E] + 0.61[U] + 1.36[T] − 0.026[M][T] − 0.007535[P][U] − 0.009594[E][U] − 0.13[U][T] |
| Failure energy, J | $[(17.405 \times (e^Y) + 2.796)/(1 + e^Y)]$ |
| HDT, °C, 1.82 MPa | 1.69[M] + 0.7[P] − 0.78[E] + 10.7[U] + 46.85[T] − 0.9[M][T] − 0.098[P][T] − 0.46[P][T] − 0.14[E][T] − 1.74[U][T] |
| Melt index, g/10 min | 0.2[M] + 0.18[P] − 0.26[E] − 3.9[U] + 4.44[T] − 0.00669[M][P] + 0.042[M][U] + 0.002669[P][E] + 0.042[P][U] − 0.092[P][T] + 0.045[E][U] |
| Mold shrinkage, % | 0.019[M] + 0.02[P] − 0.007853[E] + 0.097[U] − 0.47[T] − 0.0005556[M][P] − 0.002346[M][U] + 0.008573[M][T] − 0.001007[P][U] + 0.006448[P][T] |

[M], wt% HiMod-360 mica; [P], wt% Pro-fax 6501; [E], wt% Exact 4033; [U], wt% Unite MP-1000; [T], wt% Struktol TR-016.
*Source:* Ref. 38.

fully, this chapter will promote an increased rate of commercialization of high-performance mica-reinforced polypropylene composites.

## REFERENCES

1. FJ Meyer. Metal replacement with mica-filled polypropylene. Body Eng J 57–62, Fall 1979.
2. Loss-in-weight system eliminates bridging in plastics compounder's development lab. Powder and Bulk Engineering, Morris Plains, NJ: A Gordon Publication, June 1990.
3. J Lusis, RT Woodhams, M Xanthos. The effect of flake aspect ratio on the flexural properties of mica reinforced plastics. Polym Eng Sci 13, 1973.
4. MH Naitove. Preheating your filler makes compounding easier. Plastics Technol, 40(12), Dec. 1994.
5. MH Mack. Compounding techniques for highly filled polymers on co-rotating twin screw extruders. Dallas, TX: SPE Antec, 1990, Paper No. 351.
6. RE Godlewski. Organosilicon chemical in mica filled polyolefins. 38th Annual Conference, Reinforced Plastics/Composites Institute, Houston, TX: Society of the Plastics Industry, 1983, Session 13-E.
7. Bulletin T-6. Processing parameters for thermoplastics reinforced with Suzorite mica (phlogopite). Marietta Resources International, Ltd., 1977.
8. LA Canova, LW Ferguson, LM Parrinello, R Subramanian, HF Giles, Jr. Effect of combinations of fiber glass and mica on the physical properties and dimensional stability of injection molded polypropylene composites. Proceedings of the SPE 55th Annual Technical Conference & Exhibits, Toronto, Ontario, 1997.
9. JR Copeland. Handbook of Reinforcements for Plastics. New York: Van Nostrand Reinhold, 1987, pp. 125–136.
10. Bulletin T-5. NP surface-treated grades for use with non-polar polymers. Marietta Resources International Ltd., 1977.
11. R Shaefer, R Sherman. Diagnosing and eliminating warpage. Plastics Technol 6:80, 1991.
12. M Akay, S Ozden, T Tansey. Prediction of process-induced warpage in injection molded thermoplastics. Polymer Eng Sci 36(13):1839, 1996.
13. LFA Douven, FPT Baaijens, HEH Meijer. The computation of properties of injection-moulded products. Prog Polym Sci 20:403, 1995.
14. HF Giles, Jr. Reducing shrinkage and warpage in glass-mat thermoplastic composite. Plastic Eng LII(9):43–45, 1996.
15. U.S. Patent 4,983,647, Jan. 8, 1991.
16. U.S. Patent 4,874,809, Oct. 17, 1989.
17. U.S. Patent 4,393,153, July 12, 1983.
18. U.S. Patent 4,880,865, Nov. 14, 1989.
19. U.S. Patent 4,291,084, Sept. 12, 1981.
20. Bulletin. Mica in Polymers, Section 3. Microfine Minerals Limited, 1997.
21. M Schlack. Plastics performance tests. Plastics World 46:318–323, 1989.

## Mica Reinforcement of Polypropylene

22. LA Canova. Effects of heat treatment on mechanical and thermal properties of mica/polypropylene composites. Proceedings of the SPE 48th Annual Technical Conference and Exhibits, 1990.
23. Bulletin SUI-322. OSC Y9777/9771, A new system for filled polyolefins. Union Carbide, 1983.
24. Bulletin SC-731. UCARSIL PC organosilicon chemicals for mineral and fiberglass-reinforced polyolefins. Union Carbide, 1987.
25. RE Godlewski. Organosilicon chemicals in mica filled polyolefins. 38th Annual Conference, Houston, TX, RP/C SPI, 1983.
26. KR Marshall, RE Duncan. Study of surface-modified mica in polypropylene. 38th Annual Conference, Houston, TX, RP/C SPI, 1983.
27. Bulletin. OSi specialities PC coupling agents for fiberglass and mineral-reinforced polyolefins. OSi Specialties, Inc., July 1995.
28. LA Canova. Comparison of the relative effects of surface treatments and resin modifications on the mechanical and thermal properties of mica/polypropylene composites. Proceedings of the SPJ/CI 45th Annual Conference, Washington, DC, 1990.
29. J Griffiths, Deputy Editor. Surface modified minerals. Ind Miner 39, 1987.
30. CC Briggs, J Batchelor, Y Embu, B. Heath. Mica as a reinforcement for polyolefins, thermoplastic compcunding—innovation and specialization, Birmingham, Sept. 1985.
31. LA Canova. How to increase performance properties of mica-reinforced polyolefin composites. Polyolefins X, SPE International Conference, Houston, TX, Feb. 23–26, 1997.
32. Customer Service Report No. 87-0010-PAD. Epolene E-43 wax as a coupling additive for filled polypropylene. Eastman Chemical Products, Inc., January 1987.
33. AM Adur, RC Constable, JA Humenik. Use of acrylic acid-modified polyolefins to improve performance properties of mica filled polyolefins. 43rd Annual Conference, RP/C SPI, Cincinnati, OH, Feb. 1–5, 1988.
34. DJ Olsen, K Hyche. Anhydride based coupling agents for filled polypropylene. Proceedings of the 47th Annual Technical Conference, SPE, 1989.
35. KA Borden, RC Weil, CR Manganaro. Optimizing mica-filled polypropylene. Plastics Compounding, Sept./October, 1993.
36. KA Borden, RC Weil, CR Manganaro. The effect of polymeric coupling agent on mica-filled polypropylene: optimization of properties through statistical experimental design. Proceedings of the 51st Annual Technical Conference, SPE, New Orleans, LA, 1993.
37. JP Vander Kooi, DR Hall. Improving the processing of thermoplastic compounds with modifiers. Proceedings of the 47th Annual Technical Conference, SPE, 1989.
38. LA Canova. Mica reinforced TPO. Fourth International Conference: TPOs in Automotive '97, Novi Michigan, Oct. 27–29, 1997.
39. J Keating. Minerals used in polyolefins. Polyolefins 95 SPE Conference, Houston, TX, 1995.
40. ML Skow. U.S. Bureau of Mines Information Circular 8125, 1962.
41. Bulletin PP30496. Evaluation of mica in polyolefin composites. Franklin Industrial Minerals, 1996.

42. Bulletin P-06. Comparison of mica with other fillers. Franklin Industrial Minerals, 1997.
43. Bulletin PP31494. Properties of mica polypropylene composites. Franklin Industrial Minerals, 1994.
44. A Guillet. Organosilicon coupling systems for filled and reinforced polyolefins. SPE Technical Conference, Filler and Additives in Plastics, Goeteborg, November 1986.

# 15

# Use of Coupled Mica Systems to Enhance Properties of Polypropylene Composites

**Joseph Antonacci**
Suzorite Mica Products Inc., Boucherville, Quebec, Canada

## 15.1 INTRODUCTION

For the past 25 years, naturally occurring mica has gained use as a mineral filler in thermoplastic composites. Mica ore is crushed and suitably delaminated by either dry or wet grinding (see Section 14.1.7 in Chapter 14). The resulting thin mica flakes have high aspect ratios that impart exceptional reinforcing characteristics to thermoplastic resins. The effective use of mica flake depends on a good understanding of the mica microstructure and processing characteristics.

On the basis of relative cost (see Table 14.1 in Chapter 14), there is an incentive to replace more expensive engineering resins and many types of decorative metal parts with much cheaper polypropylene resins. By combining the inherent characteristics of mica microstructure and improved adhesive bond to the polypropylene matrix, mica-reinforced composites of polypropylene exhibit high modulus and elevated heat distortion temperature (HDT). With enhanced product performance at reduced cost, many new applications using polypropylene resins have penetrated markets that were traditionally reserved for composites of engineering resins.

Sections 15.1–15.3 apply many notions of mica technology described in Chapter 14 to make cost-effective hybrid composites with mica and glass fiber reinforcement of polypropylene resins. The ultimate purpose of this chapter is to describe a new, proprietary "SC" surface treatment for mica developed by Suzorite Mica Products, Inc. The SC coupling system provides the means to chemically bond glass fibers and mica flakes to the polypropylene matrix at lower cost than present systems. The objective of the experimental study described in Section 15.4 was to enhance the mechanical and thermal properties for both mica-filled polypropylene and hybrid mica–glass fiber-reinforced composites.

### 15.1.1 Mica Mineralogy (Phlogopite versus Muscovite)

Mica is the general name given to a group of nine different minerals: muscovite, phlogopite, vermiculite, lepidolite, biotite, roscoelite, lepidomelaane, paragonite, and zinnwaldite.

Muscovite and phlogopite types are the most used as raw materials by the mica industry for further processing to give desirable size and shape for reinforcement of plastics. Tables 14.2 and 14.3, given in Chapter 14, summarize the composition and properties of these two types of commercial mica.

Vermiculite is used for heat insulation in expanded form, whereas lepidolite is used as an ore of lithium. Other types of mica, biotite, roscoelite, lepidomelaane, paragonite, and zinnwaldite have little or no commercial importance.

### 15.1.2 Mica Flake Characteristics

Mica flake characteristics consist of mineral chemistry, color, and size/shape ratio. Chemically, mica is a complex silicate of sodium, potassium and aluminum. Most varieties of muscovite mica contain high levels of aluminum oxides; whereas the phlogopite mica is rich in magnesium oxides. There are no important end uses based solely on chemical constitution of the mica type. Individual preference for using either muscovite or phlogopite is dependent on other factors, such as visual color appearance, service temperature for the molded part, ease of delaminating mica layers, and so forth.

The higher concentration of ferrous oxide in the phlogopite mica gives it the darker golden or amber color. Muscovite mica colors range from light green to pinkish or grayish hues and is generally lighter in color.

The detailed chemical composition given in Table 15.1 includes the water content of muscovite and phlogopite mica. The higher water of constitution of the muscovite mica is driven off at 500–600°C; whereas that of the phlogopite mica is driven off at 800–1000°C. This is an important consideration when high heat applications are involved.

# Coupled Mica Systems

TABLE 15.1 Typical Composition (wt%) of Muscovite versus Phlogopite Mica

| Component | Muscovite | Phlogopite |
|---|---|---|
| $SiO_2$ | 45.3 | 40.7 |
| $Al_2O_3$ | 38.2 | 15.8 |
| MgO | 0.8 | 20.6 |
| $K_2O$ | 7.5 | 10.0 |
| Combined fluoride | — | 2.2 |
| $Fe_2O_3$ | 2.4 | 8.9 |
| $TiO_2$ | 0.1 | 0.2 |
| $Na_2O$ | 0.4 | 0.5 |
| $P_2O_5$ | 0.1 | <0.1 |
| CaO | 1.1 | Trace |
| $H_2O$ | 4.1 | 1.0 |

The crystal structure of both mica types is monoclinic. The shape is very similar and can be described as roughly hexagonal or irregular flakes with basal, eminent cleavage.

### 15.1.3 Effect of Extrusion Process on Aspect Ratio of Mica Flake

The aspect ratio $R$ is the ratio of the flake diameter $d$ to its thickness $t$:

$$R = \frac{d}{t}$$

where $d$ is diameter of an equivalent circle and $t$ is thickness of the mica flake. A larger and thinner flake having a higher aspect ratio corresponds to a composite having greater mechanical properties.

Mica flakes are generally compounded into molten polymer resin by viscous shear generated in either single- or twin-screw extruder equipment. Some degree of flake breakdown or attrition will occur during the extrusion process with a resultant decrease in aspect ratio. The size and shape of mica flakes influence the degree of attrition during the extrusion process. Larger flakes are especially prone to attrition. The extrusion process will reduce the diameter of the flakes much more than the thickness of the flake.

The magnitude of shearing stresses can be minimized by proper selection of extrusion process conditions so as to reduce attrition of mica flakes:

1. Barrel temperature settings are normally higher in compounding mica composites than in unfilled resins and should be increased with mica content.

2. A higher temperature setting in the melting zone may be advantageous for reducing flake breakdown.
3. Screw speed and torque should be minimized as much as possible for maintaining homogeneous mixing of ingredients at desired output capacity.

### 15.1.4 Correlation Between Ultimate Aspect Ratio and Composite Properties

It is possible to achieve ultrahigh aspect ratios for mica, e.g., ultimate value of $R = 300$. This is accomplished by using a specialty lab apparatus [1] for delaminating the mica platelets. By compounding under controlled processing conditions, ultrahigh-aspect-ratio mica can be incorporated into thermoset resins to give plastic composites having greatly enhanced mechanical properties [1]. The resulting composite has an unusually high degree of stiffness to plastics approaching that of specialized materials such as graphite and Kevlar fibers. Experimental evidence [2] indicates that for the same volume fraction of mica, there is a direct relationship between mica structure and composite properties. An increase in aspect ratio of the mica corresponds to an increase in composite flexural strength and modulus of the composite.

In practice, both mica and plastic processing equipment, coupled with production cost considerations, makes it impossible to achieve the ultimate aspect ratio for mica. Therefore, the compounder must resort to other means to increase the properties of plastic composites.

### 15.1.5 Choice of Mica Grade to Suit End-Use Applications

The aspect ratio measurement of mica is not an easy task and becomes more complicated as the mica flakes get smaller. An empirical correlation between aspect ratio and the loose bulk density of mica products can be used for practical determination of appropriate mica grade to use. For the same particle size product, the lower bulk density corresponds to a higher aspect ratio. Research at Suzorite Mica Products, has proven that bulk densities below 18 lb/ft$^3$ give the best reinforcing results in plastic composites.

Mica flake size can also influence the reinforcing properties of plastic composites. Test work indicates that mica flakes larger than 40 mesh or 420 µm are not suitable for plastic reinforcement. This is due to flake breakdown during the compounding extrusion process.

There are four major mica flake size products used for plastic reinforcement. Table 15.2 compares the particle size distribution for four major grades of mica with advantages and disadvantages of each in plastics. Mica with 45 µm average or median particle size is the material of choice for in manufacturing polypropylene composites having the best overall properties, appearance, and processing characteristics.

## 15.2 GENERAL REINFORCEMENT CHARACTERISTICS OF MICA-FILLED COMPOSITE

### 15.2.1 Effect on Shrinkage and Warpage of Molded Product

The aspect ratio of mineral fillers greatly affects the shrink and warp properties of molded polypropylene parts.

Spherical fillers have low aspect ratios, i.e., the diameter is practically the same as the thickness. Such grades of mica exhibit poor reinforcing properties in polypropylene and are mainly used as fillers to reduce cost.

Glass fibers offer good reinforcing characteristics in polypropylene by having a good aspect ratio along the length direction. However, uniform reinforcement is hindered by a poor aspect ratio in the width direction. This causes differential shrinkage in the flow direction versus the cross-flow direction during the filling of injection-molded parts. The glass fibers will orient themselves in the length or flow direction of the injection molding process. This directional orientation is the primary cause of warpage during mold filling.

The flat-platelet structure of mica offers unique reinforcing properties in polypropylene. This is due to the aspect ratio of mica being unity, i.e., flake length equals thickness. Orientation of the mica flakes during the injection molding or extrusion process gives consistent, equal reinforcement in the flow and cross-flow direction of the mold, unlike spherical or fibrous minerals. There is equal shrinkage in the flow and cross-flow direction during cooling of the molded part. This isotropic shrinkage results in lower part warpage in comparison with other minerals. The larger the mica flake, the better the shrink and warp properties of the molded part.

### 15.2.2 Enhancement of HDT and Flexural Modulus

Test results given in Table 15.3 demonstrate the effect of flake diameter on the stiffness and HDT-66 psi (0.455 MPa) of mica filled polypropylene. The larger mica flake increases the stiffness (flexural modulus) and HDT of polypropylene more than finer mica flake. Table 15.3 also shows that both stiffness and HDT increase with increasing mica content.

### 15.2.3 Limitations to Mechanical Strength and Impact Properties

The plastics industry uses mineral products as reinforcement to obtain specific property requirements in the finished part. In most cases, there is a compromise in

TABLE 15.2 Typical Mica Grade Selection for Plastic Reinforcement

| Grade | Advantage | Disadvantage |
| --- | --- | --- |
| −40 +100 mesh screen<br>Median particle size: 250 µm | Excellent stiffening<br>Excellent warp resistance<br>Excellent increase in HDT<br>Very low shrinkage | Decreases melt flow<br>Large flakes visible at surface<br>Moderate Izod impact loss |
| −60 +200 mesh<br>Average 170 µm | Very good reinforcement<br>Very good stiffening<br>Very good warp resistance<br>Very good HDT<br>Low shrinkage | Decreases melt flow<br>Better appearance than above<br>Smaller Izod impact loss |
| 100% thru 100 mesh<br>40–60% thru 325 mesh<br>Average 45 µm | Good stiffening<br>Good warp resistance<br>Good HDT<br>Low shrinkage | Decreases melt flow<br>Better overall appearance<br>Lower loss of Izod impact |
| >90% thru 325 mesh<br>Average 20 µm | Best Izod impact<br>Other properties slightly lower<br>Good surface appearance | Difficult to compound<br>Dusty-low bulk density |

TABLE 15.3  Effect of Mica Loading and Grade Selection on the Flexural Modulus and HDT of Polypropylene

| | Weight % | | | | |
|---|---|---|---|---|---|
| Formulation | | | | | |
| Pro-fax 6523 | 100.0 | 80.0 | 80.0 | 70.0 | 60.0 | 60.0 |
| Mica, Suzorite 60-HK | 0.0 | 20.0 | 0.0 | 30.0 | 0.0 | 40.0 | 0.0 |
| Mica, Suzorite 200-HK | 0.0 | 0.0 | 20.0 | 0.0 | 30.0 | 0.0 | 40.0 |
| Properties | | | | | |
| Flexural modulus, GPa | 1.04 | 2.10 | 2.07 | 2.74 | 2.46 | 4.11 | 3.22 |
| HDT-0.455 MPa, °C | 95.0 | 129.0 | 121.0 | 132.0 | 126.0 | 139.0 | 130.0 |

Suzorite 60-HK = (250 μm) mica.
Suzorite 200-HK = 45 (μm) mica.

the overall properties of the reinforced plastic to reach the specific property requirement for the final end-use application. Mica is used in plastics to increase the flexural modulus and HDT, and to reduce shrink and warp of plastic parts. In order to achieve these benefits, there is a trade-off in other properties.

As shown in Table 15.4, mica will typically reduce the tensile strength while increasing the flexural strength slightly. However, both tensile and flexural strength properties are reduced as the mica content is increased.

The largest trade-off in properties with mica-filled polypropylene is impact strength. Unnotched Izod impact is greatly reduced from no-break failure with virgin polypropylene resin to progressively lower impact values as the mica content is increased.

## 15.3 DESCRIPTION OF COUPLED MICA SYSTEMS

The reduction in the mechanical strength properties of untreated mica-filled polypropylene is attributed to the nonpolar nature of polypropylene resin that lacks reactive sites to bind with mica flake. Since there is poor adhesion between the polymer and the mica flakes, optimal reinforcement cannot be achieved.

The key to improved performance of mica-filled polypropylene composites is to increase the interfacial adhesion of the mica to polypropylene. Better adhesion can be obtained by modifying the surface of the mica, which is polar, and making it more compatible with the surface of polypropylene, which is nonpolar.

As described in Section 14.8 of Chapter 14, there are many types of surface treatments for mica flakes to produce coupled systems in mica-polypropylene composites. The physical and chemical interactions promote adhesion and bonding of the mica to the polypropylene matrix with an increase in tensile strength, flexural strength, and HDT for the composite material.

### 15.3.1 Design requirements for Interface of Mica-Polypropylene Matrix

In order to achieve good bonding characteristics between mica and polypropylene, a coupling system must be chosen that is compatible with both the filler and the polymer matrix.

*Need for Bifunctional Properties to Promote Compatibility.* Ideally, the additive used to promote the adhesion of mica to polypropylene would contain two functional groups: polar and nonpolar. The polar end of the bifunctional molecule attaches to the surface of the mica to form a strong bond. The nonpolar end of the additive molecule is compatible with the polymer matrix to complete a chemical bridge between the polar mica surface and the polymer molecules.

**TABLE 15.4** Effect of Mica Loading and Grade Selection on the Mechanical Strength and Impact properties of Polypropylene

| Formulation | Weight % | | | | | |
|---|---|---|---|---|---|---|
| Pro-fax 6523 | 100.0 | 80.0 | 80.0 | 70.0 | 60.0 | 60.0 |
| Mica, Suzorite 60-HK | 0.0 | 20.0 | 0.0 | 30.0 | 40.0 | 0.0 |
| Mica, Suzorite 200-HK | 0.0 | 0.0 | 20.0 | 0.0 | 0.0 | 40.0 |
| Properties | | | | | | |
| Tensile strength, MPa | 31.7 | 30.2 | 30.6 | 29.0 | 27.2 | 28.8 |
| Flexural strength, Mpa | 38.6 | 43.8 | 47.9 | 44.8 | 43.3 | 45.2 |
| Un-notched Izod impact, J/m | N.B. | 367.3 | 449.1 | 319.8 | 201.0 | 214.9 |
| Notched Izod impact, J/m | 36.4 | 79.1 | 75.9 | 59.9 | 52.4 | 54.5 |

*Adhesive Bond to Increase Interfacial Strength to Tensile Load.* The two functional groups present in an additive provide the necessary interfacial bond between mica and polypropylene. The resulting composites resist deformation under both tensile and flexural stresses, i.e., coupled systems possess load bearing characteristics.

### 15.3.2 Options to Promote Mechanical Strength

There are several ways to achieve coupling of mica to polypropylene. Mica producers prefer to pretreat the mica to produce a "value-added" product that will also facilitate subsequent compounding operations by reducing the amount of additives required. The resin suppliers prefer to modify polypropylene to increase properties. Meanwhile, compounders prefer to use proprietary formulations consisting of some type of coupling agent additive combined with a blend of other ingredients to produce finished products having improved properties.

*Surface Treatment of Mica with Silane.* Silane modification of the mica surface has been used effectively to produce mica-filled polypropylene composites to achieve higher properties for applications where polypropylene alone would not be considered.

Silanes are organofunctional coupling agents designed to form a molecular bridge by forming strong adhesive bonds between a typically organic resin and inorganic filler. It is important to understand the reactivity of the silane to obtain the proper compatibility of the filler and resin matrix so as to achieve the properties and benefits required.

In Section 11.2.1 of Chapter 11, Jancar discusses the use of silane and titanate coupling agents to produce engineered interphase layers on inorganic reinforcements. A general description of silane chemistry explains how siloxane coatings are produced on inorganic substrates. This technology is used to enhance mica reinforcement by various means of silane surface treatment.

The technique for applying silane to mica generally consists of simply spraying a solution of liquid silane onto the mica filler. A fast-evaporating solvent is used as a carrier to effectively produce thin coatings of silane layers on the mica surface.

Undiluted liquid silane can also be sprayed directly onto mica feed while being metered to a compounding extruder. An alternative method would be to separately pump silane as a separate feed stream into an injection port located in the barrel of the extruder. Then the silane would be mixed in situ with the molten mixture of polypropylene and mica ingredients. However, it is essential that no reaction occur prematurely between the silane and the polypropylene or the coupling efficiency will be reduced. Usually a larger amount of silane is required to obtain the same properties as pretreated mica with increased raw material costs.

*Addition of Maleic Anhydride Modified Polypropylene.* Modification of the polypropylene molecular chain is another way to improve the mechanical properties of mica-filled polypropylene. A polymeric coupling agent can be produced by reactive extrusion of polypropylene with polar monomers having reactive vinyl end groups such as acrylic acid and maleic anhydride. As described in Chapters 3 and 10, the grafting chemistry involves use of peroxide initiator to remove the tertiary hydrogen atoms along the polypropylene molecular chain. The free radicals that are produced are involved in a number of reactive extrusion process steps to chemically bond the functional group to the polypropylene backbone. The polar portion of the grafted adduct to the polypropylene molecular chain can then react either with the polar surface of the mica or polar interfaces formed by addition of other minerals and fibers.

Polymeric coupling agents are generally utilized in pelletized concentrate form and can be let down during the compounding extrusion process. Chapter 3 describes the use of grafted maleic anhydride in polypropylene composites. Compounders usually preblend polymeric coupling agents with virgin polypropylene at optimal concentration levels.

*Hybridization of Glass Fibers with Mica Reinforcement.* Composites of glass fiber–reinforced polypropylene require some type of chemical coupling to attain good tensile strength. Chapters 9, 12, and 13 provide details regarding glass fiber technology and chemical coupling enhancement. Generally, the coupling system consists of the addition of grafted maleic anhydride coupling agent to properly sized glass fibers at sufficient levels to maximize mechanical properties.

In many end-use applications, glass-reinforced polypropylene parts are overengineered, requiring ribbing, greater wall thickness, and special gating to meet the rigidity and HDT requirements of the end-use application. The reason for this overengineering, as described in Section 15.2.1, is the anisotropic shrinkage behavior due to unidirectional orientation of fibers in molded parts. The resulting differential shrinkage and warping of the molded parts requires expensive modification of existing molds designed for engineering plastics. All of this adds up to higher costs for the finished part.

Composites of mica–glass fiber hybrids offer a good balance between the mechanical properties as well as the warp and shrink characteristics of the part at lower cost. Mica reduces the shrink and warp problems associated with glass fibers. Then ribbing and wall thickness can be reduced while the rigidity and HDT requirements for the molded part are maintained.

*Other Additive Treatments.* Commercially available coupling agents also include titanates, zirconates, and zircoaluminates (see Table 14.29 in Chapter 14). There is insufficient evidence to prove that these products substantially enhance the mechanical properties for a mica-polypropylene matrix.

Other additive treatments include surface modifiers such as fatty acids, stearates, and other chemicals. Surface modifiers act as wetting agents on the surface of the mica and other minerals to aid in processing and compounding operations to reduce temperature and torque. This can lead to higher throughput. Surface modifiers do not necessarily interact with the filler and polymer matrix to form bonds, but rather act as lubricants improving the molten flow of the filled polypropylene.

## 15.4 SC TREATED MICA–GLASS FIBER POLYPROPYLENE COMPOSITES

In order to remedy the unidimensional shortcomings of glass fiber–reinforced polypropylene composites, a cost-effective combination of treated mica and glass fibers seems to be the logical means to produce the desired reinforcement for polypropylene with reduced shrinkage and warpage.

### 15.4.1 Design of Experiment Study to Develop New Coupled System

A preliminary investigation was undertaken by Suzorite Mica Products to develop a coupling system that would be able to bond both mica and glass fibers to polypropylene. The goal was to obtain a favorable property-to-cost ratio compared with commercial glass fiber–reinforced polypropylene composites.

The strategy of the experimental study was to characterize the effect of different coupling agents on mica-filled polypropylene, as well as different hybrid combinations of treated mica and glass fiber in polypropylene composites.

Initial testing results given in Table 15.5 describe the characteristics of two mica coupling systems of surface-treated mica that were selected as the best candidates for further investigation:

1. Mica treated with silane: 200-PP based on existing state-of-the-art technology.
2. Two-component SC coupling system: 200-SC developed by Suzorite Mica Products.

In the next step of development, these two coupling systems were combined at different ratios of glass fiber reinforcement. The goal was to determine which system was best suited to couple the mica-glass hybrid having comparable mechanical properties attained by entirely glass fiber reinforcement.

Silane-treated mica bonds very well with polypropylene to increase the mechanical properties substantially over nontreated mica. However, there are no active sites available on silane-treated mica to bond glass fibers to polypropylene,

# Coupled Mica Systems

TABLE 15.5 Mechanical Properties of Coupled and Uncoupled Mica in Polypropylene

| Materials[a] | Tensile strength (MPa) | Flexural properties Strength (MPa) | Flexural properties Modulus (GPa) | IZOD impact (J/m) Unnotched | IZOD impact (J/m) Notched | HDT (°C) @ 0.455 MPa | HDT (°C) @ 1.82 MPa |
|---|---|---|---|---|---|---|---|
| 6331 PP resin composites 40 wt% mica | | | | | | | |
| 200-HK | 28.8 | 45.2 | 3.2 | 215 | 54.5 | 130 | 80 |
| 200-PP | 32.4 | 49.1 | 3.5 | 179 | 51.2 | 134 | 84 |
| 200-SC | 37.7 | 57.5 | 3.6 | 230 | 55.5 | 138 | 86 |

Suzorite 200-HK (45 μm average size) mica. 200-PP is silane treated and 200-SC is a proprietary coupling system developed by Suzorite Mica Products, Inc.

resulting in poor mechanical properties. This leads to the conclusion that silane is not an efficient coupling system for mica-glass polypropylene composites.

The two-component SC coupling system bonds with both mica and glass to give properties similar to commercial all-glass-fiber polypropylene composites coupled with maleated polypropylene (see Table 15.6). Therefore, this coupling system was chosen as the candidate for detailed mapping of the mechanical property profile as a function of mica-fiberglass hybrid combinations [3]. Besides given hybrid combinations, the design of experiment matrix included base resin as a variable: mica homopolymer and copolymer polypropylene. Tables 15.7 and 15.8 give the entire spectrum of materials tested.

The glass fiber used in the study was PPG 3298 chopped strands with nominal 3.2-mm. length and sizing that is appropriate to composites of chemically coupled polypropylene. Polypropylene resins consisted of Montel Profax 6331 homopolymer (12 melt flow rate) and Profax SB 642 copolymer (22 melt flow rate). Compounding of mica–glass fiber composites was done using a ZSK 30 corotating twin-screw extruder.

### 15.4.2 Results of Study

The following trends in mechanical properties indicate significant flexibility of the SC coupling system for treated mica and glass fiber combinations to achieve the performance of commercial, chemically coupled, glass fiber–reinforced composites with homopolymer polypropylene. Although a similar trend in reinforcement is seen with copolymer resin, the properties of commercial glass–filled copolymer polypropylene cannot be achieved.

*Tensile Strength.* The trendline plot given in Fig. 15.1 shows that tensile strength values steadily rise for materials with incremental increase in mica and glass content. Even though the tensile strength values for uncoupled glass fibers alone are very low, as the SC mica component is increased, the composite properties increase substantially. This is due to the bifunctionality of the SC coupling system that promotes bonding at both mica and glass fiber interfaces to the polypropylene matrix.

Optimal properties are obtained at higher levels of SC mica/glass combinations, ranging from 30 MPa for virgin polypropylene to more than 80 MPa for SC mica/glass combinations, an improvement of 150%. In related data, the tensile elongation decreases at the higher filler level as expected.

*Flexural Strength.* The flexural strength (Fig. 15.2) shows the same trend as the tensile strength. The flexural strength increases from 36 MPa for virgin polypropylene to more than 108 MPa for SC mica/glass combinations, a 200% increase. Again, no increase is reported with uncoupled glass alone, but important gains are evident when SC mica is added to the same percentage of glass fibers.

## Coupled Mica Systems

**TABLE 15.6** Mechanical Properties of Coupled Mica/Glass Polypropylene Composites

| 6331 PP resin composites FG/mica[a] | Tensile strength (MPa) | Flexural properties Strength MPa | Flexural properties Modulus GPa | Izod impact (J/m) Unnotched | Izod impact (J/m) Notched | HDT (°C) @ 1.82 MPa |
|---|---|---|---|---|---|---|
| 30% FG | 80.3 | 106.6 | 3.9 | 715 | 150 | 142 |
| 20% 200-PP 20% FG | 42.8 | 57.2 | 4.5 | 219 | 53 | 138 |
| 20% SC-200-HK 20% FG | 74.5 | 57.5 | 5.1 | 603 | 91 | 144 |

FG = PPG 3298 glass fibers with 3.2 mm average chopped strand length. 200-PP = silane-treated 200-HK mica. 200-SC = two-component proprietary Suzorite mica treatment on 200-HK mica.

TABLE 15.7  Homopolymer Compositions

| Compound no. | % 6331 PP | % 200-SC mica | % 3298 FG | Compound no. | % 6331 PP | % 200-SC mica | % 3298 FG |
|---|---|---|---|---|---|---|---|
| 1  | 95 | 5  | —  | 23 | 65 | 20 | 15 |
| 2  | 90 | 10 | —  | 24 | 60 | 25 | 15 |
| 3  | 85 | 15 | —  | 25 | 55 | 30 | 15 |
| 4  | 80 | 20 | —  | 26 | 80 | —  | 20 |
| 5  | 75 | 25 | —  | 27 | 75 | 5  | 20 |
| 6  | 70 | 30 | —  | 28 | 70 | 10 | 20 |
| 7  | 65 | 35 | —  | 29 | 65 | 15 | 20 |
| 8  | 60 | 40 | —  | 30 | 60 | 20 | 20 |
| 9  | 95 | —  | 5  | 31 | 55 | 25 | 20 |
| 10 | 90 | 5  | 5  | 32 | 50 | 30 | 20 |
| 11 | 85 | 10 | 5  | 33 | 75 | —  | 25 |
| 12 | 80 | 15 | 5  | 34 | 70 | 5  | 25 |
| 13 | 90 | —  | 10 | 35 | 65 | 10 | 25 |
| 14 | 85 | 5  | 10 | 36 | 60 | 15 | 25 |
| 15 | 80 | 10 | 10 | 37 | 55 | 20 | 25 |
| 16 | 75 | 15 | 10 | 38 | 50 | 25 | 25 |
| 17 | 70 | 20 | 10 | 39 | 70 | —  | 30 |
| 18 | 65 | 25 | 10 | 40 | 65 | 5  | 30 |
| 19 | 85 | —  | 15 | 41 | 60 | 10 | 30 |
| 20 | 80 | 5  | 15 | 42 | 55 | 15 | 30 |
| 21 | 75 | 10 | 15 | 43 | 50 | 20 | 30 |
| 22 | 70 | 15 | 15 |    |    |    |    |

# Coupled Mica Systems

TABLE 15.8 Copolymer Compositions

| Compound no. | % SB642 PP | % 200-SC mica | % 3298 FG | Compound no. | % SB642 PP | % 200-SC mica | % 3298 FG |
|---|---|---|---|---|---|---|---|
| 44 | 95 | 5  | —  | 66 | 65 | 20 | 15 |
| 45 | 90 | 10 | —  | 67 | 60 | 25 | 15 |
| 46 | 85 | 15 | —  | 68 | 55 | 30 | 15 |
| 47 | 80 | 20 | —  | 69 | 80 | —  | 20 |
| 48 | 75 | 25 | —  | 70 | 75 | 5  | 20 |
| 49 | 70 | 30 | —  | 71 | 70 | 10 | 20 |
| 50 | 65 | 35 | —  | 72 | 65 | 15 | 20 |
| 51 | 60 | 40 | —  | 73 | 60 | 20 | 20 |
| 52 | 95 | —  | 5  | 74 | 55 | 25 | 20 |
| 53 | 90 | 5  | 5  | 75 | 50 | 30 | 20 |
| 54 | 85 | 10 | 5  | 76 | 75 | —  | 25 |
| 55 | 80 | 15 | 5  | 77 | 70 | 5  | 25 |
| 56 | 90 | —  | 10 | 78 | 65 | 10 | 25 |
| 57 | 85 | 5  | 10 | 79 | 60 | 15 | 25 |
| 58 | 80 | 10 | 10 | 80 | 55 | 20 | 25 |
| 59 | 75 | 15 | 10 | 81 | 50 | 25 | 25 |
| 60 | 70 | 20 | 10 | 82 | 70 | —  | 30 |
| 61 | 65 | 25 | 10 | 83 | 65 | 5  | 30 |
| 62 | 85 | —  | 15 | 84 | 60 | 10 | 30 |
| 63 | 80 | 5  | 15 | 85 | 55 | 15 | 30 |
| 64 | 75 | 10 | 15 | 86 | 50 | 20 | 30 |
| 65 | 70 | 15 | 15 |    |    |    |    |

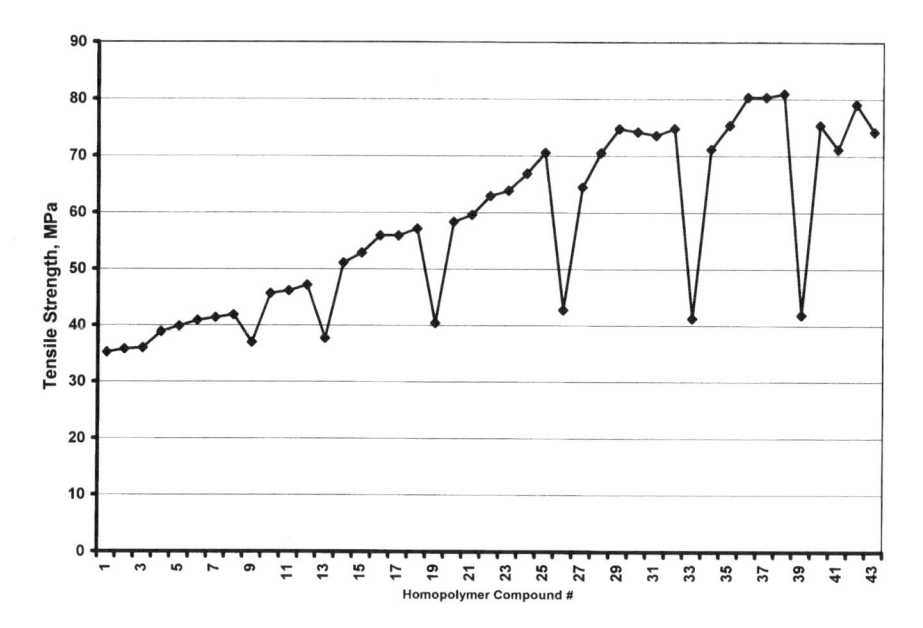

**FIGURE 15.1** Tensile strength of SC mica/glass polypropylene.

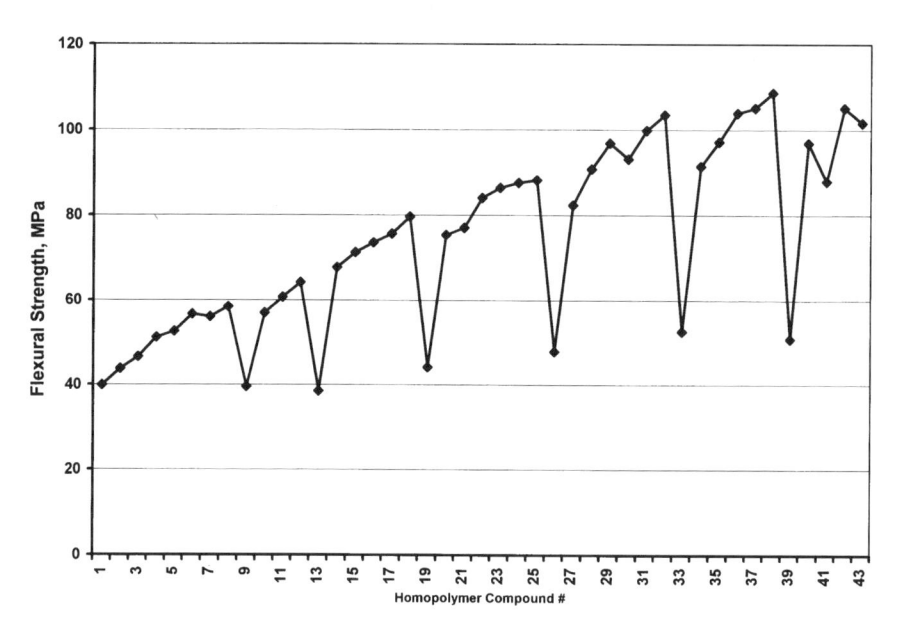

**FIGURE 15.2** Flexural strength of SC mica/glass polypropylene.

# Coupled Mica Systems

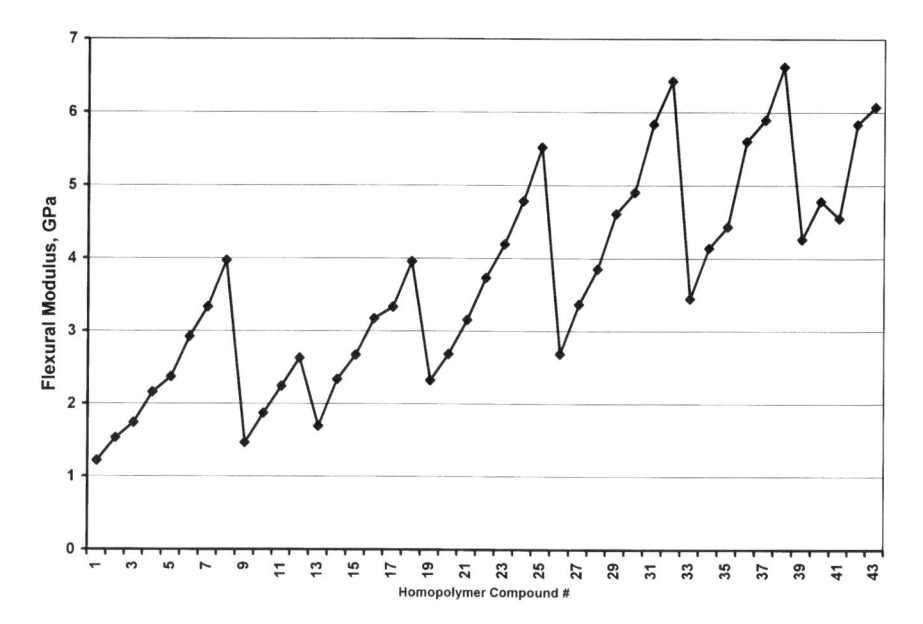

**FIGURE 15.3** Flexural modulus of SC mica/glass polypropylene.

*Flexural Modulus.* Flexural modulus (Fig. 15.3) increases substantially as the concentration of mica and glass increases. The flexural modulus jumps from 1.03 GPa for virgin polypropylene to more than 6.62 GPa for SC mica/glass polypropylene composites, a 550% improvement.

*Izod Impact Strength.* The Izod impact properties of commercial glass-reinforced polypropylene can be maintained with partial replacement of glass fibers with SC mica. Typically the replacement ratio is 2:1 by weight of SC mica to glass, e.g., a 30% glass-filled polypropylene can be replaced with 20% SC mica/20% glass with similar Izod impact.

It should be noted that glass fibers obtain superior Izod impact strength in the flow direction of the mold but poor impact strength in the cross-flow direction. This is due to the fiber's differential aspect ratio, discussed earlier in this chapter. The SC mica/glass combinations maintain the impact strength in the flow direction and improve the impact strength in the cross-flow direction of the part. This is an important factor for part design considerations.

*Heat Distortion Temperature.* The HDT of polypropylene composite increases as higher concentrations of SC mica/glass are added, reaching as high as 280% above that of virgin polypropylene (Fig. 15.4). The higher percentage of SC mica used to replace the glass fibers allows a part with the same strength values to be used in higher temperature applications.

### 15.4.3 Control of Dimensional Stability by Mica–Glass Fiber Composition

Dimensional instability of injection-molded polypropylene parts has always plagued the plastics industry. As the plastic part cools the polymer shrinks in both the flow and transverse direction of the mold. This makes it difficult to match the dimensions of the parts with other plastics or with other materials. The addition of mineral fillers, glass fibers, and fibrous minerals minimizes the shrinkage of injection-molded polypropylene. Canova et al. [4] describe post-molding shrinkage of hybrid combinations of mica and glass fibers in detail.

SC micaglass polypropylene hybrid composites exhibit reduced shrinkage as the combined concentration of mica and glass fibers is increased. Shrinkage is lower in the flow direction (positions 1, 2, and 3 in Fig. 15.5) and greater in the transverse direction (positions 4, 5, 6, and 7). There is also differential shrinkage in the transverse direction as the percent shrinkage increases away from the gate at positions 4–6 and then decreases at the far end of the plaque at position 7.

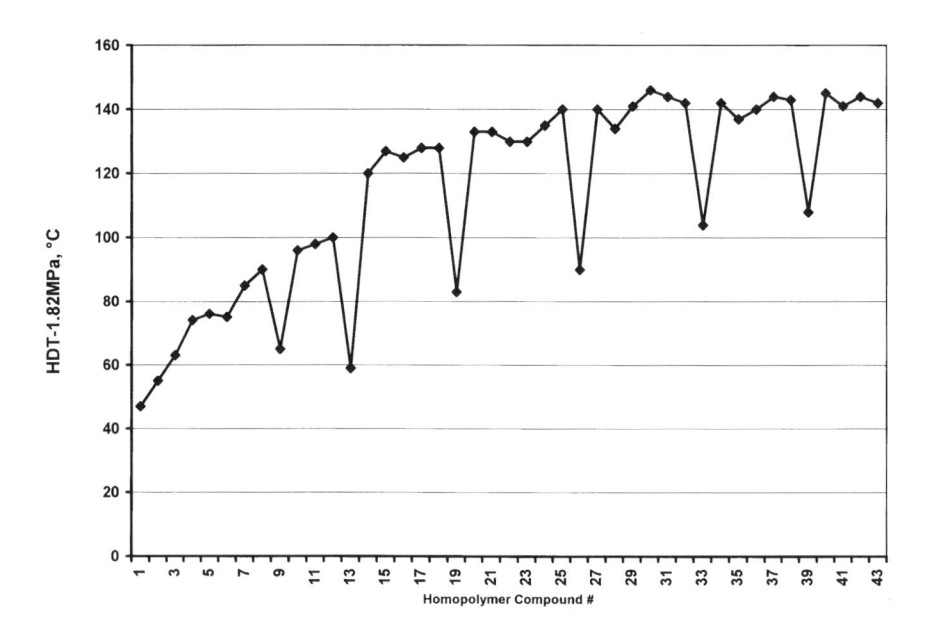

FIGURE 15.4 Heat distortion temperature of SC mica/glass polypropylene.

The glass fiber content in the composite is directly related to the differential shrinkage. An increase in glass fiber level corresponds to increased anisotropic behavior. This is due to the difference in aspect ratio of the fiber in its length and width. The shrinkage of mica alone in polypropylene is practically the same in the flow and transverse direction due to the aspect ratio of mica being essentially 1. Therefore, increasing the mica content in the hybrid composite results in decreased differential shrinkage or decreased anisotropy.

The degree of shrinkage anisotropy controls the warping of the molded plaque. Consequently, the mica content will decrease the warping of the molded plaque significantly, which allows for simpler and cheaper mold designs than entirely glass fiber–reinforced polypropylene composites.

### 15.4.4 Optimization of Cost–Mechanical Performance Balance

The SC mica coupling system allows greater flexibility to design molded parts having required mechanical properties at reduced cost than all glass fiber–reinforced polypropylene composites. The determination of optimal micaglass fiber composition ratios for hybrid composites is facilitated by using master plots of predictive properties generated from the data obtained from a series of experiments.

Table 15.9 summarizes tensile strength as a function of wt% reinforcement for an array of mica-fiberglass composites. The corresponding plots in Fig. 15.6 reflect differences in interfacial adhesion. The plot for a 50:50 hybrid ratio of mica-fiberglass content exhibits a tensile behavior that is nearly the same as chemically coupled glass fiber reinforcement. For example, a hybrid composite consisting of 20 wt% SC-200 mica and 20 wt% PPG 3298 fiberglass has a tensile strength comparable to 30 wt% chemically coupled polypropylene.

The higher stiffness and lower shrink and warp characteristics of the SC mica/glass polypropylene composites leads to simpler mold designs and reduced part thickness. The higher HDTs possible with SC mica/glass allow polypropylene to enter new markets and applications.

Reduction in glass fiber content, property and design improvements, and the cost of SC-treated 200-HK mica permits the design of a reinforced polypropylene part with optimal performance characteristics at reduced cost.

### 15.5 SUMMARY AND CONCLUSIONS

Mica can be treated with various coupling agents to improve bonding with polypropylene and thus give improvements in mechanical properties, e.g., tensile and flexural strength, flexural modulus, and HDTs. Use of SC-treated mica allows

TABLE 15.9  Tensile Strength Values (MPa) for Mica–Glass Fiber Composites

| Wt% reinforcement | Unfilled PP resin | A<br>Untreated Suzorite 200-HK mica | B<br>SC-200-HK mica | C<br>Uncoupled PPG 3298 fiberglass | D<br>50:50 SC-200-HK/ fiberglass | E<br>Chemically coupled PPG 3298 fiberglass (4) |
|---|---|---|---|---|---|---|
| 0  | 31.7 | —    | —    | —    | —    | —  |
| 10 | —    | —    | 35.8 | 37.7 | 45.7 | 50 |
| 20 | —    | 30.6 | 38.9 | 42.8 | 52.9 | 62 |
| 30 | —    | 29.8 | 40.9 | 41.9 | 62.9 | 74 |
| 40 | —    | 28.8 | 41.9 | —    | 74.5 | 83 |
| 50 | —    | —    | —    | —    | 80.9 | 90 |

# Coupled Mica Systems

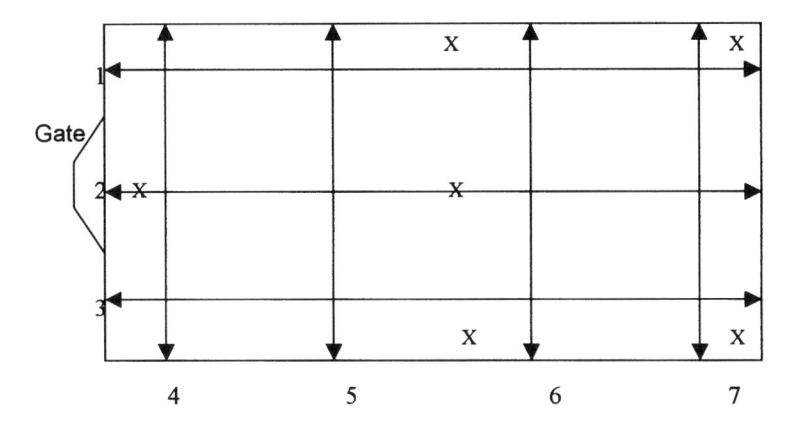

**FIGURE 15.5** Shrink and warp measurement locations.

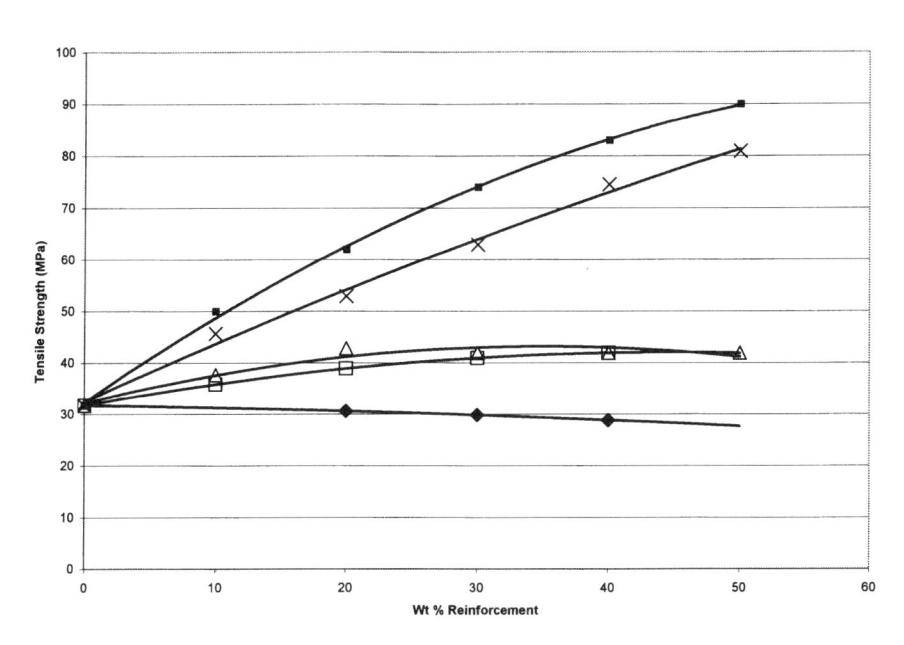

**FIGURE 15.6** Tensile behavior of mica–glass fiber composites. ◆, Untreated Suzorite 200-HK mica; □, SC-200-HK mica; △, uncoupled PPG 3298 fiberglass; ×, 50:50 SC-200-HK/fiberglass composition; ■, Chemically coupled PPG 3298 fiberglass.

hybrid composites of mica–glass fiber–reinforced polypropylene composites to compete with higher priced polymers for approval in new applications.

Mica–glass fiber hybrid systems can improve the dimensional stability and HDT resistance without sacrificing the other mechanical property requirements of the finished part.

Higher flexural modulus or stiffness is achieved as total filler level is increased, thus permitting a possible reduction in part wall thickness. Mica-fiberglass PP composites also maintain HDT levels similar to those of all fibreglass-reinforced polypropylene composites.

The design window of the SC coupling system provides a broad property response to given hybrid combinations of mica and fiberglass reinforcement. This predictive approach allows for greater flexibility in determining optimal composition of mica and glass fibers required for a particular application at lower cost than conventional coupling systems.

## REFERENCES

1. RT Woodhams, M Xanthos. HAR mica reinforced thermosets. 24th Canadian Chemical Engineering Conference, Oct. 1974.
2. M Xanthos, GC Hawley, J Antonacci. Parameters affecting the engineering properties of mica reinforcement thermoplastics. Proceedings of the SPE 35th Annual Technical Conference, 1977.
3. J Antonacci, A Khan, I Sher. "SC" coupling system enhances the properties of mica filled polypropylene and mica/glass polypropylene hybrid systems. Proceedings of the SPE 57th Annual Technical Conference, 1999, pp. 3168–3177.
4. LA Canova, LW Ferguson, LM Parrinello, R Subramanian, HF Giles Jr. Effect of combinations of fiber glass and mica on the physical properties and dimensional stability of injection molded polypropylene composites. Proceedings of the SPE 55th Annual Technical Conference, 1997, pp. 2112–2116.

# 16

# Performance of Lamellar High-Purity Submicrometer and Compacted Talc Products in Polypropylene Compounds

**Wilhelm Schober**
HiTalc Marketing and Technology GmbH, Schoconsult GmbH, Austria

**Giovanni Canalini**
Superlab S.r.l., Italy

## 16.1 DESCRIPTION OF THE MINERAL TALC AND ITS VARIATIONS

### 16.1.1 Differences in Mineralogy and Morphology: Geographic Aspects

The term "talc" summarizes a variety of minerals in nature. On the one hand, "talc" stands for pure magnesium-silicate-hydrate; on the other hand, it is a general term for a polymineral rock.

Pure talc has the chemical formula $Mg_3[Si_4O_{10}(OH)_2]$. By theoretical chemical analysis, talc composition consists of: 63.5% $SiO_2$, 31.7% MgO, and 4.8% $H_2O$.

Talc is most frequently accompanied by "chlorite," where the $Mg^{2+}$ ion has been replaced by $Al^{3+}$ or $Fe^{3+}$ ion. To compensate for the different loadings of the $Al^{3+}$ or $Fe^{3+}$ ion in comparison with the $Mg^{2+}$ ion, an additional brucite layer is added. This mineral is called chlorite; it has a lamellar structure as well, with properties similar to those of talc in most typical applications. The mineral is often green and received its name from the Greek word *chloros*, which means "green."

Pure talc is characterized by its hydrophobic properties, its slipperiness of surface, and the lowest Mohs hardness of 1. The softness of talc is attributable to the fact that the layers can easily be shifted and separated. The smooth, hydrophobic, water-repellent layers slide against one another, giving talc its slippery and fatty feel. Commercial talc grades are harder due to impurities. Pure talc remains unaltered in the presence of chemical reagents, whereas carbonate- and chlorite-containing talcs have a higher solubility in acids.

Apart from the mineralogy, talc deposits are classified by their brightness and morphology. The macrocrystalline lamellar talc is the most common modification (China; India; Mount Seabrook, Australia; France; and Italy). This type of talc is used for the full portfolio of applications, including plastics. Compactly structured talc occurs very rarely in Europe (Germany, Spain). The main deposits are in Australia (Three Springs) and in the United States (Montana). This microcrystalline type of talc is mainly used for electroceramics, paints, and paper.

Crude talc colors are gray to green, sometimes pink, rarely white. The majority of the European and North American talc deposits contain only low- and medium-brightness ore. The pure and white macrocrystalline talc is rare with primary deposits located in Australia (Mount Seabrook), China (Liaoning, Guangxi), and India (Jaipur). Limited volumes are available in France (Trimouns) and Italy (Piemont).

Hi Talc products (Table 16.1) are pure minerals with very low levels of impurities made from ore of the Australian Mount Seabrook mine. The lamellarity of the mineral is distinct. The color is slightly green as a crude ore. After grinding, the powder becomes very bright. Mount Seabrook talc has been used for many years in plastics and will continue to be a long-term source for the polypropylene industry.

### 16.1.2 Relevance of Color Consistency of Talc in Polypropylene

Talc products are usually characterized by their powder brightness and fineness. White pure talcs impart an opaque shade to talc-filled polypropylene compounds. Impurities contribute to a lower brightness. Talc powders mix very well and all levels of brightness can be achieved by blending different ores. However, in talc-filled polymer systems the use of such blends should be avoided for color sensitive applications. The shade of the talc in the compound differs significantly from that of the powder, as the impurities become visible and dominate the color.

# Lamellar High-Purity Submicrometer Talc

TABLE 16.1 Comparison Between Bulk Density of Powder and Compacted Talc Products

| Talc trade names | Type | $D_{50}$ μm | $D_{99}$ μm | Bulk density (lb/ft$^3$) | Bulk density (g/cm$^3$) |
|---|---|---|---|---|---|
| HTP1 | P | 1.8 | 11 | 16 | 0.26 |
| HTP1c | C | 1.8 | 11 | 56 | 0.90 |
| HTP05 | P | 1.4 | 10 | 16 | 0.25 |
| HTP05c | C | 1.4 | 10 | 56 | 0.90 |
| HTPultra10 | P | 1.1 | 6–7 | 13 | 0.21 |
| HTPultra10c | C | 1.1 | 6–7 | 56 | 0.90 |
| HTPultra5 | P | 0.5 | 5–6 | 11 | 0.18 |
| HTPultra5c | C | 0.5 | 5–6 | 56 | 0.90 |

P, powder; C, compacted.

Therefore, it is recommended to use talcs from suppliers only with homogenous individual talc ore bodies, instead of talcs from processors using a variety of sources.

Figure 16.1 exhibits color shades of talc filler versus compounded materials. The color of talc filled plastics can be simulated without actual compounding.

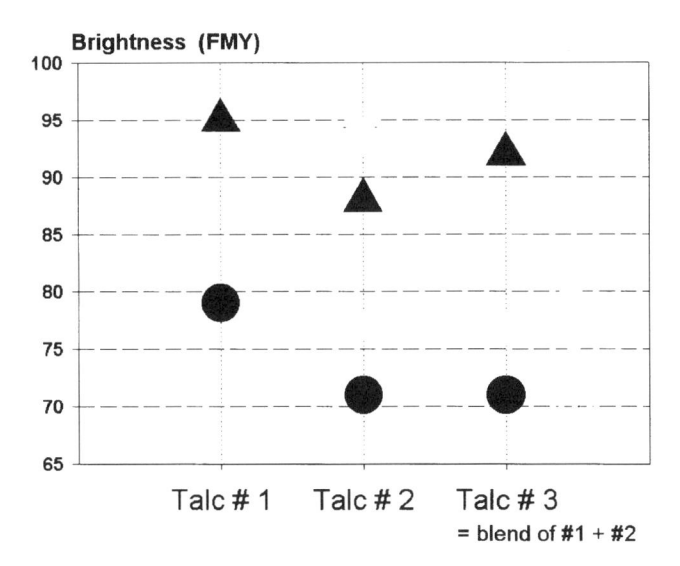

FIGURE 16.1 Talc blends and color shades in compounds. ▲, talc powder; ●, talc-filled polypropylene.

A paste obtained by mixing talc powder with oil, a plasticizer or variety of organic liquids exhibits a color of certain brightness depending on talc chemistry. The determination of "wet brightness," analogous to an oil number in the paint industry, provides an indication of final shade in a compound and might also be used for the quality control of incoming lots of talc raw material.

### 16.1.3 Delamination of Minerals and Definition of Submicrometer Talcs

The fineness of a mineral is described by its top cut and medium particle size. The top-cut $D_{98}$ or $D_{99}$ corresponds to 98% or 99% percentage of talc remaining on a defined-sieve screen that is equivalent to an upper particle size given in micrometer. Standard talcs show top cuts ($D_{99}$) of 30–70 µm, with medium particle sizes ($D_{50}$) of 5–15 µm. Micronized talcs are defined by a top cut of 25 µm or less, showing a $D_{50}$ of 4–1.5 µm. Products with a $D_{50}$ of around 1 µm and less are referred to as submicrometer talcs.

The plastics industry is primarily interested in high aspect ratio talc products because the reinforcing properties depend on this feature. The commercial talc refinement process is called *delamination*. Roller mills can produce standard talcs without destroying the platelets. A jet mill type of grinding equipment is normally used as a micronizer. In this process, standard ground products are utilized as feed for the micronizers and the talc booklets are shot against each other for further delamination. Submicrometer talcs are made by novel technologies, also securing a high aspect ratio.

Cheaper processing systems often consist of impact and/or ball mills. However, these systems are more useful for carbonates and barites than for lamellar minerals, as the talc lamina would break rather than delaminate.

In plastics compounding, only talc products produced by smooth delamination and of good aspect ratio are able to contribute to stiffness and rigidity, dimension stability, and a well-balanced impact performance.

## 16.2 DENSIFIED AND COMPACTED TALCS

Micronized and submicrometer talcs are low bulk density powders. After grinding and pneumatic transport of these products, an excess of air entrapped between particulate talc platelets causes fluidization of the material. Such fluffy materials are difficult to bag, transportation and storage occupy more volume, and more dust is created at the talc user. In addition, the compounder might encounter feeding problems and output reductions.

Solutions to this problem are densification or compaction of the filler. A densification process mainly reduces the air content in minerals or pigments for bagged materials. Volume reductions of 20–30% are possible. However, most or all of the densification effect is lost after unloading and pneumatic transport. For bulk shipments, pellets must be produced using a different compaction technology. The pellets need to be durable enough to maintain higher bulk density even after transport in bulk trucks or rail cars, after unloading at the plant silos, as well as after pneumatic transport within a compounding plant. Commercial technologies are available for the production of these pellets.

Some micronized and all submicrometer talcs must be compacted in order to make them suitable for packaging, transport, and use in plastics compounding operations. Table 16.1 makes a comparison between bulk density of powder and compacted talc products.

It is very important that pelletizing and compaction be accomplished without any addition of binders. All types of additives bear the risk of interaction with the resin or with other ingredients in the compounds. In addition, some migration might take place in due time, resulting in low paint adhesion or inhomogeneous surface properties.

Compacted talc pellets of good quality are durable enough to be transported by pneumatic systems but do not bear the risk of maintaining agglomerates during the compounding process as a consequence of being too hard to break up into fine particles of talc. A photograph of compacted talc is shown in Fig. 16.2.

## 16.3 USE OF MICRONIZED AND SUBMICROMETER TALCS FOR POLYPROPYLENE NUCLEATION

Polypropylene is a thermoplastic polymer with semicrystalline morphology. During the cooling process after manufacturing, the crystalline structure changes

FIGURE 16.2 Photograph of compacted talc.

according to the thermal conditions. The incorporation of fine particles, so-called nucleating agents, into the semicrystalline polymer matrix can induce nucleation of finer crystals at higher temperatures.

The main crystalline structures are crystallites and spherulites. Crystallites are obtained by severe cooling, as with cold water, to form microcrystalline structures. The main effects are higher transparency and higher dimensional stability after molding. Spherulites are obtained by slow cooling, as with hot water and/or air, to form larger crystalline structures characterized by opacity (low haze) and distortion, which produces warpage and low dimensional stability of the molded items.

Nucleating agents can initiate the crystallization at higher temperatures. This leads to faster development and higher rates of crystallinity in higher numbers, as well as smaller spherulites. One way to reduce negative effects on transparency and dimensional stability is to introduce some additives to the resin, which lead to the formation of microcrystals (crystallites) because of their heterogeneous nucleating effects. Nucleants have to be inert to avoid negative interactions between stabilizers and processing aids.

Such nucleated polypropylenes are addressed to applications that require

- High injection cycles (high productivity because of lower cooling time in the mold)
- Better rheology of the resin for blow molding (hollow items)
- Extrusion film production with higher output and lower thickness
- Better dimensional stability of manufactured items

Well-known additives for this purpose are salts, e.g., sodium benzoates, aluminum hydroxybenzoates. These salts have the ability to modify the kinetics of the crystallization in many semicrystalline thermoplastic polymers, such as polypropylenes, polyesters, and polyamides.

Talc has become more important as a nucleant in recent years as high-purity, micronized talcs have become available at reasonable prices. High purity talcs are important, as any disadvantageous property of the talc, such as the presence of heavy metals, leads to multiplication of errors due to fineness and a high specific surface.

The purpose of the following studies was the evaluation of the nucleating potential of some high-purity talc grades in comparison to typical nucleating agents such as benzoates. The talc products used in these studies were Hi Talc premium grades based on the Australian Mount Seabrook talc ore. These grades differed in particle size distribution.

Fine talc products are also effective as bubble nucleators for polymer foams. An important parameter is the particle size, which influences the bubble size. Purity is less important in this type of nucleation.

### 16.3.1 Talc as Nucleant in Polypropylene Homopolymer

A series of experiments were designed with different fineness of the talc products ($D_{50}$ values of 1.4–0.4 µm) and increasing amounts of micronized talc. HTP05 talc (mean diameter $D_{50}$ of 1.4 µm) in powder form was added to general-purpose polypropylene homopolymer (Valtec HS008 with MFR of 8 manufactured by Basell-Europe) at 0.2–1 wt% filler levels (0.2, 0.5, and 1.0 wt%). Mechanical property response to talc powder addition is given in Table 16.2. There is an increase in tensile strength and modulus, without any decrease in impact resistance. The tensile modulus and thermal property response to talc addition is given in Figure 16.3. In comparison with unfilled polypropylene, there is an abrupt increase in both crystallization temperature and heat distortion temperature (HDT of 455 kPa) at 0.2 wt% talc level. At higher talc levels, the thermal property values remain at a plateau level.

For compounders that have problems with dosing small amounts of talc, the use of a masterbatch (MB) is an excellent alternative. In one experiment, a 20 wt% talc-filled masterbatch was produced and added for nucleation. At the equivalent talc content of 0.6 wt% talc, we observed an even higher nucleating effect than with on-line compounding using a talc powder. This enhancement is indicated by a much higher tensile modulus value (Table 16.2) than the trendline value shown in Figure 16.3 for powdered talc addition. The assumption is that the dispersion of small percentages of additives is improved by using a masterbatch. However, using an MB with more constituents than just talc is recommended to increase the cost effectiveness.

The use of submicrometer talcs leads to further improvement, as the high specific surface seems to be directly proportional to the nucleating effect. For comparison, an aluminum benzoate hydrate was tested at the same dosage rate of 0.5 wt%. The test results given in Table 16.3 and Fig. 16.4 indicate that increasing fineness in talc (HTP Ultra 10 with $D_{50}$ of 1.1 µm) improves the HDT and crystallization temperature, as well as the tensile modulus and the notched impact properties. However, aluminum benzoate hydrate showed by far the best results in terms of mechanical properties at the same dosage rate (Table 16.3).

### 16.3.2 Talc as a Nucleant in Polypropylene Copolymer

In the past, the nucleant properties of talc were primarily used for polypropylene homopolymer. Today we can also find a wide range of polypropylene copolymer grades for which nucleation has became the standard, especially for injection molding applications.

For these studies, low- and high-ethylene ($C_2$) block polypropylene copolymer were selected. The following nucleating additives were compared: a micronized talc (HTP05), sodium benzoate (NaB), aluminium benzoate hydrate

TABLE 16.2 Talc as a Nucleant in Homopolymer Polypropylene in the Forms of Powder and Masterbatch

| Nucleating talc | HTP05 | | | | | | 0.6 added as MB |
|---|---|---|---|---|---|---|---|
| PP homopolymer | Valtec HS008 | % | 100 | 0.2 99.8 | 0.5 99.5 | 1.0 99.0 | 99.4 |
| Yield strength | ISO 527 | MPa | 26.8 | 28.5 | 29.4 | 29.4 | 29.5 |
| Elongation at yield | ISO 527 | % | 13.9 | 13.4 | 11.2 | 11.1 | 10.4 |
| Strength at 350% elongation | ISO 527 | MPa | 21.0 | 22.2 | 22.7 | 23.1 | 23.1 |
| Elongation | ISO 527 | % | 350 | 350 | 350 | 350 | 350 |
| Tensile modulus | ISO 527 | MPa | 1549 | 1545 | 1680 | 1740 | 1.860 |
| Izod impact notched | ASTM D256 | J/m | 33 | 33 | 35 | 31 | 37 |
| Izod impact unnotched | ASTM D256 | J/m | NB | NB | NB | NB | NB |

# Lamellar High-Purity Submicrometer Talc

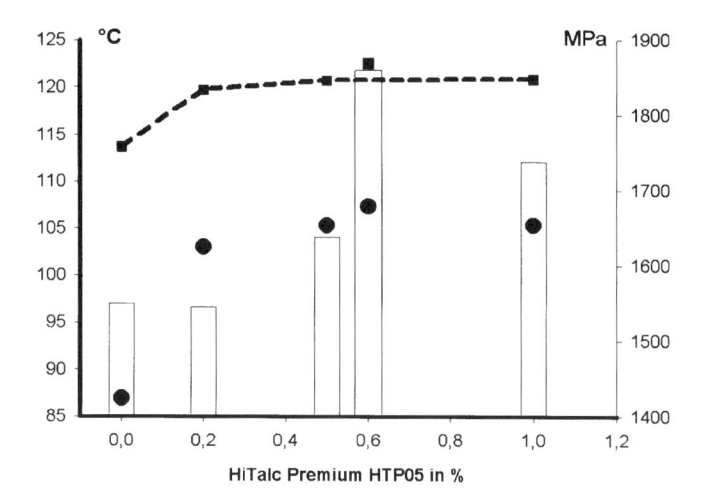

**FIGURE 16.3** Talc as a nucleant for polypropylene homopolymer. ☐, tensile modulus; ●, HDT–455 kPa; --■--, crystallization onset temperature.

(AlBH), and Sodium di-*tert*-butyl phosphate (NaDTBP). All of these products are used worldwide. Special attention was give to a combination of different nucleating agents.

In the first test series, a high ethylene–containing block polypropylene copolymer was used. Figure 16.5 describes the performance of nucleants for high-$C_2$-polypropylene block copolymers. The Na salt of DTBP alone showed the best performance at a dosage rate of 0.2 wt%; however, a blend of 0.05 wt% NaDTBP with 0.45 wt% HTP05 gave similar effects but with a better cost performance. NaB suffers by a low profile in such a system. Talc alone is used at rates of 0.5 and 1.0 wt% in practice, as even these small amounts of talc improve HDT and rigidity significantly, facilitating inexpensive nucleation.

The second series was based on a polypropylene block copolymer with lower ethylene content (Fig. 16.6). NaB and AlBH showed minor differences concerning the crystallization onset behavior. Adding micronized talc in amounts ranging from 1.0 to 1.5 wt% resulted in slightly lower crystallization temperatures but equal or even higher tensile modulus. This balanced profile is of special interest to resin producers.

## 16.4 USE OF SUBMICROMETER TALCS IN POLYPROPYLENE FOR AUTOMOTIVE APPLICATIONS

The largest single outlet for talc in filled plastics is automotive polypropylene applications for interior, exterior, and under-the-bonnet parts. The past 20 years

TABLE 16.3 Talc as a Nucleant in Homopolymer PP in Comparison with Al-benzoate

| | | Valtec HS008 | HTP1 | HTP 05 | HTPUltra 10 | Al-Benzoate hydrate |
|---|---|---|---|---|---|---|
| Medium particle size of talc (μm) | | | 1.8 | 1.4 | 1.1 | |
| Nucleating agent (%) | | 0 | 0.5 | 0.5 | 0.5 | 0.5 |
| Yield strength | ISO 527 / MPa | 26.8 | 28.8 | 29.4 | 30.3 | 31.2 |
| Elongation at yield | ISO 527 / % | 13.9 | 11.3 | 11.1 | 10.4 | 9.5 |
| Flexural modulus | ISO 527 / MPa | 1549 | 1570 | 1580 | 1880 | 1916 |
| Tensile modulus | ISO 178 / MPa | 1630 | 1850 | 1880 | 1961 | 1967 |
| Max.flex. stress | ISO 178 / MPa | 42.3 | 48.1 | 48.7 | 50.5 | 51.8 |
| Izod impact notched | ASTM D256 / J/m | 33 | 31 | 31 | 35 | 34 |
| Izod impact unnotched | ASTM D256 / J/m | NB | NB | NB | NB | 730 |

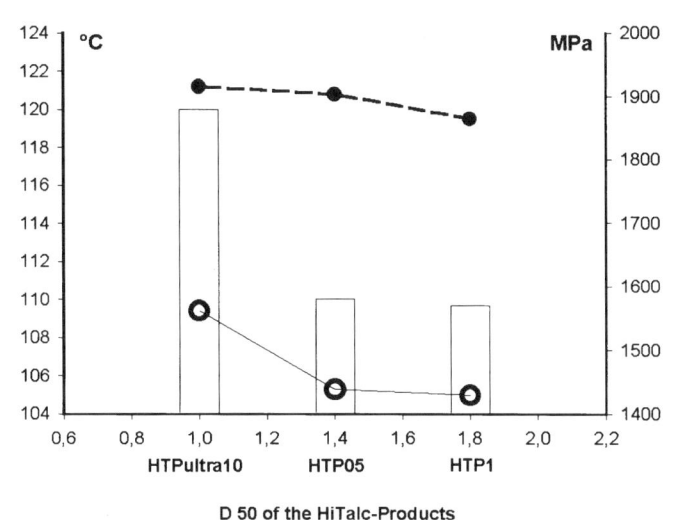

# Lamellar High-Purity Submicrometer Talc

**FIGURE 16.4** Talc as a nucleant in polypropylene homopolymer: variation of fineness. ☐, tensile modulus; ○, HDT–455 kPa; ▲, crystallization onset temperature.

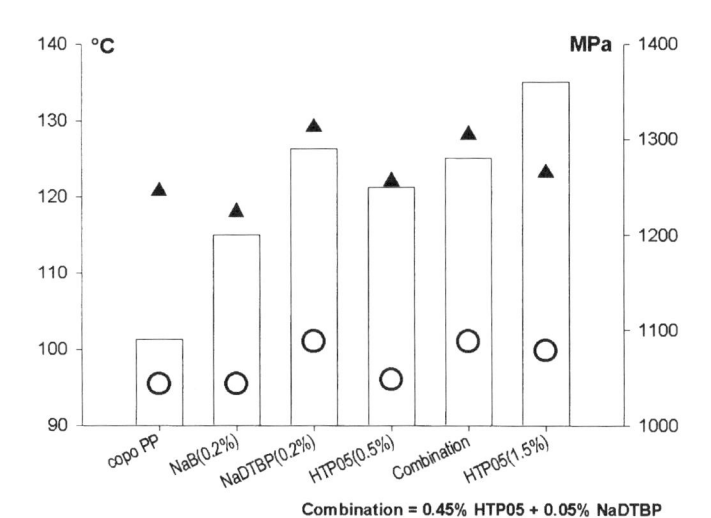

**FIGURE 16.5** Performance of nucleants for high ethylene block polypropylene copolymer. ☐, tensile modulus; ○, HDT–455 kPa; ▲, crystallization onset temperature.

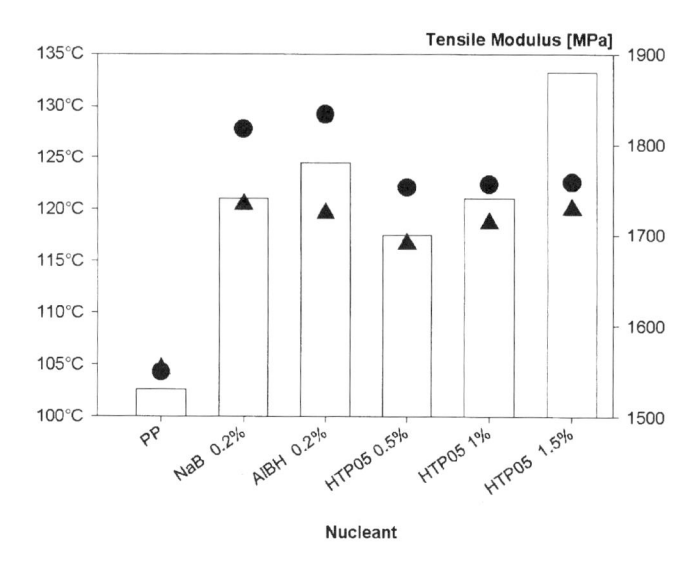

FIGURE 16.6 Performance of nucleants for low ethylene block polypropylene copolymer. ☐, tensile modulus; ●, crystallization onset temperature; ▲, HDT–455 kPa.

has seen tremendous changes in automobile design and material selection. Much of the change resulted from the demand to reduce weight and, hence, fuel consumption. The development of polypropylene/EPDM (ethylene-polypropylene-diene monomer) blends and polyolefin reactor material has facilitated this fast development and shorter model cycles.

### 16.4.1 Performance of Polypropylene Composites

Automotive polypropylene compounds normally contain 10–40 wt% talc. Talc as a filler is incorporated not only to reduce the costs of raw materials but also to impart specific properties to the polymer. Higher stiffness and rigidity, reduced shrinkage, low thermal expansion (zero gap), controlled morphology (painting), and better flow for large parts during molding are expected. Nevertheless, impact properties and scratch resistance are often poor with standard and micronized talcs. Smell reduction also calls for sophisticated formulation techniques. Moreover, talc can contribute to the reduction of cycle times and influences the surface quality of the molded parts.

*Recycling, Environment, and Safety.* In today's world, the recycling idea is strong. Polypropylene is a good example of how unification of polymers can lead to the establishment of a standard. Talc has a similar position within the functional fillers for automotive applications. This mineral is by far the most widely used functional mineral filler and contributes to an ample range of features due to its variations in fineness and shapes. Environmental and safety regulations changed the nature of the raw materials used for certain parts. Monomaterial concepts are favored in order to enable easier recycling procedures. Talc-filled polypropylene has an established position in this field. Another ecological aspect is the weight reduction of cars so as to reduce fuel consumption. Micronized and submicrometer talc-filled bumper systems are able to reduce wall thickness significantly. Increased security regulations concerning improved side impact are implemented, a problem for which talc-filled compounds offer various solutions.

*Interior Applications.* Most of the interior applications contain mineral-filled materials. Talc-filled composites are used for dashboards, door panels, pillars, scuff plates, armrests, consoles, but also for rear-end spare tire covers and seat backs, etc.

In response to the recycling demands, there are already monomaterial dashboard concepts consisting of talc-filled polypropylene carriers, polypropylene foams, and thermoplastic olefin (TPO) skins, replacing acrylonitrile-butadiene-styrene (ABS), styrene maleic anhydride (SMA), polyurethane rubber (PUR), and polyvinyl chloride (PVC). Mineral loading levels for interior applications started at 40 wt% but have been decreasing significantly over the years. Today micronized and submicrometer talcs are able to contribute much higher mechanical properties to an advanced compound compared to former standard fineness grades used during the last decade. Consequently, higher mechanical properties are achieved at lower mineral loading levels with reduced weight and improved surface properties.

Early in the automotive development of interior trim, these molded parts were colored to be as close to the final shade as possible and were painted to give the final color. This also helped to mask mold imperfections and to improve weldline strength. However, painted parts have lower durability in high-scuff areas. Today more and more interior trim parts have molded-in color, thereby avoiding the subsequent steps of painting. This technology calls for high scratch resistance surfaces and, hence, lower filler levels to reduce stress whiting on marred surfaces. Blends of minerals contribute to this feature, but the reduction of the overall filler content remains the goal.

Additional end-use requirements are high solvent and detergent resistance, as well as weatherability against thermal oxidation and medium UV resistance. Talc as an inert mineral with extremely low solubility can contribute substantially to these requirements.

*Exterior Applications.* Bumper systems and side-impact panels are the largest parts. When plastic bumpers were first introduced, designers continued to think in terms of classic metal beam concepts and the plastic shell had mainly aesthetic functions, was simply shaped, and black. Further developments integrated both optical and safety features. The new generation bumpers are designed to minimize the damages resulting from low-speed impacts. At first, TPO bumpers were made solely by compounding polypropylene with EPDM, but without any fillers. Today these polypropylene-rubber blends are still in use, but micronized talcs are implemented to improve dimension stability. Recent developments showed that reactor TPOs (heterophasic copolymers) are able to provide similar properties, eliminating the compounding step. This cost advantage exists only as long as the compound formulations do not contain functional fillers. Talc could be implemented by on-line compounding at low costs; a separate compounding step is necessary in most cases to introduce the filler.

The main challenge for plastic bumper-steel body concepts is meeting crucial in-service thermomechanical requirements. Talc can significantly reduce the coefficient of linear thermal expansion (CLTE). The content and fineness of talc, as well as the polymer matrix, play dominant roles. Standard talc products used to influence the impact properties negatively. New micronized and submicrometer talc grades improved the impact resistance, leading to the breakthrough of talc-filled bumper systems. The thin-wall bumper fascia started in Japan and recently spread to Europe and the United States. Originally wall thickness was in the range of 3.5 mm and more; today's targets are 2.3–2.8 mm. The goal is to reduce weight, cycle times and overall costs. It must be mentioned that especially notched Izod and dart impact at low temperatures ($-30°C$) are relevant.

A typical specification of such a compound is as follows:

- Flexural modulus around or above 1400 MPa.
- Sufficient dart impact resistance at $-30°C$.
- Good material flow in molds on standard injection machines; melt flow rate (MFR) targets are from 15 to 30 g/10 min.
- For a zero-gap bumper fascia, the coefficient of thermal expansion needs to be less than $70 \times 10^{-6}$ $K^{-1}$; for side sills ever tighter specifications must be reached ($5 \times 10^{-6}$ $K^{-1}$ or less).
- UV resistance for unpainted parts.

Nowadays, besides bumper systems, side body claddings and fender liners are also made of talc-filled TPO. In exterior applications, the plastic parts are often painted in the body color. Paint adhesion and surface finish are key issues. The nonpolar nature of TPO results in low adhesion to the polar water-borne paint systems. For improvement, the surface of the bumper fascia needs to be flamed (oxidized) or plasma treated in combination with primers. Moreover, a good morphology of the compound can help to improve the paint adhesion. As the result of recent developments, there are some talc-filled compounds that need neither flame treatment nor any adhesion promoter in the primer.

The surface appearance is important for painted and unpainted parts. Increasing filler content normally decreases the gloss and in some cases leads to irregularities in the surface. However, micronized and submicrometer talcs are generally able to improve this situation and to contribute significantly to reaching the desired appearance properties.

*Under-the-Bonnet Applications.* Talc-filled polypropylene composites are often used for HVAC ducts, coolant reservoirs, grill assemblies, vapor canisters, etc. The mineral reinforcement contributes stiffness and dimension stability at higher temperatures. In addition, talc filler is cost effective. For these applications, standard-fineness talc grades are used for financial reasons.

Thermal stability is an important property for under-the-hood applications. Following a common yet imprecise thesis, mineralogy and iron content are considered to be dictating factors influencing thermal stability of the talc composite. The general slogan was: "the purer the talc, the better the heat stability." Unfortunately, experience has shown that the story was not as simple as some parties thought and that scientific results were rarely in accordance with the common belief. Looking at different products with high talc content, purity was found not to be the only decisive factor. The chemical analysis alone does not allow any conclusions as to the thermal degradation behavior of different talc grades. The crystallinity of the talc and the resulting specific surface, as well as the fineness of the talc grades, also have important effects on degradation properties. Micronized talcs of the same morphology and with similar specific surface showed similar results. A higher specific surface and, hence, a higher absorption of stabilizers outweighed the differences in iron content by far. The texture of the talc product has a substantial impact; because microcrystalline talc types with high surface area significantly reduce thermal stability. A similar effect was observed with carbon black.

The heavy metal ions of the talc products can either be located within the crystal lattice of the mineral or be part of the by-mineral. Fe ions are able to form a solid solution within the dolomite and/or magnesite. In a chlorite-type talc, the majority of the Fe ions are linked to the crystal lattice of the chlorite. As long as the heavy metals are part of the crystal lattice of the chlorite-type talc, their impact on heat stability is low. Fe ions have a stronger impact on heat aging when bound to the magnesite, for example.

### 16.4.2 Performance of Submicrometer Talc in Polypropylene Copolymer

A general-purpose polypropylene block copolymer (PP/J-762 HP manufactured by Idemitsu–Japan) with MFR equal 16 g/10 min was used for this study. The ethylene content was at a level of about 5 wt%. This type of polypropylene is commonly used for automotive applications requiring impact-resistant molded parts, e.g., for Japanese cars.

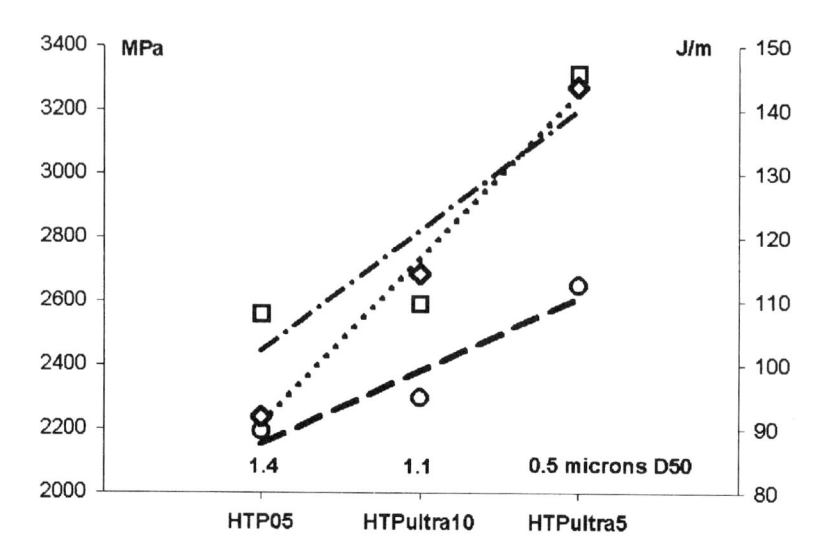

**FIGURE 16.7** Performance of a 20 wt% talc-filled polypropylene copolymer. ○, flexural modulus; □, tensile modulus; ◇, notched Izod. - - - -, trendline flexural modulus vs. talc ($D_{50}$); ·····, trendline tensile modulus vs. talc ($D_{50}$); ············, trendline notched Izod vs. talc ($D_{50}$).

The mechanical properties depended very much on the fineness of the filler material. In this study, talcs ranging from $D_{50}$ 1.4 to 0.5 µm were investigated. Increasing fineness of the talc contributed to stiffness and rigidity. Figure 16.7 portrays the performance of 20 wt% talc-filled polypropylene copolymer for various grades of micronized and submicrometer sizes (Table 16.4).

Platy fillers and those with coarser particle size distribution generally showed poor impact properties. To the contrary, submicrometer talc products were capable of significantly improving the impact resistance. Depending on the resin properties, low temperature impact properties were better in some cases. The submicron talcs were able to positively influence Izod impact, but also Dart impact properties. Full falling dart ductility at −20°C could be achieved with several formulations if using the finest submicrometer talc ($D_{50}$ equal to 0.5 µm) of the series.

### 16.4.3 Composites of Submicrometer Talc in Polypropylene Copolymer with Rubber Modification

Talc-reinforced polypropylene blends gain increased mechanical property performance by the addition of rubber as an impact modifier. The combination of

# Lamellar High-Purity Submicrometer Talc

**TABLE 16.4** Micronized and Submicronized Talc in Polypropylene Copolymer

| Formulations | Unit | Method | PP | HTP05 | HTPUltra10 | HTPUltra5 |
|---|---|---|---|---|---|---|
| Medium particle size of talc | $D_{50}$ | | | 1.4 | 1.1 | 0.5 |
| Idemitsu PP/J—762 HP | PP copolymer | | 100.0 | 79.5 | 79.5 | 79.5 |
| Talc | | | | 20.0 | 20.0 | 20.0 |
| Stabilizers | | | | 0.5 | 0.5 | 0.5 |
| Compounding Parameters | | | | | | |
| Interval temperatures | °C | | | | | |
| | 190÷220 | | | | | |
| Screw speed | rpm | | — | 240 | 240 | 240 |
| Output | kg/hr | | — | 5.5 | 5.5 | 5.0 |
| Electric power | A | | — | 11.2 | 10.8 | 10.8 |
| Physical Properties | Unit | Method | | | | |
| MVR (230°/2.16) | cm³/10' | ISO 1133 | 15.90 | 13.8 | 13.9 | 15.4 |
| Ashes (1 hr; 625°C) | % | EN 60 | 0.45 | 19.7 | 19.4 | 20.9 |
| Mould shrinkage longitudinal | % | Superlab | 1.56 | 1.01 | 0.98 | 1.06 |
| Mould shrinkage transverse | % | Superlab | 1.94 | 1.20 | 1.20 | 1.29 |
| Mechanical Properties | Unit | Method | | | | |
| Tensile stress at yield | MPa | ISO 527 | 19.9 | 20.7 | 20.8 | 21.9 |
| Tensile stress at break | MPa | ISO 527 | 15.8 | 16.1 | 16.3 | 16.2 |
| Elongation at break | % | ISO 527 | >350 | 341 | 342 | 69 |
| Tensile modulus | MPa | ISO 527 | 1291 | 2558 | 2590 | 3312 |
| Flexural modulus | MPa | ISO 178 | 1175 | 2191 | 2297 | 2648 |
| Maximal flexural stress | MPa | ISO 178 | 29.7 | 34.1 | 34.7 | 37.6 |
| Izod impact notched @ 23°C | J/m | ASTM D256 | 270 | 92 | 114 | 144 |
| Izod impact notched @ −10°C | J/m | ASTM D256 | 88 | 46 | 50 | 54 |
| Thermal properties | Unit | Method | | | | |
| HDT (1820 kPa) | °C | ASTM D648 | 92.5 | 116.9 | 118.7 | 125.3 |

homopolymer and copolymer polypropylene with rubber provides access to a wide range of properties, attainable previously by more expensive engineering resins. Compounding equipment is available in a wide range of output rates up to 10 tons/hr. Continuous mixer techniques provide additional process variables to produce desired composite properties at relatively low cost. This high degree of flexibility in process technology and raw material composition offers the compound formulator a good tool for achieving special properties. Since compounding is an additional step in the overall cost structure, TPO products are more expensive than reactor-TPO products when produced in significant volumes.

The resin used in this study was a high-impact copolymer polypropylene (Finaprop PPC 9760, Fina Chemicals, Belgium) with MFR 25. Polypropylene and impact modifier (Exxelor 703F1, ExxonMobil) were fed at the main feed port. Talc was fed downstream to the molten polymer via a corotating twin-screw extruder. Table 16.5 summarizes the formulation and composite properties of the experiment.

Submicrometer ($D_{50}$ of 0.5 µm) talcs showed only slightly better performance than micronized ($D_{50}$ of 1.8 µm) talcs in this formulation. Since the high rubber content dominates the properties, the finer talc does not induce improved performance in such a system. Rubber has the higher viscosity at increasing shear and temperature. As the rubber absorbs more energy, talc has the tendency to approach the rubber surface and to distribute well, avoiding coagulation of the rubber cells. These results could be reproduced in further investigations by varying both the filler content from 10 to 20 wt% and the mean particle size range from 1.8 to 0.3 µm.

Another experimental approach to understanding the role of talc particles in a rubber-modified block copolymer polypropylene is differential mechanical thermal analysis (DMTA). The DMTA values are given in Table 16.6. The measured difference in the rubber transition temperature between filled and unfilled rubber-polypropylene blends is the key characteristic determined by this probe of the composite microstructure. Talc lowers the transition temperature 7–10°C, which results in an improved cold impact resistance.

## 16.5 COMPOUNDING PERFORMANCE OF SUBMICROMETER TALCS

There are many variables in a compound formulation. In addition, the equipment and the processing parameters determine the dispersion of fillers and pigments, as well as the final properties of the plastic. Micronized talcs and, to an even higher degree, submicrometer talcs significantly influence the processing parameters.

# Lamellar High-Purity Submicrometer Talc

TABLE 16.5 Micronized and Submicronized Talcs in Polypropylene Copolymer + Rubber Blends

|  |  |  | HTP1 | HTPUltra5 |
|---|---|---|---|---|
| Medium particle size ($D_{50}$) | μm |  | 1.8 | 0.5 |
| Formulations |  |  |  |  |
| Finaprop PPC 9760 |  |  | 74.5 | 74.5 |
| Impact modifider |  |  | 15.0 | 15.0 |
| Talc |  |  | 10.0 | 10.0 |
| Stabilizer |  |  | 0.5 | 0.5 |
| Mechanical Properties | Method |  |  |  |
| Tensile stress at yield |  | MPa | 18.8 | 19.1 |
| Tensile strain at yield |  | % | 6.2 | 5.8 |
| Tensile stress at break | ASTM D638 | MPa | 17.6 | 17.5 |
| Tensile strain at break |  | % | >450 | >450 |
| Tensile modulus |  | MPa | 1870 | 1980 |
| Flexural modulus | ASTM D790 | MPa | 1624 | 1718 |
| Max. flexural stress | ASTM D790 | MPa | 32.4 | 34.2 |
| Izod impact notched @ 23°C | ASTM D256 | J/m | 557 | 623 |
| Izod impact notched @ −20°C | ASTM D256 | J/m | 104 | 126 |
| Physical Properties |  |  |  |  |
| MVR (230°C; 2.16 kg) | ISO 1133 | cm$^3$/10' | 19.76 | 20.24 |
| Ashes (1 hr; 625°C) | EN 60 | % | 9.7 | 9.9 |
| Thermal Properties |  |  |  |  |
| Mould Shrinkage Longitudinal | Superlab | % | 1.15 | 1.03 |
| Mould Shrinkage Transverse | Superlab | % | 1.53 | 1.44 |

## 16.5.1 Feeding of Powder and Compacted Talcs in Twin-Screw Units

There is a common opinion that melt feeding always offers advantages over feeding at the main feed hopper. The argument is that a mixture of resin granules and rubber chips with powdered talcs is not homogeneous enough in density or particle size and exhibits a tendency to segregate. Consequently, side feeding is favored and became increasingly popular, especially for twin-screw compounding extrusion. The use of talcs up to a fineness of $D_{50}$ equal to 4 μm normally does not cause any problems if these fillers are introduced at the main feed. A side feed can handle talcs as fine as 2 μm without any reduction in output. By going to finer grades of talc, these fluffy powders generate problems at both feed ports. In order to resolve this problem, compaction or densification of the talc filler is needed. The target is to achieve good feeding characteristics of fine-grade talc filler to attain enhanced composite properties at commercially meaningful output rates and higher talc loading levels.

TABLE 16.6 DMTA Characterization of Talc-Filled Materials

| DMTA | PP copolymer + rubber | HTPultra5 | HTP1 | Remarks |
|---|---|---|---|---|
| | −80°C +200°C | −100°C +150°C | −100°C +150°C | Temperature range of DMTA analysis |
| tan δ (°C) 1 | −43.0 | −40.9 | −49.9 | −49.1 | Glass transition temperature of EP rubber phase |
| 2 | 14.7 | 14.7 | 11.2 | 8.9 | Glass transition of amorphous PP (atactic morphology) |
| 3 | 74.5 | 72.1 | 82.7 | 73.9 | Secondary glass transition of EP rubber phase |
| 4 | 171.7 | | | | Crystalline transition of isotactic PP |

# Lamellar High-Purity Submicrometer Talc

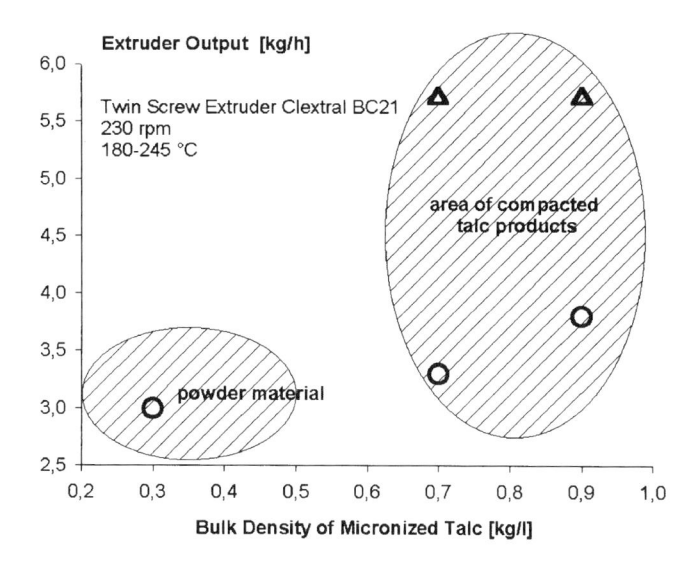

**FIGURE 16.8** Output improvement by compacted talc. △, extruder output for 40 wt% talc loading; ○, extruder output for 20 wt% talc loading.

Figure 16.8 makes a comparison between output improvement for compacted talc grades. The benefits of using compacted grades are enormous, especially if only simple equipment is available or in times of shortage in extruder capacity. Moreover, advantages in handling, as described in Sect. 16.2, have to be considered.

## 16.5.2 Performance Differences Between Main and Side Feed

Compacted talc filler can be fed effectively either with the dry blend of ingredients at the main feed hopper or to molten rubber–modified polypropylene at the side port location. Besides feeding and output characteristics, significant differences in composite properties can result from the option of feed port location for compounding twin-screw extrusion. An experimental study was conducted to investigate the performance differences between main and side feed.

The formulation consisted of polypropylene copolymer combined with a polyolefin elastomer as impact modifier in combination with a submicrometer talc with a mean particle diameter of 0.5 μm. In the case of main feeding, a preblend was made with the resin, an impact modifier, filler, and additives. In the case of side feeding, the compacted talc was fed to the melt. The polymer blend had an

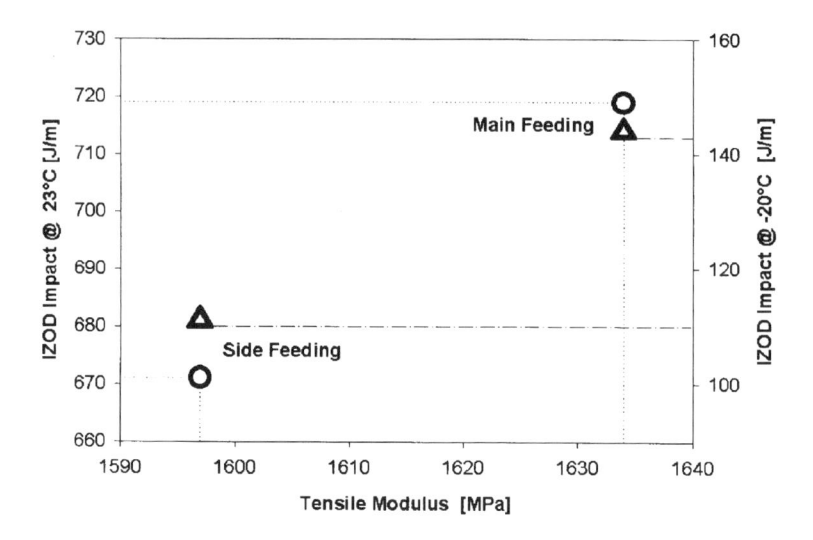

**FIGURE 16.9** Performance differences due to feed port locations. ○, Izod impact at 23°C vs. tensile modulus; △, Izod impact at −20°C versus tensile modulus.

MFR of 12–13 g/10 min. The processing temperature at the Clextral twin-screw extruder was between 170°C and 220°C.

In a second experimental series, a higher temperature range of 190–240°C was examined in order to evaluate the influence of viscosity. In this case, the dispersion of talc was obviously less at lower viscosity and, consequently, the impact values at 23°C and −20°C were slightly reduced.

The outcome of these studies was remarkable (Fig. 16.9). Feeding a compacted, submicrometer talc to the main port was superior compared to the more commonly used side feeding to the melt. Using the main feed as the filler port, the first segments of the screw mixed a dry blend of all the constituents. Compacted talc does not contain much air. This allowed the material temperature to increase quickly because of effective shearing of ingredients. The remaining screw segments are long enough to disperse the talc well within the rubber phase and the polypropylene matrix. When using the side port for feeding the talc to the melt, the remaining length of the screw is shorter, with a reduced degree of talc dispersion in the rubber-polypropylene interphase.

## 16.6 TREND FOR SUBMICROMETER TALCS

Submicrometer talcs are novel products and only a few grades are available to date. The current maximum fineness of 0.5 μm is not the end of the technological development. Besides the submicrometer minerals, we are also talking about nanominerals, which provide new ways to optimize the filler–polymer matrix interaction. One individual talc lamina has a thickness of about 20 Å, whereas the described submicrometer talcs are in the range of 300–500 Å. As we can see, talc is not far from the goal of being part of the nanomineral group of products.

The diversity of polypropylene compounds in automotive applications is assumed to grow further. Zero gap, soft nose, low wall thickness, high cold dart impact, good side impact, painted components, and the like are already part of today's car design. The area of mar resistance and the standardization of raw materials are other topics to be solved in the near future. Some automotive compounds are free of fillers today and represent a huge potential if technological solutions are found. All of these targets require special functional fillers, since they cannot be achieved using standard mineral fillers. The mineral industry needs to constantly invest in research and development work and capabilities in order to develop new generations of functional minerals, including surface treatment. New technologies facilitated a versatility of specialty talcs that will lead to a strong position of submicrometer talcs in automotive compounding.

## APPENDIX: EQUIPMENT, TEST METHODS, AND RAW MATERIALS

*Compounding Technology*

| | |
|---|---|
| Twin screw extruder | Clextral B 21 |
| Diameter (mm) | 25 |
| Length (L/D) | 36 |
| Screw speed (r.p.m.) | 50–700 |
| Specification | Corotating |
| Feeding | Two zones, variable main or side feed |
| Electric engine (KW) | 15 |
| Dosing   of dry-blend | Volumetric |
|          of separate additive | Gravimetric |
| Degassing | Forced venting |
| Die | Circular holes |
| Cooling | By water |
| Drying | By air |
| Heating | electric |

*Operative Conditions*

| | |
|---|---|
| Temperatures | 180–240°C |
| Specific energy | 0.10–0.30 kWh/kg |
| Output | 2–20 kg/hr |

*Specimen Injection*

| | |
|---|---|
| Injection equipment | Negri Bossi model V9-12 FA |
| Injection temperatures | 200–220°C/356–410°F |
| Injection speed | Medium |
| Total cycle | Optimized |
| Mould temperature | 40–50°C/104–122°F |

*Testing Methods*

| Instrument | Property | Test method |
|---|---|---|
| Instron Mod.4505 | Tensile | ASTM D638 |
|  | Flexural | ASTM D790 |
| Ceast pendulum | Izod impact, (un)notched | ASTM D256 |
| Ceast thermostatic bath | HDT | ASTM D648 |
|  | VICAT | ASTM D1525 |
| Gallenkamp oven | Thermal stability | Superlab |
| Mauser sliding gauge | Mold shrinkage | Superlab |
| Davenport grader; Ceast | Melt flow ratio | ASTM D1238 |
| Sartorius analytic balance | Density | ASTM D792 |
| Bicasa muffle furnace | Ashes content | Superlab |
| Dr. Lange gloss meter | Gloss | ASTM D523 |
| Erichsen hardness test rod 318 | Scratch resistance | Superlab |
| Perkin-Elmer DSC 7 | Differential scanning calorimeter | Superlab |

# 17

## Automotive Applications for Polypropylene and Polypropylene Composites

**Brett Flowers**
General Motors Corporation, Pontiac, Michigan, U.S.A.

### 17.1 INTRODUCTION

Polypropylene has long been used for various automotive applications. A wide variety of unfilled polypropylene and polypropylene composites can be found in the vehicle's interior, exterior, and under-the-hood. Polypropylene is often a desirable automotive material due to its low cost, colorability, chemical resistance, and UV stability. In addition, the range of potential polypropylene uses is nearly unlimited through the use of modifiers, additives, and fillers.

New developments in nanocomposites and slipping agents open the window of use for polypropylene even further. Nanocomposites exhibit good mechanical strength and stiffness with low-density parts at low reinforcement levels, whereas slip agents give improved scratch and mar resistance. Improvements in neat polypropylene as well as advancements in processing technology allow polypropylene to be the base polymer of choice for use in automotive applications previously thought to require more expensive acrylonitrile-butadiene-styrene (ABS), polycarbonate (PC), and other engineered materials.

## 17.2 TYPICAL POLYPROPYLENE APPLICATIONS

### 17.2.1 Interior

Polypropylene is one of the fastest growing interior materials for two main reasons. First, the overall material cost is lower than ABS and PC. In addition, due to the ease of UV stabilization, polypropylene parts do not require painting. This improves the recyclability of vehicle parts and greatly reduces the costs of components by eliminating the painting process.

*Trim Components.* Polypropylene is one of the more common plastics selected for trim components in automobiles today. It can often be found in molded-in-color applications for various trim items throughout the vehicle. Most pillar trim found in vehicles composed of a polypropylene copolymer. Garnish moldings are all the trim-colored body panels in the automotive interior. For example, the B-pillar garnish molding component consists of unfilled crystalline polypropylene copolymer. Copolymers with exceptional impact resistance are used to meet requirements of federal standards regarding head impact within the vehicle (FMVSS201).

The more structural parts often contain talc or another mineral to improve heat capabilities and stiffness. The upper trim component of the instrument panel may contain up to 26 wt% talc-filled polypropylene.

Many other interior components, such as glove box and console bins, as well as console housings, are molded-in-color polypropylene. Door trim consists of unfilled or talc-filled high-crystalline polypropylene with molded-in color.

*Functional Components.* Polypropylene can be compounded to meet many different component stiffness, impact, heat resistance, and appearance requirements.

Thermoplastic olefin (TPO) blends of polypropylene with metallocene plastomers are described in Chapter 7. In order to make automotive components having both toughness and processability, high melt flow resistance (low molecular weight) polypropylene resins are important ingredients in TPO formulations. This requirement poses a challenge to the material engineer because of the dual need to disperse high molecular weight elastomer into low molecular weight polymer matrix (high viscosity ratio condition) in order to attain effective impact modification.

Glass fiber- and/or mineral-reinforced polypropylene materials can often be found in structural parts throughout the vehicle. Armrest substrates and console substrates are often injection molded with 20–30% glass-reinforced polypropylene. These products offer a good combination of stiffness and impact while maintaining the lower specific gravity characteristics of polypropylene. In addition, components that utilize a TPO skin or polyolefin based skin are highly recyclable.

# Automotive Applications for Polypropylene

Pillar trim uppers consist of a wide variety of polypropylene materials depending on air bags that are in the vehicles. In the case where no side impact airbags are installed, highly crystalline polypropylene resin with a foam energy absorber or a polypropylene energy absorber will suffice. Another alternative to material of construction is medium-impact copolymer with molded-in ribs. If there are side-impact airbags, an unfilled high-impact copolymer with a notched impact value of greater than 10 ft-lb°F/in. is required. If a roof-rail airbag system exists, an interior-grade TPO is used to suit impact resistance over a wide temperature range between $-40°F$ in the extremely cold winter to over $100°F$ on hot summer days.

When impact resistance is not critical, it may be possible to use less expensive filler, such as talc or mica. Talc is inherently better for use in color-critical applications. The use of mica is reserved for black, non-show-type parts.

## 17.2.2 Exterior

*Weather-Resistant, Molded-In Color.* Polypropylene use is somewhat limited in the exterior of automobiles to air inlet panels and wheel house liners. However, TPOs, which are closely related to polypropylene, are frequently used for automotive exterior parts. Both polypropylene and TPOs offer excellent weatherability properties. Air inlet panels often require substantial weather resistance as these parts are subjected to extreme temperatures (high and low), sunlight, and rain or snow. These parts are often polypropylene copolymer with 10–20 wt% talc to offer improved stiffness and good impact. Many molded-in-color bumper fascias and body-side moldings are made with TPO. TPO offers excellent impact at a low cost and a low specific gravity. Bumper fascias are typically molded from an unfilled TPO whereas body side moldings typically contain anywhere from 5 to 20 wt% mineral filler. The filler helps improve stiffness while reducing the coefficient of linear thermal expansion.

*Painted Parts.* A significant reason for the choice of TPO over polypropylene resin is for paintability of the molded part. TPO can be made paintable through the use of adhesion promotors and special primers. Polypropylene is much more difficult to paint due to its superior chemical resistance. TPO is often the material of choice for bumper fascias due to its lower weight, lower cost, and impact resistance. In many cases, the same material is used for both painted parts and molded-in-color parts. These parts contain a special weathering package that improves paint adhesion.

## 17.2.3 Under-the-Hood

A variety of polypropylene compounds can be found under the hood. Figure 17.1 shows the location of the fuse block cover (A), radiator or fan shroud (B), and

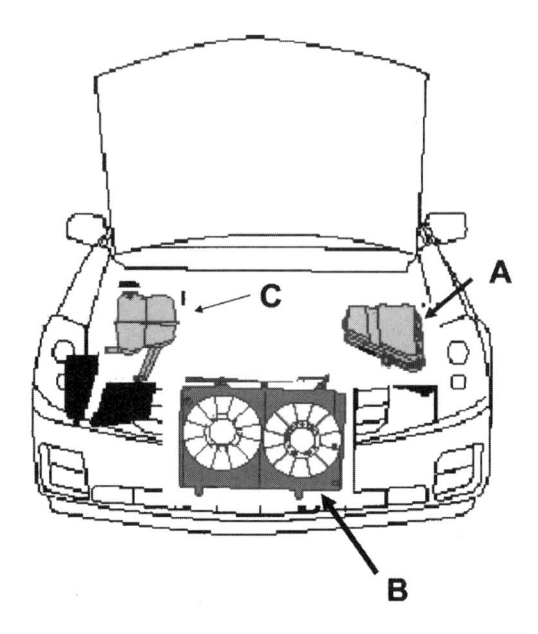

**FIGURE 17.1** Under-the-hood automotive parts.

coolant reservoir (C). Some of the close-out panels that are located further from the engine are molded-in-color unfilled copolymer. Sheet, injection-molded, and blow-molded grades of polypropylene are available, making it a very versatile material for under-the-hood applications.

Most under-the-hood parts consist of glass-reinforced or mineral-filled thermoplastic composites. These choices of material design maintain the required stiffness and dimensional stability of molded parts subjected to extremes of ambient temperature and fluctuating stress loads. For example, a radiator shroud component has a highly structural part design to maintain mechanical integrity under high heat loads and vibrations due to electric fan motor attachments. Besides static flexural stress loads, this under-the-hood component needs to be resistant to time-dependent fatigue and creep deformations at elevated temperatures (150–170°C).

In the case of trucks having fans that are clutch driven, there is much less heat load to contend with. Depending on the specific material requirements, materials used in these components range from 25 wt% mica-reinforced polypropylene copolymer to 20–30 wt% chemically coupled glass fiber-reinforced polypropylene homopolymer.

In more remote locations from the engine block, there is much less requirement for mechanical stiffness at elevated temperatures. The fuse block cover consists of 12 wt% mica-filled polypropylene copolymer. The coolant reservoir consists of only unfilled copolymer resin.

## 17.3 BENEFITS OF POLYPROPYLENE

### 17.3.1 Cost Reduction

Polypropylene resin often allows for substantial cost reduction of parts over most other plastic materials due to its low cost per pound, coupled to its low specific gravity. Many engineered materials can be replaced with a polypropylene compound at a lower cost per pound. In addition, since polypropylene offers excellent weatherability, many parts can have molded-in color versus applied paint. For example, for interior trim components, it is possible to replace a painted ABS with a molded-in color polypropylene compound. This will typically offer a savings in material costs but, more importantly, a substantial savings due to elimination of the painting operation.

### 17.3.2 Processing Advantages

Automotive applications using composites based on polypropylene resin provide a wide spectrum of molded parts that meet even more stringent mechanical and thermal specifications. Because of relatively low polymer cost and the narrow range of process temperatures required to injection-mold a variety of component parts, both the compounder and molder have an opportunity to make cost-effective products with minimal variation of process control and material design parameters. Hence, the increasing expectation for consistent product manufacture within ever tightening variation constraints can be attained. Furthermore, the recyclability of polypropylene-based materials helps to provide a reliable source of feedstock in an effort to reduce manufacturing cost and eliminate waste.

Thermoplastic composites based on polypropylene resin can be injection molded into a wide variety of automotive components having different wall thicknesses. With development of improved thermal and processing stabilizers, molded parts can be made with relatively tight constraints on melt flow rate in order to maintain consistent compounding and molding process conditions. Furthermore, the utilization of added peroxide for in situ controlled rheology of polypropylene resin during the compounding process (Chapter 10) permits a wider window for effective control on finished product flow characteristics.

Advances in the compounding technology for incorporating appropriate fillers and reinforcement fibers into polypropylene resin allow one to attain desired reinforcement while maintaining a good cost–performance balance. Development of chemical coupling additives (Chapter 3) enhances the mechanical integrity of composites produced with less raw material cost. Hence, the inherent incompatibility of polypropylene resin with polar ingredients is overcome by this improved chemistry at the interphase of the filler particle and the polymer matrix.

### 17.3.3 Specific Gravity or Part Weight

As a consequence of low specific gravity for polypropylene resin and metallocene elastomers, trim components made from TPOs have experienced significant growth in automotive applications. Furthermore, as a result of the advent of reinforced composites based on nanocomposite technology (see Chapters 3 and 20) to make structural parts having a combination of mechanical strength-stiffness, thermal resistance, and low specific gravity, there are possibilities for reducing cost while maintaining product performance.

## 17.4 NEW AUTOMOTIVE POLYPROPYLENE TECHNOLOGY

### 17.4.1 Scratch- and Mar-Resistant Polypropylene

In Section 18.5.2 in Chapter 18, scratch and mar resistance was described as an important characteristic for good visual appearance. Hence, various additives are required to improve molded surface properties affected by the incorporation of mineral fillers to attain desired part stiffness in automotive applications.

In Section 8.6.3 of Chapter 8, there is discussion regarding the surface appearance problems associated with replacement of engineering thermoplastics by TPO-type materials. Stress-whitening results from the use of untreated talc filler in TPOs (see also Chapter 7). Development of a new generation of modified talcs containing reactive groups might provide a new route for improving scratch and mar resistance. Development of proprietary surface modifiers will continue in the automotive area in order to reduce material costs.

An alternative route to improvement of surface hardness is partial replacement of talc filler by wollastonite filler, as described in Chapter 18.

### 17.4.2 Polypropylene Nanocomposites

Chapter 20 and Section 3.18 in Chapter 3 provide detailed description of developments in nanocomposite technology based on compounding of exfoliated clay layers into polypropylene resin.

Nonpolar polypropylene resin requires addition of chemical coupling to form mechanically strong nanocomposites of clay platelets. As discussed in Chapters 3 and 11, chemical coupling agents consist of bifunctional polar/nonpolar molecular groups (maleated polypropylene and organosilane) to compatibilize dissimilar ingredients. The improved adhesion between nonpolar polypropylene molecules and ionic clay platelet layers promotes the same type of stress transfer mechanism needed in glass fiber reinforcement (Chapter 9).

# Automotive Applications for Polypropylene

Applied stress-load to the polypropylene matrix is thereby transferred to the load-bearing exfoliated clay. In spite of low concentrations of clay platelets, mechanical strength/stiffness is greatly enhanced by particulates having very high aspect ratio (clay platelet diameter of micrometer size/very thin nanometer thickness).

The cooperative effort between GM, Basell Polyolefins, and Southern Clay has led to commercial application of polypropylene nanocomposites. As a result of continued interest by both academia and industry, polymer scientists and automotive engineers are pooling their resources and talents to invent even more unique materials. Besides the economic benefits of utilizing nanocomposites in automotive applications, a further understanding of the nano-domain of nature can lead to a whole series of technological breakthroughs in material science.

## 17.4.3 Long Glass Fiber–Reinforced Polypropylene

In Section 9.2 in Chapter 9, there is a detailed description of the manufacturing process for long glass fiber–reinforced polypropylene long glass fiber–reinforced polypropylene composites. Automotive applications involving structural parts (like bumper beams) made from long glass fiber–reinforced polypropylene certainly have possibilities in regard to mechanical strength and stiffness. However, certain cost constraints exist for this type of structural material in comparison with short glass fiber–reinforced composites.

## 17.4.4 Polypropylene/Fiberglass Sheet and Roving

In Section 9.3 in Chapter 9, the series of development of compression molding glass mat (GMT) composites for bumper beam construction is described as beginning with the Azdel process.

Azdel Plus was a subsequently developed material that combines GMT with unidirectional roving. This innovation provides additional strength in the longitudinal direction (ref. 212 in Chapter 9).

Following that, GMT manufactured by the Taffen process was described as utilizing continuous fiberglass strand mat yielding superior impact strength. The major automotive application was a structural battery tray.

## 17.4.5 Continuous Fiber–Reinforced Thermoplastic Composites

In Section 9.4 of Chapter 9, the Twintex process, developed by Vetrotex (Europe) in 1995, is described as "the most exciting and revolutionary development of glass fiber reinforced polypropylene composite." Continuous fiber–reinforced polypropylene (CFRP) composites made by this process combine glass and polypropylene fibers "as they are being fiberized into commingle rovings."

TABLE 17.1 Polypropylene-Based Materials for Automotive Applications {2}

| Automotive location | Component | Trade name | Type PP resin/composite | MFR (°/min) | Outstanding property |
|---|---|---|---|---|---|
| Interior | Pillar and trim | Marlex AGN-380 | Impact copolymer, nucleated | 38 | Balance stiffness-impact |
| | | Marlex AGN-450 | Impact copolymer, nucleated | 45 | High stiffness and high flow |
| | | Marlex AGN-250 | Talc-filled impact copolymer | 25 | High modulus for stationary trim |
| | Door panels | Marlex BPC-45 | Talc-filled impact copolymer | 45 | Talc-filled for injection/low-pressure molding |
| | | Marlex BPC-70 | Talc-filled impact copolymer | 80 | Talc-filled for low-pressure molding |
| | Head impact/ side impact | Marlex AGM-250 | Impact copolymer, nucleated | 25 | Ultra high impact |
| | | Marlex AGM-350 | Impact copolymer, nucleated | 35 | Ultra high impact |
| | Instrument panels | Marlex CHP-261 | Talc-filled modified-impact copolymer | 26 | High stiffness and low-temperature impact |
| | | Marlex CT9917 | Talc-filled scratch-resistant and impact copolymer | 37 | High impact/high modulus for injection molding and scratch resistance |
| | | Marlex CDB-240 | | 19 | |

# Automotive Applications for Polypropylene

| | | | | |
|---|---|---|---|---|
| Exterior | Filled bumper fascias | Marlex AGM-110 | | Engineered fascia grade, good balance of impact and flexural properties and excellent paintability |
| | Splash shields, fender liners | Marlex AHX-120 Black Color | 12 | Unfilled black TPO, excellent impact properties |
| Under-the-hood | Blow-molded ducts | Marlex AMN-010 | 1.0 | Heat stabilized, excellent melt strength |
| | Reservoirs | Marlex AHN-030 | 3.0 | Impact copolymer for brake reservoir |
| | | Marlex HHX-007 | 0.65 | Heat stabilized, cooling bottle application |

Notes (Description column):
- Marlex AHX-120 Black Color: Impact copolymer, cold temperature impact, heat stabilized
- Marlex AMN-010: Impact copolymer, cold temperature impact, heat stabilized and nucleated
- Marlex AHN-030: Impact copolymer, heat stabilized and nucleated
- Marlex HHX-007: Fractional melt flow, good impact resistance, heat stabilized

Recently, Knox [1] published a technical paper describing automotive applications of continuous fiber–reinforced thermoplastic composites made by the Twintex process. This in-line process for producing dry prepregs [1] of single-end roving by intimate "commingling" of polymer and 60–80 wt% reinforcing fibers is described as a high-volume, low-cost method.

The intimate contact of individual polymer–glass fiber filaments is on the micrometer scale. Weaving of the filaments into fabric enhances efficient wet-out of glass fibers by molten polymer during various fabrication processes, e.g., hybrid thermoforming or diaphragm forming process. Light-weight, dent-free, and corrosive-free automotive underbody protection (shield) competes with metals. CFRP shields are tougher and much more processable than thermoset materials.

New GMT composites exhibit the high strength requirement dominated by steel construction of bumper beams. The thermoplastic alternative consists of 60 wt% CFRP skins combined with random chopped glass core. Compared to steel, the bumper beam has 40% reduction in weight, increased energy absorption and elastic recovery after impact deformation [1].

The new CFRP technology has been utilized in sandwich panels for load floors with reduced weight and other fabrication attributes.

## 17.5 CONCLUSIONS

In this chapter, we have described different types of automotive applications based on polypropylene resins and polypropylene composites. In comparison with engineering plastics, metal components, and thermoset materials, we have provided many reasons why polypropylene-based components are the preferred option for material selection. The combination of low cost, ease of processing, flexibility of reactor catalysts to tailor-make desired properties, and recyclability are just a few reasons for this market growth in the automotive area.

Table 17.1 lists the array of polypropylene-based products [2] that are used to make various component parts for interior, exterior, and under-the-hood automotive applications. Furthermore, innovative nanocomposite and CFRP composite technologies exemplify how reinforcement of polypropylene extends the scope of applications to an ever-growing list of possibilities.

## REFERENCES

1. MP Knox. Continuous fiber reinforced thermoplastic composites in the automotive industry. Automotive Composites Conference, Troy, Michigan, Sept. 2001.
2. Phillips Sumika Polypropylene Co. literature.

# 18

## Wollastonite-Reinforced Polypropylene

**Roland Beck, Dick Columbo, and Gary Phillips**
Nyco Sales, Calgary, Alberta, Canada

## 18.1 INTRODUCTION

### 18.1.1 History and Overview

Although it is a uniquely functional mineral, wollastonite has a relatively short history of industrial use. Deposits of sufficient size and purity to be of commercial interest are unusual, even though it is also a relatively common mineral in a geological sense. Wollastonite was probably first mined for industrial purposes in California in 1933, when it was used in production trials in the manufacture of mineral wool.

Significant commercial production did not start until 1953, at what is now NYCO's Willsboro, New York deposit. Early research beginning in 1947 by the New York State College of Ceramics, Alfred University (Alfred, New York) led to many of the early commercial uses of wollastonite [1]. These early studies also identified many potential uses where wollastonite is still in common use today. Early applications were in the ceramic industry where it was and is still used as an additive in many types of glazes and ceramic body formulations. Other early uses were in the production of paints, concrete, asphalt, welding fluxes, metallurgical fluxes, glass, mineral wool, filters, insulation, plastics, elastomers, and abrasives. Wollastonite is still used in most if not all of these applications today.

Until the early 1960s, commercial production of wollastonite was carried out mainly by two U.S. producers: NYCO (Willsboro, New York) and R.T. Vanderbilt (Gouverneur, New York). NYCO has always been the larger of these two producers. During the 1960s, wollastonite became widely known as an industrial mineral with a variety of potential uses. Production in Finland and in Mexico began commercially in 1967, in India in 1970, in Africa in 1975, and in China in 1983. Worldwide production has more than doubled in the last 10 years with an average annual growth rate of more than 10% per year. Throughout the years, NYCO has dominated the market. With the recent addition of production from Minera NYCO in Mexico, it is likely that NYCO will lead the market for years to come.

Geologically, wollastonite can be formed in nature in a variety of ways. For commercially useful deposits, it is generally accepted that there are two formation routes. Both routes involve metamorphism (heat and pressure) of limestones (calcite) in the presence of silica-bearing material. The simple metamorphic reaction between silica and calcium carbonate to form wollastonite occurs at about 600°C at shallow depths.

### 18.1.2 Chemistry

Wollastonite is a mineral comprised of calcium, silicon, and oxygen. Its molecular formula can be expressed as $CaSiO_3$ or as $CaO \cdot SiO_2$. It is commonly referred to as calcium silicate and has a theoretical composition of 48.28% $CaO$ and 51.72% $SiO_2$.

However, natural wollastonite may contain trace or minor amounts of various metal ions such as aluminum, iron, magnesium, manganese, potassium, and sodium. Any of these ions may partially substitute for calcium. Wollastonite is rarely found by itself. Ore zones in major deposits contain 20–98 wt% wollastonite. Table 18.1 describes typical chemical compositions for wollastonite mined by NYCO and other producers at geographic locations around the world.

### 18.1.3 Size and Shape

One of the most interesting properties of crushed and ground wollastonite is its cleavage. Fragments of crushed and ground wollastonite tend to be in the form of blades and needles. Crystals are usually in fibrous, somewhat splintery masses of elongated crystals flattened parallel to the base and to the front pinacoid, giving the impression of slender prismatic needles.

Wollastonite can generally be made to cleave into very fine microfibers with high nominal aspect ratio (length/diameter, L/D). The aspect ratio of the ground

TABLE 18.1  Typical Chemical Composition of Commercial Wollastonite Products

| Component (wt%) | NYCO minerals (New York) | Minera NYCO (Mexico) | Finland | India | China | Africa | Synthetic grade |
|---|---|---|---|---|---|---|---|
| CaO | 47.5 | 46.1 | 44.5 | 47.0 | 45.3 | 44.6 | 45.7 |
| $SiO_2$ | 51.0 | 51.4 | 51.8 | 49.5 | 50.6 | 50.5 | 52.6 |
| $Fe_2O_3$ | 0.40 | 0.25 | 0.22 | 0.43 | 0.34 | 0.42 | 0.26 |
| $Al_2O_3$ | 0.20 | 0.66 | 0.44 | 0.60 | 0.68 | 0.80 | 0.47 |
| MnO | 0.10 | 0.05 | 0.01 | 0.29 | 0.06 | | 0.04 |
| MgO | 0.10 | 0.50 | 0.56 | 0.20 | 0.70 | 0.50 | 0.55 |
| $TiO_2$ | 0.02 | 0.02 | 0.05 | 0.01 | 0.02 | 0.09 | 0.13 |
| $K_2O$ | 0.05 | 0.32 | 0.01 | 0.11 | 0.12 | 0.13 | 0.38 |
| Loss on ignition | 0.68 | 0.65 | 2.20 | 1.79 | 1.60 | 2.32 | 0.36 |

| Commercial Wollastonite Products | Median Diameter D(micron) | Median Length L(micron) | Aspect Ratio L/D | CHART OF RELATIVE WOLLASTONITE PARTICLE SIZE Diameter X Length Cross-Section |
|---|---|---|---|---|
| NYGLOS® M3 | 3.0 | 15 | 5/1 | — |
| NYGLOS® 4 | 3.8 | 40 | 11/1 | — |
| NYGLOS® 5 | 4.7 | 60 | 13/1 | — |
| NYGLOS® 8 | 8.0 | 150 | 19/1 | — |
| NYGLOS® 12 | 12 | 160 | 13/1 | — |
| NYGLOS® M-15 | 15 | 120 | 8/1 | — |
| NYGLOS® 20 | 20 | 260 | 13/1 | — |
| NYAD® G | 40 | 600 | 15/1 | — |
| NYAD® M 1250 | 3.0 | 10 | 3/1 | — |
| NYAD® M 400 | 7.0 | 20 | 3/1 | — |
| NYAD® M 325 | 9.0 | 35 | 4/1 | — |
| NYAD® M 200 | 15 | 75 | 5/1 | — |

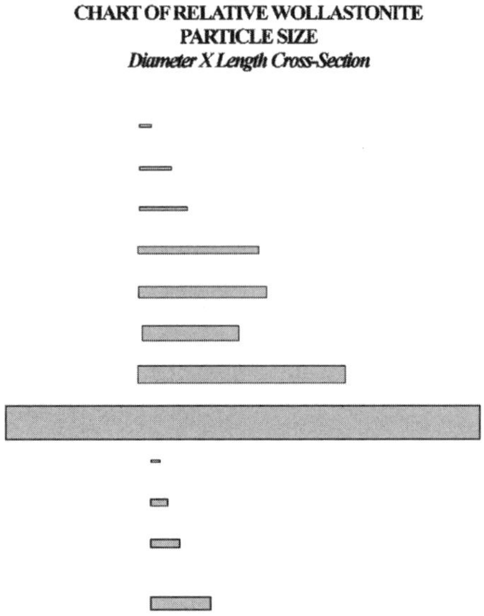

FIGURE 18.1  Relative size chart.

particles is commonly between 5:1 and 10:1 and may be more than 20:1 depending on the milling techniques.

Commercial grades available from one producer are characterized in Fig. 18.1. They are divided into low nominal aspect ratio products (generally with aspect ratios from 3:1 to 5:1) and high nominal aspect ratio products (generally with aspect ratios from 8:1 to 20:1). High nominal aspect ratio is important in many filler applications due to its mechanical reinforcing effect in both thermoplastic and thermoset polymer composites.

The optimal particle size for wollastonite products is determined primarily by the application. In general, coarse particle sizes are considered undesirable when used in polymer composites. They detract from mechanical reinforcement, segregate and settle quickly, affect the processing and quality of end-use products, lead to higher abrasion, and affect surface finish. On the other hand, excessive amounts of fines can lead to ineffective mechanical reinforcement and problems associated with materials handling.

### 18.1.4  Physical Properties

Table 18.2 summarizes the physical properties of wollastonite. The following description of properties helps to explain the unique characteristics of wollastonite as a reinforcement fiber for polymeric composites.

# Wollastonite-Reinforced Polypropylene

TABLE 18.2  Physical Properties of Wollastonite

| Property | Value |
|---|---|
| Morphology | Acicular |
| Nominal aspect ratio | 20:1 to 3:1 |
| Particle size | Products vary from −10 mesh to −1250 mesh |
| Loose bulk density, kg/m$^3$(lb/ft$^3$) | Products vary from 220 to 1360 (14 to 85) |
| Tapped bulk density, kg/m$^3$(lb/ft$^3$) | Products vary from 420 to 1440 (26 to 90) |
| Specific gravity of solids | 2.87–3.09 |
| pH (10 wt% slurry) | 8–10 |
| Molecular formula | CaSiO$_3$ or alternativelyCaO·SiO$_2$ |
| Color | Brilliant white to cream |
| Melting point (°C) | 1540 |
| Water solubility (g/100 cm$^3$) | 0.0095 |

*Specific Gravity.* The specific gravity of pure wollastonite (triclinic) can be calculated based on unit cell parameters to be 2.96. Measured specific gravities typically fall in the range of 2.87–3.09. This variation is due to trace or minor amounts of various impurity ions such as aluminum, iron, magnesium, manganese, potassium, and sodium, which substitute for calcium and distort the crystal lattice. The specific gravity (s.g.) of commercial wollastonite products is also affected by the content of impurity minerals such as calcite (s.g. 2.70–2.95), garnet (s.g. 3.5–3.8), diopside (s.g. 3.2–3.3), and so forth.

*Bulk Density.* Mixing, compounding, storing, and shipping of ground materials requires knowledge of their apparent bulk densities. The bulk density of commercial wollastonite products depends primarily on their fineness and nominal aspect ratio. In addition, specific gravity, moisture content, and test method can also play a role. However, wollastonite is not hygroscopic and is normally provided with less than 0.1 wt% moisture.

Measurement is done in the "loose" condition (aerated) and in the "tapped" condition (compacted). Typical measured bulk densities for commercially available products are as follows. For loose bulk density, products vary from 220 to 1360 kg/m$^3$ (14 to 85 lb/ft$^3$). In comparison, for tapped bulk density, products vary from 420 to 1440 kg/m$^3$ (26 to 90 lb/ft$^3$).

*Color.* When pure, the mineral is brilliantly white, but impurities even in trace amounts may color it cream, gray, pink, brown, or red. This color change is related to the presence of iron and other coloring ions. Color may be imparted by impurities on the crystal surface (deposited by the passage of groundwater through the deposit) or by impurities contained in the crystal structure.

The luster is glassy to silky (vitreous to pearly). Luster is important in applications such as plastics, paints, and coatings as it in turn imparts luster to surface finishes.

*Loss on Ignition.* Loss on ignition (LOI) is the amount of volatile matter driven off when the mineral is heated to 1000°C. Commercial wollastonite products have an ignition loss ranging from 0.4 to 2.0 wt%. The LOI for higher quality wollastonite products are typically less than 0.7 wt%.

Loss on ignition can be attributed primarily to the decomposition of carbonates (mainly calcite). However, such things as water of hydration, organic materials (roots, wood, plants, etc.), and sulfur-containing minerals may also contribute in part for some wollastonites.

*Brightness.* The dry brightness and whiteness of wollastonite are important in determining its suitability for certain filler applications. Brightness is determined by measuring the reflectance of finely ground powder against a standard that is assigned a brightness of 100. Magnesium oxide and barium sulfate are the two standards used. GE brightness, a term used in North America, refers to brightness measured with a General Electric reflectometer. Commercial wollastonite products usually have a GE brightness ranging from 80 to 95.

*Melting Point.* The melting point for pure wollastonite is generally accepted as being 1540°C. The fluid temperature for commercially produced wollastonites is generally somewhat lower than this value. Certain producers have found that the fluid temperature can be as low as 1380°C.

*Thermal Conductivity.* Thermal conductivity measurements are specific to the application. Wollastonite is normally considered to have low thermal conductivity. Functional fillers for plastics typically also have low thermal conductivity. However, wollastonite may be advantageous in these applications relative to other fillers. It can increase thermal conductivity of the polymer matrix without deterioration of electrical insulating properties.

*Thermal Expansion.* A characteristically low coefficient of thermal expansion combined with aspect ratio imparts high thermal shock resistance and dimensional stability in high-temperature applications such as fire-resistant board or refractory linings. The coefficient of linear thermal expansion is generally accepted as being $6.5 \times 10^{-6}$ mm/mm/°C.

*Electrical Properties.* Wollastonite can be considered to be an insulator or nonconductor of electricity. For pure wollastonite, the commonly accepted value for DC electrical conductivity is $1.5 \times 10^{-11}$ mho/m. The dielectric permittivity of pure wollastonite is generally accepted as 8.60 at 1 MHz.

## 18.1.5 Health and Safety Aspects

The benign health aspects of wollastonite provide a compelling reason for its use in place of other industrial minerals and fibers. Since wollastonite products are considered nontoxic and noncarcinogenic, health and safety concerns during handling are limited to those associated with nuisance dusts. Standard industry practice with regard to dust containment and collection should be observed.

Wollastonite products are composed of very stable oxides and dusts generated during handling can be considered nonconductive, nonflammable, and nonexplosive.

## 18.2 PRODUCTION OVERVIEW

### 18.2.1 Mining

Wollastonite ore is generally mined using surface methods, e.g., open pits, quarries. The wollastonite is separated from its associated minerals by three common methods as follows:

1. Crushing followed by dry magnetic separation to remove minerals such as garnet, diopside, ilmenite, hematite, and sphene. These minerals are weakly magnetic or paramagnetic and generally require high-intensity magnetic separation. Thermal drying of the ore is generally required for effective magnetic separation.
2. Crushing and grinding followed by wet processing (flotation) to remove minerals such as calcite, diopside, and feldspars. Filtration and thermal drying of the resulting wollastonite concentrate is required after wet processing.
3. Manual separation (hand sorting) from blasted and/or crushed ore. Optical sorting has been proposed to replace or augment this method.

The methods used depend on the associated minerals present, product requirements, mineral liberation size, selling price, and operating cost.

### 18.2.2 Production Process

Initial size reduction is generally accomplished by conventional crushing and ores are generally fed to a jaw crusher to produce at least a $-150$-mm ($-6$-in.) product. This setting corresponds to a maximum crushed ore with diameter less than 6 in. Additional size reduction, commonly by cone crushers, is effected to reduce the ore to a size suitable for grinding or beneficiation. Hand sorting is generally carried out on jaw-crushed or finely blasted ore.

After the ore has been beneficiated, grinding to preserve nominal aspect ratio is carried out by a variety of specialized grinding mills and techniques.

Specialized air classification also plays an integral part in these processes. The additional costs incurred by these types of processes generally result in higher prices for the resultant high aspect ratio products. The requirement for cost effective products is development of in-house technology to retain high nominal aspect ratio to the greatest extent possible with minimal manufacturing costs.

Grinding of beneficiated feed to produce low nominal aspect ratio products is generally carried out in conventional grinding equipment, including pebble mills and roller mills. The requirements are generally to minimize operating costs and minimize the introduction of contaminants. The nominal aspect ratio of the ground particles is commonly between 3:1 and 5:1. These products are generally destined for ceramic coatings or metallurgical applications.

## 18.3 COMPOUNDING WITH WOLLASTONITE

Twin-screw compounding is the preferred process for compounding nominal high aspect ratio (HAR) wollastonite. Typically HAR values are between 8:1 and 19:1. It is normally not recommended to add wollastonite products into the main feed throat of the extruder due to the possibility of aspect ratio attrition, especially for HAR products, and because of possible abrasive wear to the equipment. Downstream feeding into the melt via a side feeder or open-top barrel is generally more successful (Fig. 18.2).

A good general rule of thumb is to process HAR wollastonite in the same manner as chopped fiberglass strands.

When feeding more than 30 wt% wollastonite, it may be necessary to "back-vent" the barrel immediately upstream of the point of addition. This is especially useful when processing polymers that may contain excess moisture. Also, when processing a high loading of more than 30 wt%, it may be beneficial to split-feed the wollastonite into two feed zones. Proper feeding and screw design allows for retention of aspect ratio throughout compounding and results in a superior end product (Fig. 18.3). Even with 100 wt% rework, HAR wollastonite maintains its integrity on a properly designed twin-screw compounding line (Fig. 18.4).

Wollastonite products having lower nominal aspect ratios (i.e., 3:1 to 5:1 powder grades) have been successfully compounded on continuous mixer, Banbury, and single-screw compounding lines. Higher nominal aspect ratio products can also be successfully compounded on single-screw compounding lines if feed and screw designs are properly optimized. However, compounding HAR wollastonite in a Banbury mixer, for example, causes attrition of aspect ratio with little resultant increase in end-product properties compared with using powder grade wollastonite having low aspect ratio.

As with all twin-screw compounding, proper screw design is required to achieve optimal performance and properties. This is especially important for

## Wollastonite-Reinforced Polypropylene 659

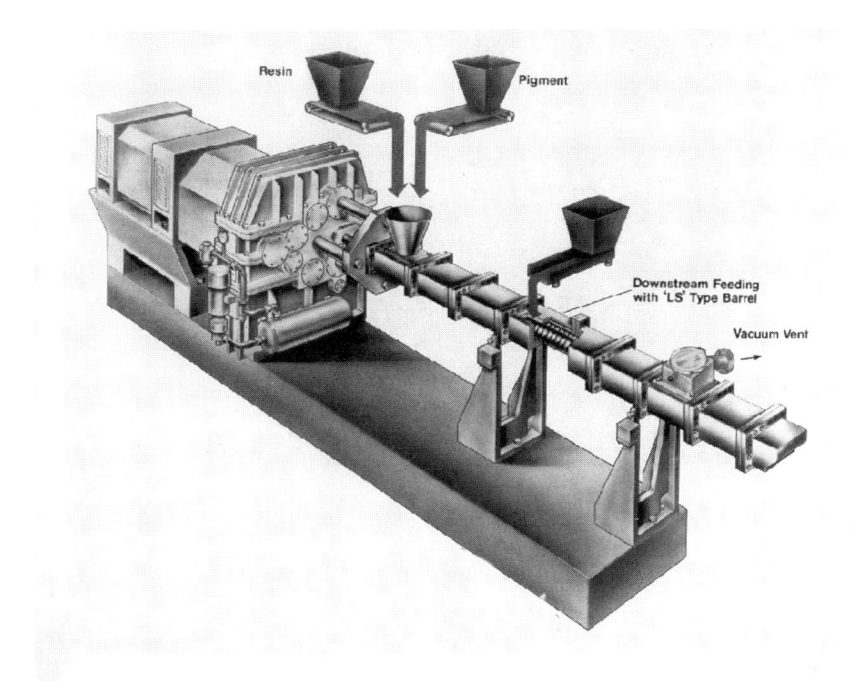

**FIGURE 18.2** W&P twin screw.

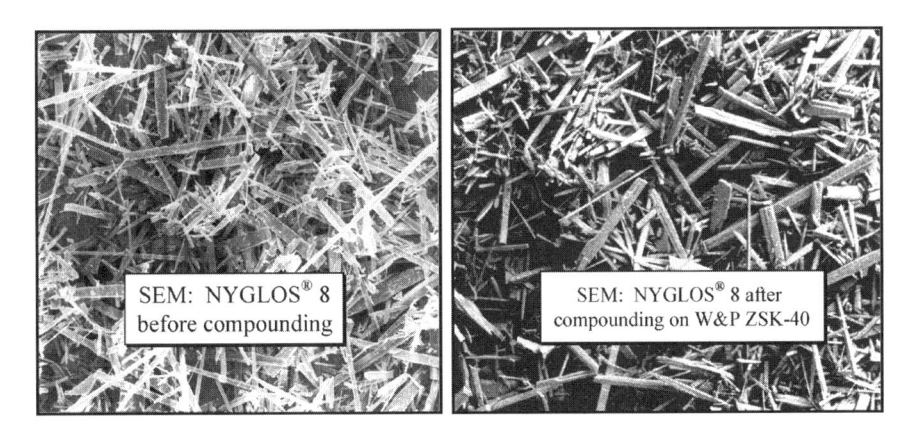

**FIGURE 18.3** Scanning electron micrograph NYGLOS 8 before and after compounding.

SEM: NYGLOS® 8 after
2 passes on W&P ZSK-40,
100 % re-work

X 250

**FIGURE 18.4** Scanning electron micrograph of NYGLOS 8 after two passes thru extruder.

HAR products. For instance, bilobed ("two-flighted") screws are preferred to trilobed ("three-flighted") screws as there is more free space in the restricted area within the extruder with a bilobed screw, and hence less attrition of the HAR wollastonite.

In order to minimize aspect ratio attrition, the arrangement of the kneading elements in the corotating twin-screw extruder must be optimized to give low-shear dispersive mixing of ingredients. As a result, the compounding process will yield good finished product properties and superior end-use performance.

## 18.4 HANDLING WOLLASTONITE

### 18.4.1 Conveying Issues

Wollastonite products can be moved horizontally or vertically with a variety of conveying equipment, including:

- Vibrating conveyors
- Screw conveyors
- Bucket elevators
- Airslides
- Pneumatic conveyors

The selection of the appropriate conveyor depends on the product, situation, and application.

*Vibrating Conveyors.* Vibrating conveyors commonly referred to as "shakers" are efficient for transferring the coarser wollastonite products (200 mesh or coarser) in the horizontal direction. A design factor of 1.25 on capacity is desirable to allow for flow surges. These conveyors are generally coupled with a bucket elevator if a vertical lift is required.

*Screw Conveyors.* Screw conveyors are commonly used for transferring the finer wollastonite products (325 mesh or finer) in the horizontal direction. These conveyors are generally coupled to a bucket elevator if a vertical lift is required. However, inclined screw conveyors have been successfully used with wollastonite products.

Capacity is limited to 45% of total volume. A design factor of 1.25 on capacity is desirable to allow for flow surges. Screw speed is limited to 60 revolutions per minute to satisfy wear concerns.

*Bucket Elevators.* Bucket elevators are commonly used for vertical transfers of wollastonite products. These conveyors are generally coupled with either a vibrating conveyor or a screw conveyor if a horizontal transfer is also required. A design factor of 1.25 on capacity is desirable to allow for flow surges. Use of belt-style bucket elevators is recommended.

*Airslide Conveyors.* Airslides are used in some situations for transferring wollastonite products along downward slopes. Airslide gravity conveyors operate on the principle of fluidization. When air is introduced into wollastonite products that can be fluidized, they can be made to flow like a liquid, so that it is possible to move wollastonite along low-angle downward slopes. Air pressures of 35 kPag (5 psig) are typically required.

*Pneumatic Conveyors.* Pneumatic conveyors are used for transferring any wollastonite product in both the horizontal and vertical directions. Coupled with a diverter, a pneumatic conveyor can send material to a variety of locations.

Two-phase systems are commonly employed, although dense phase and dilute phase and vacuum have also been used. Full dense phase systems are prone to plugging, and full dilute phase systems can result in particle degradation. A design factor of 1.25 on capacity is desirable to allow for flow surges.

### 18.4.2 Transfer Issues

Wollastonite products can be successfully handled at transfer points using the following equipment:

- Chutes and hoppers
- Rotary airlocks

- Screw feeders
- Bin activators
- Loss in weight feeders

*Chutes and Hoppers.* Gravity chutes and hoppers are employed for simple transfers between conveyors and between storage and conveying equipment. Chutes and hoppers are to be designed with a minimal angle of 60°. Valley angles are avoided if at all possible. Otherwise, valley angles are the highest possible (minimum 60°).

*Rotary Airlocks.* Rotary airlocks are utilised to transfer wollastonite products and at the same time isolate one system from another. Rotary airlocks are also useful in preventing uncontrolled flow with products prone to "flooding."

*Screw Feeders.* Screw feeders are utilized to transfer wollastonite products and at the same time provide volumetric metering. While screw feeders can be effective, they are limited in holding back wollastonite products that have a high degree of floodability.

A variable-pitch screw is recommended to produce a uniform draw of material across the entire hopper opening. The extended section of screw that projects beyond the hopper at the discharge provides some degree of "sealing" against flooding and uncontrollable discharge. Double-screw feeders are utilized in applications requiring more precise metering. A large-diameter screw provides full feed rate to 80–90% full, and a smaller diameter screw provides a controlled dribble feed rate to 100 wt% full.

*Bin Activators.* Bin activators are utilized to transfer wollastonite products from storage bins. Bin activators are capable of maintaining a controlled uninterrupted continuous flow.

*Loss-in-Weight Feeders.* Loss-in-weight feeders are utilized to transfer wollastonite products and at the same time provide gravimetric metering. Loss-in-weight feeders are designed to control the mass rate of discharge to ±1 wt% accuracy. The feeder is generally designed for a 7- to 9-min retention time with a 1-min refill cycle every 5 min.

## 18.5 CHARACTERISTICS OF WOLLASTONITE IN POLYPROPYLENE

### 18.5.1 Physical Property Enhancement

The physical properties of copolymer and homopolymer polypropylenes can be changed substantially by the addition of mineral fillers or reinforcement to make composite materials. Chopped fiberglass, talc, mica, and calcium carbonate are commonly used to make a variety of polypropylene composites. Each mineral

# Wollastonite-Reinforced Polypropylene

TABLE 18.3 Effect of Wollastonite Characteristics on Stiffness and Impact of Reinforced Polypropylene

| Material properties | Neat polypropylene resin | 20 wt% wollastonite | 20 wt% wollastonite | 20 wt% wollastonite |
|---|---|---|---|---|
| Nominal aspect ratio ($L/D$) | — | 5:1 | 8:1 | 19:1 |
| Diameter $D_{50}$ (μm) | — | 3 | 15 | 8 |
| Flexural modulus | 1.00 | 1.65 | 1.80 | 2.25 |
| Gardner impact @ room temperature | 1.00 | 0.73 | 0.60 | 0.65 |

type has a different balance of stiffness and impact characteristics depending on its geometry and other features.

Wollastonite has a needle-like or rod-shaped structure with an aspect ratio that can be varied by different grinding and milling techniques. This is unique among naturally occurring minerals. The effect of modifying the size and shape of the wollastonite on stiffness and impact properties of a polypropylene copolymer are shown in Table 18.3. A relative comparison is made between properties of three commercially available wollastonite products from NYCO and unfilled polypropylene resin. Measured properties of filled composites are normalized by dividing by values obtained for the unfilled polymer resin. This comparison features variation in particle size diameters and nominal aspect ratios.

Flexural modulus increased predictably and substantially with increased nominal aspect ratio: 65% with low aspect ratio wollastonite (5:1) and up to 125% with high aspect ratio wollastonite (19:1). Likewise, the impact resistances decreased with increased particle size (i.e., increased $D_{50}$ or median diameter), as is the case with most additives, with the possible exception being chopped glass. These property values are likely the result of a combination of both the particle diameter and the nominal aspect ratio. Stiffness is primarily a function of nominal aspect ratio, whereas impact strength is a function of particle size.

The raw material cost of wollastonite is a key variable in formulating cost-effective polypropylene composites, e.g., coarser minerals are less expensive than finer grades. Therefore, the balance of properties may include stiffness and impact strength as well as cost as primary concerns. Appearance, mar and scratch resistance, and other factors are additional concerns. Table 18.3 demonstrates that wollastonite as a reinforcement for polypropylene can be tailored to meet a wide range of performance criteria.

TABLE 18.4 Relative Comparison Between HAR Wollastonite and Fine-Course Talc

| Filler type | Unfilled polypropylene | Wollastonite | Wollastonite | Talc | Talc |
|---|---|---|---|---|---|
| Raw material | Basell SG-702 | NYCO NYGLOS 8 | NYCO NYGLOS M3 | IMI Fabi B2207 | IMI Fabi B2202 |
| Filler loading wt% | 0 | 20 | 20 | 20 | 20 |
| Diameter $D_{50}$ (μm) | — | 8 | 3 | 7 | 2 |
| Aspect ratio $L/D$ | — | 19.1 | 5.1 | — | — |
| Tensile strength | 1.00 | 1.07 | 1.00 | 1.00 | 1.07 |
| Flexural modulus | 1.00 | 2.25 | 1.65 | 1.47 | 1.85 |
| Notched Izod | 1.00 | 0.65 | 0.74 | 0.53 | 0.65 |
| HDT, 0.5 MPa | 1.00 | 1.53 | 1.25 | 1.39 | 1.45 |
| Melt flow index (MFI) | 1.00 | 1.00 | 1.00 | 0.95 | 0.93 |

# Wollastonite-Reinforced Polypropylene

Comparing the performance of wollastonite directly with competing talc mineral yields data outlined in Table 18.4. The table shows the balance of physical properties attainable with both a "low" (5:1) and a "high" (19:1) nominal aspect ratio wollastonite compared with fine (2 μm) and coarse (7 μm) talc (needle versus platelet). The baseline is "neat" resin only (Basell SG-702) and the mineral loading for all other data sets (other than the baseline) is 20 wt% in all samples.

It is notable that tensile strength at yield is relatively unaffected by the addition of talc or wollastonite minerals. With the addition of 20 wt% high nominal aspect ratio wollastonite, the flexural modulus is substantially increased (225%) in comparison with the neat resin as also shown in Table 18.3. The coarse and fine talc increased the flexural modulus by 47% and 85%, respectively. This difference is lower than that of the wollastonite.

Heat distortion temperature (HDT) is increased by more than 50% for the low-diameter, high nominal aspect wollastonite. HDT values are higher than that attainable by talc-filled polypropylene, whereas the melt flow index (MFI) of the polypropylene composite is unaffected by the addition of either grade of wollastonite.

As shown in Table 18.4, impact properties are reduced with the addition of the minerals. Compared to the neat resin, the finer, lower nominal aspect wollastonite (3 μm, 5:1 $L/D$) showed approximately 26% reduction, less than the coarse and fine talcs (47% and 35%, respectively). The high nominal aspect ratio, low-diameter wollastonite (8 μm, 19:1 $L/D$) for NYGLOS 8 wollastonite also gave a 35% reduction.

Table 18.5 provides data for the physical properties for polypropylene composites of NYGLOS 8 wollastonite as a function of loading level. Figures 18.5–18.7 portray smooth trendline plots for flexural modulus, Gardner impact, and HDT temperatures, respectively. Flexural modulus (Fig. 18.5) increases with wollastonite loading level; while Gardner impact, drops off in an inverse manner (Fig. 18.6). The corresponding increase in HDT parallels the increase in stiffness with loading level (Fig. 18.7).

TABLE 18.5 Effect of NYGLOS 8 Wollastonite Loading on Composite Physical Properties

| Wt% wollastonite | 0% | 5% | 10% | 15% | 20% | 30% |
|---|---|---|---|---|---|---|
| Flexural modulus | 1.00 | 1.20 | 1.45 | 1.70 | 2.25 | 2.65 |
| Gardner impact @ RT | 1.00 | 0.84 | 0.79 | 0.71 | 0.65 | 0.50 |
| HDT @ 0.5 MPa°C | 1.00 | 1.16 | 1.35 | 1.46 | 1.53 | 1.65 |
| MFI (g/10 min) | 1.00 | 1.02 | 1.05 | 0.93 | 1.00 | 0.85 |

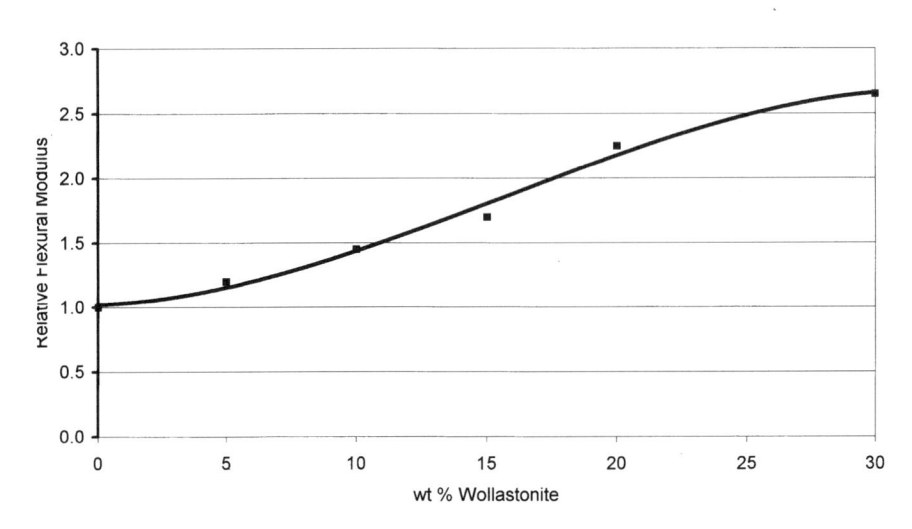

**FIGURE 18.5** Flexural modulus versus wollastonite loading level.

To mitigate the effect of this reduction in impact strength, Fig. 18.6 demonstrates that lowering the loading level of wollastonite increases impact resistance. This attribute can be used to offer a different balance of properties compared with commonly used talc and other mineral additives while achieving equal stiffness and higher impact properties.

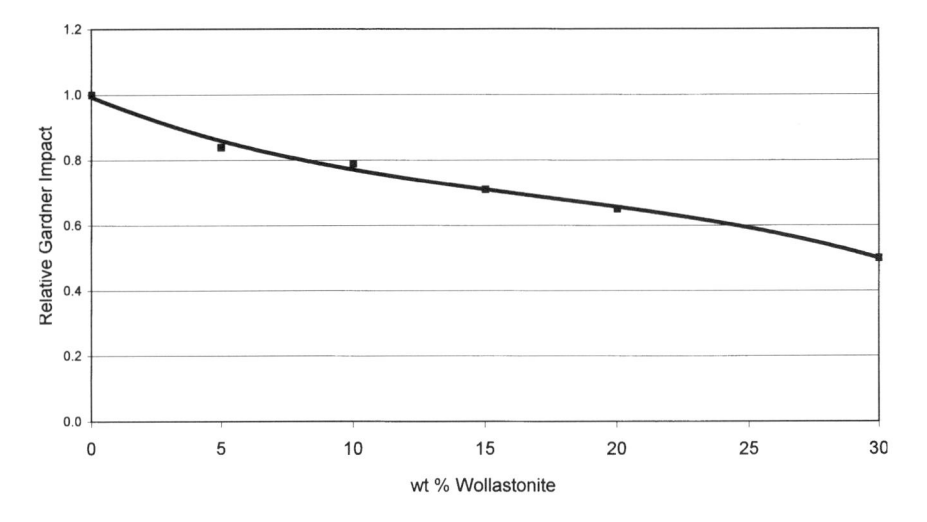

**FIGURE 18.6** Gardner impact versus wollastonite loading level.

# Wollastonite-Reinforced Polypropylene

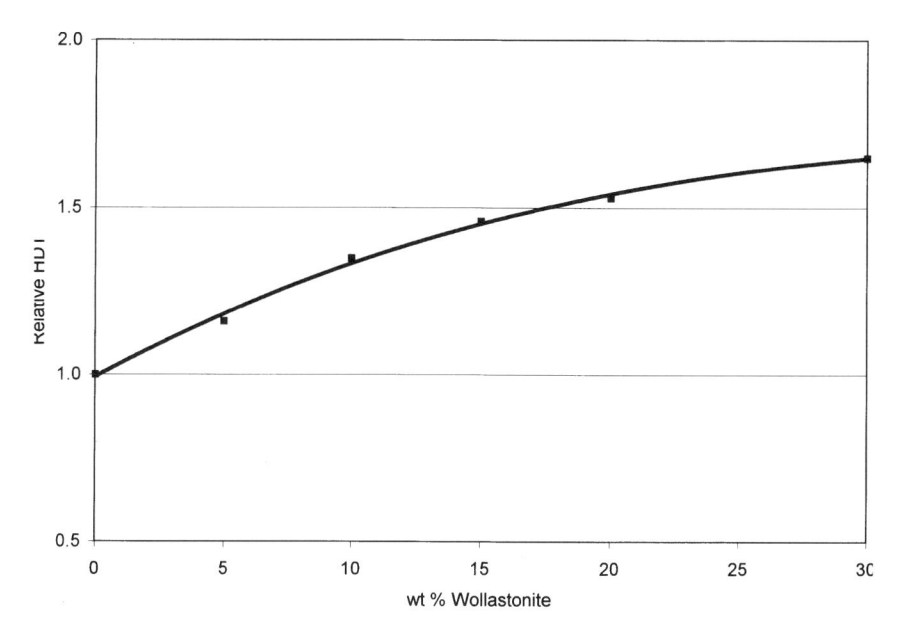

**FIGURE 18.7** Heat distortion temperature versus wollastonite level.

## 18.5.2 Mar and Scratch

Resistance to mar and scratch, as measured by the Visteon Five Finger Scratch Test [2], is one of the most important benefits derived from using wollastonite compared with other minerals.

Scratch and mar resistance is a primary concern in automotive and other appearance parts. As described in [2,3], there are many characteristics that affect scratch visibility. These include polymer type, type of reinforcement mineral, mineral filler loading level, addition of impact modifiers, surface morphology, etc. Unfilled polypropylene has inherently good scratch resistance, which is better than the acrylonitrile-butadiene-styrene (ABS) replaced in automotive interior parts. However, the need to improve stiffness of the polypropylene by the addition of talc resulted in a severe deterioration in scratch resistance, as shown in Table 18.6.

Scratch depth at 7 Newton force increased from 1.5 μm for the unfilled polypropylene to 8.8 μm with the addition of 13 wt% talc. Replacing half of the talc with wollastonite reduced the scratch depth to 3.7 μm or about 60% improvement. Additional improvements to 3.4 μm were obtained by replacing

TABLE 18.6 Mar and Scratch Resistance of Wollastonite Compared with Talc

| Materials | Apparent scratch depth ($\mu$m) at 7 N applied force [2] |
|---|---|
| Unfilled polypropylene copolymer | 1.5 |
| 13% talc | 8.8 |
| Blend of 6.5% talc + 6.5% wollastonite | 3.7 |
| 13% wollastonite | 3.4 |
| General-purpose ABS | 3.6 |

all of the talc with wollastonite. As shown in Table 18.6, this is equal to or better than the ABS that the polypropylene originally replaced.

### 18.5.3 Surface Appearance

The gloss and depth or distinctness of image (DOI) of reinforced parts are influenced by the substrate formulation, type, and concentration of the reinforcement, and primer–top coat type, paint thickness, along with paint cure conditions. The automotive industry continues to strive for improved part quality for all applications. A new generation of short-fiber high aspect ratio wollastonite grades has been developed and is commercially available for new polymer applications. Exterior composites reinforced with short-fiber HAR wollastonite have provided the automotive industry improved part quality when employed in most polymers as exemplified by high DOI thermoset and thermoplastic formulations reinforced with these grades. In the past, certain polymer formulations have been limited by deficiencies in surface quality (DOI) and impact strength when reinforced with minerals and/or milled-glass fiber. Short-fiber wollastonite, depending on grade, has overcome or has now been found to meet new and existing original equipment manufacturer (OEM) class A painted surface requirements.

The new generation of short-fiber HAR wollastonite combines surface quality, essentially equivalent to that of painted sheet metal, with an extremely uniform physical property profile. These new wollastonite grades combine superior surface appearance properties with excellent dimensional stability, processing characteristics, and multiple improvements in overall properties to make class A surface composites now achievable by application.

### 18.5.4 Effect on Coefficient of Linear Thermal Expansion

One of the principal benefits that wollastonite brings to any composite is in controlling that composite's ability to withstand variations in temperature. In

# Wollastonite-Reinforced Polypropylene

TABLE 18.7  Coefficient of Linear Thermal Expansion (CLTE)

| CLTE Test Method: ASTM E831-93<br>Temperature range: −40 to 100°C | Cross-flow direction ($\times 10^{-6}/°C$) | Parallel flow direction ($\times 10^{-6}/°C$) | Cross/parallel ratio |
|---|---|---|---|
| Neat polypropylene copolymer, Equistar 8762 | 103 | 75 | 1.37 |
| Wollastonite (8:1 $L/D$) at 20% loading, NYGLOS M15 | 89 | 53 | 1.68 |
| Wollastonite (19:1 $L/D$) at 20% Loading, NYGLOS 8 | 85 | 44 | 1.93 |

effect, the high melt temperature of wollastonite acts as a heat sink, thereby stabilizing the effect of elevated temperature on the composite as a whole. This benefit is highly prized in automotive applications especially, where use of wollastonite in polypropylene and other resin systems allows design engineers to match the shrinkage of wollastonite-containing plastic parts with those of steel and aluminum. Table 18.7 describes the effect of loading level on coefficient of linear thermal expansion (CLTE).

Adding 20 wt% by weight of wollastonite having an intermediate nominal aspect ratio and $D_{50}$ of 15 µm reduces the CLTE by 15% in the flow direction and by 21% using high nominal aspect ratio wollastonite, $D_{50}$ of 8 µm. As with shrinkage, the difference between cross and flow direction CLTE must be considered in tool design.

## 18.5.5  Effect of Surface Treated Wollastonite

Surface chemistry is important in applications where wollastonite is added as a reinforcement in other materials. Coupling agents or surface modifiers enhance the performance of wollastonite in the matrix by enabling a superior bond with the resin. Stronger bonding with the matrix enhances mechanical properties, reduces shrinkage, increases weather resistance, and reduces or eliminates surface or internal defects. In addition, surface-treated wollastonite permits higher filler loadings, improves dispersion, and improves powder flow characteristics, as well as flow during mixing and molding in compounding systems.

The three main groups of chemical agents used for treating plastic fillers are: acids, organosilanes, and organometallics. Both silanes and titanates are used to treat wollastonite. Of the two, the organotitanates have relatively small and specialized markets. The surface treatment method used plays a major part in the success of a treated wollastonite product. After treatment, proper handling of

TABLE 18.8 Effect of Surface Modification on Impact Properties

| Test values for filled system are divided by neat polymer values | Neat polymer | 20% Untreated wollastonite (NYGLOS 8) | 20% Treated wollastonite (NYGLOS 8 + 10992) |
|---|---|---|---|
| Notched Izod (J/m) | 1.00 | 0.65 | 0.91 |
| Flexural modulus | 1.00 | 2.25 | 1.90 |

wollastonite is important as broken particles expose untreated mineral surface that can result in potential weak points in the finished composite.

Table 18.8 outlines the effect of surface treating one grade of wollastonite by way of example. This demonstrates that impact properties of the polypropylene system can be balanced with increased flexural modulus to achieve superior performance in the system, while also achieving favorable economics through the use of a filler.

### 18.5.6 Shrinkage in Parallel vs. Perpendicular Directions

A relatively high-impact polypropylene copolymer (Equistar 8762) was used as the base polymer for an array of composites of wollastonite and talc at 20 wt% filler loading. Table 18.9 summarizes the measured postmold shrinkage along two directions for an edge-gated molded plaque: change in length in the parallel direction to flow and change in plaque width in the perpendicular or cross-flow

TABLE 18.9 Shrinkage Data for Polypropylene Composites with 20 wt% Filler Loading

| Measured values after 48 hr conditioning at 23°C | Perpendicular direction (%) | Parallel direction (%) | Perpendicular/parallel shrinkage ratio |
|---|---|---|---|
| Neat resin-Equistar 8762HS polypropylene copolymer | 1.16 | 1.08 | 1.07 |
| Wollastonite Intermediate $L/D$ NYGLOS M15 | 0.90 | 0.63 | 1.43 |
| Wollastonite High $L/D$ NYGLOS 8 | 0.85 | 0.49 | 1.73 |
| 2-μm Talc | 0.68 | 0.63 | 1.08 |
| 7-μm Talc | 0.75 | 0.68 | 1.10 |

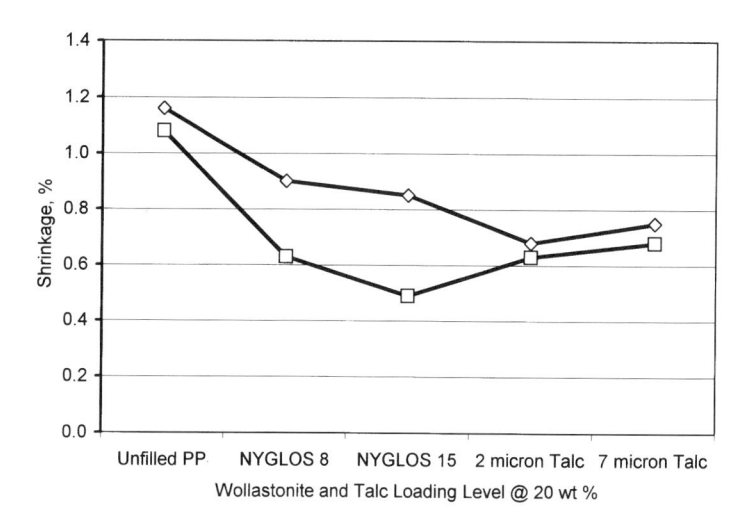

FIGURE 18.8 Shrinkage data, wollastonite and talc-filled polypropylene composites. ◇, perpendicular flow direction; □, parallel flow direction.

direction. The corresponding plots in Fig. 18.8 provide useful information regarding the effects of different mineral types in the polypropylene composites.

The cross-flow shrinkage of the neat resin is about 5–10% higher than in the flow direction. Adding 20 wt% wollastonite or talc substantially reduces the shrinkage in both directions, but not equally. As would be expected, the "needle-shaped" wollastonite shrinks differently than talc that has a platelet shape. The wollastonite shrinks less in the flow direction and more in the cross-flow direction than the talc. This means that tooling modifications are required to replace a talc-reinforced compound with one containing wollastonite if the tooling was originally designed for talc. New parts with new tools, designed with wollastonite in mind from the start, would not face this hurdle. However, wollastonite's shrinkage characteristic may even be favored for certain parts, such as long, thin body strips on automobiles.

One leading wollastonite producer has recently commercialized a new grade of wollastonite (NYGLOS M3) which has parallel as opposed to perpendicular flow characteristics that match those of fine talc (IMI Fabi 2202 talc with 2-μm median diameter), allowing for the "drop-in" use of wollastonite in low-modulus applications where talc had been used. The drop-in would have benefited the compounder and end-use customer in terms of superior part flatness, excellent stiffness, and good mar and scratch performance that wollastonite gives in applications like interior pillars and rub strips, while still using tooling originally cut for fine talc.

### 18.5.7 Reworkability

The discussion in Section 18.3 emphasized the importance of compounding extrusion with low-shear dispersive mixing of high nominal aspect ratio, fine diameter wollastonite to minimize aspect ratio attrition. The obvious question, then, is "can processed wollastonite be reworked under normal conditions?"

As shown in Table 18.10, reworking of both intermediate and high nominal aspect ratio wollastonite into the extruder feed throat at a typical 25% rate shows essentially no degradation of key physical properties. It is obvious that the reworked wollastonite, having been protected by the polymer during the first pass on a correctly optimized extruder and thus not exposed to the grinding effect of unmelted polymer and dry wollastonite under high shear on the second pass, shows very little attrition of aspect ratio. This means that the compounder can rework wollastonite-containing blends with no fear of loss in properties in the end-use material.

## 18.6 ECONOMICS OF USING WOLLASTONITE IN POLYPROPYLENE

The improved performance found to date in polypropylene in comparison with competitive minerals has promoted a whole new generation for wollastonite-reinforced polypropylene composites. These improvements have provided for the replacement or extension of more costly reinforcements, such as chopped and milled fibreglass, and of more costly engineering polymer systems.

For example, the replacement of a popular grade of talc with the appropriate and comparably priced high aspect ratio wollastonite provides substantially higher flexural modulus. Consequently, this allows the design engineer to downsize the wall thickness or other dimensions of the part by 25% or more. Furthermore, relatively stiffer composites of wollastonite at reduced loading levels can drive the overall material savings for the end user. In particular, this results in overall weight savings for automotive applications.

Key factors that are improved are CLTE and mar and scratch resistance. These improvements are accompanied by an excellent balance of stiffness and impact. Injection molding of wollastonite-reinforced composites yields consistently good quality parts with reduced scrap and rework.

Prices for wollastonite are competitive with other reinforcements, including talc, mica, clay, and glass fiber, but its superior performance characteristics allow for a "greater bang for the buck" for composites makers and their customers, such as automotive companies.

# Wollastonite-Reinforced Polypropylene

TABLE 18.10 Rework Blends of Wollastonite Composites at 20 wt% Loading

| Measured values for rework blends are divided by values of virgin material | 100% Intermediate aspect wollastonite[a] | Intermediate aspect wollastonite + 25% rework | 100% High-aspect wollastonite[b] | High-aspect wollastonite + 25% rework |
|---|---|---|---|---|
| Tensile | 1.00 | 1.07 | 1.00 | 1.02 |
| Flexural modulus | 1.00 | 0.95 | 1.00 | 0.99 |
| Notched Izod | 1.00 | 1.00 | 1.00 | 1.00 |
| HDT @ 0.5 MPa | 1.00 | 1.02 | 1.00 | 0.99 |
| MFI | 1.00 | 1.01 | 1.00 | 1.02 |

[a]NYCO Minerals, Inc. NYGLOS M15.
[b]NYCO Minerals, Inc. NYGLOS 8.

## 18.7 FUTURE DEVELOPMENTS AND PROSPECTS

There are several fronts of technological development by the leading wollastonite producers that relate to polypropylene. New short-fiber HAR technology for wollastonite has positioned these emerging products as superior mineral reinforcement in polymers in both existing and new-application development. Major activity is seen in polypropylene and engineering alloys to replace talc, mica, and calcined clay.

This development is driven by short-fiber HAR wollastonite providing a better balance in stiffness and impact, which allows the compounder a broader window in formulation to achieve the requested physical properties sought by the end user by application. Short-fiber HAR wollastonite also improves surface appearance comparable to the neat resin and/or painted composites.

Based on the above, wollastonite is becoming a critical mineral reinforcement for future applications such as the replacement of fine grades of talc in fluffy or densified form for OEM interior and exterior polypropylene composites. Applications will encompass automotive instrument panels, pillars, and fascia.

Other key developments to watch for are in the area of densified-prilled wollastonite for improved handling and feeding during compounding accompanied by elimination of air-borne dust. Also, conductive (metallized) wollastonite is being developed for OEM fascia composites to improve physical properties with improved paint adhesion.

The prospects for these and other new developments are virtually unlimited, implying that wollastonite will be used in polypropylene composites for many years to come.

## 18.8 SUMMARY AND CONCLUSION

The production of wollastonite continues to be expanded worldwide. This growth is driven in part by increased demand for enhanced physical properties and surface appearance in various polymer composites. These improvements are typically met with high aspect ratio wollastonite grades, both uncoated and surface modified. The improved performance found to date in polypropylene when compared with competitive minerals has promoted a whole new generation for wollastonite-reinforced polypropylene composites.

## REFERENCES

1. CR Amberg, JF McMahon. Wollastonite, an industrial mineral. Bulletin No. 4, Ceramic Experiment Station, New York State College of Ceramics, 1948.
2. J Chu, L Rumao, B Coleman. Scratch and mar resistance of mineral-filled polypropylene materials. SAE Tech Paper 970659, Detroit, 1997, pp. 75–77.
3. J Chu, L Rumao, B Coleman. Scratch and mar resistance of filled polypropylene materials. Polym Eng Sci 38:1906–1914, 1998.

# 19

# Part Shrinkage Behavior of Polypropylene Resins and Polypropylene Composites

**Harutun George Karian**
RheTech, Inc., Whitmore Lake, Michigan, U.S.A.

## 19.1 INTRODUCTION

In the preceding 18 chapters of this second edition, we have discussed a wide variety of filled and reinforced polypropylene composites that can be utilized for new product development. For example, conventional fillers like talc take on a new meaning in the domain of submicrometer particle sizes (less than 0.5 µm described in Chapter 16). The compaction of submicrometer talc to offset inherent light bulk density provides the required feeding characteristics for cost-effective compounding extrusion processes. Furthermore, the development of nanocomposite technology makes it possible to attain both low density and stiff composites to meet increasingly sophisticated requirements for light-weight automotive parts. Hence, the material engineer has many options in tailoring interphase design for truly engineered plastic applications.

However, there remains one bottleneck that hinders the rapid deployment of new technology into the market area, i.e., accurate prediction of mold shrinkage using polypropylene-based composites. The difficulty lies in the inherent crystallinity of semicrystalline polypropylene homopolymer that can produce significant shrinkage for molded parts in spite of ever tightening tolerances for control of molded part size and shape.

In order to better appreciate the need for better control of mold shrinkage for material designs, let us trace the development of a typical new product from its inception to final approval by the customer. The systematic reduction of candidate materials generally proceeds in the following stepwise manner [1,2]:

1. Material cost (raw material + manufacturing) is always an important first consideration for preliminary sorting of possible candidates in new product development. Immediate elimination is made for materials that do not suit cost constraints.
2. Depending on the part geometry and end-use application, an experimental matrix can be generated for possible combinations of polymer resin–type melt flow characteristics and filler-type loading.
3. The material engineer can then utilize design of experiment (DOE) methodology to make benchmark comparisons between candidate materials versus criteria for acceptable performance in the field.
4. Mechanical and thermal properties are compared versus customer specifications based on structural needs at specific stress load temperature conditions. Standardized laboratory tests are conducted to develop trendline plots of mechanical data versus material design parameters, e.g., tensile strength or flexural modulus versus filler loading.
5. Final material selection is decided by characterizing long-term durability by accelerated testing, e.g., LTHA, UV weatherability, creep, fatigue resistance, abrasion resistance to continuous wear applications, etc. This determination can take about 1000 hours or 42 days.
6. Because of tooling costs of replacing the existing mold, there is a mold trial at the customer's plant to test out the moldability of the new product. For example, a mold originally designed for nylon-66 might be used to make parts of polypropylene as a cost-down effort to reduce product cost.
7. Most likely there may be a need to fine tune an existing mold design in an effort to meet ever tightening dimensional constraints for the molded part.
8. If the part geometry has never been molded, an entirely new mold design is required. Herein lies the existing problem. Even with computer-aided engineering (CAE) to optimize mold design parameters, some aspects of input data are *inadequate* to make computations giving sufficiently accurate results.
9. Quite often, the mold designer has to make a series of trial-and-error corrections to a number of prototype mold designs before obtaining an acceptable production mold.

## Part Shrinkage Behavior of Polypropylene

Computers now have tremendous speed and memory to handle complex tasks of simulating the injection molding process via finite element methodology. As we will see later in the chapter, simulation of the injection molding process requires computational models that accurately predict the response of specific volume and shrinkage to changes in pressure and temperature during melt–solid transitions. However, characterization of pressure-volume-temperature (PVT) and shrinkage have certain deficiencies that make CAE calculation of mold design difficult:

1. PVT measurements are typically conducted under equilibrium conditions maintained at low cooling rates, e.g., 1–3 °C/min. Injection molding processes have significantly higher cooling rates that cause nonequilibrium PVT behavior.
2. Particularly for semicrystalline resins, accurate computation of volumetric shrinkage requires inclusion of expressions for crystallization kinetics. Therefore, empirical values for linear shrinkage measurements need to be built into the computer model in order to obtain accurate mold design information. However, measured values of linear shrinkage, using standardized test specimens via ASTM D955 method, are not sensitive enough to variations of molding parameters within the operating range of the injection molding process.

In the following sections of this chapter, we will tie together fundamental concepts of linear-volumetric shrinkage and empirical trendline relationships between shrinkage and injection molding parameters. The resulting semirigorous model can then be used to quantitatively predict shrinkage as a function of material characteristics within the operating window for the injection molding process conditions and mold design parameters.

### 19.2 FUNDAMENTAL CONCEPTS OF PART SHRINKAGE

During the injection molding process, molten polymer is initially compressed in the mold cavity at a given applied pressure. Upon decompression, the volume of the frozen part begins to contract as the material begins to cool. After a period of cooling, the mold cycle is terminated by part ejection from the mold. The ejected part continues to contract as the material equilibrates to room temperature.

Generally, mold shrinkage is volumetric in nature. One can visualize volumetric shrinkage as a composite of linear contraction components aligned parallel or perpendicular with respect to the direction of molten flow streamlines within the mold cavity. The linear shrinkage component along the longitudinal (parallel) flow direction is defined as $S_L$. The corresponding shrinkage component that is transverse (perpendicular) to mold flow is $S_T$.

### 19.2.1 Definition of Linear–Volumetric Shrinkage

The ASTM D955 method uses a molded flexural bar having a gate at one end. Linear shrinkage $S_L$ is defined as a ratio of the measured change in the flexural bar length divided by the initial length determined by the linear dimension of the corresponding mold cavity [3,4]:

$$S_L = \text{linear shrinkage} = \frac{L_{\text{mold}}(T_{\text{mold}}) - L_{\text{part}}(T_{\text{part}})}{L_{\text{mold}}(T_{\text{mold}})} \quad (19.1)$$

Mold length ($L_{\text{mold}}$) and part length ($L_{\text{part}}$) measurements are temperature dependent. The choice of mold and part temperatures differs between engineering shrinkage determined by ASTM D955 method and true linear shrinkage defined from a more fundamental viewpoint [3]:

Engineering shrinkage : $T_{\text{mold}} = T_{\text{part}} = 23 \pm 2\,°C$

True shrinkage : $T_{\text{mold}}$ = steady-state temperature for hot mold during molding, and

$$T_{\text{part}} = 23° \pm 2\,°C \quad (19.2)$$

The corresponding volumetric shrinkage $S_V$ is calculated using specific volumes of the part at the transition temperature $T_t$ at which the gate freezes (point at which mold cavity is fully packed) and the final part temperature ($T_{\text{room}}$):

$$S_V = \text{volumetric shrinkage} = \frac{V_{\text{freeze-off}}(T_t) - V_{\text{room}}(T_{\text{room}})}{V_{\text{freeze-off}}(T_t)} \quad (19.3)$$

### 19.2.2 Dependence of Shrinkage on System Parameters

Shrinkage is dependent on three system properties [3]:

1. *Material.* Degree of crystallinity, ratio of amorphous-crystalline composition, thermal properties, PVT diagram plots, and melt rheology.
2. *Process conditions.* Melt and mold temperatures, cavity pressure, injection fill time, and cooling cycle time.
3. *Part geometry.* Mold dimensions (length, width, thickness), molded part tolerance, gate position, and design.

Volumetric shrinkage is related to changes in specific volume as a function of temperature and pressure. Without compressive forces of applied pressure to fully pack molten material into a mold cavity and hold the shape of the part, there would be significant volumetric shrinkage or contraction of about 25% [4] for all plastic materials undergoing cooling into the frozen state.

Molded part shrinkage depends on many other process and mold design factors besides cavity pressure. Mold design parameters include location of gate positions and flow lengths that influence streamline flow patterns of molten polymer within the mold cavity. Cooling rate is affected by a combination of part thickness and temperature gradients of molten layers of material within the thickness of the part. The combined effects of material properties and process variables funnel down into factors that determine the degree of crystallinity, which is particularly responsible for shrinkage of semicrystalline resins.

### 19.2.3 Need for Functional Relationships Between Part Shrinkage and System Parameters

PVT diagrams characterize the many changes in material properties that occur during transition from solid to molten phases of plastic material. Specific volume for each phase is an individual function of pressure and temperature. The shape of specific volume versus temperature curve in the transition region between melt and frozen solid is dependent on changes in polymer morphology due to the degree of crystallinity for amorphous and semicrystalline polymers [5]. Furthermore, the mold cavity pressure is dependent on changes in melt viscosity of the cooling molten mass, i.e., a function of shear rate and temperature.

Tabulated mold shrinkage values given on most material property sheets for unfilled and filled polypropylene composites fail to reflect the quantitative influence of process parameters on the crystallization temperature, e.g., the effect of cooling rate on the location of freeze-off transition temperature. Since polypropylene resin is a semicrystalline material, a reliable model [6] is needed to bracket the influence of compressibility and crystallinity on shrinkage within the operational window for process parameter settings for injection molding.

## 19.3 SHRINKAGE DEPENDENCE ON MATERIAL PROPERTIES

### 19.3.1 Effect of Fillers

Even though the magnitude of shrinkage for unfilled polymers indicates a strong sensitivity of resin density or specific volume to changes in pressure and temperature, the addition of filler to the polymer matrix can drastically alter shrinkage behavior to give reduced values [5,6]. The magnitude of actual shrinkage reduction due to introduction of filler into base resin is dependent on the filler type and filler content [2,6]. Since fillers are incompressible and generally have a higher solid density than the unfilled polypropylene resins [7], the addition of filler to the polymer matrix in effect decreases or dulls the sensitivity of the composite density to variations in pressure and temperature.

Fillers generally fall into two groups [6]:

1. *Particulate fillers*: PTFE, carbon black, milled glass, glass beads, talc, and calcium carbonate
2. *Fiber reinforcements*: glass fiber, carbon fiber, and some grades of wollastonite

Since most particulate fillers generally have a low aspect ratio (1 : 1), the corresponding polypropylene composites tend to shrink equally in the longitudinal flow (parallel) and transverse flow (perpendicular) directions to give isotropic shrinkage behavior: $S_T/S_L = 1$ for the shrinkage ratio of transverse (*T*) to longitudinal (*L*) flow directions.

Glass fibers have high aspect ratios (length/diameter > 20 : 1). As a consequence of orientation in the longitudinal flow direction, glass fiber–reinforced polypropylene composites exhibit differential shrinkage or anisotropic behavior. Compared with flexible polypropylene molecules, each glass fiber has a rigid structure that takes up a certain volume. With less polymer resin to linearly contract in the direction parallel to molten flow, shrinkage values $S_L$ for the composite are significantly reduced in comparison with unfilled polymer resin. With less replacement of polymer resin volume by glass fibers in the transverse direction, shrinkage $S_T$ perpendicular to flow is more dependent on the contraction of polymer molecules [5,8]. Thus, $S_T$ values are greater than $S_L$ values. Consequently, the anisotropic ratio $S_T/S_L$ is much greater than unity.

The differences between shrinkage behavior of glass reinforcement and unfilled resin can be exemplified by the following data [5] obtained at a cavity pressure of 300 bar pressure. The shrinkage values for the 30 wt% glass composite are 0.3% ($S_L$) and 1.1% ($S_T$). The corresponding anisotropic ratio is $S_T/S_L = 3.7$. In comparison, unfilled polypropylene resin has shrinkage values of 1.6% ($S_L$) and 1.9% ($S_T$) to give nearly isotropic behavior with $S_T/S_L = 1.2$.

### 19.3.2 Effect of Mineral Filler Type and Loading Level

Table 19.1 provides a comparison of linear shrinkage values for mineral-filled composites of polypropylene homopolymer resin with either talc or calcium carbonate with loading levels at 10–50 wt%. For given filler loading, the range of shrinkage values due to different filler types is narrow. With an increase in filler level from 10 to 50 wt%, the average shrinkage value for the set of data decreases linearly from about 1.6% (0.0160 in./in.) to 0.7% (0.0070 in./in.).

### 19.3.3 Effect of Polypropylene Resin Type

Shrinkage data given in Chapters 16 and 18 are summarized in Table 19.2 for mineral-filled composites of polypropylene copolymer (random copolymer and heterophasic blends of impact copolymer) composites. Comparison is made for shrinkage values between filled and unfilled copolymer resins. Shrinkage values

TABLE 19.1  Shrinkage Data for Polypropylene Homopolymer Materials

| Filler content (wt%) | Talc-filled material | Shrinkage flow direction (%) | Calcium carbonate–filled material | Shrinkage flow direction (%) |
|---|---|---|---|---|
| 10 | P-10TC-1156 | 1.6 | P-10CC-Y488 | 1.6 |
|  | T10P100-01 | 1.2–1.5 | CPP1010 | 1.1–1.5 |
| 20 | P-20TC-5100 | 1.4 | CPP-1020 | 1.0–1.4 |
|  | EA-14 | 0.8–1.2 | EZ-14 | 1.2–1.6 |
|  | T20P100-01 | 1.2–1.5 | CC20P100-00 | 1.2–1.5 |
| 30 | T30P100-00 | 0.9–1.2 | CPP-1030 | 0.9–1.3 |
|  | EA-16 | 0.6–1.0 | EZ-16 | 1.0–1.4 |
|  |  |  | CC30P100-00 | 1.0–1.4 |
| 40 | P-40TC-8100 | 1.0 | P-40CC-1100 | 1.0 |
|  | EA-18 | 0.6–1.0 | CPP-1040 | 0.8–1.0 |
|  | T40P100-01 | 0.9–1.2 | EZ-18 | 0.8–1.2 |
|  |  |  | CC40P100-01 | 1.0–1.3 |
| 50 |  |  | CPP-1050 | 0.7–0.9 |

*Polypropylene Products from Compounder's Literature:*
Thermofil, Inc.: P-10TC-1156, P20TC-5100, P-40TC-8100, P10CC-Y488, P-40CC-1100.
RheTech, Inc.: T10P100-01, T20P100-01, T40P100-01, CC20P100-00, CC30P100-00, CC40P100-01.
Singapore Polymers Corp.: CPP-1010, CPP-1020, CPP-1030, CPP-1040, CPP-1050.
Adell Plastics, Inc.: EA-14, EA-16, EA-18, EZ-14, EZ-16, EZ-18.

are given for $S_L$ (parallel-longitudinal flow direction) and $S_T$ (perpendicular-transverse direction). The ratio of shrinkage values for $S_T/S_L$ (transverse/longitudinal flows) describes a measure of anisotropic behavior.

Compared to shrinkage for polypropylene homopolymer–based composites (Table 19.1), the magnitude of shrinkage values given in Table 19.2 for polypropylene copolymer composites having the same filler content are significantly lower. This reduction in shrinkage is attributed to decreased volume fraction of crystalline molecular chains by increasing levels of randomly distributed and loosely entangled amorphous molecular chains.

The material engineer may want to design a polymer composite having desired stiffness, enhanced impact resistance, and limited shrinkage. One can see by the following property–ingredient relationships that the optimization of material formulation is hindered by opposing interactions:

> Stiffness is enhanced by increasing filler loading level and favors high crystallinity in the polymer matrix. Higher filler level causes both reduced impact and shrinkage.

TABLE 19.2 Shrinkage Data for Polypropylene Copolymer–Based Materials

| Material | Type filler | Filler content (wt%) | $S_L$ parallel direction | $S_T$ perpendicular direction | Anisotropy ratio $S_T/S_L$ |
|---|---|---|---|---|---|
| Equistar 8762-HS | Neat resin | 0 | 1.08 | 1.16 | 1.07 |
| Super impact copolymer (Table 18.9) | Wollastonite intermediate aspect ratio | 20 | 0.63 | 0.90 | 1.43 |
|  | Wollastonite high aspect ratio | 20 | 0.49 | 0.85 | 1.73 |
|  | Talc, 2 μm | 20 | 0.63 | 0.68 | 1.08 |
|  | Talc, 7 μm | 20 | 0.68 | 0.75 | 1.10 |
| Finaprop PPC 9760 | Talc, 0.5 μm | 10 | 1.15 | 1.53 | 1.33 |
| high-impact copolymer +15 wt% impact modifier (Table 16.5) | Talc, 1.8 μm | 10 | 1.03 | 1.44 | 1.40 |
| Idemitsu J-762HP | Neat resin | 0 | 1.56 | 1.94 | 1.24 |
| PP block copolymer | Talc, 0.5 μm | 20 | 1.06 | 1.29 | 1.22 |
| (Table 16.4) | Talc, 1.1 μm | 20 | 0.98 | 1.20 | 1.22 |
|  | Talc, 1.4 μm | 20 | 1.01 | 1.20 | 1.19 |
| Dow C703-35U 35 MFR impact copolymer RG-1 (Ref. 14) | Neat resin | 0 | 1.49 | 1.59 | 1.07 |
| Dow C702-20 20 MFR impact copolymer RG-3 (Ref. 14) | Neat resin | 0 | 1.48 | 1.61 | 1.09 |

# Part Shrinkage Behavior of Polypropylene

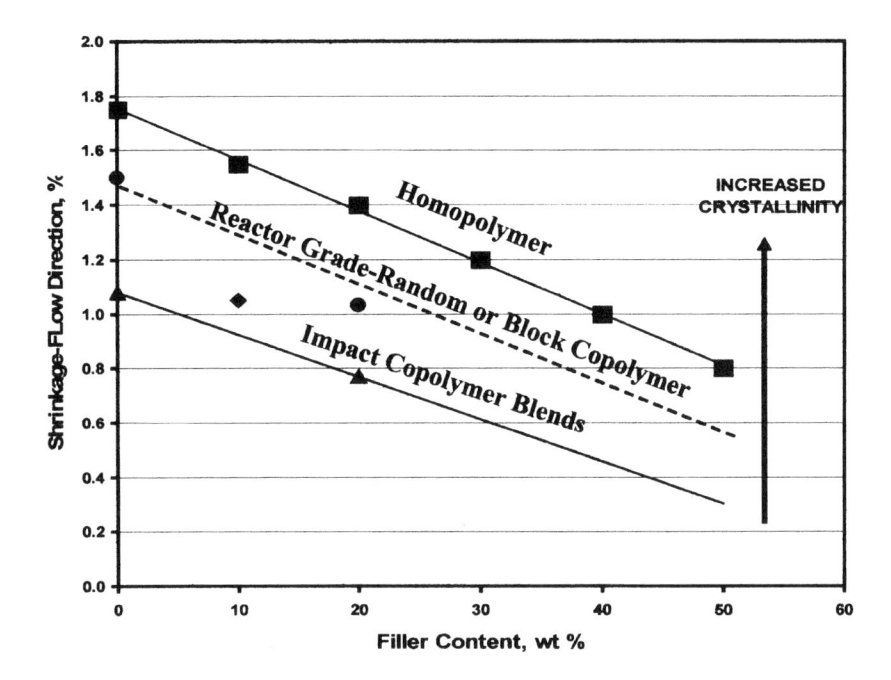

FIGURE 19.1 Spectrum of shrinkage values for polypropylene resin type and filler content. ■, Homopolymer values from Table 19.1; ●, unfilled and 20 wt% Talc in Idemitsu J 76 HP; , 20 wt% talc in Finaprop PPC 9760; ▲, 20 wt% talc-filled Equistar 8762-HS. (Data from Tables 19.1 and 19.2.)

Impact resistance is enhanced by EPR or plastomer addition to random copolymers (Chapter 7). The modified polypropylene resin has a higher amorphous/crystalline ratio and lower stiffness with reduced shrinkage. Shrinkage trendline plots shown in Fig. 19.1 describe interrelationships between filler content and degree of crystallinity.

## 19.3.4 Comparison of Shrinkage Between Semicrystalline and Amorphous Polymers

In order to obtain a better handle on quantitative prediction of shrinkage, there is a need to determine the effects of crystalline and amorphous molecular structure on specific volume as a function of pressure, temperature, and cooling rate. Shrinkage data given in Table 19.3 and Fig. 19.2 make a comparison between two groups of thermoplastic resins: amorphous and semicrystalline plastic resins.

TABLE 19.3  Percent Shrinkage in Flow Direction

|  | Bolurpe [20] | Green [21] | GE polymerland [22] | PP user guide [5] |
|---|---|---|---|---|
| AMORPHOUS POLYMER |  |  |  |  |
| Rigid PVC | 0.5–0.7 | 0.3–0.8 | 0.2–0.4 |  |
| Noryl (PPO+HIPS) |  |  | 0.5–0.7 |  |
| Polystyrene (PS) | 0.5–0.7 | 0.2–0.6 | 0.2–0.8 | 0.4–0.7 |
| High-impact Polystyrene (HIPS) |  |  | 0.4–0.6 |  |
| Polymethyl Methacrylate (PMMA) |  |  |  | 0.3–0.8 |
| ABS | 0.4–0.6 | 0.5–0.7 | 0.5–0.7 | 0.4–0.7 |
| Polycarbonate (PC) | 0.5–0.7 | 0.5–0.7 | 0.5–0.7 | 0.6–0.8 |
| PPO (Polyphenylene Oxide) |  |  |  | 0.5–0.7 |
| SAN | 0.4–0.6 |  |  | 0.4–0.6 |
| Acrylic | 0.3–0.6 | 0.2–0.8 | 0.2–0.9 |  |
| SEMICRYSTALLINE POLYMER |  |  |  |  |
| LDPE | 1.5–3.0 | 1.5–5.0 | 1.5–3.5 | 1.0–5.0 |
| HDPE | 2.0–3.0 | 1.5–5.0 | 1.5–3.0 | 1.5–3.0 |
| Polypropylene (PP) | 1.2–2.0 | 1.0–2.5 | 1.0–3.0 | 1.2–2.5 |
| PET |  | 0.5–1.2 |  | 1.6–2.0 |
| PBT |  |  |  | 1.0–2.0 |
| Nylon-6 (PA-6) | 1.0–1.5 | 0.6–1.4 | 0.7–1.5 | 0.2–1.2 |
| Nylon-66 (PA-66) | 1.0–2.0 | 1.2–1.8 | 1.0–2.5 | 0.8–2.0 |
| Polyamide-12 (PA-12) |  |  |  | 1.0–2.0 |
| Acetal |  | 1.8–2.5 | 2.0–3.5 | 1.5–2.5 |
| Polyamide-6/10 (PA-6/10) |  |  |  | 0.8–2.0 |

Amorphous resins exhibit a lower range of shrinkage values ($S_L = 0.5$–$0.8\%$) than semicrystalline polymer materials ($S_L = 1.5$–$3.0\%$).

The distinction between amorphous and crystalline type polymers is outlined in Table 19.4. The degree of shrinkage depends on how tightly molecules are arranged in the polymer matrix during the upstream polymerization process. Semicrystalline plastics consist of mostly dense crystalline regions surrounded by amorphous domains having loosely entangled molecular chains. Upon melting, both groups of molecules coalesce into one phase of disentangled and expanded chains. Upon cooling, molecules having amorphous or crystalline structures have a tendency to reassume their original molecular configuration and density. However, injection molding process conditions can alter or slow down the recrystallization of cooling molten mass. Consequently, the shrinkage of semicrystalline polymeric materials is complicated by a greater sensitivity to

## Part Shrinkage Behavior of Polypropylene

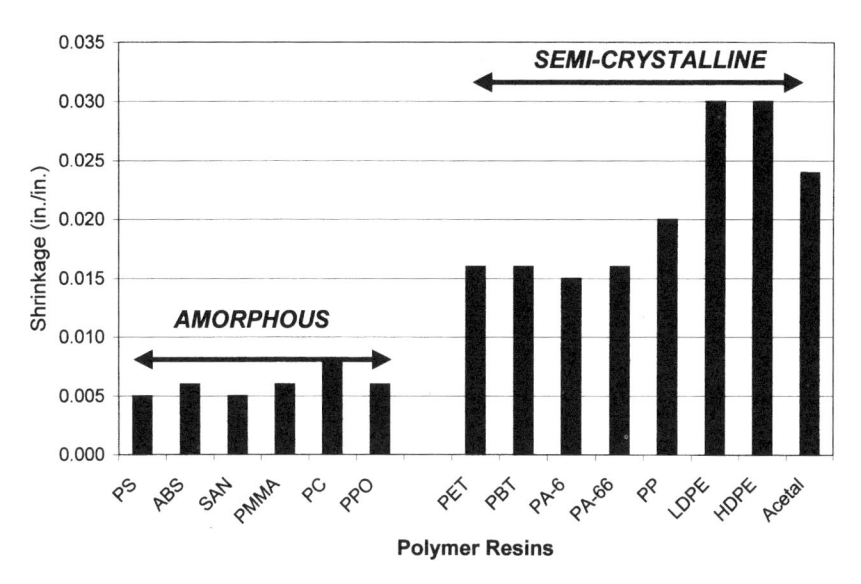

**FIGURE 19.2** Shrinkage spectrum of amorphous versus semicrystalline polymeric resins.

temperature and pressure. Therefore, accurate prediction of shrinkage requires an in-depth PVT analysis of interactions between system properties.

### 19.3.5 PVT Diagrams of Amorphous and Semicrystalline Polymers

In both molten and frozen states of matter, polymers can be defined as compressible fluids with specific volume directly dependent on temperature and inversely dependent on applied pressure [9,10]. This behavior can be represented by the ideal gas law equation:

$$PV = [\text{constant}]T$$

or

$$V = [\text{constant}]T/P$$

where $P$ is the pressure of molten material in mold cavity which can be expressed in psi, MPa, atmospheres, or bar units. The specific volume $V$ (cm$^3$/g) of the polymer matrix is directly dependent on temperature and inversely varies as pressure. The melt temperature $T$ is given in kelvins (k).

The PVT diagrams given in Figs. 19.3 and 19.4 characterize pressure-volume-temperature interrelationships for polycarbonate (amorphous resin) and polypropylene homopolymer (semicrystalline resin), respectively. The differences

TABLE 19.4 Comparison Between Amorphous and Semicrystalline Polymers

| Characteristic | Amorphous polymer | Semicrystalline polymer |
|---|---|---|
| Molecular arrangement in solid state | Random chain orientation in both solid/molten phases. Molecular structure is incapable of forming regular order. Molecules are randomly entangled, twisted, kinked, and coiled. | Molecular structure allows polymer chains to pack into an ordered arrangement as molecules cool. Decrease in free volume caused by formation of crystals. |
| Melting point | No sharp melting point with broad softening range. | Relatively sharp melting point. The crystal domains remain unchanged until the polymer reaches its melting point. |
| Change in molecular arrangement upon heating. | Tangled molecular chains become mobile and begin unravelling into a fluid mass. Don't flow as easily as semicrystalline material. | Upon melt transition, polymer behaves just like an amorphous polymer with generation of freely moving molecular chains at temperatures above the melting point. |
| Change in molecular arrangement upon cooling. | Rigidity returns to the polymer and molecular movement decreases. | Onset of crystallization as molten mass cools at a given temperature depending on the degree of crystallinity and cooling rate. |
| Visual appearance of molded part | Usually transparent. | Molded part is usually opaque due to crystal structure. |

## Part Shrinkage Behavior of Polypropylene

| | | |
|---|---|---|
| Shrinkage | Relatively low shrinkage when cooling from liquid to solid. | High shrinkage due to decrease in free volume due to formation of crystals in a packed and orderly arrangement. A greater degree of crystallinity leads to more shrinkage of the molded part. |
| Temperature/pressure dependence of density | Less | Greater |
| Effects of reinforcement | Fiberglass and/or mineral filler only slightly improve heat distortion temperature (HDT). | Increase load bearing capabilities and HDT. |
| Warpage in molded parts | Shrink isotropically in flow/transverse directions with little warpage of molded part. Uniform properties in all directions. | Depending on material composition and type of reinforcement, shrinkage tends to be quite anisotropic making molded part prone to warpage. |

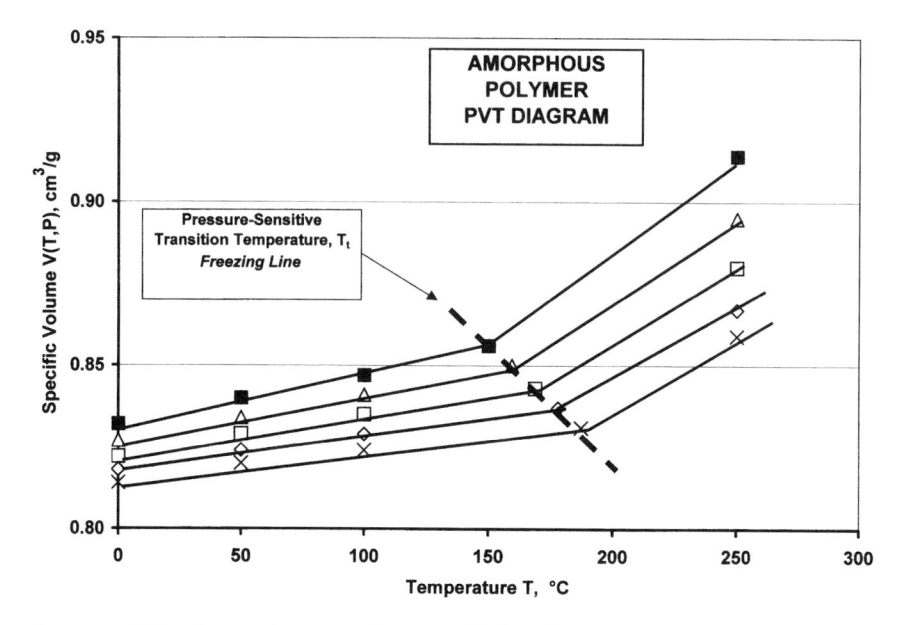

**FIGURE 19.3** Amorphous polymer PVT diagram. ■, $P = 0$ MPa; ◇, $P = 90$ MPa; △, $P = 30$ MPa; X, $P = 120$ MPa; □, $P = 60$ MPa.

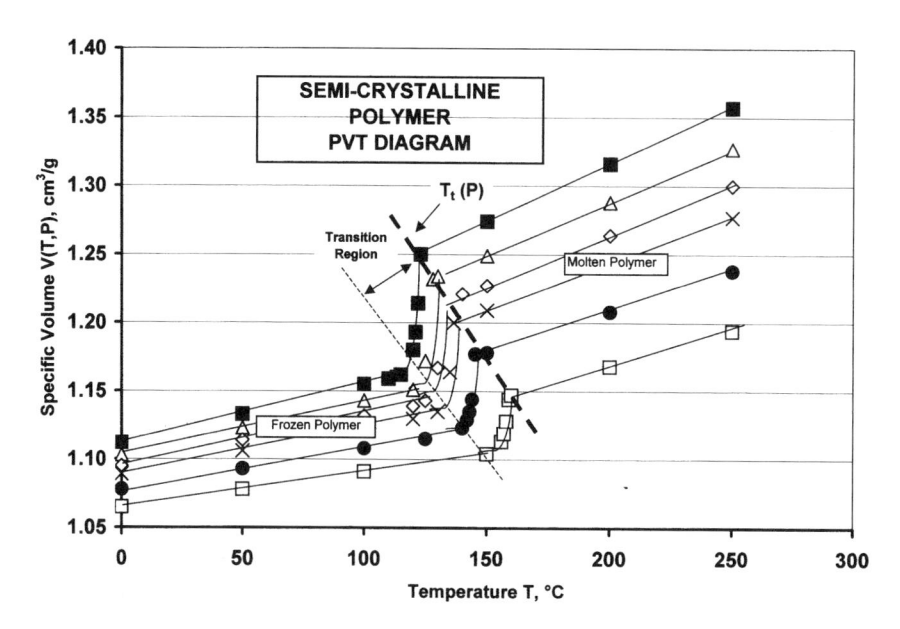

**FIGURE 19.4** Semicrystalline polymer PVT diagram. ■, $P = 0.1$ MPa; X, $P = 60$ MPa; △, $P = 20$ MPa; ●, $P = 100$ MPa; ◇, $P = 40$ MPa; □, $P = 160$ MPa.

## Part Shrinkage Behavior of Polypropylene

in the magnitude and slope of the corresponding specific volume versus temperature plots (under isobaric conditions) can be used to explain relative volumetric shrinkage behavior for the packing of molecular chains into cooling molten mass confined in the shape of the mold cavity.

Specific volume plots for amorphous resins, such as polycarbonate shown in Fig. 19.3, feature the following characteristics [7,11,12]:

1. $V(T, P)$ plots for so-called solid and molten states are both linear at a constant pressure condition (isobaric) with an abrupt change in slope at just one point that corresponds to freeze-off or solidification.
2. This solidification corresponds to a transition temperature $T_t$ under isobaric conditions.
3. The transition temperature $T_t$ is linearly dependent on pressure, i.e., $T_t = T_t(P) = b_5 + b_6 P$ where $b_5$ and $b_6$ parameters are determined from analysis of measured data.
4. At atmospheric pressure, $T_t$ is defined as the glass transition temperature $T_g$.
5. A tieline connects all of the $T_t(P)$ points.

Various analytical expressions for $V(T, P)$ have been derived that fit measured PVT data for different polymer resins. For polycarbonate, the PVT diagram shown in Fig. 19.3 is obtained using the following equation for the double-domain modified Tait model [11]:

$$V(T, P) = V_0(T)\left[1 - C\log_e\left(1 + \frac{P}{B(T)}\right)\right] \tag{19.5}$$

where

$C = 0.0894 =$ universal constant.
$V_0(T) = b_1 + b_2(T - b_5)$
$B(T) = b_3 \exp[-b_4(T - b_5)]$.

The units for $V$, $T$, $P$ are cm$^3$/g, °C, and MPa, respectively. Table 19.5 gives the values for the parameters used in Eq. (19.5).

The corresponding set of PVT plots for semicrystalline polypropylene homopolymer (Profax 6523, 4 melt flow rate) are given in Fig. 19.4. The PVT diagram is generated from equations and parameters derived by Guo et al. [12]. The linear portions of the PVT data for specific volume versus temperature were fit to an expression based on the Spencer and Gilmore model [9,10]. This model is similar to Eq. (19.4) but is modified for nonideal gases using the following van der Waals equation to better describe linear volumetric changes in molten and frozen polymers:

$$(P + P_0)[V(T, P) - V_0] = R_0 T \tag{19.6}$$

TABLE 19.5 Parameters for Fitting PVT Plots

| Polymer | Parameter | Units | Solid State | Molten State | Transition |
|---|---|---|---|---|---|
| Polycarbonate (amorphous) [11] | $b_1$ | cm$^3$/g | 0.856 | 0.856 | — |
| | $b_2$ | cm$^3$/g °C | $1.62 \times 10^{-4}$ | $5.84 \times 10^{-4}$ | — |
| | $b_3$ | MPa | 273.15 | 177.78 | — |
| | $b_4$ | (°C)$^{-1}$ | $3.384 \times 10^{-3}$ | $4.411 \times 10^{-3}$ | — |
| | $b_5$ | °C | — | — | 150 |
| | $b_6$ | °C(MPa)$^{-1}$ | — | — | 0.311 |
| Polypropylene (semicrystalline) [12] | $P_0$ | MPa | 240.2 | 265.0 | — |
| | $V_0$ | cm$^3$/g | 0.9950 | 0.9925 | — |
| | $R_0$ | MPa cm$^3$ °Kg | 0.1029 | 0.2203 | — |
| | $b_5$ | °K | — | — | 396 |
| | $b_6$ | °K(MPa)$^{-1}$ | — | — | 0.225 |
| | $b_7$ | cm$^3$/g | — | — | $8.7 \times 10^{-2}$ |
| | $b_8$ | °K$^{-1}$ | — | — | 0.537 |

## Part Shrinkage Behavior of Polypropylene

Upon rearrangement of Eq. (19.6), the specific volume is a linear function of temperature at a given pressure:

$$V(T,P) = \frac{R_0 T}{P + P_0} + V_0 \tag{19.7}$$

The parameters $R_0$, $P_0$, and $V_0$ are given in Table 19.5 for the solid and molten states.

Equation (19.7) was used [12] to fit the linear portion of PVT plots for temperatures greater than the transition temperature $(T > T_t)$. The pressure-dependent transition temperature is expressed as:

$$T_t(P) = b_5 + b_6 P \tag{19.8}$$

The parameter values for $b_5$ and $b_6$ are given in Table 19.5 for polypropylene homopolymer.

The knee part of the PVT curve for Fig. 19.4 was expressed by an exponential term [12] that is added to the linear expression for the *solid* phase for temperatures *less than* the transition temperature $[T < T_t(P)]$:

$$V(T,P) = \frac{R_0 T}{P + P_0} + b_7 \exp[b_8(T - b_5) - b_9 P] \tag{19.9}$$

### 19.4 EFFECTS OF INJECTION MOLDING PROCESS AND MOLD DESIGN ON VOLUMETRIC SHRINKAGE

The PVT diagram (Fig. 19.3) for polycarbonate can be used to make an estimate of volumetric shrinkage in terms of an actual representation of the injection molding process. In addition, the transformation of volumetric shrinkage values into linear shrinkage for amorphous resins (like polycarbonate) is rather simple.

#### 19.4.1 PVT Diagram for Injection Molding Process Steps

The PVT diagram shown in Fig. 19.5 describes the entire injection molding process for an amorphous polymer. The successive steps shown on the diagram are as follows:

1. The polymer resin pellets are heated to a temperature of 250 °C at atmospheric pressure.
2. Upon injection into the mold cavity, the applied cavity pressure is set at about 60 MPa with little change in the melt temperature. The compression leads to reduction of specific volume from about 0.91 to 0.88 cm³/g.
3. While maintaining cavity pressure for a sufficient period of time, the molten mass is cooled by heat transfer with additional material packed

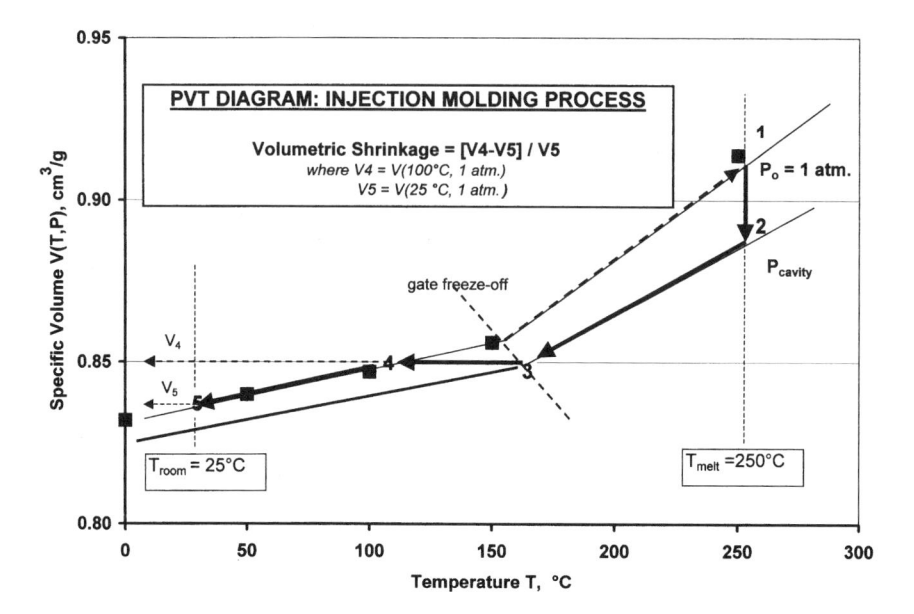

**FIGURE 19.5** PVT diagram of the injection molding process. ■, Specific volume versus temperature plot at $P = 1$ atm. 1–5: Process steps for PVT changes. 1, Molten polymer at 250 °C and $P = 1$ atm (0.1 MPa). 2, Increase pressure to 60 MPa at $T = 250$ °C. 3, Melt temperature cools to gate freeze-off @170 °C with $P = 60$ MPa. 4, Pressure drops rapidly to $P = 0.1$ MPa at $T = 110$ °C. 5, Ejected part at $P = 1$ atm. and $T = 25$ °C (room temperature).

into the mold cavity to fill the available volume of the mold until freeze-off occurs at 170 °C.

4. At freeze-off, the material pressure drops quickly to atmospheric pressure for the fully packed cavity ($V_4$), while the material temperature continues to cool with onset of shrinkage.
5. The part is ejected and cools down to room temperature with part shrinkage to $V_5$.

The volumetric shrinkage $S_V$ is expressed as:

$$S_V = \frac{V_4 - V_5}{V_5} \tag{19.10}$$

Since the amorphous polymer is essentially isotropic, the linear shrinkage $S_L$ is estimated as [3]:

$$S_L = \frac{S_V}{3} \tag{19.11}$$

## Part Shrinkage Behavior of Polypropylene

This determination of linear shrinkage in terms of process and material properties has much greater fundamental meaning than just a single measurement of shrinkage by the ASTM D955 methodology.

### 19.4.2 Determining Factors for Mold Shrinkage

Let us summarize what determines the magnitude of shrinkage for a molded part. Shrinkage results when polymer molecules relax after molding. The direction and magnitude of shrinkage depends on the three system properties listed in Table 19.6. The various factors that influence shrinkage can be rationalized by similar analysis of PVT diagrams as discussed in the previous section.

### 19.4.3 DOE Studies of Molding Process–Mold Design Variables.

Due to high crystallinity, semicrystalline resin-based polymer composites are much more sensitive to any variation in process conditions [6]. Furthermore, differential shrinkage in parallel and perpendicular mold flow directions, which gives anisotropic behavior, can lead to undesirable warpage [4] of the molded part's shape Therefore, shrinkage (decrease in size) and warpage (change in shape) combine, resulting in a serious challenge to the mold designer and material engineer alike.

Based on the above application of the PVT diagram for amorphous resins, it is evident that shrinkage values given on material data sheets are generally inadequate for design purposes. Similar treatment of PVT diagrams for semicrystalline-based composites is much more difficult due to the complex effect of crystallinity on the magnitude of shrinkage. Regardless, specific volume defined as a function of pressure and temperature continues to have fundamental meaning in developing more quantitative computational models for predicting mold shrinkage in general.

Particularly for semicrystalline materials, there is a need to develop *empirical* expressions based on DOE studies that characterize interrelationships between measured shrinkage values and injection molding process variables as suggested by PVT diagrams.

Herbert [13] and, more recently, Gipson et al. [14] studied the effects of key system variables (melt/mold temperature, part thickness, variation in mold length, average cavity pressure, etc.) on shrinkage values for polypropylene copolymer resins in the parallel and perpendicular flow directions. By bracketing the high–low variable limits at a sufficiently wide range, the resulting correlation plots define the entire operating window of injection molding process. Instead of obtaining averaged single-valued shrinkage numbers for a given material, such as that shown in typical material data sheets, shrinkage is now given a functional meaning in terms of key system parameters.

TABLE 19.6  Determining Factors for Mold Shrinkage

| System property | Decrease in shrinkage | Increase in shrinkage |
|---|---|---|
| Material | Amorphous microstructure<br>Particulate filler reinforcement<br>Fiber reinforcement in flow direction<br>Low-crystallinity polymer matrix | Crystalline microstructure<br><br><br>High-crystallinity polymer matrix |
| Injection molding process conditions | High cavity pressure<br>High injection rate<br>Low mold temperature<br>Fast cooling time<br>High-conductivity steel | Low holding-injection pressure<br>Insufficient gate freeze-off time<br>High mold temperature<br>Short pack-hold time or cooling time |
| Mold design | | Increase in wall thickness<br>Long flow path<br>Small gate area |

## Part Shrinkage Behavior of Polypropylene

The results for the two DOE studies are summarized in Table 19.7. There is good agreement of mold/melt temperature and cavity pressure trendlines between the two works. Furthermore, plots of shrinkage versus the range of variable values are linear. These charts provide guidelines for mold/process optimization.

The key process variable for both investigations [14,15] is mold cavity pressure. It is particularly encouraging that the linear plot for cavity pressure versus shrinkage is statistically *identical* in these DOE results and related data in the literature [15,16]. The constant negative slope of $-1.0 \times 10^{-6}$ (in./in.)/psi of the linear plot of shrinkage versus cavity pressure is applicable for a very wide range of pressures [250–4500 psi] and for a wide variety of mold designs.

Gipson et al. [14] interpreted the observed trends in shrinkage behavior as a function of individual system parameters:

1. *Increase in cavity pressure—shrinkage decreases.* There is more packing of polymer molecules leading to less shrinkage. This is visualized in Fig. 19.5.
2. *Part thickness increases—shrinkage increases.* Thicker molded parts cool more slowly to permit polymer molecules to recrystallize into a more ordered state with a higher degree of crystallinity and correspondingly higher shrinkage.
3. *Increase in mold temperature–shrinkage increases.* Higher mold surface temperature in the cavity leads to slower cooling rates with increasing crystallinity and increased shrinkage.
4. *Increase in melt temperature–shrinkage decreases.* This behavior was somewhat of a surprise considering the above mold temperature–shrinkage behavior. The increased freedom for polymer molecules to move due to decreased viscosity gives lower cavity pressure and longer time to gate seal.

### 19.4.4 Effect of Cooling Rate on Crystallization Temperature

Injection molding processes have very high or "supercooling" rates of 500–5000 °C/min [7,17,18]. It is apparent that shrinkage trends obtained by DOE studies are strongly influenced by the effective cooling rates during the very brief time (20–60 sec) that the molten mass is confined in the mold cavity.

Because semicrystalline resins have a higher degree of crystallinity than amorphous resins, the change in specific volume is highly dependent on the cooling rate during the crystallization process. The cooling rate depends on part wall thickness and temperature gradient between the molten polymer and colder mold surface.

In the laboratory, we can simulate the effects of cooling rate on crystallization temperature to a very limited degree by using differential scanning calorimetry (DSC). Figures 19.6–19.8 depict the effects of 3, 10, and 50 °C/min

TABLE 19.7  Effect of Molding Process Conditions and Mold Design on Shrinkage of Polypropylene Copolymer

| Parameter | Mold design $W$ = width (mm) $L$ = length (mm) $t$ = thickness (mm) $W \times L \times t$ | Bracket parameter $P$ range values Low — High | Incremental increase parameter value $\Delta P$ | Length-wise shrinkage values $S_L$ (in./in.) | Change in length-wise shrinkage $\Delta S_L$ (in./in.) | Guideline for predicting shrinkage per incremental increase in parameter value $\Delta S_L / \Delta P$ |
|---|---|---|---|---|---|---|
| Melt temp. | $66.7 \times 66.7 \times 2.5$ | 390 to 460°F | 70°F | 0.0159 to 0.0145 | −0.0014 | −2.00E-05 (in./in.)/°F |
|  | $101 \times 305 \times 2.5$ | 390 to 460°F | 70°F | 0.0150 to 0.0140 | −0.0010 | −1.43E-05 (in./in.)/°F |
|  | $127 \times 127 \times 2.3$ | 399 to 500°F | 101°F | 0.0159 to 0.0153 | −0.0006 | −0.60E-05 (in./in.)/°F |
| Mold temp. | $66.7 \times 66.7 \times 2.5$ | 100° to 160°F | 60°F | 0.0144 to 0.0160 | 0.0016 | 2.66E-05 in./in.)/°F |
|  | $101 \times 305 \times 2.5$ | 80° to 140°F | 60°F | 0.0151 to 0.0162 | 0.0011 | 1.85E-05 in./in.)/°F |
|  | $127 \times 127 \times 2.3$ | 80° to 140°F | 60°F | 0.0144 to 0.0150 | 0.0006 | 1.00E-05 in./in.)/°F |
| Part thickness | $66.7 \times 66.7 \times t$ | 2.0 to 3.5 mm | 1.5 mm | 0.0140 to 0.0175 | 0.0035 | 2.33E-03 (in./in.)/mm |
| Average cavity pressure | $66.7 \times 66.7 \times 2.5$ | 250 to 3000 psi | 2750 psi | 0.0174 to 0.0140 | −0.0034 | −1.24E-06 (in./in.)/psi |
|  | $101 \times 305 \times 2.5$ | 1500 to 4000 psi | 2500 psi | 0.0160 to 0.0134 | −0.0026 | −1.04E-06 (in./in.)/psi |
|  | $127 \times 127 \times 2.3$ | 1450 to 4495 psi | 3045 psi | 0.0173 to 0.0142 | −0.0031 | −1.02E-06 (in./in.)/psi |

Note: $66.7 \times 66.7 \times t$ and $101 \times 305 \times 2.5$ molds were used in [14] and $127 \times 127 \times 2.3$ was used in [13].

cooling rate at atmospheric pressure for DSC analysis of a polypropylene homopolymer resin. In comparison, PVT measurements are generally done at a low cooling rate of 3 °C/min in order to attain desired equilibrium results.

Table 19.8 indicates that for just a range of 3 °C/min to 50 °C/min, there is a profound effect of cooling rate on crystallization temperature and degree of crystallization (inferred by changes in the heat of crystallization). Shay et al. [7] plotted crystallization temperature versus cooling rate on semilog plot to extrapolate DSC results to determine the effect of supercooling encountered in the injection molding process. By using the same extrapolation technique to data given in Table 19.8, Fig. 19.9 was generated. This analysis predicts that the crystallization temperature is greatly reduced to 60 °C at 1000 °C/min cooling rate.

In 1997, R. Y. Chang et al. [17] developed a computational model using a modified Tait equation [much like Eq. (19.5)]. They introduced the cooling rate effect into the model by describing nonequilibrium PVT behavior and crystallization kinetics of semicrystalline polymer. The crystallization kinetic model was developed using DSC data obtained at different cooling rates (1, 2.5, and 5 °C/min) for polypropylene homopolymer.

More recently, Lobo and Bethard [18] utilized a similar hybrid technique to correct PVT measurements for cooling rate effects. DSC data were used to determine shifts in crystallization temperature for nylon-66 resin over a broad range of cooling rates (3, 20, 40, 100 °C/min) at atmospheric pressure. The linear portion of the specific volume–temperature plot for the molten polymer was extrapolated to the onset crystallization temperature determined by DSC at the given cooling rate. PVT measurements were made at the same cooling rates and for a range of pressures (10, 40, and 80 MPa). By combining the DSC and PVT measurements into a hybrid model, they were able to obtain specific volume–temperature plots as a function of cooling rate. This permitted more accurate prediction of specific volume as a function of temperature and pressure for use in CAE programs for simulating injection molding processes.

### 19.4.5 Shift in Specific Volume vs. Temperature Due to High Cooling Rate

A typical crystallization plot obtained via DSC features the onset of a crystallization exotherm peak at a given temperature during cooling of the molten mass. Using data given in Table 19.8, the values of onset-peak-end crystallization temperatures for 3 °C/min and 50 °C/min cooling rates were superposed onto the PVT diagram given in Fig. 19.10. The figure exhibits a shift in the original specific volume–temperature curve to demonstrate the effects of the higher cooling rate. The modified PVT diagram shown in Fig. 19.10 is a practical application of the above hybrid method [17].

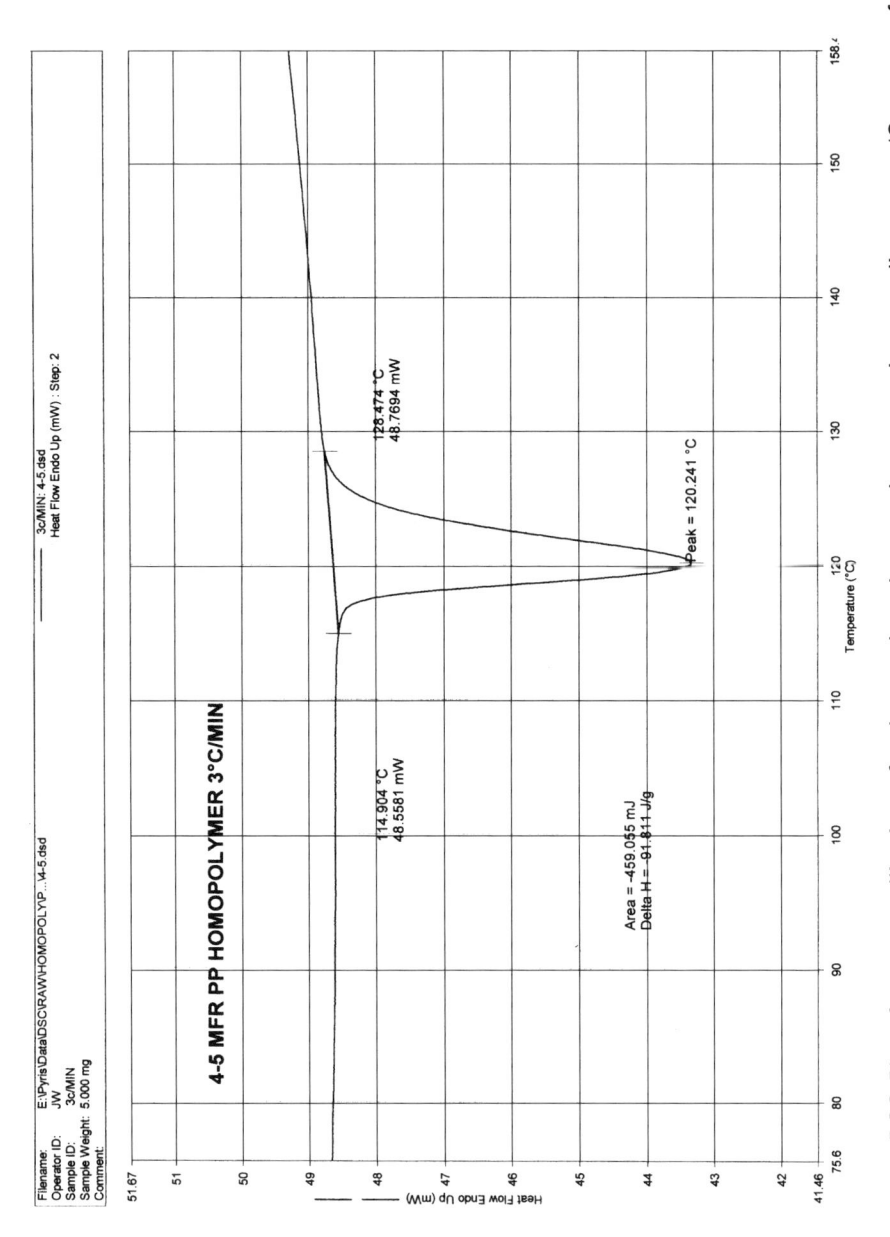

FIGURE 19.6 DSC Plots for recrystallization of polypropylene homopolymer at given cooling rates. (Courtesy of Mr. Joseph Westermeier, RheTech, Inc.) Cooling rate at 3 °C/min.

# Part Shrinkage Behavior of Polypropylene

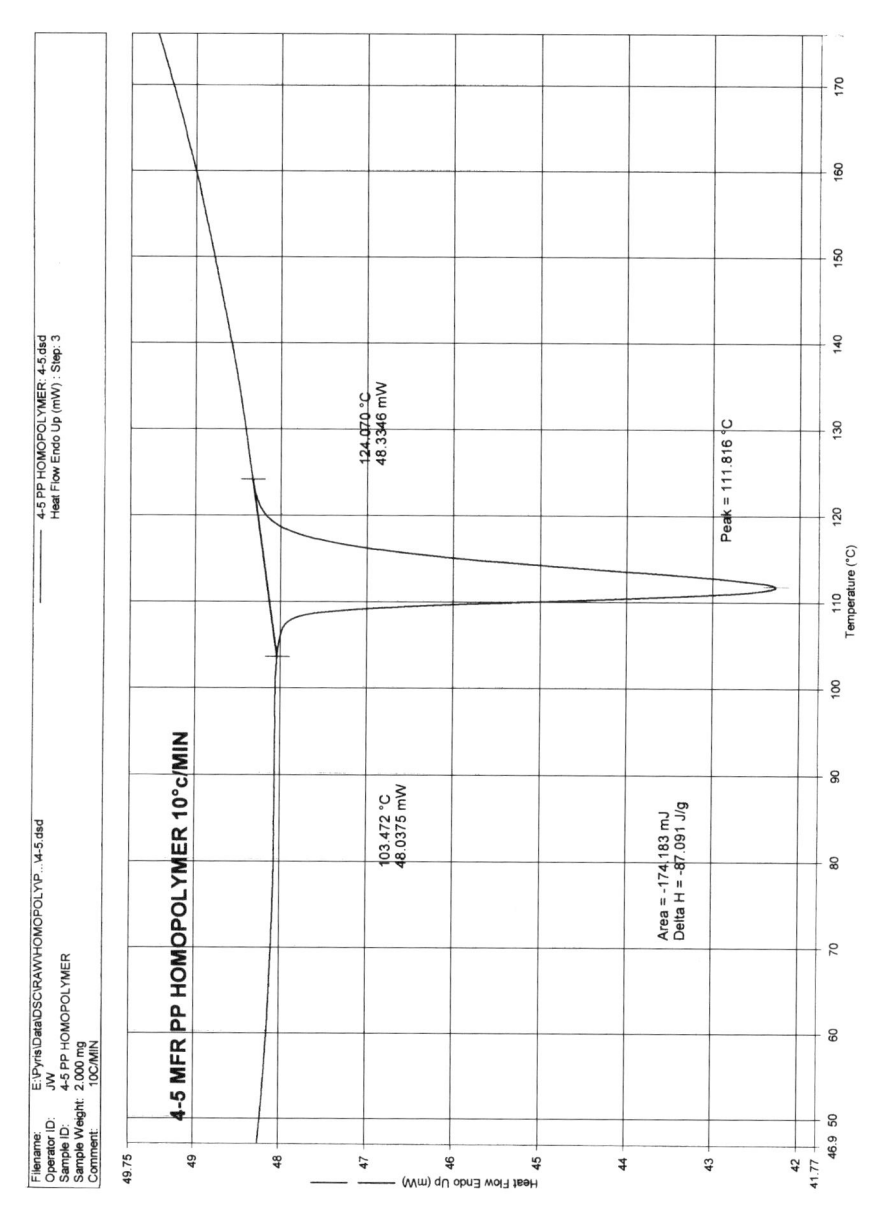

FIGURE 19.7 DSC Plots for recrystallization of polypropylene homopolymer at given cooling rates. (Courtesy of Mr. Joseph Westermeier, RheTech, Inc.) Cooling rate at 10 °C/min.

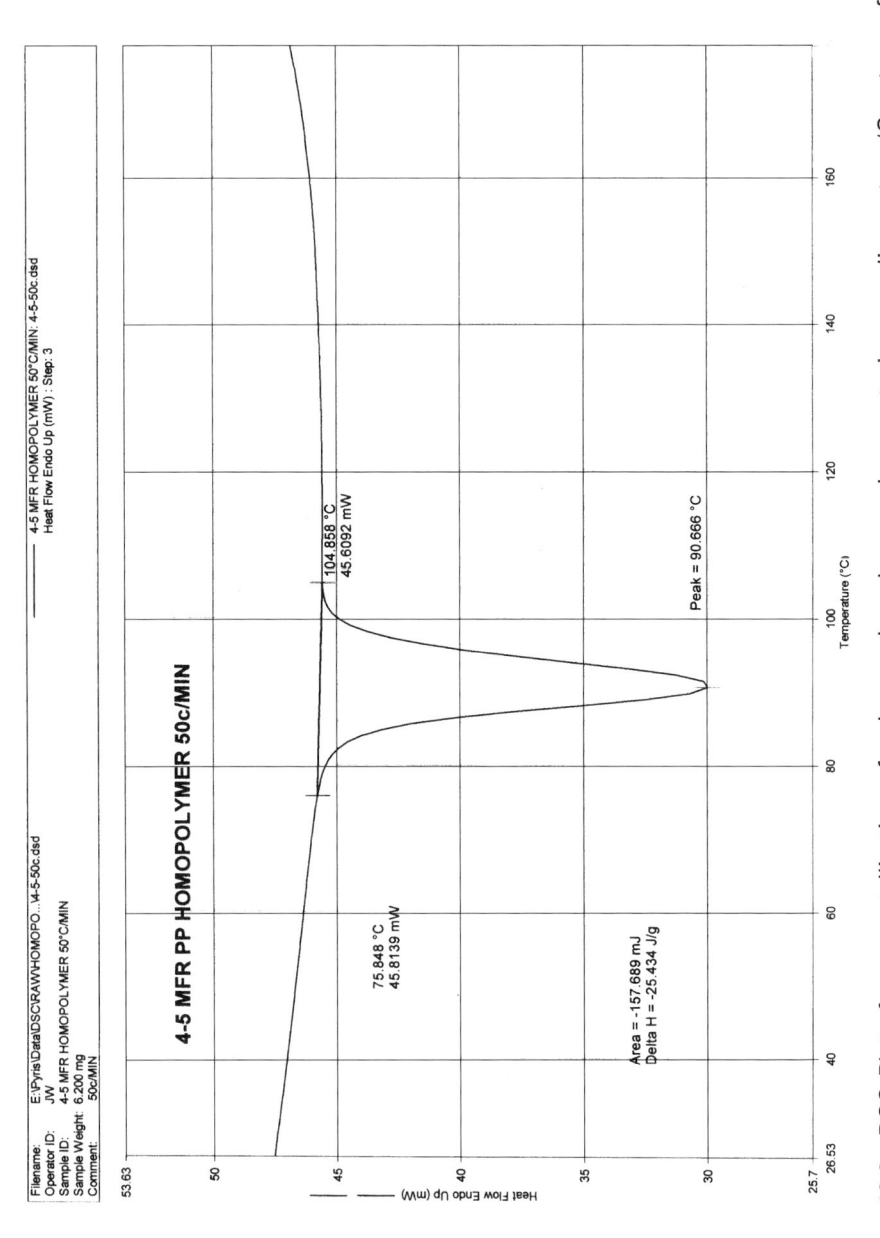

**FIGURE 19.8** DSC Plots for recrystallization of polypropylene homopolymer at given cooling rates. (Courtesy of Mr. Joseph Westermeier, RheTech, Inc.) Cooling rate at 50 °C/min.

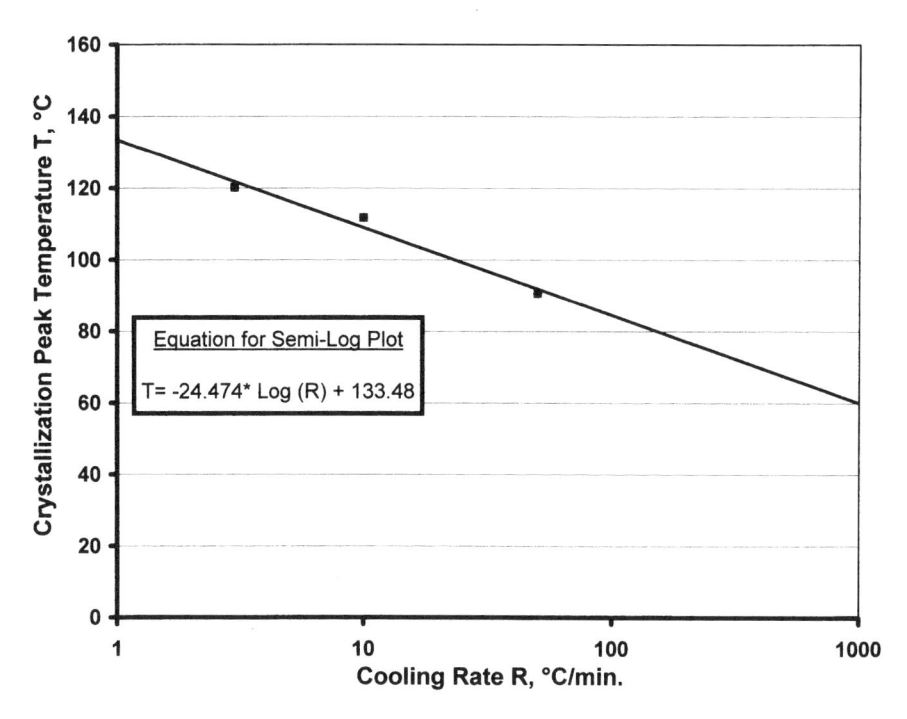

FIGURE 19.9  Semilog plot of crystallization peak temperature versus cooling rate. ■, Data obtained from DSC plots given in Figs. 19.6–19.8.

The 3 °C/min crystallization temperature (onset, end) falls precisely on the original PVT diagram as one would expect. Then the linear portion of the specific volume–temperature plot for the molten polypropylene was extrapolated to the onset crystallization temperature at 104.9 °C that corresponds to a 50 °C/min cooling rate. With the transition curve ending at 75.9 °C for the end of crystallization, the resulting dashed line in Fig. 19.10 traces out the shift in the PVT plot due to a higher cooling rate of 50 °C/min compared with the original 3 °C/min rate for actual PVT measurement.

## 19.5  CONCLUSIONS

Baine et al. [19] pointed out many shortcomings associated with taking a narrow range of shrinkage values for polymer composites to design molds for a given part's geometry. They described the consequences of having inadequate input data for computer simulation of shrinkage behavior due to the injection molding

TABLE 19.8 Effect of Cooling Rate on Crystallization of Polypropylene via Differential Scanning Calorimetry

| Cooling rate (°C/min) | Temp. for onset of crystallization (°C) | Temp. at maximal peak height (°C) | Temp. at end of crystallization (°C) | Heat of crystallization (J/g) |
|---|---|---|---|---|
| 3 | 128.5 | 120.2 | 114.9 | 91.81 |
| 10 | 124.1 | 111.8 | 103.5 | 87.09 |
| 50 | 104.9 | 90.7 | 75.9 | 25.40 |

# Part Shrinkage Behavior of Polypropylene

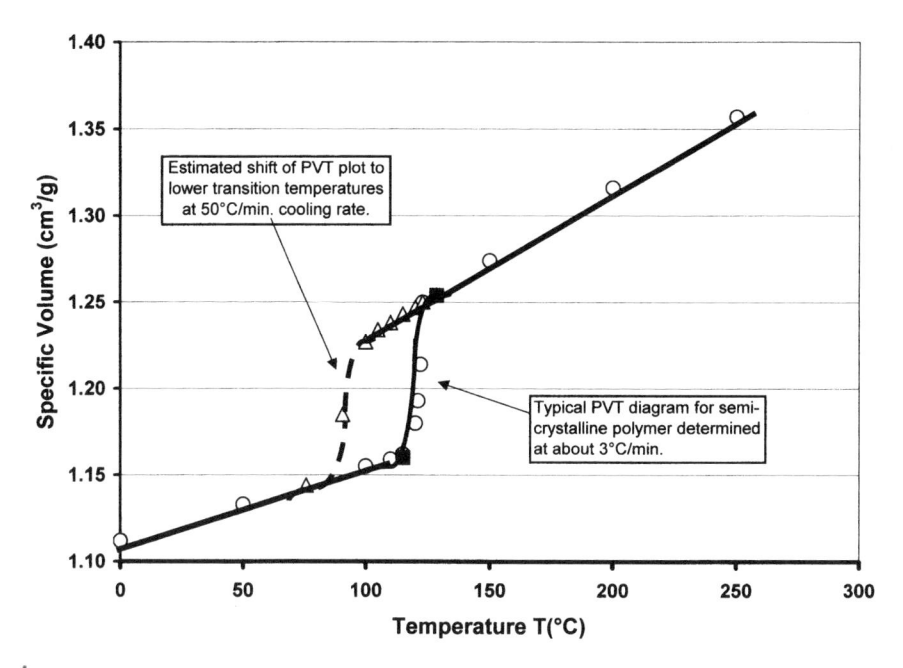

**FIGURE 19.10** Effect of cooling rate on PVT diagram. ○, Original PVT plot measured at about 3 °C/min. [12]. △, Shift in transition curve estimated from DSC data (Fig. 19.8).

process. It results in prolonged and expensive cutting of many prototype molds before finally obtaining a suitable mold design.

An assortment of "correction factors" are required to reduce the errors due to insufficient input data for computerized mold design. Hence, current methods for attaining final mold design are semiempirical at best. Even in the development of a hybrid method, Lobo and Bethard [18] emphasized the need for implementing proven crystallization kinetics into the hybrid model in order to better describe the effects of cooling rate.

The question raised in 1992 by Baine et al. (19) remains the same today: "Why isn't there more specific injection molding shrinkage data available?" [19]. By combining hybrid methodology of DSC-PVT measurements [17,18] with DOE investigation of system parameters (material, process conditions, and part geometry), we can finally begin to answer this almost elusive query. There is a need for more rigorous analytical techniques based on sound fundamental principles of polymer morphology, rheology, and thermal properties of polymer composites under compressive conditions.

The degree of crystallinity of the base polymer resin and the type of filler/fiber reinforcement remain the most dominant system parameters that determine the size (shrinkage) and shape (warpage) of molded plastic parts. Semicrystalline materials require explicit mathematical expressions for specific volume change as a function of cooling time, temperature gradients in the mold cavity, and crystallization kinetics.

Cavity pressure of molten plastic is clearly the most influential process parameter that determines the magnitude of shrinkage. There is a need to better define polymer viscosity as a function of melt temperature and melt pressure in order to accurately compute cavity pressure.

## REFERENCES

1. HR Sheth, RE Nunn. Temperature based adaptive holding control, using injection molding machine generated pvt data. SPE ANTEC Preprints, Toronto, 1997, pp. 551–555.
2. KA Beiter, K Ishii. Incorporating dimensional requirements into material section and design of injection molded parts. SPE ANTEC Preprints, Toronto, 1997, pp. 3295–3299.
3. Intelligent Systems Laboratory. Michigan State University, 1999, E. Lansing, Michigan.
4. Warpage Design Principles—Making Accurate Plastic Parts, Moldflow Pty Ltd, 1991.
5. C Maier, T Calafut, eds. Polypropylene: The Definitive User's Guide and Databook. New York: Plastics Design Library, 1998, pp. 151–152, 171–172.
6. LNP literature. Understanding Shrinkage and Warpage of Reinforced and Filled Thermoplastics.
7. RM Shay Jr., AJ Poslinski, Y Fakhreddine. Estimating linear shrinkage of semicrystalline resins from pressure–volume–temperature (pvt) data. SPE ANTEC Preprints, Atlanta, 1998, pp. 579–583.
8. DA Velarde, MJ Yeagley. Linear shrinkage differences in injection molded parts. Plastics Eng 56(12):60–64, 2000.
9. PS Spencer, GD Gilmore. Equation of state for high polymers. J Appl Phys 21:523, 1950.
10. GN Foster III, N Waldman, RG Griskey. Pressure–volume–temperature behavior. Polym Eng Sci 6(4):131–134, 1966.
11. WC Bushko, UK Stokes. Shrinkage of packed injection-molded parts: simulation of gate freeze-off effects. SPE ANTEC Preprints, San Francisco, 1994, pp. 559–564.
12. X Guo, AI Isayev, M Demiray. Crystallinity and microstructure in injection moldings of isotactic polypropylene. II: Simulation and Experiment. J Polym Sci 39(11):2132–2149, 1999.
13. G Herbert. High modulus impact copolymer polypropylene. Effects of polymer and process variables on the physical properties of injection molded articles. SPE ANTEC Preprints, San Francisco, 1994, pp. 579–586.

14. PM Gipson, PF Grelle, BA Salamon. SPE Automotive TPO Global Conference '99, Northfield Hilton, Troy, Michigan, Sept. 20–22, 1999, pp. 355–363.
15. DC Paulson, Injection Molding Technology, video series and workbook, 1983, Paulson Training Programs, Inc.
16. PA Tres, Polypropylene Product Design and Processing, 1996–1997.
17. RY Chang, YC Hsien, FH Lin, CH Hsu. Modifying the Tait equation with cooling rate effect to predict the pressure–volume–temperature behavior of semi-crystalline polymers: model and experiment. SPE ANTEC Preprints, Toronto, 1997, pp. 2081–2085.
18. H Lobo, T Bethard. Capturing pvt behavior of injection molded parts using hybrid methods. SPE ANTEC Preprints, New York, 1999, pp. 788–793.
19. MF Baine Jr, SL Janicki, AS Ulmer, LS Thomas. Mold shrinkage: not a single data point. SPE ANTEC Preprints, Detroit, 1992, pp. 977–980.
20. PC Bolurpe. A Guide to Injection Molding of Plastics. Chennai, India: Allied Publishers, Ltd., 2000.
21. JP Green, Assoc Prof, Dept ME/Manufacturing, California State University, Chico, CA.
22. GE Polymerland literature. Thermoplastic Material Properties—Physical Properties.

# 20

## Polypropylene Nanocomposite

**Guoqiang Qian and Tie Lan**
Nanocor, Inc., Arlington Heights, Illinois, U.S.A.

### 20.1  INTRODUCTION

Polypropylene is one of the fastest growing classes of thermoplastics. This growth is attributed to its attractive combination of low cost, low density, and high heat distortion temperature (HDT). The extraordinary versatility of unfilled virgin resin and reinforced polypropylene suits a wide spectrum of end-use applications for fibers, films, and molded parts. However, there always exist certain shortcomings in physical and chemical properties that can limit universal use of any given polymer resin. In packaging, for example, polypropylene resins have poor oxygen barriers, while low dimensional and thermal stability limits the scope of polypropylene composites in automotive applications. Most schemes to improve polypropylene gas barrier properties involve either addition of higher barrier plastics via a multilayer structure or surface coatings. Although effective, the increased cost of these approaches negates one big attraction for using polypropylene in the first place—economy. Currently, automotive and appliance applications employ glass or mineral-filled systems with loading levels ranging from 15 to 50 wt%. This approach improves most mechanical properties, but polypropylene's ease of processing is somewhat compromised. Furthermore, the need for higher filler loading leads to greater molded part weight.

Polymer-clay nanocomposites have advanced significantly in recent years [1–4]. New methods and novel applications are evolving rapidly, as evidenced by a global explosion of issued patents, especially in the United States, Europe, and Asia. Nanocomposites are filled or reinforced polymer resins in which at least one dimension of the dispersed phase is in the nanometer range (< 100 nm).

Based on the interaction between polymer resin and clay, three structures can be formed upon mixing: immiscible, intercalated, and exfoliated. Ideally the structure of the nanocomposite is defined as intercalated or exfoliated. Intercalated structures are formed by incorporation of a polymer chain between clay layered silicate regions to form a well-ordered multilayer structure with alternating polymer/inorganic layers. Exfoliated hybrids are materials in which the clay layers are exfoliated and uniformly dispersed in a continuous polymer matrix. The most preferred structure is exfoliated structure, even though most of the intercalated nanocomposites also demonstrate excellent properties. Finally, if the components are immiscible, undesirable microcomposites having poor properties will result.

Practically, it is difficult to define nanocomposite structures by a single measurement. The most common method to obtain structural information is X-ray diffraction (XRD). XRD can be used to measure the interlayer spacing distance ($d$) between clay layers. A given range of $d$-spacing values is associated with certain material properties:

1. When the $d$ spacing of the composite is similar to that of the starting layered silicate, the composite is normally defined as immiscible.
2. When the $d$ spacing of the composite is greater than that of the initial layered silicate, the composite is normally defined as intercalated nanocomposite.
3. If the $d$ spacing cannot be detected by XRD, the composite is normally defined as exfoliated nanocomposite.

In most cases, immiscible/intercalated or intercalated/exfoliated structure coexists in the same nanocomposite materials. Therefore, a simple determination of nanocomposite structure by XRD alone is not conclusive. It is recommended to combine different analytical methods to obtain a comprehensive picture of the nanocomposite structure, e.g., XRD combined with transmission electron microscopy (TEM).

When nanocomposites are formed, they exhibit significant improvements in physical, chemical, and mechanical properties. Usually at very low layered silicate loading, nanocomposites exhibit a greatly improved tensile strength, stiffness, better dimensional stability, decreased thermal expansion coefficient, and reduced gas barrier properties in comparison with neat or unfilled polymer

## Polypropylene Nanocomposite

resin. In both academic and industrial locations, the study of polypropylene nanocomposites is an intense area of current interest and investigation. The driving force for such efforts is attributed to huge commercial opportunities in both automotive and packaging applications. Material design at relatively low clay loading addresses the inherent shortcomings of polypropylene resin by itself and does so with favorable cost, processing, and reduced molded-part weight profiles.

### 20.2 STRUCTURE OF MONTMORILLONITE CLAY

The structure of montmorillonite belongs to the general family of 2:1 layered silicates [5]. Their crystal structure consists of layers made up of two silica tetrahedra fused to an edge-shared octahedral sheet of aluminum hydroxide (Fig. 20.1). The stacking of the silicate layers leads to a gap between layers called the *interlayer* or *gallery*. Typical van der Waals forces are active in this domain. Isomorphic substitution within the layers generates negative charges that are normally counterbalanced by cations residing in the interlayer space. In naturally occurring montmorillonite clay, the interlayer cations are usually hydrated $Na^+$, $K^+$, or $Ca^{2+}$. This structure is highly hydrophilic, and it is not miscible with most engineering polymers. However, ion exchange reactions of interlayer cations with various organic cations (e.g., alkylammonium cations) can render the clay surface organophilic. The organic cations lower the surface energy of the clay surface and

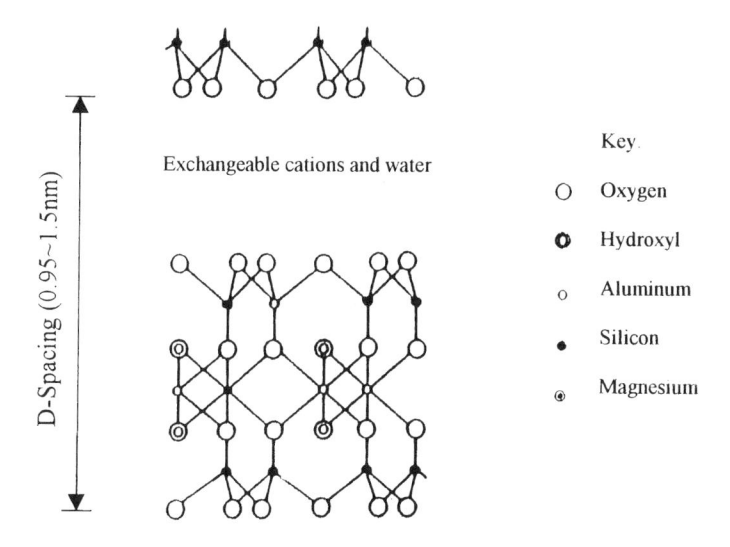

**FIGURE 20.1** Structure of montmorillonite clay.

improve wetting with the polymer matrix. In addition, the organic cations may contain various functional groups that react with the polymer to improve adhesion between the inorganic phase and the polymer matrix. In most cases, polymer-clay nanocomposites are made from surface-modified montmorillonite clay.

## 20.3 FORMATION OF POLYPROPYLENE NANOCOMPOSITES

In general, there are three ways to make polymer-clay nanocomposites: in situ polymerization, solution intercalation, and melt compounding [6]. Solution intercalation usually requires dissolving the polymer resin in an organic solvent and then intercalating into clay layers. This method is not feasible for large-scale production of polypropylene nanocomposite. In situ polymerization and melt compounding, on the other hand, have been used successfully to form polypropylene nanocomposite.

### 20.3.1 In Situ Polymerization

The successful synthesis of nylon-clay nanocomposite by in situ polymerization [7] has drawn tremendous interest in making high-performance nanocomposite with montmorillonite clay. Bayer AG and Honeywell have announced commercialization of nylon-6 nanocomposite. By introducing only a few percent of layered silicate into nylon-6 matrix via in situ polymerization, the tensile strength and modulus were greatly improved. The HDT increased from 65°C for unfilled nylon 6 resin to 152°C for the corresponding nanocomposite. The oxygen barrier properties also improve dramatically. However, the success of making nylon nanocomposite by in situ polymerization method has not been extended to making polypropylene nanocomposites at commercial scale. The major impediment is the chemical sensitivity of current polymerization catalysts for polypropylene resin.

Ziegler-Natta catalysts are the most common commercial catalysts for polypropylene [8]. The original Ziegler-Natta catalyst chemistry consisted of a complex of transition metal halide (titanium trichloride or $TiCl_3$) combined with organometallic compound (typically triethylaluminum) as cocatalyst to initiate polymerization. Since the surface area of $TiCl_3$ ingredient was low and only titanium atoms on the catalyst surface were accessible to the organometallic compound, few active sites were formed. Consequently, the amount of polypropylene produced per gram of catalyst used was low.

Supported Ziegler-Natta catalysts were then developed using magnesium chloride ($MgCl_2$) as the inert support material. Addition of Lewis base as an internal electron donor to a Lewis base as an external donor to the $MgCl_2$-supported catalyst increased catalyst activity and stereospecificity. This combina-

# Polypropylene Nanocomposite

tion of Lewis acid-base eliminated the necessity for postreactor removal of catalyst residues.

More recently, metallocene catalysts have been utilized to tailor make polyolefin and plastomer resins (see Chapter 7) having desired end-use properties. Metallocenes are organometallic compounds with a sandwich-like arrangement, consisting of a transition metal situated between two cyclic organic compounds. Current metallocene catalyst systems commonly use zirconium chloride as the transition metal complex, with a cyclopentadiene as the organic compound and an aluminoxane such as methylaluminoxane (MAO) as cocatalyst. Polypropylene resins with different morphologies, molecular weights, and other properties can be produced by varying the transition metal and organic compound used [8].

Both Ziegler-Natta catalysts and metallocene catalysts are chemically sensitive. However, by carefully controlling the reaction conditions, polypropylene-clay nanocomposites can be synthesized by in situ polymerization using both kinds of catalysts, at least at the laboratory scale. Ma and coworkers [9] used hexadecyloctodecyltrimethylammonium to modify the clay. The organoclay was then mixed with $MgCl_2$ and dispersed in toluene. A certain amount of internal electron donor and $TiCl_4$ was added to the slurry and activate the organoclay. During the polymerization reaction, toluene, triethylaluminum, and external electron donor were injected into the reactor. Different organoclay loading levels (2.5–8.1 wt%) were obtained by controlling the polymerization time. The resulting polypropylene-clay nanocomposite exhibited much higher storage modulus ($E'$) than pure polypropylene, especially at temperatures higher than $T_g$ (Table 20.1). For example, at 8.1% organoclay loading, the storage modulus tripled at 120°C. The corresponding HDT of the nanocomposite also increased significantly with increasing clay content (Table 20.1).

Metallocene catalyst has also been used to make polypropylene nanocomposite by Sun and coworkers [10,11]. In one case, 4-tetradecylanilinium-modified clay was dispersed in toluene. Then tripropylaluminum and metallocene poly-

TABLE 20.1 Effect of the Clay Content on $E'$, $T_g$, and HDT of PP-Clay Nanocomposites

| Samples | $E'$ (GPa) ||||  $T_g$ (°C) | HDT (°C) |
| --- | --- | --- | --- | --- | --- | --- |
|  | −40°C | 20°C | 80°C | 120°C |  |  |
| PP | 1.42 | 0.38 | 0.21 | 0.12 | 6.2 | 110 |
| PP/MMT1 (2.5 wt%) | 2.08 | 0.76 | 0.36 | 0.24 | 12.1 | 138 |
| PP/MMT2 (4.6 wt%) | 2.22 | 0.82 | 0.46 | 0.28 | 9.0 | 144 |
| PP/MMT3 (8.1 wt%) | 2.43 | 0.98 | 0.54 | 0.36 | 8.0 | 151 |

merization catalyst [dimethylsilysbis(2-methyl-4-phenylindenys)zirconium(II)-1,4-diphenyl-1,3-butadiene] was added to the slurry. In the next synthesis step, propylene gas was introduced at ambient temperature to form polypropylene nanocomposite. XRD analysis of the reaction mass as polymerization progressed showed a change in microstructure. The diffraction peak of organoclay decreased and then eventually disappeared, indicating that the nanocomposite had achieved exfoliation/disordered structure. High-resolution TEM revealed that some of the initial particles were exfoliated to a single layer or a few layers. However, optical microscopy indicated that there were still some microsized organoclay particles that were not dispersed during the polymerization reaction. The resulting polypropylene nanocomposite had greatly improved mechanical properties in comparison with neat resin. For example, the nanocomposite has up to 600 kpsi Young's modulus at about 10 wt% organoclay loading, compared with 220–260 kpsi for neat resin.

### 20.3.2 Melt Compounding Process

So far, most of the research activities have focused on preparation of polypropylene nanocomposite by melt compounding. This preference is due to easy processing of the base resin. Because polypropylene is a low-polarity polymer, it is extremely challenging to make well-dispersed polypropylene nanocomposite. Numerous parameters can influence nanocomposite formation in a melt compounding process. Figure 20.2 lists some of the most important ones. These factors will be discussed in more detail below.

*Effect of Clay Surface Treatment.* The most common method to modify clay surface is by ion-exchange reaction using alkylammonium. The molecular chain length of alkylammonium has a pronounced effect on clay surface polarity, and therefore affects the interaction between the polymer matrix and organoclay reinforcement. Reichert and coworkers [12] modified inorganic clay using protonated $C_4$ to $C_{18}$ alkylamine. Polypropylene nanocomposites were then compounded using different organoclay with (or without) maleic anhydride–modified polypropylene (MA-PP) as compatibilizer. In the absence of compatibilizer, Young's modulus of polypropylene composites was very similar to that of neat resin. However, with the addition of MA-PP, there was significant improvement in Young's modulus when the chain length was 12 carbon atoms or higher. This influence of alkyl chain length on mechanical properties was attributed to the changes of superstructures. In the absence of compatibilizer, the interaction between polymer matrix and organoclay was limited. No matter what alkyl chain length there is in the organoclay, a macrocomposite formed that behaved as conventional filler. On the other hand, when the alkyl chain length was 12 and higher, partially exfoliated nanocomposites were formed in the presence of

# Polypropylene Nanocomposite

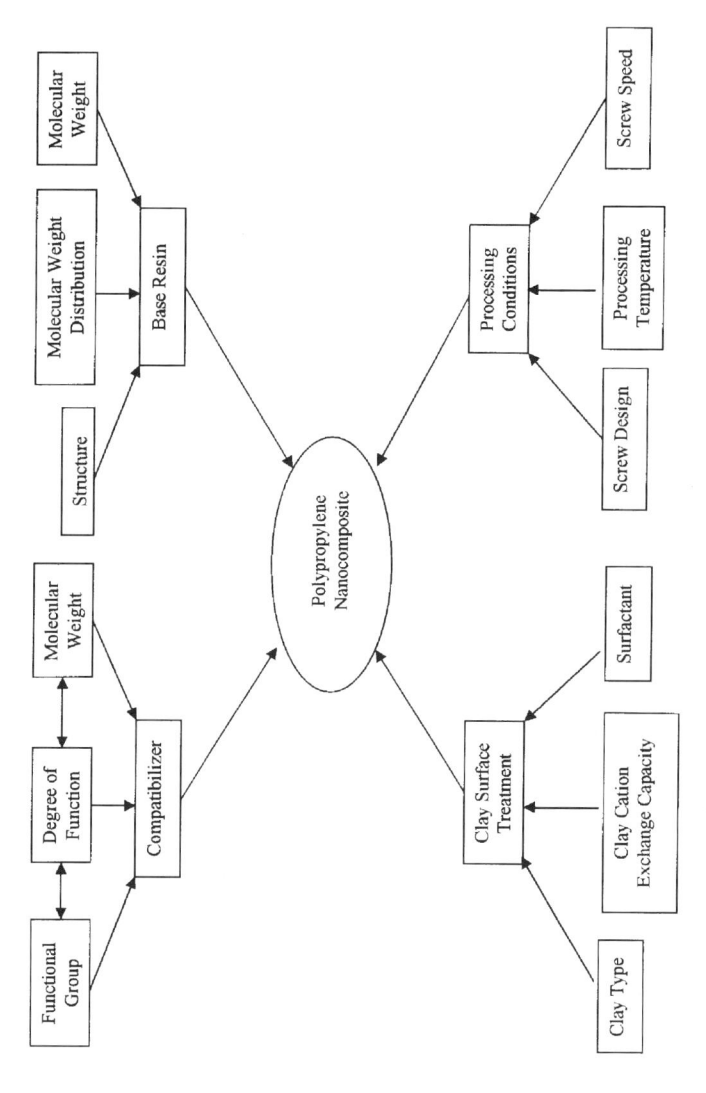

**FIGURE 20.2** Factors affecting melt compounding of polypropylene nanocomposites.

compatibilizer. The exfoliated structure was detected by a broadened peak from wide-angle X-ray scattering (WAXS) and TEM [12]. As a result, the mechanical properties of these nanocomposites are greatly improved.

Theoretically, an increase in carbon chain length would give a more hydrophilic organoclay. However, in the case of making polypropylene nanocomposite, longer chain length doesn't necessarily translate into better properties. This lack of property improvement is due to a higher organic loading in an organoclay when the alkyl chain length is longer. At the same organoclay loading level, the inorganic part of organoclay is smaller, and the reinforcement effect of organoclay will be decreased. Consequently, there is an optimal organic loading level that is determined by both alkyl chain length and clay cation exchange capacity (CEC). Typically, an increase in chain length of alkylammonium corresponds to an increase in interlayer spacing.

On the basis of clay CEC and carbon atom chain length, the alkylammonium confined between two clay layers adopts a monolayer, bilayer, or trilayer structure [13]. Let us consider the clay CEC is 0.8 meq/g. The following structures are generated: the alkylammonium adopts a single layer at chain lengths of $C_7$–$C_{11}$. A bilayer is formed at $C_{12}$ and higher. However, when the clay CEC is 1.5 meq/g, alkylammonium forms a bilayer structure at $C_{10}$ to $C_{12}$, "two and a half" layers for $C_{13}$ to $C_{15}$, and trilayer for $C_{16}$ to $C_{19}$. This means at a certain chain length, the CEC of particular clay determines the number of molecular chains per unit area.

Nanocor developed two surface-modified montmorillonite products for compounding polypropylene nanocomposite: Nanomer I.30P and I.44PA, respectively [14]. These two products are formulated differently, with a focus for I.30P on film, engineering, and flame retardancy applications, and I.44PA on engineering and flame retardancy applications. In Table 20.2 are compared the mechanical properties of polypropylene nanocomposite prepared from Nanomer I.30P and

TABLE 20.2 Mechanical Properties of Polypropylene Nanocomposite Containing Nanomer I.30P and I.44PA

| Material | Additive | Nanomer loading (%) | Tensile strength (MPa) | Flexural modulus (MPa) | Notched Izod (ft-lb/in.) | HDT (°C) |
|---|---|---|---|---|---|---|
| Neat polypropylene | None | 0 | 34.8 | 1310 | 0.6 | 98 |
| Polypropylene nanocomposite | Nanomer I.30P | 6 | 39.8 | 2240 | 0.6 | 122 |
| Polypropylene nanocomposite | Nanomer I.44PA | 6 | 39.4 | 2160 | 0.8 | 124 |

# Polypropylene Nanocomposite

I.44PA. From the table we can see that both nanocomposites achieved significantly improved mechanical properties and HDT without impact loss.

*Effect of Compatibilizer.* The effect of reinforcement by layered silicate in polypropylene is determined by at least two important factors: clay dispersion (exfoliation or intercalation) and interaction between clay and polymer. When the clay surface is covered by a sufficient amount of organic surfactant (such as alkylammonium), organoclay can be well dispersed in the polypropylene matrix by using appropriate compounding techniques. However, in many cases, the mechanical properties remain quite poor due to inadequate stress transfer from the polymer matrix under applied stress to load-bearing clay reinforcement. The cause for this weakness is attributed to poor compatibility in the interphase region that exists between nonpolar polypropylene molecules and polar clay layers. To overcome this problem, the most effective approach is to add bifunctional compatibilizer to the nanocomposite formulation to enhance adhesion at the polymer–clay interface (see Chapters 3, 11, and 12).

Yurokawa [15–18] modified organoclay by blending polydiacetone acrylamide, synthesized by free radical polymerization, with maleic acid dissolved in solvent carrier to make a masterbatch. This masterbatch was further blended with polypropylene to make the final nanocomposite. Although effective, this procedure was too laborious and not suitable for large-scale production. A slightly simpler approach has been proposed by using polyolefin diol as compatibilizer [19]; however, this approach also required toluene as the solvent.

The introduction of MA-PP as compatibilizer made it feasible to compound polypropylene nanocomposite without using solvent [20–22]. Commercially available MA-PP products (see Chapters 3 and 12) have a wide variety of grades, including different molecular weight and maleic anhydride grafting content. Optimization of the type and amount of compatibilizer is crucial to make polypropylene nanocomposite successfully. The most effective compatibilizer must be compatible with both the layered silicate structures and polypropylene matrix.

*Effect of Compounding Conditions.* Melt extrusion processing conditions are one of the most important factors to disperse layered silicate. Unfortunately, not enough research has been done in this area. Thus far, the most common compounding method is direct compounding in a single-extrusion process. By using this method, a small amount of clay is blended into base polypropylene using a twin-screw extruder or other mixing device to form the polypropylene nanocomposite. However, in most cases the clay dispersion is not optimized.

Recently, a masterbatch compounding method was developed by Nanocor [23]. By using this method, optimized clay dispersion and property improvement can be achieved. A typical masterbatch compound consists of 40–50 wt% Nanomer combined with polypropylene resin and compatibilizer. Polypropylene

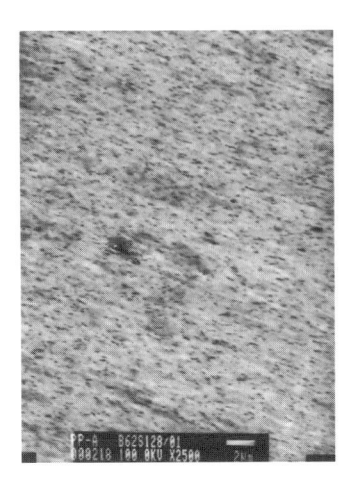

FIGURE 20.3 TEM images of polypropylene nanocomposite. (a) ×2,500, (b) ×30,000.

masterbatch can be prepared by using high-shear compounding devices, such as twin-screw extruder or internal mixer. Nanocor's masterbatch products can be further let down into polypropylene to make the final nanocomposite at the desired loading level. Optimized Nanomer dispersion and performance properties have been achieved through this masterbatch process.

TEM photomicrographs of polypropylene nanocomposite prepared by the masterbatch method are shown in Fig. 20.3. These pictures indicate that optimal layered silicate dispersion has been attained. This fine degree of dispersion is responsible for the observed property improvement for polypropylene nanocomposite. Table 20.3 compares the mechanical properties of polypropylene nanocomposites prepared from masterbatch compounding. We can definitely conclude

TABLE 20.3 Effect of Processing Conditions on Mechanical Properties of Polypropylene Nanocomposite

| Processing method | Additive | Nanomer loading (%) | Tensile strength (MPa) | Flexural modulus (MPa) | Notched Izod (ft-lb/in.) | HDT (°C) |
|---|---|---|---|---|---|---|
| Control | None | 0 | 31.6 | 1020 | 0.76 | 89 |
| Direct compounding | Nanomer I.30P | 6 | 34.2 | 1586 | 0.59 | 110 |
| Masterbatch let-down | Nanomer I.30P | 6 | 35.7 | 1753 | 0.63 | 111 |

## Polypropylene Nanocomposite

that tensile strength and flexural modulus of polypropylene nanocomposite prepared by masterbatch compounding is higher than direct compounding.

Nanocor's masterbatch products can also be let down using a single-screw extruder [24]. Under certain conditions, finished nanocomposite product can even be made directly by injection molding or other fabrication methods without using twin-screw extrusion let-down. In order to disperse masterbatch effectively by single-screw extrusion, there is generally a need to have some type of high-shear mixing elements before the diehead. New Castle Industrial, a single-screw extruder manufacturer, has developed a patented technology called NanoMixer. The NanoMixer is effective in both dispersive and distributive mixing. When the Nanomer masterbatch is let down in the NanoMixer, comparable dispersion and mechanical property has been achieved compared with twin-screw extrusion.

By using a modified twin-screw extruder, Nanocor was also able to selectively add ingredients to make nanocomposite by a sequence of resin additions at different feed ports. In this process, Nanomer, compatibilizer, and base resin are fed in the first hopper of twin screw extruder to make the masterbatch, and at the same time, base polypropylene is fed in the downstream port to let down the masterbatch. As a result, well-dispersed polypropylene nanocomposite is obtained in a single processing step.

*Effect of Base Resin.* In both masterbatch and let-down processes, the characteristic properties of base resin (mainly molecular weight and molecular weight distribution) play an important role in making polypropylene nanocomposite. During twin-screw extrusion, the length of the polymer molecular backbone not only affects intercalation of the clay but also affects the melt viscosity. Melt viscosity directly affects shear stress applied to the polymer melt. For example, the same masterbatch was let down into a twin-screw extruder with various polypropylene resins having different melt flow rates. The mechanical properties for the array of nanocomposites produced are listed in Table 20.4. At the low melt flow range, good improvement in mechanical properties is achieved.

TABLE 20.4 Effect of Melt Flow on Mechanical Properties of Polypropylene Nanocomposite

| Melt flow (g/10 min) | Additive | Nanomer loading (%) | Tensile strength (MPa) | Flexural modulus (MPa) | Notched Izod (ft-lb/in.) | HDT (°C) |
|---|---|---|---|---|---|---|
| 4 | None | 0 | 32 | 1150 | 0.7 | 86 |
|  | Nanomer I.30P | 6 | 38 | 2040 | 0.7 | 115 |
| 35 | None | 0 | 35.2 | 1590 | 0.4 | 113 |
|  | Nanomer I.30P | 6 | 39.9 | 2310 | 0.4 | 121 |

However, for high melt flow polypropylene, the improvement in mechanical properties is relatively small. The enhanced properties for low melt flow polypropylene are attributed to the corresponding higher melt viscosity with increased shear deformation of ingredients during the extrusion process. Also, the melt viscosity of low melt flow polypropylene has a better match with the melt viscosity of masterbatch to enhance clay dispersion.

## 20.4 PROPERTIES OF POLYPROPYLENE NANOCOMPOSITE

### 20.4.1 Mechanical Properties

When well-dispersed polypropylene nanocomposites are formed, greatly improved mechanical properties can be achieved at relatively low organoclay loading (4–6 wt%). Table 20.5 lists the mechanical properties of polypropylene homopolymer, copolymer, and TPO nanocomposite. From the table we can see that nanocomposites exhibited much higher tensile strength, flexural modulus, and HDT without any sacrifice impact strength. Compared with traditional talc or glass fiber–reinforced polypropylene, nanocomposite can achieve similar stiffness at lower inorganic loading level (Fig. 20.4). As a result, the same mechanical properties can be obtained by reduced wall thickness. This down-gauging opportunity can help maintain cost-effective utility of polypropylene nanocomposite. Furthermore, low inorganic loading level is combined with this down-gauging opportunity; easy processing and recycling make polypropylene nanocomposite well suited for automotive, appliance, industrial, and other applications.

TABLE 20.5 Mechanical Properties of Polypropylene Nanocomposite

| Resin | Additive | Nanomer loading (%) | Tensile strength (MPa) | Flexural modulus (MPa) | Notched Izod (ft-lb/in.) | HDT (°C) |
|---|---|---|---|---|---|---|
| Homopolymer polypropylene | None | 0 | 32.0 | 1150 | 0.7 | 86 |
|  | Nanomer I.30P | 6 | 38.0 | 2040 | 0.7 | 115 |
| Copolymer polypropylene | None | 0 | 23.2 | 995 | 2.2 | 86.3 |
|  | Nanomer I.30P | 6 | 27.1 | 1600 | 2.0 | 102 |
| TPO | None | 0 | 19.5 | 780 | 9.8 | 71 |
|  | Nanomer I.30P | 6 | 21.8 | 1230 | 9.8 | 85 |

# Polypropylene Nanocomposite

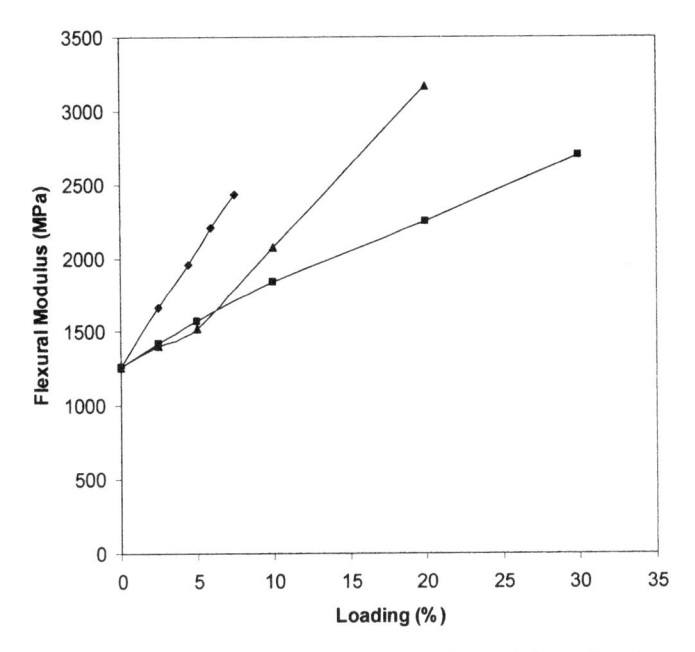

**FIGURE 20.4** Effect of filler loading on flexural modulus of polypropylene. ◆, Nanomer I.30P; ▲, glass fiber; ■, talc.

## 20.4.2 Barrier Properties

For gas and moisture barrier properties, the tortuous path theory has been well accepted for filled polymer systems [26]. According to this theory, the impermeable inorganic filler in the relatively permeable polymer matrix will create a "tortuous path" for the diffusing molecules. As a result, the filled polymer demonstrates reduced gas permeability, and the minimal permeability can be determined using the following equation:

$$P_f/P_u = \Phi_p/[1 + (L/2W)\Phi_f]$$

where $P_f$ and $P_u$ are the permeability of the filled and unfilled polymer, $\Phi_p$ and $\Phi_f$ are the volume fraction of the polymer and filler, and $L/W$ is the aspect ratio of the filler. According to this theory, there is an inverse relationship between aspect ratio of the filler and permeability of the filled polymer. An increase in aspect ratio of the filler corresponds to a decrease in the permeability of the filled polymer.

Compared with most inorganic fillers, silicate layers of montmorillonite have very high aspect ratio. The thickness of the single layer is 1 nm, while the length is in the range 200–500 nm. When the layered silicate is fully exfoliated in

TABLE 20.6 Oxygen Barrier Properties of Polypropylene Nanocomposite

| Resin | Product form | Additive | Nanomer loading (%) | OTR (cm$^3$/100 in.$^2$ day) |
|---|---|---|---|---|
| Random copolymer polypropylene | Cast film | None | 0 | 205 |
| | | Nanomer I.30P | 6 | 108 |
| TPO | Cast film | None | 0 | 1820 |
| | | Nanomer I.30P | 7 | 1270 |
| TPO | Blown film | None | 0 | 2270 |
| | | Nanomer I.30P | 7 | 1020 |

the polymer matrix, greatly reduced gas permeability can be achieved. So in order to reduce the gas permeability of polypropylene nanocomposite, it is important to fully disperse the silicate layers into polypropylene matrix. Sometimes, if the dispersion of the silicate is not good, there is almost no improvement in permeability; even increased permeability has been observed in some extreme cases. Table 20.6 shows the oxygen transmission rate (OTR) of polypropylene and TPO nanocomposite containing Nanomer I.30P. When the Nanomer is well dispersed, the OTR of the nanocomposite can decrease from 30% to 50%. It is interesting to see that processing conditions also have a big effect on barrier property. From Table 20.6 we can see for the same TPO nanocomposite, if it is processed through the cast film method, the OTR was reduced by 30%. When it is processed through the blown film method, the OTR was reduced by around 55%. This is because in the blown film process the silicate layers are more aligned parallel to the film surface. This generates a more tortuous path to the diffusing gas.

## 20.4.3 Flame Retardancy

The reduced flammability of polymer nanocomposite has been demonstrated by Gilman and coworkers [27,28]. In their cone calorimetry study, thermoplastic nanocomposites showed significantly reduced peak heat release rate (PHRR), a change in the char structure, and a decrease in the rate of mass loss. According to the authors, the mechanism is a consequence of high-performance carbonaceous silicate char buildup on the surface during burning. This charred surface insulates the underlying material and slows down the mass loss rate of decomposition products [29]. This residue layer forms as the polymer burns away and the clay layer reassembles. It is interesting to note that in some cases 40% reduction in PHRR can be achieved by adding as little as 0.1 wt% organoclay [30]. This

## Polypropylene Nanocomposite

TABLE 20.7 Nanomer Synergistic with DBDPO in Polypropylene

| Components | Bromine Flame Retardant Formulations | | | |
|---|---|---|---|---|
| Polypropylene (%) | 73.3 | 80 | 77 | 74 |
| DBDPO (%) | 20 | 15 | 15 | 15 |
| $Sb_2O_3$ (wt%) | 6.7 | 5 | 5 | 5 |
| Nanomer I.44PA (wt%) | 0 | 0 | 3 | 6 |
| UL-94 rating | V-2 | Fail | V-2 | V-0 |

attribute indicates that other mechanisms of reduced nanocomposite flammability are possible.

Even though PHRR decreases with an increase in silicate loading level, there is little benefit in promoting self-extinguishing of a highly filled polymer nanocomposite that has been ignited by a flame. Sometimes polymer nanocomposites are more effective in combination with other flame-retardant additive as a means to reduce dripping and improve mechanical properties. A synergistic effect between layered silicate and halogenated flame-retardant additives has been demonstrated in polypropylene matrix [31]. Decabromodiphenyl oxide (DBDPO) has been widely used in filled polypropylene to provide flame retardancy properties (see Chapter 4). Normally the loading level of DBDPO has to be at least 20 wt% to pass UL-94 testing. When combined with 3–6 wt% Nanomer I.44PA, the DBDPO level can be reduced to 15 wt% while maintaining the same UL-94 rating (see Table 20.7). Another example of Nanomer synergistic effect occurs with metal oxide (Table 20.8). Compared with traditional FR formulation, nanocomposite formulations burn at a noticeably reduced burning rate with a hard char formed on the surface. They not only reduce the loading of traditional FR package but also exhibit minimal dripping and fire sparkling. The UL-94 test results indicate that 60 wt% addition of $Mg(OH)_2$ is required to maintain FR rating for EVA system. Inclusion of 3–6 wt% Nanomer I.30P maintains the V-0 rating while lowering the $Mg(OH)_2$ to 50–55 wt%. With lowering the amount of $Mg(OH)_2$, the formulations containing organoclay

TABLE 20.8 Nanomer Synergistic with $Mg(OH)_2$ in EVA

| Components | Magnesium Hydroxide Flame Retardant Formulations | | | | | |
|---|---|---|---|---|---|---|
| EVA (%) | 40 | 45 | 42 | 39 | 50 | 47 |
| $Mg(OH)_2$ (%) | 60 | 55 | 55 | 55 | 50 | 50 |
| Nanomer I.30P (wt%) | 0 | 0 | 3 | 6 | 0 | 3 |
| UL-94 rating | V-0 | Fail | V-0 | V-0 | Fail | V-0 |

provide comparable FR rating, but improved processing and mechanical properties.

Another benefit of using nanocomposite in a FR formulation is the observed "antiblooming" effect. Some traditional FR additives, such as DBDPO, will bloom to the material surface under certain conditions. When this happens, it not only affects the appearance and moisture absorption but also affects the electrical properties. Sometimes people have to use more expensive, nonblooming-type FR additive to avoid this blooming problem. Nanocomposite, on the other hand, has a strong antiblooming effect. This effect is related to the high barrier properties of nanocomposite. When layered silicate is combined with traditional FR additive, it can prevent FR additive from blooming or at least greatly slow down the blooming rate. For example, when DBDPO and antimony trioxide (ATO) are incorporated into a polypropylene formulation, they can easily bloom to the plaque surface when placed in oven at 80°C for several days. The blooming can be detected by wiping the plaque surface using a black cloth. However, when 3 wt% Nanomer I.30P is incorporated in the system, almost no blooming effect can be detected. This clearly demonstrated that cost savings can be achieved by avoiding the use of expensive, nonblooming-type FR additive in the formulation.

### 20.4.4 Other Properties

*Crystallization Behavior.* Polypropylene is semicrystalline polymer, and its properties depend significantly on its crystallization behaviors. When polypropylene nanocomposites are formed, the presence of organoclay in polypropylene matrix affects its crystallization. With a fundamental relationship between crystallinity and mechanical properties (see Chapter 12), it is important to understand the crystallization behavior of polypropylene nanocomposite.

Ma and coworkers reported that during isothermal crystallization organoclay layers act as heterogeneous nuclei of polypropylene [32]. This microstructure leads to an increase in the crystallization rate of polypropylene nanocomposite. Typically, pure polypropylene crystallizes via both heterogeneous and homogeneous nucleation mechanisms. Homogeneous nucleation starts spontaneously by chain aggregation below the melting point and requires more time, where heterogeneous nuclei form simultaneously as soon as the sample reaches the crystallization temperature. In the case of polypropylene nanocomposite, a large number of dispersed organoclay layers act as heterogeneous nuclei, so that the crystallization rate is much faster. Also, higher organoclay loading corresponds to a faster crystallization rate. In fact, when organoclay is present in the polypropylene matrix, not only is the crystallization rate of polypropylene nanocomposite faster, but also the crystallization temperature of nanocomposite is increased. By using a polarizing optical microscope, there appears to be small

## Polypropylene Nanocomposite

crystallites formed in the polypropylene nanocomposite with a much higher concentration of spherulites in the limited space [32].

The nucleating effect of organoclay in polypropylene is also observed during nonisothermal crystallization [33]. At a given cooling rate, polypropylene nanocomposite has a faster crystallization rate. Thus, polypropylene nanocomposite can be easily fabricated using traditional injection molding processes with much faster cycle times than neat polymer.

*Coefficient of Linear Thermal Expansion.* An additional benefit from the addition of layered silicate to polypropylene is reduced thermal expansion. In Table 20.9, the coefficient of linear thermal expansion (CLTE) of polypropylene nanocomposite is compared with neat polypropylene. At 6 wt% loading, Nanomer reduces the thermal expansion by 20%. Greater dimensional stability will allow the use of polypropylene nanocomposites in combination with other low CLTE materials.

*Mold Shrinkage.* The addition of layered silicate to polypropylene also reduces mold shrinkage. In Table 20.9, the mold shrinkage of polypropylene homopolymer and TPO nanocomposite is compared with that of neat resin. As can be seen, the mold shrinkage was reduced by about 10–20% in both homopolymer and TPO. This reduction in mold shrinkage is beneficial since it minimizes mold dimension changes caused by changing polymer systems.

*Ultraviolet Stability.* Ultraviolet (UV) radiation is very destructive to polymeric materials. The energy in UV radiation is strong enough to break molecular bonds. This activity in the polymer brings about thermal oxidative degradation, which results in embrittlement, discoloration, and overall reduction in physical properties. By introducing organoclay in polypropylene, the damage of the polymer chain by UV radiation is reduced. As a result, the property reductions of nanocomposite are better than those of neat resin. Figure 20.5 compares the flexural modulus of polypropylene neat resin and nanocomposite

TABLE 20.9  CLTE and Mold Shrinkage of Polypropylene Nanocomposites

| Resin | Additive | Organoclay loading (%) | CLTE (X10E6) | Mold shrinkage (in./in.) |
|---|---|---|---|---|
| Polypropylene | None | 0 | 133 | 0.016 |
| | Nanomer I.30P | 6 | 106 | 0.014 |
| TPO | None | 0 | 86 | 0.013 |
| | Nanomer I.30P | 6 | 65 | 0.010 |

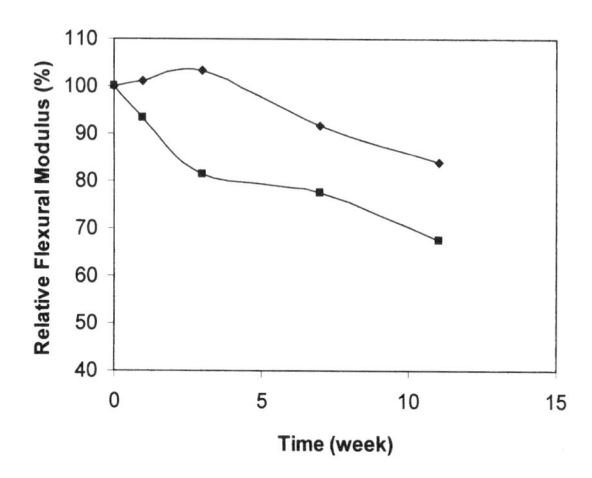

**FIGURE 20.5** Ultraviolet stability of polypropylene nanocomposite. ■, Neat polypropylene; ▲, polypropylene nanocomposite.

exposed to UV radiation in a QUV unit. The flexural modulus of neat polypropylene decreased rapidly with exposure to ultraviolet, while that of nanocomposite decreased at a much lower rate.

*Thermal Stability.* The thermal oxidation of polypropylene follows the reactions shown in Figure 20.6 [34]. Below 200°C, oxygen addition occurs at the carbon radical created within the polymer chain by H abstraction (process A). Above the temperature range 200–250°C, hydrogen abstraction resulting in oxidative dehydrogenation (process B). As the temperature increases, the concentration of chain end radicals II increases because of β scission of radicals I (process C). When nanocomposites are formed, weight loss in thermogravimetric analysis (TGA) begins 40°C higher than neat polymer. Also, the maximal weight loss shifted 100°C higher [34]. This drastically improved thermal stability is attributed to the barrier effect of organoclay. On the one hand, the dispersed layered silicate slows down the oxygen diffusion into the polymer matrix; on the other hand, it also slows down the release of volatiles by either the barrier effect or intercalation of polar volatile into silicate layers.

## 20.5 CONCLUSIONS

Polypropylene nanocomposite is a fast-growing area, and new developments and commercial applications are emerging rapidly. When well-dispersed polypropylene nanocomposites with relatively low clay loading are compounded, pelletized,

# Polypropylene Nanocomposite

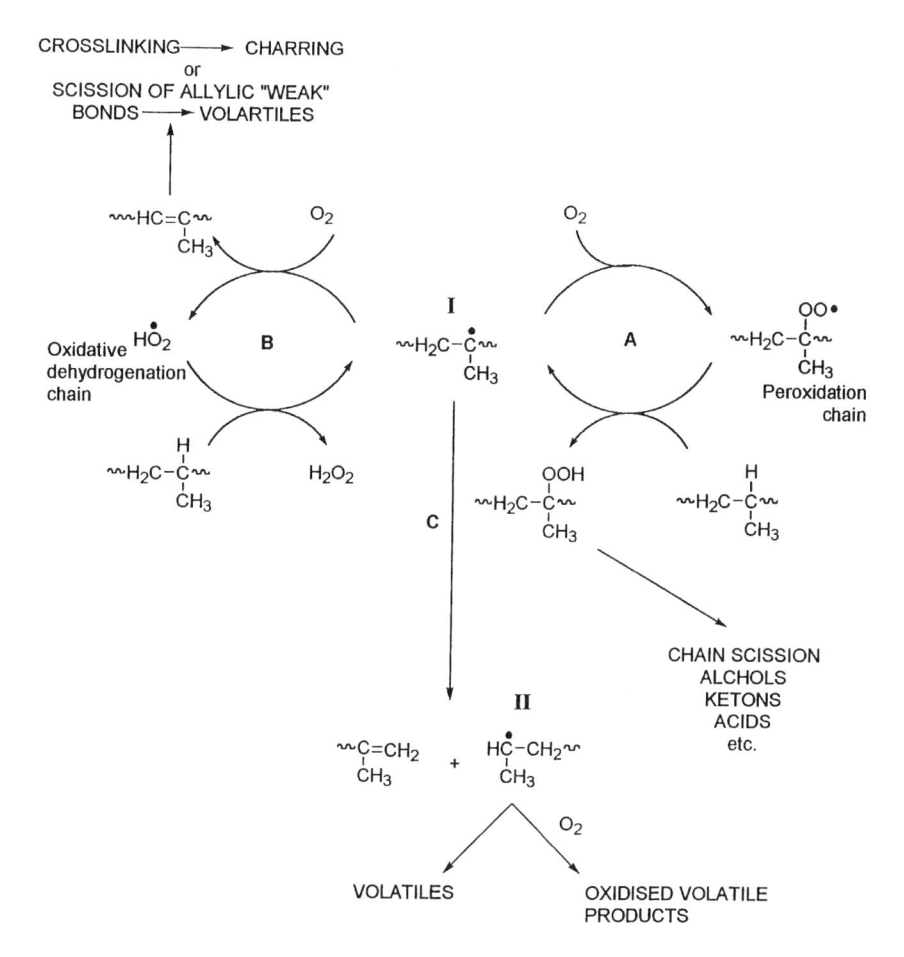

**FIGURE 20.6** Reaction paths for thermal oxidative degradation of polypropylene [34].

and molded into fabricated parts, there are many end-use benefits that enhance product performance:

1. Greatly improved mechanical properties
2. Barrier properties
3. Flame retardancy
4. Thermal stability

Cost-effective use of polypropylene nanocomposite can be achieved by down-gauging or by replacing more expensive materials.

This chapter has provided an overview of recent development in the polypropylene nanocomposite area. Hopefully, insights into the underlying fundamentals involved in compounding methodology and microstructure determination will promote even more commercial applications of high-performance polypropylene nanocomposite.

## ACKNOWLEDGMENT

The authors thank AMCOL International/Nanocor, Inc. management for their support of polyolefin nanocomposite program. The authors also want to thank the contributions from other Nanocor technical staff: Jae Whan Cho, Ying Liang, Sarah Crummy, Jason Logsdon, Scott Omachinski, and Vasiliki Psihogios.

## REFERENCES

1. EP Giannelis. Polymer layered silicate nanocomposites. Adv Mater 8:29–35, 1996.
2. EP Giannelis. Polymer-layered silicate nanocomposite: synthesis, propertics, and applications. Appl Organometal Chem 12:675–680, 1998.
3. B Singh, RN Jagtap. Polymer clay nanocomposite: a review. Pop. Plast. Packaging 47:51–59, 2002.
4. M Biswas, SS Ray. Recent progress in synthesis and evaluation of polymer-montmorillonite nanocomposites. Adv Polym Sci 155:167–221, 2001.
5. BKG Theng. The Chemistry of Clay–Organic Reactions. New York: John Wiley and Sons, 1974, pp. 9–13.
6. RA Vaia, H Ishii, EP Giannelis. Synthesis and properties of two-dimensional nanostructures by direct intercalation of polymer melt in layered silicates. Chem Mater 5:1694–1696, 1993.
7. Y Kojima, A Usuki, M Kawasumi, A Okada, T Kurauchi, O Kamigaito. Synthesis of nylon 6–clay hybride by montmorillonite intercalated with ε-caprolactam. J Polym Sci A Polym Chem 31:983–986, 1993.
8. C Maier, T Calafut. Polypropylene. Norwich, NY: Plastics Design Library, 1998, pp. 7–8.
9. J Ma, Z Qi, Y Hu. Synthesis and characterization of polypropylene-clay nanocomposites. J Appl Polym Sci 82:3611–3617, 2001.
10. T Sun, JM Garcés. High-performance polypropylene-clay nanocomposites by in-situ polymerization with metallocene/clay catalysts. Adv Mater 14:128–130, 2002.
11. T Sun, JM Garcés, ZR Jovanovic. Nanocomposite polymers. WO 01/30864 A2.
12. P Reichert, H Nitz, S Klinke, R Brandsch, R Thomann, R Mülhaupt. Polypropylene/organoclay nanocomposite formation: influence of compatibilizer functionality and organoclay modification. Macromol Mater Eng 275:8–17, 2000.
13. E Hackett, E Manias, EP Giannelis. Molecular dynamics simulations of organically modified layered silicates. J Chem Phys 108:7410–7415, 1998.

## Polypropylene Nanocomposite

14. G Qian, JW Cho, T Lan. Preparation and properties of polyolefin nanocomposites. International Conference on Polyolefins, Houston, 2001, pp. 553–559.
15. Y Kurokawa, H Yasuda, A Oya. Preparation of a nanocomposite of polypropylene and smectite. J Mater Sci Lett 15:1481–1483, 1996.
16. N Furuichi, Y Kurokawa, F Fujita, A Oya, H Yasuda, M Kiso. Preparation and properties of polypropylene reinforced by smectite. J Mater Sci 31:4307–4310, 1996.
17. Y Kurokawa, H Yasuda, M Kashiwagi, A Oyo. Structure and properties of a montmorillonite/polypropylene nanocomposite. J Mater Sci Lett 16:1670–1672, 1997.
18. A Oya, Y Kurokawa, H Yasuda. Factors controlling mechanical properties of clay mineral/polypropylene nanocomposites. J Mater Sci 35:1045–1050, 2000.
19. A Uauki, M Kato, A Okada, T Kurauchi. Synthesis of polypropylene-clay hybrid. J Appl Polym Sci 63:137–139, 1977.
20. M Kato, A Usuki, A Okada. Synthesis of polypropylene oligomer-clay intercalation compounds. J Appl Polym Sci 66:1781–1785, 1997.
21. M Kawasumi, N Hasegawa, M Kato, A Usuki, A Okada. Preparation and mechanical properties of polypropylene-clay hybrids. Macromolecule 30:6333–6338, 1997.
22. N Hasegawa, M Kawasumi, M Kato, A Usuki, A Okada. Preparation and mechanical properties of polypropylene-clay hybrids using a maleic anhydride–modified polypropylene oligomer. J Appl Polym Sci 67:87–92, 1998.
23. T Lan, G Qian. Preparation of high performance polypropylene nanocomposites. Additive 2000, Clearwater Beach, FL, 2000.
24. JW Cho, J Logsdon, S Omachinski, G Qian, T Lan, TW Womer, WS Smith. Nanocomposite: a single screw mixing study of nanoclay-filled polypropylene. Society of Plastics Engineers Annual Technical Conference, San Francisco, 2002.
25. Guidelines and setup parameters for polypropylene nanocomposite. Nanocor Technical Data Sheet P-801.
26. LE Nielsen. Models for the permeability of filled polymer systems. J Macromol Sci (Chem) A1 5:929–942, 1967.
27. JW Gilman, T Kashiwagi, JET Brown, S Lomakin, EP Giannelis, E Manias. Flammability studies of polymer layered silicate nanocomposite. 43rd International SAMPE Symposium, 1998, pp. 1053–1066.
28. JW Gilman, CL Jackson, AB Morgan, R Harris, E Manias, EP Giannelis, M Wuthenow, D Hilton, SH Phillips. Flammability properties of polymer-layered silicate nanocomposite. Polypropylene and polystyrene nanocomposites. Chem Mater 12:1866–1873, 2000.
29. JW Gilman, T Kashiwagi, M Nyden, JET Brown, CL Jackson, S Lomakin, EP Giannelis, E Manias. Flammability studies of polymer layered silicate nanocomposites: polyolefin, epoxy, and vinyl ester resins. In: S Al-Malaika, A Golovoy, CA Wilkie, eds. Chemistry and Technology of Polymer Additives. Oxford, UK: Blackwell Science, 1999, pp. 249–265.
30. J Zhu, CA Wilkie. Thermal and fire studies on polystyrene-clay nanocomposites. Polym Int 49:1158–1163, 2000.
31. T Lan, G Qian, Y Liang, JW Cho. FR applications of plastics nanocomposites. Fire Retardant Chemicals Association Spring Conference, 2002, pp. 115–119.

32. J Ma, S Zhang, Z Qi, G Li, Y Hu. Crystallization behavior of polypropylene/montmorillonite nanocomposites. J Appl Polym Sci 83:1978–1985, 2002.
33. W Xu, M Ge, P He. Nonisothermal crystallization kinetics of polypropylene/montmorillonite nanocomposite. J Polym Sci B Polym Sci 40:408–414, 2002.
34. M Zanetti, G Camino, P Reichert, R Mülhaupt. Thermal behavior of polypropylene layered silicate nanocomposites. Macromol Rapid Commun 22:176–180, 2001.

# Index

Acrylic acid (AA) grafted PP, 37–40
   chemical structure, 51
   homopolymerization of AA monomer, 38,40
   melt extrusion process, 38–40, 387
   reactions mechanisms for, 39
   thermal properties, 48–49
American Society of Test Methods (ASTM):
   accelerated UV aging [D4459], 98
   fracture toughness [E399], 201
   horizontal flammability [D635], 100
   standard Charpy test [D256], 182
   vertical flammability [D3801], 100
Amorphous domains, 428, 437, 439–440
Anisotropic behavior, 162, 429, 498
Antioxidants:
   butylated hydroxytoluene [BHT], 84–87
   dibutylnonylphenol [DBNP], 85–87

[Antioxidants]
   polyphenolic-multi-site free radical trap, 87
   primary-radical terminating, 82, 87
   secondary-phosphites, 88
   secondary-thiosynergist, 84, 297
   volatility and migration, 86
ASEAN free trade area, 5
Average annual growth rate (AAGR):
   of polypropylene resin, 1
   of TPO compounds, 222

Blow molding, 32
   HDPE fuel tank, 360
   injection, 247–248
   melt strength of HDPE in, 32
   shredder housing, 360
Blush resistance to stress whitening, 246–247
Brittle failure, 293

Brittle fracture, 157, 452–454
Bruceton staircase procedure, 183
Bumper fascia, 222, 255

Calcium carbonate-filled PP properties, 200, 204–207
    effect of untreated filler on, 445
    optimum filler level for, 204
Carpet backing, 25
Catalyst, 17
    metallocene, 17, 223–224
    Ziegler-Natta, 17, 224, 710
CEAST impact test equipment, 169–170
    Fractovis falling weight tester, 170, 186–187
    Resil 5.5 tester, 167, 169
Cellulose reinforced PP, 73–76, 471
    interphase thickness of cellulose fibers in, 54
Charpy notched impact strength (CNIS), 183, 204–215
Clarifying agents (see nucleating agents)
Class types of fracture modes, 190–191
Clear random copolymer, 249–251
    by refractive index matching, 249–250
    using clarifying agents, 249
Co-crystallization, 51, 54
Coefficient of linear thermal expansion (CLTE), 294, 630, 668–669, 723
Compatibilizer, 120–121
    bridge between glass fiber-polymer matrix, 467
    Kraton FG 1901X rubber as, 128–129
    maleated SEBS (styrene-ethylene-butene-styrene) triblock as, 128
Compounding extrusion (*see also* Twin screw extruders),
    of commingled plastics, 119, 131
    of compacted talc, 635–638
    of compatibilized, recycled HDPE-PP blends, 128–129
    effect of mixing intensity on dispersion of plastomer, 241–243

[Compounding extrusion]
    effect of talc bulk density on, 308–309
    of GFRP, 350–354
    for interphase design study, 477–479
    of mica reinforced PP, 549–551, 595–596
    of organosilane crosslinked TPV by single pass, 273
    of powdered PVC resins, 396
    of PP-plastomer blends, 242–243
Continuous fiber reinforced composites, 367, 647–650
    by Twintex process, 367–368, 650
Controlled rheology of PP, 17, 384–385
    by beta scission, 384
    closed loop regulation, 385
    extruder screw design for, 386
    free radical mechanism for, 385
    by peroxide-induced, 385
    by thermally-induced, 385–386
    for vis breaking, 17, 384
Core-shell inclusions, 445
Coupling agent, 36–37
    carboxylated amorphous PP, 336
    comparison between types, 56–57
    glicidyl methacrylate grafted PP, 389
    Hercuprime G, 129, 336, 471
    for interfacial adhesion, 53, 466
    organofunctional groups, 36
    organosilane, 36, 55, 419–422
    PPgAA, 36
    PPgMAH, 36
    SC coupling system, 604
    Titanates, 419–420
    Ucarsil (peroxide-organosilane blend), 37, 569–570, 575
Crack propagation, 194
Creep deformation behavior, 530
Creep rupture strength-temperature, 540
Creep rupture time, 532–533
    analytical expression of load-temperature, 533
    effect of glass fiber content, 533
    for mega-coupled GFRP, 534

# Index

Critical strain energy release rate, 172, 198
Crystalline domains, 163–164, 428, 439–440
Crystallinity, 12–13, 158, 160
  level of, 17
  by nucleating agent addition, 18, 339
Crystallites, 12–13, 437, 622
  lamella morphology of, 13, 339
  as physical cross-links in amorphous phase, 17
  thickness of, 13
Crystallization temperature, 18, 297
  effect on mold cooling time, 251–253, 695–697

Design of experiment (DOE) methodology:
  of Box-Behnken statistical design, 132–133
  for compatibilized PP-HDPE blends, 135
  of Design-Expert software, 584
  to develop SC-mica coupling system, 604–606
  effects of TPO composition on properties by, 255–262
  full cubic three-component design for, 256–257
  mechanical property-response factors by, 136
  for mold shrinkage response, 693–695
  for optimizing stiffness-impact strength balance, 237
  predictive equations from, 137–141, 589
  process variable input ranges for, 134
  response surface analysis by, 140–143, 258–268
Detergent resistance of GFRP:
  definition of suds resistance or, 64
  effect of glass fiber type on, 65
  effect of PPgMAH type on, 66
  effect of water contact on, 323, 327

Differential scanning calorimetry (DSC), 18
  crystallization temperature by, 18, 297
  determination of $T_g$ by, 18
  endothermic peak in, 18
  heat capacity determination by, 484
  heat capacity jump at $T_g$ by, 484–485
  heat of fusion by, 18
  of metallocene plastomers, 225
Dimensional stability of mica-glass fiber hybrid composite, 612
Dispersive mixing, 119–120, 409–410
  effect of shear rate, 141–143
  HDPE dispersed phase, 143
  Taylor model of, 120
  viscosity matching for, 129–130
Dissipation of mechanical energy, 157–158
  by crazing mechanisms, 157, 160–162, 438, 449
  by crack pinning, 158
  via inclusion cavitation, 158, 160
  by interfacial cavitation, 158
  by particle deformation, 158
  by shear yielding and banding, 158–160, 454–455
Distributive mixing, 395
  for kneading block design, 407–410
Ductile-brittle transition (DBT), 158, 189–194, 454
Ductile failure, 294, 448
Ductile fracture, 151
Ductility index, 235

Elastic moduli of interphase, 426–433
  upper and lower limits for, 418–419
Elastomers (*see also* Metallocene plastomers):
  EPDM, 221, 416, 423
  ethylene-propylene rubber (EPR), 221, 416, 423
  natural rubber, 221
  styrene-butadiene-styrene (SBS) rubber, 221

[Elastomers]
   styrene-ethylene-butene-styrene
      (SEBS) tri-block, 221
   very low density polyethylene
      (VLDPE), 221
Engineered Interphase Design:
   control of polarity of EIL, 416–417
   definition of concepts for, 414
      engineered interphase layers
         (EILs), 414
      interface, 414
      interphase, 414
      mesophase, 414
   elastomeric EILs of, 423–424
   for elastomer interphase layer,
      441–442
   fracture toughness for, 372, 458–459
   by grafting oligomers onto glass fiber
      surface, 422–423
   localized plastic deformation of, 446
   for non-spherical reinforcement,
      431–433
   optimum stiffness-toughness balance
      via, 416–419
   of silane and titanate EILs, 419–422
   of spherical inclusions, 429–431
   transcrystalline EILs of, 424–426

Flame retardant additives:
   alkyl organohalogen, 103
   alumina trihydrate (ATH), 94–95, 416
   ammonium polyphosphate salts, 97,
      108
   antimony oxide, 98–99
   aromatic organohalogen, 103
   decabromodiphenyloxide, 90, 721
   halogen flame retardant, 96, 99
   with low dust form, 108
   magnesium hydroxide, 35–36,
      416–418
   non-halogen intumescent, 92
   in PP compound formulations, 104
Flame retardant mechanisms:
   for antimony oxide synergy, 96
   with chain-branching, 83

[Flame retardant mechanisms]
   in combustion, 93
   for condensed phase-intumescent,
      92, 96
   fire triangle of, 93
   flame-branching reaction in, 96
   for heat sink-endothermic processes,
      94–95
   in nanocomposite compounds,
      720–722
   in vapor phase-halogen FR, 92, 96
Flexural behavior of GFR PP:
   effect of moisture on, 496
   flexural creep of mega-coupled,
      490–491, 511–512
   flexural fatigue of, 490, 493–494
Flexural modulus, 16
   versus glass content, 352
Fourier transform infrared (FTIR)
      analysis, 43
   determination of MAH content by,
      46–47
Fracture failure process, 156
   Real-life testing of, 184
Fracture toughness, 196, 204
   of particulate-filled PP, 458–459
   for processes at molecular level,
      455–458

Gas phase reactor design, 21–23, 222
   for horizontal stirred bed, 23, 222
   for vertical fluidized bed, 22–23,
      222
Gel permeation chromatography (GPC),
   15
Glass fiber characteristics:
   of filament diameter, 63, 320–321,
      347, 520–521
   of filament length, 346
   for fuzz generation, 55, 334
   for SEM micrograph of virgin E-glass
      fiber, 323
   sizing chemistry of, 55–56, 328–339,
      334–339, 420–421, 477,
      526–527

# Index

[Glass fiber characteristics]
  TEX, 321
  for types of short chopped fibers, 61–64
Glass fiber manufacture, 317–318
  fabrication of product forms in, 315
  film forming process of, 312–315, 334–336
  glass composition of, 312–314
  in-line chopped strand in, 523
  off-line, two-step process for, 523–524
Glass fiber-reinforced PP(GFR PP):
  AA grafting reaction-glass fiber addition, 390–391
  chemical coupling mechanism, 48–54
  detergent resistance of, 64–66
  effects of glass fiber length-orientation on molded, 498
  interphase microstructure of, 414, 467–468
  market-applications of, 356–369
  physical properties of, 318–319
Glass flake-reinforced PP, 333
Glass mat thermoplastic (GMT), 36
Glass transition temperature (Tg), 437
Glicidyl methacrylate grafted PP, 389
Global trends of polypropylene growth:
  for regional issues, 3–5
  by application type, 5
  in recycling, 6–7
  or specialty grades, 7–8
Griffith-Orowan criterion, 165–166, 452–453

Head-to-head monomer linkage, 14
Heat capacity behavior:
  due to fiber glass content, 482–486
  for interphase design, 485
  at glass transition temperature, 485
Heat capacity jump, 485
Heat distortion temperature, 156, 289, 556–557, 597, 599, 665
  dependence on glass sizing, 470

Hybrid mica-glass fiber composites, 66–67, 603, 607–611
Hydroperoxide, 82

Impact copolymer (ICP), 13–14, 222
  rubber content of, 13
Impact modification of PP:
  filled impact copolymer, 255
  homopolymer, 236–247
  impact copolymer, 253–263
  impact resistance by, 13
  random copolymer, 247–253
Impact modifiers, 233
Impact tests:
  charpy, 182–184
  clamping force for Izod, 184
  Gardner, 234, 561–562, 666, 579–580
  instrumented impact tester, 171, 185–189, 234
    bouncing effect of, 175
    inertia effects, 178
  Izod notched, 58, 184–185, 234
  Izod notched radius, 178–181
  Izod reverse notched, 57, 168
  test geometry of, 168
Injection molding (*see also* Two dimensional mapping), 30–32
  of GFR PP composite, 354
  generation of weldline, 335–356
  process variables for, 678, 694
Instrument panel surface, 274
Interfacial adhesion, 52, 211–214, 451, 466
Interfacial strength of PP composites, 333, 339
  definition of, 339
  test methods for determining, 341–343
Interphase Design (*see also* Engineered Interphase Design), 471–479
  compatibilizer bridge in, 471–472
  glass fiber content of, 351–352
  glass fiber diameter of, 347–350
  glass fiber sizing chemistry in, 55–56, 471

[Interphase Design]
    glass fiber type for, 61–64
    importance of, 415–416
    interphase thickness, 459, 467,
        479–480, 486
    Lipitov model of, 481–482
    optimum design attributes, 476–479
    physical entanglement, 36–37,
        472–473
    polymer brush model, 473
    PP-g-MA type-level in, 58–59
    tensile creep behavior, 490, 493
Interphase microstructure, 467–468
    geometry of, 468
    methods to probe, 479–480
    thermodynamic probe of, 480
Interphase thickness, 450, 467, 480, 486
    correlation tensile strength, 489–492
Intumescent chars, 96, 721

Linear elastic fracture mechanics
    (LEFM), 164–165
Lipitov model, 481–482
Living hinge, 31
Localized plastic deformation,
    446–452
    with crack initiation in, 446–447
    by crack propagation in, 447–448
    crazing of, 449
    effect on composite fracture toughness, 452
    finite element analysis (FEA), 443
    fracture toughness of, 458
    by shear banding of, 450
Long glass fiber-reinforced PP, 36,
    360–363, 647
    attrition of glass fiber length, 341,
        353–354, 362
    melt impregnation process of,
        360–361
Loss tangent tan $\delta$, 181
Low temperature instrumented impact
    behavior, 254–255
    effect of plastomer addition level on,
        254–255

Macrocrystalline talc, 288–289
Magnesium hydroxide-PP composite,
    72–73
    flame retardancy (FR), 35–36,
        416–417
    low smoke for non-halogen, 35–36,
        416–417
    mechanical properties of, 73
    with reduced fracture toughness, 416
Maleic anhydride content:
    by $^{13}$C-NMR method, 43
    by FTIR method, 46–47
    by titration method, 47, 471–472
Maleic anhydride grafted PP, 40–43, 475
    chemical structure of, 43, 51
    control of homopolymerization for, 43
    determination of MAH level, 43,
        46–48
    effect of MAH level on melt flow rate
        of, 45–46
    effect of MAH content on efficiency
        of, 58–60
    effect of PP MW on performance of,
        60–61
    grafting reaction mechanisms of,
        41–42, 44–45
    homopolymerization of MAH
        monomer, 43
    via melt extrusion, 40, 388
    peroxide selection for, 43, 45
    via solution method, 41
    via thermal grafting, 40
    thermal properties of, 48
Manufacturing processes for PP resins,
    19–25
    Amoco gas phase technology for, 21,
        24
    by BASF-Novelen, 21–22
    in bulk (slurry), 19–20
    by Montell-Spheripol, 21
    by Union Carbide-Unipol, 21
Mega coupled GFRP(*see also* Interphase
    design), 54, 488–497
    comparision with engineering resins,
        494–495

# Index

[Mega coupled GFRP]
  effects of impact modification for, 512–514
  interphase thickness - fiberglass content, 491
  mechanical properties of, 492
  perfect coupling limit, 496–497
Melt flow rate (MFR):
  for bulked continuous filament, 26
  for cast film, 28
  for continuous filament, 26
  controlled rheology, 17
  increase by vis breaking, 17
  measure of molecular weight, 15, 17, 473, 475
  for melt-blown fibers, 27
  for monofilament, 28
  range for interior trim applications, 34
  range for thermoforming, 29–30
  for rigid packaging containers, 31
  for slit film, 25
  test method, 15
Mercosur market countries, 5
Metallocene catalysts, 223–224
Metallocene plastomers, 222–231
  composition distribution of methylene groups, 228–231
    by CRYSTAF method, 228
    by SITS method, 228, 230–232
    by TREF method, 228
  of ethylene-butene (EB), 227
  of ethylene-hexene (EH), 233
  of ethylene-octene (EO), 233
  melting point via DSC, 230
  molecular weight distribution of, 229
  morphology, 243–246
    correlation impact resistance, 243–246
    by LVSEM method, 236, 251
    peroxide crosslinked TPE, 272
  physical properties of, 226
  thermal transitions in, 225–227
    determined by DMA methodology, 227

[Metallocene plastomers]
  $\beta$-transition in DMA spectrum, 225
  $\gamma$-transition in DMA spectrum, 225
Mica:
  applications market, 543–544
  aspect ratio, 546, 595–596
  characteristics of, 594
  chemical resistance, 547
  coupled systems, 600–604
  dry-wet grinding of, 548–549
  enhanced weatherability, 547
  manufacture, 548–549
  mineralogy, 594
  morphology, 545–546
  muscovite type, 544
  phlogopite type, 544
  scratch resistance of, 547
  surface treated, 569–576
    by maleic anhydride grafted PP, 603
    by SC coupling system, 604
    by silane, 602
  type color variations,
Mica reinforced-PP composites, 68–70, 543–589, 593–616
  additives use in, 571–574, 577–583
  comparison with other filled-GFR PP composites, 553–554
  compounding extrusion process, 549–551, 595–596
  control of warpage in, 564
  effect of loading level on MFR of, 553
  factors affecting mechanical properties of:
    by annealing, 565–569
    for mica particle size, 69, 562–564
    for mica type-level, 69, 555–562
    by PPgAA chemical coupling, 68–70
    by surface treatment of mica, 68, 569–576
  injection molding of, 551–552
  shrinkage-warpage of, 597
Microcrystalline talc, 288–289

Modular twin screw design, 397–410 (see also Twin screw extruders)
  axial pressure buildup in, 403–404
  building block concept of, 393–395
  distributive mixing by, 407–410
  drag-pressure flow characteristics, 398–404
  flat plate model for, 399
  flow patterns in channels, 400–402
  Newtonian flow model for, 397
  staggered arrays of kneading discs, 398
Mold cooling time, 251–253
Mold shrinkage, 597, 670–671, 675–705,
  dependence on material properties, 679
  dependence on system parameters, 678–679
  effect of cooling rate, 697–701, 723
  effect of crystallinity, 683–689
  effect of fillers, 670–671, 679–680
  effect of injection molding process conditions, 693–695
  effect of PP resin type, 680–693
  exhibited by PVT diagram, 677, 685–691
  fundamentals, 677–679
  relationship linear-volumetric shrinkage, 678
  for talc filled PP, 293
Molecular tailored blends, 257–263
Molecular weight,
  number-average, 15, 17
  weight-average of, 15, 17
  z-average of, 15, 17
Molecular weight distribution (MWD), 15–17
  broad by two-reactor configuration, 17
  effect on melt viscosity, 17
  measured by GPC, 15, 228
  using metallocene PP catalysts, 17
Montmorillonite clay, 233
  chemical structure of, 709
  surface treatment of, 712

Morphology (*see also* SEM), 17
  of crystallites in the PP matrix, 17
  dispersed phase domain size, 141–142
  homopolymer-plastomer blend, 244–246
  phase contrast optical microscopy, 142

Nanocomposite (see Polypropylene nanocomposite)
Nucleating agents, 18, 249, 339
Nucleation of PP, 248–249, 623–628
  by aluminum hydroxybenzoate, 622–625
  by fine grade talc particles, 296–297, 623
  by micronized talc, 623
  by sodium benzoate, 622–625
  by sodium di-tert-butyl phosphate, 625

Paintability-automotive TPOs, 263–268, 630, 643
  characterized by test methods, 266–268
Particulate-filled and rubber modified composites, 202–203, 433
Perfect coupling limit (*see* Mega coupled GFR PP)
Peroxide (*see* Controlled degradation of PP)
Peroxide selection for grafting of polyolefins, 43, 45
Plane stress - strain regions, 198
Plastomers (*see* Metallocene plastomers)
Poisson's ratio, 199, 201–202, 206, 430
Polypropylene applications (*see also* PP conversion), 25–34
  automotive, 33–34, 629–631, 641–650
    air inlet panels using talc filled PP, 643
    bumper fascia, 631
    exterior, 33, 643
    fuse box cover, 643
    interior, 33–34, 642–643

# Index

[Polypropylene applications]
  molded in color, 3, 642–643
  painted parts, 33, 643
  trim components, 33, 642
  under-hood, 631, 644
 blow molding, 32–33
  extrusion, 32
  injection, 33
  injection-stretch, 33
 fibers and fabrics, 25–28
  bulked continuous filament, 26
  carpets, 26
  continuous filament fibers, 26
  melt-blown fiber/fabric, 27
  monofilament, 28
  nonwoven fabrics, 27
  slit-film, 25
  spunbonded fabric, 5
  staple fibers, 27
 film, 28–29
  biaxially oriented, 29
  cast, 28
 injection molding, 30–32
  applications and hand tools, 31–32
  caps and closures, 31
  consumer-industrial, 32
  medical, 32
  rigid packaging containers and housewares, 31
 stampable PP composites, 363–367
 strapping, 28
Polypropylene blends:
 economic analysis of, 143–144
 Hivalloy-PET, 122
 PP-EPDM, 125
 PP-HDPE, 125
 PP-PET, 123
Polypropylene conversion:
 benefits of, 645–646
  by cost reduction, 645
  for processing advantage, 645–646
  with reduced part weight, 646
 from engineering resins, 5, 305

[Polypropylene conversion]
 for automotive fan shroud, 643–644
 in automotive instrument panels, 274, 648
 due to lower cost/volume, 302
 global homopolymer production for, 7
 growth and uses for, 5
 from PVC, 268–270, 304
 from styrenic-base materials, 5
 from thermoset resins by TPV, 272–273
 tailored to needs of:
  by structural integrity of GFR PP composites, 647–648
  for under-the-hood applications, 643–644
  using functional components, 642
  for weather resistant exterior parts, 643
Polypropylene nanocomposite, 76–78, 646–647, 707–726
 barrier properties, 719
 crystalline behavior, 722–723
 flammability, 234, 720–722
 manufacture:
  by in situ polymerization, 710–712
  by melt compounding process, 712–718
   effect of base polymer resin, 717–718
   effect of compatibilizer, 715
   effect of process conditions, 715–717
  by solution intercalation, 710
 mechanical properties, 718
 thermal properties of, 234, 724–725
 UV stability, 723–724
Polypropylene resins:
 alpha crystalline (monoclinic) form, 17, 339
 atactic, 14
 beta crystalline (hexagonal) form of, 17–18, 339
 consumption of, 5

[Polypropylene resins
  copolymer, 13
  high crystalline, 73–76
  homopolymer, 12–13
  impact copolymer, 13–14
  monomer, 2, 14
  morphology, 17
  random copolymer, 12–13, 247–253
  semicrystalline, 11
  super impact copolymer, 263
  syndiotactic, 14
Prediction of service life:
  by accelerated tensile creep, 533–540
  for ductile-brittle failure, 157
  by UL 746B for long term properties, 102
  for UV aging under Florida sunlight exposure, 99
PVT diagrams of amorphous-semicrystalline polymers, 685–688
  for injection molding process steps, 691–692

Random copolymers, 12, 247–253
Recycling of plastics:
  from automotive parts, 115
  from battery casings, 7, 115
  from bumpers, 115
  economics of, 116
  from painted TPO, 7
  from postconsumer sources, 7, 114–115
  from postindustrial fiber reclaim, 7
  from waste stream sources, 117
Recycling technology:
  for GFRP, 367
  of incineration, 116
  for monomer recovery-depolymerization, 118
  for postreactor compatibilization, 121, 124
  for pyrolysis processes, 118
  for selective dissolution processes, 118–119
Reference state, 204

Refractive index matching (see Clear RCP)
Rheology:
  definition of, 16
  of HDPE and PP, 130
  melt elasticity
  melt viscosity, 17
    relation to molecular weight, 18
    shear sensitivity of, 18
  of molten fiber extrusion, 16–17

Scanning electron microscopy (SEM):
  of brittle-ductile failure for talc filled PP, 292–294
  for chemically coupled-GFR PP composites, 337–338
  of fracture surfaces of PP composites, 52–53, 467–469
  of muscovite mica, 546
  virgin E-glass fiber, 323
Scratch-mar resistant PP, 646, 667–668, 672
Shear yielding at crack tip, 454
Specialty PP products, 7–8
Spherulites, 162, 622
Splay on molded parts, 304
Stampable polypropylene composites:
  Attributes of, 511
  Azdel process for, 363–366
  Taffen process for, 366
  Twintex process for, 367
Stiffness (see also Flexural modulus), 16
  definition of, 16
  effect of crystallinity on, 16
  in thermoforming, 30, 274–275
Stiffness-toughness balance, 156, 366, 416–417
  for automotive TPO, 237
  by elastomer addition, 417
  by incorporation rigid fillers, 416
  for mica filled-PP composites, 585
  molecular tailored blends for, 257–263
  by tailoring PP-HDPE blends, 150–151

# Index

[Stiffness-toughness balance]
    for talc filled PP composites, 255, 289–293, 296, 306
Strength (see also Tensile behavior of GFR PP):
    definition of, 16
Stress concentration, 159
Stress state, 167
Super olefin polymer, 263
Surface treatment of fillers and glass fibers:
    by azide silane, 69, 570
    by filler deactivator, 298
    glass fiber sizing, 55–56, 328–329, 334–339, 420–421, 477, 526–527
    by organosilanes, 37, 328–331, 576
    by stearic acid, 36
    thickness of silane layer of, 331–332
    by titanites, 36
    by zirconates, 36

Tacticity, 14–15
    atactic, 11, 14–15
    isotactic, 11, 14
    level of, 17
    syndiotactic, 14–15
Tail-to-tail monomer linkage, 14
Talc filled-PP, 70–71, 287–289
    effect of chemical coupling, 70
    effect of HDT, 282
    effect of loading level on properties, 288
    effect on shrinkage and melt flow, 294
    long term heat aging [LTHA], 291, 296
        effect of iron content, 298
    molded in color, 290, 299, 307–308
    UV stability for composites of, 299–301
Talc filler:
    chemical composition, 282–283
    color consistency, 299, 618–620
    compacted form, 303, 620–621
        by zero force technology, 308–309

[Talc filler]
    densified, 620–621
    geography of deposits, 283, 618
    geology of, 281
    loose bulk density, 286, 303
    macro-versus micro-crystalline, 288–289
    mining/processing steps for, 283, 285
    mineralogy-morphology of, 617–618
    R-Talc, 305–306
    splay on molded parts, 304
    surface characteristics of, 283, 285
    surface modifications of, 304–306
Talc particle size:
    effect on impact strength, 291
    effect on reinforcement, 289
    effect on tensile strength, 290
    effect on thermal stability, 291
    grind measurement of, 286
    median, 286
    micronized, 233
    submicrometer, 620
    top size, 296
Tensile behavior of GFR PP(see also Tensile Creep):
    anisotropic (see also Two-dimensional mapping), 498
    contributing factors, 466
    creep-fatigue correlation, 528–529
    effect of chemical coupling level on, 474
Tensile creep:
    analysis of GFRP composite, 533–540
    analytical expression, 532–533
    correlation with tensile fatigue, 528–529
    deformation stages, 530
    effect of chopped-glass fiber chopped strand process,
    effect of sizing type, 526–528
    mega coupled, 534
    method, 530
    rupture time, 532–533
    test equipment, 519, 531

Tensile fatigue:
  effect of glass content, 522
  effect of glass fiber sizing,
Tensile strength:
  angular dependence of, 499
  chopped glass fiber strand process, 523–524
  correlation interphase thickness, 492
  effect of sizing type, 526–528
  versus glass fiber content, 351
Thermal analysis:
  differential scanning calorimetry (DSC), 18
  differential thermal analysis (DTA), 18
  dynamic mechanical analysis (DMA), 18
  heat space gas chromatography(GC)-mass spectra (MS), 48
  thermal mechanical analyzer (TMA), 18
  thermogravimetric analysis (TGA), 18
    weight loss of talc via, 284
  TGA-FTIR combination in, 48
Thermal stabilization:
  by decomposition of hydroperoxides in, 82
  by free radial termination in, 82–84
  by radical trapping mechanism for BHT, 84
Thermodynamic probe of interphase microstructure (see Interphase design)
Thermoforming technology, 29–30, 274–276
  negative vacuum forming processes, 274
  solid-phase pressure forming of, 30
  stiffness requirement for, 30
Thermogravimetric analysis (TGA):
  assessment of volatility-migration stabilizers by, 84–86
  of ATH decomposition, 95
  for PPgAA, 49
  for PPgMAH, 50
  of types of phenolic antioxidants, 85

Thermo-oxidative degradation:
  accelerated by trace metals, 82
  free radical chain mechanism of, 82–83, 725
Thermoplastic elastomer (TPE), 270–273
  dynamic vulcanization of, 270, 273
  melt rheology of, 270
  organosilane-moisture crosslinked, 272
  peroxide initiated crosslinked, 271–272
Thermoplastic olefin (TPO), 221
  bumper fascia application, 34
  effect of mixing process on,
  flame retardant
  flexible compounds to replace PVC, 268–270
  impact improvement of mica filled, 584–588
  micropellets for rotomolding, 276
  paintability of, 263–268
  reactor grade (RXTPO), 222
  sheeting, 274–276
  slushing molding process, 276
Thermoplastic vulcanizate (TPV), 273
Thermotropic liquide crystallization polymer (TLCP), 163
Transcrystallinity (see also Crystallinity):
  in engineered interphase design, 424–426
  glass fiber surface, 54, 344
  growth of alpha-beta PP crystallites, 339
  morphology for wood fiber reinforced PP composite, 54
  SEM view of transcrystallinity, 340
  stress-induced transcrystallization, 345
Twin-screw extruders (see also Modular screw design):
  comparison between types, 393–394
  counter-rotating, 393
  high volume compounding by, 396–397

# Index

[Twin-screw extruders]
  intermeshing corotating, 393–394
  technology, 392–397
Two-dimensional mapping of tensile strength:
  anisotropic tensile behavior in, 498
  experimental method to determine, 499–500
  load angle-fiber orientation of, 499
  longitudinal-transverse values for, 509–510
  toughness in transverse direction, 355
  tripartite model of Stowell-Liu for, 501–505

Ultimate tensile strength (UTS), 339
Ultraviolet durability:
  bicomponent FR fiber for, 110
  effect of Florida aging on, 99
  effect of processing on, 88
  of flame-retarded PP, 81, 88
  of muscovite mica, 547
  retention of tensile strength of, 98
  standard failure criteria for, 99
  surface coating for, 89
Ultraviolet stabilizer:
  carbon black, 89
  cyclic mechanism for HALS radical trapping, 90–91
  hindered amine light stabilizer (HALS), 90
  monomeric grafted stabilizer, 109
  secondary HALS, 91
  silicone based HALS, 109–110
UV light absorber:
  benzophenone type, 89
  benzotriazoles, 89–90
UV reflectance spectra-brominated flame retardants,

Underwriters Laboratories (UL) protocol:
  for flammability (UL 94), 101–102
  for relative thermal index (RTI), 102
  of UL 746A-short term properties, 102
  of UL 746B-long term properties, 102

Vis breaking (*see* Controlled degradation of PP)
Warpage of molded parts, 564–565, 597, 613
  causes for, 564
  of mica-glass fiber composites, 564–565
Weldline strength, 355–356
Wollastonite fiber:
  chemistry, 652
  handling, 660–661
  manufacturing process, 657–658
  mining, 657
  physical properties, 654–656
  size and shape, 652–654
Wollastonite reinforced PP, 651–674
  CLTE, 668–669
  compounding extrusion, 658–660
  effect of surface treated fiber, 669–670
  economics, 672
  mar-scratch resistance, 667–668
  properties of, 662–666
  shrinkage, 670–671
Wood flour-filled PP, 73–76
Wool fiber-glass fiber hybrid composite, lapinus wood fiber filler, 66
  mechanical properties of, 67
  with high crystalline base PP, 74–76
World scale PP technology, 22–23

Yielding of composites with engineered interphases, 433–445